REACTIVE OXYGEN SPECIES in BIOLOGY and HUMAN HEALTH

Edited by

SHAMIM I. AHMAD
Nottingham Trent University
United Kingdom

CRC Press
Taylor & Francis Group
Boca Raton London New York

CRC Press is an imprint of the
Taylor & Francis Group, an **informa** business

CRC Press
Taylor & Francis Group
6000 Broken Sound Parkway NW, Suite 300
Boca Raton, FL 33487-2742

First issued in paperback 2022

ISBN-13: 978-1-498-73545-2 (hbk)
ISBN-13: 978-1-03-233997-9 (pbk)
DOI: 10.1201/b20228

Library of Congress Cataloging-in-Publication Data

Names: Ahmad, Shamim I., editor.
Title: Reactive oxygen species in biology and human health / [edited by] Shamim I. Ahmad.
Description: Boca Raton : Taylor & Francis, 2016. | Includes bibliographical references and index.
Identifiers: LCCN 2016010468 | ISBN 9781498735452 (alk. paper)
Subjects: | MESH: Reactive Oxygen Species--adverse effects | Oxidative Stress--drug effects
Classification: LCC RB170 | NLM QV 312 | DDC 571.9/453--dc23
LC record available at http://lccn.loc.gov/2016010468

REACTIVE OXYGEN SPECIES in BIOLOGY and HUMAN HEALTH

Contents

SECTION I   Introduction, Detection, and Production of ROS

SECTION II   Importance of Accumulation of Iron in the Brain

SECTION III   Neurodegenerative Diseases Caused by ROS

SECTION IV   Autoimmune Diseases Caused by ROS

SECTION V   Cancers Induced by ROS

SECTION VI   Cardiovascular Diseases Induced by ROS

SECTION VII   Other Uncommon Diseases Induced by Oxidative Stress

Preface

Reactive oxygen species (ROS) have been identified to play major roles in our lives. This is evident from the fact that since their first documented publication in 1945 by Stuffins and Weatherall [1] and in 1946 by Mann and Quastel [2], 161,400 research papers have been documented (PubMed, November 6, 2015). Indeed, ROS production, their interactions with a large number of biomolecules and the resulting damage and consequences, and their roles in inducing a large number of human diseases, also positively affecting certain metabolic processes and playing roles in aging, are highly complex areas of human biochemical metabolism and require much more studies to obtain a more complete picture.

ROS are unstable reactive molecules, have one or more unpaired electron(s), and are able to oxidize nearby molecules to gain an electron to enter the ground states. A number of ROS in biological systems include lipid peroxides, nitric oxide (NO), singlet oxygen ($^1O_2$), ozone (O_3), and hypochlorous acid (HOCl); however, superoxide anions ($O_2{}^{\bullet-}$), hydroxyl radicals ($^\bullet OH$), and hydrogen peroxide (H_2O_2) are the three most important ROS playing dual roles in biological systems. Interestingly, these three ROS are interconvertible, in that the oxygen molecule can be converted to $O_2{}^{\bullet-}$ by accepting an electron. In turn, the dismutation reaction by superoxide dismutase (SOD) can convert $O_2{}^{\bullet-}$ to H_2O_2, and this finally may be partially reduced to $^\bullet OH$ or fully reduced to water.

ROS can be produced endogenously or exogenously. Endogenously, they are produced naturally and continuously in biological systems as a result of leakage of electrons from the electron transport chain in mitochondria. Additionally, they are produced by various enzyme reactions in the systems such as those carried out by xanthine oxidase and by cytochrome P450. Also, autoxidation of small molecules (e.g., catecholamine), response to xenobiotics and exogenous environmental exposures, ischemia, and inflammatory stimuli are also responsible for the production of ROS. Endogenously, ROS are also produced through the ROS NADPH oxidase (NOX) complex in the cell membrane, mitochondria, peroxisomes, and endoplasmic reticulum.

A number of exogenous agents playing roles in ROS production include xenobiotics, pollutants, drugs, smoke, tobacco, and radiation. A good example of production of ROS by ionizing radiation is a set of reactions in which water is converted to $^\bullet OH$, then to H_2O_2 and to $O_2{}^{\bullet-}$, and ultimately to oxygen.

ROS being highly reactive can cause damages to biological systems, including DNA strand breaks, base modifications leading to mutations, inhibition of RNA and protein synthesis, protein damage including disruption of amino acid bonds and also their cross-linking, oxidation of membrane phospholipids, lipid peroxidation, disruption of membrane ion gradients, and depletion of cellular levels of ATP leading to cellular dysfunction. Mitochondria, having the highest turnover of oxygen, involving enzymes of the respiratory chain, are the specific targets of ROS. Out of the three ROS, $^\bullet OH$ is the most reactive and can immediately interact with any molecule in its vicinity and can remove electron, turning that molecule into a free radical, giving rise to chain reactions. $^\bullet OH$ specifically induces hydroxylation of deoxyguanosine in DNA forming 8-OH-dG, which can be a site for mutagenesis and, possibly, cancer. $O_2{}^{\bullet-}$ in comparison to $^\bullet OH$ is not that much reactive by itself but can initiate lipid peroxidation in its protonated form or can inactivate certain specific enzymes. H_2O_2 has low reactivity, is more stable, and hence can travel into the nucleus and can react with important components such as nucleic acids and nuclear proteins besides other cellular components such as lipids.

ROS are beneficial at physiological levels and assist in maintaining normal cell function; it is therefore essential that they be maintained at homeostatic levels. The beneficial roles of ROS include induction of host defense against a variety of pathogens. Defense against viral infections is carried out by potentiating RIG-like helicase-1 and mitochondrial antiviral signaling protein. ROS have also been implicated with mobilization of the ion transport system and a variety of inflammatory responses, including cardiovascular disease (CVD) and ischemic injury and also in hearing impairment due to ototoxicity by cisplatin and congenital deafness. Interestingly, people suffering from granulomatous disease, in which ROS production is reduced or absent, become more susceptible to infection by a range of pathogenic bacterial species.

Nature has endowed living systems with a number of protective mechanisms mostly involving enzymes whose roles are to scavenge excessive and unwanted ROS present in the system. ROS scavenging enzymes include catalase and peroxidase (for H_2O_2) and SOD that catalyze the dismutation of $O_2^{\bullet-}$ into oxygen and H_2O_2. Catalase converts H_2O_2 to oxygen and $2H_2O$. Also in the system, low-molecular-weight antioxidants such as ascorbic acid, α-tocopherol or vitamin E, and glutathione are present. In a recent article, it has been shown that mitochondria are the major source of MnSOD, and this enzyme constitutes an essential defense against $O_2^{\bullet-}$; further, this and other nitrogen species are responsible for various redox-related diseases and aging. A number of neurodegenerative diseases (NDs) associated with ROS include amyotrophic lateral sclerosis (ALS), Parkinson's disease, Alzheimer's disease (AD), and Huntington's disease (HD) (for more details, see [3]), as well as age-related macular degeneration, atherosclerosis, and various types of cancers.

In this book, attempts have been made to cover as much information about ROS as possible with the exception of their role in aging, which is still at the hypothesis stage.

For the readers' easy reading the contents of the chapter has been sectionalized below.

Section I includes Chapters 1 through 6 in which the introduction, detection, and production of ROS are described. In Chapter 5, special emphasis is given to phenolic compounds and vitamin E for their abilities to act as antioxidants; however, due to their overall low concentrations, doubt has been casted on their *in vivo* effectiveness. Chapter 6 highlights the role of lipoprotein-associated oxidative stress and suggests that its assessment in plasma may be used as the best biomarkers of oxidative stress.

Section II embraces Chapter 7 in which Carmen et al. have intricately addressed the importance of accumulation of iron in the brain resulting in the development of a group of neurodegenerative disorders. Several causative genes for neurodegeneration with brain iron accumulation (NBIA) have been identified, which are associated with Parkinsonism-related disorders. It is suggested that increased knowledge of NBIA genes and their functions should help to better understand the clinical picture, MRI findings, and disease mechanisms.

Section III discusses a number of NDs. In Chapter 8, Perry addresses one of the most important diseases, ALS, also known as motor neuron disease. Despite a large number of studies being carried out, its exact mechanism still remains unclear. One important finding linked with the disease is the mutation in SOD-1 (coding for Cu, Zn SOD), which may be responsible for the disease pathology. Chapter 9 presents the complexity of AD and its therapeutic strategies. HD, another ND, which is associated with mitochondrial dysfunction, has been addressed comprehensively in Chapter 10, and readers may gain considerable insight from it. In Chapter 11, epilepsy, another ND caused by ROS and reactive nitrogen species (RNS) associated with mitochondrial dysfunction, is described. Mitochondrial DNA and phospholipids have specifically been identified as the targets of ROS and RNS. Interestingly, certain psychological disorders have also been associated with ROS-induced brain injury, which is critically addressed in Chapter 12. In Chapter 13, studies on another frequently found neurological disease, multiple sclerosis, and the latest treatment regimen worthwhile to be studied are presented.

Section IV addresses autoimmune diseases caused by ROS. In Chapters 14 and 15, two important diseases are included: asthma, caused by the dysregulation of the oxidant–antioxidant balance, and autoimmune liver diseases, caused by oxidative stress from cigarette smoking, wherein the smoke contains a large number of toxic, carcinogenic, and mutagenic chemicals. Chapter 15 also includes other autoimmune diseases such as rheumatoid arthritis, multiple sclerosis, thyroid disease, and primary biliary cirrhosis. The use of immunosuppressants has been recommended as a possible regimen. Systemic lupus described in Chapter 27 is another ROS-activated autoimmune disease that can affect several organs, including the kidneys, skin, lungs, brain, and heart. Oxidative stress, leading to oxidative modifications of functional proteins, lipid, and DNA, ultimately results in the breakage of immune tolerance, apoptosis, necrosis, autophagy, and increased tissue damage.

Section V includes Chapters 16 through 19 that address a number of different types of cancers, including lung cancer, breast cancer, and melanoma. Detailed analyses of these cancers alongside their possible treatment regimens are intricately presented and deserve priority in the reading list.

Section VI discusses CVDs induced by ROS. They too are a highly important set of diseases. In Chapters 20 and 21, Johnson and Shimokawa, respectively, present their in-depth knowledge of this subject. CVDs remain the leading cause of human death and much information about their causes, mechanisms, and treatments is available. Mangge in Chapter 22 intricately presents the ROS-associated complex biochemical processes inducing inflammation as an important cause of CVDs. Chapter 23 explains the roles carotenoids play in preventing CVDs and how this agent can improve cardiovascular health. These authors recommend more studies to be carried out to minimize the risk of CVDs.

Section VII includes Chapters 24 through 39 addressing other human diseases induced by oxidative stress. These include sickle cell disease (Chapter 24), which is a group of complex genetic hemolytic disorders associated with high morbidity and mortality and which still remains to be fully understood. Nonalcoholic steatohepatitis (Chapter 25) addressed by Yamamoto is more commonly found in patients suffering from diabetes. ROS have been accepted as the main inducers of this disease in which liver fat deposition commonly prevails. In Chapter 26 on retinopathy, one of the several diabetic complications, the authors have ascertained that oxidative stress is the central factor involved in its pathology and the polyol pathway plays key roles in the production of ROS. Fibromyalgia, addressed by Fatima in Chapter 28, is one of the least understood human diseases caused by imbalance between malondialdehyde and SOD, and patients suffering from this disease sustain persistent and widespread pain and tingling sensations, muscle spasms, limb weakness, nerve pain, and muscle twitching. Unfortunately, the etiology of this disease is not well known, and only limited treatments are available. Chronic obstructive pulmonary disease, asthma, and pulmonary hypertension are shown, by Zuo in Chapter 29, to be induced by excessive ROS production. Further studies have been suggested to unravel the complete picture and successful treatment. Chapters 30 through 32 are dedicated to human fertility, and its awareness may assist those suffering from infertility. Chapter 33 presents a detailed study on the aging of human skin by ROS, and, more importantly, a good number of color figures have been added to enrich the chapter. Also, awareness has been included on how to avoid or minimize skin aging specially when induced by exogenous agents. Ataxia telangiectasia, an inborn genetic disease, has been shown, by Nakajima in Chapter 34, to be due to a mutation in the ATM gene, and the regulation of this gene expression is carried out by oxidative stress. It is intriguing to learn in Chapter 35 by Hong that oxidative stress can have profound effects on viral pathogenesis and on the host, and this is due to an imbalance between pro- and antioxidants leading to inflammatory response, viral replication, and apoptosis. Hargreaves in Chapter 36 highlights the importance of organophosphates, the highly toxic group of chemicals that can affect the nervous system by exerting its toxicity via oxidative stress. Conclusion is drawn on the basis of antioxidants attenuating some of the toxic effects. Chapters 37 through 39 address the roles of ROS in morphine addiction, exercise, nitric oxide, and epigenetics, respectively.

It is hoped that the material presented in this book will stimulate both experts and novice researchers in the field with excellent overviews of the current status of research and pointers to future research goals. Clinicians, nurses, carers, and families should also benefit from the information presented in handling and treating their specific patients. Also, the insights obtained should prove valuable for further understanding such a large number of diseases at the molecular level and allow the development of new biomarkers, novel diagnostic tools, and highly effective therapeutic drugs to treat patients suffering from these devastating diseases.

Shamim I. Ahmad, BSc, MSc, PhD
School of Science and Technology
Nottingham Trent University
Nottingham, United Kingdom

REFERENCES

1. Stuffins, C.B. and Weatherall, H., Determination of the peroxide value of oil and fat, *Analyst*, 70, 403, 1945.
2. Mann, P.J.G. and Quastel, J.H., Toxic effects of oxygen and hydrogen peroxide on brain metabolism, *Biochemistry*, 40, 139, 1946.
3. Ahmad, S.I. (ed.), *Neurodegenerative Diseases*, Springer-Verlag, New York, 2012.

Acknowledgments

The editor cordially acknowledges the various authors of this book for their contribution of the chapters with in-depth knowledge and highly skilled presentation. Without their input, it would not have been possible to produce a book on such an important subject. The editor also acknowledges the hard work, friendly approach, and patience of Taylor & Francis Group staff, especially Ashley Weinstein, Chuck Crumly, and Hilary LaFoe, for their efficient and highly professional handling of this publication.

Editor

Shamim I. Ahmad, after earning his master's degree in botany from Patna University, Bihar, India, and his PhD in molecular genetics from the University of Leicester, England, joined Nottingham Polytechnic as a grade 1 lecturer and was subsequently promoted to the senior lecturer post. Nottingham Polytechnic subsequently became Nottingham Trent University, where, after serving for about 35 years, he took early retirement, yet continued serving as a part-time senior lecturer. Dr. Ahmad is now spending much of his time producing/writing medical books. For more than three decades, he researched on different areas of molecular biology/genetics, including thymineless death in bacteria, genetic control of nucleotide catabolism, development of anti-AIDs drug, control of microbial infection of burns, phages of thermophilic bacteria, and microbial flora of Chernobyl after the accident at the nuclear power station. But his main interest, which started about 30 years ago, is DNA damage and repair specifically by near-ultraviolet light, especially through the photolysis of biological compounds, production of reactive oxygen species, and their implications on human health, including skin cancer. He is also investigating near-ultraviolet photolysis of nonbiological compounds such as 8-metoxypsoralen and mitomycin C and their importance in psoriasis treatment and in Fanconi's anemia. In collaboration with the University of Osaka, Japan, in his recent research publication, Dr. Ahmad and his colleagues were able to show that a number of naturally occurring enzymes were able to scavenge the reactive oxygen species.

In 2003, Dr. Ahmad received the prestigious "Asian Jewel Award" in Britain for "Excellence in Education." He has been editor for the following books published by Landes Bioscience/Springer: *Molecular Mechanisms of Fanconi Anemia* (2006), of *Xeroderma Pigmentosum* (2009), of *Cockayne Syndrome* (2009), of *Ataxia Telangiectasia* (2009), *Diseases of DNA Repair* (2010), *Neurodegenerative Diseases* (2012), *Diabetes: An Old Disease, a New Insight* (2013), and *Obesity: A Practical Guide* (2016).

Contributors

Oluwasesan Adegoke
Research Institute of Green Science and
    Technology
Shizuoka University
Shizuoka, Japan

and

Department of Chemistry
University of Pretoria
Pretoria, South Africa

Ashok Agarwal
American Center for Reproductive Medicine
Cleveland Clinic
Cleveland, Ohio

and

Department of Physiology
University of Health Sciences
Lahore, Pakistan

Ashish Aggarwal
Department of Biochemistry
Panjab University
and
Centre of Excellence
Department of Internal Medicine
Postgraduate Institute of Medical Education
    and Research
Chandigarh, India

Gulfam Ahmad
American Center for Reproductive Medicine
Cleveland Clinic
Cleveland, Ohio

and

Department of Physiology
University of Health Sciences
Lahore, Pakistan

Shamim I. Ahmad
School of Science and Technology
Nottingham Trent University
Nottingham, United Kingdom

Sheikh F. Ahmad
Department of Pharmacology and Toxicology
King Saud University
Riyadh, Saudi Arabia

Haseeb Ahsan
Department of Biochemistry
Faculty of Dentistry
Jamia Millia Islamia
New Delhi, India

Anami Ahuja
Department of Biotechnology
Meerut Institute of Engineering and
    Technology
Meerut, India
and
Division of Regenerative Medicine
National Innovative Research Academy
Meerut, India

Mohammed M. Al-Harbi
Department of Pharmacology and Toxicology
King Saud University
Riyadh, Saudi Arabia

Finley J. Allgaier
Department of Biochemistry
University of California
Riverside, California

Eduardo Alves de Almeida
Department of Chemistry and Environmental
    Sciences
Sao Paulo State University
Sao Jose do Rio Preto, Brazil

Riccardo Amorati
Dipartimento di Chimica "G. Ciamician"
Università di Bologna
Bologna, Italy

Ana Carolina de Andrade
Laboratory of Inflammatory Mediators
State University of West Paraná
Unioeste, Brazil

António Araújo
Medical Oncology Service of Centro
    Hospitalar do Porto (CHP)
Unit of Oncobiology Research
    Unit for Multidisciplinar Research In
    Biomedicine (UNIO-UMIB)
and
Abel Salazar Institute for the Biomedical
    Sciences (ICBAS)
University of Porto
Porto, Portugal

Luiz Roberto G. Bechara
School of Physical Education and Sport
University of São Paulo
São Paulo, Brazil

Edis Belini Junior
Department of Biology
Sao Paulo State University
Sao Jose do Rio Preto, Brazil

R. Boguen
Centre of Reproductive Biotechnology
University of La Frontera
Temuco, Chile

Claudia Regina Bonini-Domingos
Department of Biology
Sao Paulo State University
Sao Jose do Rio Preto, Brazil

Tanima Bose
Department of Molecular Physiology
Leibniz Institute for Neurobiology
Magdeburg, Germany

Michaël Boyer-Guittaut
Laboratoire de Biochimie
EA3922 Research Team
Université Bourgogne Franche-Comté
Besançon, France

Abhishek Chandra
Feil Family Brain and Mind Research Institute
Weill Cornell Medical College
New York, New York

Chia-Chen Chuang
Radiologic Sciences and Respiratory Therapy
    Division
Wexner Medical Center
and
Interdisciplinary Biophysics Graduate Program
The Ohio State University
Columbus, Ohio

Ana Coelho
Molecular Oncology and Viral Pathology
    Group–CI
Portuguese Institute of Oncology
and
Unit of Oncobiology Research
Unit for Multidisciplinar Research In
    Biomedicine (UNIO-UMIB)
and
LPCC Research Department
Portuguese League Against Cancer (NRNorte)
and
Faculty of Medicine
University of Porto
Porto, Portugal

Alejandra Darling
Centro de Investigación Príncipe Felipe
Valencia, Spain

Kapil Dev
Department of Cytokinetics
Institute of Biophysics
Brno, Czech Republic

Stefan S. Du Plessis
Division of Medical Physiology
Stellenbosch University
Stellenbosch, South Africa
and
Center for Reproductive Medicine
Cleveland Clinic
Cleveland, Ohio

Abraham Eisenstark
Division of Biological Sciences
University of Missouri
Columbia, Missouri

Ahmed M. El-Sherbeeny
Industrial Engineering Department
King Saud University
Riyadh, Saudi Arabia

Carmen Espinós
Hospital Sant Joan de Déu
Barcelona, Spain
and
Centro de Investigación Biomédica en Red
Valencia, Spain

Ghizal Fatima
Department of Biochemistry
King George Medical University
Lucknow, India

Jaroslava Folbergrová
Department of Developmental Epileptology
Institute of Physiology of the Czech Academy
  of Sciences
Prague, Czech Republic

Patricia B.C. Forbes
Department of Chemistry
University of Pretoria
Pretoria, South Africa

Mario C. Foti
Consiglio Nazionale delle Ricerche
Istituto di Chimica Biomolecolare
Catania, Italy
Istituto di Chimica Biomolecolare del CNR
Catania, Italy

Maria Alessandra Gammone
Department of Medical Oral Biotechnological
  Science
University G.d'Annunzio
Chieti, Italy

Andrzej Głąbiński
Department of Neurology and Stroke
Medical University of Lodz
Lodz, Poland

Mónica Gomes
Molecular Oncology and Viral Pathology
  Group–CI
Portuguese Institute of Oncology
and
Unit of Oncobiology Research
Unit for Multidisciplinar Research In
  Biomedicine (UNIO-UMIB)
and
Abel Salazar Institute for the Biomedical
  Sciences (ICBAS)
University of Porto
and
LPCC Research Department
Portuguese League Against Cancer (NRNorte)
Porto, Portugal

Johanna M. Gostner
Division of Medical Biochemistry
Biocenter of Innsbruck Medical University
Innsbruck, Austria

Alan J. Hargreaves
School of Science and Technology
Nottingham Trent University
Nottingham, United Kingdom

Eric Hervouet
Laboratoire de Biochimie
EA3922 Research Team
Université Bourgogne Franche-Comté
Besançon, France

James Hoerter
Department of Biological Sciences
Ferris State University
Big Rapids, Michigan

Jiann-Ruey Hong
Laboratory of Molecular Virology and
  Biotechnology
Institute of Biotechnology
National Cheng Kung University
Tainan, Taiwan, Republic of China

Eric L. Johnson
Department of Family and Community
    Medicine
University of North Dakota
and
School of Medicine and Health Sciences
Altru Diabetes Center
Grand Forks, North Dakota

Ashu Johri
Feil Family Brain and Mind Research Institute
Weill Cornell Medical College
New York, New York

Ma Jun
Binzhou medical University
Yantai, People's Republic of China

Dominika Książek-Winiarek
Department of Neurology and Stroke
Medical University of Lodz
Lodz, Poland

Ngoc-Anh Le
Biomarker Core Laboratory
Atlanta Research and Education Foundation
Atlanta, Georgia

Hyoung-gon Lee
Department of Pathology
Case Western Reserve University
Cleveland, Ohio

Feng Liu-Smith
Department of Epidemiology
and
Department of Medicine
University of California, Irvine
Irvine, California

Vincenzo Lupo
Hospital Sant Joan de Déu
Barcelona, Spain
and
Centro de Investigación Biomédica en Red
Valencia, Spain

Nidhi Mahajan
Department of Biochemistry
Panjab University
Chandigarh, India

Harald Mangge
Clinical Institute of Medical and Chemical
    Laboratory Diagnosis
Medical University of Graz
Graz, Austria

Rui Medeiros
Molecular Oncology and Viral Pathology
    Group–CI
Portuguese Institute of Oncology
and
Abel Salazar Institute for the Biomedical
    Sciences
University of Porto
and
LPCC Research Department
Portuguese League Against Cancer (NRNorte)
and
Faculty of Health Sciences
Fernando Pessoa University
Porto, Portugal

Elżbieta Miller
Department of Physical Medicine
Medical University of Lodz
Lodz, Poland
and
Neurorehabilitation Ward
III General Hospital in Lodz
Lodz, Poland

Nobutaka Motohashi
Department of Neuropsychiatry
University of Yamanashi
Chuo, Japan

Ahmed Nadeem
Department of Pharmacology and Toxicology
King Saud University
Riyadh, Saudi Arabia

Tetsuo Nakajima
Research Center for Radiation Protection
National Institute of Radiological Sciences
Chiba, Japan

Swapan K. Nath
Arthritis and Clinical Immunology Research
    Program
Oklahoma Medical Research Foundation
Oklahoma City, Oklahoma

Akihiko Nunomura
Department of Neuropsychiatry
University of Yamanashi
Chuo, Japan

Mohammad Shamsul Ola
Department of Biochemistry
King Saud University
Riyadh, Saudi Arabia

Ibukun P. Oyeyipo
Division of Medical Physiology
Stellenbosch University
Stellenbosch, South Africa
and
Department of Physiology
Osun State University
Osogbo, Nigeria

Carolina Panis
Laboratory of Inflammatory Mediators
State University of West Paraná
Unioeste, Brazil

Bishnuhari Paudyal
Department of Radiology
Thomas Jefferson University
Philadelphia, Pennsylvania

Belén Pérez-Dueñas
Centro de Investigación Biomédica en Red
and
Centro de Investigación Príncipe Felipe
Valencia, Spain

George Perry
Department of Pathology
Case Western Reserve University
Cleveland, Ohio
and
Neurosciences Institute
and
Department of Biology
University of Texas at San Antonio
San Antonio, Texas

J. Jefferson P. Perry
Department of Biochemistry
University of California, Riverside
Riverside, California

Bruno Ricardo Barreto Pires
Laboratory of Stem Cells
National Cancer Institute
Inca, Brazil

Lenora Ann Pluchino
Department of Biomedical and Diagnostic
    Sciences
University of Tennessee
Knoxville, Tennessee

Zheng Qiusheng
Binzhou Medical University
Yantai, People's Republic of China

Paulo R. Ramires
School of Physical Education and Sport
University of São Paulo
São Paulo, Brazil

Vibha Rani
Department of Biotechnology
Jaypee Institute of Information Technology
Noida, India

Mohammad Latif Reshi
Laboratory of Molecular Virology and
    Biotechnology
and
Department of Life Sciences
National Cheng Kung University
Tainan, Taiwan, Republic of China

Kimio Satoh
Department of Cardiovascular Medicine
Tohoku University
Sendai, Japan

Dilip Shah
Center for Translational Medicine
Thomas Jefferson University
Philadelphia, Pennsylvania

Hiroaki Shimokawa
Department of Cardiovascular Medicine
Tohoku University
Sendai, Japan

Danilo Grünig Humberto da Silva
Department of Chemistry and Environmental
    Sciences
Sao Paulo State University
Sao Jose do Rio Preto, Brazil

Mateus Batista Silva
Laboratory of Inflammatory Mediators
State University of West Paraná
Unioeste, Brazil

Bongekile Skosana
Division of Medical Physiology
Stellenbosch University
Stellenbosch, South Africa

Mansoor Ali Syed
Department of Pediatrics
Drexel University
Philadelphia, Pennsylvania

Akinobu Takaki
Department of Gastroenterology and
    Hepatology
Okayama University
Okayama, Japan

Toshio Tamaoki
Department of Neuropsychiatry
University of Yamanashi
Chuo, Japan

Leonardo Y. Tanaka
School of Physical Education and Sport
and
Vascular Biology Laboratory
University of São Paulo
São Paulo, Brazil

Ana Luísa Teixeira
Molecular Oncology and Viral Pathology
    Group–CI
Portuguese Institute of Oncology
and
Unit of Oncobiology Research
Unit for Multidisciplinar Research In
    Biomedicine (UNIO-UMIB)
and
Abel Salazar Institute for the Biomedical
    Sciences (ICBAS)
University of Porto
and
LPCC Research Department
Portuguese League Against Cancer (NRNorte)
Porto, Portugal

Cristina Tello
Hospital Sant Joan de Déu
Barcelona, Spain

F. Treulen
Centre of Reproductive Biotechnology
University of La Frontera
Temuco, Chile

Pankaj K. Tyagi
Department of Biotechnology
Meerut Institute of Engineering and
    Technology
Meerut, India

Daisuke Uchida
Department of Gastroenterology and
    Hepatology
Okayama University
Okayama, Japan

P. Uribe
Centre of Reproductive Biotechnology
University of La Frontera
Temuco, Chile

Vanessa Jacob Victorino
Faculty of Medicine
University of São Paulo
São Paulo, Brazil

J.V. Villegas
Department of Internal Medicine
University of La Frontera
Temuco, Chile

Hwa-Chain Robert Wang
Department of Biomedical and Diagnostic
    Sciences
University of Tennessee
Knoxville, Tennessee

Tomasz Włodarczyk
Ophthalmology Department
Warminski Hospital
Bydgoszcz, Poland

Kazuhide Yamamoto
Department of Gastroenterology and
    Hepatology
Okayama University
Okayama, Japan

Jianhua Zhang
Department of Pathology
University of Alabama at Birmingham
and
Department of Veterans Affairs
Birmingham VA Medical Center
Birmingham, Alabama

Tingyang Zhou
Radiologic Sciences and Respiratory Therapy
    Division
Wexner Medical Center
and
Interdisciplinary Biophysics Graduate
    Program
The Ohio State University
Columbus, Ohio

Xiongwei Zhu
Department of Pathology
Case Western Reserve University
Cleveland, Ohio

Li Zuo
Radiologic Sciences and Respiratory Therapy
    Division
Wexner Medical Center
and
Interdisciplinary Biophysics Graduate
    Program
The Ohio State University
Columbus, Ohio

Section I

Introduction, Detection, and Production of ROS

1 Introduction to Reactive Oxygen Species
Emphasizing Their Importance in the Male Reproductive System

Bongekile Skosana, Ibukun P. Oyeyipo,
and Stefan S. Du Plessis

CONTENTS

ABSTRACT

Oxygen is life sustaining, but its metabolism by organisms is not without consequences. Aerobic metabolism provides energy predominantly through mitochondrial oxidative phosphorylation. This process leads to the formation of by-products known as oxygen free radicals, which are toxic in high quantities. Oxygen toxicity is an intrinsic challenge to aerobic life that also affects spermatozoa, cells essential for the propagation of a species.

Free radicals are "molecules with one or more unpaired electron(s)" (Halliwell and Gutteridge, 1990). They are unstable, reactive molecules and are able to oxidize nearby molecules to gain an electron, damaging adjacent cells. At physiological levels, free radicals are beneficial and assist in maintaining normal cell function. It is therefore essential that they be maintained at homeostatic levels.

This chapter introduces free radicals present within the male reproductive system and their formation, sources, and antioxidant systems in place to maintain them within physiological levels.

1.1 INTRODUCTION

Oxygen is essential for life on Earth. Our ability to metabolize and utilize oxygen is dependent on aerobic processes that use it to regulate the oxidation of carbon-containing molecules, resulting in energy production (Tvrda et al., 2011). The energy produced by aerobic metabolism is predominantly created through mitochondrial oxidative phosphorylation, a process resulting in the reduction of oxygen to water. This metabolic process leads to the formation of by-products known as oxygen free radicals (Tremellen, 2008).

Free radicals are "molecules with one or more unpaired electron(s)" (Halliwell and Gutteridge, 1990). They are unstable and extremely reactive molecules with the ability to oxidize and modify nearby molecules to gain an electron in order to return to a ground state (Agarwal et al., 2006a). These oxidized molecules become free radicals themselves, perpetuating a cycle that results in the destruction of neighboring cells and tissues.

At physiological levels, free radicals are beneficial and assist in the maintenance of normal cell function. However, these molecules have the potential to be harmful to cell survival. Aerobic cells are therefore continually confronted by an "oxygen paradox" (Sies, 1993), necessitating the need for the regulation of these molecules at a homeostatic level.

Free radicals come in two main forms: reactive oxygen species (ROS) and reactive nitrogen species (RNS) (Sharma and Agarwal, 1996; Tvrda et al., 2011). The superoxide anion (O_2^-), hydroxyl radical (OH^-), peroxyl radical (HO_2^-), and hydrogen peroxide (H_2O_2) are all subsets of ROS (Sikka, 2001). RNS include nitric oxide (NO), nitrous oxide (N_2O), peroxynitrite ($ONOO^-$), peroxynitrous acid (HNO_3), and the nitroxyl anion (HNO) (Sharma and Agarwal, 1996; Tvrda et al., 2011). RNS are considered to be a subdivision of ROS (Sikka, 2001).

Antioxidants are substances that inhibit oxidation. They scavenge free radicals and help maintain them at physiological levels, while staving off the pathological effects that may occur. When ROS levels are in excess and antioxidants are under strain to control them, an imbalance occurs, resulting in a state known as oxidative stress (OS) (Agarwal et al., 2004). OS results from a reduction in antioxidant capacity, an increase in free radical concentrations, or a combination of the two (Kothari et al., 2010). In this state, the production of peroxides is amplified. These molecules then have the ability to damage all components of a cell, including lipids, proteins, and DNA (Schafer and Buettner, 2001).

ROS leads to the pathogenesis of numerous diseases and conditions. However, the effects exerted by ROS are dependent on the type and amount of ROS involved (De Lamirande and Gagnon, 1992a,b; Aitken et al., 1993a; Griveau et al., 1995) and the length and the moment of exposure to ROS (De Lamirande and Gagnon, 1992a).

The negative consequences brought on by the presence of ROS do not negate the positive roles they play within an organism. In fact, at controlled concentrations, ROS have been shown to act as regulatory mediators involved in signal transduction processes (Griveau and Le Lannou, 1997). This is among several other physiological roles that free radicals play in regulating reproductive processes.

This chapter presents basic information on free radicals, including their production, regulation, and mechanisms of action.

1.2 FORMATION OF ROS

Diatomic oxygen (O_2), known more simply as oxygen, is naturally found in a ground state. The two unpaired, complementary electrons in its outer shell spin in opposite directions, stabilizing the atom. This makes oxygen inert, preventing it from reacting with and damaging surrounding structures. When O_2 is metabolized, a reduction reaction (used for cellular respiration) occurs, resulting in the creation of the O_2^- radical. It is formed by the transfer of an electron to the O_2 atom as it moves from complex I to III of the electron transport chain (Aitken et al., 1989; Koppers et al., 2008). O_2^- is the main ROS formed by respiring cells (Griveau and Le Lannou, 1997; Agarwal and Said, 2005). It is somewhat unreactive but, in the presence of H^+, it is converted to form the more reactive and membrane permeable free radical H_2O_2. This interaction results in the spontaneous or enzyme-catalyzed dismutation of $O_2^{\bullet-}$ into H_2O_2 (Halliwell and Cross, 1992). The reaction is catalyzed by superoxide dismutase (SOD), an enzyme commonly found on spermatozoa membranes, in the epididymis and in seminal plasma (Storey, 1997; Vernet et al., 2004).

The interaction of H_2O_2 with transitional metals results in its decomposition via the reaction of Fenton. This reaction occurs in two steps: First, $O_2^{\bullet-}$ reacts with ferric iron (Fe^{3+}), a reaction that yields ferrous iron (Fe^{2+}) and oxygen. Second, Fe^{2+}, in the presence of H_2O_2, is reduced to Fe^{3+} with H_2O_2 itself being converted into OH^-, an extremely reactive free radical that has been shown to instantly react with surrounding biological molecules (Griveau and Le Lannou, 1997). More infamously, it is known to be a potent initiator of lipid peroxidation (Halliwel and Gutteridge, 2004). The hydroxyl reaction may also be created through the interaction of H_2O_2 with O_2^- in the presence of Fe^{2+} using the Haber–Weiss reaction.

Reaction of Fenton:

$$Fe^{2+} + H_2O_2 \rightarrow Fe^{3+} + {}^\circ OH + OH^-$$

$$H_2O_2 + Fe^{3+} \rightarrow Fe^{2+} + H^+ + {}^\circ OOH$$

Haber–Weiss balance:

$$H_2O_2 + O_2^- \xrightarrow{Fe^{2+}} O_2 + {}^\circ OH + {}^- OH$$

The reaction of Fenton readily occurs within plasma as both copper and iron exist in a free state within the fluid (Kwenang et al., 1987), which freely react with O_2^- anions and H_2O_2 to propagate OS (Griveau and Le Lannou, 1997).

RNS are also created through the interaction of cellular compounds with ROS. NO production is catalyzed by nitric oxide synthase (NOS) isoenzymes (eNOS, nNOS, iNOS), some of which have been localized within tissues of the reproductive system. With NOS as a catalyst, nicotinamide adenine dinucleotide phosphate (NADPH) initiates a reaction between L-arginine and O_2, resulting in a redox reaction that produces NO and forms L-citrulline as a by-product. NO also reacts with O_2^- to form $ONOO^-$, which is highly toxic (Sikka, 2001).

$$\text{L-Arginine} + \text{NADPH} \xrightarrow{NOS} \text{L-Citrulline} + NO$$

$$O_2^- + NO \rightarrow ONOO^-$$

The presence of NO may contribute to OS as its reaction with O_2^- is three times faster than SOD's reaction with O_2^-. When NO levels increase by only nanomolar increments, it may compete with SOD in the scavenging of O_2^-. However, homeostatic levels of NO have demonstrated a beneficial role in the transduction of signals required in motility, capacitation, and acrosome reaction (AR), especially in spermatozoa (Zini et al., 1995; Harrero and Gagnon, 2001).

1.3 NATURE AND SOURCE OF ROS

As ROS originating within (endogenous) the reproductive system and from external factors (Kothari et al., 2010) play important roles in this system, this section will describe ROS with a focus on human sperm.

1.4 ENDOGENOUS (SEMINAL) SOURCES OF ROS

Spermatozoa travel within a nourishing fluid medium called seminal fluid (or seminal plasma) as they migrate from the epididymis to the female genital tract. The constituents of this fluid originate from secretions produced by the testes, prostate, seminal vesicles, and Cowper's glands. Seminal fluid also contains various other cell types: immature sperm cells, leukocytes, macrophages, and Sertoli cells (Smith et al., 1989; Fisher and Aitken, 1997). Of these cell types, immature spermatozoa and leukocytes are the main sources of endogenous ROS production (Kothari et al., 2010).

1.4.1 IMMATURE SPERMATOZOA

Sperm undergo a vital maturation step within the epididymis. The cytoplasm within the head of the sperm migrates from the proximal to the distal end and is ultimately extruded from spermatozoa during or immediately after ejaculation (Kato et al., 1996). Impaired sperm maturation does not allow for proper extrusion and may lead to the retention of cytoplasmic droplets. This excess residual cytoplasm contains high levels of ROS-generating enzymes as it traps enzymes within the head of the sperm that would otherwise not be present. Enzymes such as G6PD, lactic acid dehydrogenase, and creatine phosphokinase can fuel ROS production by generating NADPH through the hexose monophosphate shunt (Casano et al., 1991; Aitken et al., 1994; Huszar and Vigue, 1994). Therefore, spermatozoa with excess residual cytoplasm are said to be immature and functionally defective (Huszar et al., 1997). Surpluses of these enzymes create NADPH, thereby activating the NADPH system, leading to excessive ROS production and sperm damage. Semen with high concentrations of immature spermatozoa has been shown to have lower sperm quality (Kefer et al., 2009). As concentrations increase, so does the number of mature spermatozoa with damaged DNA (Gil-Guzman et al., 2001).

1.4.2 LEUKOCYTES AND ROS

Leukocyte populations within semen consist mainly of polymorphonuclear (PMN) leukocytes and macrophages (Wolff, 1995). These originate from the prostate gland and seminal vesicles (Kothari et al., 2010). Leukocytes become activated spontaneously or during disease states. When activated by a disease state or infection, leukocytes induce high levels of ROS as a cytotoxic reaction against the invading material. In this state, leukocytes facilitate ROS production through the NADPH system. NADPH oxidase then catalyzes the formation of free radicals (Du Plessis et al., 2015) and this results in an oxidative burst and overproduction of ROS (Ochsendorf, 1999), thereby killing nearby pathogens and protecting the male reproductive tract from invasion.

Infections result in the activation and chemotaxis of more leukocytes to the site of infection. This may result in excessive ROS production, which is detrimental to spermatozoa. Leukocytes are a more predominant source of ROS. They have been shown to produce a 1000-fold more ROS in semen than immature spermatozoa (Du Plessis et al., 2015).

According to the World Health Organization (WHO), seminal leukocyte concentrations of 1×10^6 per milliliter constitute leukocytospermia (Cooper et al., 2010; WHO, 2010). The clinical significance of leukocytospermia is yet to be fully defined as there are conflicting data on its effect on fertilizing potential. It is, however, considered to be a marker of a systemic or urological infection and of possible spermatozoal impairments (Kefer et al., 2009).

1.4.3 SERTOLI CELLS AND ROS

Sertoli cells have the ability to produce ROS. Scavestrogens are free radical scavengers derived from the structure of 17α-estradiol, which have the ability to inhibit ROS production and iron-induced cellular injuries in vitro. Scavestrogens J811 and J861 have been demonstrated to inhibit ROS production by Sertoli cells, leading to the hypothesis that Sertoli cells mediate their spermatogenic functions through the production of ROS (Hipler et al., 2000).

1.4.4 VARICOCELE AND ROS

Varicocele is described as the "uncharacteristic dilatation and tortuosity of veins in the pampiniform plexus around the human male spermatic cord" (Agarwal et al., 2009). It has consistently been shown to be associated with pathological levels of ROS and impaired sperm function (Hendin et al., 1999). Varicocele appears to increase ROS production through increasing levels of NO and xanthine oxidase, a source for O_2^- production (Mitropoulos et al., 1996). Higher NO levels were observed in infertile men with varicocele when compared to fertile and infertile men without varicocele (Mehraban et al., 2009). Increased ROS levels in men have also been linked to higher grades of varicocele (Allamaneni et al., 2004). Varicocele patients exhibit markers of lipid peroxidation, DNA damage, and OS (Saleh et al., 2003; Agarwal et al., 2006). Patients who have undergone varicocelectomy show increased total antioxidant capacity and a reduction in ROS levels, leading to an improvement in sperm parameters (Mostafa et al., 2001).

1.5 EXOGENOUS SOURCES OF ROS

Various external factors influence ROS production within the male reproductive system. These include chemical compounds (e.g., industrial compounds), lifestyle factors (e.g., smoking and alcohol consumption), and bodily insults (e.g., spinal cord injury) (Kothari et al., 2010).

1.5.1 INDUSTRIAL POLLUTANTS

Industrial waste and by-products impact male fertility directly and indirectly. Manufacturing by-products in the environment such as lead, mercury, sulfur dioxide, and cadmium (Slivkova et al., 2009) and industrial waste products such as phthalate (commonly found in plastics and cosmetic products), polychlorinated biphenyls, dioxins, and nonylphenol (Chitra et al., 2002; Latchoumycandane et al., 2003; Krishnamoorthy et al., 2007) have been demonstrated to stimulate the production of ROS in male testicular tissue, specifically O_2^- and H_2O_2. This pathological increase in ROS levels may lead to impaired spermatogenesis and DNA damage (Kothari et al., 2010). Industrial workers constantly exposed to heavy metals are prone to reductions in sperm volume, concentration, and quality (Jurasović et al., 2004).

1.5.2 CIGARETTE SMOKING

There are more than 4000 compounds in a cigarette. These include alkaloids, inorganic molecules, and nitrosamines (Lavranos et al., 2012). Smoking increases leukocyte concentrations within seminal plasma, resulting in an increased ROS generation (Saleh et al., 2002). Specific constituents

of cigarettes, and especially nicotine, have been correlated with increases in free radical production (Traber et al., 2000). These compounds disturb the homeostatic balance between ROS and
antioxidants in semen by increasing ROS generation and reducing seminal plasma antioxidants,
placing spermatozoa at risk of oxidative damage (Lavranos et al., 2012). This has been seen in seminal plasma samples of smokers, where a substantial elevation of the oxidative damage biomarker
8-OHdG was observed (Estevez, 2002). The spermatozoa of cigarette smokers are considerably
more sensitive to DNA denaturation and have a higher incidence of DNA strand breaks (Jarow,
2003). Prolonged exposure to cigarette smoke brings about spermatozoal DNA damage and apoptosis, which may lead to infertility. Smoking has been correlated with lower sperm parameters,
specifically motility, morphology, and concentration (Vine, 1996; Saleh et al., 2002).

1.5.3 ALCOHOL CONSUMPTION AND ROS

Acetaldehyde is a breakdown product of ethanol metabolism. It interacts with proteins and lipids
to induce ROS production, while also interfering with the body's antioxidant capacity. Ethanol-
induced ROS production, in excess, may result in the molecular degradation of proteins, lipids, and
DNA. This has been shown to lead to a decrease in normal spermatozoa (Agarwal and Prabakaran,
2005; Saalu, 2010).

1.5.4 SPINAL CORD INJURY

Of the men who have incurred spinal cord injuries, more than 90% have subsequently become infertile (Linsenmeyer and Perkash, 1991). They have also been found to have excessively high counts
of PMN leukocytes within their semen, leading to a mechanism of ROS-induced sperm damage to
be postulated (Sharma and Agarwal, 1996). These pathological ROS concentrations are associated
with reductions in sperm motility and morphology.

1.5.5 RADIATION

Radiation is a natural source of energy. It was found to have significant clinical implications since
radiation emitted from mobile phones was linked to increased ROS production in seminal plasma
and impairments in sperm function (Agarwal et al., 2005, 2008). Radio frequency electromagnetic
waves may disturb the flow of electrons along the inner membranes of cells due to an abundance of
charged molecules within the cytoplasm, causing a disruption to normal cellular function (Lavranos
et al., 2012). Any resultant DNA damage may cause a further decline in concentration, motility, and
viability of spermatozoa, depending on the length of exposure to radiation (De Iuliis et al., 2009).

1.6   REGULATION OF ROS LEVELS AND THEIR MEASUREMENT

The effects of endogenous and exogenous ROS are limited and controlled by regulatory systems
within spermatozoa and seminal plasma. To thwart and encourage the physiological effects of ROS,
defense mechanisms are required to control the production and metabolism of ROS, maintaining
their levels within certain ranges. These mechanisms maintain a homeostatic ROS balance through
the collaboration of various systems, each with their own localization and targets. These reduce the
damaging effects of ROS by intervening in the various steps of the peroxidative process (Nissen
and Kreysel, 1983).

Spermatozoa and seminal plasma contain antioxidants that defend the body against oxidation-
induced damage. Antioxidants are "compounds or enzymes that scavenge and inhibit the formation
of ROS, or oppose their actions" (Sikka et al., 1995). When ROS is elevated to pathological levels,
antioxidants scavenge them and protect cells using three methods: prevention, interception, and
repair (Agarwal et al., 2004).

Antioxidants can be divided into two main categories:

1. Enzymatic—SOD, catalase (CAT), and glutathione peroxidases (GPx).
2. Nonenzymatic—Vitamins A, C, and E; carotenoids; glutathione; pyruvate; albumin; uric acid; ascorbate; taurine; and hypotaurine (Agarwal and Prabakaran, 2005; Agarwal et al., 2007).

Spermatozoa have restricted antioxidant defense capabilities due to their limited cytoplasmic volume and low antioxidant levels. Mammalian sperm largely employ enzymatic antioxidants, which are predominantly concentrated in the midpiece. The plasma membrane provides some assistance with nonenzymatic antioxidants, which provide the sperm with a preventive role against ROS damage.

Seminal plasma is the essential protectant of spermatozoa, providing both enzymatic and nonenzymatic antioxidants for protection against free radical formation. These antioxidants scavenge ROS produced by leukocytes and abnormal spermatozoa to improve semen quality, prevent DNA damage, inhibit premature sperm maturation, and reduce cryodamage in assisted reproduction (Agarwal et al., 2007).

1.7 ENZYMATIC ANTIOXIDANTS

1.7.1 SUPEROXIDE DISMUTASE

Superoxide dismutase are "metal-containing enzymes that catalyze the conversion of two O_2^- anions into O_2 and H_2O_2," protecting spermatozoa from oxygen toxicity and lipid peroxidation (Alvarez et al., 1987). Three types of SODs reside in different compartments within a cell: Mn-SOD in mitochondrial matrix, Cu/Zn-SOD in the cytosol and intermembrane space, and secretory extracellular SOD (ECSOD) (Alvarez and Storey, 1989; Peeker et al., 1997; Dandekar et al., 2002). The dimeric SOD, Cu/Zn-SOD, is the dominant SOD enzyme in both sperm and seminal plasma (Alvarez and Storey, 1989).

1.7.2 CATALASE

Catalase is an enzyme mainly located within peroxisomes. Its function is to catalyze the degradation of H_2O_2 into O_2 and water. Therefore, it works together with SOD to remove O_2^-, decreasing lipid peroxidation (Sikka et al., 1995).

1.7.3 GLUTATHIONE PEROXIDASES

This is a family of enzymes with varying properties that are dependent on the tissue they are found in. Their main function is to remove peroxyl radicals from peroxides, including H_2O_2 (Sikka et al., 1995). Thus, they catalyze the reduction of lipid peroxides and H_2O_2 using reduced glutathione (Dandekar et al., 2002). GPx1, the classic intracellular glutathione peroxidase, can be located in sperm and within the male genital tract. A direct correlation has been made between this enzyme and sperm motility (Foresta et al., 2002). Its inhibition leads to sperm immobilization and an inability to undergo acrosome reaction (Alvarez and Storey, 1989). Phospholipid hydroperoxide glutathione peroxidase (PHGPx), or GPx4, is a selenoprotein vastly expressed in testicular tissue. A strong relationship has been established between GPx4 and male fertility (Idriss et al., 2008).

Catalase activity is weak in spermatozoa but is primarily present in seminal fluid. SOD and GPx activity appear to play the essential role in sperm survival (Sikka et al., 1995). Other enzymes (glutathione-S-transferases, heme oxygenase-1, ceruplasmin) have also been implicated in the enzymatic control of ROS and their products (Idriss et al., 2008; Tremellen, 2008).

1.8 NONENZYMATIC ANTIOXIDANTS

Free radical scavengers, also present in spermatozoa and seminal plasma, neutralize ROS and retard the extent of oxidation reactions (Griveau and Le Lannou, 1997).

1.8.1 GLUTATHIONE

Glutathione is a non-thiol protein and is the most abundant protein of this kind within mammalian cells (Irvine, 1996). It has a cysteine subunit with a sulfhydryl (SH) group that instantly scavenges free radicals. SH is oxidized in this process, forming glutathione disulfide (GSSG), which is then regenerated through its reduction by glutathione reductase (Nistiar, 2009). This cycle buffers high levels of ROS to protect the body against their damaging effects.

1.8.2 VITAMINS

Vitamins A, C, and E are valuable antioxidants. Vitamin C is a "water-soluble, chain-breaking antioxidant" that counteracts O_2^-, H_2O_2, and hydroxyl radicals (Agarwal et al., 2004) and singlet oxygen. It also prevents spermatozoa from agglutinating. Its actions result in the prevention of lipid peroxidation and a reduction in hydrogen peroxide–induced DNA damage (Argawal et al., 2007).

Vitamin E, also a chain-breaking antioxidant, is found in spermatozoal membranes and has a dose-dependent effect (Hull et al., 2000). Its position allows it to protect sperm cell membranes from oxidative damage, where it traps and scavenges O_2^-, H_2O_2, and hydroxyl radicals. It also counteracts lipid radicals (Chow, 1991). Vitamin E is recycled by vitamin C and coenzyme Q-10.

1.8.3 COENZYME Q-10

Coenzyme Q-10, located in the sperm midpiece related to low-density lipoproteins, guards sperm against peroxidative damage (Frei et al., 1990). As mentioned, it recycles vitamin E, but also inhibits its pro-oxidant capacity (Karbownik et al., 2001). Furthermore, it performs an energy-promoting function, where it assists to increase sperm motility (Lewin and Lavon, 1997).

Other substances aid in the protection of sperm against OS by their presence in the reproductive system. Resveratrol is an exogenous, powerful lipid-soluble antioxidant, demonstrated to aid in the reduction of lipid peroxidation (Sarlós et al., 2002). The presence of some endogenous compounds does not prevent against the actions of ROS or decrease their production but reduce the risk of the development of OS (Tvrda et al., 2011). These are uric acid (Ciereszko et al., 1991), bilirubin (Sedlak et al., 2009), albumin (Roche et al., 2008), and taurine or hypotaurine (Aruoma et al., 1988). The actions of these antioxidants are summarized in Table 1.1.

Attempts have been made to define the values within which ROS need to exist in order for their physiological roles to take effect and their pathological effects to be kept at bay. Using fertilization and pregnancy outcomes as parameters, it was determined that an upper threshold of 0.075×10^6 counts per minute (cpm) per million spermatozoa was the highest level of ROS required (Das et al., 2008). Above this value, ROS was found to cause a significant reduction in fertilization and clinical pregnancy rates. Using ROS levels between fertile and infertile men, this number was changed to 0.0185×10^6 cpm/million spermatozoa (Desai et al., 2009). With this cutoff value, two groups were divided and observed, and differences in motility, concentration, and viability were detected between the groups. This value was found to have a higher accuracy, specificity, and sensitivity in determining fertilizing potential. However, these reference values did not correspond when the results from different studies were compared. This might probably be due to the small sample size per study, pathogenesis of patients, or the use of nonstandardized assays to determine ROS (Chen et al., 2013). Cutoff values for upper and lower ROS limits are laboratory dependent and individually determined but are currently not universally accepted. Individual clinical and fertility

TABLE 1.1

Antioxidants Beneficial to Spermatozoa and Their Function

Antioxidants	Function
Enzymatic	
SODs	Catalyze the degradation of superoxide anions
	Protect spermatozoa from oxygen toxicity and lipid peroxidation
Catalase	Catalyzes the degradation of hydrogen peroxide
	Decreases lipid peroxidation
Glutathione peroxidases	Remove peroxyl radicals from peroxides
	Decrease lipid peroxidation and enhance sperm mobility and acrosome reaction
Nonenzymatic	
Glutathione	Scavenges free radicals to buffer against high levels of ROS
Vitamin C	Counteracts superoxide, hydrogen peroxide, and hydroxyl radicals and singlet oxygen actions
	Prevents lipid peroxidation and reduces hydrogen peroxide–induced DNA damage
	Prevents spermatozoa agglutination
	Recycles vitamin E
Vitamin E	Traps and scavenges superoxide, hydrogen peroxide, and hydroxyl radicals
	Protects sperm cell membranes from oxidative damage
Coenzyme Q-10	Related to low-density lipoproteins
	Guards sperm against peroxidative damage
	Energy-promoting function—assists to increase sperm motility
	Recycles vitamin E and inhibits its pro-oxidant capacity
Albumin, bilirubin, taurine, hypotaurine	Do not prevent against actions of ROS or decrease their production
	Reduce the risk of the development of oxidative stress

laboratories effectively standardize seminal ROS reference ranges from their own patients and use these to determine high and low ROS values. Cutoff values of 0.0125×10^6 cpm/20×10^6 sperm and 0.0145×10^6 cpm/20×10^6 sperm have been reported by Agarwal et al. (2009) and Allamaneni et al. (2005), respectively, and a study by Homa et al. (2015) has revealed an ideal reference value of <24.1 relative light units (RLU)/s/10^6 sperm for ROS, a number they found to be both highly specific and sensitive in differentiating between samples with normal semen parameters and those with high PMN in their semen. Kashou and his colleagues (2013) found a normal ROS range of <20 RLU/s/10^6, a very similar and corroborative cutoff value to the one established by Homa.

Values less than the specified cutoff values are generally considered to be low or basal levels of ROS, while levels above these values are considered high. More research needs to be carried out to determine the exact values within which ROS need to be maintained in order to inhibit their negative effects.

1.9 CONCLUSION

The presence of oxygen is essential to life on Earth but presents organisms with an "oxygen paradox." Oxygen toxicity is an intrinsic challenge to aerobic life that also affects spermatozoa, which are essential to the propagation of a species (Sikka et al., 1995).

When spermatozoa were first discovered to produce ROS, the negative association between ROS and male fertility came into focus. Excessive ROS production was shown to result in reduced sperm motility (Eskenazi et al., 2003), increased DNA fragmentation, membrane lipid peroxidation, and apoptosis (Aitken and Baker, 1995; Armstrong et al., 1999), as well as a reduction in successful sperm–oocyte fusion (Agarwal et al., 2007). The perception that ROS is damaging to male gametes became dominant, and it was not until the work carried out by Aitken and his colleagues in 1989 that more light was shed on their existence in the male reproductive system.

ROS have now been shown to act as regulatory mediators involved in signal transduction processes (Griveau and Le Lannou, 1997). At physiological levels, ROS have been implicated in the activation of intracellular pathways leading to sperm maturation, hyperactivation, capacitation, acrosome reaction, and chemotactic processes important in localizing the oocyte and sperm–oocyte fusion, all of which are functions vital to the success of fertilization.

In order to acquire fertilization capacity, a balance must be maintained for the formation and degradation of ROS acting at certain instances during the existence of spermatozoa (Du Plessis et al., 2015). Their levels within the body are kept at homeostatic or physiological levels by numerous antioxidants to prevent the state of OS. This state may lead to the damage of cellular components and to the pathogenesis of disease and infertility.

The positive roles played by ROS do not negate the negative consequences brought on by their presence within an organism. ROS need to be maintained at suitable levels to ensure proper physiological function, while also making sure that pathological damage is averted (Kothari et al., 2010).

REFERENCES

Agarwal A, Deepinder F, Sharma RK, Ranga G, Li J. Effect of cell phone usage on semen analysis in men attending infertility clinic: An observational study. *Fertility and Sterility*, 2008;89:124–128.

Agarwal A, Nallella KP, Allamaneni SS, Said TM. Role of antioxidants in treatment of male infertility: An overview of the literature. *Reproductive Biomedicine Online*, 2004;8(6):616–627.

Agarwal A, Prabakaran SA. Mechanism, measurement, and prevention of oxidative stress in male reproductive physiology. *Indian Journal of Experimental Biology*, 2005;43:963–974.

Agarwal A, Prabakaran S, Allamaneni S. What an andrologist/urologist should know about free radicals and why. *Urology*, 2006a;67(1):2–8.

Agarwal A, Prabakaran S, Allamaneni SS. Relationship between oxidative stress, varicocele and infertility: A meta-analysis. *Reproductive Biomedicine Online*, 2006b;12(5):630–633.

Agarwal A, Prabakaran SA, Sikka SC. Clinical relevance of oxidative stress in patients with male factor infertility: Evidence-based analysis. *American Andrology Association Update Series*, 2007;26:1–12.

Agarwal A, Said TM. Oxidative stress, DNA damage and apoptosis in male infertility: A clinical approach. *British Journal of Urology International*, 2005;95:503–507.

Agarwal A, Sharma RK, Desai NR, Prabakaran S, Tavares A, Sabanegh E. Role of oxidative stress in pathogenesis of varicocele and infertility. *Urology*, 2009;73(3):461–469.

Aitken RJ. Molecular mechanisms regulating human sperm function. *Molecular Human Reproduction*, 1997;3(3):169–173.

Aitken RJ, Baker HG. Andrology: Seminal leukocytes: Passengers, terrorists or good Samaritans? *Human Reproduction*, 1995;10(7):1736–1739.

Aitken RJ, Buckingham DW, Harkiss D. Use of a xanthine oxidase free radical generative system to investigate the cytotoxic effects of reactive oxygen species on human spermatozoa. *Journal of Reproduction and Fertility*, 1993;97:441–450.

Aitken RJ, Clarkson JS. Cellular basis of defective sperm function and its association with the genesis of reactive oxygen species by human spermatozoa. *Journal of Reproduction and Fertility*, 1987;81:459–469.

Aitken RJ, Clarkson JS, Fischel S. Generation of reactive oxygen species, lipid peroxidation and human sperm function. *Biology of Reproduction*, 1989a;40:183–197.

Aitken RJ, Clarkson JS, Hargreave TB, Irvine DS, Wu FCW. Analysis of the relationship between defective sperm function and the generation of reactive oxygen species in cases of oligozoospermia. *Journal of Andrology*, 1989b;10:214–220.

Aitken RJ, Fisher HM, Fulton N, Gomez E, Knox W, Lewis B, et al. Reactive oxygen species generation by human spermatozoa is induced by exogenous NADPH and inhibited by the flavoprotein inhibitors diphenylene iodonium and quinacrine. *Molecular Reproduction and Development* 1997;47(4):468–482.

Aitken RJ, Harkiss D, Knox W, Paterson M, Irvine DS. A novel signal transduction cascade in capacitating human spermatozoa characterised by a redox-regulated, cAMP-mediated induction of tyrosine phosphorylation. *Journal of Cell Science*, 1998a;111(Pt 5):645–656.

Aitken RJ, Harkiss D, Knox W, Paterson M, Irvine S. On the cellular mechanisms by which the bicarbonate ion mediates the extragenomic action of progesterone on human spermatozoa. *Biology of Reproduction*, 1998b;58(1):186–196.

Aitken RJ, Krausz C, Buckingham DW. Relationship between biochemical markers for residual sperm cytoplasm, reactive oxygen species generation and the presence of leucocytes and precursor germ cells in human sperm suspensions. *Molecular Reproduction and Development*, 1994;39:268–279.

Allamaneni SS, Agarwal A, Nallella KP, Sharma RK, Thomas AJ Jr, Sikka SC. Characterization of oxidative stress status by evaluation of reactive oxygen species levels in whole semen and isolated spermatozoa. *Fertility and Sterility* 2005;83:800–803.

Allamaneni SS, Naughton CK, Sharma RK, Thomas AJ, Agarwal A. Increased seminal reactive oxygen species levels in patients with varicoceles correlate with varicocele grade but not with testis size. *Fertility and Sterility*, 2004;82(6):1684–1686.

Alvarez JG, Storey BT. Assessment of cell damage caused by spontaneous lipid peroxidation in rabbit spermatozoa. *Biology of Reproduction*, 1984;30:323–331.

Alvarez JG, Storey BT. Role of glutathione peroxides in protecting mammalian spermatozoa from loss of motility caused by spontaneous lipid. *Gamete Research*, 1989;23:77–90.

Alvarez JG, Touchstone JC, Blasco I, Storey BT. Spontaneous lipid peroxidation and production of hydrogen peroxide and superoxide in human spermatozoa. *Journal of Andrology*, 1987;8:338–348.

Armstrong JS, Rajasekaran M, Chamulitrat W, Gatti P, Hellstrom WJ, Sikka SC. Characterization of reactive oxygen species induced effects on human spermatozoa movement and energy metabolism. *Free Radical Biology and Medicine*, 1999;26(7):869–880.

Aruoma OI, Halliwell B, Hoey BM, Butler J. The antioxidant action of taurine, hypotaurine and their metabolic precursors. *Biochemical Journal*, 1988;25:251–255.

Babior BM. NADPH oxidase: An update. *Blood*, 1999;93(5):1464–1476.

Casano R, Orlando C, Serio M, Forti G. LDH and LDH-X activity in sperm from normospermic and oligozoospermic men. *International Journal of Andrology*, 1991;14:257–263.

Chitra K, Latchoumycandane C, Mathur, P. Effect of nonylphenol on the antioxidant system in epididymal sperm of rats. *Archives of Toxicology*, 2002;76:545–551.

Chow CK. Vitamin E and oxidative stress. *Free Radicals Biology and Medicine*, 1991;11:215–232.

Ciereszko A, Dabrowski K, Kucharczyk D, Dobosz S, Goryczko K, Glogowski J. The presence of uric acid, an antioxidative substance, in fish seminal plasma. *Fish Physiology and Biochemistry*, 1991;21:313–315.

Cooper TG, Noonan E, von Eckardstein S, Auger J, Baker HW, Behre HM et al. World Health Organization reference values for human semen characteristics. *Human Reproduction Update*, 2010;16(3):231–245.

Dandekar SP, Nadkarni GD, Kulkarni VS, Punekar S. Lipid peroxidation and antioxidant enzymes in male infertility. *Journal of Postgraduate Medicine*, 2002;48:186–189.

Das S, Chattopadhyay R, Jana SK, Narendra BK, Chakraborty C, Chakravarty B et al. Cut-off value of reactive oxygen species for predicting semen quality and fertilization outcome. *Systems Biology in Reproductive Medicine*, 2008;54(1):47–54.

De Iuliis GN, Newey RJ, King BV, Aitken RJ. Mobile phone radiation induces reactive oxygen species production and DNA damage in human spermatozoa in vitro. *PLoS One*, 2009;4:e6446.

De Lamirande E, Gagnon C. Reactive oxygen species and human spermatozoa I—Effects on the motility of intact spermatozoa and on sperm axonemes. *Journal of Andrology*, 1992a;13:5368–5378.

De Lamirande E, Gagnon C. Reactive oxygen species and human spermatozoa II. Depletion of adenosine triphosphate plays an important role in the inhibition of sperm motility. *Journal of Andrology*, 1992b;13:5379–5386.

De Lamirande E, Gagnon C. A positive role for the superoxide anion in the triggering of human sperm hyperactivation and capacitation. *International Journal of Andrology*, 1993a;16:21–25.

Desai N, Sharma R, Makker K, Sabanegh E, Agarwal A. Physiologic and pathologic levels of reactive oxygen species in neat semen of infertile men. *Fertility and Sterility*, 2009;92(5):1626–1631.

Du Plessis SS, Agarwal A, Halabi J, Tvrda E. Contemporary evidence on the physiological role of reactive oxygen species in human sperm function. *Journal of Reproductive Genetics*, 2015;32(4):509–520.

Eskenazi B, Wyrobek AJ, Sloter E, Kidd SA, Moore L, Young S, Moore D. The association of age and semen quality in healthy men. *Human Reproduction*, 2003;18(2):447–454.

Esteves SC. Effect of cigarette smoking on levels of seminal oxidative stress in infertile men: A prospective study. *International Brazilian Journal of Urology*, 2002;28:484–485.

Fisher HM, Aitken RJ. Comparative analysis of the ability of precursor germ cells and epididymal spermatozoa to generate reactive oxygen metabolites. *Journal of Experimental Zoology*, 1997;277(5):390–400.

Foresta C, Flohe C, Garolla A, Roveri A, Ursini F, Maiorino M. Male fertility is linked to the selenoprotein phospholipid hydroperoxide glutathione peroxidase. *Biology of Reproduction*, 2002;67:967–971.

Frei B, Kim MC, Ames BN. Ubiquinol-10 is an effective lipid soluble antioxidant at physiological concentrations. *Proceedings of the National Academy of Sciences of the United States of America*, 1990;87:4879–4883.

Gil-Guzman E, Ollero M, Lopez MC, Sharma RK, Alvarez JG, Thomas AJ Jr, Agarwal A. Differential production of reactive oxygen species by subsets of human spermatozoa at different stages of maturation. *Human Reproduction*, 2001;16(9):1922–1930.

Griveau JF, Dumont E, Renard P, Le Lannou D. Reactive oxygen species, lipid peroxidation and enzymatic defense systems in human spermatozoa. *Journal of Reproduction and Fertility*, 1995;103:17–26.

Griveau JF, Le Lannou D. Reactive oxygen species and human spermatozoa: Physiology and pathology. *International Journal of Andrology*, 1997;20:61–69.

Halliwell B, Cross CE. Oxygen-derived species: Their relation to human disease and environmental stress. *Environmental Health Perspectives*, 1992;102(Suppl. 10):5–12.

Halliwell B, Gutteridge JM. Role of free radicals and catalytic metal ions in human disease: an overview. *Methods in Enzymology,* 1990;186:1–85.

Harrero MB, Gagnon C. Nitric oxide: A novel mediator of sperm function. *Journal of Andrology*, 2001;22(3):349–356.

Hendin BN, Kolettis PN, Sharma RK, Thomas AJ, Agarwal A. Varicocele is associated with elevated spermatozoal reactive oxygen species production and diminished seminal plasma antioxidant capacity. *The Journal of Urology*, 1999;161(6):1831–1834.

Hipler UC, Görnig M, Hipler B, Römer W, Schreiber G. Stimulation and scavestrogen-induced inhibition of reactive oxygen species generated by rat sertoli cells. *Archives of Andrology*, 2000;44(2):147–154.

Homa ST, Vessey W, Perez-Miranda A, Riyait T, Agarwal A. Reactive Oxygen Species (ROS) in human semen: Determination of a reference range. *Journal of Assisted Reproduction and Genetics*, 2015;32:757–764.

Hull MG, North K, Taylor H, Farrow A, Ford WC. Delayed conception and active and passive smoking. The avon longitudinal study of pregnancy and childhood study team. *Fertility and Sterility*, 2000;74:725–733.

Huszar G, Sbracia M, Vigue L, Miller DJ, Shur BD. Sperm plasma membrane remodeling during spermiogenetic maturation in men: Relationship among plasma membrane beta 1,4-galactosyltransferase, cytoplasmic creatine phosphokinase, and creatine phosphokinase isoform rations. *Biology of Reproduction*, 1997;56:1020–1024.

Huszar G, Vigue L. Correlation between the rate of lipid peroxidation and cellular maturity as measured by creatine kinase activity in human spermatozoa. *Journal of Andrology*, 1994;15:71–77.

Idriss NK, Blann AD, Lip GYH. Hemoxygenase-1 in cardiovascular disease. *Journal of the American College of Cardioliology*, 2008;52:971–978.

Irvine DS. Glutathione as a treatment for male infertility. *Reviews of Reproduction*, 1996;1:6–12.

Jarow JP. Semen quality of male smokers and non-smokers in infertile couples. *Journal of Urology*, 2003;170:675–676.

Jurasović J, Cvitković P, Pizent A, Colak B, Telisman S. Semen quality and reproductive endocrine function with regard to blood cadmium in Croatian male subjects. *Biometals*, 2004;17:735–743.

Karbownik M, Gitto E, Lewinski A, Reiter RJ. Induction of lipid peroxidation in hamster organs by the carcinogen cadmium: Melioration by melatonin. *Cell Biology and Toxicology*, 2001;17:33–40.

Kashou AH, Sharma R, Agarwal A. Assessment of oxidative stress in sperm and semen. *Methods in Molecular Biology*, 2013;927:351–361.

Kato S, Shibukawa T, Harayama H, Kanna Y. Timing of shedding and disintegration of cytoplasmic droplets from boar and goat spermatozoa. *Journal of Reproduction and Development* (Tokyo), 1996;42:237–241.

Kefer JC, Agarwal A, Sabanegh E. Role of antioxidants in the treatment of male infertility. *International Journal of Urology*, 2009;16:449–457.

Koppers AJ, De luliis GN, Finnie JM, McLaughlin EA, Aitken RJ. Significance of mitochondrial reactive oxygen species in the generation of oxidative stress in spermatozoa. *Journal of Clinical Endocrinology and Metabolism*, 2008;93:3199–3207.

Kothari S, Thompson A, Agarwal A, Du Plessis SS. Free radicals: Their beneficial and detrimental effects on sperm function. *Indian Journal of Experimental Biology*, 2010;48:425–435.

Krishnamoorthy G, Venkataraman P, Arunkumar A, Vignesh RC, Aruldhas MM, Arunakaran J. Ameliorative effect of vitamins (alpha-tocopherol and ascorbic acid) on PCB (Aroclor 1254) induced oxidative stress in rat epididymal sperm. *Reproductive Toxicology*, 2007;23:239–245.

Kwenang A, Krous MJ, Koster JF, Van Eijk HG. Iron, ferritin and copper in seminal plasma. *Human Reproduction*, 1987;2:387–388.

Latchoumycandane C, Chitra KC, Mathur PP. 2,3,7,8-Tetrachlorodibenzo-pdioxin (TCDD) induces oxidative stress in the epididymis and epididymal sperm of adult rats. *Molecular Toxicology*, 2003;77:280–284.

Lavranos G, Balla M, Tzortzopoulou A, Syriou V, Angelopoulou R. Investigating ROS sources in male infertility: A common end for numerous pathways. *Reproductive Toxicology*, 2012;34:298–307.

Lewin A, Lavon H. The effect of coenzyme Q10 on sperm motility and function. *Molecular Aspects of Medicine*, 1997;18(Suppl.):213–219.

Linsenmeyer TA, Perkash I. Infertility in men with spinal cord injury. *Archives of Physical Medicine and Rehabilitation*, 1991;72(10):747–754.

Mehraban D, Ansari M, Keyhan H, Sedighi Gilani M, Naderi G, Esfehani F. Comparison of nitric oxide concentration in seminal fluid between infertile patients with and without varicocele and normal fertile men. *Urology Journal*, 2009;2(2):106–110.

Mitropoulos D, Deliconstantinos G, Zervas A, Villiotou V, Dimopoulos C, Stavrides J. Nitric oxide synthase and xanthine oxidase activities in the spermatic vein of patients with varicocele: A potential role for nitric oxide and peroxynitrite in sperm dysfunction. *The Journal of Urology*, 1996;156(6):1952–1958.

Mostafa T, Anis TH, El-Nashar A, Imam H, Othman IA. Varicocelectomy reduces reactive oxygen species levels and increases antioxidant activity of seminal plasma from infertile men with varicocele. *International Journal of Andrology*, 2001;24(5):261–265.

Nissen HP, Kreysel HW. Superoxide dismutase in human semen. *Klinische Wochenschrift*, 1983;61:63–65.

Nistiar F, Racz O, Benacka R, Lukacinova A. Effect of bioflavonoids on type 1 diabetes onset and antioxidant status in spontaneously diabetic rats of BB strain. *Journal of Central European Agriculture*, 2009;11:113–140.

Ochsendorf FR. Infections in male genital tract and reactive oxygen species. *Human Reproduction Update*, 1999;5:399–420.

Peeker R, Abramsson L, Marklund SL. Superoxide dismutase isoenzymes in human seminal plasma and spermatozoa. *Molecular Human Reproduction*, 1997;13:1061–1066.

Rivlin J, Mendel J, Rubinstein S, Etkovitz N, Breitbart H. Role of hydrogen peroxide in sperm capacitation and acrosome reaction. *Biology of Reproduction*, 2004;70:518–522.

Roche M, Rondeau P, Singh NR, Tarnus E, Bourdon E. The antioxidant properties of serum albumin. *FEBS Letters*, 2008;582:1783–1787.

Rousseaux J, Rousseaux-Prevost R. Molecular localization of free thiols in human sperm chromatin. *Biology of Reproduction*, 1995;52:1066–1072.

Roveri A, Ursini F, Flohe L, Maiorino M. PHGPx and spermatogenesis. *BioFactors*, 2001;14:213–222.

Saalu LC. The incriminating role of reactive oxygen species in idiopathic male infertility: An evidence based evaluation. *Pakistan Journal of Biological Science*, 2010;13:413–422.

Saleh RA, Agarwal A, Sharma RK, Nelson DR, Thomas AJ. Effect of cigarette smoking on levels of seminal oxidative stress in infertile men: A prospective study. *Fertility and Sterility*, 2002;78(3):491–499.

Saleh RA, Agarwal A, Sharma RK, Said TM, Sikka SC, Thomas AJ. Evaluation of nuclear DNA damage in spermatozoa from infertile men with varicocele. *Fertility and Sterility*, 2003;80(6):1431–1436.

Sarlós P, Molnár A, Kókai M, Gábor Gy, Rátky J. Comparative evaluation of the effect of antioxidants in the conservation of ram semen. *Acta Veterinaria Hungarica*, 2002;50:235–245.

Schafer FQ, Buettner GR. Redox environment of the cell as viewed through the redox state of the glutathione disulfide/glutathione couple. *Free Radical Biology and Medicine*, 2001;30:1191–1212.

Sedlak TW, Saleh M, Higginson DS, Paul BD, Juluri KR, Snyder SH. Bilirubin and glutathione have complementary antioxidant and cytoprotective roles. *Proceedings of the National Academy of Sciences of the United States of America*, 2009;106:5171–5176.

Sharma RK, Agarwal A. Role of reactive oxygen species in male infertility. *Urology*, 1996;48(6):835–850.

Sies H. Strategies of antioxidant defence. *European Journal of Biochemistry*, 1993;215:213–219.

Sikka SC. Relative impact of oxidative stress on male reproductive function. *Current Medicinal Chemistry*, 2001;8(7):851–862.

Sikka SC, Rajasekaran M, Hellstrom WJ. Role of oxidative stress and antioxidants in male infertility. *Journal of Andrology*, 1995;16:464–468.

Slivkova J, Popelkova M, Massanyi P, Toporcerova S, Stawarz R, Formicki G, Lukac N, Putała A, Guzik M. Concentration of trace elements in human semen and relation to spermatozoa quality. *Journal of Environmental Science and Health. Part A: Toxic/Hazardous Substances & Environmental Engineering*, 2009;44:370–375.

Smith DC, Barratt CL, Williams MA. The characterisation of nonsperm cells in the ejaculates of fertile men using transmission electron microscopy. *Andrologia*, 1989;21(4):319–333.

Storey BT. Biochemistry of the induction and prevention of lipoperoxidative damage in human spermatozoa. *Molecular Human Reproduction*, 1997;3(3):203–213.

Tanghe S, Soom AV, Mehrzad J, Maes D, Duchateau L, Kruif Ad. Cumulus contributions during bovine fertilization in vitro. *Theriogenology*, 2003;60:135–149.

Traber MG, van der Vliet A, Reznick AZ, Cross CE. Tobacco-related diseases: Is there a role for antioxidant micronutrient supplementation?. *Clinics in Chest Medicine*, 2000;21(1):173–187.

Tremellen K. Oxidative stress and male infertility: A clinical perspective. *Human Reproduction Update*, 2008;14:243–258.

Tvrda E, Kňažická Z, Bárdos L, Massányi P, Lukáč N. Impact of oxidative stress on male fertility—A review. *Acta Veterinaria Hungarica*, 2011;59(4):465–484.

Vernet P, Aitken RJ, Drevet JR. Antioxidant strategies in the epididymis. *Molecular and Cellular Endocrinology*, 2004;216:31–39.

Vine MF. Smoking and male reproduction: A review. *International Journal of Andrology*, 1996;19(6):323–337.

Wolff H. The biologic significance of white blood cells in semen. *Fertility and Sterility*, 1995;63(6):1143–1157.

World Health Organisation. *WHO Laboratory Manual for the Examination and Processing of Human Semen*, 5th edn. World Health Organization, Geneva, Switzerland, 2010.

Zini A, De Lamirande E, Gagnon C. Low levels of nitric oxide promote human sperm capacitation in vitro. *Journal of Andrology*, 1995;16(5):424–431.

2 Detection of Reactive Oxygen Species

Oluwasesan Adegoke and Patricia B.C. Forbes

CONTENTS

ABSTRACT

Semiconductor quantum dot (QD) nanocrystals and organic fluorophore–based probes are efficient detection tools for the fluorescent monitoring of reactive oxygen species in both *in vivo* and *in vitro* systems. Organic fluorophore dyes, which are currently the most commonly used probes, provide chemical information with respect to changes in the redox environment of the sample system. Chemical reactions of these probes with the target analyte are known to generate fluorescent products that correspond to a fluorescence signal enhancement that is proportional to the concentration of the analyte. QDs, on the other hand, have superior optical properties than those of organic fluorophore dyes. However, their development as fluorescent probes for reactive oxygen species (ROS) has been limited. Generally, each of these classes of probes has their merits and demerits. In this chapter, we provide a brief summary of probes for ROS that combine the requirements of specificity and a clear reaction mechanism needed for any efficient fluorescent probe.

2.1 INTRODUCTION

2.1.1 MONITORING OF REACTIVE OXYGEN SPECIES IN BIOLOGICAL SYSTEMS

The major problem in fully elucidating the roles of reactive oxygen species (ROS) in biological systems is the lack of suitable detection systems, which is challenging due to the presence of various antioxidants within cell systems and the short lifetimes exhibited by ROS species. Among various probes developed for ROS detection, synthetic fluorescent probes have emerged as a powerful tool due to their simple design, high sensitivities, surface modification strategies, and low cost [1–4]. These probes also allow for the sharp visualization of intracellular functions, and they may be used as labels on biomolecules of interest as fluorescent reporters.

In order to unravel the unique properties of ROS within biological systems, there are several different types of fluorescent probes that have been developed [5–8]. In this chapter, we simply introduce the different types of fluorescent probes that have been developed for ROS, specifically hydrogen peroxide (H_2O_2), the hydroxyl radical ($HO^{\bullet}$), the superoxide radical ($O_2^{\bullet-}$), and hypochlorous acid (HOCl). Examples of synthetic fluorescent probes for the detection of ROS in both *in vivo* and *in vitro* systems are organic fluorophore dyes, encoded fluorescent proteins, nanoparticle-based fluorescent probes, and small molecule probes. In this chapter, however, our emphasis is on the use of organic fluorophore dyes and nanocrystal quantum dots (QDs) as fluorescent probes for ROS due to their extensive use and potential over other types of synthetic fluorescent probes. It is important to note that several probes that have been developed for ROS have pitfalls such as the lack of a clear reaction mechanism or specificity. For detail, readers are advised to consult the articles that cover the disadvantages of organic fluorescent–based probes [9] and QD-based probes [10] for ROS; while here we focus on those fluorescent probes that produce appropriate specificity for the target ROS and a clear reaction mechanism. It should be noted that analytical methods for ROS are not only limited to synthetic fluorescent probes, but conventional techniques such as electron paramagnetic resonance, mass spectrometry, and high-performance liquid chromatography have also been employed to detect specific ROS either via the oxidation of proteins, DNA, exogenous probes, or other biomolecules [11–15]. However, these conventional techniques have limitations in terms of real-time monitoring; hence, they are not covered in this chapter.

2.1.2 SEMICONDUCTOR QDs AS FLUOROPHORE PROBES

Semiconductor QDs are nanocrystals that exhibit electronic and optical properties that depend on their size, quantum confinement property, and surface to volume ratio [16]. The quantum confinement phenomenon exhibited by QDs is due to the confinement of energetic levels to discrete values, which results from a decrease in the nanocrystal dimensions to a few nanometers and an increase in the bandgap. Hence, the mobility of the exciton (pair of electron and hole) is restricted to the nanoscale dimension [17]. QDs are composed of a core that is usually surrounded by a shell layer. Fluorescent QDs that have been used for the detection of ROS are typically composed of cadmium chalcogenide materials. QD nanocrystals have been exploited as fluorescent probes for the detection of ROS due to their unique properties that have advantages over traditional organic fluorophore dyes, such as their broad absorption and narrow emission spectra, size-tunable fluorescence emission spectra, excellent photostability, high resistance to photodegradation, bright fluorescence, and multiplex detection potentials [18,19]. The success of QDs as fluorescent probes for ROS has been limited. This is due to the difficulty in finding an appropriate molecule that can bind to their surface for the detection of the target ROS. Nonetheless, they still hold great promise for the detection of ROS due to their unique properties.

2.1.3 ORGANIC DYES AS FLUOROPHORE PROBES

The most commonly used fluorescent probes for the monitoring of ROS are organic fluorophore dyes. However, several questions have been raised about the erroneous and over-interpretation of experimental data emanating from these probes [9]. They can be classified as redox- and non-redox-sensitive probes, respectively. Redox-sensitive probes are the most common, and their mode of interaction involves their ability to penetrate cells in which they are oxidized by the ROS to fluorescent products. Hence, their esterified derivative undergoes hydrolysis [9]. However, these probes lack selectivity for any specific type of ROS.

The nonredox organic fluorescent probes are a class of new generation probes that were developed to circumvent the problems associated with redox-sensitive probes. In this case, the chemical principle involves using a blocking group to mask the fluorescence of the probe, and a nucleophilic attack by the reactive species releases the blocking group. The oxidized form of the fluorophore then fluoresces and the oxidized form of the blocking group is released [9]. The limitations of this class of probe are their unclear reaction kinetics in cellular media and their lack of specificity.

In general, the detection of ROS using organic fluorophore dyes still remains a huge challenge.

2.2 PROBES FOR HYDROGEN PEROXIDE

2.2.1 QDs

QD-based fluorescent probes for the detection of H_2O_2 in cellular environments have not been developed to date. However, progress has been made for their detection in aqueous solution systems. A QD-enzyme hybrid system was developed by Yuan et al. for the detection of H_2O_2 [20]; for example, horseradish peroxidase (HRP) was catalytically coupled onto the surface of the QD and hydroquinone (H_2Q) was added to form a H_2Q-HRP-QD hybrid sensor. HRP, a redox enzyme, was employed due to its catalytic activity that chemically reduced H_2O_2. Due to this, Fe^{2+} present in the heme of HRP was converted to Fe^{3+} by the transfer of an electron [21]. Hence, the chemical coupling of HRP onto the surface of the QD enhanced the specificity for H_2O_2 detection.

In a related development, a Förster resonance energy transfer (FRET) sensor for H_2O_2 was reported [22] with high sensitivity, specificity, and a well-elucidated reaction mechanism. Tyramide-labeled Cytochrome 5 (Cy5) was coated on the surface of the QD via the catalytic activity of HRP. The FRET sensor was based on the energy transfer of the QDs to Cy5, which resulted in the decrease of the QD emission with simultaneous increase in the emission of Cy5 as the concentration of H_2O_2 was increased. Although it was not demonstrated if other radical scavengers could intercept the HRP-QD assay, the clear reaction mechanism suggested that the probe showed some level of specificity for H_2O_2.

2.2.2 ORGANIC FLUOROPHORES

There are many organic fluorophore–based probes for H_2O_2 that have been reported in the literature, but the majority of these are not specific for H_2O_2 because the probe displayed a similar response to other competing species such as $^\bullet OH$, peroxynitrite, and nitric oxide [23]. The limited number of fluorescent probes that are specific for H_2O_2 include the one developed by Kim et al. [24] that employed a diacetyl derivative of 2′,7′-dichlorofluorescin (Probe **1**) (Figure 2.1) combined with organically modified silane (Ormosil) particles that enhanced selectively toward H_2O_2 over other types of reactive oxygen and nitrogen species. The developed probe proved efficient for the bioimaging of H_2O_2.

In vivo and *in vitro* monitoring of H_2O_2 has been performed using boronate-based luminescent probes. These probes are highly efficient because H_2O_2 can selectively hydrolyze boronates. An excellent development of such probe was reported by Lo et al. [25]; in their work, a H_2O_2-specific boronate probe (Probe **2**) was developed in which the p-dihydroxyborylbenzyloxycarbonyl blocking group protected the latent aminocoumarin fluorophore (Figure 2.2). A fluorescent aminocoumarin was generated by the 1,6-elimination reaction that promoted the formation of the intermediate complex from H_2O_2-based hydrolysis.

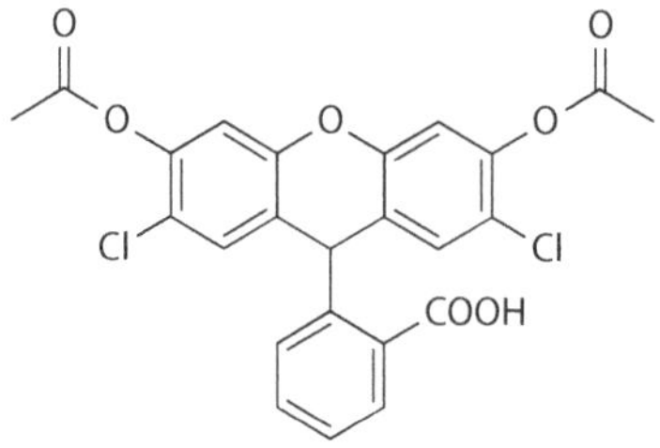

FIGURE 2.1   Diacetyl derivative of 2',7'-dichlorofluorescein Probe **1** for H$_2$O$_2$ detection.

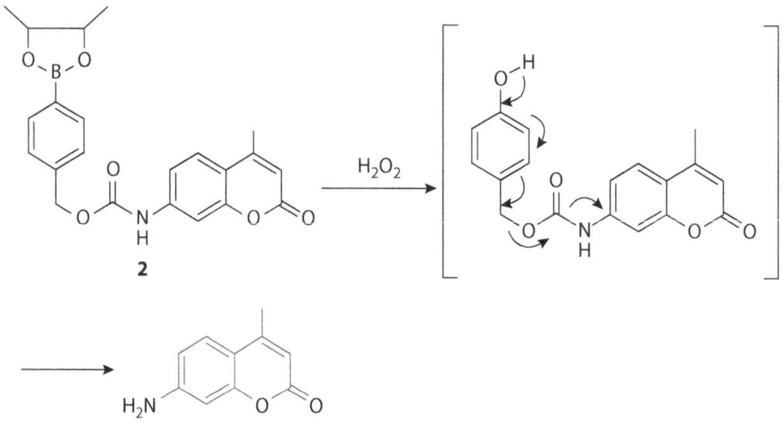

FIGURE 2.2   Working principle of Probe **2** for H$_2$O$_2$.

Other types of organic-based fluorescent probes, specific to H$_2$O$_2$, include H$_2$O$_2$-induced conversion of arylboronates to phenols, organic dye-polymer-conjugated fluorescent probes, and xanthone and resorufin probes for H$_2$O$_2$ in living cells [26–28].

2.3   PROBES FOR THE HYDROXYL RADICAL

2.3.1   QDs

To date, there are no reported QD-based probes for ˙OH that combine the requirements of specificity and a clear reaction mechanism. However, readers are referred to a review article that further highlights these limitations for ˙OH detection [10].

2.3.2   ORGANIC FLUOROPHORES

More than two decades ago, Pou et al. developed a conjugated nitroxide fluorescent probe (Probe **3**) that was specific to HO˙ in dimethyl sulfoxide (DMSO) [29]. The probe exhibited a weak fluorescence in the absence of HO˙ as a result of intramolecular quenching induced by the singlet excited state of the nitroxide moiety of the fluorophore. The reaction of ˙OH in DMSO converted the fluorophore (Probe **3**) to its reaction product (Figure 2.3) via methyl radicals; hence, the fluorescence of the probe was enhanced. Following this successful development, several studies have demonstrated a similar use of the nitroxide moiety to develop probes for ˙OH [30–34].

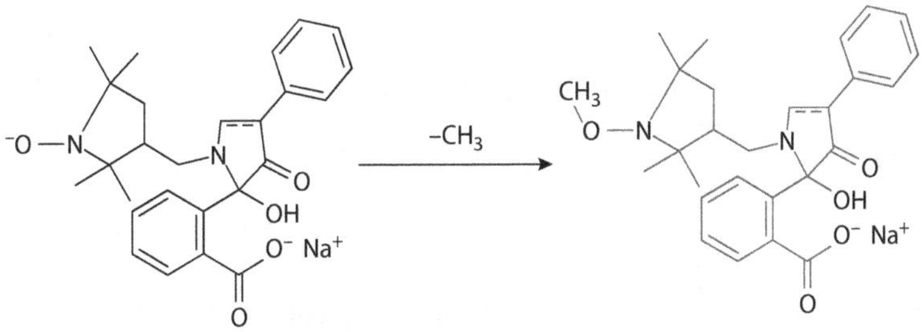

FIGURE 2.3 Conjugated nitroxide fluorophore for the detection of HO$^•$ (Probe **3**).

2.4 PROBES FOR THE SUPEROXIDE RADICAL

2.4.1 QDs

A QD bioconjugate probe that incorporates the binding of cytochrome c to CdSe/ZnS QDs was developed by Li et al. [35]. In their work, cytochrome c, a heme redox protein, was used as a recognition biomolecule for the specific detection of $O_2^{•-}$. Both aqueous phase and *in vivo* detection of $O_2^{•-}$ was conducted with success. The working principle for the aqueous phase detection was based on the quenching of the fluorescence of the negatively charged QDs by the oxidized cytochrome c, while the $O_2^{•-}$ generated from the pyrogallol assay reduced the oxidized cytochrome c and consequently switched on the fluorescence of the QDs in a concentration-dependent manner. The assay proved to be highly selective to $O_2^{•-}$ over other types of biological molecules, and the sensing of $O_2^{•-}$ in biological systems was also conducted in HeLa and HL-7702 cells [35].

2.4.2 ORGANIC FLUOROPHORES

Intracellular detection of $O_2^{•-}$ has been widely conducted using Probe **4**. This probe produces a product called 2-hydroxyethidium (Figure 2.4) that binds to DNA leading to a fluorescence enhancement effect [36]. Results have shown that low levels of $O_2^{•-}$ (1.5 pmol) in 5 mg of biological sample can be quantitatively detected by this probe [37]. In a related study, Robinson et al. modified the structure of 2-hydroxyethidium with a triphenylphosphonium moiety to form Probe **5** (Figure 2.4), which was then employed to target mitochondria by facilitating transport through the phospholipid bilayers [38]. It was proposed that the detection mechanisms of both Probes **4** and **5** were via two routes. The first route involved nonsuperoxide, cellular-dependent, and autooxidation processes that led to the formation of fluorescent ethidium species. The second route involved the formation of 2-hydroxyl ethidium products via the oxidation of $O_2^{•-}$ when excited at 396 and 510 nm [38].

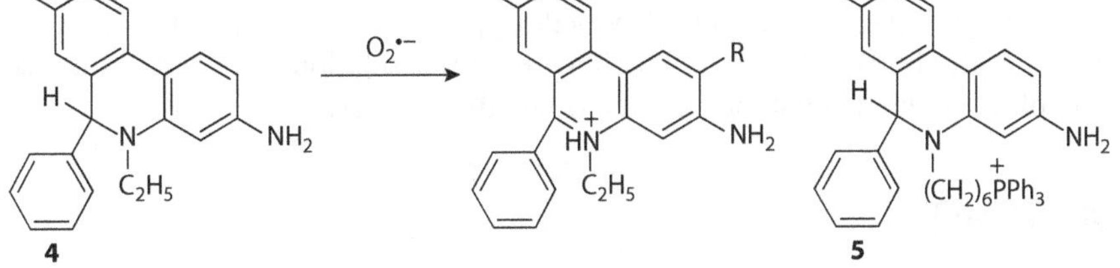

FIGURE 2.4 Detection mechanism of Probe **4** for $O_2^{•-}$ and the structure of a similar probe (Probe **5**). R = H = ethidium.

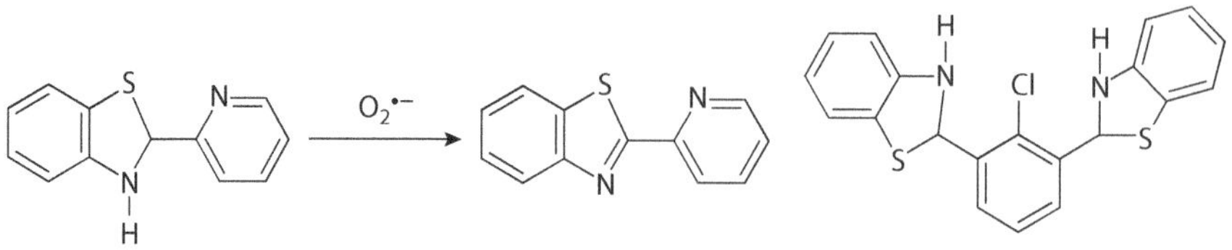

FIGURE 2.5   Detection principle of Probe **6** for O$_2$$^{\bullet-}$ and the structure of Probe **7**.

Tang et al. developed Probes **6** and **7** in separate studies for the detection of O$_2$$^{\bullet-}$ [39,40]. Oxidation by O$_2$$^{\bullet-}$ on Probe **6** led to a strongly fluorescent product (Figure 2.5) that selectively detected O$_2$$^{\bullet-}$. Similarly, the faster and selective fluorescence response from Probe **7** was successfully employed to detect O$_2$$^{\bullet-}$ based on the generation of a strongly fluorescent product.

2.5   PROBES FOR HYPOCHLOROUS ACID

2.5.1   QDs

Aqueous and cellular detection of HOCl was demonstrated by using QDs capped with certain organic ligands [41]. CdSe/ZnS QDs were coated with four different types of organic compounds containing carboxylate, sulfide, hydrocarbon, and methylamino functional groups, respectively, and their sensitivity toward HOCl detection was investigated. It was found that QDs capped with the carboxylate group exhibited higher sensitivity than the rest of the functionalized QDs. The working principle of the probe was based on the ability of HOCl to induce oxidative etching on the QD surface through its diffusion through the negative layer of the coated polymer. Intracellular sensing of HOCl was performed in HL60 cells.

An excellent study to quantify endogenous HOCl that was generated via leukocytes was reported using a polymer microbead–coated CdSe QD probe [42]. Microbeads were employed to localize the QDs through phagocytosis into the leukocyte; hence, intraphagosomal HOCl generated via neutrophils was quantified cellularly to be $6.5 \pm 0.9 \times 10^8$ molecules. This work was the first reported quantification of HOCl in cells by any fluorescent probe.

2.5.2   ORGANIC FLUOROPHORES

The development of a tetramethylrhodamine-based fluorescent probe that is specific for HOCl has been reported by Kenmoku et al. [43]. The working principle of this probe involves the reaction of HOCl with the thioether group to generate a sulfonate product. This reaction led to the enhancement of the fluorescence of the probe based on the ring opening of the spiro system. Other rhodamine-based probes for HOCl have also been reported [44–46].

BODIPY (boron-dipyromethene)-based fluorescent probes that are specific for HOCl have been developed by Sun et al. [47]. Oxidation of p-methoxyphenol to benzoquinone was accomplished by the reaction of the BODIPY probe with HOCl in buffer solution. This led to the fluorescence enhancement of the probe without interference from other species.

2.6   CONCLUSION

Generally, organic fluorophore probes have been more widely employed to date for the detection of ROS than those of inorganic QD nanocrystals. However, many of the organic fluorophore probes have limitations with respect to an unclear reaction mechanism and lack of analyte specificity.

Inorganic QD probes, on the other hand, are more difficult to develop due to the need to functionalize their surfaces with molecules that can render specificity to the target analyte. Therefore, it will be of research interest to develop analyte-specific molecules that can bind to the surface of QDs and thus enable targeted detection. Further work on the coating of shells of low toxicity on core semiconductor QDs is also important to allow for their use *in vivo*. A key ongoing challenge regarding all ROS detection systems is target analyte selectivity.

REFERENCES

1. Ko, S.-K., Chen, X., Yoon, J., and Shin, J. 2011. Zebrafish as a good vertebrate model for molecular imaging using fluorescent probes. *Chem. Soc. Rev.* 40:2120–2130.
2. Chen, X., Zhou, Y., Peng, X., and Yoon, J. 2010. Fluorescent and colorimetric probes for detection of thiols. *Chem. Soc. Rev.* 39:2120–2135.
3. Kikuchi, K. 2010. Design, synthesis and biological application of chemical probes for bio-imaging. *Chem. Soc. Rev.* 39:2048–2053.
4. Kim, J.S. and Quang, D.T. 2007. Calixarene-derived fluorescent probes. *Chem. Rev.* 107:3780–3799.
5. Nagano, T. 2002. Bioimaging of nitric oxide. *Chem. Rev.* 102:1235–1270.
6. McQuade, L.E. and Lippard, S.J. 2010. Fluorescent probes to investigate nitric oxide and other reactive nitrogen species in biology (truncated form: fluorescent probes of reactive nitrogen species). *Curr. Opin. Chem. Biol.* 14:43–49.
7. Dickinson, B.C., Srikun, D., and Chang, C.J. 2010. Mitochondrial-targeted fluorescent probes for reactive oxygen species. *Curr. Opin. Chem. Biol.* 14:50–56.
8. Gomes, A., Fernandes, E., and Lima, J.L.F.C. 2005. Fluorescence probes used for detection of reactive oxygen species. *J. Biochem. Biophys. Methods* 65:45–80.
9. Winterbourn, C.C. 2014. The challenges of using fluorescent probes to detect and quantify specific reactive oxygen species in living cells. *Biochim. Biophys. Acta Gen. Subj.* 1840:730–738.
10. Adegoke, O. and Forbes, P.B.C. 2015. Challenges and advances in quantum dot fluorescent probes to detect reactive oxygen and nitrogen species: A review. *Anal. Chim. Acta* 862:1–3.
11. Leichert, L.I., Gehrke, F., Gudiseva, H.V., Blackwell, T., Ilbert, M., Walker, A.K., Strahler, J.R., Andrews, P.C., and Jakob, U. 2008. Quantifying changes in the thiol redox proteome upon oxidative stress in vivo. *Proc. Natl. Acad. Sci. U.S.A.* 105:8197–8202.
12. Niki, E. 2014. Biomarkers of lipid peroxidation in clinical material. *Biochim. Biophys. Acta* 1840:809–817.
13. Cadet, J., Douki, T., Gasparutto, D., and Ravanat, J.L. 2003. Oxidative damage to DNA: Formation, measurement and biochemical features. *Mutat. Res.* 531:5–23.
14. Thornalley, P.J. and Rabbani, N. 2014. Detection of oxidized and glycated proteins in clinical samples using mass spectrometry—A user's perspective. *Biochim. Biophys. Acta* 1840:818–829.
15. Zielonka, J., Zielonka, M., Sikora, A., Adamus, J., Joseph, J., Hardy, M., Ouari, O., Dranka, B.P., and Kalyanaraman B. 2012. Global profiling of reactive oxygen and nitrogen species in biological systems: High-throughput real-time analyses. *J. Biol. Chem.* 287:2984–2995.
16. Efros, A.L. and Efros, A.L. 1982. Interband absorption of light in a semiconductor sphere. *Sov. Phys. Semicond.* 16:772–775.
17. Bawendi, M.G., Steigerwald, M.L., and Brus, L.E. 1990. The quantum mechanics of larger semiconductor clusters ("quantum dots"). *Annu. Rev. Phys. Chem.* 41:477–496.
18. Chang, E., Thekkek, N., Yu, W.W., Colvin, V.L., and Drezek, R. 2006. Evaluation of quantum dot cytotoxicity based on intracellular uptake. *Small* 2:1412–1417.
19. Tortiglione, C., Quarta, A., Tino, A., Manna, L., Cingolani, R., and Pellegrino, T. 2007. Synthesis and biological assay of GSH functionalized fluorescent quantum dots for staining *Hydra vulgaris*. *Bioconjug. Chem.* 18:829–835.
20. Yuan, J., Guo, W., and Wang, E. 2008. Utilizing a CdTe quantum dots-enzyme hybrid system for the determination of both phenolic compounds and hydrogen peroxide. *Anal. Chem.* 80:1141–1145.
21. Zhao, J.G., Henkens, R.W., Stonehuerner, J., Daly, J.P.O., and Crumbliss, A.L. 1992. Direct electron transfer at horseradish peroxidase–colloidal gold modified electrodes. *J. Electroanal. Chem.* 327:109–119.
22. Huang, X., Wang, J., Liu, H., Lan, T., and Ren, J. 2013. Quantum dot-based FRET for sensitive determination of hydrogen peroxide and glucose using tyramide reaction. *Talanta* 106:79–84.
23. Chen, X., Tian, X., Shion, I., and Yoon, J. 2011. Fluorescent and luminescent probes for detection of reactive oxygen and nitrogen species. *Chem. Soc. Rev.* 40:4783–4804.

24. Kim, G., Lee, Y.-E. K., Xu, H., Philbert, M.A., and Kopelman, R. 2010. Nano-encapsulation method for high selectivity sensing of hydrogen peroxide inside live cells. *Anal. Chem.* 82:2165–2169.

25. Lo, L.-C. and Chu, C.-Y. 2003. Development of highly selective and sensitive probes for hydrogen peroxide. *Chem. Commun.* 2728–2729.

26. Chang, M.C.Y., Pralle, A., Isacoff, E.Y., and Chang, C.J. 2004. A selective, cell-permeable optical probe for hydrogen peroxide in living cells. *J. Am. Chem. Soc.* 126:15392–15393.

27. He, F., Tang, Y., Yu, M., Wang, S., Li, Y., and Zhu, D. 2006. Fluorescence-amplifying detection of hydrogen peroxide with cationic conjugated polymers, and its application to glucose sensing. *Adv. Funct. Mater.* 16:91–94.

28. Miller, E.W., Albers, A.E., Pralle, A., Isacoff, E.Y., and Chang, C.J. 2005. Boronate-based fluorescent probes for imaging cellular hydrogen peroxide. *J. Am. Chem. Soc.* 127:16652–16659.

29. Pou, S., Huang, Y.-I., Bhan, A., Bhadti, V.S., Hosmane, R.S., Wu, S.Y., Cao, G.-L., and Rosen, G.M. 1993. A fluorophore-containing nitroxide as a probe to detect superoxide and hydroxyl radical generated by stimulated neutrophils. *Anal. Biochem.* 212:85–90.

30. Yang, X.-F. and Guo, X.-Q. 2001. Study of nitroxide-linked naphthalene as a fluorescence probe for hydroxyl radicals. *Anal. Chim. Acta* 434:169–177.

31. Yang, X.-F. and Guo, X.-Q. 2001. Investigation of the anthracene–nitroxide hybrid molecule as a probe for hydroxyl radicals. *Analyst* 126:1800–1804.

32. Li, P., Xie, T., Duan, X., Yu, F., Wang, X., and Tang, B. 2010. A new highly selective and sensitive assay for fluorescence imaging of ˙OH in living cells: Effectively avoiding the interference of peroxynitrite. *Chem. Eur. J.* 16:1834–1840.

33. Maki, T., Soh, N., Fukaminato, T., Nakajima, H., Nakano, K., and Imato, T. 2009. Perylenebisimide-linked nitroxide for the detection of hydroxyl radicals. *Anal. Chim. Acta* 639:78–82.

34. Soh, N., Makihara, K., Sakoda, E., and Imato, T. 2004. A ratiometric fluorescent probe for imaging hydroxyl radicals in living cells. *Chem. Commun.* 496–497.

35. Li, D.-W., Qin, L.-X., Li, Y., Nia, R.P., Long, Y.-T., and Chen, H.-Y. 2011. CdSe/ZnS quantum dot—cytochrome c bioconjugates for selective intracellular O_2 sensing. *Chem. Commun.* 47:8539–8541.

36. Zhao, H., Kalivendi, S., Zhang, H., Joseph, J., Nithipatikom, K., Vasquez-Vivar, J., and Kalyanaraman, B. 2003. Superoxide reacts with hydroethidine but forms a fluorescent product that is distinctly different from ethidium: Potential implications in intracellular fluorescence detection of superoxide. *Free Radic. Biol. Med.* 34:1359–1368.

37. Georgiou, C.D., Papapostolou, I., Patsoukis, N., Tsegenidis, T., and Sideris, T. 2005. An ultrasensitive fluorescent assay for the in vivo quantification of superoxide radical in organisms. *Anal. Biochem.* 347:144–151.

38. Robinson, K.M., Janes, M.S., Pehar, M., Monette, J.S., Ross, M.F., Hagen, T.M., Murphy, M.P., and Beckman, J.S. 2006. Selective fluorescent imaging of superoxide in vivo using ethidium-based probes. *Proc. Natl. Acad. Sci. U.S.A.* 103:15038–15043.

39. Tang, B., Zhang, L., and Zhang, L.L. 2004. Study and application of flow injection spectrofluorimetry with a fluorescent probe of 2-(2-pyridil)-benzothiazoline for superoxide anion radicals. *Anal. Biochem.* 326:176–182.

40. Gao, J.J., Xu, K.H., Tang, B., Yin, L.L., Yang, G.W., and An, L.G. 2007. Selective detection of superoxide anion radicals generated from macrophages by using a novel fluorescent probe. *FEBS J.* 274:1725–1733.

41. Yan, Y., Wang, S., Liu, Z., Wang, H., and Huang, D. 2010. CdSe–ZnS quantum dots for selective and sensitive detection and quantification of hypochlorite. *Anal. Chem.* 82:9775–9781.

42. Yang, Y.-C., Lu, H.-H., Wang, W.-T., and Liau, I. 2011. Selective and absolute quantification of endogenous hypochlorous acid with quantum-dot conjugated microbeads. *Anal. Chem.* 83:8267–8272.

43. Kenmoku, S., Urano, Y., Kojima, H., and Nagano, T. 2007. Development of a highly specific rhodamine-based fluorescence probe for hypochlorous acid and its application to real-time imaging of phagocytosis. *J. Am. Chem. Soc.* 129:7313–7318.

44. Yang, Y.-K., Cho, H.J., Lee, J., Shin, I., and Tae, J. 2009. A rhodamine–hydroxamic acid-based fluorescent probe for hypochlorous acid and its applications to biological imaging. *Org. Lett.* 11:859–861.

45. Kim, H.N., Kim, M.J., Cho, C.-S., Lee, S.Y., and Lee, W.J. 2011. A specific and sensitive method for detection of hypochlorous acid for the imaging of microbe-induced HOCl production. *Chem. Commun.* 47:4373–4775.

46. Chen, X., Wang, X., Wang, S., Shi, W., Wang, K., and Ma, H. 2008. A highly selective and sensitive fluorescence probe for the hypochlorite anion. *Chem. Eur. J.* 14:4719–4724.

47. Sun, Z.-N., Liu, F.-Q., Chen, Y., Tam, P.K.H., and Yang, D. 2008. A highly specific BODIPY-based fluorescent probe for the detection of hypochlorous acid. *Org. Lett.* 10:2171–2174.

3 Ultraviolet Light, Chromophores, Reactive Oxygen Species, and Human Health

Shamim I. Ahmad

CONTENTS

ABSTRACT

Since the time living organisms came into existence, ultraviolet (UV) light, specifically from the sun, has been playing important roles for almost all forms of life, including plants. Out of the three major forms of solar UV light—UVA (320–400 nm), UVB (280–320 nm), and UVC (180–280 nm)—only UVA and some UVB can reach the earth, and all UVC and most UVB is absorbed by the stratospheric ozone layer. All three forms play dual roles in that they can not only impose damage but also provide beneficial effects to humans.

Effects of UV light can be direct in that the radiation can penetrate the skin and damage various cellular components, most importantly DNA. In laboratory conditions, UVC light has been shown to be the most damaging to DNA, and, if not properly repaired, can lead to various forms of skin diseases, most commonly skin cancers, including basal cell carcinoma, squamous cell carcinoma, and melanoma. The induction of skin cancers is carried out by both types of UV lights, UVB and UVA, although their mechanisms are different. UVB can induce cancer through the same mechanisms as those of UVC. On the other hand, UVA induces cancer if combined with one of the large number of man-made chemicals or naturally synthesized biological compounds acting as photosensitizers or chromophores. The damage to DNA, leading to mutation and cancer, occurs via the production of reactive oxygen species (ROS).

The most important beneficial effect reaped from UV light is the synthesis of vitamin D, which in recent studies has been proven to be a highly important compound for a healthy life. People with vitamin D deficiency (estimated to be around 50% of the global population) can suffer from a number of ailments. UV light with or without sensitizers has also been in use as a therapeutic agent to cure a number of human skin diseases.

This chapter focuses on various aspects of UVA and its interaction with sensitizers, and the production of ROS affecting human life. Other forms of UV are included where necessary.

3.1 INTRODUCTION

The two major sources of ultraviolet (UV) light are the sun and man-made UV lamps. Solar UV light contains a number of different wavelengths: from 180 to 280 nm is UVC, 280 to 320 is UVB, and 320 to 400 is UVA. Initially, experiments with UV light were carried out mostly employing UVC lamps, and they were presented as UV radiation, UV rays, or just photo-irradiation. Later, when different kinds of filters were developed to stop unwanted UV rays to go through, various types of UV lamps were produced, including UVA, B, and C types. Also, a TL-01 lamp geared to generate UV of the specific wavelength of 311 nm (Barth and Pinzer, 1990) has been produced. In the beginning, UVA light was often referred to as "black light" or "near UV light."

Nowadays, almost all kinds of tailor-made UV lamps are available, and the most commonly used are the UVA and UVB lamps employed for human skin tanning, in research, in industries to carry out chemical reactions, and in phototherapies. UVA and UVB lamps have been used to treat psoriasis and neonate jaundice (Ring et al., 2015; Woodgate and Jardine, 2015).

UV light from the sun is the most difficult to control and can penetrate the stratosphere to reach the earth. However, out of the three forms of solar UV light, the UVC cannot reach the earth as it is mostly absorbed by the stratospheric ozone layer surrounding the earth. This is fortunate, as UVC is a potent damaging agent to biological systems, causing principally cyclobutane pyrimidine dimers (CPD) and (6–4) pyrimidine–pyrimidone photoproducts in the DNA, resulting in damage, mutation, and cancer. Skin erythema is also induced by UVC light and by UVB. When it was realized that UVC cannot reach the earth, research interest on this light affecting biological systems declined. On the other hand, some UVB (about 5%–10% of the total UV) and all UVA (90%–95% of total) can reach us.

Later, more interest developed in UVA and UVB research and was found that even though UVA was quite weak to cause direct damaging effects to biological systems, it can be a powerful damaging agent via generating reactive oxygen species (ROS) when presented with a photosensitive agent—this reaction was named "synergistic effect." This chapter mostly focuses on information pooled with UVA and UVB band of light, and UVC studies will only be presented where appropriate. In order to maintain consistency, the terms "black light" and "near UV light" will be replaced by "UVA light."

3.2 CLASSIFICATION OF UV LIGHT

UVA/B light (290–400 nm) was later classified into UVB (290–320 nm), narrowband UVB (311–313 nm), man-made excimer laser (380 nm), UVA-2 (320–340 nm), and UVA-1

(340–400 nm) (Situm et al., 2014). Interestingly, UVB shows certain properties associated with UVC and UVA; hence, wherever needed, information on UVB is also included.

3.3 BACKGROUND OF UV STUDIES

Research on UV light started when the UVC lamp became available and its effects studied from irradiating a number of chemical agents (too many to mention) and subsequently on biological materials including DNA, RNA, amino acids, and lipids, as well on whole organisms. A number of viruses and bacterial species were employed for in vivo studies to measure the effects of UVC light. One such bacterial species was *Escherichia coli*, and studies with this microbe demonstrated that DNA was the most important cellular target affected by UVC. This light was shown to cause two kinds of products: (1) a link between two adjacent pyrimidine residues on the same strand of DNA, known as CPD, which normally occurs between two thymine residues or thymine to cytosine residues on the same strand to link, and (2) (6–4) pyrimidine–pyrimidone photoproducts (Beukers et al., 2008). The importance of DNA damage and its repair became apparent when wild-type *E. coli* and their mutant derivatives, deficient in DNA repair, were studied under UVC exposure (Webb and Brown, 1976).

By comparing the effects of the bacterial killing by UVC versus UVA, it was found that the latter showed little significant effect when equivalent dose of UVC light was given (Webb and Brown, 1979). However, later experiments revealed that although long-time exposure of microbes to UVA had some effects, the effect was significantly enhanced when this light was presented with a sensitizer (details presented later).

3.4 UVA AND BIOLOGICAL SYSTEMS: CHRONOLOGICAL DISCOVERIES

The earliest recorded study with UV light on biological system was when Gamlen (1903) carried out a treatment of lupus by x-rays and UV rays. Subsequently, in 1922, the importance of UV light on living organisms was shown on *Drosophila* (Lutz and Richtmyer, 1922). Encouraged by the success of these studies, Nicholls (1928) wrote an article on the application of UV rays in therapeutics. All these experiments used the term "UV light," and there is little doubt that most researchers had employed the lamps (or sources) emitting UVC (ironically, abstracts for these citations are not available). Then in 1946, the method of making filters, transmitting the UVA light (most likely UVA and UVB), and absorbing visual light was applied (Smith, 1946). Subsequently, it was shown that certain chemicals such as monoalkyl-substituted benzene can absorb UVA light and give rise to hyperconjugation and the Baker–Nathan effect (Matsen et al., 1947). This was followed by a study on photoreactivation of glyceraldehyde-3-phosphate dehydrogenase (Shugard, 1951). Studies employing *E. coli* mutants lacking manganese or iron superoxide dismutase (Mn-SOD and Fe-SOD) confirmed the role of $O_2^{\cdot-}$ in DNA damage leading to a ninefold increased mutagenesis in bacteria when exposed to UVA light (Hoerter and Eisenstark, 1989).

3.5 UVA: ITS SENSITIZERS AND ROS PRODUCTION

A breakthrough came in 1976 when the group of Eisenstark reported that one of the UVA photoproducts of tryptophan was hydrogen peroxide (H_2O_2). Its production was determined by spectrophotometric, chromatographic, and biological activities (McCormick et al., 1976). Further studies in this laboratory showed that when bacteriophage T7 was exposed to a combination of H_2O_2 and UVA, it was inactivated synergistically, which means the inactivation of phage occurred at doses at which the individual agent showed little or no effect (Ananthaswamy and Eisenstark, 1976). The experiment was repeated with a number of repair-deficient mutants, *lex A*, *recA*, *recB*, *recC*, *polA*, and *uvrA*, and their parent strain, *E. coli*. Results showed that all strains were inactivated synergistically. It was interpreted that as all the mutant strains were unable to repair the DNA damage they were highly sensitive to this photoactivation (Hartman and Eisenstark, 1978). Single-strand DNA breaks in human cells were also demonstrated to be synergistically affected (Wang et al., 1980).

At this stage, I joined the team of Eisenstark to look for the photolytic product(s) of H_2O_2 and the reaction mechanisms involved. A number of experimental data using the reduction reaction of nitroblue tetrazolium (NBT) to formazan blue (Auclair et al., 1978) and the oxidation of NADPH showed that indeed superoxide anion ($O_2^{\cdot-}$) was produced when H_2O_2 was photolyzed by UVA. Research for the generation of hydroxyl radical ($^{\cdot}$OH) failed, suggesting that either this radical was not produced or, if it was, our detection techniques could not pick it up. Based on these results, the following reaction was proposed:

$$NBT + NUV \rightarrow NBT^*$$

$$NBT^* + H_2O_2 \rightarrow NBTH^{\cdot} + H^+ + O_2^{\cdot-}$$

$$NBTH^{\cdot} + O_2 \rightarrow NBT + H^+ + O_2^{\cdot-}$$

In this study, it was also revealed that the production of $O_2^{\cdot-}$ occurred maximally at 310 nm of the UVA light. Furthermore, pH had an effect on the reduction of NBT in that as the pH increased from 7 to 9, the OD_{560} of the reaction mixture increased 10-fold. Additional evidence of $O_2^{\cdot-}$ came when phage T7 inactivation was found to be significantly reduced in the presence of SOD (the enzymes specifically involved in scavenging $O_2^{\cdot-}$). But when $^{\cdot}$OH scavengers (sodium benzoate, potassium iodide, mannitol, and isopropyl alcohol) were used, no effects were observed, suggesting that most likely this radical was not produced from UVA photolysis of H_2O_2 (Ahmad, 1981). Ironically, soon after the Elsevier's publications of the *Photobiochemistry and Photobiophysics* had stopped, and possibly for this reason the paper could not be cited in PubMed.

Subsequent exploration of other agents able to produce ROS showed that a number of antibiotics, including lincomycin, cephalothin, and erythromycin, also generated $O_2^{\cdot-}$ when exposed to sunlight, UVA (320–400 nm), or UVB (290–320). Moreover, it was suggested that precautions ought to be taken when treating patients with these drugs to be exposed to UV radiations especially in tropical and subtropical countries (Ray et al., 2001). Other chemicals able to produce ROS from UV photolysis were the polycyclic aromatic hydrocarbons (PAHs, a class of genotoxic environmental contaminants). It was also found that ROS produced through this photolysis caused (1) cytotoxicity and DNA damage, (2) single-strand DNA breaks, (3) the formation of 8-oxo-deoxyguanosine (8-OXO-dG), and (4) lipid peroxidation (Yu et al., 2006). These results may implicate $^{\cdot}$OH production in the process. In another study, nitropolycyclic aromatic hydrocarbons were added to the list (Xia et al., 2013).

Naproxen, a nonsteroidal anti-inflammatory drug, can also induce photocytotoxicity via ROS production when photolyzed by UVA (Bracchitta et al., 2013). Protein *S*-glutathionylation, the oxidation of the proliferating cell nuclear antigen, the oxidation of cellular tryptophan, the phosphorylation of Chk1, and the inhibition of DNA replication were also recorded. Also, the immunosuppressant, azathioprine, and the fluoroquinolone antibiotics, ciprofloxacin and ofloxacin, produced ROS, and this was found to diminish the Nucleotide Excision Repair system of DNA. This diminution has been associated with the damage to one or more proteins involved in the repair pathway (Guven et al., 2015). Nano-titanium dioxide, a potent inducer of apoptosis able to produce ROS under UVA photolysis, was also added to the list (Xue et al., 2015). UVA photolytic product of quinine in human skin cells was shown to be 6-methoxy-quinolin-4-ylmethyl-oxonium leading to the production of $^1O_2, O_2^{\cdot-}$, and $^{\cdot}$OH. The ROS thus produced was responsible for the degradation of 2-oxyguanosine, single-strand DNA breaks, arresting of G2 phase of cell cycle, and induced apoptosis. It also caused the upregulation of p21 and p53 gene expressions (Yadav et al., 2013). The implication of the data is that precautions need to be taken for excessive exposure to sunlight when treating with such agents.

Biologically available iron plays a key role in both oxidative stress and photoinduced skin damage, and the main causes of oxidative stress include ROS produced in the skin by UVA and biologically available iron (Wright et al., 2014). The recently reported UV sensitizer, benzophenone (used as sunscreen protector), has been shown to produce $^1O_2$, $O_2^{\cdot-}$, and $^{\cdot}OH$ through type I and type II photodynamic mechanisms. It has been suggested that this agent may be replaced by a safer agent in the cosmetic preparation of topical applications (Amar et al., 2015). Interestingly, polystyrene was also added to the list of UVA sensitizers; it is hard to envisage what effect it may have on human health as this material is commonly used in daily life (Hoerter and Eisenstark, 1988).

3.6 BIOLOGICAL SENSITIZERS OF UVA

Attempts to discover new UVA sensitizers were also based on the NBT detection system. Upon screening a large number of biochemical compounds, four of them—phenylalanine, tyrosine, tryptophan, and histidine—gave positive results (Craggs et al., 1994; Paretzoglou et al., 1998). Subsequent attempts to find more of these used electro-paramagnetic resonance (EPR) analysis besides NBT reaction; results confirmed that L-mandelate and β-phenyl pyruvic acid also produced $^{\cdot}OH$ and H_2O_2 (Ahmad et al., 2004; Hargreaves et al., 2007). Additionally, all these compounds were able to synergistically inactivate phage T7 with high efficiency, implying that either the ROS detected in these studies were involved in T7 hypersensitivity or that there may be other reactive molecules, such as $^1O_2$ produced, during the photolysis of these compounds that could not be detected by the methods applied.

3.7 COMPARATIVE STUDIES OF VARIOUS WAVEBANDS OF UV LIGHT

A comparative study employing cultured adult rat liver cells, H_2O_2 alone, and H_2O_2 plus UVA or UVB revealed that H_2O_2 alone increased the production of 8-OXO-dG by 42% compared to its presence in nontreated cultures. UVB exposure alone, on the other hand, raised this level to 8.4% only, whereas in the presence of both these agents a synergistic action increased the level of 8-OXO-dG to 155%. In contrast, UVA light alone did not increase the 8-OXO-dG level, but when combined with H_2O_2, the level increased to 310%. It was concluded that both UVA and UVB were able to generate ROS-oxidizing deoxyguanosine in the DNA (Rosen et al., 1996).

Studies on the wavelength dependence of UV radiation were also carried out on a number of bacterial isolates, and the results were as expected in that the inactivation of all isolates was the highest for UVC and the lowest for UVA, suggesting a weaker action of the latter. Furthermore, the generation of ROS and protein and lipid oxidation followed the same pattern, although the DNA double-strand breaks (DSB) showed an inverse trend. The conclusion was that each type of UV was responsible in inducing DSB and lipid oxidation, most likely generated by ROS (Santos et al., 2013).

3.8 UVA AND MICROORGANISMS

An early study on the effects of UVA on *E. coli* and *Streptomyces gricius* showed that photoactivation of the latter species occurred in the region between 365 and 500 nm with the most effective near 436 nm. On the other hand, the activation of *E. coli* occurred in the region between 365 and 470 nm with the most effective wavelength lying near 375 nm (Kelner, 1951). This is intriguing as the two different microbial species are showing two different sensitivities to the UV radiation. One explanation is the different genetic makeup of the microorganisms used. Another study using a mercury arc lamp showed that *E. coli* sustained three effects—division delay, growth delay, and photoprotection—all induced by a common type of critical event (Phillips et al., 1967). Then, by using a variety of DNA repair–deficient mutants of *E. coli*, it was shown that the UVA photoproduct of L-tryptophan was inhibiting DNA replication (Yoakum et al., 1974).

Subsequent studies on the importance of fluorescent and photo lamps demonstrated that the inactivation of bacteria and phage was repairable in these microorganisms although recombination-deficient mutants of *Salmonella typhimurium* (most likely *recA*, Ferron et al., 1972) were more sensitive to visible and UVA light (Eisenstark, 1970a,b). These results presented additional information that not only UVA but also visible light from fluorescent and photo lamps is harmful to biological systems.

3.9   UVA AND EUKARYOTES

UVA exposure in mouse embryonic fibroblasts and fertilized sea urchin eggs in the presence of tryptophan resulted in marked inhibition of the cells to incorporate labeled precursors of protein, RNA, and DNA, implying that the presence of this amino acid plays an important role in cell activities (Zigman and Hare, 1976). The generation of ROS, including $^{\bullet}OH$ by UVA photolysis, was confirmed when HaCaT keratinocytes were exposed to this light. H_2O_2 production was emphasized when no photoreactive agents were included in the reaction mixture. Hence, it was established that H_2O_2 played a central role in damaging DNA, causing single-strand breaks. $O_2^{\bullet-}$, if any produced, was suggested to be a likely substrate for H_2O_2 production (Petersen et al., 2000). In another study, mouse ocular tissues were exposed to this light for a period of 19 weeks. Unexpected results were observed in that no histological changes were found in the cornea and it was interpreted that a repair mechanism may have played a role in preventing any damage (Zigman, 1975).

Irradiation of human lenses from an aged group (55–75 years) to UVA light for 1 h resulted in 70% loss of glutathione reductase (GR)-specific activity and a 24% reduction in glyceraldehyde-3-phosphate dehydrogenase (G3PD) activities. Thus, ROS have been implicated in these inactivations (Linetsky et al., 2003) and possibly certain eye diseases.

3.10   UVA, ROS, AND HUMAN DISEASES

A large number of human diseases have been found to be induced via oxidative stress, and this book attempts to cover as much of them as possible. Skin cancer is an important and major disease. This is a vast subject (requiring probably an entire book) and is presented in this chapter in its highly abridged form. Certain other diseases, not covered in this book, are photokeratitis and photo-conjunctivitis, cataract, pterygium, and squamous cell carcinoma (SCC) of the cornea and conjunctiva. Also ocular melanoma, age related macular degeneration, ageing, and circadian clock, all are associated with the UV exposure (Lucas, 2011; Lai et al., 2012; Sohal and Orr, 2012).

3.10.1   ROS AND SKIN CANCER

Melanoma and nonmelanoma skin cancers, including SCC, actinic keratosis, and basal cell carcinoma (BCC), are on the rise globally, specially among the people of Caucasian origin; the reason being the inadequate protection from solar radiation, specifically UVA, UVA plus endogenous chromophores, and UVB. Studies have shown that the occurrence of BCC is strongly correlated with intermittent UV exposure (particularly those at childhood or adolescence) while SCC is more strongly linked to constant or cumulative sun exposure (Calzavara-Pinton et al., 2015).

A review by Epstein (1978) suggested that sun is the primary stimulus for most human skin cancer formation. Yet, it has been mostly animal experiments that gave rise to data on action spectra, time–dose relationship, energy level requirements, etc. UV light between 280 and 320 nm was primarily responsible for tumor formation. Environmental factors such as heat wind and moisture also played a role in the stimulation of tumor.

Later studies to determine the type and the mechanism of inducing skin cancers used UVA (315–400 nm) and UVA-1 (340–400 nm) lights and found that UVA-1 (340–400 nm) induced SCC mediated through ROS largely without the characteristic point mutation in P53, whereas both UVB (280–315 nm) and UVA (315–400 nm) gave rise to ROS-related point mutation G to T in P53 and crude genomic alterations such as deletions in both SCC and BCC (de Gruijl, 2002).

Human 8-oxoguanine-DNA glycosylase (hOGG1) is the enzyme that participates in the repair of DNA affected by UV and ROS that induce the development of BCC. The mechanism is as follows: if the damaged DNA with 8-OXO-dG is not repaired prior to cell division, then the gene mutation may persist in daughter cells that may culminate into BCC. Furthermore, the dermis with a low level of hOGG1 protein was more susceptible to BCC development than the dermis with a normal level of hOGG1. It was interpreted that the low-level expression of hOGG1 and inefficient DNA repair may be the prime reasons for the development of BCC via the accumulation of ROS, which generated and induced the production of 8-OXO-dG (Huang et al., 2012).

With the growing worldwide incidence of malignant melanoma caused mostly due to solar exposure, an in vitro study was carried out to understand the role of melanin in the skin. The effects of UVA/UVB plus H_2O_2 have been studied in human melanoma cells. Three types of cells—nonpigmented (CHL-1, A375), moderately pigmented (FM55, SKmel23), and highly pigmented (FM49, hyperpigmented FM55)—were employed. Using UVA, UVB as cellular stressors, it was observed that cellular melanin played a role like a double-edged sword in that it acted as a photoprotector and as a photosensitizer. A positive effect of melanin action was that it protected cells against mitochondrial $O_2{}^{\bullet-}$ generation, thereby preventing mtDNA damage; however, when melanin binded directly to DNA it acted as a direct sensitizer of mtDNA damage during UVA irradiation (Swalwell et al., 2012).

In terms of endogenous ROS production, the complex II in mitochondrial respiratory chain (RC) has been suggested as a major generator of ROS, which is thought to be an underlying cause of carcinogenesis in many tissues and the general aging process. Skin cells, as they are more exposed to sunlight, especially UVA, are supposed to be most affected by oxidative stress. Tests on dermal and epidermal skin cells indicated that the complex II inhibitor, TTFA (thenoyltrifluoroacetone), was the only respiratory chain inhibitor to significantly increase UVA-induced ROS production in skin cells. Furthermore, the activities of RC enzymes decreased with increasing age, and telomere length is correlated with aging (Anderson et al., 2014).

3.11 BENEFICIAL ASPECTS OF UV RADIATION

Although UV light has been branded as an enemy of human health (as discussed earlier), a large number of benefits can also be reaped from it. These include the treatment of a large number of skin diseases, such as psoriasis, vitiligo, atopic dermatitis, and localized scleroderma, either by solar UV (heliotherapy) or by artificial UV radiation (phototherapy). UV can also suppress the clinical symptoms of multiple sclerosis and can generate nitric oxide (NO), which can act as antimicrobial agent, reduce blood pressure, and improve the function of the cardiovascular system. NO can also act as a neurotransmitter and improve mood due to the release of endorphins (Juzeniene and Moan, 2012). Another important use of UV is the curing of jaundice in neonates. Mostly, preterm babies develop jaundice 2–4 days after birth in which bilirubin deposition in skin becomes apparent. Jaundice, normally, is the result of an increase in red blood cell breakdown into bilirubin and accumulation due to reduction in excretion (Woodgate and Jardine, 2015). Most neonates with jaundice are treated with UVA or UVB lamps. For successful treatment, the exposure should be for more than 48 h.

3.12 PSORALEN PLUS UVA IN THERAPY

Furocoumarins, which include psoralen compounds, have been in use for centuries in the form of plant extracts to treat skin disorders such as vitiligo. They are present in many kinds of plants and fruits, including parsnips, celery, parsley, fig, and citrus fruits. Although a large number of furocoumarin derivatives have been synthesized, including trioxalen (Isaacs et al., 1977), and tested as therapeutic agents for a number of diseases, 8-methoxypsoralen (MOP) is used most commonly. MOP on its own has little therapeutic activity but when combined with UVA, its activity is significantly enhanced. This combination is commonly known as PUVA treatment.

PUVA principally induces DNA damage, causing interstrand cross-links. The three-stage reaction starts with MOP intercalating in the DNA grove and forming a noncovalent bond with a pyrimidine base in the DNA. In stage 2, upon receiving a UVA photon, a monoadduct forms between the psoralen and the base. The base involved usually is a thymine base, although photoaddition of cytosine can also occur (Calvin and Hanawalt, 1987). In stage 3, the absorption of a second photon by the thymine monoadduct causes it to form a covalent adduct with another pyrimidine base in the complementary DNA strand, causing interstrand cross-links. Further studies have shown that this additional reaction involves the 4,5 double bond of the pyrone ring at one end at 3,4 double bond with furan ring at the other (Dall'Acqua et al., 1968; Hearst et al., 1984).

Later studies on the photolysis of MOP by UVA revealed that this reaction also generated $^{\cdot}OH$, $O_2^{\cdot -}$, and $^1O_2$ (Decuyper et al., 1983; Aboul-Enein et al., 2003; Orimo et al., 2006). Confirmation of $^{\cdot}OH$ was also carried out from the generation of 8-oxo-dG in DNA (Liu et al., 1999). The DNA damage if not effectively repaired can lead to mutagenesis, which was recorded fairly early (Igali et al., 1970).

In a recent study, it was shown that unlike MOP, which upon UVA photolysis produces ROS and damages DNA by causing interstrand cross-links, trioxalen (another form of methoxypsoralen) does not produce ROS but damage DNA just by causing interstrand cross-links (Ahmad et al., 2012).

Although research shows that UV light has been in use as a therapeutic agent for long (Nicholls, 1928), in recent years, PUVA treatment was tried for a large number of skin diseases, including localized morphoea (Usmani et al., 2008); cutaneous T-cell lymphoma (Geskin, 2007); palmoplantar psoriasis (Sezer et al., 2007); anetodermic mastocytosis (Del Pozo et al., 2007); systemic sclerosis (Sakakibara et al., 2008); Schamberg disease (Seckin et al., 2008), multiple melanoma skin cancer, especially squamous cell carcinoma (Sarnoff et al., 2008); Woringer–Kolopp disease (Lee et al., 2008); folliculotropic mycosis fungoidis (Gerami et al., 2008); mycosis fungoidis (Trautinger, 2011); and vitiligo (Bruckner, 1979). Interestingly, in Spain in a study during 1982 and 1996, 41 skin diseases, including psoriasis and cutaneous T-cell lymphomas, were attempted to be treated with PUVA (Grau-Salvat et al., 2007). Also, in one study, the effect of PUVA on the apoptosis of HL-60 leukemia cells was tested, and it was found that the growth of HL-60 cells was inhibited by PUVA in a time- and concentration-dependent manner by inducing cell apoptosis (Chen et al., 2008).

Additionally, proposals have been put forward that HUVA (H_2O_2 + UVA) treatment may be tried to cure leprosy (Ahmad, 2001). Another work proposed a combination of ferrous sulfate and UVA (known to generate ROS) to treat a variety of post-burn bacterial infections, including the notorious *Pseudomonas*, which is resistant to many antibiotics (Ahmad and Iranzo, 2003). Also, H_2O_2 (0.6%–3%) may be tried to cure (or reduce) certain solid tumors by injecting it directly using a multithronged injection (Symons et al., 2001).

Azelaic acid (AzA), used to treat inflammatory skin disease such as acne and rosacea, has been tested on the effect left on the skin after the PUVA treatment. AzA appeared to counteract stress-induced premature cell senescence. Its ability to activate PPAR-gamma (a nuclear receptor that plays a relevant role in inflammation and aging) has been postulated to impose reduction in senescence (Briganti et al., 2013).

3.12.1 PUVA THERAPY OF PSORIASIS

PUVA has been successfully used for a long time for the treatment of psoriasis. The common practice is to administer psoralen orally (or as a bath) and subsequently expose the affected area to UVA light (Wolf, 2013). Pathak et al. (1981) presented a comprehensive report on this treatment. Another study reported to achieve good to excellent result (9 out of 10) on 15 patients tried upon. Two forms of psoralen, MOP + UVA and trimethylpsoralen (TMP) + UVA, were employed, and better results were obtained by the former agent (Ros and Wennersten, 1987). Oxidative damage by PUVA has been reported when calf thymus DNA was exposed to these agents; a sixfold increased production of 8-OHdG was recorded in comparison with nontreated DNA. The addition of ROS scavengers such as sodium azide and genistein significantly reduced PUVA-induced 8-OHdG formation, whereas catalase, SOD, and mannitol exhibited no effect (Liu et al., 1999). The implication is that the ˙OH formed could attack DNA, generating 8-OHdG.

3.13   UV AND VITAMIN D

UV's role to synthesize vitamin D was discovered in 1987 when it was shown that 7-dehydrocholesterol is transformed to vitamin D in two stages: in stage 1, this chemical is photolyzed by UV light in a 6-electron conrotatory ring-opening electrocyclic reaction (possibly involving 7-dehydrocholesterol as referred by Wacker and Holick, 2013), and the product is spontaneously isomerized product is previtamin D3. In stage 2, the products is spontaneously isomerized to vitamin D3 or cholecalciferol in an antrafacial sigmatropic hydride shift (Holick, 1987). Previtamin D3 and vitamin D3, after absorbing UVB, are converted to a variety of photoproducts including 1,25-dihydroxyvitamin D3 (calcitriol) which plays an important role in regulating calcium and phosphate metabolism for the maintenance of metabolic functions. The product, 1,25-hydroxyvitamin D is a ligand for Vitamin D receptor, also a transcription factor, binding to sites in the DNA called Vitamin D response elements (VDREs) is then comes to the Vitamin D receptors, present in thousands of the binding sites regulating hundreds of genes in a cell specific manner (Bikle, 2014).

Calcitriol influences a large number of biological pathways and plays a role in the reduction of chronic diseases such as autoimmune diseases, some cancers, cardiovascular diseases, infectious diseases, schizophrenia, and type 2 diabetes. Its analogs have been suggested to be used for the treatment of immunological, inflammatory, and infectious skin diseases; also, it can reduce UV-induced DNA damage (Lehmann and Meurer, 2010; Wacker and Holick, 2013). Later, it was conformed that it is UVB in the range of 280–320 nm that carries out the transformation reaction (Bendik et al., 2014). Vitamin D3 can be either synthesized endogenously or obtained from food such as fatty fish, fish liver oils, and hen's eggs, especially those fed on vitamin D (Bendik et al., 2014).

3.13.1   Diseases of Vitamin D Deficiency

It is estimated that approximately one billion people worldwide are vitamin D deficient (Villacis et al., 2014), and this can lead to a number of diseases, including the bone disease osteoporosis; certain autoimmune diseases due to poor immune function such as psoriasis, multiple sclerosis, asthma; different types of cancers, and cardiac problems such as hypertension (Holick, 2014). Measurement of vitamin D in the human body is carried out by determining the 25-hydroxy-vitamin D in plasma, a level of around 75 nmol/L is considered normal and 25 nmol/L is considered deficiency (Bendik et al., 2014).

In an alarming report, it was shown that in 1978 in the United States alone there were 1000–2000 suntan saloons that used a type of fluorescent UVB source that emitted highly injurious sunburn rays with wavelengths between 290 and 320 nm. These rays can cause acute cellular injury and erythema and at least at the experimental conditions are the most carcinogenic (Epstein, 1981). On the other hand, inappropriate amounts of vitamin D can damage the kidneys; hence, this must only be given under strict control.

For this reason UV light, on the one hand, has been branded as a foe and, on the other, a friend. There is controversial debate among the scientific community as to how much UV light is appropriate to bring a balance between the positive and negative effects of the solar and man-made UV generators. The debate has reached to the recommendation stage that in case of vitamin D deficiency, oral supplements should be prescribed, and as UV radiation has been confirmed to be highly injurious and carcinogenic (specially for skin), exposure to natural or artificial UV radiation should not be prescribed. Also, the population must be warned against misleading advertising from the tanning industry (Leccia, 2013).

3.13.2 Diseases Affected by UV

Children's genetic diseases in which UV plays important roles are xeroderma pigmentosum and Fanconi anemia.

3.13.2.1 Xeroderma Pigmentosum

The most important clinical feature of xeroderma pigmentosum (XP) is the extreme photosensitivity of patients to UV light (especially to solar UV, hence they are protected as much as possible from exposure to the sun as soon as the disease is diagnosed). Other skin injuries include cutaneous atrophy, actinic keratosis, sunburn with blistering, erythema, marked freckle-like pigmentation of the face, freckles, letigen, and nontender ulcerated nodular lesion on the nose. Nonskin diseases include ocular neoplasms, photophobia, impaired vision, and corneal and conjunctival abnormalities. Neurological dysfunctions that include acquired microcephaly diminished or absent deep tendon stretch reflexes, progressive sensorineural hearing loss, and progressive cognitive impairment. According to Wikipedia less than 40% of XP patients survive beyond the age of 20 years and some with less severe cases can live into their 40. A large number of XP genes have now been identified (Ahmad and Hanaoka, 2008).

3.13.2.2 Fanconi Anemia

Unlike XP, Fanconi anemia (FA) in children is not directly linked to UV but involves oxidative stress, and cells suffering from FA are highly sensitive to PUVA and other DNA-intercalating agents, including mitomycin C, diepoxybutane, *cis*-platin, cyclophosphamide, and hexavalent chromium compounds (Fendrick and Hallick, 1984). DNA repair deficiency is the reason for this disease, and a large number of genes have been identified associated with a set of redox abnormalities and sensitivity. Also, mitochondrial dysfunction (MD) has been associated with the disease. Further studies employing PUVA to treat fibroblast from patients with FA revealed that their mitochondria, due to loss of the cellular defense system against ROS, suffer from mitochondrial matrix densification (Pagano et al., 2012). The result was interpreted as some kind of membrane damage caused this densification (Rousset et al., 2002). It is likely that ROS may have caused lipid peroxidation, resulting in the densification. Genetic data show the involvement of FANCA, C, D2, and G genes in the phenotype (Pagano et al., 2013) alongside many other genes discovered for this disease (Ahmad and Kirk, 2006).

The clinical features of this disease include hematological abnormalities; leukemia; solid tumors in the liver, head, and neck; oesophageal and vulvar carcinoma; Wilms tumors; and brain tumor, including medulla blastoma and astrocytoma. Other features include congenital abnormalities such as microcephaly and abnormalities of skin pigmentation, slower growth both in utero and after birth, and lower birth rate. Upper limb abnormalities, absent radius and thumb anomalies, renal abnormalities (e.g., horseshoe kidney), cardiopulmonary defects, deafness, and neurological abnormalities were also observed (Ahmad and Kirk, 2006).

3.14 SCAVENGERS OF ROS

Although ROS including H_2O_2, $O_2^{\cdot-}$, $^{\cdot}OH$, and certain organic and lipid peroxides have potentials to damage nucleic acids, proteins, and membranes, mechanisms have evolved in living systems including in man to scavenge these damaging agents. A number of antioxidants have been identified both endogenously produced and others may be taken from exogenous sources; these include uric acid, ascorbic acid or vitamin C, glutathione, melatonin and tocopherol or vitamin E. Also included in the scavengers are the enzymes superoxide dismutase, catalase, peroxiredoxins, thioredoxin, glutathione reductase, glutathione peroxidase, and glutathione transferase.

Studies on the regulatory mechanisms involved in ROS regulons, in *E. coli*, show that the soxRS regulon controls the synthesis of Mn-SOD, oxyR controls catalase HPI, and rpoS positively regulates HPII. On the other hand, *delta fur* (ferric uptake regulator) plays an important role in defense against UVA radiation: in addition to regulating all the genes directly involved in iron acquisition it also regulates the expression of Mn-SOD and Fe-SOD, the key enzymes against oxygen toxicity in the bacterium. Mutation in *fur* gene makes the strain hypersensitive to a number of oxidative agents including UVA (Eisenstark et al., 1995; Hoerter et al., 2005). In another study, it was revealed that *rpos* (Sigma-38 factor of RNA polymerase) mutant of *E. coli* is hypersensitive to near UV light plus H_2O_2 and is the major regulator of genes for their survival in their stationary phase, whereas *oxyR* gene product (a transcriptional regulator) responds to H_2O_2-induced stress in exponential phase (Becker-Hapak and Eisenstark, 1995; Eisenstark et al., 1996). Subsequent studies added glutathione reductase, exonuclease III, and DNA glycosylase (Eisenstark, 1998), the enzymes involved in DNA repair in *E. coli*.

Polyphenols are another set of compounds that are known to scavenge ROS. As green tea has been found to be rich in polyphenol, studies carried out to determine their values in human health, 10 healthy volunteres were given green tea (540 mL). Samples of whole blood were obtained 30, 60, and 90 min after drinking the tea. From the blood, the leukocytes were obtained and exposed to increasing period of UVA/VIS light. Results showed that drinking green tea significantly reduced the genotoxic effects of radiation. It was concluded that polyphenols had acted as ROS scavengers present in the green tea (Malhomme de la Roche et al., 2010).

Azelaic acid has recently been added in the list of antioxidant (Briganti et al., 2013). Also, chromene sargachromanol E, isolated from the marine brown algae, *Sargassum horneri*, has been found to inhibit UVA-induced aging of skin in human dermal fibroblasts (Kim et al., 2015).

A study conducted to determine the protective effect of antioxidants, MitoQ and tiron, on mitochondrial DNA (mtDNA) damage in human dermal fibroblast showed that tiron (EC50 10 mM) gave complete protection against both UVA- and H_2O_2-induced mtDNA damage, whereas MitoQ (EC50 750 nM) provided less protection. Furthermore, the tiron's protective effect against H_2O_2-induced nuclear DNA damage was greater than the cellular antioxidant and MitoQ (Oyewole et al., 2014). Resveratrol, a component of red grapes, has also been found to exert protective effect against the damage by UVA onto the retinal pigment epithelial (RPE) cells. Age-related macular degeneration (AMD) has been associated with the oxidative injury to the retinal pigment epithelium. As resveratrol has been found to provide protective effects to several other cell types, the study may be extended with an aim to use it as a chemoprotective agent for the prevention of AMD (Chan et al., 2015). No wonder why red wine and green teas have gained public awareness for healthy life.

Benzophenone (BP) has been used in the sunscreen to protect the skin from the harmful sun exposure. In a recent study, it has been claimed that this chemical agent on one hand may be providing protection, and on the other it produces ROS of various types including $^1O_2$, $O_2^{\cdot-}$, and $^\cdot OH$ through type 1 and type 2 photodynamic mechanisms. The phototoxicity was shown to induce pyrimidine dimers and single-strand DNA breaks in DNA. Also, lipid peroxidation and LDH leakage were enhanced by BP. Based on these adverse effects, it is proposed that its replacement should be found from the cosmetic preparations of topical application (Amar et al., 2015).

In a recent study, a number of naturally occurring ROS scavengers have been discovered and as the method employed is unique and intriguing it is worth briefing it here. The work started from the isolation of a PUVA hyper-resistant mutant of *E. coli*. A 1D PAGE analysis of this mutant identified a protein of ~55 kDa in higher concentration than its parent strain (Holland et al., 1991). Next, a heavier spot was seen on the 2D PAGE analysis from the cell extract in comparison to the lighter spot at the same position of the extract from the parent strain. An end terminal amino acid sequence of the heavy spot identified it as malate dehydrogenase. Next, tests on *E. coli* mutant of malate dehydrogenase showed that it was highly sensitive to 8-methoxypsoralen + ultraviolet A (PUVA). From the results, the possible involvement of this enzyme was pointed out in determining the hyper-resistance, and it was suggested that most likely the enzyme was a ROS scavenger. The experiment was extended with a large number of other putative *E. coli* mutants, and it was found that succinate dehydrogenase as well as NADH: ubiquinone oxidoreductase (respiratory gene complex) was sensitive. Another unidentified mutation gene (*zhd*) in *E. coli* was also found to be sensitive to PUVA. Additional confirmatory tests were carried out for confirmation and conclusion drawn (Ahmad et al., 2012).

3.15 UV DOSIMETER

With the awareness of the harmful effects of UVA and UVB light, especially induction of the vigorous type skin cancer, melanoma, attempts have been made to produce an economical and reliable dosimeter, which can be used by the sunseekers and outdoor workers to avoid over UV exposure and subsequent skin cancer. A miniaturized UVA dosimeter using polyphenylene was developed previously, but its commercial success remained to be questionable. However, recently, a UVA dosimeter has been developed, which uses 8-MOP to measure UVA (320–400 nm) dose over a long period. A Mylar as a UVB filter has been added to it to cut off the UVB light going through the sensor (Wainwright et al., 2015).

3.16 CONCLUSION

Solar light including the heat play the most important roles in our life. Also the solar UV components are equally important as provider of the vitamin D. They also play important roles in the supply of oxygen through the photosynthetic chain reactions. On the other hand, a number of human diseases are also induced by UV light, most importantly the skin cancers. If we balance out the two contradictory roles of UV, we can see that the balance tilts to more for pros than cons. Further tilting of the balance toward pros can be considered simply by reducing the exposure of our body to the sunlight, which ironically is not taken into consideration nowadays—skin tanning for cosmetic reason is on increase so are the skin cancers. Also, the misleading information that we should get UV from the solaria (and sun) for compromising the vitamin D requirements for our good health ought to be discouraged. The advice is that the body's requirement of vitamin D ought to be provided through food containing this vitamin.

Nature has provided this huge wealth (the sun) to living organisms and plants and it is our duty to make the best use of this treasure in the best possible ways, and for this we need more knowledge on this subject via research as well dispersing the knowledge in public in the best possible ways.

REFERENCES

Aboul-Enein HY, Klanda A, Kruk I et al. 2003. Effect of psoralen on Fenton-like reaction generating reactive oxygen species. *Biopolymers* 72:59–68.

Ahmad SI. 1981. Synergistic action of near ultraviolet radiation and hydrogen peroxide on the killing of coliphage T7: Possible role of superoxide radical. *Photobiochem Photobiophys* 2:173–180.

Ahmad SI. 2001. Control of skin infections by a combined action of ultraviolet A (from sun or UVA lamp) and hydrogen peroxide (HUVA therapy), with special emphasis on leprosy. *Med Hypothesis* 57(4):484–486.

Ahmad SI, Hanaoka F (Eds.). 2008. *Molecular Mechanisms of Xeroderma Pigmentosum Advances in Experimental Medicine and Biology.* Landes Bioscience Publication, Austin, TX.

Ahmad SI, Hargreaves A, Taiwo FA. 2004. Near UV photolysis of L-mandelate, formation of reactive oxygen species, inactivation of phage T7 and implication on human health. *J Photochem Photobiol B* 77(1–3):55–62.

Ahmad SI, Iranzo OG. 2003. Treatment of bacterial infections by Fenton reagent, particularly the ubiquitous multiple drug resistant *Pseudomonas* spp. *Med Hypothesis* 61(4):431–434.

Ahmad SI, Kirk SH (Eds.). 2006. *Molecular Mechanisms of Fanconi Anemia.* Landes Bioscience Publication, Austin, TX.

Ahmad SI, Yokoi M, Hanaoka F. 2012. Identification of new scavengers for hydroxyl radicals and superoxide by utilizing ultraviolet A photoreaction of 8-methoxypsoralen and a variety of mutants of *Escherichia coli*: Implications on certain diseases of DNA repair deficiency. *J Photochem Photobiol B* 116:30–36.

Amar SK, Goyal S, Mujtaba SF et al. 2015. Role of type I and type II reactions in DNA damage and activation of caspase 3 via mitochondrial pathway induced by photosensitized benzophenone. *Toxicocol Lett* 235(2):84–95.

Ananthaswamy HN, Eisenstark A. 1976. Near-UV-induced breaks in phage DNA: Sensitization by hydrogen peroxide (a tryptophan photoproduct) *Photochem Photobiol* 24(5):439–442.

Anderson A, Bowman A, Boulton SJ et al. 2014. A role for human mitochondrial complex II in the production of reactive oxygen species in human skin. *Redox Biol* 2C:1016–1022.

Auclair C, Torres M, Hakim J. 1978. Superoxide anion involvement in NBT reduction catalyzed by NADPH-cytochrome P-450 reductase: A pitfall *FEBS Lett* 89(1):26–28.

Barth J, Pinzer B. 1990. Therapy of psoriasis with the Phillips TL01 ultraviolet lamp. *Dermatol Monatsschr* 176(11):707–710.

Becker-Hapak M, Eisenstark A. 1995. Role of rpoS in the regulation of glutathione oxidoreductase (gor) in *Escherichia coli*. *FEMS Microbiol Lett* 134(1):39–44.

Bendik I, Friedel A, Roos FF et al. 2014. Vitamin D: A critical and essential micronutrient for human health. *Front Physiol* 5:248.

Beukers R, Eker AP, Lohman PH. 2008. 50 years thymine dimers. *DNA Repair (Amst)* 7(3):530–543.

Bikle DD. 2014. Vitamin D metabolism, mechanism of action, and clinical applications. *Chem Biol* 21(3):319–329.

Bracchitta G, Catalfo A, Martineau S et al. 2013. Investigation of the phototoxicity of naproxen, a non-steroidal anti-inflammatory drug, in human fibroblasts. *Photochem Photobiol Sci* 12(5):911–922.

Briganti S, Flori E, Mastrofrancesco A, Kovacs D. 2013. Azelaic acid reduced senescence-like phenotype in photo-irradiated human dermal fibroblasts: Possible implication of PPARy. *Exp Dermatol* 22:41–47.

Bruckner V. 1979. Photosensitizers and radiosensitizers in dermatology and oncology. *S Afr Med J* 56(13):528–531.

Calvin NM, Hanawalt PC. 1987. Photoadducts of 8-methoxypsoralen to cytosine in DNA. *Photochem Photobiol* 45(3):323–330.

Calzavara-Pinton P, Ortel B, Venturini M. 2015. Non-melanoma skin cancer, sun exposure and sun protection. *G Ital Dermatol Venereol* 150(4):369–378.

Chan CM, Huang CH, Li HJ et al. 2015. Protective effects of resveratrol against UVA-induced damage in ARPE19 cells. *Int J Mol Sci* 16(3):5789–5802.

Chen NN, Huang SL, Xiang Y et al. 2008. Effect of psoralen plus long wave ultraviolet-A on apoptosis of HL60 leukemia cell. *Zhong Xi Yi Jie He Xue Bao* 6:852–855.

Craggs J, Kirk SH, Ahmad SI. 1994. Synergistic action of near-UV and phenylalanine, tyrosine or tryptophan on the inactivation of phage T7: Role of superoxide radicals and hydrogen peroxide. *J Photochem Photobiol B* 24(2):123–128.

Dall'Acqua F, Marciani S, Bordin F. 1968. Studies on the photoreaction (365 nm) between psoralen and thymine. *Ric Sci* 38(11):1094–1099.

Decuyper J, Piette A, Van de Vorst. 1983. Activated oxygen species produced by photoexcited furocoumarin derivatives. *Arch Int Physiol Biochim* 91:471–476

De Gruijl FR. 2002. Photocarcinogenesis: UVA vs UVB radiation. *Skin Pharmacol Appl Skin Physiol* 15(5):316–320.

Del Pozo J, Pimentel MT, Paradela S et al. 2007. Anetodermic mastocytosis: Response to PUVA therapy. *J Dermatolog Treat* 18:184–187.

Eisenstark A. 1970a. Sensitivity of *Salmonella typhimurium* recombinationless (rec) mutants to visible and near-visible light. *Mutat Res* 10(1):1–6.

Eisenstark A. 1970b. Repair in phage and bacteria inactivated by light from fluorescent and photo lamps. *Biochem Biophys Res Commun* 38(2):244–248.

Eisenstark A. 1998. Bacterial gene products in response to near-ultraviolet radiation. *Mutat Res* 422(1):85–95.

Eisenstark A, Calcutt MJ, Becker-Hapak et al. 1996. Role of *Escherichia coli* rpoS and associated genes in defense against oxidative damage. *Free Radic Biol Med* 21(7):975–973.

Eisenstark A, Yallaly P, Ivanova A et al. 1995. Genetic mechanisms involved in cellular recovery from oxidative stress. *Arch Insect Biochem Physiol* 29(2):159–173.

Epstein JH. 1978. Photocarcinogenesis: A review. *Natl Cancer Inst Monogr* 50:13–25.

Epstein JH. 1981. Suntan salon and the American skin. *South Med J* 74(7):837–840.

Fendrick JL, Hallick LM. 1984. Psoralen photoactivation of herpes simplex virus: Monoadduct and cross-link repair by Xeroderma pigmentosum and Fanconi's anemia cells. *J Invest Dermatol* 83(1 Suppl.):96s–101s.

Ferron WL, Eisenstark A, Mackay D. 1972. Distinction between far- and near-ultraviolet light killing of recombinationless (recA) *Salmonella typhimurium*. *Biochim Biophys Acta* 277(3):651–658.

Gamlen HE. 1903. Treatment of lupus by X-rays and ultraviolet rays. *Br Med J* 1(2214):1310–1313.

Gerami P, Rosen S, Kuzel T et al. 2008. Folliculotropic mycosis fungoidis: An aggressive variant of cutaneous T-cell lymphoma. *Arch Dermatol* 144:738–746.

Geskin L. 2007. ECP versus PUVA for the treatment of cutaneous T-cell lymphoma. *Skin Therapy Lett* 12:1–4.

Grau-Salvat C, Vilata-Corell JJ, Azon-Massoliver A et al. 2007. Use of psoralen plus UV-A therapy in the autonomous community in Valencia, Spain. *Actas Dermosifilogr* 98:611–616.

Guven M, Brem R, Macpherson P et al. 2015. Oxidative damage to RPA limits the nucleotide excision repair capacity of human cells. *J Invest Dermatol* 135(11):2834–2841.

Hargreaves A, Taiwo FA, Duggan O et al. 2007. Near-ultraviolet photolysis of beta-phenylpyruvic acid generates free radicals and results in DNA damage. *J Photochem Photobiol B* 89(2–3):110–116.

Hartman PS, Eisenstark A. 1978. Synergistic killing of *Escherichia coli* by near-UV radiation and hydrogen peroxide: Distinction between recA-repairable and recA nonrepairable damage. *J Bacteriol* 133(2):769–774.

Hearst JE, Isaacc ST, Kanne D. 1984. The reaction of the psoralen with deoxyribonucleic acid. *Q Rev Biophys* 17:1–44.

Hoerter J, Eisenstark A. 1988. Synergistic killing of bacteria and phage by polystyrene and ultraviolet radiation. *Environ Mol Mutagen* 12(2):261–264.

Hoerter J, Eisenstark A, Touati. 1989. Mutation by near-ultraviolet radiation in *Escherichia coli* strains lacking superoxide dismutase. *Mutat Res* 215(2):161–165.

Hoerter JD, Arnold AA, Ward CS et al. 2005. Reduced hydroperoxidases (HPI and HPII) activity in the Deltafur mutant contributes to increased sensitivity to UVA radiation in *Escherichia coli*. *J Photochem Photobiol B* 79(2):151–157.

Holland IB, Holland J, Ahmad SI. 1991. DNA damage by 8-methoxypsoralen plus near ultraviolet light (PUVA) and its repair in *Escherichia coli*: Genetic analysis. *Mutat Res* 254:289–298.

Holick MF. 1987. Photosynthesis of vitamin D in the skin: Effect of environmental and life style variables. *Fed Proc* 46(5):1876–1882.

Holick MF. 2014. Sunlight, ultraviolet radiation, vitamin D and skin cancer: How much sunlight do we need. *Adv Exp Med Biol* 810:1–16.

Huang XX, Scolver RA, Abubakar A et al. 2012. Human 8-oxoguanine-DNA glycosylase-1 is downregulated in human basal cell carcinoma. *Mol Genet Metab* 106(1):127–130.

Igali S, Bridges BA, Ashwood-Smith MJ et al. 1970. Mutagenesis in *Escherichia coli*. IV. Photosensitization to near ultraviolet light by 8-methoxypsoralen. *Mutat Res* 9(1):21–30.

Isaacs ST, Shen CK, Hearst JE et al. 1977. Synthesis and characterization of new psoralen derivatives with superior photoreactivity with DNA and RNA. *Biochemistry* 16(6):1058–1064.

Juzeniene A, Moan J. 2012. Beneficial effects of UV radiation other than via vitamin D production. *Dermatoendocrinology* 4(2):109–117.

Kelner A. 1951. Action spectra for photoreactivation of ultraviolated irradiated *Escherichia coli* and *Streptomyces griseus*. *J Gen Physiol* 34(6):835–852.

Kim ME, Jung YC, Jung I et al. 2015. Anti-inflammatory effects of ethanolic extract from *Sargassum horneri* (Turner) C. Agardh on lipopolysaccharide-stimulate macrophage activation via NF-KB pathway regulation. *Immunol Invest* 44(2):137–146.

Lai AG, Doherty CJ, Mueller-Roeber B et al. 2012. Circadian clock-associated 1 regulates ROS homeostasis and oxidative stress responses. *Proc Natl Acad Sci USA* 109(42):17129–17134.

Leccia MT. 2013. Skin, sun exposure and vitamin D: Facts and controversies. *Ann Dermatol Venereol* 140(3):176–182.

Lee J, Viakhireva N, Cesca C et al. 2008. Clinicopathologic features and treatment outcomes in Woringer-Kolopp disease. *J Am Acad Dermatol* 59:706–712.

Lehmann B, Meurer M. 2010. Vitamin D metabolism. *Dermatol Ther* 23(1):2–12.

Linetsky M, Chemoganskiy VG, Hu F et al. 2003. Effect of UVA light on the activity of several aged human lense enzymes. *Invest Opthalmol Vis Sci* 44(1):264–274.

Liu Z, Lu Y, Lebwohl M et al. 1999. PUVA (8-methoxypsoralen plus ultraviolet A) induces the formation of 8-hydroxy-2′-deoxyguanosine and DNA fragmentation in calf thymus DNA and human epidermoid carcinoma cells. *Free Radic Biol Med* 27:127–133.

Lucas RM. 2011. An epidemiological perspective of ultraviolet exposure-public health concern. *Eye Contact Lens* 37(4):168–175.

Lutz FE, Richtmyer FK. 1922. The reaction of *Drosophila* to ultraviolet. *Science* 55(1428):519.

Malhomme de la Roche H, Seagrove S, Mehta A et al. 2010. Using natural dietary source of antioxidants to protect against ultraviolet and visible radiation-induced DNA damage: An investigation of human green tea injestion. *J Photochem Photobiol B* 101(2):169–173.

Matsen FA, Robertson WW, Chuoke RL. 1947. The near-ultraviolet absorption spectra of monoalkyl-substituted benzene: Hyperconjugation and Baker-Nathan effect. *Chem Rev* 41(2):273–279.

McCormick JP, Fischer JR, Pachlatko et al. 1976. Characterization of a cell-lethal product from the photooxidation of tryptophan: Hydrogen peroxide. *Science* 191(4226):468–469.

Nicholls AG. 1928. The ultraviolet rays in therapeutics. *Can Med Assoc J* 18(3):321–322.

Orimo H, Tokura Y, Hino R et al. 2006. Formation of 8-hydroxy 2′-deoxyguanosine in the DNA of cultured human keratinocytes by clinically used doses of narrowband and broad band ultraviolet B and psoralen plus ultraviolet A. *Cancer Sci* 97:99–105.

Oyewole AO, Wilmot MC, Fowler M et al. 2014. Comparing the effects of mitochondrial targeted and localized antioxidants with cellular antioxidants in human skin cells exposed to UVA and hydrogen peroxide. *FASEB J* 28(1):485–494.

Pagano G, Talarnaca AA, Castello G. 2012. Oxidative stress in Fanconi anaemia: From cells and molecules towards prospects in clinical management. *Biol Chem* 393(1–2):11–21.

Pagano G, Talamanca AA, Castello G et al. 2013. From clinical description, to in vitro and animal studies, and backward to patients: Oxidative stress and mitochondrial dysfunction in Fanconi anemia. *Free Radic Biol Med* 58:118–125.

Paretzoglou A, Stockenhuber C, Kirk S et al. 1998. Generation of reactive oxygen species from the photolysis of histidine by near-ultraviolet light: Effects on T7 as a model biological system. *J Photochem Photobiol B* 43:101–105.

Pathak MA, Parrish JA, Fitzpatrick TB. 1981. Psoralens in photochemotherapy of skin diseases. *Farmaco Sci* 36(7):479–491.

Petersen AB, Gniadecki R, Vicanova J et al. 2000. Hydrogen peroxide is responsible for UVA-induced DNA damage measured by alkaline comet assay in HaCaT keratinocytes. *J Photochem Photobiol B* 59(1–3):123–131.

Phillips SL, Person S, Jagger J. 1967. Division delay in *Escherichia coli* by near-ultraviolet radiation. *J Bacteriol* 94(1):165–170.

Ray RS, Mehrotra S, Shakar U et al. 2001. Evaluation of UV-induced superoxide radical generation potential of some common antibiotics. *Drug Chem Toxicol* 24(2):191–200.

Ring HC, Randskov VG, Miller IM. 2015. Time spent per delta PASI (TSdP) among psoriasis patients undergoing UVB-therapy—A pilot study. *J Dermatolog Treat* 20:1–3.

Ros AM, Wennersten G. 1987. PUVA therapy for photosensitive psoriasis. *Acta Derm Venereol* 67(6):501–505.

Rosen JE, Prahalad AK, William GM. 1996. 8-oxodeoxyguanosine formation in the DNA of cultured cells after exposure to H_2O_2 alone or with UVB or UVA irradiation. *Photochem Photobiol* 64(1):117–122.

Rousset S, Nocentini S, Rouillard D et al. 2002. Mitochondrial alterations in Fanconi anemia fibroblasts following ultraviolet A or psoralen photoactivation. *Photochem Photobiol* 75(2):159–166.

Sakakibara N, Sugano S, Morita A et al. 2008. Ultrastructural changes induced in cutaneous collagen by ultraviolet A-1 and psoralen plus ultraviolet A therapy in systemic sclerosis. *J Dermatol* 35:63–69.

Santos AL, Oliviera V, Baptista et al. 2013. Wavelength dependence of biological damage induced by UV radiation on bacteria. *Arch Microbiol* 195(1):63–74.

Sarnoff DS, Saini R. 2008. Multiple nonmelanoma skin cancer in a patient with epidermolytic hyperkeratosis on long standing retinoid therapy. *J Drugs Dermatol* 7:475–478.

Seckin Z, Yazici Z, Senol A et al. 2008. A case of Schamberg's disease responding dramatically to PUVA treatment. *Photodermal Photoimmunol Photomed* 24:95–96.

Sezer E, Erbil AH, Kurumlu Z et al. 2007. Comparison of the efficacy of local narrowband ultraviolet B (NB-UVB) phototherapy versus psoralen plus ultraviolet A (PUVA) paint for palmoplantar psoriasis. *J Dermatol* 34:435–440.

Shugard D. 1951. Photoreactivation in the near ultraviolet of D-glyceraldehyde-3-phosphate dehydrogenase. *Experientia* 7(1):26–28.

Situm M, Bulat V, Majcen K et al. 2014. Benefit of controlled ultraviolet radiation in the treatment of dermatological diseases. *Coll Antropol* 38(4):1249–1253.

Smith B. 1946. Method of making filters transmitting the near ultraviolet and absorbing visual light. *Science* 104(2708):490–491.

Sohal RS, Orr WC. 2012. The redox stress hypothesis of aging. *Free Radic Biol Med* 52(3):539–555.

Swalwell H, Latimer J, Haywood RM et al. 2012. Investigating the role of melanin in UVA/UVB-and hydrogen peroxide-induced cellular and mitochondrial ROS production and mitochondrial DNA damage in human melanoma. *Free Radic Biol Med* 52(3):626–634.

Symons MRC, Rusakiewicz S, Rees RC et al. 2001. Hydrogen peroxide: A potent cytotoxic agent effective in causing cellular damage and used in the possible treatment for certain tumors. *Med Hypothesis* 57(1):56–58.

Trautinger F. 2011. Phototherapy of mycosis fungoidis. Photodermatol *Photoimmunol Photomed* 27:68–74.

Usmani A, Murphy D, Veale V et al. 2008. Photochemotherapy for localized morphoea: Effect on clinical and molecular markers. *Clin Exp Dermatol* 33:698–704.

Villacis D, Yi A, Jahn R et al. 2014. Prevalence of abnormal vitamin D levels among division I NCAA athletes. *Sports Health* 6(4):340–347.

Wacker M, Holick MF. 2013. Sunlight and vitamin D: A global perspective for health. *Dermatoendocrinol* 5(1):51–108.

Wainwright L, Parisi AV, Downs N. 2015. Dosimeter based on 8-methoxypsoralen for UVA exposure over extended period. *Photochem Photobiol B* 148:246–251.

Wang RJ, Anantahswamy HN, Nixon BT et al. 1980. Induction of single strand DNA breaks in human cells by H_2O_2 formed in near-UV (black light) irradiated medium. *Radiat Res* 82(2):269–276.

Webb RB, Brown MS. 1976. Sensitivity of strains of *Escherichia coli* differing in repair capability to far UV, near UV and visible radiations. *Photochem Photobiol* 24(5):425–432.

Webb RB, Brown MS. 1979. Action spectra for oxygen-dependent and independent inactivation of *Escherichia coli* WP2S from 254 to 460 nm. *Photochem Photobiol* 29(2):407–409.

Wolf P. 2013. Bath vs. oral psoralen plus ultraviolet A: Is one more effective than the other? *Br J Dermatol* 169(3):492–493.

Woodgate P, Jardine LA. 2015. Neonatal jaundice: Phototherapy. *BMJ Clin Evid* 2015:pii:0319.

Wright JA, Richards T, Srai SK. 2014. The role of iron in the skin and cutaneous wound healing. *Front Pharmacol* 5:156.

Xia Q, Yin JJ, Zhao Y et al. 2013. UVA photoirradiation of nitro-polycyclic aromatic hydrocarbons-induction of reactive oxygen species and formation of lipid peroxides. *Int J Environ Res Public Health* 10(3):1062–1084.

Xue C, Luo W, Yang XL. 2015. A mechanism for nano-titanium dioxide-induced cytotoxicity in HaCaT cells under UVA irradiation. *BioSci Biotechnol Biochem* 79(8):1384–1390.

Yadav N, Dwivedi A, Muitaba SF et al. 2013. Ambient UV-induced expression of p53 and apoptosis in human skin melanoma A375 cell line by quinine. *Photochem Photobiol* 89(3):655–664.

Yoakum G, Ferron W, Eisenstark A et al. 1974. Inhibition of replication gap closure in *Escherichia coli* by near ultraviolet light photoproducts of L-tryptophan. *J Bacteriol* 119(1):62–69.

Yu H, Xiao Q, Yan J et al. 2006. Photo irradiation of polycyclic aromatic hydrocarbons with UVA light—A pathway leading to the generation of reactive oxygen species, lipid peroxidation, and DNA damage *Int J Environ Res Public Health* 3(4):348–354.

Zigman S, Groff J, Yulo T et al. 1975. The response of mouse ocular tissues to continuous near-UV light exposure. *Invest Opthalmol* 14(9):710–713.

Zigman S, Hare JD. 1976. Inhibition of cell growth by near ultraviolet light photoproducts of tryptophan. *Mol Cell Biochem* 10(3):131–135.

4 Ultraviolet A Radiation and ROS

Observations from Studies with Bacteria

Abraham Eisenstark and James Hoerter

CONTENTS

ABSTRACT

Understanding the mechanisms involved in cellular damage and repair caused by reactive oxygen species (ROS) has evolved, in part, from studies with appropriate bacterial mutants. The need for this information arose from scientific discoveries of x-ray and radioactive isotopes. Unfortunately, the earlier users suffered from these radiations before the mechanisms of biological damage were fully understood. Knowledge and understanding progressed rapidly mainly due to extensive collections of metabolic and regulatory genetic mutants of bacteria and libraries of publications on gene damage and repair mechanisms.

4.1 INTRODUCTION

Decades of epidemiological studies have linked exposure to solar ultraviolet (UV) radiation to human skin cancers [1]. Most adverse biological consequences of exposure to solar UV light can be attributed to the generation of increased levels of reactive oxygen species (ROS). Solar UV light has been classified into three categories of wavelengths: UVC (100–290 nm), UVB (290–320 nm), and UVA (300–400 nm). The ozone layer under normal atmospheric conditions absorbs all UVC and over 90% of UVB, shielding the earth from most of the harmful effects of solar UV. As laboratory UV experiments have shown this light to be most mutagenic and carcinogenic, it is almost certain that exposure to solar light also exerts cancer.

UVA radiation penetrates the ozone layer and is the most abundant portion of the solar electromagnetic spectrum that reaches the earth's surface. Interestingly, UVA radiation is able to penetrate more deeply in the human skin up to the basal layer of the epidermis than UVB radiation [2,3]. Long-term effects such as persistent genomic instability and bystander effects have been observed in mammalian cells following UVA exposure [4–6]. Thus, UV radiation has at least the same potential to damage cells as UVB. In the past, UVA had often been addressed as NUV (near UV) and black light, but from now on in this chapter, it will be addressed as UVA.

4.2 BACTERIAL MUTANTS TO PINPOINT WAYS THAT CELLS DEAL WITH ROS DAMAGE

Insights into mechanisms that govern how cells protect and defend themselves from ROS generated from UVA irradiation as well as repair of the resulting oxidative damage have come from the study of bacterial mutants that are deficient in these functions [7]. This includes studies of bacterial mutants deficient in superoxide dismutases (SODs), hydroperoxidases, and the capacity of the mutants to repair DNA damage as a result of ROS-induced damage [8–14]. Variables influencing the biological response to UVA radiation include flunk, irradiance, and wavelength distribution within the UVA spectrum. These variables are significant when evaluating the biological responses. For example, variables affecting the fluence rate reaching skin cells *in situ* include differences in UVA- and UVB-penetrating ability, presence or absence of sunscreens, atmospheric conditions, season, and geographical location. All these variables affect how biological systems react to a given dose of UVA radiation [15,16].

UVA exerts toxic effects via ROS, which include the superoxide anion radical ($O_2^{\bullet-}$), hydroxyl radical ($^\bullet OH$), hydrogen peroxide (H_2O_2), and singlet oxygen ($^1O_2$). ROS production leads to oxidative damage to nucleic acids, proteins, enzymes, and lipids [17–19]. In DNA, ROS induce cyclobutane pyrimidine dimers (CPDs), pyrimidine (6-4), pyrimidone photoproducts (PDs), and single-strand breaks [20]. Generation of these photoproducts is wavelength dependent and UVB radiation is 1000-fold as detrimental as UVA for these products. These photolesions can block DNA replication and transcription and affect protein–DNA interactions. If the photoproducts are improperly repaired or escape repair processes, they can cause, in human, mutations contributing to the development of skin cancer such as melanoma depending on the types of cells affected [21]. In humans, CPDs and (6-4) PDs are normally removed from DNA by nucleotide excision repair; however, if not repaired, they can contribute to the impairment of important biological pathways [22,23].

4.3 TRYPTOPHAN PHOTOPRODUCT

UVA-induced lethality is mostly oxygen dependent, and several active oxygen intermediates are generated by this irradiation in biological systems [24]. For example, UVA irradiation of tryptophan is known to produce H_2O_2, which can then generate mutagenic $^\bullet OH$ via the Fenton reaction [16,25,26]. Absorption of UVA light by endogenous photosensitizers, such as vitamins, generates $^1O_2$, which contributes to cell and tissue damage [27]. Also, certain amino acids have been shown to be activated by UVA generating $O_2^{\bullet-}$ [79]. The types of damage induced in these molecules can lead to cytotoxicity, mutations, and alterations in signaling pathways in cells [28].

4.4 IMPORTANCE OF FLUENCE RATE

Irradiance (fluence rate) of a given dose of UVA light has significant effect to the extent on oxidative stress. Irradiance levels over that normally found in solar radiation increase the oxidative stress indicators (ROS, protein oxidation, glutathione, heme oxygenase-1, etc.) in human dermal fibroblasts [15]. This has important implications for assessing the harmful effects of artificial sources of UVA/UVB radiation including tanning beds, which often have irradiance levels more than 10 times that of the normal sunlight.

Solar UVA irradiation promotes the formation of ROS and induces oxidative stress by the production of highly deleterious $^{\bullet}OH$ according to the Fenton and the Haber–Weiss reactions

$$H_2O_2 + O_2^{\bullet-} \rightarrow O_2 + OH^{\bullet} + OH^- \quad \text{(Haber–Weiss reaction)}$$

$$H_2O_2 + Fe^{2+} \rightarrow Fe^{3+} + OH^{\bullet} + OH^- \quad \text{(Fenton reaction)}$$

Although ROS are produced by all aerobic cells during metabolism under normal (21%) oxygen conditions, their levels of activity are elevated following exposure to UVA radiation. H_2O_2, a photoproduct of UVA, can react with naturally occurring iron complexes to generate highly reactive $^{\bullet}OH$ [29–31]. By generating the $O_2^{\bullet-}$, UVA irradiation initiates the release of iron from naturally occurring iron complexes [32]. Superoxide-mediated iron reduction creates favorable conditions for oxidative stress and DNA damage [12]. For this reason, iron carriers and mechanisms regulating intracellular iron pools are thought to have roles in the production of ROS from solar UVA radiation [33].

4.5 DETOXIFICATION BY CELL ENZYMES

Enzymes, including catalases, SODs, and glutathione reductase, have major roles to play in detoxifying ROS generated by UVA irradiation. Other endogenous radical scavengers, such as glutathione, also play significant roles. During UVA stress, these enzymes and radical scavengers act in a highly sophisticated manner to remove ROS when cells are exposed to UVA [9,34–36].

Following UVB irradiation, significant increases in Cu-Zn SOD activity have been recorded, which occur in keratinocytes as a cutaneous antioxidant defense mechanism that protects against cytotoxicity [37,38]. The role of SOD in defending against UVA irradiation is supported by studies also in bacteria showing increased UVA sensitivity and mutation rate in *E. coli* strains deficient in SOD. MnSOD activity in skin is induced fivefold by repetitive UVA irradiation [39] and chronic UVB radiation [40].

Under certain irradiation conditions (higher dosage and irradiance), the activity and level of antioxidant enzymes are lowered and restrict the protective effects of these enzymes against oxidative stress. For example, exposure to high fluence (300 mJ/cm²) of UVA radiation (higher than 280 nm) in hairless mice caused a rapid and statistically significant inhibition in the activity of glutathione reductase and catalase [41]. Also, repeated irradiation of the cornea of albino rats with UVB evoked a deficiency in antioxidant enzymes (SOD, glutathione peroxidase, and catalase) in the corneal epithelium, which most probably contribute to the damage of the cornea (also possibly the deeper parts of the eye) from UVB rays and the reactive oxygen products generated from them [42].

Catalase may also have a direct role to play in generating oxidants by mediating the production of ROS. This occurs when DNA-damaging UVB light is absorbed by the catalase and is converted to ROS [43]. The catalytic activity of hydroxyperoxidase II (HPII) is rapidly eliminated by lethal UVA irradiation. This suggests that HPII is more susceptible than hydroperoxidases I (HPI) and plays a less protective role in preventing the accumulation of harmful ROS through the Harber–Weiss and Fenton reactions during UVA irradiation. Electrophoretic movement of HPI and HPII is progressively altered on gels after sublethal and lethal doses of UVA irradiation, suggesting that structural changes to HPI and HPII are contributing to the reduced enzymatic activities. The structural changes to HPII are most likely due to destruction of the heme present as a core in the enzyme, which absorbs the UVA and is the active site of catalase [44,45].

4.6 GLUTATHIONE REDUCTASE

Studies with *E. coli* have shown that glutathione reductase (GR) has an important role in cellular defense against ROS by sustaining reduced glutathione, a potent endogenous antioxidant [46].

Glutathione, encoded by *gorA*, is regulated by OxyR. Both OxyR and glutathione are also involved in the regulation of *katG* [47,48]. Induction of GR is an adaptive response to sublethal UVA irradiation in *E. coli* and may play an important role in the complex regulatory network that controls defense against oxidative stress induced by UVA irradiation [45]. The induction of GR may be even greater than that reflected in assays of activity levels because GR contains a UVA-absorbing flavin, and the enzyme activity may be reduced under conditions of UVA stress [49].

4.7 CELLULAR ADAPTIVE RESPONSES IN *E. COLI*

Four adaptive responses (induction of three regulons and one sigma factor) are involved in the cellular response to oxidative stress initiated by UVA irradiation: OxyR (oxidative stress response regulon), SoxRS (superoxide stress regulon), Fur (ferric uptake regulator), and the stationary phase sigma factor, RpoS. OxyR directly detects oxidative stress and is a transcriptional activator of stress response genes including *katG*, encoding HPI [50,51]. SoxRS regulates *fur* by activating the expression of a transcript encoding both flavodoxin and Fur protein [52]. RpoS is a stress response protein that functions as an alternative sigma factor (σ^{38}) for RNA polymerase, thereby affecting global switching of gene expression when cells undergo stress or enter the stationary phase [53]. The RpoS protein is regulated at transcription, translation, and protein stabilization levels, and its stability is dependent on other proteins, including Fur protein.

Fur regulates iron superoxide dismutase (FeSOD) and manganese superoxide dismutase (MnSOD), the key enzymes in protecting cells against oxygen toxicity [52,54]. Fur is activated by OxyR and SoxRS, regulators of the antioxidant defense response in *E. coli*, thereby reducing the production of highly reactive $^•$OH and other toxic oxygen species through the Fenton reaction [55]. Strains lacking the Fur protein exhibit a high sensitivity to various oxidative agents and are highly sensitive to UVA irradiation [56]. The sensitivity of Δfur mutant to UVA irradiation is associated with a reduction in HPI and HPII activities, which may be due to a decrease in the transcription of the *katE* and *katG* genes. The *fur* gene product can mediate this effect by decreasing the transcription of *rpoS* (σ^{38}) [57].

Figure 4.1 provides a summary of the signaling pathways in the antioxidant response to UVA radiation. Of special interest is the cascade of signal pathways that are released.

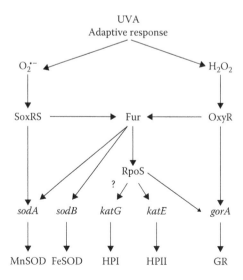

FIGURE 4.1 Signaling pathways in the antioxidant response to NUV irradiation.

4.8 CONCLUSION

Our awareness of biological radiation effects arose from our observations of the blockage of the *E. coli* cell division [58]. As research progressed, details of genetic, metabolic, and regulatory pathways were chronicled in periodic reviews [8,59–68].

Studies, using appropriate bacterial mutants, have revealed the important role in unraveling the mechanisms and metabolic pathways involved in radiation damage and repair along the electromagnetic spectrum. The role of ROS following solar radiation has increased our understanding of the molecular mechanisms of damage and repair both in prokaryotes and in eukaryotes [68–79]. This knowledge has influenced the use of sunscreen, the use of antioxidants in foods and pharmaceuticals, and protection against excess sunlight (wearing wide brim hats, using sun umbrellas and sunglasses, etc.).

ACKNOWLEDGMENTS

We thank Jacki Kian Mehr and Alycia McGee, for the tedious tasks of assembling references and repeated rewrites of the manuscript, and Professors Olen Brown (University of Missouri–Columbia) and Shamim Ahmad (Nottingham University).

REFERENCES

1. A. Besaratinia and G.P. Pfeifer, Sunlight ultraviolet irradiation and BRAF V600 mutagenesis in human melanoma, *Hum. Mutat.* 29 (2008), 983–991.
2. W.A. Bruls, H. Slaper, J.C. van der Leun, and L. Berrens, Transmission of human epidermis and stratum corneum as a function of thickness in the ultraviolet and visible wavelengths, *Photochem. Photobiol.* 40 (1984), 485–494.
3. W.A. Bruls, W.H. van, and J.C. van der Leun, Transmission of UV-radiation through human epidermal layers as a factor influencing the minimal erythema dose, *Photochem. Photobiol.* 39 (1984), 63–67.
4. J. Dahle, E. Angell-Petersen, H.B. Steen, and J. Moan, Bystander effects in cell death induced by photodynamic treatment UVA radiation and inhibitors of ATP synthesis, *Photochem. Photobiol.* 73 (2001), 378–387.
5. J. Dahle and E. Kvam, Induction of delayed mutations and chromosomal instability in fibroblasts after UVA-, UVB-, and X-radiation, *Cancer Res.* 63 (2003), 1464–1469.
6. J. Dahle, E. Kvam, and T. Stokke, Bystander effects in UV-induced genomic instability: Antioxidants inhibit delayed mutagenesis induced by ultraviolet A and B radiation, *J. Carcinog.* 4 (2005), 11.
7. S.I. Ahmad, A. Atkinson, and A. Eisenstark, Isolation and characterization of a mutant of *Escherichia coli* K12 synthesizing DNA polymerase I and endonuclease I constitutively, *J. Gen. Microbiol.* 117 (1980), 419–422.
8. I. Compan and D. Touati, Interaction of six global transcription regulators in expression of manganese superoxide dismutase in *Escherichia coli* K-12, *J. Bacteriol.* 175 (1993), 1687–1696.
9. A. Eisenstark, M.J. Calcutt, M. Becker-Hapak, and A. Ivanova, Role of *Escherichia coli* rpoS and associated genes in defense against oxidative damage, *Free Radic. Biol. Med.* 21 (1996), 975–993.
10. R.L. Knowles and A. Eisenstark, Near-ultraviolet mutagenesis in superoxide dismutase-deficient strains of *Escherichia coli*, *Environ. Health Perspect.* 102 (1994), 88–94.
11. B.D. Sak, A. Eisenstark, and D. Touati, Exonuclease III and the catalase hydroperoxidase II in *Escherichia coli* are both regulated by the katF gene product, *Proc. Natl. Acad. Sci. U.S.A.* 86 (1989), 3271–3275.
12. D. Touati, M. Jacques, B. Tardat, L. Bouchard, and S. Despied, Lethal oxidative damage and mutagenesis are generated by iron in delta fur mutants of *Escherichia coli*: Protective role of superoxide dismutase, *J. Bacteriol.* 177 (1995), 2305–2314.
13. R.W. Tuveson, The interaction of a gene (nur) controlling near-UV sensitivity and the polA1 gene in strains of *E. coli* K12, *Photochem. Photobiol.* 33 (1981), 919–923.
14. P.S. Hartman and A. Eisenstark, Synergistic killing of *Escherichia coli* by near-UV radiation and hydrogen peroxide: Distinction between recA-repairable and recA-nonrepairable damage, *J. Bacteriol.* 133 (1978), 769–774.

15. J.D. Hoerter, C.S. Ward et al., Effect of UVA fluence rate on indicators of oxidative stress in human dermal fibroblasts, *Int. J. Biol. Sci.* 4 (2008), 63–70.

16. L. Glatzer, D.A. Rincon, and A. Eisenstark, The binding of near-ultraviolet light-induced tryptophan photoproduct(s) to DNA, *Biochim. Biophys. Acta* 418 (1976), 137–145.

17. E. Smyk-Randall, O.R. Brown, A. Wilke, A. Eisenstark, and D.H. Flint, Near ultraviolet light inactivation of dihydroxyacid dehydratase in *Escherichia coli, Free Radic. Biol. Med.* 14 (1993), 609–613.

18. S. Leven, A. Heimberger, and A. Eisenstark, Catalase HPI influences membrane permeability in *Escherichia coli* following near-UV stress, *Biochem. Biophys. Res. Commun.* 171 (1990), 1224–1228.

19. A. Eisenstark, S. Kovacs, and J. Terry, Cell damage by near ultraviolet radiation: Role of DNA-protein cross-links, *J. Natl. Cancer Inst.* 69 (1982), 177–181.

20. R.J. Wang, H.N. Ananthaswamy, B.T. Nixon, P.S. Hartman, and A. Eisenstark, Induction of single-strand DNA breaks in human cells by H_2O_2 formed in near-UV (black light) irradiated medium, *Radiat. Res.* 82 (1980), 269–276.

21. P. Caimi and A. Eisenstark, Sensitivity of *Deinococcus radiodurans* to near-ultraviolet radiation, *Mutat. Res.* 162 (1986), 145–151.

22. E. Kvam and R.M. Tyrrell, Induction of oxidative DNA base damage in human skin cells by UV and near visible radiation, *Carcinogenesis* 18 (1997), 2379–2384.

23. S. Mouret, C. Baudouin, M. Charveron, A. Favier, J. Cadet, and T. Douki, Cyclobutane pyrimidine dimers are predominant DNA lesions in whole human skin exposed to UVA radiation, *Proc. Natl. Acad. Sci. U.S.A.* 103 (2006), 13765–13770.

24. R.B. Webb and J.R. Lorenz, Oxygen dependence and repair of lethal effects of near ultraviolet and visible light, *Photochem. Photobiol.* 12 (1970), 283–289.

25. J.A. Imlay and S. Linn, DNA damage and oxygen radical toxicity, *Science* 240 (1988), 1302–1309.

26. J.P. McCormick, J.R. Fischer, J.P. Pachlatko, and A. Eisenstark, Characterization of a cell-lethal product from the photooxidation of tryptophan: Hydrogen peroxide, *Science* 191 (1976), 468–469.

27. A. Knak, J. Regensburger, T. Maisch, and W. Baumler, Exposure of vitamins to UVB and UVA radiation generates singlet oxygen, *Photochem. Photobiol. Sci.* 13 (2014), 820–829.

28. T.J. McMillan, E. Leatherman, A. Ridley, J. Shorrocks, S.E. Tobi, and J.R. Whiteside, Cellular effects of long wavelength UV light (UVA) in mammalian cells, *J. Pharm. Pharmacol.* 60 (2008), 969–976.

29. J.A. Imlay, S.M. Chin, and S. Linn, Toxic DNA damage by hydrogen peroxide through the Fenton reaction in vivo and in vitro, *Science* 240 (1988), 640–642.

30. P. Tachon, Ferric and cupric ions requirement for DNA single-strand breakage by H_2O_2, *Free Radic. Res. Commun.* 7 (1989), 1–10.

31. H.N. Ananthaswamy and A. Eisenstark, Near-UV-induced breaks in phage DNA: Sensitization by hydrogen peroxide (a tryptophan photoproduct), *Photochem. Photobiol.* 24 (1976), 439–442.

32. J. Hoerter, A. Eisenstark, and D. Touati, Mutations by near-ultraviolet radiation in *Escherichia coli* strains lacking superoxide dismutase, *Mutat. Res.* 215 (1989), 161–165.

33. R.A. Larson, R.E. Lloyd, K.A. Marley, and R.W. Tuveson, Ferric-ion-photosensitized damage to DNA by hydroxyl and non-hydroxyl radical mechanisms, *J. Photochem. Photobiol. B* 14 (1992), 345–357.

34. A. Eisenstark, Bacterial gene products in response to near-ultraviolet radiation, *Mutat. Res.* 422 (1998), 85–95.

35. A. Heimberger and A. Eisenstark, Compartmentalization of catalases in *Escherichia coli, Biochem. Biophys. Res. Commun.* 154 (1988), 392–397.

36. A. Eisenstark and G. Perrot, Catalase has only a minor role in protection against near-ultraviolet radiation damage in bacteria, *Mol. Gen. Genet.* 207 (1987), 68–72.

37. H. Sasaki, H. Akamatsu, and T.Horio, Effects of a single exposure to UVB radiation on the activities and protein levels of copper-zinc and manganese superoxide dismutase in cultured human keratinocytes, *Photochem. Photobiol.* 65 (1997), 707–713.

38. H. Sasaki, H. Akamatsu, and T. Horio, Protective role of copper, zinc superoxide dismutase against UVB-induced injury of the human keratinocyte cell line HaCaT, *J. Invest Dermatol.* 114 (2000), 502–507.

39. A. Poswig, J. Wenk et al., Adaptive antioxidant response of manganese-superoxide dismutase following repetitive UVA irradiation, *J. Invest. Dermatol.* 112 (1999), 13–18.

40. K. Punnonen, K. Lehtola, P. Autio, U. Kiistala, and M. Ahotupa, Chronic UVB irradiation induces superoxide dismutase activity in human epidermis in vivo, *J. Photochem. Photobiol. B* 30 (1995), 43–48.

41. J. Fuchs, M.E. Huflejt, L.M. Rothfuss, D.S. Wilson, G. Carcamo, and L. Packer, Impairment of enzymic and nonenzymic antioxidants in skin by UVB irradiation, *J. Invest. Dermatol.* 93 (1989), 769–773.

42. J. Cejkova, S. Stipek, J. Crkovska, and T. Ardan, Changes of superoxide dismutase, catalase and glutathione peroxidase in the corneal epithelium after UVB rays. Histochemical and biochemical study, *Histol. Histopathol.* 15 (2000), 1043–1050.

43. D.E. Heck, A.M. Vetrano, T.M. Mariano, and J.D. Laskin, UVB light stimulates production of reactive oxygen species: Unexpected role for catalase, *J. Biol. Chem.* 278 (2003), 22432–22436.

44. Y. Shindo and T. Hashimoto, Time course of changes in antioxidant enzymes in human skin fibroblasts after UVA irradiation, *J. Dermatol. Sci.* 14 (1997), 225–232.

45. J.D. Hoerter, A.A. Arnold et al., Effects of sublethal UVA irradiation on activity levels of oxidative defense enzymes and protein oxidation in *Escherichia coli*, *J. Photochem. Photobiol. B* 81 (2005), 171–180.

46. H.S. Yoon, I.A. Lee, H. Lee, B.H. Lee, and J. Jo, Overexpression of a eukaryotic glutathione reductase gene from *Brassica campestris* improved resistance to oxidative stress in *Escherichia coli*, *Biochem. Biophys. Res. Commun.* 326 (2005), 618–623.

47. M.F. Christman, R.W. Morgan, F.S. Jacobson, and B.N. Ames, Positive control of a regulon for defenses against oxidative stress and some heat-shock proteins in *Salmonella typhimurium*, *Cell* 41 (1985), 753–762.

48. G. Storz and J.A. Imlay, Oxidative stress, *Curr. Opin. Microbiol.* 2 (1999), 188–194.

49. J. Fuchs, M.E. Huflejt, L.M. Rothfuss, D.S. Wilson, G. Carcamo, and L. Packer, Acute effects of near ultraviolet and visible light on the cutaneous antioxidant defense system, *Photochem. Photobiol.* 50 (1989), 739–744.

50. S. Altuvia, M. Almiron, G. Huisman, R. Kolter, and G. Storz, The dps promoter is activated by OxyR during growth and by IHF and sigma S in stationary phase, *Mol. Microbiol.* 13 (1994), 265–272.

51. A. Eisenstark, P. Yallaly, A. Ivanova, and C. Miller, Genetic mechanisms involved in cellular recovery from oxidative stress, *Arch. Insect Biochem. Physiol.* 29 (1995), 159–173.

52. M. Zheng, B. Doan, T.D. Schneider, and G. Storz, OxyR and SoxRS regulation of fur, *J. Bacteriol.* 181 (1999), 4639–4643.

53. R. Hengge-Aronis, Signal transduction and regulatory mechanisms involved in control of the sigma(S) (RpoS) subunit of RNA polymerase, *Microbiol. Mol. Biol. Rev.* 66 (2002), 373–395.

54. B. Tardat and D. Touati, Two global regulators repress the anaerobic expression of MnSOD in *Escherichia coli*::Fur (ferric uptake regulation) and Arc (aerobic respiration control), *Mol. Microbiol.* 5 (1991), 455–465.

55. D.J. Hassett, P.A. Sokol, M.L. Howell, J.F. Ma, H.T. Schweizer, U. Ochsner, and M.L. Vasil, Ferric uptake regulator (Fur) mutants of *Pseudomonas aeruginosa* demonstrate defective siderophore-mediated iron uptake, altered aerobic growth, and decreased superoxide dismutase and catalase activities, *J. Bacteriol.* 178 (1996), 3996–4003.

56. J. Hoerter, A. Pierce, C. Troupe, J. Epperson, and A. Eisenstark, Role of enterobactin and intracellular iron in cell lethality during near-UV irradiation in *Escherichia coli*, *Photochem. Photobiol.* 64 (1996), 537–541.

57. J.D. Hoerter, A.A. Arnold, C.S. Ward, M. Sauer, S. Johnson, T. Fleming, and A. Eisenstark, Reduced hydroperoxidase (HPI and HPII) activity in the Deltafur mutant contributes to increased sensitivity to UVA radiation in *Escherichia coli*. *J. Photochem. Photobiol. B* 79 (2005), 151–157.

58. A. Eisenstark, R. Eisenstark et al., Radiation—Sensitive and recombinationless mutants of *Salmonella typhimurium*, *Mutat. Res.* 8(3) (1969), 497–504.

59. A. Eisenstark, Sensitivity of *Salmonella typhimurium* recombinationless (rec) mutants to visible and near-visible light, *Mutat. Res.* 10 (1970), 1–6.

60. A. Eisenstark and D. Ruff, Repair in phage and bacteria inactivated by light from fluorescent and photo lamps, *Biochem. Biophys. Res. Commun.* 38(2) (1970), 244–248.

61. A. Eisenstark, Mutagenic and lethal effects of visible and near-ultraviolet light on bacterial cells, *Adv. Genet.* 16 (1971), 167–198.

62. G. Yoakum and A. Eisenstark, Toxicity of L-tryptophan photoproduct on recombinationless (rec) mutants of *Salmonella typhimurium*, *J. Bacteriol.* 112(1) (1972), 653–655.

63. A. Eisenstark, Tryptophan photoproduct as a genetic probe: Effects on bacteria. *Stadler Symposium*, Vol. 5, University of Missouri, Columbia, MO, 1973, pp. 49–60.

64. G. Yoakum, W. Ferron et al., Inhibition of replication gap closure in *Escherichia coli* by near-ultraviolet light photoproducts of L-tryptophan, *J. Bacteriol.* 119(1) (1974), 62–69.

65. G. Yoakum, A. Eisenstark et al., Near-UV photoproduct(s) of L-typtophan: An inhibitor of medium-dependent repair of x-ray-induced single-strand breaks in DNA which also inhibits replication-gap closure in *Escherichia coli* DNA, *Basic Life Sci.* 5B (1975), 453–458.

66. G.H.Yoakum, A. Eisenstark, Trytophan photoproduct(s): Sensitized induction of strand breaks (or alkali-labile bonds) in bacterial deoxyribonucleic acid during near-ultraviolet irradiation, *J. Bacteriol.* 122(1) (1975), 199–205.

67. H.N. Ananthaswamy and A. Eisenstark, Repair of hydrogen peroxide-induced single-strand breaks in *Escherichia coli* deoxyribonucleic acid, *J. Bacteriol.* 130(1) (1977), 187–191.

68. S.I. Ahmad and A. Eisenstark, Thymidine sensitivity of certain strains of *Escherichia coli* K12, *Mol. Gen. Genet.* 172(2) (1979), 229–231.

69. P.S. Hartman, A. Eisenstark et al., Inactivation of phage T7 by near-ultraviolet radiation plus hydrogen peroxide: DNA-protein crosslinks prevent DNA injection, *Proc. Natl. Acad. Sci. U.S.A.* 76(7) (1979), 3228–3232.

70. P.S. Hartman and A. Eisenstark, Killing of *Escherichia coli* K-12 by near-ultraviolet radiation in the presence of hydrogen peroxide: Role of double-strand DNA breaks in absence of recombinational repair, *Mutat. Res.* 72(1) (1980), 31–42.

71. J.P. McCormick, S. Klita, J. Terry, M. Schrodt, and A. Eisenstark, *Formation by Hydrogen Peroxide or 254 nm Radiation of a Near-UV Chromophore from Peptite-Bound Cysteine.* DC and DBS, University of Missouri, Columbia, MO, 1982.

72. A. Eisenstark, Genetic damage in *Salmonella typhimurium* by near-ultraviolet radiation. Lack of repair by plasmid pKM101, *Mutat. Res.* 122(3–4) (1983), 267–272.

73. M.A. Turner and A. Eisenstark, Near-ultraviolet radiation blocks SOS responses to DNA damage in *Escherichia coli. Mol. Gen. Genet.* 193(1) (1984), 33–37.

74. M. Schaechter, et al., The molecular biology of bacterial growth. A symposium held in honor of aole Maaløe. Jones and Bartlett Publ., Inc., Boston, MA, pp. 243–255.

75. A. Eisenstark, Mutagenic and lethal effects of near-ultraviolet radiation (290–400 nm) on bacteria and phage, *Environ. Mol. Mutagen.* 10(3) (1987), 317–337.

76. J. Hoerter and A. Eisenstark, Synergistic killing of bacteria and phage by polystyrene and ultraviolet radiation, *Environ. Mol. Mutagen.* 12(2) (1988), 261–264.

77. J. Hoerter and A. Eisenstark, Patterns of protein synthesis in a growth delay mutant (nuv) of *Escherichia coli* after treatment by near-UV radiation or hydrogen peroxide, *J. Photochem. Photobiol. B* 6(3) (1990), 283–289.

78. M. Becker-Hapak and A. Eisenstark, Role of rpoS in the regulation of glutathione oxidoreductase (gor) in *Escherichia coli, FEMS Microbiol. Lett.* 134(1) (1995), 39–44.

79. J. Craggs, S.H. Kirk, and S.I. Ahmad, Synergistic action of near-UV and phenylalanine, tyrosine or tryptophan on the inactivation of phage T7: Role of superoxide radicals and hydrogen peroxide, *J. Photochem. Photobiol. B* 24 (1994), 123–128.

5 ROS and Phenolic Compounds

Mario C. Foti and Riccardo Amorati

CONTENTS

ABSTRACT

Reactive oxygen and nitrogen species (ROS and RNS) are detrimental to human health because they initiate free radical–catalyzed oxidations of fundamental biomolecules such as DNA, proteins, lipids in low-density lipoprotein (LDL) and cell membranes, polysaccharides, etc. Molecular oxygen in its triplet ground state (the oxygen we breathe), $^3O_2$, is the oxidant species in these processes called autoxidation or peroxidation. Cells are equipped with defensive systems able to quench most of the radicals responsible for initiating or propagating autoxidation in organic matter. Enzymes (superoxide dismutases, catalases, glutathione peroxidases, and peroxiredoxins) destroy radicals such as $O_2^{\cdot-}$ or non-radical species such as H_2O_2 and ROOH. Other small molecules, mainly phenols, present in the diet are able to react with radicals and hence may cooperate with the enzymes in keeping "oxidative stress" at bay. However, the physiological concentrations of phenols are frequently low, and this has cast doubt on their effectiveness *in vivo*. On the other hand, there is evidence that vitamin E is an effective *peroxyl* radical (ROO$^\cdot$) scavenger both *in vitro* and *in vivo* with rate constants of ~10^6 M^{-1} s^{-1}. Its effectiveness *in vivo* against other radicals and non-radical oxidative species (HO$^\cdot$, R$^\cdot$, NO$_2^\cdot$, RS$^\cdot$, $^1O_2$, O_3, and HOCl) is however uncertain. Vitamin C and ubiquinol-10 are able to regenerate vitamin E from its radicals.

5.1 INTRODUCTION

Organic matter exposed to air is thermodynamically unstable because molecular oxygen can oxidize it, releasing large amounts of energy [1]. Fortunately, this spontaneous process is slow at room temperature; and thus, the aerobic organisms, even though immersed in an atmosphere of oxygen, can thrive. Without this chemical inertia, aerobes could not exist because they would burn as a piece of paper in a flame. Yet, an *imperceptible* process of oxidation called "peroxidation" or "autoxidation" does happen and, in humans, seems to be implicated in the onset of many degenerative diseases such as atherosclerosis, arthritis, inflammation, heart disease, cancer, and aging [2].

Cellular metabolism continuously and inevitably generates small quantities of various types of radicals, which are able to promote a (relatively slow) peroxidation process [3]. These initiators of peroxidation are apparently a potential danger for life [2]. That is why living organisms have developed, over time, mechanisms to scavenge these species, in order to contain the biological damage caused by oxygen [2–5]. The molecules subjected to peroxidation are all cell components, that is, phospholipids in membranes, proteins, polysaccharides, and nucleic acids [2,3]. The destructive potential of oxygen can better be evaluated at high temperatures when peroxidation becomes very fast and destructive (combustion), and the complex chemical architecture of cells is *dismounted* carbon-by-carbon, by oxygen producing essentially carbon dioxide, water, and other simple molecules (NO_x, SO_x, etc.).

5.2 REACTIVE OXYGEN AND NITROGEN SPECIES

The radicals mentioned earlier are commonly oxygen-centered radicals such as hydroxyl radical (HO$^\bullet$), alkoxyl radicals (RO$^\bullet$), peroxyl radicals (ROO$^\bullet$), and superoxide radical anion ($O_2^{\bullet-}$), which is the most abundant radical in the human body [2]. In healthy subjects, approximately 2–17 kg per year of inhaled oxygen is converted to $O_2^{\bullet-}$ [2]. In the presence of traces of transition metals (such as Fe and Cu), non-radical species such as lipid hydroperoxides, ROOH, hydrogen peroxide (H_2O_2), etc., act as peroxidation initiators [6]. All these species are collectively called reactive oxygen species (ROS). Together with these, reactive nitrogen species (RNS), such as NO$^\bullet$, NO$_2^\bullet$, and ONO$_2^-$, are also present in living organisms [3–5]. ROS and RNS are involved in the catalysis of unwanted free radical processes that lead to peroxidation of biological substrates. The primary products of these reactions (ROOH) accelerate oxidation, which (in the absence of ROOH removal) becomes autocatalytic [1]. Toxic secondary products of lipid peroxidation are mostly short-chain aldehydes, for example, acrolein (CH_2=CH–CHO), glyoxal (OHC–CHO), methylglyoxal (OHC–CO–CH_3), epoxides, ketones, and carboxylic acids [7,8]. Reactive carbonyl species (RCS) are also produced from *in vivo* oxidation of carbohydrates and amino acids [7,8]. Generally, RCS are mutagenic and cytotoxic as they are implicated in chemical modifications of proteins, nucleic acids, and aminophospholipids [7,8]. These effects are believed to be responsible for aging and the development of a wide range of human diseases [7,8].

5.3 AUTOXIDATION OF ORGANIC MATTER AND ANTIOXIDANT ACTION

Figure 5.1 shows the mechanism of oxidative conversion of a hydrocarbon or a lipid chain, RH, into a hydroperoxide, ROOH. As inferred from the sign of $\Delta_r G_1^0$, which is significantly negative

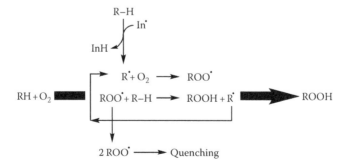

FIGURE 5.1 Conversion of a hydrocarbon RH into a hydroperoxide ROOH by molecular oxygen. The process is initiated by In$^\bullet$ radicals and is sustained by catalytic amounts of peroxyl radicals ROO$^\bullet$, which abstract H atoms from RH to give ROOH and then regenerate. The peroxidation rate of RH is proportional to the stationary concentration of ROO$^\bullet$.

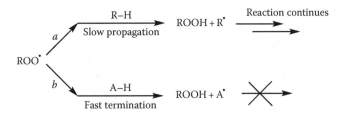

FIGURE 5.2 Concurrent reactions of ROO˙ radicals with RH (propagation), path *a*, and with antioxidants AH (termination), path *b*. While in the case of path *a* there is the formation of a chain-carrying radical, R˙, in the case of path *b* there is no further propagation and peroxidation stops (dead-end route). The radical A˙ reacts with ROO˙ or another A˙ to give non-radical products.

(ca. $-10\,\text{kcal mol}^{-1}$ of O_2), peroxidation is a spontaneous process [1]. However, it requires ROO˙ radical catalysis in order to occur. In fact, peroxidation generally involves a free radical chain mechanism in which (Figure 5.1) In˙ is an ROS/RNS or an ROS-/RNS-derived radical that abstracts a H atom from weak C–H bonds in RH to give a carbon-centered radical R˙ (initiation step) [1,9,10]. Afterward, the R˙ radical adds quickly ($k_{O_2} \sim 10^9\,\text{M}^{-1}\,\text{s}^{-1}$) oxygen to give the chain-carrying peroxyl, ROO˙ [1,9,10]. This radical restarts the chain (propagation step) by reacting with RH ($k_p \leq 100\,\text{M}^{-1}\,\text{s}^{-1}$) to yield R˙ and ROOH [1,9,10]. Finally, when two ROO˙ radicals encounter and quench by a radical/radical recombination ($2k_t \sim 10^3 - 10^8\,\text{M}^{-1}\,\text{s}^{-1}$), two chains are terminated (termination step) [11]. Under normal partial pressure of oxygen ($P_{O_2} \sim 210\,\text{mmHg}$), the stationary concentration of R˙ is very low because most of it is transformed into ROO˙. On the other hand, the concentration of the latter increases until a steady-state condition is attained as the rate of ROO˙ quenching reaches the rate of formation of In˙. In connection with the [ROO˙], it should be noted that the overall rate of peroxidation, that is, the rate at which oxygen alters and hence damages essential biomolecules, increases with [ROO˙] [1]. In several pathological conditions (inflammation, for instance) and because of the use of certain drugs, cosmetics, etc., the formation of initiating radicals (In˙) can increase, causing in turn an increase in "oxygen toxicity" [12–14].

As explained, the stationary concentration of ROO˙ controls the rate of oxidative injury. Two possible strategies for cells to reduce [ROO˙] and prevent damage are (1) the removal of a portion of initiating radicals, In˙, by cellular enzymes (superoxide dismutases, catalases, glutathione peroxidases, peroxiredoxins) that are capable of scavenging (preventive antioxidants) the O_2˙⁻, ROOH, and H_2O_2 [4,5,10], and (2) the use of simple antioxidant molecules, AH (chain-breaking antioxidants) that act as H-atom donors to the chain-carrying radicals ROO˙, ROO˙ + AH → ROOH + A˙, in competition with the substrate RH (Figure 5.2) [1,4,5,9]. The rate constant of ROO˙ + AH for effective antioxidants is orders of magnitude greater than that of ROO˙ + RH [1,9,10]. This means that a significant amount of ROO˙ reacts and disappears through path *b* (Figure 5.2), in competition with path *a*, even if [AH] ≪ [RH]. On the other hand, it must be noted that in path *b* the radical ROO˙ exchanges with an antioxidant-derived radical, A˙. Again, if this radical is derived from effective antioxidants, generally it is unable to start new peroxidation chains, making, therefore, path *b* a dead end (Figure 5.2) [9]. In this case, A˙ decays by self-quenching, that is, A˙ + A˙ → products, or by quickly reacting with another ROO˙ radical, ROO˙ + A˙ → products. Nonprotein antioxidants can have an exogenous (diet) or endogenous origin (uric acid, bilirubin, glutathione, ferritin, α-lipoic acid, etc.) [10,15]. Typical examples of effective antioxidants derived from the diet include vitamin E, vitamin C, and polyphenols [15].

5.3.1 PHENOLICS AS ANTIOXIDANTS

Phenols (ArOH) are among the best antioxidants in nature because they are able to quickly donate H atoms to ROO˙ (path *b*, Figure 5.2) from their OH group [1,9,16–18]. The aryloxyl radical ArO$^\bullet$ generated is stabilized by resonance and is unable to react with oxygen to form peroxyl radicals [9]. Thereby, the ArO$^\bullet$ radical is incapable of continuing the peroxidation chain (for exceptions, see Section 5.4.1). At low temperatures (<100°C), it usually disappears by quenching another ROO$^\bullet$ radical [17]. The stoichiometric factor, n, for the reaction ArOH + n ROO$^\bullet$ → products is therefore 2.0, indicating that two chains are terminated per single ArOH moiety. At higher temperatures, however, $n \to 0$ and ArOH becomes increasingly ineffective in retarding peroxidation [17].

The efficiency of phenols in the protection of living organisms from oxygen toxicity will be shown in Sections 5.4.1 and 5.4.2. In commercial terms, phenols *and* amines have a huge relevance since these compounds are essential to prolong the shelf-life and durability of man-made products [9,17]. Similar to phenols, amines are also powerful antioxidants but their chemistry is far more complex [17]. At higher temperatures, for instance, the stoichiometric factor of diarylamine antioxidants can reach a substantial value of 40 because of a mechanism in which the amine is regenerated from its radical [17].

The antioxidant ability of phenols depends essentially on the nature of the substituents present on the ring [9,18,19]. Later, it will be shown that electron-donating groups reinforce in phenols their ability to quench ROO$^\bullet$ and other radicals [9,19]. On the other hand, it is important to realize that the antioxidant ability (AA) of ArOH is a relative kinetic property because the level of inhibition of peroxidation, namely, the AA, is directly related to the rate of path b in relation to path a. This "ability" can therefore be expressed by the rate constant of path b, that is, the rate constant of ArOH + ROO$^\bullet$ (hereafter indicated as k_H). It should be noted that the effectiveness of a given ArOH depends on the "oxidizability" and concentration of RH (see Section 5.4.1) [1]. Other parameters that can be found in the literature as a measure of AA (DPPH$^\bullet$ EC$_{50}$, for instance) are groundless [20]. Interestingly, thermodynamics can help in assessing the antioxidant ability of phenols because the swiftness at which ArOH donates its H atom to ROO$^\bullet$ (k_H) depends on the ArO–H bond dissociation enthalpy (BDE) [19]. Generally, phenols with a weaker O–H bond are better antioxidants, ArO–H BDEs $\cong$ 77–87 kcal mol^{-1}. However, this obvious relation can, in a few cases, lead to a paradox since ArOHs with very weak O–H bonds might behave as strong pro-oxidants rather than as potent antioxidants! [9]. For instance, dialkylamino groups cannot be used to increase the AA of phenols [9,21]. This is because they are strong electron donors (much stronger than alkoxy groups), and phenols with these groups have low ionization potentials to the point that they can react via electron transfer to molecular oxygen generating free radicals [9].

5.4 CLASSES OF BIOLOGICALLY ACTIVE PHENOLS

5.4.1 TOCOPHEROLS AND TOCOTRIENOLS

Vitamin E is a family of liposoluble phenols called tocopherols and tocotrienols, which is found in vegetable oils, nuts, seeds, and green leafy vegetables [15,18]. These phenols are composed of a relatively polar 6-hydroxychromane head and a C$_{16}$ hydrocarbon tail at 2-position, which facilitates the passage into biological membranes (Figure 5.3) [22]. Tocopherols and tocotrienols differ in the presence of unsaturation on the hydrocarbon side chain. These phenols are further classified on the basis of the number of methyl groups present on the 6-hydroxychromane nucleus so that 8 natural vitamin E components exist: α, β, γ, and δ-tocopherol and α, β, γ, and δ-tocotrienol, as shown in Figure 5.3. Tocopherols have also three chiral centers at positions 2 and 4′ and 8′ in the chain, while tocotrienols have only one chiral center at position 2 on the chromane ring. In natural tocopherols, only RRR enantiomers are present, whereas synthetic α-tocopherol contains a mixture of enantiomers and diastereoisomers, all-*rac*-α-tocopherol. In humans, tocopherols are absorbed and delivered to the liver by chylomicrons, but only *RRR*-α-tocopherol is retained by the α-tocopherol transfer

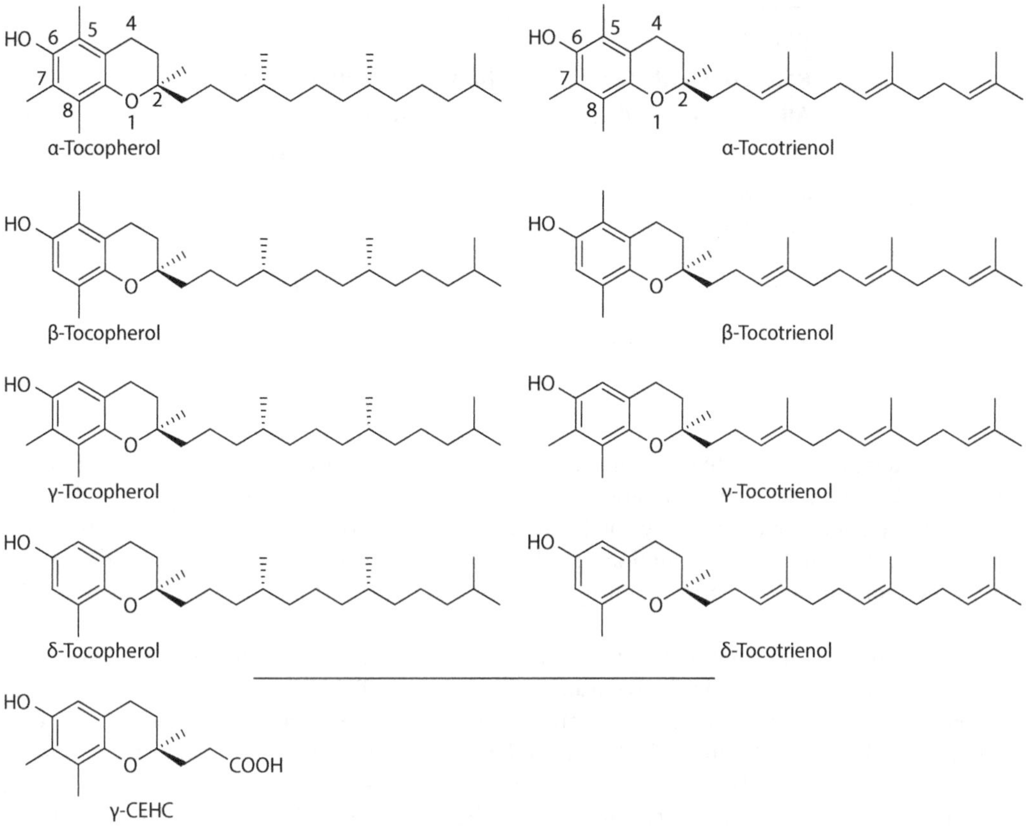

FIGURE 5.3 Structures of tocopherols and tocotrienols present in vitamin E and of the carboxyethyl hydroxychroman metabolite.

TABLE 5.1

Rates of Reaction between α-Tocopherol with Alkoxyl and Peroxyl Radicals

ROS	Solvent	k_H (M^{-1} s^{-1})	Reference
ROO•	Chlorobenzene	3.2×10^6	[21]
ROO•	MeCN	6.8×10^5	[23]
ROO•	LDL pH 7.4	5.9×10^5	[24]
RO•	Benzene	3.1×10^9	[25]
RO•	MeCN	2.9×10^8	[25]

protein (α-TTP), which specifically binds the natural isomer while other isoforms are degraded and excreted in urine or bile [22].

The phenolic head present in all members of the vitamin E family is able to trap free radicals. The rate constants k_H for the reaction of α-tocopherol with two oxygen-centered radicals are reported in Table 5.1. Multiple ring methylation has a positive effect on the AA of tocopherols and tocotrienols, the rate constant for the reaction with ROO• being about one order of magnitude larger for α-tocopherol relative to δ-tocopherol (Table 5.2) [21]. This is because alkyl groups are electron donors and stabilize the tocopheroxyl radical. On the other hand, stereochemistry and the nature of the C-2 hydrocarbon tail have no effect on the AA of these phenols.

TABLE 5.2
Rate Constants for the Reaction of Tocopherols with
Alkylperoxyl Radicals in Chlorobenzene at 30°C

	k_H (M^{-1} s^{-1})
α-tocopherol	3.2×10^6
β-tocopherol	1.3×10^6
γ-tocopherol	1.4×10^6
δ-tocopherol	4.4×10^5

Source: Burton, G.W. et al., *J. Am. Chem. Soc.*, 107, 7053, 1985.

As reported in Table 5.1, the k_H's in acetonitrile and low-density lipoprotein (LDL) at pH 7.4 are lower than those determined in apolar solvents irrespective of the ROS employed. This is a general effect that polar solvents have on H-atom transfer reactions from phenols (and other polar compounds). It is due to the formation of a H bond between ArOH and the polar solvent that makes the removal of H atom difficult by the attacking radical (kinetic solvent effects, KSEs) [25,26].

α-Tocopherol (and the other components of vitamin E) appears to be structurally designed for the protection of cell membrane's polyunsaturated fatty acids (PUFAs) against ROO$^•$. As stated in Section 5.1, the AA of phenols is a kinetic property that is properly expressed by the ratio between the rates of paths *b* and *a* (Figure 5.2). In connection with this argument, Niki interestingly noticed that for α-tocopherol (α-TOH), an effective antioxidant of PUFAs, the rate constant for α-TOH + ROS has to be at least 10,000 times larger than that for PUFA + ROS [27]. This estimation was based on the physiological concentration of PUFAs and α-TOH, whose molar ratio [PUFA]/[α-TOH] is about 10^3. On the basis of these estimations and of the rate constants $k_{H,ROS}$ in the literature, the only ROS that can be scavenged by α-tocopherol in a biological environment seems to be ROO$^•$ since HO$^•$, R$^•$, NO$_2^•$, RS$^•$, singlet oxygen ($^1O_2$), O$_3$, and HOCl are expected to react faster with PUFA than with α-tocopherol [27].

An important feature of the tocopheroxyl radical, α-TO$^•$, formed in the reaction with ROO$^•$ is that it can be reduced back to the phenolic form by reaction with ascorbate (vitamin C) or ubiquinol-10 [28–30]. This process is important since the permanence of α-TO$^•$ in human LDL is deleterious because of a process called "tocopherol-mediated peroxidation" [28,29]. This process consists in a slow peroxidation of the LDL initiated by the α-TO$^•$ radical. The inability of α-TO$^•$ to escape from an LDL particle allows it to propagate a radical chain via its reaction with PUFA within the particle ($k_H = 0.1$ M^{-1} s^{-1}); see Figure 5.4. However, the ascorbate anion in the aqueous phase or ubiquinol-10 in the LDL particle can both quench α-TO$^•$ by reducing it back to α-TOH; see Figure 5.4.

This section previously focused its attention on the properties of vitamin E as an antioxidant of PUFAs in a lipid milieu [28–30]. The physiological role of α-tocopherol in cells embraces, however, other important functions [31]. Also, the other constituents of the vitamin E family and their metabolites may play a role in the prevention of disease. γ-Tocopherol, for instance, is abundantly present in the diet but only in low amounts in blood plasma because it is quickly metabolized to 2,7,8-trimethyl-2-(β-carboxyethyl)-6-hydroxychroman (γ-CEHC) in the liver (Figure 5.3) [32]. This metabolite has anti-inflammatory and natriuretic properties, which may explain why epidemiologic studies show an inverse correlation between γ-tocopherol intake and cardiovascular diseases and cancer [32]. Tocotrienols are also gaining increasing interest by the scientific community because they display promising effects in cancer, bone resorption, diabetes, and in cardiovascular and neurological diseases [33].

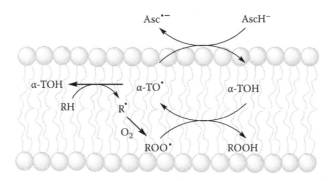

FIGURE 5.4 Mechanism of α-tocopherol-mediated peroxidation of PUFA lipids in the double layer of a cell membrane. This slow peroxidation, $k_H = 0.1$ M^{-1} s^{-1} for α-TO$^•$+RH, is due to the inability of α-TO$^•$ to escape from the apolar environment into the water phase. The polar –O$^•$ head can now and then be in contact with the aqueous phase where it can be reduced by ascorbate (AscH$^-$) with the formation of the Asc$^{•-}$ radical, which can be further oxidized to dehydroascorbic acid (30). This "radical export" from the membrane into the aqueous phase prevents the α-tocopherol-mediated peroxidation of PUFAs.

5.4.2 UBIQUINOL

Ubiquinol (UQH$_2$) is the reduced form of coenzyme Q. It acts as an electron carrier in mitochondrial respiratory chain and as an antioxidant in cell membranes (Figure 5.5) [3]. Unlike vitamin E, which is absorbed from food, ubiquinol is produced endogenously in all cells and tissues in variable amounts [34]. It is composed of a 1,4-diphenolic (hydroquinone) moiety with a hydrocarbon tail of 6–10 isoprene units (10 in humans), (Figure 5.5) [3]. Ubiquinol-10 is therefore very lipophilic and is located in the interior of membranes despite the presence of two hydroxyl groups that are H bonded to the adjacent OMe groups [35]. In lipid bilayers, UQH$_2$ is oxidized by radicals to the semiquinone radical (UQH$^•$). This species quickly deprotonates in aqueous solution at physiologic pH values to give the radical anion UQ$^{•-}$ as the pK_a of semiquinones is about 5 (Figure 5.5) [29,36].

The presence of intramolecular H bonds influences the reactivity of the phenolic OHs with radicals. Abstraction of a H atom from ArOH groups engaged in a H bond is more difficult than from *free* ArOHs [37]. However, the abstraction of a H atom from one of the OH groups in UQH$_2$ is facilitated by the concerted strengthening of the H bond in the other one (remote H-bond effect) [38,39].

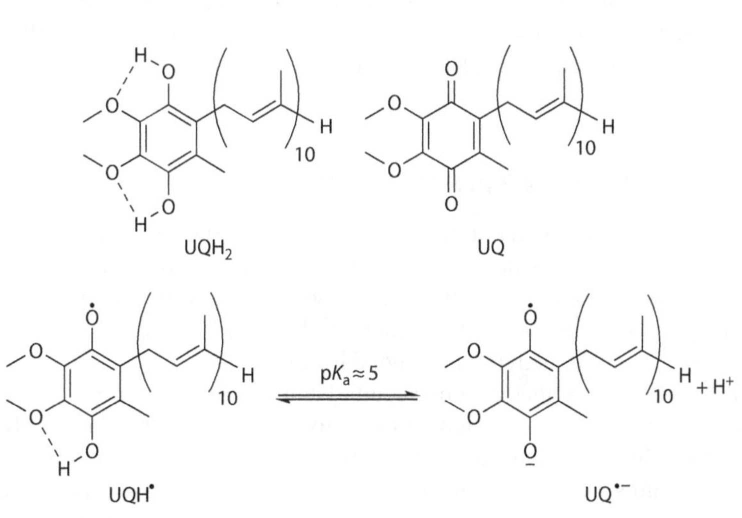

FIGURE 5.5 Structures of coenzyme Q10 in its various oxidation states.

TABLE 5.3

Rate Constants for the Reaction of Ubiquinol with Various ROS at Room Temperature

Radical	k_H (M^{-1} s^{-1})	Solvent	Reference
ROO$^{\bullet}$[a]	4.4×10^5	Chlorobenzene	[40]
RO$^{\bullet}$[a]	2.2×10^9	Benzene	[41]
RO$^{\bullet}$[a]	4.3×10^8	Acetonitrile	[41]
NO$^{\bullet}$	1.6×10^4	Water, pH 6	[42]

[a] Data for ubiquinol-0 in which there is no isoprene tail.

UQH$_2$ reacts rapidly with ROS, in particular with peroxyl [40], alkoxyl [41], and NO$^{\bullet}$ [42] radicals (Table 5.3). The reactivity toward RO$^{\bullet}$ radicals reduces in acetonitrile relative to that in benzene because of the kinetic solvent effects mentioned earlier [26].

The concentration of coenzyme Q10 in cell membranes is usually 5–20 times larger than that of α-tocopherol except in LDLs, where it is about 10 times smaller [34]. Interestingly, ubiquinol is able to spare not only α-tocopherol in PUFA liposomes through the reaction of UQH$_2$ but also UQH$^{\bullet}$ with α-TO$^{\bullet}$ to give α-TOH [35].

As reported earlier, ubiquinol-10 and ascorbic acid inhibit the tocopherol-mediated peroxidation in human LDLs [28,29]. This inhibition is evident for ascorbic acid because there is an apparent "radical export" into the aqueous phase through the formation of Asc$^{\bullet-}$ (Figure 5.4). In the case of UQH$_2$ this is less evident since the α-TO$^{\bullet}$ radical is converted into UQH$^{\bullet}$, which still resides in the LDL particle and thereby might start a new peroxidation chain similar to the α-TO$^{\bullet}$ radical [28,29]. However, an important feature of the semiquinone radical UQH$^{\bullet}$ is that it reacts with O$_2$ to give the hydroperoxyl radical HOO$^{\bullet}$ (or O$_2^{\bullet-}$ formed by electron transfer from UQ$^{\bullet-}$; the two radical species are a Brønsted–Lowry acid–base couple, HOO$^{\bullet} \leftrightarrows$ H$^+$ + O$_2^{\bullet-}$) [28]. The formation of HOO$^{\bullet}$ has different kinetic consequences depending on the system being homogeneous or heterogeneous. In a homogeneous solution, the experiment shows that the reaction, UQH$^{\bullet}$ + O$_2 \rightarrow$ UQ + HOO$^{\bullet}$, reduces markedly the number of ROO$^{\bullet}$ radicals trapped by one UQH$_2$ [43]. On the contrary, in the case of peroxidation involving vesicles and LDL, the formation of HOO$^{\bullet}$ decreases the rate of peroxidation since HOO$^{\bullet}$ is a small and polar radical that can easily diffuse outside the particle into the aqueous phase ("radical export"). In water, at pH 7.4, the predominant form of the radical is O$_2^{\bullet-}$ (p$K_a \approx 4.5$) [44], which is unable to abstract H atoms from PUFA [36]. The migration of O$_2^{\bullet-}$/HOO$^{\bullet}$ in the aqueous phase is the reason, unlike α-tocopherol, UQH$_2$ is not associated with *pro-oxidant* effects in LDL [28].

5.5   DIETARY PHENOLS AND POLYPHENOLS

Epidemiological evidence shows that a diet rich in fruits, vegetables, and in plant-derived commodities such as tea, coffee, red wine, and chocolate determines a lower incidence of cancer, diabetes, and cardiovascular diseases [45]. What all these foods have in common is the presence of comparatively large amounts of phenolics, compounds that the scientific community is increasingly considering to be beneficial for maintaining a good health [18].

Polyphenols are secondary metabolites accumulated in high concentrations in many edible plants even though these compounds have little role in their growth. Flavonoids are the most typical polyphenols [46]. They have a C6–C3–C6 carbon core and often are glycosylated with one to several sugar units [46,47]. Flavonoids are classified according to their aglycone structure [46]. The main structures of dietary interest are illustrated in Figure 5.6. Other important phenolics include phenolic acids and hydroxyl derivatives of stilbene; see Figure 5.7. Oligomerization

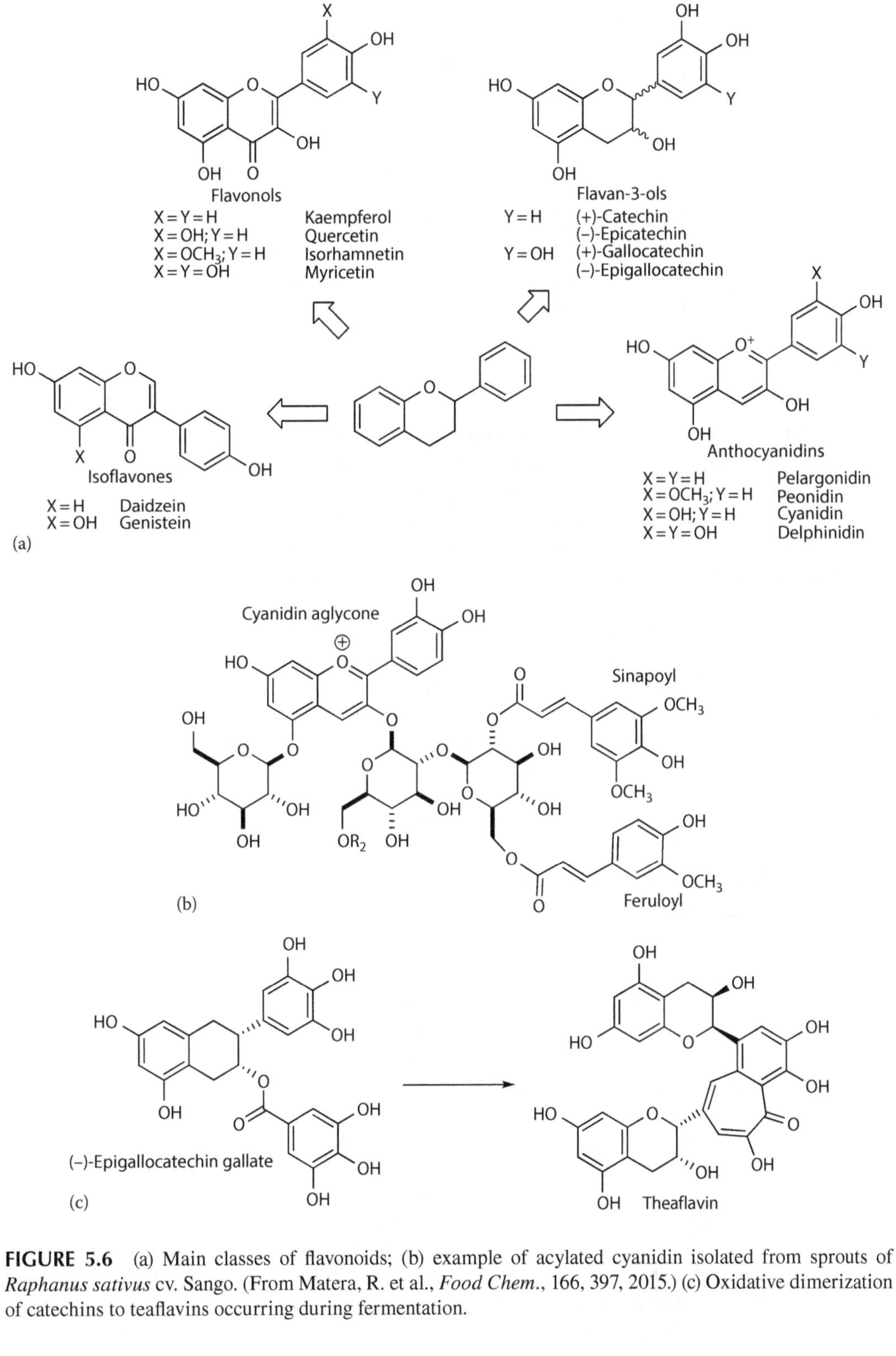

FIGURE 5.6 (a) Main classes of flavonoids; (b) example of acylated cyanidin isolated from sprouts of *Raphanus sativus* cv. Sango. (From Matera, R. et al., *Food Chem.*, 166, 397, 2015.) (c) Oxidative dimerization of catechins to teaflavins occurring during fermentation.

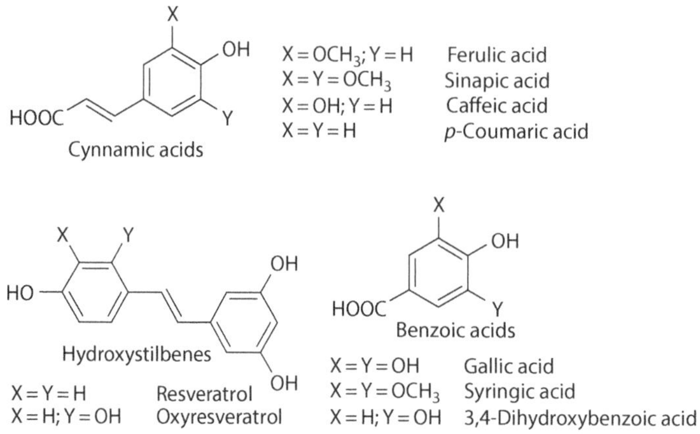

FIGURE 5.7 Phenolic acids and hydroxylstilbenes.

and partial oxidation of flavonoids, hydroxystilbenes, cinnamic, and gallic acids lead to the formation of a wide array of polyphenolic derivatives such as theaflavins, which are formed during the fermentation of tea leaves (Figure 5.6). In general, the procedures employed in the preparation of coffee and tea (roasting and fermentation) can yield many new phenolics (Figure 5.6) [45].

The presence of numerous phenolic OH groups in polyphenols makes these compounds good antioxidants [48]. The available rate constants for the reaction of dietary phenols with peroxyl radicals are reported in Table 5.4 [49–54].

The possibility that the antioxidant activity of polyphenols is the reason for the *in vivo* positive health effects of a diet rich in fruits and vegetables, although suggestive, has not been proven yet, and indeed several lines of evidence suggest a much more complex scenario. Some dietary phenols are preferentially adsorbed in the small intestine, while others reach the large intestine where they are extensively degraded by the colonic microflora. The adsorbed fraction is metabolized by

TABLE 5.4

Rates of Reactions of Selected Polyphenols with Peroxyl Radicals from Autoxidation Studies

Phenol	k_H (M^{-1} s^{-1})	Solvent	Reference
Epicatechin	4.2×10^5	PhCl, 50°C	[49]
	1.7×10^4	*t*-BuOH, 50°C	[49]
Quercetin	5.0×10^5	PhCl, 30°C	[50]
	2.1×10^4	*t*-BuOH, 50°C	[49]
Hydroxytyrosol	8.0×10^5	PhCl, 30°C	[50]
	3.0×10^4	MeCN, 30°C	[50]
Caffeic acid	2.9×10^5	PhCl, 1% MeOH, 30°C	[51]
Caffeic acid phenetyl ester	6.8×10^5	PhCl, 30°C	[52]
	1.3×10^4	MeCN, 30°C	[52]
Resveratrol	1.8×10^5	PhCl, 30°C	[53]
	1.0×10^4	MeCN, 30°C	[53]
3,4-Dihydroxybenzoic acid	6.5×10^4	PhCl, 1% MeOH, 30°C	[51]
Epigallocatechin gallate	1.3×10^4	SDS micelles, pH 7.4, 37°C	[54]

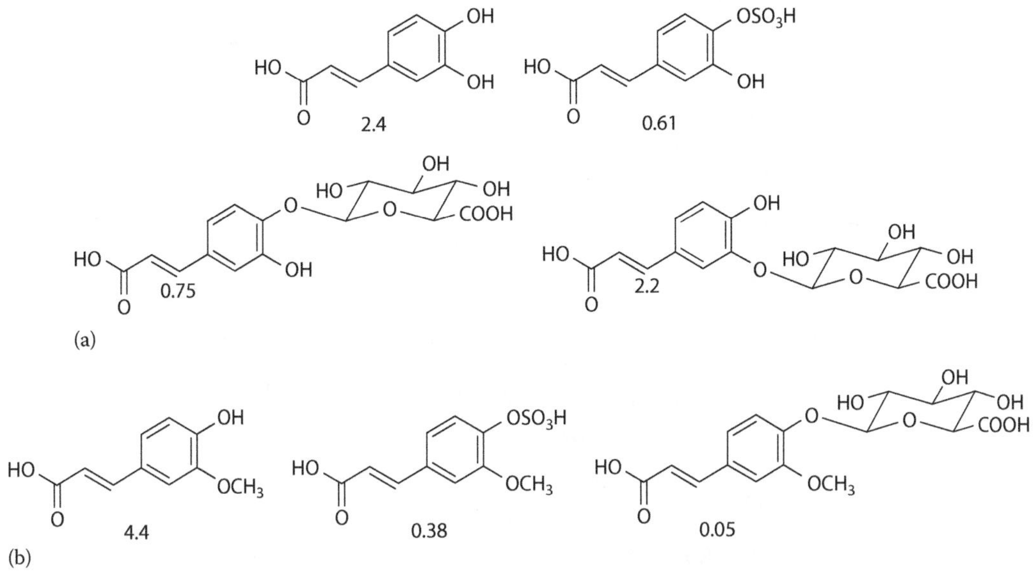

FIGURE 5.8 Radical-trapping abilities of (a) caffeic acid and ferulic acid (b) and of their sulfate or glucuronide metabolites measured by the ABTS assay. The numbers represent the slope of the plot, concentration of quenched ABTS *versus* concentration of antioxidant. Higher slopes denote higher antiradical activities. (From Piazzon, A. et al., *J. Agric. Food Chem.*, 60, 12312, 2012.)

the intestinal cells to glucuronides and sulfate derivatives, which are usually found in plasma in concentration in the range of 1–1000 nM. The half-life of such metabolites is usually short (<5 h), and this suggests that further metabolic modifications and a rapid excretion in urines occur [45]. The actual plasmatic concentrations of the polyphenol metabolites seem not to be consistent with the *classical* antioxidant theory of the health effects of vegetables and fruits, particularly if these concentrations are compared with those of antioxidants such as vitamin C or endogenous uric acid [55]. On the other hand, the AAs of polyphenol metabolites are not often known since the scientific interest has mainly been concentrated on their precursors (polyphenols). Recent studies, however, pointed out that these metabolic modifications (glucuronidation and sulfation) of the (phenolic) OH groups may decrease or even nullify the antioxidant activity of polyphenols [10,56]. In this respect, Figure 5.8 shows how the antiradical abilities of caffeic and ferulic acids decline as their structure is subjected to the metabolic transformations we mentioned. Also, the data clearly show that the antiradical abilities of these phenolic acids rely essentially on the phenolic OH *para* to the propenoic acid group.

A recent hypothesis that has been put forward to explain the health protective effects of polyphenols is that these compounds, rather than acting as antioxidants, behave as mild electrophilic stressors that upregulate the natural defenses against oxidative stress. According to this hypothesis, polyphenols are oxidized *in vivo* to the corresponding quinones, which in turn react with activated cysteines present in proteins that control the expression of protective genes [57] (Figure 5.9). After the reaction, *ortho* quinones are converted to phenols with a thioalkyl substituent (Figure 5.9). Interestingly, it has been observed that these alkylthiophenols are still reactive toward peroxyl radicals and hence are able to undergo further oxidation processes. For instance, 5-lipoylhydroxytyrosol (Figure 5.9), formed by oxidizing hydroxytyrosol in the presence of lipoic acid, has a k_H value of 2.0×10^5 M^{-1} s^{-1} in PhCl at 30°C, only four times lower than that of hydroxytyrosol (Table 5.4) [50].

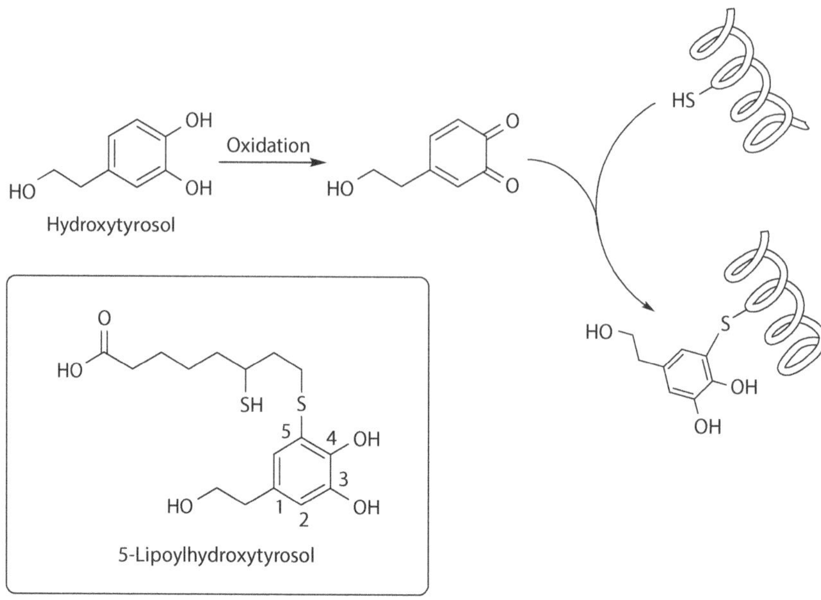

FIGURE 5.9 Adduct formation between orthoquinones and activated thiols in proteins. The inset shows the adduct formed from the reaction between the orthoquinone of hydroxytyrosol and lipoic acid.

5.6 CHEMICAL METHODS TO MEASURE THE ANTIOXIDANT ABILITY OF PHENOLS

The antioxidant ability of compounds can be assessed by using *in vivo* and *in vitro* assays [58]. For *in vivo* evaluation, specific markers of oxidative stress, univocally related to the oxidation of biomolecules (e.g., isoprostanes in plasma and 8-hydroxy-2′-deoxyguanosine [8-OHdG] in urine) are measured [59]. On the other hand, *in vitro* assays have often been described as "surrogates" or preliminary screening tests. In these assays, biochemical and chemical methods are used in order to evaluate the antioxidant effects of a compound. In the first group, the antioxidant activity of a given compound is evaluated from its ability to reduce the oxidation of a probe inside the cells when the latter are exposed to an oxidant such as H_2O_2 [58,60]. Assays based on chemical methods measure instead the ability of a given compound to slow down the autoxidation of oxidizable compounds. Chemical methods also provide mechanistic insights for nonmedical applications (e.g., stabilization of food or of pharmaceutical formulations). Furthermore, these methods, being simpler than the biochemical ones, are amenable to being used for screening purposes [58]. Unfortunately, this has generated a plethora of assays with low chemical soundness as shown by us in the case of the antioxidant activity of essential oils [61].

5.6.1 FIRST CHOICE METHODS: AUTOXIDATION-BASED METHODS

The best way to measure the AA of chemicals is to follow the autoxidation of a substrate in the presence and absence of antioxidants (Figure 5.10). The two rates of oxidation give the protecting ability of the chemical a protecting ability. However, spontaneous autoxidation is poorly controllable because the formation of initiating radicals depends on traces of peroxides and transition metal ions present as impurities in the substrate and hence in variable amounts. Thereby, spontaneous autoxidation is usually slow and may take hours or even days to convert a small fraction of the substrate [58]. In order to overcome this problem and the scarce reproducibility of the initiation process, several radical initiators are commonly used. Azoinitiators are

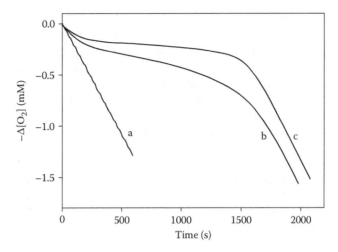

FIGURE 5.10 Oxygen consumption measured during the autoxidation of styrene (4.3 M) in chlorobenzene initiated by 0.05 M AIBN (Figure 5.11) at 30°C in the absence (a) or in the presence of 5 μM of (b) Trolox or (c) 2,2,5,7,8-pentamethyl-6-chromanol.

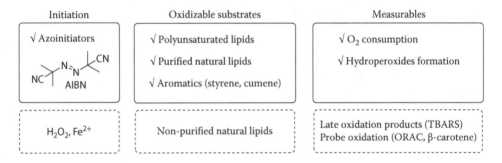

FIGURE 5.11 Recommended (on top) and not-recommended procedures for measuring antioxidant activity in autoxidation studies. AIBN: azobisisobutyronitrile, a liposoluble azoinitiator.

the best ones because they decompose at a constant rate that depends on the temperature only. Hydrophilic and lipophilic azoinitiators are both commercially available (Figure 5.11). Other popular methods include the addition of Fe^{2+} and H_2O_2, which generate HO˙ radicals through the Fenton reaction [6]. These initiators should be used with caution because iron salts and hydrogen peroxide interfere with the radicals that sustain the autoxidation chain [62] and because the hydroperoxides formed during the autoxidation of the substrate may react with Fe^{2+}, giving an autocatalytic acceleration of the autoxidation [58].

In order to follow the autoxidation of a substrate, different procedures can be used. The simplest is based on the measurement of oxygen consumption in the sample. This can be made (1) by using a sensitive pressure gauge (best choice for experiments in organic solvents) or (2) by using dissolved oxygen sensors (particularly in aqueous solutions) [63]. Alternative methods include the spectrophotometric measurement of conjugated-diene hydroperoxides formed during the autoxidation of unsaturated lipids. These compounds have a characteristic UV–vis absorption spectrum at about 232 nm [64]. Determination by HPLC–UV analysis is, however, preferred in order to exclude interferences by other absorbing species that may be formed during the autoxidation process [65]. Hydroperoxides are early oxidation products, and their concentration is directly linked to the rate of autoxidation. In this respect, the analysis of the regioisomers of the hydroperoxides formed by methyl linoleate autoxidation also provides information about the kinetics of antioxidants [66]. On the other hand, various methods based on the detection of *advanced* oxidation products have imprudently become very popular. These methods include the TBARS (thiobarbituric acid reactive species) assay that detects malondialdehyde, which is formed after the breakdown of hydroperoxides at high temperatures. This assay does not afford a quantitative measure of the rate of autoxidation and should be used to get a qualitative ranking of the antioxidant potency [58,61].

The choice of the oxidizable substrate in this autoxidation is another important point. Pure polyunsaturated lipids (e.g., methyl linoleate) are best suited because of their high oxidation rate. However, natural mixtures such as egg yolk lecithin or seed oils can also be used, providing that the natural antioxidants present in the lipids are removed in advance. Synthetic substrates such as styrene or cumene are extensively used because they are readily available and can be easily purified. In this context, it should be noted that the ORAC (oxygen radical antioxidant capacity) assay, a very popular method for nutritional studies, cannot be considered as an autoxidation-based method because no oxidizable substrate is present. In the ORAC assay, the antioxidant competes with a fluorescent probe (e.g., phycoerythrin or fluorescein) for quenching peroxyl radicals generated from AAPH (2,2′-azobis(2-amidinopropane) dihydrochloride), a water-soluble thermal azo initiator [58].

Finally, the chemistry behind many assays is often very different from the chemistry involved in the free and inhibited autoxidation of an organic substrate (see Figures 5.1 and 5.2). The alleged "antioxidant activity" measured or evaluated by means of many of these tests is actually (in simple cases) an antiradical activity. This is the very case with the quenching of stable or persistent colored radicals, the most common being DPPH$^{\bullet}$ (2,2-diphenyl-1-picrylhydrazyl) and ABTS (2,2′-azinobis(3-ethylbenzothiazoline-6-sulfonic acid), which give a radical trapping ability only! Often, the information gained from these tests is not sufficient to establish whether or not a given compound is able to inhibit the autoxidation of organic matter. For instance, H_2O_2 is able to quench the DPPH$^{\bullet}$ radical but certainly it is not an antioxidant! [67] On the other hand, tests that are based on the reduction of metal ions, such as FRAP (ferric reducing antioxidant power), CUPRAC (cupric reducing antioxidant capacity), and Folin–Ciocalteu tests, provide the stoichiometry of any reducing molecules present in the sample and which are necessarily not antioxidants [58].

5.7 CONCLUSIONS

The *in vitro* autoxidation of organic matter is a good model for the real radical processes that can occur in cell compartments. Oxidative damage to membranes, mitochondria, and DNA is increasingly considered to be implicated in the decline of physiologic functions that occur during aging. Cells have, however, an elaborate defense system to keep ROS and RNS at bay, including enzymes and small molecules (particularly phenols) able to convert these species to harmless products. Vitamin E is one of the major lipid antioxidants present in the human body along with ubiquinol-10 and vitamin C, which are notably involved in its regeneration. New kinetic interpretation of available data supports the antioxidant role of vitamin E *in vivo*, particularly against peroxyl radicals in LDL. Other *in vitro* antioxidant compounds introduced with the diet (flavonoids, phenolic acids, hydroxystilbenes, etc.) are molecularly equipped to have similar roles. However, the *physiological* concentration of these compounds is often very low and thus they are unable to trap ROS and RNS. Thereby, the health benefits attributed to these phenolic compounds must not simply rely on their antioxidant properties. The regulation of oxidative stress presumably relies on a more complex scenario (the activation of the NF-E2-related factor 2 signaling pathway) on which further experimentation is desirable.

REFERENCES

1. Foti MC. Antioxidant properties of phenols. *Journal of Pharmacy and Pharmacology* 2007;59:1673–1685.
2. Halliwell B and Gutteridge JMC. *Free Radicals in Biology and Medicine*, 4th edn., 2006. Oxford University Press, Oxford, U.K.
3. Nelson DL and Cox MM. *Lehninger Principles of Biochemistry*, 4th edn., 2005. W.H. Freeman and Company, New York.
4. Halliwell B. Oxidative stress and neurodegeneration: Where are we now? *Journal of Neurochemistry* 2006;97:1634–1658.
5. Halliwell B. Reactive species and antioxidants. Redox biology is a fundamental theme of aerobic life. *Plant Physiology* 2006;141:312–322.
6. Goldstein S, Meyerstein D, and Czapski G. The Fenton reagents. *Free Radical Biology and Medicine* 1993;15:435–445.
7. Semchyshyn HM. Reactive carbonyl species in vivo: Generation and dual biological effects. *The Scientific World Journal* 2014;2014:417842.
8. Chetyrkin S, Mathis M, Pedchenko V et al. Glucose autoxidation induces functional damage to proteins via modification of critical arginine residues. *Biochemistry* 2011;50:6102–6112.
9. Ingold KU and Pratt DA. Advances in radical-trapping antioxidant chemistry in the 21st century: A kinetics and mechanisms perspective. *Chemical Reviews* 2014;114:9022–9046.
10. Foti MC and Amorati R. Non-phenolic radical-trapping antioxidants. *Journal of Pharmacy and Pharmacology* 2009;61:1435–1448.
11. Foti MC and Ingold KU. Mechanism of inhibition of lipid peroxidation by γ-terpinene, an unusual and potentially useful hydrocarbon antioxidant. *Journal of Agricultural and Food Chemistry* 2003;51:2758–2765.
12. Conner EM and Grisham MB. Inflammation, free radicals, and antioxidants. *Nutrition* 1996;12(4):274–281.
13. Khansari N, Shakiba Y, and Mahmoudi M. Chronic inflammation and oxidative stress as a major cause of age-related diseases and cancer. *Recent Patents on Inflammation & Allergy Drug Discovery* 2009;3(1):73–80.
14. Sinha BK. Free radicals in anticancer drug pharmacology. *Chemico-Biological Interactions* 1989;69(4):293–317.
15. Li Y. *Antioxidants in Biology and Medicine: Essentials, Advances, and Clinical Applications*, 2011. Nova Science Publishers, Inc., New York.
16. Burton GW and Ingold KU. Autoxidation of biological molecules. 1. The antioxidant activity of vitamin E and related chain-breaking phenolic antioxidants in vitro. *Journal of the American Chemical Society* 1981;103:6472–6477.
17. Gryn'ova G, Ingold KU, and Coote ML. New insights into the mechanism of amine/nitroxide cycling during the hindered amine light stabilizer inhibited oxidative degradation of polymers. *Journal of the American Chemical Society* 2012;134:12979–12988.
18. Rappoport Z. *The Chemistry of Phenols*, 2003. John Wiley & Sons, Ltd., Chichester, U.K.
19. Foti MC, Daquino C, Mackie ID et al. Reaction of phenols with the 2,2-diphenyl-1-picrylhydrazyl radical. Kinetics and DFT calculations applied to determine ArO-H bond dissociation enthalpies and reaction mechanism. *Journal of Organic Chemistry* 2008;73(23):9270–9282.
20. Foti MC. The use and abuse of the DPPH• radical. A review. *Journal of Agricultural and Food Chemistry* 2015;63(40):8765–8776.
21. Burton GW, Doba T, Gabe E et al. Autoxidation of biological molecules. 4. Maximizing the antioxidant activity of phenols. *Journal of the American Chemical Society* 1985;107:7053–7065.
22. Cardenas E and Ghosh R. Vitamin E: A dark horse at the crossroad of cancer management. *Biochemical Pharmacology* 2013;86:845–852.
23. Amorati R, Pedulli GF, and Valgimigli L. Kinetic and thermodynamic aspects of the chain-breaking antioxidant activity of ascorbic acid derivatives in non-aqueous media. *Organic and Biomolecular Chemistry* 2011;9:3792–3800.
24. Culbertson SM, Antunes F, Havrilla CM et al. Determination of the α-tocopherol inhibition rate constant for peroxidation in low-density lipoprotein. *Chemical Research in Toxicology* 2002;15:870–876.
25. Valgimigli L, Banks JT, Lusztyk J et al. Solvent effects on the antioxidant activity of vitamin E. *Journal of Organic Chemistry* 1999;64:3381–3383.

26. Snelgrove DW, Lusztyk J, Banks JT et al. Kinetic solvent effects on hydrogen-atom abstractions: Reliable, quantitative predictions via a single empirical equation. *Journal of the American Chemical Society* 2001;123(3):469–477.

27. Niki E. Role of vitamin E as a lipid-soluble peroxyl radical scavenger: In vitro and in vivo evidence. *Free Radical Biology and Medicine* 2014;66:3–12.

28. Bowry VW and Stocker R. Tocopherol-mediated peroxidation. The prooxidant effect of vitamin E on the radical-initiated oxidation of human low-density lipoprotein. *Journal of the American Chemical Society* 1993;115(14):6029–6044.

29. Bowry VW and Ingold KU. The unexpected role of vitamin E (α-tocopherol) in the peroxidation of human low-density lipoprotein. *Accounts of Chemical Research* 1999;32:27–34.

30. Traber MG and Stevens JF. Vitamins C and E: Beneficial effects from a mechanistic perspective. *Free Radical Biology and Medicine* 2011;51:1000–1013.

31. Azzi A. Molecular mechanism of α-tocopherol action. *Free Radical Biology and Medicine* 2007;43: 16–21.

32. Jiang Q, Christen S, Shigenaga MK et al. γ-Tocopherol, the major form of vitamin E in the US diet, deserves more attention. *The American Journal of Clinical Nutrition* 2001;74:714–722.

33. Aggarwal BB, Sundaram C, Prasad S et al. Tocotrienols, the vitamin E of the 21st century: Its potential against cancer and other chronic diseases. *Biochemical Pharmacology* 2010;80:1613–1631.

34. Bentinger M, Brismar K, and Dallner G. The antioxidant role of coenzyme Q. *Mitochondrion* 2007;7(Suppl.):S41–S50.

35. James AM, Smith RAJ, and Murphy MP. Antioxidant and prooxidant properties of mitochondrial coenzyme Q. *Archives of Biochemistry and Biophysics* 2004;423:47–56.

36. Warren JJ, Tronic TA, and Mayer JM. Thermochemistry of proton-coupled electron transfer reagents and its implications. *Chemical Reviews* 2010;110:6961–7001.

37. Amorati R, Menichetti S, Mileo E et al. Hydrogen-atom transfer reactions from ortho-alkoxy-substituted phenols: An experimental approach. *Chemistry: A European Journal* 2009;15:4402–4410.

38. Amorati R, Franchi P, and Pedulli GF. Intermolecular hydrogen bonding modulates the hydrogen-atom-donating ability of hydroquinones. *Angewandte Chemie, International Edition* 2007;46:6336–6338.

39. Foti MC, Amorati R, Pedulli GF et al. Influence of "remote" intramolecular hydrogen bonds on the stabilities of phenoxyl radicals and benzyl cations. *Journal of Organic Chemistry* 2010;75:4434–4440.

40. Roginsky V, Barsukova T, Loshadkin D et al. Substituted p-hydroquinones as inhibitors of lipid peroxidation. *Chemistry and Physics of Lipids* 2003;125:49–58.

41. de Heer MI, Mulder P, Korth H-G et al. Hydrogen atom abstraction kinetics from intramolecularly hydrogen bonded ubiquinol-0 and other (poly)methoxy phenols. *Journal of the American Chemical Society* 2000;122:2355–2360.

42. Poderoso JJ, Carreras MC, Schöpfer F et al. The reaction of nitric oxide with ubiquinol: Kinetic properties and biological significance. *Free Radical Biology and Medicine* 1999;26(7):925–935.

43. Valgimigli L, Amorati R, Fumo MG et al. The Unusual reaction of semiquinone radicals with molecular oxygen. *Journal of Organic Chemistry* 2008;73:1830–1841.

44. Foti MC, Sortino S, and Ingold KU. New insight into solvent effects on the formal HOO + HOO reaction. *Chemistry a European Journal* 2005;11:1942–1948.

45. Crozier A, Jaganath IB, and Clifford MN. Dietary phenolics: Chemistry, bioavailability and effects on health. *Natural Product Reports* 2009;26:1001–1043.

46. Grotewold E. *The Science of Flavonoids*, 2006. Springer Science & Business Media, Inc., New York.

47. Matera R, Gabbanini S, Berretti S et al. Acylated anthocyanins from sprouts of *Raphanus sativus* cv. Sango: Isolation, structure elucidation and antioxidant activity. *Food Chemistry* 2015;166:397–406.

48. Amorati R and Valgimigli L. Modulation of the antioxidant activity of phenols by non-covalent interactions. *Organic and Biomolecular Chemistry* 2012;10:4147–4158.

49. Pedrielli P, Pedulli GF, and Skibsted LH. Antioxidant mechanism of flavonoids. solvent effect on rate constant for chain-breaking reaction of quercetin and epicatechin in autoxidation of methyl linoleate. *Journal of Agricultural and Food Chemistry* 2001;49:3034–3040.

50. Amorati R, Valgimigli L, Panzella L et al. 5-*S*-Lipoylhydroxytyrosol, a multidefense antioxidant featuring a solvent-tunable peroxyl radical-scavenging 3-thio-1,2-dihydroxybenzene motif. *Journal of Organic Chemistry* 2013;78:9857–9864.

51. Amorati R, Pedulli GF, Cabrini L et al. Solvent and pH effects on the antioxidant activity of caffeic and other phenolic acids. *Journal of Agricultural and Food Chemistry* 2006;54:2932–2937.

52. Spatafora C, Daquino C, Tringali C et al. Reaction of benzoxanthene lignans with peroxyl radicals in polar and non-polar media: Cooperative behaviour of OH groups. *Organic and Biomolecular Chemistry* 2013;11:4291–4294.
53. Tanini D, Panzella L, Amorati R et al. Resveratrol-based benzoselenophenes with an enhanced antioxidant and chain breaking capacity. *Organic and Biomolecular Chemistry* 2015;13:5757–5764.
54. Zhou B, Wu LM, Yang L et al. Evidence for α-tocopherol regeneration reaction of green tea polyphenols in SDS micelles. *Free Radical Biology and Medicine* 2005;38:78–84.
55. Hollmann PCH. Unravelling of the health effects of polyphenols is a complex puzzle complicated by metabolism. *Archives of Biochemistry and Biophysics* 2014;559:100–105.
56. Piazzon A, Vrhovsek U, Masuero D et al. Antioxidant activity of phenolic acids and their metabolites: Synthesis and antioxidant properties of the sulfate derivatives of ferulic and caffeic acids and of the acyl glucuronide of ferulic acid. *Journal of Agricultural and Food Chemistry* 2012;60:12312–12323.
57. Forman HJ, Davies KJA, and Ursini F. How do nutritional antioxidants really work: Nucleophilic tone and para-hormesis versus free radical scavenging in vivo. *Free Radical Biology and Medicine* 2014;66:24–35.
58. Amorati R and Valgimigli L. Advantages and limitations of common testing methods for antioxidants. *Free Radical Research* 2015;49:633–649.
59. Seet RCS, Lee CYJ, Lim ECH et al. Oxidative damage in Parkinson disease: Measurement using accurate biomarkers. *Free Radical Biology and Medicine* 2010;48:560–566.
60. Viglianisi C, Di Pilla V, Menichetti S et al. Linking an α-tocopherol derivative to cobalt(0) nanomagnets: Magnetically responsive antioxidants with superior radical trapping activity and reduced cytotoxicity. *Chemistry: A European Journal* 2014;20:6857–6860.
61. Amorati R, Foti MC, and Valgimigli L. Antioxidant activity of essential oils. *Journal of Agricultural and Food Chemistry* 2013;61:10835–10847.
62. Foti MC and Ingold KU. Unexpected superoxide dismutase antioxidant activity of ferric chloride in acetonitrile. *Journal of Organic Chemistry* 2003;68(23):9162–9165.
63. Viglianisi C, Bartolozzi MG, Pedulli GF et al. Optimization of the antioxidant activity of hydroxy substituted 4-thiaflavanes: A proof-of-concept study. *Chemistry: A European Journal* 2011;17:12396–12404.
64. Foti MC and Ruberto G. Kinetic solvent effects on phenolic antioxidants determined by spectrophotometric measurements. *Journal of Agricultural and Food Chemistry* 2001;49:342–348.
65. Amorati R, Valgimigli L, Diner P et al. Multi-faceted reactivity of alkyltellurophenols towards peroxyl radicals: Catalytic antioxidant versus thiol-depletion effect. *Chemistry: A European Journal* 2013;19:7510–7522.
66. Tallman KA, Pratt DA, and Porter NA. Kinetic products of linoleate peroxidation: Rapid β-fragmentation of nonconjugated peroxyls. *Journal of the American Chemical Society* 2001;123:11827–11828.
67. Ionita P. Is DPPH stable free radical a good scavenger for oxygen active species? *Chemical Papers* 2005;59:11–16.

6 Lipoprotein-Associated Oxidative Stress

Ngoc-Anh Le

CONTENTS

ABSTRACT

Oxidative stress is implicated in the initiation and progression of different chronic diseases from diabetes to cardiovascular and neurological diseases. Reactive oxygen species (ROS) that contribute to oxidative stress are continuously being generated as part of the normal cellular metabolism. Oxidative stress occurs when there is excess generation of ROS that cannot be compensated by available antioxidant processes. Due to the short biological half-lives of these metabolites, one of the main challenges in biomarker research is to validate reliable and efficient markers that could be used to identify individuals who are at risk of developing oxidative stress and who would benefit the most of an early preventive management program. We propose that plasma lipoproteins may serve as natural biosensors of early oxidative stress in the arterial wall. Furthermore, indices of lipoprotein-associated oxidative stress assessed in plasma in the fasting state and after a physiologic bout of oxidative challenge might potentially be among the best biomarkers of early oxidative stress.

6.1   INTRODUCTION

Oxidative stress is recognized as one of the primary processes underlying the initiation and progression of atherosclerotic vascular disease. Generation of reactive oxygen species (ROS) is part of normal cellular metabolism and is counterbalanced by a multitiered antioxidant defense system. Excess generation of ROS that could not be compensated by available antioxidant capacity is responsible for a number of pathological conditions. Due to their short biological half-lives, direct measurement of ROS levels is not reliable, and a number of different surrogate measures have been suggested and used. In this chapter, we will present data to support the use of plasma lipoproteins as a natural biosensor of oxidative stress in the arterial wall.

Plasma lipoproteins represent a dynamic system for the transport of lipids, which is characterized by a spectrum of spherical particles ranging in diameter from 5 to 1200 nm. Their primary role is to deliver triglycerides (TGs) from both the intestine (dietary origin) and the liver (endogenous origin) to peripheral tissues for storage as sources of energy. They also contribute to cholesterol homeostasis. Impaired lipoprotein metabolism has been demonstrated to result in the accumulation of cholesterol in the arterial wall leading to endothelial dysfunction and atherosclerotic vascular diseases. A brief review of lipoprotein physiology will be presented here with the emphasis on a potential new function of the triglyceride-rich lipoproteins (TRLs) as biosensors of oxidative stress in the arterial wall. We will review in vitro and in vivo evidence, which suggest that in view of the high contents of polyunsaturated fatty acids (PUFAs) that are susceptible to oxidative modification, plasma lipoproteins could readily be seeded with ROS generated by the inflamed arterial wall. Depending on the level of antioxidative defense available on plasma lipoproteins and in the circulation, the oxidative modification process could be quenched or allowed to propagate leading to further inflammation and oxidative damage. Furthermore, we will present data that suggest that this process may be significantly enhanced acutely during postprandial lipemia. This is of special importance in obesity as the metabolism of postprandial lipoproteins is known to be impaired.

6.2   LIPOPROTEIN METABOLISM

Lipids, in particular TGs, cholesterol, and phospholipids, are both source of energy and structural components of cells. Cholesterol and phospholipids are essential for the structure of cell membranes. Cholesterol serves as a precursor of bile acids, vitamin D, and other steroids. TGs, with their 3 fatty acids, are a high-energy metabolic fuel and the most efficient form of energy storage. The main sites of fatty acid metabolism are the liver and muscle, with adipose tissue serving as the principal storage depot.

6.2.1   PLASMA LIPOPROTEINS

There have been numerous rationales for the classification of plasma lipoproteins, from size, density, floatation constant, and electrophoretic mobility to nuclear magnetic resonance spectroscopy (Dominiczak, 2000). For the purpose of this chapter, we will characterize plasma lipoproteins as two major classes, A-lipoproteins and B-lipoproteins (Figure 6.1). The primary protein component of the A-lipoproteins is apolipoprotein A-I (apoA-I), while the primary protein component of the B-lipoproteins is apolipoprotein B (apoB). The A-lipoproteins are responsible primarily for the reverse cholesterol transport pathway and are represented by high-density lipoprotein (HDL). The B-lipoproteins that are responsible for the transport of the *bad* cholesterol can be identified as the exogenous and the endogenous pathways. The exogenous pathway is responsible for the transport of dietary lipids from the intestine to peripheral tissues and the liver in the form of chylomicrons. The endogenous pathway is responsible for the transport of lipids synthesized in the liver (very low-density lipoproteins, VLDL) to peripheral tissues with the remnant particles (remnants and

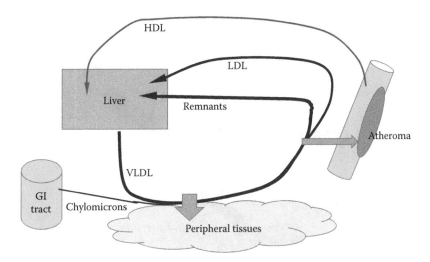

FIGURE 6.1 **(See color insert.)** Basic lipid transport in plasma. The two major classes of lipoproteins are the A-lipoproteins (good cholesterol), represented by high-density lipoproteins or HDL, and the B-lipoproteins (bad cholesterol), represented by chylomicrons, very low-density lipoproteins or VLDL, remnants, and low-density lipoproteins or LDL.

low-density lipoproteins, LDL) being returned to the liver. The B-lipoproteins are considered the bad lipoproteins because they contribute to the accumulation of cholesterol in the atheroma. In contrast, the A-lipoproteins are the good lipoproteins as their primary functions are to facilitate the efflux of excess cholesterol from peripheral tissues for transport back to the liver and to protect the B-lipoproteins from oxidative and inflammatory modifications.

Plasma lipoproteins are traditionally assessed in fasting plasma as stable entities, for their lipid contents, their protein contents, or the number of particles. This is simply a snapshot of an extremely dynamic system of particles that are continuously undergoing modifications and transformations. Lipid and protein moieties can be exchanged, converted, or exchanged among different classes of lipoproteins as modulated by a panel of plasma enzymes and cell surface receptors. First among the enzymes are the lipolytic enzymes that are responsible for the hydrolysis of lipoprotein TGs, a necessary step for their delivery to peripheral tissues for storage as sources of energy. Lipoprotein lipase (LPL) is anchored to heparan sulfate proteoglycans lining the arterial wall and is the first step in the metabolism of the TRLs, intestinal chylomicrons carrying dietary TGs, and hepatic VLDL carrying de novo TGs. Several excellent reviews on the biology and regulation of LPL are available (Young and Zechner, 2013; Kersten, 2014). The released free fatty acids (FFA) and monoglycerides readily move across the endothelial barrier and are resynthesized as TGs for storage. The residual particles that are partially depleted of their TG cargo can undergo further hydrolysis via the action of hepatic triglyceride lipase (HTGL or HL, hepatic lipase), leading to the formation of chylomicron remnants or LDL, in the case of the hepatic VLDL.

6.2.2 Lipid Transfer Proteins

While this hydrolytic process is taking place, several lipid transfer proteins are actively and continually modifying the composition of plasma lipoproteins (Figure 6.2; Stein and Stein, 2005; Quintao and Cazita, 2010). Cholesteryl ester transfer protein (CETP) is actively exchanging cholesteryl esters from plasma HDL for TGs in chylomicrons and VLDL (Lagrost et al., 1993). The net result is TG-poor, cholesterol-rich remnant particles that require recognition by specific surface receptors for irreversible removal. In diseased conditions, saturation of the removal process leads to extended

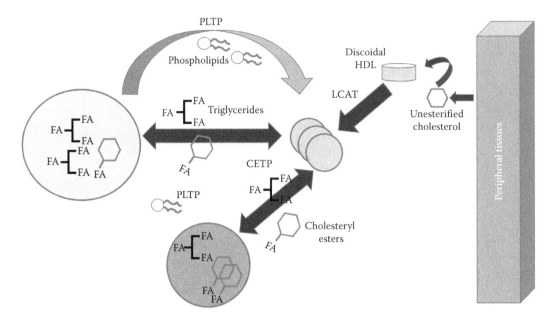

FIGURE 6.2 **(See color insert.)** Key lipid transfer enzymes responsible for the remodeling of plasma lipoproteins. PLTP, phospholipid transfer protein; CETP, cholesteryl ester transfer protein; LCAT, lecithin cholesterol acyl transferase.

residence of the remnants in the vascular space and increased probability for these cholesterol-rich remnants to become *damaged*. Available in vitro and in vivo evidence would suggest that native remnants and LDL do not contribute to cholesterol accumulation in the atheroma but the modified particles do (Brown et al., 1979; Fogelman et al., 1980).

In addition to CETP, there are two other lipid transfer proteins that are involved in the modulation of cholesteryl ester synthesis and transport, acyl-CoA:cholesterol acyltransferase (ACAT) and lecithin cholesterol acyltransferase (LCAT). ACAT is responsible for the esterification of intracellular cholesterol. With the inhibition of ACAT, most of the intracellular pool of cholesterol is unesterified and can thus freely diffuse in and out of the cells. Under certain conditions, ACAT inhibitors have been reported to reduce atherosclerotic lesions even when only a modest reduction in plasma could be demonstrated (Delsing et al., 2001). LCAT, on the other hand, is responsible for the esterification of cholesterol in the vascular space. It plays a key role in the modification and maturation of nascent HDL (Francone et al., 1989; Zannis et al., 2015). In brief, lipid-poor nascent HDL particles are avid acceptor of free cholesterol from cell membranes as facilitated by ATP-binding cassette transporter A1 (ABCA1) and scavenger receptor B1 (SR-B1), resulting in the formation of discoidal HDL–containing phospholipid and unesterified cholesterol. With the increase in hydrophobicity of cholesteryl esters formed by the action of LCAT and free cholesterol, discoidal HDL is transformed into its spherical mature form. Subsequent interactions with CETP result in a spectrum of HDL particles varying in size, TGs, and cholesteryl ester contents.

Phospholipid transfer protein (PLTP) is the fourth of the plasma lipid transfer proteins and is getting increasing attention for its role in lipoprotein metabolism (Tzotzas et al., 2009; Jiang et al., 2012). In addition to its role in the exchange of phospholipids among plasma lipoproteins, PLTP is essential for the net transfer of phospholipids from the B-lipoproteins to HDL (Tall et al., 1978; Tall et al., 1985). The importance of this process in lipoprotein oxidation will be discussed in a later section.

Figure 6.3 illustrates the physical changes in plasma lipoproteins that could occur in vitro in the presence of CETP, LCAT, and PLTP using nondenaturing gradient polyacrylamide gel

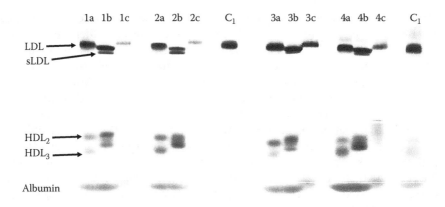

FIGURE 6.3 **(See color insert.)** Effects of lipid transfer enzymes of physical characteristics of plasma lipoproteins: Changes in lipoprotein particle diameters and electrophoretic mobility as assessed by non-denaturing polyacrylamide gradient gel electrophoresis. Whole plasma was incubated at 37°C for 0 h (designated as "a"), 2 h (designated as "b"), and 6 h (designated as "c") for four different individuals (designated as 1–4). C1 designates the control plasma.

electrophoresis (Li et al., 1997). After 2 h at 37°C, the net transfer of TGs and phospholipids from the B-lipoproteins results in the shifting of HDL toward larger size and LDL toward smaller size. Depending on the relative activities of the different enzymes, distinct subpopulations of LDL could be observed in some individuals (Subjects 1 and 2) but not in others (Subjects 3 and 4). With extended incubations (6 h or greater), the integrity of the HDL particles is no longer maintained and LDL becomes a more homogeneous subpopulation, which is larger in size than native LDL, suggestive of increase in core lipids.

6.3 OXIDATIVE HYPOTHESIS OF ATHEROSCLEROSIS

One of the landmarks of atherosclerotic disease is the accumulation of cholesterol in macrophage-derived foam cells, which constitute the plaques that occlude normal blood flow. While LDL and its cholesterol contents have been implicated in the initiation, formation, and progression of atherosclerotic plaques, cell culture studies from several laboratories have conclusively demonstrated that human monocyte–derived macrophages do not transform lipid-laden foam cells when exposed to normal LDL (Goldstein et al., 1979; Fogelman et al., 1980), even at very high concentrations. However, in the presence of small amount of chemically altered LDLs, the macrophages can become foam cells within a short incubation period. These observations culminated into the oxidative hypothesis of atherosclerosis (Steinberg et al., 1989; Witztum and Steinberg, 1991; Chisolm and Steinberg, 2000; Steinberg and Witztum, 2010).

6.3.1 EVIDENCE FROM AUTOANTIBODIES TO OXIDIZED LDL

Support for this hypothesis is available from a number of different lines of research. While several chemically altered forms of LDL have been examined with respect to their ability to promote cholesterol accumulation and formation of foam cells, oxidatively modified LDL is one form of modification that has been demonstrated to be present in vivo (Palinski et al., 1989; Palinski et al., 1996), specifically in regions of active atherosclerotic lesion growth (Avogaro et al., 1988; Yla-Herttuala et al., 1989).

The presence of circulating antibodies that recognize oxidatively modified LDL has also been reported (Salonen et al., 1992; Puurunen et al., 1994; Bui et al., 1996; van de Vivjer et al., 1996). However, the significance of antibody titers and disease severity remains unclear. Some studies reported strong relation between antibody titers and disease status (Salonen et al., 1992; Puurunen et al., 1994; Bui et al., 1996), while other studies noted no difference in titers between cases and controls (van de Vivjer, 1996; Uusitupa et al., 1996). In a small case–control study, our group has reported that autoantibody titers were not different between individuals with documented coronary artery disease (CAD) and young healthy adults with normal lipids and no history of heart disease (Le et al., 2000). However, meal-induced changes in antibody titers were dramatically different when these subjects were challenged with a 600 kcal mixed meal. There were no changes in autoantibody titers among the controls, while patients with documented CAD exhibited a 25% reduction in autoantibody titers (Le et al., 2000). We suggested that the reduction in antibody titers observed in CAD patients reflects an excess production of oxidatively modified epitopes during postprandial lipemia (Le et al., 2000). In fact, we subsequently demonstrated that, in patients with vascular disease, this meal-induced reduction in autoantibody titers is specific for a test meal enriched in PUFAs and was not observed following isocaloric meal challenge containing saturated or monounsaturated fatty acids (MUFAs) that are less susceptible to oxidative modification (Gradek et al., 2004). This in vivo observation is consistent with an earlier report from Mabile and coworkers on the relationship between oxidizability of chylomicrons and dietary fatty acid composition (Mabile et al., 1995).

6.3.2 Pathogenesis of Oxidized LDL

While the impact of oxidized LDL (oxLDL) in atherogenesis is well accepted, the exact process for the in vivo modification of plasma LDL remains unclear. Given the endogenous antioxidant defense system present in human blood (Stocker and Frei, 1991), it has been suggested that plasma LDL must be trapped in a microenvironment in the presence of excess oxidants for some extended period of time in order for the modification to be completed. According to the oxidation hypothesis, LDL, in particular small dense LDL, is trapped in the subendothelium surrounded by oxidants secreted

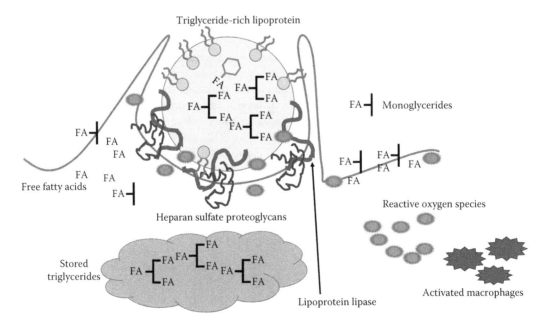

FIGURE 6.4   (See color insert.) Schematic representation of the interactions of triglyceride-rich lipoproteins (TRL) with lipoprotein lipase (LPL) anchored to the arterial wall via heparan sulfate proteoglycans. Triglycerides are hydrolyzed to free fatty acids (FA) and monoglycerides that can move across the endothelium and are reconstituted as triglycerides for storage. Reactive oxygen species (ROS) that are generated from activated macrophages can diffuse through the endothelium and seed the fatty acids on plasma lipoproteins.

by activated macrophages (Steinberg et al., 1989). Once modified, oxLDL contributes to the atherosclerotic process by (1) promoting the recruitment of circulating monocytes into the intimal surface via the secretion of monocyte chemotactic protein-1 (MCP-1) (Cushing et al., 1990), (2) retaining resident macrophages within the intima (Quinn et al., 1987), (3) upregulating the expression of SRs on macrophages leading to the uptake and accumulation of oxLDL and subsequent formation of foam cells (Shechter et al., 1981), and (4) being cytotoxic to endothelial cells leading to the loss of endothelial integrity (Hessler et al., 1979; Cathcart et al., 1991; Mabile et al., 1995).

Figure 6.4 schematizes an alternate scenario for the oxidative modification of plasma lipoproteins in vivo. In this scheme, as TRLs come into contact with LPL that is attached to heparan sulfate proteoglycan lining the arterial wall, the movement of various metabolites across the endothelial barrier may be possible. From the hydrolysis of lipoprotein TGs, the released FFAs and monoglycerides can diffuse across the endothelium and are reconstituted as TGs in the underlying cells for storage. We are postulating that during this interaction, ROS that are generated by activated macrophages in the subendothelium can diffuse through the endothelial layer and seed the surface lipids of plasma lipoproteins. Thus, excess radicals are readily transferred from the arterial wall to circulating plasma lipoproteins. Further evidence in support of the scheme is presented in later sections.

6.4 ASSESSMENT OF OXIDATIVE STATUS

Oxidative stress normally occurs when available antioxidant defense processes are inadequate to inactivate ROS (Stadtman and Bertlett, 1997) and reactive nitrogen species (RNS) (Dedon and Tannenbaum, 2004). This may result from either excess production of ROS/RNS or a dysfunctional antioxidant system. Even with the availability of special equipment that can capture the electron spin resonance of reactive radicals (Tarpey et al., 2004), reliable detection is a challenge in view of the inherent instability of these radicals. This is the case even when various surrogate markers of oxidation have often been used including stable metabolites such as nitrate or nitrite, target end products such as lipid peroxidation end products (Draper et al., 2000; Cracowski et al., 2002), and oxidized proteins (Dean et al., 1997; Requena et al., 2003).

6.4.1 LIPID PEROXIDATION PRODUCTS

First, let us focus on lipid peroxidation products. Lipids can be oxidized by three different mechanisms (1) nonenzymatic, nonradical oxidation, (2) nonenzymatic, free radical–mediated oxidation, and (3) enzymatic oxidation (Niki et al., 2005). The most significant nonradical molecules involved in nonenzymatic lipid oxidation are singlet oxygen, ozone, and molecular chlorine (Iuliano, 2011). While this process has been overlooked for the other, more familiar process, in vitro oxidation of LDL by exposure to ozone has been reported (Horl et al., 2014). By virtue of the multiple bonds, PUFAs are highly susceptible to oxidation via enzymatic and nonenzymatic free radical–mediated pathways (Yoshida et al., 2015). Lipid peroxidation is a chain reaction process characterized by repetitive abstraction by hydroxyl radicals ($^\bullet$OH) and RO$^\bullet$, followed by the addition of O_2 to the alkyl radical R$^\bullet$ with subsequent generation of peroxyl radical ROO$^\bullet$ and ultimately the oxidative destruction of PUFA. Reactive species that contribute to the oxidative toxicity include the intermediates of the partial reduction of oxygen, superoxide radical ($O_2^{\bullet-}$), hydrogen peroxide (H_2O_2), $^\bullet$OH, peroxyl radicals (ROO$^\bullet$), nitric oxide (NO), peroxynitrite (ONOO$^-$), and singlet oxygen ($^1O_2$).

Among enzymatic oxidation pathways, the key enzymes in LDL oxidation are lipoxygenase (LPO) and myeloperoxidase (MPO) (Yoshida and Kisugi, 2010). Lipoxygenase (15-LPO) from endothelial cells (Parthasarathy et al., 1989) and fibroblasts (Ezaki et al., 1995) have been reported to oxidatively modify human LDL and convert it into a cytotoxic metabolite (Cathcart et al., 1991). It has also been demonstrated to be expressed in human atherosclerotic plaques with co-localization of LPO mRNA and immunohistochemical evidence of protein with oxidized epitopes (Yla-Herttuala et al., 1990, 1991; Folcik et al., 1995; Kuhn et al., 1997). MPO is a key enzyme expressed in response to inflammation and stress by neutrophils and monocytes (Klebanoff, 2005; Schindhelm et al., 2009). It catalyzes the formation of HOCl from H_2O_2 and chloride. HOCl acts as a potent cytotoxin as part of the host defense against pathogens. MPO has been implicated in the development and progression of atherosclerosis (Daugherty et al., 1994; Hazell et al., 1996; Hazen and Heinecke, 1997). In addition to these experimental evidences, epidemiologic studies have shown that higher concentrations of plasma MPO are associated with an increased risk of cardiovascular disease (CVD) (Stocker and Keaney, 2005; Schinfhelm et al., 2009). In addition to these data, it should also be noted that in vivo observations in transgenic animals have been confusing with more severe atherosclerotic lesions have been confusing with the observation that MPO knockout mice actually have more severe atherosclerotic lesions (Brennan et al., 2001).

Lipid peroxidation generates a number of stable end products such as hydroperoxides; reactive aldehydes, for example, malondialdehyde (MDA), 4-hydroxy-2-nonenal (HNE), and acrolein (Esterbauer et al., 1991); prostanes (Montuschi et al., 2004), for example, F2 isoprostanes; and hydroxyoctadecanoic acid (HODE) (Yoshida et al., 2015).

The reaction of MDA with 2-thiobarbituric acid (TBA) is the most widely used method for assessing oxidative stress. While spectroscopic method is very convenient and commonly used (Ohkawa et al., 1979), it lacks specificity since TBA could react with other oxidant products in plasma. Thus, it must be emphasized that, while direct measurements of MDA by gas chromatography/mass spectrometry (GC/MS) are strongly associated with TBARS (thiobarbituric acid reactive substances), the two measurements are not the same (Liu et al., 1997). In the longitudinal analysis of the PREVENT study, TBAR levels were predictive of future cardiovascular events in patients with stable CAD (Walter et al., 2004). Increased levels of MDA have been reported in plasma and atherosclerotic plaques of patients with type 2 diabetes mellitus (T2DM) (Slatter et al., 2000). In plasma, MDA can react with lysine residues of proteins. MDA-modified proteins are characteristically immunogenic, and auto-antibodies against MDA-modified proteins have been detected in vivo (Stocker and Keaney, 2004).

HNE is generated from free radical modification of n-6 PUFA and is a toxic second messenger of oxygen-free radicals (Esterbauer et al., 1991). It undergoes reactions with proteins, phospholipids, and nucleic acids, resulting in changes in a variety of biological activities (Parola et al., 1999).

Acrolein is present in environmental sources, specifically cigarette smoke. Acrolein reacts with lysine residues of the major protein in HDLs, resulting in the impairment of the function of HDL in reverse cholesterol transport. Acrolein-modified HDL has been identified in human atherosclerotic plaques (Shao et al., 2005).

F2-isoprostanes are a family of prostaglandin-like compounds generated in vivo by free radical–catalyzed nonenzymatic peroxidation of esterified arachidonic acid. They are released into the circulation by the action of phospholipases and subsequently excreted in the urine (Cracowski et al., 2002; Montuschi et al., 2004). While F2-isoprostanes are chemically stable products, they are rapidly metabolized and excreted as free acids in the urine.

6.4.2 Protein Oxidation Products

As with lipid peroxidation, plasma proteins can also undergo oxidative modification via a number of processes. Under conditions of moderate oxidative stress, cysteine residues can undergo oxidative modification resulting in the formation of mixed sulfides with glutathione, a process known as S-glutathionylation. The ratio of GSSG:GSH (glutathione disulfide:reduced glutathione) in blood has been used as an indicator of oxidative status in humans (Dalle-Donne et al., 2006). Glutathionylated hemoglobin has been reported to increase in patients with increased oxidative stress, including patients with diabetes and hyperlipidemia, as well as those on hemodialysis or peritoneal dialysis (Giustarini et al., 2004).

Tyrosine moieties of plasma proteins can be modified resulting in the formation of 3-nitrotyrosine (NO_2-Tyr), 3-chlorotyrosine (Cl-Tyr), or 3-bromotyrosine (Gaut et al., 2001; Brennan et al., 2002).

High levels of HDL proteins modified with NO_2-Tyr and Cl-Tyr have been reported (Zheng et al., 2004) as isolated from whole plasma or from human atherosclerotic plaques (Bergt et al., 2004; Pennathur et al., 2004).

In the context of atherosclerosis, plasma levels of oxLDL have been suggested to be a good marker for lipoprotein-associated oxidation. Several antibodies specific for different epitopes on oxLDL have been described, and elevated levels of oxLDL have been linked to disease severity. Well-characterized antibodies include DLH3 specific for oxidized phosphotidylcholine (Itabe et al., 1996), 4E6 specific for Cu^{++}-oxidized LDL (Holvoet et al., 1996), MDA2 specific for MDA-modified LDL (Boyd et al., 1989), and E06 specific for oxidized phospholipids (oxPLs) (Palinski et al., 1994; Tsimikas, 2006). A review of the different assays for oxidatively modified LDL is available (Le, 2009).

The commercially available enzyme-linked immunoassay (ELISA) kit using 4E6 has been approved by the U.S. Food and Drug Administration for diagnostic purposes. In the Atherosclerosis and Insulin Resistance study that is based on individuals with subclinical atherosclerosis, higher levels of oxLDL were reported in patients with carotid and femoral plaques than in subjects with no plaques (Hulthe and Fagerberg, 2002). Using this assay in a prospective study of clinically healthy men, plasma level of oxLDL was reported to be a prognostic marker of subclinical atherosclerosis development (Wallenfeldt et al., 2004). In the "Health, Aging and Body Composition trial," subjects with metabolic syndrome were found to have higher plasma levels of oxLDL using this same assay (Holvoet et al., 2004). In the Multi-Ethnic Study of Atherosclerosis (MESA), individuals with subclinical atherosclerosis had higher levels of oxLDL than subjects without subclinical CVD (Holvoet et al., 2007).

The other monoclonal antibody that has received a lot of attention is the E06 initially characterized in the apoE knockout mouse model of atherosclerosis. Data using this antibody, specific for oxPL, are typically expressed as the ratio of oxPL to apoB. In patients undergoing coronary angiography, the oxPL:apoB ratio was found to be strongly correlated with the extent of CAD (Tsimikas et al., 2005). In nonhuman primate model of atherosclerotic progression, oxPL:apoB ratio has been reported to be increased as plasma LDL levels are increased (Tsimikas, 2007). When the atherogenic diet was replaced by normal chow to induce disease regression in this same animal model, plasma oxPL levels were reduced concomitant with reduction in plasma LDL. However, during regression, an unexpected increase in the oxPL:apoB ratio was found (Tsimikas et al., 2007). In other words, with reduction in the number of LDL particles during regression, there was a net reduction in the levels of oxPL in plasma, but the number of oxidized epitopes per LDL particle was actually increased. A similar unexpected increase in oxPL:apoB ratio was also observed in the MIRACL study in patients with acute coronary syndromes (Tsimikas et al., 2004). In this trial, patients were aggressively treated with high-dose atorvastatin, and significant reductions in both LDL levels and plasma oxPL levels were achieved. In the REVERSAL (Reversal of Atherosclerosis with Aggressive Lipid Lowering) trial, in spite of the reduction in LDL cholesterol to below 79 mg/dL and reduction in total atheroma volume after 18 months, both measures of oxidatively modified LDL were increased, oxPL:apoB ratio and MDA-modified apoB as detected with the MDA2 antibody (Nissen et al., 2004; Nissen, 2005).

A possible explanation for these unexpected increases in oxPL:apoB ratio may be available from the data of Nishi et al. (2002), which suggest that a significantly higher fraction of LDL in atherosclerotic plaques exhibited oxidative epitopes as compared to circulating plasma LDL. This is true in patients undergoing carotid endarterectomy, 11.9 versus 0.18 ng of oxLDL per μg of apoB, as well as in autopsy samples from age-matched healthy controls, 1.86 versus 0.13 ng of oxLDL per μg of apoB (Nishi et al., 2002). It is possible that during regression, oxidatively modified LDL may be released from the atheroma. Consequently, while the total number of LDL particles may be reduced, the contribution of this subpopulation of plasma LDL derived from the atheroma could account for the higher oxPL:apoB ratio observed in these regression studies (Le, 2009).

6.5 OXIDATIVE SUSCEPTIBILITY OF PLASMA LIPOPROTEINS

PUFAs attached to core lipids, TGs and cholesteryl esters, and surface lipids, phospholipids of plasma lipoproteins are the prime candidate for lipid peroxidation. Indeed, while the focus has been on oxidatively modified LDL, all plasma lipoproteins are subject to oxidative modification, including chylomicrons (Mabile et al., 1995), VLDL (Rabini et al., 1999; McEneny et al., 2000; Le, 2015), and HDL (Le, 2015).

6.5.1 EFFECT OF LIPID-SOLUBLE ANTIOXIDANTS

In the typical oxidative susceptibility assay, isolated plasma lipoproteins are exposed to a catalyst such as Cu^{++}, and the rate of formation conjugated dienes, as detected at 234 nm, is monitored for 5–6 h (Esterbauer et al., 1989). This process can be characterized into three phases: initiation, propagation, and degradation. The duration of each of these phases depends on the temperature of the reaction, the concentration of Cu^{++}, the amount of PUFA, and the amount of antioxidants present on the lipoprotein particle. During the initiation phase, all exogenous antioxidants present on the lipoprotein particle are being consumed; these include vitamin E, vitamin C, β-carotene, resveratrol, and other similar compounds (Tribble et al., 1994; Burkitt, 2001). Higher antioxidant protection is associated with longer lag times. Oral supplementation with D-α-tocopherol has been reported to significantly increase LDL lag times (Dieber-Rotheneder et al., 1991). There was also a strong independent correlation between percent change in LDL lag time and percent change in plasma α-tocopherol ($r = 0.47$, $p < 0.01$; Abbey et al., 1993). Combined supplementation with ascorbate (vitamin C), β-carotene, and α-tocopherol was found not to be superior to high-dose α-tocopherol alone in inhibiting LDL oxidation (Jialal and Grundy, 1993).

6.5.2 EFFECT OF FATTY ACID SATURATION

Once the oxidative process is initiated, another key determinant of susceptibility to oxidation is the fatty acid composition. LDL particles rich in PUFA, presenting multiple double bonds accessible for oxidative attack, are more readily oxidatively modified. In New Zealand white (NZW) rabbits, substituting a special oil enriched in oleate for 6 weeks resulted in the presence of LDL with an extremely short lag time as compared to LDL isolated from animals maintained on conventional sunflower oil enriched in linoleic acid (Parthasarathy et al., 1990). Indeed, controlled trials with saturated fatty acids (SFAs), MUFAs, w6-PUFA, and w3-PUFA in healthy normolipidemic subjects demonstrated that LDL lag time was the longest, indicative of high resistance of oxidative modification, during MUFA (mean ± SD 55.1 ± 7.3 min) and shortest with SFA and both w6 and w3-PUFA (45.3 ± 6.4, 47.1 ± 8.4, and 45.3 ± 7.0, respectively) (Mata et al., 1996). Plasma TBARS levels were also significantly higher during the 2 PUFA periods (1.51 ± 0.50 and 1.69 ± 0.48 nmol MDA/mg LDL protein) as compared to either the MUFA (1.15 ± 0.35) or SFA (1.15 ± 0.57) period (Mata et al., 1996). Thus, diets enriched in PUFA have two opposing effects. While diets in which PUFAs are substituted for SFA are considered beneficial in which they lower plasma cholesterol levels (Czernichow et al., 2010) and reduce the risk of CAD (Michas et al., 2014), it is also well documented that LDL isolated from individuals maintained on a high PUFA diet is enriched with PUFA and is more susceptible to oxidative modification. One possible explanation is that the reductions in LDL levels with a PUFA-rich diet have greater beneficial effect than the deleterious effect of increased oxidative susceptibility (Kaikkonen et al., 2014).

These observations would support the concept that diets enriched in MUFA, such as Mediterranean-type diet, may confer additional protection by reducing both plasma LDL levels and LDL oxidative susceptibility (Aronis et al., 2007). Dietary oil composition has also been reported to affect in vitro susceptibility to oxidation of TRL and LDL isolated from both fasting plasma and postprandial plasma (Nielsen et al., 2002).

6.5.3 LIPOPROTEIN OXIDIZABILITY AND DISEASE

The key question remains: Is there a relationship between lipoprotein oxidizability and disease? Increased oxidative susceptibility of isolated VLDL and isolated LDL has been reported in patients with T2DM (Rabini et al., 1999; McEneny et al., 2000) and in patients with ischemic stroke (Ryglewicz et al., 2002). The increased oxidative susceptibility of LDL in some patients has been suggested to be associated with the predominance of small, dense LDL, phenotype B (Tribble et al., 1992, 2001). In a case–control study of a subset of participants in the Kupio Atherosclerosis Prevention Study, among the strongest predictors of a 3-year increase in carotid wall thickness were lipid hydroperoxides in LDL (determined fluorometrically as TBARS) and oxidative susceptibility of the B-lipoproteins (VLDL + LDL) in the presence of hemin and H_2O_2 (Salonen et al., 1997). In a case–control study of men between the age of 40 and 83 years, measures of oxidative susceptibility, including lag time, oxidation rate, and maximal change in oxidation, were strongly predictive of the presence of disease by logistic regression analysis, even after adjustment for the traditional covariates (age, LDL-C, HDL-C, apoA-I, and smoking status) (Hendrickson et al., 2005). The susceptibility of LDL to oxidation was reported to predict new carotid artery atherosclerosis over a 5-year period, and higher susceptibility was associated with a higher incidence of new carotid artery atherosclerosis (Aoki et al., 2012).

6.6 LIPOPROTEINS AS BIOSENSORS OF OXIDATIVE STRESS

In view of the close physical association of plasma TRL with the endothelium during TG hydrolysis, we have hypothesized that plasma lipoproteins may serve as a natural biosensor of oxidative stress in the arterial wall. Figure 6.5 illustrates this process for the normal, healthy arterial wall (Figure 6.5a) and the inflamed arterial wall (Figure 6.5b).

6.6.1 OXIDATIVE MODIFICATION OF LIPOPROTEINS IN HEALTH AND DISEASE

In the healthy arterial wall, there are few monocytes trapped in the subendothelium, and most of them remain inactive. ROS generation is minimal. As TRLs attach themselves to LPL anchored along the arterial wall, the hydrolytic process is efficient and the contact period is brief, resulting in minimal seeding of the plasma lipoproteins with ROS. Depending on the contents of PUFA on the TRL and antioxidants, there would be a lag phase before oxidative modification can propagate from the surface lipids to the core lipids and ultimately to the protein moieties. However, upon their release into the circulation, some lipid molecules from the partially hydrolyzed TRL may be exchanged or transferred into plasma HDL with their panel of antioxidant enzymes, including paraoxonase-1 (PON1), lipoprotein-associated phospholipase A_2 (Lp-PLA$_2$), and glutathione peroxidase (GPx-1) (Mackness et al., 1993; Ballantyne et al., 2004; Kontush and Chapman, 2006; Ansell et al., 2007). The end product is a cholesterol-rich, TG-depleted particle that has minimal or no oxidative epitopes. If the TRL particles are intestinally derived chylomicrons secreted during postprandial lipemia, the end products would be chylomicron remnants, and if the TRL particles are hepatic VLDL present throughout the day, the end products would be LDL.

In the diseased endothelium with chronic inflammation, we would expect not only to have more monocytes trapped in the subendothelium but also to find more activated macrophages secreting MCP-1 to attract more monocytes and excess quantities of ROS. In this instance, as TRL becomes attached to LPL, they are exposed to a higher concentration of ROS and are seeded with excessive amount of ROS. Even though PLTP and CETP may remain effective in the transfer of oxidized lipids to HDL, the antioxidant capacity of HDL is overwhelmed by either the excess amount of ROS transferred or low activities of the antioxidant enzymes or both (Kontush and Chapman, 2006; Ansell et al., 2007). In consequence, the resulting end products still contain high levels of

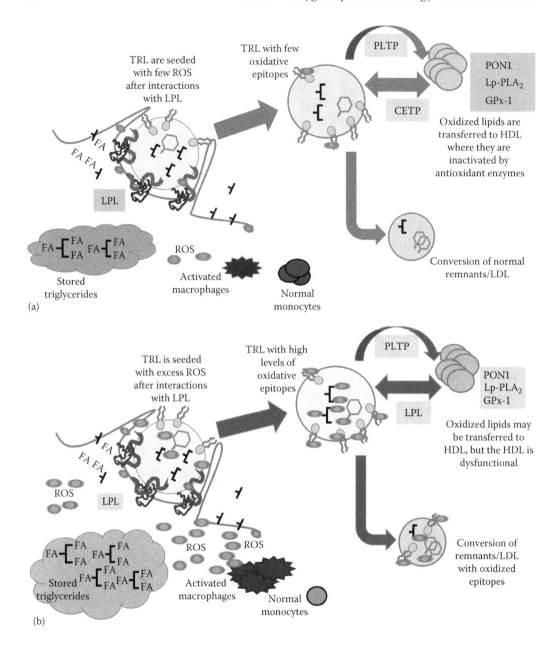

FIGURE 6.5 (See color insert.) Schematic representation of the interactions between the arterial wall, triglyceride-rich lipoproteins (TRL), and high-density lipoproteins (HDL) under normal and diseased conditions. (a) Under normal conditions, only a few of the monocytes trapped in the subendothelium are activated with the generation of low quantity of ROS. During interactions of TRL with LPL, there are few ROS to seed the plasma lipoproteins. These ROS are efficiently transferred to circulating HDL and are rapidly inactivated by the antioxidant enzymes on HDL, including paraoxonase-1 (PON-1), lipoprotein-associated phospholipase A$_2$ (Lp-PLA$_2$), and glutathione peroxidase (GPx-1). The final product is a normal remnant/LDL particle. (b) Under diseased conditions, the arterial wall is inflamed, and most of the monocytes have been transformed to activated macrophages generating excessive quantities of ROS. Interactions of TRL with LPL along the arterial wall result in the seeding of plasma lipoproteins with excess ROS. The excess ROS overwhelm the capability of the HDL-associated antioxidant enzymes and proceed to propagate the oxidative modification process, especially if the highly oxidizable polyunsaturated fatty acids (PUFAs) are present. The net result is the generation of remnant/LDL particles with oxidatively modified epitopes.

ROS that have now propagated to other PUFA and to the protein moiety, a process that cannot be reversed. The net effect is a high level of oxidatively modified proteins in plasma that could be detected as oxLDL.

6.6.2 IN VIVO METABOLIC FATE OF MODIFIED LIPOPROTEINS

Figure 6.6 illustrates the potential problem why plasma levels of oxidatively modified epitopes may not be the best parameter to monitor. In plasma, the presence of autoantibodies against oxidatively modified epitopes has been well established. Depending on the concentrations and titers of the antibodies, the LDL immune complexes could be rapidly removed and thus biasing the levels of oxidized epitopes that could be detected in plasma. Furthermore, a few particles with multiple oxidized epitopes may bind multiple antibody molecules forming a larger complex that may be cleared preferentially as compared to many particles, each with only a single oxidized epitope and binding to a single antibody molecule.

The presence of autoantibodies against oxidatively modified epitopes also plays a role in the metabolism of oxidatively modified lipoproteins and in disease progression. In the LDL receptor–deficient Watanabe heritable hyperlipidemic (WHHL) rabbits, less atherosclerotic lesions were observed when they were preimmunized with MDA-modified LDL prior to being placed on the high-fat, high-cholesterol atherogenic diet (Palinski et al., 1995). In contrast, NZW rabbits that were started on the atherogenic diet prior to being immunized with oxLDL failed to be protected from atherosclerosis (Ameli et al., 1996). Immunization with apoB peptide sequences or direct injection with recombinant antibodies to oxDL epitope in a mouse model resulted in the inhibition of atherosclerosis (Fredrikson et al., 2003; Schiopu et al., 2007). However, in LDL receptor–deficient mice, increased autoantibody titers against oxLDL were associated with increased atherosclerosis (Palinski et al., 1995). As schematized in Figure 6.7, we have two different scenarios depending on

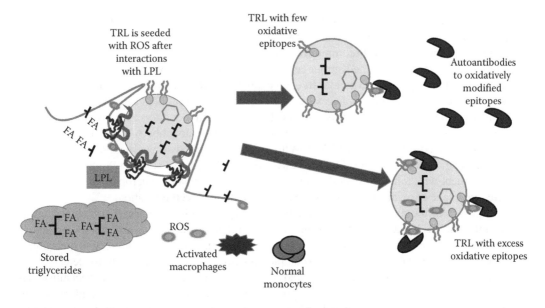

FIGURE 6.6 (See color insert.) Interactions of TRL with autoantibodies against oxidized epitopes: During postprandial lipemia, interactions of TRL with LPL along the arterial wall will generate partially hydrolyzed particles seeded by either few ROS or excess ROS depending on the levels of ROS produced by the macrophages. If there are only few oxidized epitopes and there are adequate circulating autoantibodies, the impact of postprandial lipemia on autoantibody titers is negligible. If, however, there are excess oxidized epitopes, the pool of circulating autoantibodies can be acutely depleted resulting in a transient reduction in autoantibody levels.

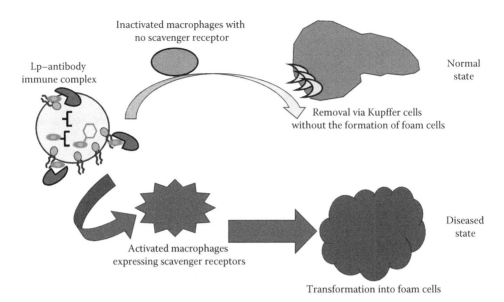

FIGURE 6.7 (**See color insert.**) Metabolic fates of lipoprotein–autoantibody immune complexes. The association of oxidized lipoproteins and autoantibodies is an ongoing process, and the immune complexes are normally taken up by Kupffer cells in the liver and degraded. However, in diseased states with a large number of activated macrophages scattered through the arterial tree and expressing scavenger receptors, there would be higher probability that the immune complexes are taken up by macrophages prior to reaching the liver for normal degradation. In this instance, excess autoantibodies and immune complexes might be expected to accelerate disease progression.

the activation state of the macrophages. In the normal state with low expression of the SRs, immune complexes formed by the binding of oxLDL to autoantibodies are directly toward the Kupffer cells for degradation and excretion. Preimmunization with oxLDL would increase autoantibody titers and enhance rapid removal via the Kupffer cells. In the diseased state with highly activated macrophages scattered throughout the vascular tree, all expressing SRs, the immune complexes are more likely to be taken up by the macrophages leading to the formation of foam cells than to be cleared by the liver via the Kupffer cells. In this instance, higher autoantibody titers actually would accelerate disease progression and severity. According to this scheme, autoantibody titers alone would not be sufficient to determine the risk of CVD. While high autoantibody titers in a healthy individual would enhance the efficient removal of any newly generated oxidative epitopes, high autoantibody titers in a patient with inflamed endothelium populated with activated macrophages may actually promote the uptake of oxidized epitopes via SRs leading to foam cell formation.

As discussed in the preceding sections, measurements of oxidative markers in fasting plasma have an inherent bias because of the contribution of several processes for the generation of oxidative epitopes as well as for the removal of modified epitopes, both as part of a self-protective mechanism and as part of disease progression. In such a complex system, when abnormal levels are detected, it is often the case that the disease has reached an advanced state and any compensatory mechanism has already failed. Analogous to the use of the response to an oral glucose tolerance test to identify individuals who are at risk of developing diabetes, we propose to use assess the response to a physiologic bout of oxidative stress to determine whether or not an individual is at the edge of pro-oxidant state. We have chosen a standardized mixed meal enriched in PUFA as the metabolic challenge (Cortner et al., 1987, Cortner et al., 1992, Le et al., 2000). With the meal challenge (600 kcal with 30% fat, 55% carbohydrate, and 15% protein), newly absorbed fatty acids are transported in intestinal chylomicrons and interact with LPL throughout the body for the delivery of TGs to peripheral tissues.

During TG hydrolysis by LPL, these particles enriched in PUFA serve as acceptors for ROS generated in the subendothelium. The acute, meal-induced changes in plasma levels of various indices of oxidation reflect the net effect of pro- and antioxidant processes in any one individual.

6.6.3 MEAL-INDUCED CHANGES IN LIPOPROTEIN-ASSOCIATED OXIDATION

Based on the observations of other investigators, we looked at autoantibodies against MDA-modified LDL in fasting and postprandial plasma. We reported that in obese individuals with documented CAD, postprandial lipemia was associated with a transient 10%–20% reduction in autoantibody levels, with the greatest reduction being observed at 4 h after meal consumption (Le et al., 2000). In contrast, young normolipidemic individuals with no family history of premature heart disease, there was no meal-induced change in autoantibody titers. After 6 months of management with caloric restriction and modest supervised physical activity but no lipid-lowering medication, this meal-induced reduction was reproducible, in spite of significant reductions in plasma lipids (Le et al., 2000). We subsequently demonstrated that this acute reduction in autoantibody titers was specific for test meals enriched in PUFA and not with test meals containing MUFA or SFA (Gradek et al., 2004). There was no difference in the fasting level of autoantibodies in either study. We interpret these findings to indicate that, during postprandial lipemia, there is excess generation of oxidative epitopes in plasma of patients with diseased endothelium. Any amount of circulating autoantibodies available in these patients were overwhelmed by the excessive levels of oxidized epitopes in plasma lipoproteins resulting in a reduction in autoantibodies, which typically return to fasting levels by 6–8 h after the meal.

Another approach that we have investigated is to assess the changes in oxidative susceptibility of individual lipoprotein subfractions as well as a mixture of lipoproteins isolated in fasting and postprandial plasma. In a small study of patients with metabolic syndrome, we have reported that isolated LDL from some patients can undergo spontaneous oxidative modification at room temperature without requiring the addition of Cu^{++} as the catalyst (Le et al., 2013). We would consider these individuals to have the highest risk. For most patients, LDL isolated from postprandial plasma after a mixed meal enriched in PUFA is more susceptible to oxidative medication. This would be the expected response, with the greater reduction in lag time being associated with higher risk. When oxidative susceptibility was examined in a mixture of autologous LDL and HDL, at a fixed ratio of 3:1 by cholesterol content, we noted that in the majority of the patients with metabolic syndrome, the addition of HDL actually increased LDL oxidative susceptibility (Le et al., 2013). After 8 weeks of therapy with fenofibric acid, oxidative susceptibility of fasting LDL was reduced, even when LDL cholesterol was not changed. The key lipid changes were fasting and postprandial TG (Le et al., 2013). HDL from some patients had antioxidant protection properties after therapy, but this was not observed in all patients. These results would suggest that (1) even when HDL levels may be normal in some of the patients with metabolic syndrome, they are dysfunctional and failed to protect LDL from Cu^{++}-induced oxidative modification; (2) oxidative susceptibility of LDL can be significantly and transiently increased during postprandial lipemia; and (3) improvement in postprandial hypertriglyceridemia achieved with fenofibric acid is associated with partial improvement in LDL oxidative susceptibility, possibly as a result of more efficient hydrolysis and less contact of TRL with the endothelium.

6.7 IMPAIRED LIPOPROTEIN METABOLISM IN OBESITY

Obesity is a chronic condition associated with impaired fat metabolism as characterized by excess accumulation of TG in adipose tissue. A consequence of the impaired metabolism is the excess production of adipokines such as plasminogen activator-1 (PAI-1), tumor necrosis factor α (TNF-α), interleukin-6 (IL-6), resistin, leptin, and adiponectin (Fonseca-Alaniz et al., 2007; Ouchi et al., 2011). Excess production of some of these bioactive substances is responsible for the underlying chronic proinflammatory state (Bruun et al., 2006) and the excess production of ROS associated with obesity. Due in part to the poor regulation of these processes, the presence of obesity contributes directly to the increase in mortality and in prevalence of other chronic health problems such as CVDs, diabetes, and colon cancer.

During mobilization of fat from adipose tissue, TGs stored in adipocytes are converted to monoglycerides and FFA by hormone-sensitive lipase (HSL) for release into the circulation. During the postprandial period, with meal-induced increase in plasma levels of insulin, HSL is inhibited and dietary fats carried on chylomicrons are deposited in adipose tissue. During the fasting state, plasma insulin is low and HSL is activated resulting in the efflux of stored FFA that would be used by the liver to synthesize VLDL. Adipocytes from obese individuals have a lower insulin receptor density, which accounts in part for the insulin resistance associated with obesity and, more importantly, for the impaired regulation of HSL. The net result is an increase in the lipolysis rate and excess mobilization of FFA from adipocytes for delivery to the liver. With the increased flux of FFA precursors, hepatic production of the TG-rich VLDL is stimulated causing the saturation of the LPL system, which translates to a hypertriglyceridemic state. Key consequences of hypertriglyceridemia are a reduction in the rate of TG hydrolysis and prolongation of the attachment of TRL to the arterial wall, a process that enhances the seeding of plasma lipoproteins with ROS secreted by activated macrophages (Figure 6.4).

Obesity is associated with excess secretion of proinflammatory cytokines such as TNF-α, IL-6, and IL-1β. In fact, inflammation has been considered as the link connecting obesity, metabolic syndrome, and T2DM (Esser et al., 2014). Obesity is also associated with increased production of MCP-1, a key factor in promoting the infiltration of monocytes into the subendothelium and thus providing additional fuel to the high inflammatory and oxidative state (Di Gregorio et al., 2005). The net effect of all these processes is a vicious circle that continually feeds upon itself by promoting the infiltration of adipose tissue by macrophages, inducing the secretion of proinflammatory cytokines, stimulating the production of ROS, enhancing the efflux of FFA from peripheral tissues to the liver, increasing the production of hepatic VLDL, prolonging the exposure of TRL to the inflamed arterial wall, and resulting ultimately in the excess formation of oxidized lipoproteins. Obesity accelerates the progression of CVD by enhancing all of these processes (Figure 6.8).

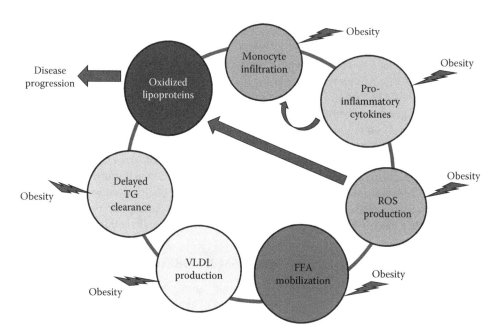

FIGURE 6.8   Impact of obesity on multiple processes leading to increased oxidative modification of plasma lipoproteins.

6.8 DIETARY AND MANAGEMENT IMPLICATIONS

We have presented evidence in support of the use of indices of lipoprotein-associated oxidation as markers of oxidative stress. The evidence is strong for these indices in case–control studies and some prospective studies. However, the impact of lipid-reduction therapies and regression interventions on these biomarkers remains controversial. According to the hypothesis proposed here, by the time abnormal levels are detected in fasting plasma, the disease has already progressed to an advanced stage. In order to identify individuals who might be at increased risk in the early stages of the disease process, we would need to examine the changes in these oxidative indices as induced by specific, physiologic, and standardized oxidative stress challenge. Meal consumption and physical activity (Fisher-Wellman et al., 2009; Nikolaidis et al., 2011) are examples of such metabolic challenges. Both are associated with high oxidative stress, in particular meal enriched in highly oxidizable PUFA. Both have also been demonstrated to be beneficial in reducing the cardiovascular risk. We would expect that healthy individuals with adequate antioxidant protection should be able to compensate for any excess oxidative stress triggered by these physiologic challenges.

If we accept the premises set forth in this hypothesis with respect to the effect of meal consumption, the most effective intervention would be to reduce meal-induced oxidative stress. This can be achieved by reducing the opportunity for TRL to be seeded with ROS generated in the arterial wall by enhancing LPL activities and by reducing postprandial hypertriglyceridemia. Therapies that increase LPL activities would be able to reduce postprandial lipemia. Reduction of the amount of fat available with each meal would also improve postprandial lipemia, that is, more frequent consumption of smaller meals. Substitution of the highly oxidizable PUFA with MUFA, which has comparable lipid-lowering effect, could also reduce the oxidative susceptibility of plasma lipoproteins.

Another approach would be to increase the antioxidant properties of TRL by supplementing the meal with exogenous antioxidants that could be incorporated directly into the newly formed intestinal lipoproteins. Probucol is an example of a very potent lipid-soluble, anti-atherogenic agent that is carried by atherogenic plasma lipoproteins to the arterial wall (Simon et al., 1991; Kuzuya and Kuzuya, 1993). AGI-1067 is another example of an anti-atherogenic agent that is most efficacious when administered with meals in spite of minimal effect of plasma lipid levels (Tardiff et al., 2003; Wasserman et al., 2003). However, clinical trial studies with oral supplementation of the lipid-soluble vitamin E have been less conclusive. It may be that variability in factors such as intestinal absorption and incorporation into intestinal chylomicrons, exchange with other plasma lipoproteins, and efficiency of cell uptake have not been taken into account in evaluating the impact of vitamin E supplementation (Hacquebard and Carpentier, 2005; Ozkanlar and Akcay, 2012; Tinkel et al., 2012). A better understanding of these processes may allow a better strategy for delivering antioxidants to the appropriate sites.

6.9 CONCLUSION

We have presented evidence that plasma lipoproteins, in particular the B-lipoproteins, can readily undergo oxidative modifications *in vivo*. In general, the assessment of various indices of oxidative modification in fasting plasma has been demonstrated to adequately discriminate between individuals with documented CVD and healthy controls. The usefulness of these markers in intervention studies, however, has been mixed and, in some cases, counterintuitive. This is probably due in part to the complexity in the metabolism of the oxidatively modified lipoproteins, which is still poorly understood. We propose that the indices of lipoprotein oxidation may best be assessed as acute changes induced by a specific, standardized, and physiologic oxidative challenge. According to the proposed hypothesis, individuals at risk of developing oxidative stress would respond poorly to the challenge with excessive increase in oxidative indices, while healthy individuals would have sufficient antioxidant defense to compensate for the challenge. In conclusion, the assessment of lipoprotein oxidative status after a fat tolerance challenge may prove to be the counterpart of the oral glucose tolerance test in the identification of individuals at risk of developing CVD.

REFERENCES

Abbey M, Nestel PJ, and Baghurst PA (1993). Antioxidant vitamins and low-density lipoprotein oxidation. *Am J Clin Nutr* 58: 525–532.

Ameli S, Hultgardh-Nilsson A, Rengstrom J et al. (1996). Effect of immunization with homologous LDL and oxidized LDL on early atherosclerosis in hypercholesterolemic rabbits. *Arterioscler Thromb Vasc Biol* 16: 1074–1079.

Ansell BJ, Fonarow GC, and Fogelman AM (2007). The paradox of dysfunctional HDL. *Curr Opin Lipidol* 18: 427–434.

Aoki T, Abe T, Yamada E, Matsuto T, and Okada M (2012). Increased LDL susceptibility to oxidation accelerates future carotid artery atherosclerosis. *Lipids Health Dis* 11: 4–11.

Aronis P, Antonopoulou S, Karantonis HC et al. (2007). Effect of fast-food Mediterranean-type diet on human plasma oxidation. *J Med Food* 10: 511–520.

Avogaro P, Bon GB, and Cazzolato G (1988). Presence of a modified LDL in humans. *Arteriosclerosis* 8: 79–87.

Ballantyne CM, Hoogeveen RC, Bang H et al. (2004). Lipoprotein-associated phospholipase A2, high-sensitivity C-reactive protein, and risk for incident coronary heart disease in middle-aged men and women in the Atherosclerosis Risk in Communities (ARIC) Study. *Circulation* 109: 837–842.

Bergt C, Pennathur S, Fu X et al. (2004). The myeloperoxidase product hypochlorous acid oxidizes HDL in the human artery wall and impairs ABCA1 cholesterol transport. *Proc Natl Acad Sci USA* 101: 13032–13037.

Boyd HC, Down AM, Wolfbauer G, and Chait A (1989). Direct evidence for a protein recognized by a monoclonal antibody against oxidatively modified LDL in atherosclerotic lesions from a Watanabe heritable hyperlipidemic rabbit. *Am J Pathol* 135: 815–825.

Brennan ML, Anderson MM, Shih DM, Qu XD, Wang X, Mehta AC, Lim LL, Shi W, Hazen SL, Jacob JS, Crowley JR, Heinecke JW, and Lusis AJ (2001). Increased atherosclerosis in myeloperoxidase-deficient mice. *J Clin Invest* 107: 419–430.

Brennan ML, Wu W, Fu X et al. (2002). A tale of two controversies: Defining both the role of peroxidases in nitrotyrosine formation in vivo using eosinophil peroxidase in myeloperoxidase-deficient mice and the nature of peroxidase-generated reactive nitrogen species. *J Biol Chem* 277: 17415–17422.

Brown MS, Goldstein JL, Krieger M, Ho YK, and Anderson RG (1979). Reversible accumulation of cholesteryl esters in macrophages incubated with acetylated lipoproteins. *J Cell Biol* 82: 597–613.

Bruun JM, Helge JW, Richelsen B, and Stallnecht B (2006). Diet and exercise reduce low-grade inflammation and macrophage infiltration in adipose tissue but not in skeletal muscle in severely obese subjects. *Am J Physiol Endocrinol Metab* 290: E961–E967.

Bui MN, Sack MN, Moutsatsos G et al. (1996). Autoantibody titers to oxLDL in patients with coronary atherosclerosis. *Am Heart J* 131: 663–667.

Burkitt MJ (2001). A critical overview of the chemistry of copper-dependent LDL oxidation: Roles of lipid hydroperoxides, α-tocopherol, thiols, and ceruloplasmin. *Arch Biochem Biophys* 394: 117–135.

Cathcart MK, McNally AK, and Chisolm GM III (1991). Lipoxygenas-mediated transformation of human LDL to an oxidized and cytotoxic complex. *J Lipid Res* 32: 63–70.

Chisolm GM and Steinberg D (2000). The oxidative modification hypothesis of atherogenesis: An overview. *Free Radic Biol Med* 28: 1815–1826.

Cortner JA, Coates PM, Le N-A et al. (1987). Chylomicron remnant clearance studies in normal and hypertriglyceridemic subjects. *J Lipid Res* 28: 195–206.

Cortner JA, Le N-A, Coates PM, Bennett MJ, and Cryer DR (1992). Determinants of fasting plasma TG levels: Metabolism of hepatic and intestinal lipoproteins. *Eur J Clin Invest* 22: 158–165.

Cracowski JL, Durand T, and Bessard G (2002). Isoprostanes as a biomarker of lipid peroxidation in humans: Physiology, pharmacology, and clinical implications. *Trends Pharmacol Sci* 23: 230–236.

Cushing SD, Berliner JA, Valente AJ et al. (1990). Minimally modified LDL induces MCP-1 in human endothelial cells and smooth muscle cells. *Proc Natl Acad Sci USA* 87: 5134–5138.

Czernichow S, Thomas D, and Bruckert E (2010). n-6 Fatty acids and cardiovascular health: A review of the evidence for dietary intake recommendations. *Br J Nutr* 104: 788–796.

Dalle-Donne I, Rossi R, Colombo R et al. (2006). Biomarkers of oxidative damage in human disease. *Clin Chem* 52: 601–623.

Daugherty A, Dunn JL, Rateri DL et al. (1994). Myeloperoxidase, a catalyst for lipoprotein oxidation, is expressed in human atherosclerotic lesions. *J Clin Invest* 94: 437–444.

Dean RT, Fu S, Stocker R, and Davies MJ (1997). Biochemsitry and pathology of radical-mediated protein oxidation. *Biochem J* 324: 1–18.

Dedon PC and Tannebaum SR (2004). Reactive nitrogen species in the chemical biology of inflammation. *Arch Biochem Biophys* 433: 12–22.

Delsing DJ, Offerman EH, and van Duyvenvoorde W (2001). Acyl-CoA:cholesterol acyltransferase inhibitor avasimibe reduces atherosclerosis in addition to its cholesterol-lowering effect in apoE*3-Leiden mice. *Circulation* 103: 1778–1786.

Dieber-Rotheneder M, Puhl H, Waeg G, Striegl G, and Esterbauer H (1991). Effect of oral supplementation of D-α-tocopherol on the vitamin E content of human LDL and resistance to oxidation. *J Lipid Res* 32: 1325–1332.

Di Gregorio GB, Yao-Borengasser A, Rasouli N et al. (2005). Expression of CD68 and MCP-1 genes in human adipose and muscle tissue: Association with cytokine expression, insulin resistance, and reduction by pioglitazone. *Diabetes* 54: 2305–2313.

Dominiczak MH (2000). Apolipoproteins and lipoproteins in human plasma in *Handbook of Lipoprotein Testing*. N Rifai, GR Warnick, and MH Dominiczak, Eds. Chapter 1, 2nd Edition, AACC Press, Washington DC, pp. 1–29.

Draper HH, Csallany AS, and Hadley M (2000). Urinary aldehydes as indicators of lipid peroxidation in vivo. *Free Radic Biol Med* 29: 1071–1077.

Esser N, Legrand-Poels S, Piette J et al. (2014). Inflammation as a link between obesity, metabolic syndrome and type 2 diabetes. *Diabetes Res Clin Pract* 105: 141–150.

Esterbauer H, Schaur R, and Zollner H (1991). Chemistry and biochemistry of 4-hydroxynonenal, malondialdehyde and related aldehydes. *Free Radic Biol Med* 11: 81–128.

Esterbauer H, Striegl G, Puhl H, and Rotheneder M (1989). Continuous monitoring of in vitro oxidation of human LDL. *Free Radic Biol Med* 6: 67–75.

Ezaki M, Witztum JL, and Steinberg D (1995). Lipoperoxides in LDL incubated with fibroblasts that overexpress 15-lipoxygenase. *J Lipid Res* 36: 1996–2004.

Fisher-Wellman K, Bell HK, and Bloomer RJ (2009). Oxidative stress and antioxidant defense mechanisms linked to exercise during cardiopulmonary and metabolic disorders. *Oxid Med Cell Longev* 2: 43–51.

Fogelman AM, Schechter I, Seager J, Hokom M, Child JS, and Edwards PA (1980). Malondialdehyde alteration of low density lipoproteins leads to cholesteryl ester accumulation in human monocyte-macrophages. *Proc Natl Acad Sci USA* 77: 2214–2218.

Folcik VA, Nivar-Aristy RA, Krajewski LP, and Cathcart MK (1995). Lipoxygenase contributes to the oxidation of lipids in human atherosclerotic plaques. *J Clin Invest* 96: 504–510.

Fonseca-Alaniz MH, Takada J, Alonso-Vale MI, and Lima FB (2007). Adipose tissue as an endocrine organ: From theory to practice. *J Pediatr* 83 (Suppl 5): S192–S203.

Francone OL, Gurakar A, and Fielding C (1989). Distribution and functions of lecithin: Cholesterol acyltransferase and cholesteryl ester transfer protein in plasma lipoproteins. Evidence for a functional unit containing these activities together with apolipoproteins A-I and D that catalyzes the esterification and transfer of cell-derived cholesterol. *J Biol Chem* 264: 7066–7072.

Fredrikson GN, Soderberg I, Lindholm M, Dimayuga P, Chyu KY, Shah PK, and Nilsson J (2003). Inhibition of atherosclerosis in apoE-null mouse by immunization with apoB-100 peptide sequences. *Arterioscl Thromb Vasc Biol* 23: 879–884.

Gaut J, Yeh G, Tran H et al. (2001). Neutrophils employ the myeloperoxidase system to generate antimicrobial brominating and chlorinating oxidants during sepsis. *Proc Natl Acad Sci USA* 98: 11961–11966.

Giustarini D, Rossi R, Milzani A et al. (2004). S-Glutathonylation: From redox regulation of protein functions to human diseases. *J Cell Mol Med* 8: 201–212.

Goldstein JL, Ho YK, Basu SK, and Brown MS (1979). Binding site on macrophages that mediates uptake and degradation of acetylated low density lipoprotein, producing massive cholesterol deposition. *Proc Natl Acad Sci USA* 76: 333–337.

Gradek Q, Harris M, Yahia N, Davis W, Le N-A, and Brown WV (2004). Polyunsaturated fatty acids acutely suppress antibodies to malondialdehyde-modified LDL in patients with vascular disease. *Am J Cardiol* 93: 881–885.

Hacquebard M and Carpentier YA (2005). Vitamin E: Absorption, plasma transport and cell uptake. *Curr Opin Clin Nutr Metab Care* 8: 133–138.

Hazell LJ, Arnold L, Flowers D et al. (1996). Presence of hypochlorite-modified proteins in human atherosclerotic lesions. *J Clin Invest* 97: 1535–1544.

Hazen SL and Heinecke JW (1997). 3-Chlorotyrosine, a specific marker of myeloperoxidase-catalyzed oxidation, is markedly elevated in low-density lipoproteins isolated from human atherosclerotic intima. *J Clin Invest* 99: 2075–2081.

Hendrickson A, McKinstry LA, Lewis JK et al. (2005). Ex vivo measures of LDL oxidative susceptibility predict coronary artery disease. *Atherosclerosis* 179: 147–153.

Hessler JR, Robertson AL Jr, and Chisolm GM III (1979). LDL-induced cytotoxicity and its inhibition by HDL in human vascular smooth muscle and endothelial cells in culture. *Atherosclerosis* 32: 213–229.

Holvoet P, Donck J, Landeloos M et al. (1996). Correlation between oxLDL and von Willebrand factor in chronic renal failure. *Thromb Haemost* 76: 663–669.

Holvoet P, Jenny NS, Schreiner PJ, Tracy RP, and Jacobs DR (2007). The relationship between oxidized LDL and other cardiovascular risk factors and subclinical CVD in different ethnic groups: The Multi-Ethnic Study of Atherosclerosis (MESA). *Atherosclerosis* 194: 245–252.

Holvoet P, Kritchesky SB, Tracy RP et al. (2004). The metabolic syndrome, circulating oxidized LDL, and risk of myocardial infarction in well-functioning elderly people in the Health, Aging, and Body Composition cohort. *Diabetes* 53: 1068–1073.

Horl G, Ledinski G, Kager G et al. (2014). In vitro oxidation of LDL by ozone. *Chem Phys Lipids* 183: 18–21.

Hulthe J and Fagerberg B (2002). Circulating oxidized LDL is associated with subclinical atherosclerosis development and inflammatory cytokines (AIR study). *Arterioscler Thromb Vasc Biol* 22: 1162–1167.

Itabe H, Yamamoto H, Imanaka T et al. (1996). Sensitive detection of oxidatively modified LDL using a monoclonal antibody. *J Lipid Res* 37: 45–53.

Iuliano L (2011). Pathways of cholesterol oxidation via non-enzymatic mechanisms. *Chem Phys Lipids* 164: 457–468.

Jialal I and Grundy SM (1993). Effect of combined supplementation with alpha-tocopherol, ascorbate, and beta carotene on LDL oxidation. *Circulation* 88: 2780–2786.

Jiang X-C, Jin W, and Hussain MM (2012). The impact of phospholipid transfer protein (PLTP) on lipoprotein metabolism. *Nutr Metab* 9: 75–81.

Kaikkonen JE, Kresanov P, Ahotupa P et al. (2014). High serum n6 fatty acid proportion is associated with lowered LDL oxidation and inflammation. The Cardiovascular Risk in Young Finns Study. *Free Rad Res* 48: 420–426.

Kersten S (2014). Physiological regulation of lipoprotein lipase. *Biochim Biophys Acta* 1841: 919–933.

Klebanoff SJ (2005). Myeloperoxidase: Friend or foe. *J Leukoc Biol* 77: 598–625.

Kontush A and Chapman MJ (2006). Functionally defective HDL: A new therapeutic target at the crossroads of dyslipidemia, inflammation and atherosclerosis. *Pharmacol Rev* 58: 342–374.

Kuhn H, Heydeck D, Hugou I, and Gniwotta C (1997). In vivo action of 15-lipoxygenase in early stages of human atherogenesis. *J Clin Invest* 99: 888–893.

Kuzuya M and Kuzuya F (1993). Probucol as an antioxidant and antiatherogenic drug. *Free Rad Biol Med* 14: 67–77.

Lagrost L, Gandjini H, Athias A et al. (1993). Influence of plasma CETP activity on the LDL and HDL distribution profiles in normolipidemic subjects. *Arterioscler Thromb* 13: 815–825.

Le N-A (2009). Oxidized lipids and lipoproteins: Indices of risk or targets of management. *Clin Lipidol* 4: 41–54.

Le N-A (2015). Lipoprotein-associated oxidative stress: A new twist to the postprandial hypothesis. *Int J Mol Sci* 16: 401–419.

Le N-A, Farkas-Epperson M, Sweeney ME, WIlson PWF, and Brown WV (2013). Effect of ABT-335 (fenofibric acid) on meal-induced oxidative stress in patients with metabolic syndrome. *Atherosclerosis* 231: 268–273.

Le N-A, Li X, Kyung S, and Brown WV (2000). Evidence for the in vivo generation of oxidatively modified epitopes in patients with documented CAD. *Metabolism* 49: 1271–1277.

Li X, Innis-Whitehouse W, Le N-A, and Brown WV (1997). Protocol for the preparation of segmental linear polyacrylamide gradient gel: Simultaneous determination of Lp(a), LDL, and HDL particle sizes. *J Lipid Res* 38: 2603–2614.

Liu J, Yeo HC, Doniger SJ, and Ames BN (1997). Assays of aldehydes from lipid peroxidation: GC/MS compared to thiobarbituric acid. *Anal Biochem* 245: 161–166.

Mabile L, Salvayre R, Bonnafe MJ, and Negre-Salvayre A (1995). Oxidizability and subsequent cytotoxicity of chylomicrons to monocytic U937 and endothelial cells are dependent on dietary fatty acid composition. *Free Rad Biol Med* 19: 599–607.

Mackness MI. Arrol S, Abbott CA, and Durrington PN (1993). Protection of LDL against oxidative modification by HDL-associated paraoxonase. *Atherosclerosis* 104: 129–135.

Mata P, Alonso R, Lopez-Farre A et al. (1996). Effect of dietary fat saturation on LDL oxidation and monocyte adhesion to human endothelial cells in vitro. *Arterioscler Thromb Vasc Biol* 16: 1347–1355.

McEneny J, O'Kane MJ, Moles KW, McMaster C, McMaster D, Mercer C, Trimble ER, Young IS (2000). Very low density lipoprotein subfractions in Type II diabetes mellitus: Alterations in composition and susceptibility to oxidation. *Diabetologia* 43: 485–493.

Michas G, Micha R, and Zampelas A (2014). Dietary fats and cardiovascular disease: Putting together the pieces of a complicated puzzle. *Atherosclerosis* 234: 320–328.

Montuschi P, Barnes PJ, and Roberts LJ (2004). Isoprostanes: Markers and mediators of oxidative stress. *FASEB J* 84: 1381–1478.

Nielsen NS, Pedersen A, Sandstrom B, Marckman P, and Hoy CE (2002). Different effects of diets rich in olive oil, rapeseed oil and sunflower oil on postprandial lipid and lipoprotein concentrations and on lipoprotein oxidation susceptibility. *Br J Nutr* 87: 489–499.

Niki E, Yoshida Y, Saito Y, and Noguchi N (2005). Lipid peroxidation: Mechanisms, inhibition, and biological effects. *Biochem Biophys Res Commun* 338: 668–676.

Nikolaidis MG, Kyparos A, and Vrabas IS (2011). F2-isoprostane formation, measurement, and interpretation: The role of exercise. *Prog Lipid Res* 50: 89–103.

Nishi K, Itabe H, Uno M et al. (2002). Oxidized LDL in carotid plaques and plasma associates with plaque instability. *Arterioscler Thromb Vasc Biol* 22: 1649–1654.

Nissen SE (2005). Halting the progression of atherosclerosis with intensive lipid lowering: Results from the Reversal of Atherosclerosis with Aggressive Lipid Lowring (REVERSAL) trial. *Am J Med* 118 (Suppl. 12A): 22–27.

Nissen SE, Tuzcu EM, Schoenhagen P et al. (2004). Effect of intensive compared with moderate lipid-lowering therapy on progression of coronary atherosclerosis: A randomized controlled trial. *JAMA* 291: 1071–1080.

Ohkawa H, Ohishi N, and Yagi K (1979). Assay for lipid peroxides in animal tissues by thiobarbituric acid reaction. *Anal Biochem* 95: 351–358.

Ouchi N, Parker JL, Lugus JL, and Walsh K (2011). Adipokines in inflammation and metabolic disease. *Nat Rev Immunol* 11: 85–97.

Ozkanlar S and Akcay F (2012). Antioxidant vitamins in atherosclerosis—Animal experiments and clinical studies. *Adv Clin Exp Med* 21: 115–123.

Palinski W, Horkko S, Miller E et al. (1996). Cloning of monoclonal antibodies to epitopes of oxidized lipoproteins from apoE-deficient mice. Demonstration of epitopes of oxidized LDL in human plasma. *J Clin Invest* 98: 800–814.

Palinski W, Miller E, and Witztum JL (1995). Immunization of LDL receptor-deficient rabbits with homologous MDA-modified LDL reduces atherosclerosis. *Proc Natl Acad Sci USA* 92: 821–825.

Palinski W, Ord VA, Plump AS et al. (1994). ApoE-deficient mice are a model of lipoprotein oxidation in atherogenesis. Demonstration of oxidation specific epitopes in lesions and high titers of antibodies to malondialdehyde-lysine in serum. *Arterioscl Thromb* 14: 605–616.

Palinski W, Rosenfeld ME, Yla-Herttuala S et al. (1989). LDL undergoes oxidative modification in vivo. *Proc Natl Acad Sci USA* 86: 1372–1376.

Parola M, Bellamo G, Robino G et al. (1999). 4-hydroxynonenal as a biological signal: Molecular basis and pathological implications. *Antioxid Redox Signal* 1: 255–284.

Parthasarathy S, Khoo JC, Miller E et al. (1990). LDL rich in oleic acid is protected against oxidative modification: Implications for dietary prevention of atherosclerosis. *Proc Natl Acad Sci USA* 87: 3894–3898.

Parthasarathy S, Wieland E, and Steinberg D (1989). A role of endothelial cell lipoxygenase in the oxidative modification of LDL. *Proc Natl Acad Sci USA* 86: 1046–1050.

Pennathur S, Bergt C, Shao B et al. (2004). Human atherosclerotic intima and blood of patients with established CAD contain HDL damaged by reactive nitrogen species. *J Biol Chem* 279: 42977–42983.

Puurunen M, Manttari M, Manninen V et al. (1994). Antibody against oxidized LDL predicting myocardial infarction. *Arch Intern Med* 154: 2605–2609.

Quintao EC and Cazita PM (2010). Lipid transfer proteins: Past, present, and perspectives. *Atherosclerosis* 209: 1–9.

Quinn MT, Parthasarathy S, Fong LG, and Steinberg D (1987). Oxidatively modified LDL: A potential role in recruitment and retention of monocyte-macrophages during atherogenesis. *Proc Natl Acad Sci USA* 84: 2995–2998.

Rabini RA, Tesei M, Galeazzi T, Dousset N, Ferretti G, and Mazzanti L (1999). Increased susceptiility to peroxidation of VLDL from non-insulin-dependent diabetic patients: A possible correlation with fatty acid composition. *Mol Cell Biochem* 199: 63–67.

Requena JR, Levine RL, and Stadtman ER (2003). Recent advances in the analysis of oxidized proteins. *Amino Acids* 25: 221–226.

Ryglewicz D, Rodo M, Roszczynko M, Baranska-Gieruszczak M, Szirkowiec W, SwiderskaM and Wehr H (2002). Dynamics of LDL oxidation in ischemic stroke patients. *Acta Neurol Scand* 105: 185–188.

Salonen JT, Nyyssonen K, Salonen R et al. (1997). Lipoprotein oxidation and progression of carotid atherosclerosis. *Circulation* 95: 840–845.

Salonen JT, Yla-Herttuala S, Yamamoto R et al. (1992). Autoantibody against LDL and progression of carotid atherosclerosis. *Lancet* 339: 883–887.

Schindhelm RK, van der Zwan LP, Teerlink T, and Scheffer PG (2009). Myeloperoxidase: A useful biomarker for cardiovascular disease risk stratification? *Clin Chem* 55: 1462–1470.

Schiopu A, Frendeus B, Jansson B, Soderberg I, Ljungcrantz I, Araya Z, Shah PK, Carlsson R, Nilsson J, and Fredrickson GN (2007). Recombinant antibodies to an oxidized low-density lipoprotein epitope induce rapid regression of atherosclerosis in apobec-1(-/-)/low-density lipoprotein receptor (-/-) mice. *J Am Coll Cardiol* 50: 2313–1318.

Shao B, O-Brien KD, McDonald TO et al. (2005). Acrolein modifies apolipoprotein A-I in the human arterial wall. *Ann NY Acad Sci* 1043: 396–403.

Shechter I, Fogelman AM, Haberland ME et al. (1981). The metabolism of native and MDA-altered LDL by human monocyte-macrophages. *J Lipid Res* 22: 63–71.

Simon SJT, Yates MT, Rechtin AE, Jackson RL, and van Sickle WA (1991). Antioxidant activity of probucol and its analogues in hypercholesterolemic Watanabe rabbits. *J Med Chem* 34: 298–302.

Slatter DA, Bolton CH, and Bailey AJ (2000). The importance of lipid-derived malondialdehyde in diabetes mellitus. *Diabetologia* 43: 550–557.

Stadtman E and Berlett BS (1997). Reactive oxygen-mediated protein oxidation in ageing and disease. *Chem Res Toxicol* 10: 485–494.

Stein O and Stein Y (2005). Lipid transfer proteins (LTP) and atherosclerosis. *Atherosclerosis* 178: 217–230.

Steinberg D, Parthasarathy S, Carew TE, Khoo JC, and Witztum JL (1989). Beyond cholesterol: Modifications of low density lipoproteins that increase it atherogenicity. *N Engl J Med* 320: 915–924.

Steinberg D and Witztum JL (2010). Oxidized low-density lipoproteins and atherosclerosis. *Arterioscler Thromb Vasc Biol* 30: 2311–2316.

Stocker R and Frei B (1991). Endogenous antioxidant defenses in human blood plasma. In *Oxidative Stress: Oxidants and Antioxidants*. H Sies, Ed., Academic Press, London, U.K., pp. 213–243.

Stocker R and Keaney JF Jr (2004). Role of oxidative modification in atherosclerosis. *Physiol Rev* 84: 1381–1478.

Stocker R and Keaney JF Jr (2005). New insights on oxidative stress in the artery wall. *J Thromb Haemost* 3: 1825–1834.

Tall AR, Hogan V, Askinazi L, and Small D (1978). Interactions of plasma HDL with dimyristoyl lecithin multilamellar liposomes. *Biochemsitry* 17: 322–326.

Tall AR, Krumholz S, Olivecrona T, and Deckelbaum RJ (1985). Plasma phospholipid transfer protein enhances transfer and exchange of phospholipids between VLDL and HDL during lipolysis. *J Lipid Res* 26: 842–851.

Tardiff JC, Gregoire J, Schwartz L et al. (2003). Canadian Antioxidant Restenosis Trial (CART-1). Effects of AGI-1067 and probucol after percutaneous coronary interventions. *Circulation* 107: 552–558.

Tarpey MM, Wink DA, and Grisham MB (2004). Methods for detection of reactive metabolites of oxygen and nitrogen: in vitro and in vivo considerations. *Am J Physiol Regul Integr Comp Physiol* 286: R431–R444.

Tinkel J, Hassanain H, and Khouri SJ (2012). Cardiovascular antioxidant therapy: A review of supplements, pharmacotherapies, and mechanisms. *Cardiol Rev* 20: 77–83.

Tribble DL, Holl LG, Wood PD, and Krauss RM (1992). Variations in oxidative susceptibility among six LDL subfractions of different density and particle size. *Atherosclerosis* 93: 189–199.

Tribble DL, Rizzo M, Chait A et al. (2001). Enhanced oxidative susceptibility and reduced antioxidant content of metabolic precursors of small, dense LDL. *Am J Med* 110: 103–110.

Tribble DL, van der Berg JJM, Motchnik PA et al. (1994). Oxidative susceptibility of LDL subfractions is related to their ubiquinal-10 and a-tocopherol content. *Proc Natl Acad Sci USA* 91: 1183–1187.

Tsimikas S (2006). Oxidative biomarkers in the diagnosis and prognosis of cardiovascular disease. *Am J Cardiol* 98: 9P–17P.

Tsimikas S, Aikawa M, Miller FJ, Miller ER, Torzewski M, Lentz SR, Bergmark C, Heistad DD, Libby P, and Witztum JL. (2007). Increased plasma oxidized phospholipid:apoB-100 ratio with concomitant deple-tion of oxidized phospholipids from atherosclerotic lesions after dietary lipid lowering. *Arterioscler Thromb Vasc Biol* 27: 175–181.

Tsimikas S, Brikakis ES, Miller ER, McConnell JP, Lennon RJ, Kornman KS, Witztum JL, and Berger PB. (2005). Oxidized phospholipids, Lp(a) lipoprotein and coronary artery disease. *N Engl J Med* 353: 46–57.

Tsimikas S, Witztum JL, Miller ER, Sasiela WJ, Szarek M, Olsson AG, and Schwartz GG. (2004). High-dose atorvastatin reduces total plasma levels of oxidized ohospholipids and immune complexes present on apoB in patients with acute coronary syndromes in the MIRACL trial. *Circulation* 110: 1406–1412.

Tzotzas T, Desrumaux C, and Lagrost L (2009). Plasma phospholipid transfer protein (PLTP): Review of an emerging cardiometabolic risk factor. *Obes Rev* 10: 403–411.

Uusitupa MI, Niskanen L, Luoma J, Vilja P, Mercuri M, Rauramaa R, and Yla-Hertutuala S. (1996). Autoantibodies against oxidized LDL do not predict atherosclerotic vascular disease in non-insulin-dependent diabetes mellitus. *Arterioscler Thromb Vasc Biol* 16: 1236–1242.

van de Vivjer LP, Steyger R, van Poppel G, Boer JM, Kruijssen DA, Seidell JC and Princen HM. (1996). Autoantibodies against MDA-LDL in subjects with severe and minor atherosclerosis and healthy popu-lation controls. *Atherosclerosis* 122: 245–253.

Wallenfeldt K, Fagerberg B, Wikstrand J, and Hulthe J (2004). Oxidized low-density lipoproteins in plasma is a prognostic marker of subclinical atherosclerosis development in clinically healthy men. *J Intern Med* 256: 413–420.

Walter MW, Jacob RF, Jeffers B et al. (2004). Serum levels of TVARS predict cardiovascular events in patients with stable CAD: A longitudinal analysis of the PREVENT study. *J Am Coll Cardiol* 44: 1996–2002.

Wasserman MA, Sundell CL, Kunsch C et al. (2003). Chemistry and pharmacology of vascular protectants: A Novel approach to the treatment of atherosclerosis and CAD. *Am J Cardiol* 91(Suppl.): 34A–40A.

Witztum JL and Steinberg D (1991). Role of oxidized low density lipoproteins in atherogenesis. *J Clin Invest* 88: 1785–1792.

Yla-Herttuala S, Palinski W, Rosenfeld ME et al. (1989). Evidence for the presence of oxidatively modified LDL in atherosclerotic lesions of rabbit and man. *J Clin Invest* 84: 1086–1095.

Yla-Herttuala S, Rosenfeld ME, Parthasarathy S et al. (1990). Colocalization of 15-lipoxygenase mRNA and protein with epitopes of oxidized LDL in macrophage-rich areas of atherosclerotic lesions. *Proc Natl Acad Sci USA* 87: 6959–6963.

Yla-Herttuala S, Rosenfeld ME, Parthasarathy S et al. (1991). Gene expression in macrophage-rich human atherosclerotic lesions, 15-lipoxygenase and acetyl LDL receptor messenger colocalize with oxidation specific lipid-protein adducts. *J Clin Invest* 87: 1146–1152.

Yoshida H and Kisugi R (2010). Mechanisms of LDL oxidation. *Clin Chim Acta* 411: 1875–1882.

Yoshida Y, Umeno A, Akazawa Y et al. (2015). Chemistry of lipid peroxidation products and their use as biomarkers in early detection of diseases. *J Oleo Sci* 64: 347–356.

Young SG and Zechner R (2013). Biochemistry and pathophysiology of intravascular and intracellular lipolysis. *Genes Dev* 27: 459–484.

Zannis VI, Fotakis P, Koukos G, Kardassis D, Ehnholm C, Jauhiainen M, and Chroni A (2015). HDL biogenesis, remodeling, and catabolism. *Handb Exp Pharmacol* 224: 53–111.

Zheng L, Nukuna B, Brennan M et al. (2004). Apolipoprotein A-I is a selective target for myeloperoxidase-catalyzed oxidation and functional impairment in subjects with cardiovascular disease. *J Clin Invest* 114: 529–541.

Section II

*Importance of Accumulation
of Iron in the Brain*

7 Role of Oxidative Damage in Neurodegeneration with Brain Iron Accumulation Disorders

Vincenzo Lupo, Alejandra Darling, Cristina Tello,
Belén Pérez-Dueñas*, and Carmen Espinós*

CONTENTS

ABSTRACT

Neurodegeneration with brain iron accumulation (NBIA) is a heterogeneous group of inherited neurologic disorders characterized by progressive movement disorders and abnormal accumulation of iron in the basal ganglia (Gregory and Hayflick 2011, Schneider et al. 2012).

The two core syndromes are pantothenate kinase–associated neurodegeneration (PKAN) and PLA2G6-associated neurodegeneration (PLAN). However, different approaches have allowed the identification of several genes causing NBIA that encode for proteins involved in processes such as iron metabolism, mitochondrial dynamics, reactive oxygen species (ROS)-induced damage, lipid metabolism, and autophagy. A better knowledge of the NBIA genes and their functions will help to link clinical picture, magnetic resonance imaging findings, and disease mechanisms.

---

* These authors have contributed equally.

7.1 CLINICAL MANIFESTATIONS

Our aim is to provide a summary of the main clinical features of the neurodegeneration with brain iron accumulation (NBIA) syndromes (Table 7.1) and highlight the broad clinical spectrum of these complex disorders.

7.1.1 PANTOTHENATE KINASE–ASSOCIATED NEURODEGENERATION

Mutations in the *PANK2* gene lead to pantothenate kinase–associated neurodegeneration (PKAN) (Zhou et al. 2001), which is the most frequent NBIA form, accounting for 50% of patients. The phenotypic continuum includes classic PKAN and atypical PKAN: *Classic PKAN* is characterized by an early onset (1st decade) with severe extrapyramidal signs and progresses rapidly with loss of ambulation within 15 years from onset (Schneider et al. 2012). Dystonia is the predominant feature and involves limbs, the face, and the oromandibular region. Oculomotor abnormalities suggestive of midbrain degeneration are common. *Atypical PKAN* has an onset in the second to third decade with less severe extrapyramidal signs. Individuals are likely to present with speech difficulty and psychiatric symptoms. The phenotype may be less severe, and overall progression is slower compared with the classic form (Hayflick et al. 2003, Gregory et al. 2009).

7.1.2 PLA2G6-ASSOCIATED NEURODEGENERATION

Mutations in the *PLA2G6* gene are linked to two diseases: PLA2G6-associated neurodegeneration (PLAN) and Parkinson disease-14 (PARK14) (Morgan et al. 2006, Paisán-Ruiz et al. 2012). PLAN is the second more frequent NBIA form and occurs in around 20% of patients. Three phenotypes are observed in this disease: infantile neuroaxonal dystrophy (INAD), atypical NAD, and PLA2G6-related dystonia–parkinsonism (Morgan et al. 2006, Paisán Ruiz et al. 2012). INAD is characterized by early-onset manifestations that begin with developmental regression, truncal hypotonia, early visual disturbances due to optic atrophy, cerebellar ataxia, and progressive pyramidal signs (Morgan et al. 2006, Kurian et al. 2008). Atypical NAD manifestations begin in childhood with slower progression; dystonia and spastic tetraparesis are common. Patients often present with speech delay and diminished social interactions. *PLA2G6-related dystonia–parkinsonism* occurs in late adolescence or early adulthood with subacute onset of dystonia–parkinsonism.

7.1.3 MITOCHONDRIAL MEMBRANE PROTEIN–ASSOCIATED NEURODEGENERATION

Mitochondrial membrane protein–associated neurodegeneration (MPAN) represents approximately 6%–10% of NBIA patients. The MPAN patients harbor mutations in the *C19orf12* gene (Hartig et al. 2011). Onset typically occurs in childhood to early adulthood, but it has also been reported to occur as late as age 30 years. The progression of MPAN is usually slow with good survival into adulthood (Hartig et al. 2011, Hogarth et al. 2013). The most common presenting feature is impaired gait, followed by progressive spastic paresis, lower motor neuron signs, and visual loss due to optic atrophy. Progressive cognitive decline is the rule in MPAN. Neuropsychiatric changes as well as parkinsonism, dystonia, dysarthria, and dysphagia are frequent.

7.1.4 β-PROPELLER PROTEIN–ASSOCIATED NEURODEGENERATION

β-propeller protein–associated neurodegeneration (BPAN) is an X-linked form caused by mutations in the *WDR45* gene (Haack et al. 2012, Hayflick et al., 2013). BPAN accounts for 1%–2% of NBIA patients, but this percentage is probably underestimated since the *WDR45* gene has recently been identified. Affected individuals are mostly females, with seizures and global developmental delay during childhood with slow motor and cognitive gains; however, during adolescence or adulthood,

TABLE 7.1

NBIA Disorders: Clinical Features

NBIA Subtype Gene Frequency		Onset	Main Clinical Features
PKAN PANK2 35%–50%	Classical form	Childhood	Extrapyramidal features: dystonia and dysarthria
			Prominent oromandibular involvement
			Additional findings: oculomotor abnormalities, chorea, parkinsonism, pyramidal features, cognitive features, behavioral changes
			Retinal degeneration
	Atypical form	Juvenile–Adulthood	Speech difficulty
			Psychiatric symptoms
			Extrapyramidal signs less severe
PLAN PLA2G6 20%	Infantile neuroaxonal dystrophy (INAD)	6 months–3 years	Developmental regression
			Hypotonia
			Cerebellar ataxia
			Psychomotor delay
			Spastic tetraparesis
			Optic atrophy
	Atypical NAD	Later in childhood	Dystonia
			Spastic tetraparesis
	PLA2G6-related dystonia–parkinsonism	Late adolescence–early adulthood	Dystonia–parkinsonism of subcacute onset
MPAN C19orf12 6%–10%		Childhood to early adulthood	Spastic paresis (more prominent)
			Dystonia
			Neuropsychiatric abnormalities
			Cognitive decline
BPAN WDR45 1%–2%		Childhood	Developmental delay
			Dystonia
			Parkinsonism
			Dementia
			Mainly females
FAHN FA2H Rare		Childhood	Dystonia
			Spasticity
			Progressive intellectual decline
Kufor–Rakeb syndrome ATP13A2 Rare		Juvenile	Parkinsonism
			Pyramidal signs
			Dementia
Neuroferritinopathy FTL Rare		Adulthood	Chorea
			Dystonia
			Cognitive changes

(Continued)

TABLE 7.1 (*Continued*)
NBIA Disorders: Clinical Features

NBIA Subtype

Gene

Frequency	Onset	Main Clinical Features
Aceruloplasminemia	Adulthood	Clinical triad
CP		Movement disorders: dystonia,
Rare		tremor, chorea
		Ataxia
		Dementia
Woodhouse–Sakati syndrome	Juvenile to adult onset	Progressive extrapyramidal disorder
DCAF17		Cognitive decline
Rare		Endocrine abnormalities
CoPAN	Childhood	Spastic–dystonic paraparesis
COASY		Oromandibular dystonia and
Rare		dysarthria
		Axonal neuropathy
		Parkinsonism
		Cognitive impairment and
		obsessive–compulsive behavior

they exhibit a relatively sudden onset of progressive dystonia–parkinsonism and dementia (Haack et al. 2012).

7.1.5 Fatty Acid Hydroxylase–Associated Neurodegeneration

In fatty acid hydroxylase–associated neurodegeneration (FAHN), *FA2H* gene is mutated in this rare NBIA type (Kruer et al. 2010) and also in leukodystrophy and spastic paraplegia-35 (SPG35) (Edvardson et al. 2008, Dick et al. 2010). Symptoms usually begin in childhood and are slowly progressive with gait disturbance, due to lower limb dystonia and corticospinal tract involvement. Later patients develop ataxia, dysarthria, progressive tetraparesis, optic atrophy, and visual loss. Progressive intellectual decline is reported in most affected individuals (Kruer et al. 2010).

7.1.6 Kufor–Rakeb Syndrome

Mutations in the *ATP13A2* gene lead to Parkinson disease-9 (PARK9), neuronal ceroid lipofuscinosis, and Kufor–Rakeb syndrome (Di Fonzo et al. 2007, Schneider et al. 2010, Bras et al. 2012). The limited number of affected individuals previously reported is characterized by juvenile-onset parkinsonism, dementia, pyramidal signs, supranuclear gaze palsy, facial–faucial–finger myoclonus, visual hallucinations, and oculogyric dystonic spasms (Schneider et al. 2010).

7.1.7 Neuroferritinopathy

Mutations in the gene encoding ferritin light chain (*FTL*) are responsible for this type of NBIA (Curtis et al. 2001). The onset is in midlife with chorea and dystonia. Neuroferritinopathy progresses from extremity involvement to a more generalized movement disorder; most affected individuals develop a characteristic orofacial action–specific dystonia related to speech. Serum ferritin concentration may be low (Chinnery et al. 2007).

7.1.8 Aceruloplasminemia

Patients affected by aceruloplasminemia are carriers of mutations in the *CP* gene, encoding ceruloplasmin (CP) (Harris et al. 1995). Its main hallmark is the iron accumulation in the brain and viscera. The clinical trial of retinal degeneration, diabetes mellitus (DM), and neurologic disease is seen in individuals ranging from 25 years to older than 60 years. The neurologic findings include movement disorders and ataxia. Psychiatric disturbance includes depression and cognitive dysfunction. Laboratory investigations reveal the absence of serum CP, low serum copper or iron concentration, high serum ferritin concentration, and increased hepatic iron concentration (Kono et al. 2006).

7.1.9 Woodhouse–Sakati Syndrome

This rare syndrome is due to mutations in the *DCAF17* gene (Alazami et al. 2008). Neurologic findings include progressive extrapyramidal signs, generalized and focal dystonia, dysarthria, and cognitive decline. Endocrine abnormalities comprise hypogonadism, alopecia, and DM (Woodhouse and Sakati 1983, Al-Semari and Bohlega 2007).

7.1.10 COASY Protein–Associated Neurodegeneration

Mutations in the *COASY* gene have been recently published in COASY protein–associated neurodegeneration (CoPAN) patients (Dusi et al. 2014). The subjects described to date presented with early-onset spastic–dystonic paraparesis and later development of oromandibular dystonia, dysarthria, axonal neuropathy, parkinsonism, cognitive impairment, and obsessive–compulsive behavior. Affected individuals had slow progression and were alive in their third decades (Dusi et al. 2014).

7.2 NEUROIMAGING IN NBIA

Brain magnetic resonance imaging (MRI) is a standard component for the diagnosis of NBIA, as the majority of patients show abnormal iron accumulation in the basal ganglia. This is usually appreciated as hypointense lesions in the globus pallidus and substantia nigra on T2-weighted images and susceptibility-weighted imaging (SWI) (Figure 7.1). Iron accumulation is sometimes observed in the red nucleus, dentate nucleus, putamen, or caudate (Gregory et al. 2009). MRI is also useful in the distinction of the different forms of NBIA. There is a high correlation between the presence of a central region of hyperintensity in the globus pallidus with surrounding hypointensity on T2 weighted, a pattern called the eye-of-the-tiger sign, and the presence of mutations in *PANK2* (Hayflick et al. 2003; Figure 7.1). Low signal in the globus pallidus and substantia nigra on T2-weighted images, indicating iron accumulation, may be present in around half of INAD cases. In both INAD and atypical NAD, cerebellar and optic atrophy are hallmark features (Kurian et al. 2008). Patients with BPAN show a T1-hyperintense hallo surrounding the substantia nigra in the mesencephalon.

7.3 PATHOPHYSIOLOGICAL MECHANISMS IN NBIA

Despite the clear association of NBIA disorder with iron accumulation, only *FTL* and *CP* genes are directly involved in iron metabolism. The remaining known genes encode for proteins that play a role in multiple biological processes such as lipid metabolism, coenzyme A (CoA) biosynthesis, mitochondrial function, and autophagy. Figure 7.2 shows the localization of NBIA genes in the context of lipid and iron metabolism and autophagy. Iron, mitochondrial involvement, reactive oxygen species (ROS) damage, lipid metabolism, and autophagy represent the wide spectrum of mechanisms related to the NBIA genes.

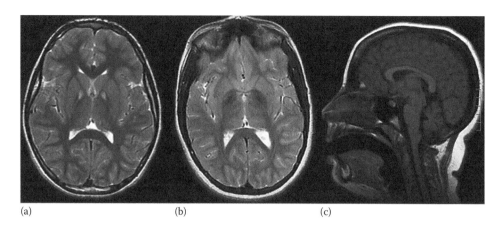

FIGURE 7.1 MRI features of PKAN and PLAN forms. (a) PANK2 mutation–positive patient shows hypointensity with a central region of hyperintensity in the globus pallidus (eye-of-the-tiger sign). (b and c) PLA2G6 mutation–positive patient shows globus pallidus hypointensity and cerebellar atrophy.

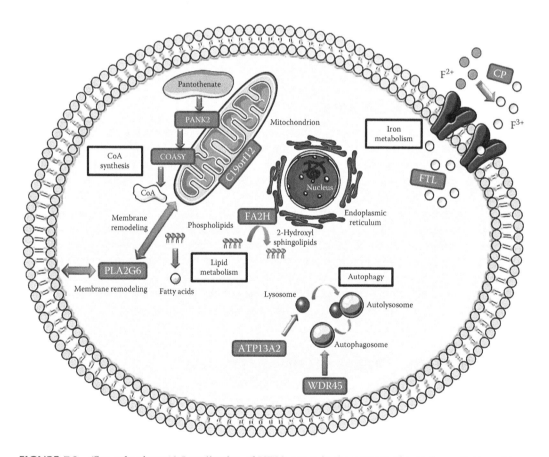

FIGURE 7.2 **(See color insert.)** Localization of NBIA genes in the context of autophagy, lipid, and iron metabolism.

Iron can exist in two redox states, and this property is useful for electron donation, which makes iron essential for mammalian cell metabolism in the brain, acting as a key factor in numerous processes, including energy production, DNA synthesis and repair, phospholipid metabolism, oxidative phosphorylation, myelin synthesis, and neurotransmitter production (Crichton et al. 2011). Iron accumulation can be destructive to the cell via formation of ROS (Crichton et al. 2002), which can be triggered by a direct reaction between ferrous iron and oxygen, and further reaction with hydrogen peroxide produced, and so generating hydroxyl radicals (·OH) that are highly reactive. Therefore, "free iron" can lead to the increase of oxidative stress and neuronal toxicity. For this reason, it is extremely important that homeostatic mechanisms of iron have to be tightly controlled in both peripheral and central nervous systems (CNSs) (Andrews and Schmidt 2007, Rouault 2013). In the CNS, extracellular iron is bound to transferrin and internalized by transferrin receptor (TFRC) through the protein channel DMT1 (divalent metal transporter 1). Ferritin probably sequesters iron in the neuron cell body and subsequently promotes storage of iron into the organelles, such as mitochondria. Ferroportin (FPN1) is responsible for iron export from the cell body, which is stabilized at the plasma membrane by the CP. Mice lacking CP show an increased lipid peroxidation in some regions of the CNS, implicating oxidative stress and damage in the brain (Patel et al. 2002). Based on other in vitro and in vivo experiments and according to the patient assessment, it seems that ROS production by high levels of iron, cellular oxidation process, and also mitochondrial dysfunction are key aceruloplasminemia-causing factors. Experiments carried out in *FTL* mutant mouse model as well as in HeLa and SH-SY5Y cells have confirmed that ferritin impairment results in iron accumulation with enhanced ROS production and increased oxidized protein levels (Cozzi et al. 2010). Additionally, an increased expression of the mitochondrial superoxide dismutase 2 (SOD2) is reported, a protein that is important for cell protection against mitochondrial superoxide (Barbeito et al. 2009). All together, these findings implicate the iron imbalance in the production of free radicals that cause cell damage and subsequent neurodegeneration.

A second aspect of some NBIA mutant conditions is the evidence of a morphology disruption of mitochondrial cristae. PLA2G6 plays a fundamental role in phospholipid homeostasis, and a loss of this protein impairs the mitochondrial membrane, suggesting that respiratory chain super structures might be disrupted as a result of lipid insufficiencies (Shinzawa et al. 2008). In fact, mitochondrial dysfunction, such as reduced mitochondrial membrane potential and swollen mitochondria with altered cristae, has been described in mouse and fruit fly *Pank2* knockout models (Bosveld et al. 2008, Brunetti et al. 2012). In PKAN and PLAN pathogenesis, degenerative and swelled mitochondria are transported through the axons and release ROS and proapoptotic factors (Seleznev et al. 2006), which progressively lead to neuronal death. In Kufor–Rakeb syndrome, the mutated form of ATP13A2 protein, a lysosomal ATPase, is retained in the endoplasmic reticulum (ER), leading to ER stress. Lysosomal vesicles with fragmented mitochondria result in an increase in mitochondria-derived ROS. Mitochondria represent a major storage of iron inside the cell, and therefore, the accumulation of this element in some NBIA forms could extremely exacerbate oxidative stress toxicity.

The catalytic activity of PANK2 is crucial for the phosphorylation of pantothenate to 4-phosphopantothenate, required for the biosynthesis of CoA, which in turn plays a role in β-oxidation, the citric acid cycle, and the metabolism of fatty acids and amino acids. It has been theorized that the enzymatic inactivation of PANK2 increases N-pantothenyl cysteine and free cysteine, which can chelate physiological iron and indirectly cause iron accumulation and ROS-derived damage (Perry et al. 1985).

Autophagy has also been involved in the pathological mechanisms of NBIA disorders, as reported by Chen et al. (2013), who showed that in rat brains, increased iron levels induce autophagy pathway activation and subsequent cell death. An important hypothesis in BPAN pathological mechanism is based on a possible disruption of autophagosome maturation, which maybe responsible for leading to the aggregation of damaged cellular components that can have toxic effects on the cell. The BPAN-causing gene *WDR45* encodes a β-propeller scaffolding protein that is involved in the autophagy pathway. Mutation in this gene results in a dramatic decrease of WDR45 protein expression, which is associated with the accumulation of aberrant early autophagic structures.

7.4   CONCLUSIONS

NBIA disorders are a group of devastating and life-threatening rare diseases that encompass a wide spectrum of adult and pediatric forms. Genetic analysis provides an accurate diagnosis and provides information useful for prognosis. Genetic spectrum of NBIA disorders is continuously expanding. Unfortunately, a considerable proportion of patients are molecularly undiagnosed (idiopathic NBIA), suggesting that other genes are still unknown. To date, the reported NBIA genes have been related to different processes including lipid and iron metabolism, autophagy, mitochondrial involvement, and ROS damage. No treatment for NBIA disorders is currently known, and management strategies focus on the medical and surgical palliation of symptoms with the aim of improving the quality of life. The research focused on the underlying disease mechanisms will provide new clues to better understand the link between mitochondrial functions, membrane homeostasis, and iron metabolism in the development of NBIA disorders and will hopefully lead to the development of novel therapeutic interventions in the future in order to improve the clinical treatment of NBIA patients.

ACKNOWLEDGMENT

This work was supported by Fundació La Marató de TV3 [Grant No. 20143130-31].

REFERENCES

Alazami, A.M., Al-Saif, A., Al-Semari, A. et al. 2008. Mutations in C2orf37, enconding a nucleolar protein, cause hypogonadism, alopecia, diabetes mellitus, mental retardation, and extrapyramidal syndrome. *Am J Hum Genet* 83: 684–691.

Al-Semari, A., Bohlega, S. 2007. Autosomal-recessive syndrome with alopecia, hypogonadism, progressive extra-pyramidal disorder, white matter disease, sensory neural deafness, diabetes mellitus, and low IGF1. *Am J Med Genet A* 143: 149–160.

Andrews, N.C., Schmidt, P.J. 2007. Iron homeostasis. *Annu Rev Physiol* 69: 69–85.

Barbeito, A.G., Garringer, H.J., Baraibar, M.A. et al. 2009. Abnormal iron metabolism and oxidative stress in mice expressing a mutant form of the ferritin light polypeptide gene. *J Neurochem* 109: 1067–1078.

Bosveld, F., Rana, A., van der Wouden, P.E. et al. 2008. De novo CoA biosynthesis is required to maintain DNA integrity during development of the *Drosophila* nervous system. *Hum Mol Genet* 17: 2058–2069.

Bras, J., Verloes, A., Scheneider, S.A., Mole, S.E., Guerreiro, R.J. 2012. Mutation of the parkinsonism gene ATP13A2 causes neuronal ceroid-lipofuscinosis. *Hum Mol Genet* 21: 2646–2650.

Brunetti, D., Dusi, S., Morbin, M. et al. 2012. Pantothenate kinase-associated neurodegeneration: Altered mitochondria membrane potential and defective respiration in Pank2 knock-out mouse model. *Hum Mol Genet* 21: 5294–5305.

Chen, G., Jing, C.H., Liu, P.P., Ruan, D., Wang, L. 2013. Induction of autophagic cell death in the rat brain-caused by iron. *Am J Med Sci* 345: 369–374.

Chinnery, P.F., Crompton, D.E., Birchall, D. et al. 2007. Clinical features and natural history of neuroferritinopathy caused by the FTL1 460InsA mutation. *Brain* 130: 110–119.

Cozzi, A., Rovelli, E., Frizzale, G. et al. 2010. Oxidative stress and cell death in cells expressing L-ferritin variants causing neuroferritinopathy. *Neurobiol Dis* 37: 77–85.

Crichton, R.R., Dexter, D.T., Ward, R.J. 2011. Brain iron metabolism and its perturbation in neurological diseases. *J Neural Transm* 118: 301–314.

Crichton, R.R., Wilmet, S., Legssyer, R., Ward, R.J. 2002. Molecular and cellular mechanisms of iron homeostasis and toxicity in mammalian cells. *J Inorg Biochem* 91: 9–18.

Curtis, A.J., Fey, C., Morris, C.M. et al. 2001. Mutation in the gene encoding ferritin light chain polypeptide causes adult onset autosomal dominant basal ganglia disease. *Nat Genet* 28: 350–354.

Dick, K.J., Eckhardt, M., Paisán-Ruiz, C. et al. 2010. Mutation of FA2H underlies a complicated form of hereditary spastic paraplegia (SPG35). *Hum Mutat* 31: 1251–1260.

Di Fonzo, A., Chien, H.F., Socal, M. et al. 2007. ATP13A2 missense mutations in juvenile parkinsonism and young onset Parkinson disease. *Neurology* 68: 1557–1562.

Dusi, S., Valletta, L., Haack, T.B. et al. 2014. Exome sequence reveals mutations in CoA synthase as a cause of neurodegeneration with brain iron accumulation. *Am J Hum Genet* 94: 11–22.

Edvardson, S., Hama, H., Shaag, A. et al. 2008. Mutations in the fatty acid 2-hydroxylase gene are associated with leukodystrophy with spastic paraparesis and dystonia. *Am J Hum Genet* 83: 643–648.

Gregory, A., Hayflick, S.J. 2011. Genetics of neurodegeneration with brain iron accumulation. *Curr Neurol Neurosci Rep* 11: 254–261.

Gregory, A., Polster, B.J., Hayflick, S.J. 2009. Clinical and genetic delineation of neurodegeneration with brain iron accumulation. *J Med Genet* 46: 73–80.

Haack, T.B., Hogarth, P., Kruer, M.C. et al. 2012. Exome sequencing reveals de novo WDR45 mutations causing a phenotypically distinct, X-linked dominant form of NBIA. *Am J Hum Genet* 91: 1144–1149.

Harris, Z.L., Takahashi, Y., Miyajima, H., Serizawa, M., MacGillivray, R.T.A., Gitlin, J.D. 1995. Aceruloplasminemia: Molecular characterization of this disorder of iron metabolism. *Proc Natl Acad Sci USA* 92: 2539–2543.

Hartig, M.B., Iuso, A., Haack, T. et al. 2011. Absence of an orphan mitochondrial protein, c19orf12, causes a distinct clinical subtype of neurodegeneration with brain iron accumulation. *Am J Hum Genet* 89: 543–550.

Hayflick, S.J., Kruer, M.C., Gregory, A. et al. 2013. β-propeller protein-associated neurodegeneration: A new X-linked dominant disorder with brain iron accumulation. *Brain* 136: 1708–1717.

Hayflick, S.J., Westaway, S.K., Levinson, B. et al. 2003. Genetic, clinical, and radiographic delineation of Hallervorden-Spatz syndrome. *N Engl J Med* 348: 33–40.

Hogarth, P., Gregory, A., Kruer, M.C. et al. 2013. New NBIA subtype: Genetic, clinical, pathologic, and radiographic features of MPAN. *Neurology* 80: 268–75.

Kono, S., Suzuki, H. Oda, T. et al. 2006. Biochemical features of ceruloplasmin gene mutations linked to aceruloplasminemia. *Neuromolecular Med* 8: 361–374.

Kruer, M.C., Paisán-Ruiz, C., Boddaert, N. et al. 2010. Defective FA2H leads to a novel form of neurodegeneration with brain iron accumulation (NBIA). *Ann Neurol* 68: 611–618.

Kurian, M.A., Morgan, N.V., MacPherson, L. et al. 2008. Phenotypic spectrum of neurodegeneration associated with mutations in the PLA2G6 gene (PLAN). *Neurology* 70: 1623–1629.

Morgan, N.V., Westaway, S.K., Morton, J.E. et al. 2006. PLA2G6, encoding a phospholipase A2, is mutated in neurodegenerative disorders with high brain iron. *Nat Genet* 38: 752–754.

Paisán-Ruiz, C., Li, A., Schneider, S.A. et al. 2012. Widespread Lewy body and tau accumulation in childhood and adult onset dystonia-parkinsonism cases with PLA2G6 mutations. *Neurobiol Aging* 33: 814–823.

Patel, B.N., Dunn, R.J., Jeong, S.Y. et al. 2002. Ceruloplasmin regulates iron levels in the CNS and prevents free radical injury. *Neuroscience* 22: 6578–6586.

Perry, T.L., Norman, M.G., Yong, V.W. et al. 1985. Hallervorden-Spatz disease: Cysteine accumulation and cysteine dioxygenase deficiency in the globuspallidus. *Ann Neurol* 18: 482–489.

Rouault, T. 2013. Iron metabolism in the CNS: Implications for neurodegenerative diseases. *Nat Rev Neurosci* 14(8): 551–564.

Schneider, S.A., Hardy, J., Bhatia, K.P. 2012. Syndromes of neurodegeneration with brain iron accumulation (NBIA): An update on clinical presentations, histological and genetic underpinnings, and treatment considerations. *Mov Disord* 27: 42–53.

Schneider, S.A., Paisán-Ruiz, C., Quinn, N.P. et al. 2010. ATP13A2 mutations (PARK9) cause neurodegeneration with brain iron accumulation. *Mov Disord* 25: 979–984.

Seleznev, K., Zhao, C., Zhang, X.H. et al. 2006. Calcium-independent phospholipase A2 localizes in and protects mitochondria during apoptotic induction by staurosporine. *J Biol Chem* 281: 22275–22288.

Shinzawa, K., Sumi, H., Ikawa, M. et al. 2008. Neuroaxonal dystrophy caused bygroup VIA phospholipase A2 deficiency in mice: A model of human neurodegenerative disease. *J Neurosci* 28: 2212–2220.

Woodhouse, N.J., Sakati, N.A. 1983. A syndrome of hypogonadism, alopecia, diabetes mellitus, mental retardation, deafness, and ECG abnormalities. *J Med Genet* 20: 216–219.

Zhou, B., Westaway, S.K., Levinson, B., Johnson, M.A., Gitschier, J., Hayflick, S.J. 2001. A novel pantothenate kinase gene (PANK2) is defective in Hallervorden-Spatz syndrome. *Nat Genet* 28: 345–349.

Section III

*Neurodegenerative Diseases
Caused by ROS*

8 Reactive Oxygen Species in Amyotrophic Lateral Sclerosis

Finley J. Allgaier and J. Jefferson P. Perry

CONTENTS

ABSTRACT

Amyotrophic lateral sclerosis (ALS) is a neurodegenerative motor neuron disease, for which there is no treatment. Amyotrophic refers to the lack (A-) of muscle (-myo-) nourishment (-trophic), resulting in the wasting of fibers, lateral refers to the lateral corticospinal tract of affected neurons between the brain and the spinal cord, and sclerosis is the resultant hardening of the tissue. ALS has also been termed "Charcot's disease" (maladie de Charcot) after the French clinician Jean Martin Charcot who first described its pathology in 1869, or in the United States, it is also known as Lou Gehrig's disease, after the New York Yankees baseball player who was afflicted by it. In Europe, it is also known as motor neuron disease, reflecting the cells chiefly affected, but this term may also refer to a wider group of motor neuron diseases in general. Despite the identification of disorder close to 150 years ago, and a modern understanding of its pathology, ALS remains an ill-defined and fatal disease with few medicinal options and a typical life expectancy of only a few years after diagnosis. New molecular-based understandings of ALS are being elucidated, and one important factor is a strong link to reactive oxygen species (ROS) in disease progression. This link between ROS and disease progression includes that mutant products of the SOD1 gene, which encodes a Cu,Zn superoxide dismutase (Cu,ZnSOD) with important functions in removing oxygen free radicals from the cell, can result in ALS. Here, we describe the disease, the links between ROS and ALS, in addition to the current understanding of how Cu,ZnSOD mutations may give rise to the disease pathology.

8.1 INTRODUCTION

Amyotrophic lateral sclerosis (ALS) is primarily characterized by the degeneration of motor neurons and atrophy of associated muscle tissues and supporting cells. This loss of neurological function is caused by the degeneration of both the upper and lower motor neurons (UMN and LMN, respectively),[1–5] while other motor neuron diseases only affect a single subgroup of neurons. The brain stem, cerebral cortex, and the spinal cord are primarily affected, so while progression of this disease is characterized by muscular failure in relation to limb movement and upper body control (lungs and throat), cognitive function is left intact in most cases.[1–6] General symptoms of the disease include, but are not limited to, muscle weakness and wasting, that is primarily in the limbs, speech difficulty, spasticity, paralysis, and eventually death typically within 3–5 years of developing the symptoms.[7,8] Lack of voluntary control over breathing, talking, and swallowing (dysphagia) develops during the middle stages of the disease, and control over these functions is severely compromised during the final stages,[9] meaning patients usually succumb to the disease because of respiratory failure.[5]

ALS develops later in life with a 1 in 400 to 1 in 1000 chance of developing symptoms at 70 years.[6] It is the most widespread type of motor neuron diseases with a probability of the disease occurring at 1–3 out of 100,000 people[3,4] and a prevalence of about 4–6 out of 100,000 people. Riluzole is the only commercially available drug approved to treat ALS, but life expectancy is only prolonged by 2–3 months[10] and treatment results in many side effects.[10,11] Therefore, there is an urgent need for advances in the diagnostics of this disease for the development of alternative treatments and for personalized medicine that targets a patient's distinctive symptoms.

The etiology of the disease has yet to be fully defined, but the roles of several genes that have been determined to be involved in the emergence and advancement of this disease are being investigated. Copper–zinc superoxide dismutase (Cu,ZnSOD) encoded by SOD1 was the first to be discovered,[12,13] and mutations of this gene are common in both forms of ALS, familial ALS (FALS) and sporadic ALS (SALS). SOD1 is involved in removing superoxide from cells. Other genes causing this disease include UBQLN2[16,17] that is involved in proteolytic functions and the open reading frame C9ORF72 that was observed to have disease-causing nucleotide repeat expansions[18–20] and is the most common mutation identified in ALS to date. This is in addition to mutations that lead to ALS can occur in the OPTN,[14] TARDBP, fused in sarcoma (FUS), and ANG[15] genes that are all involved in RNA metabolism. Elucidating how such a multifactorial disease leads to a common ALS disease phenotype remains a major challenge for the scientific community.

8.2 DIAGNOSIS OF ALS

8.2.1 FALS AND SALS

Excluding the age of onset, FALS and SALS have clinical and pathological presentations that are largely indistinguishable from each other.[4,24] SALS comprises the majority of cases, and this indicates a potential role of environmental factors, which may include heavy metals, pesticides, and/or excitotoxins.[7] However, much remains to be defined in terms of the molecular mechanisms of SALS development. A majority of FALS that occur in middle age arise from autosomal dominant transmission,[21] but the more rare juvenile cases are associated with recessive transmission.[3]

The most common form of FALS, at about 20% of cases, results from a gain-of-function mutation of the SOD1 gene,[13] whose disruption leads to neuronal death.[4,5,7,8]

8.2.2 Epidemiology of ALS, FALS, and SALS

FALS overall constitutes 5%–10% of all cases,[25] with a range of 2%–15% in different populations,[26] with regional and ethnic variations in incidence[27,28] and penetrance[29] complicating this estimation.[30] The onset of ALS is mainly age dependent, which for SALS is 58–63 years,[5] while the average for FALS is 47–52 years.[5] A mutation in SOD1 is linked to a mean earlier age of onset compared to FALS cases that do not have the mutation.[22] Juvenile cases have also been documented but are rare,[8] and younger patients (of both FALS and SALS) survive much longer than patients who develop the disease later in life. Between men and women, the average onset ratio is between 1.5:1 and 2:1, but this ratio becomes closer with an increase in age, where under the age of 40 the ratio is 3:1, but over the age of 70, this ratio becomes 1:1.[23]

8.2.3 Symptoms

Partly due to the multifactorial nature of ALS, no single reliable test for diagnosing the disease is currently available. Most cases are characterized based on a combination of symptom presentation, progression, and tests to eliminate overlapping conditions.[31] Since ALS is defined by the degradation of both UMNs and LMNs, doctors look for a combination of symptoms that affect both motor neuron groups. Symptoms associated with the degradation of UMNs include weakness, difficulty in controlling speech, overactive reflexes, spasticity, and inappropriate emotionality. LMN symptoms include weakness, and patients may also have decreased reflexes, cramps, twitching, and muscle wasting.[32,33] Spinal onset, where the disease starts affecting the limbs, is the more common form of onset, while a quarter of patients have a bulbar onset,[5] where the disease affects the facial, mouth, jaw, and tongue muscles controlled by the lower brain stem (or "bulb"). One defining trait of ALS is rapid progression, so over time most patients will eventually display features from both spinal and bulbar onsets.

Many other diseases share several clinical features with ALS and, therefore, need to be ruled out as mimics before proper diagnosis, which include spinal injuries, cervical spondylosis, metabolic problems such as enzyme/vitamin deficiency (vitamin B-12, etc.), thyroid problems, stroke, myopathies or neuropathies, inclusion body myositis, infections such as Lyme or HIV, or diseases such as myasthenia gravis, syringomyelia, cancer, Kennedy's disease, Tay–Sachs disease, or multiple sclerosis.[24,32–36,40] As such, the diagnosis of ALS is through a process of elimination, with family history proving helpful, but misdiagnosis is common[24,37] with a rate of about 10% of patients.[38,39]

8.2.4 Pathophysiology

Postmortem examinations of patient brain and spinal cord sections have shown that neuronal atrophy and the presence of cellular inclusions are a shared trait among patients suffering from ALS. These inclusions include cystatin C and transferrin-immunoreactive Bunina bodies,[41] as well as skein-like or round Lewy body–like hyaline inclusions from ubiquitination,[42] which implies that defects in the ubiquitin proteasome system may be a more generalized feature of ALS. Irregular glutamate metabolism is also highly associated with ALS,[16] which is targeted by the drug riluzole,[43] because elevated synaptic glutamate can overstimulate glutamate receptors and cause nerve damage and death through excitotoxicity. An increase in p53-mediated apoptosis, impaired axonal transport, and cytoskeletal and mitochondrial dysfunction[44–47] are also associated with progression of the disease. The hallmark of ALS is rapid disease progression especially later in life, and thus, cumulative damage occurring through increased levels of oxidative stress may be a significant contributor to the disease.[48] Interestingly, a recent study analyzing the cerebrospinal fluid of ALS patients revealed that there were significant differences in the metabolite profiles of patients with FALS (non-SOD1), SALS, and patients carrying a mutation in the SOD1 gene. This indicated that the neurodegenerative disease process in these distinct subtypes could be at least partially dissimilar.[49]

8.3 ROS AND SOD

8.3.1 REACTIVE OXYGEN SPECIES

Reactive oxygen species (ROS) are a class of oxygen-containing molecules that are either chemically reactive themselves or are precursors of other chemically reactive species. They can be produced by both exogenous sources (e.g., pollutants, smoke, and radiation) and endogenous sources (e.g., aerobic metabolism and oxidative phosphorylation). An imbalance of ROS may cause significant damage to the cell and DNA, which if not dealt with by repair mechanisms can lead to lethal mutations and genomic instability, as well as inflammation and DNA damage. Aging, cancer, cardiovascular disease, hereditary diseases, and neurodegenerative diseases have all been linked to the deleterious effects of ROS. Proteins that control cellular levels of ROS and DNA damage responses are essential for the cell to operate correctly and especially for the nervous system, which is particularly vulnerable. Examples of ROS are superoxide radicals ($O_2^{\bullet-}$), hydroxyl radicals ($^{\bullet}OH$), hydroxyl ions (OH^-), and hydrogen peroxide (H_2O_2).

8.3.2 SUPEROXIDE DISMUTASE

Superoxide dismutases (SODs) are a class of enzyme that catalyzes the disproportionation (a redox reaction) of $O_2^{\bullet-}$ to molecular oxygen and H_2O_2 at very high enzyme rates. SODs are critical for regulating cellular ROS levels and may have potential uses in therapeutics for treating oxidative stress–related diseases. There are three distinct classes of SOD that have discrete protein folds and catalytic metal ions: the Cu,ZnSODs, MnSOD/FeSODs, and NiSODs.[50] Cu,ZnSODs, with SOD1, present in the cytoplasm, mitochondrial intermembrane space, and extracellular SOD3 have been observed in eukaryotes and certain prokaryotes. Point mutations in human Cu,ZnSOD (SOD1 and HsCu,ZnSOD) may cause ALS. FeSOD and MnSOD (SOD2 in humans) have a common ancestral gene but have diverged significantly enough to the point where they cannot functionally substitute for each other.[51] The FeSOD gene has been observed in primitive eukaryotes, the plastids of plants, and in bacteria, and MnSOD occurs in all the major domains of life.[51] NiSOD is the newest discovered class but has only been observed in bacteria.

8.4 Cu,ZnSOD (SOD1)

8.4.1 STRUCTURE

The structure[62] of human Cu,ZnSOD (SOD1, PDB code: 1PU0) reveals that the enzyme's fold and domain organization are highly conserved in eukaryotes from the primary to quaternary structure[51,52] and that this structure allows for high stability and optimally fast catalysis. Human Cu,ZnSOD is an extremely stable homodimer that is composed of two identical subunits related by a twofold symmetry (Figure 8.1). Each of the two subunits consists of a Greek key β-barrel composed of eight antiparallel β-strands with a core tightly packed with hydrophobic residues.[52,53] The active site channel is formed by two major external loops, β4/β5 and β7/β8. The β4/β5 loop confers stability to the structure by tethering the dimer interface with the active site zinc with the help of an intervening disulfide bond, which stabilizes both the subunit fold and the dimer interface.[50,51,53–55] The β7/β8 loop guides and accelerates the substrate into the active site.[51,54] Overall, the stable Greek key scaffold supports elements for electrostatic guidance, dimer formation, and active site metallochemistry, where steric exclusion allows superoxide but not larger anions into the active site. The Cu

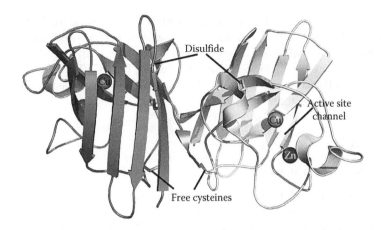

FIGURE 8.1 (See color insert.) The human Cu,ZnSOD crystal structure (PDB ID 1PU0). hCu,ZnSOD is a homodimer (green and cyan), with each subunit consisting of a Greek key motif that is a barrel-like domain flanked by extended loops that form the active site cleft containing the Cu and Zn ions (large spheres). A conserved disulfide bond in each subunit provides structural rigidity, while the free cysteine residues are implicated in irreversible unfolding.

and Zn sites tie structural elements together and provide catalytic roles to the protein. Their respective sites are positioned outside the β-barrel in the active site channel where hydrophobic anchors tie the Cu ion to the β-barrel. Each active site contains one Cu ion ligated to three histidine residues and one Zn ion also ligated to three histidine residues and to one aspartic acid residue. When in the oxidized state, a bridging histidine links the Zn and Cu ions. The Cu is exposed and coordinated to a water molecule, and the Cu ion can be found occupying two positions in a single crystal, which may be important mechanistically.[56-59]

8.4.2 Mechanism

Around 20% of FALS cases can be linked to mutations in the SOD1 gene, which are usually single amino acid substitutions and are found widely dispersed throughout the polypeptide. In earlier studies, it was not clear how all these different mutation sites could give rise to the same FALS pathology. Further clarification was provided by combined structural, biochemical, and biophysical characterizations[68] of two FALS mutant Cu,ZnSOD proteins, H43R (which disrupts the protein's hydrophobic packing) and A4V (which disrupts the dimer interface).[67] Point mutations cause an overall architectural destabilization of these proteins to promote the formation of fibril-like aggregates,[68] the severity of which is likely enhanced by the presence of free cysteine residues. The formation of disulfide bonds by these residues covalently locks the fibrous aggregates, which have been found to resemble those found in postmortem studies of FALS patients and have an amyloid-like structure.

Evidence for a molecular mechanism for initiating FALS toxicity has been elucidated from structural biochemistry data. Protein misfolding or impaired protein degradation may result in framework destabilization and loss of dimer assembly, which promotes fibril formation, and therefore, the accumulation of toxic intracellular human Cu,ZnSOD aggregates in the brain tissue. These cytoplasmic aggregates, which are common in many neurodegenerative diseases, are strongly immunoreactive to human Cu,ZnSOD antibodies and cannot be dissociated with detergents or reducing agents.

In order for Cu,ZnSOD to perform its primary function of catalyzing the disproportionation reaction, the enzyme must have extreme specificity for superoxide instead of other molecules that are very close in size (molecular oxygen, nitric oxide). After the substrate is guided into the active site by electrostatic recognition and positively charged residues, for example, human Cu,ZnSOD

Lys136 and Arg143, the first half-reaction begins with the O_2^- binding to Cu(II). Cu(II) is then reduced to Cu(I) and O_2^- is oxidized to molecular oxygen. The bond between the Cu ion and bridging histidine (human Cu,ZnSOD His63) is broken, which leaves His63 Nε1 protonated. In the second half-reaction, a proton from this Nε1 and an electron from Cu(I) are donated to $O_2^{\cdot-}$. Cu(I) is then oxidized to Cu(II) and $O_2^{\cdot-}$ is reduced to H_2O_2 or a HO^-, and the bond connecting the Cu ion to the bridging histidine is restored.

8.5 ROS IN ALS

8.5.1 DYING-BACK HYPOTHESIS

ROS have been found to inhibit transmitter release and neuromuscular junction (NMJ) function by inducing oxidative stress, thus hindering synaptic transmission. An increase in ROS levels in the presence of already established and elevated levels of oxidative stress has shown that oxidative damage could start in peripheral tissues and proceed backward to neurons.[70] Initial damage to the NMJ has been found to occur without altering their fundamental morphology. Oxidative stress, dysfunctional mitochondria, and increased intracellular Ca^{2+} augment presynaptic decline in NMJ and are necessary for the underlying mechanisms of ALS.[71–74] During the later stage of ALS, the main damage results from the accumulation of ROS and inflammatory factors as well as an absence of neurotrophic factors. Oxidative stress can be produced in various ways: by ROS, abnormal metabolism, and by ALS-linked proteins such as mutant TAR DNA-binding protein 43 (TDP-43).[75] This stress promotes tissue damage by escalating other events that promote motor neuron degeneration (e.g., inflammation, proinflammatory cytokines, and enzymes such as interleukin-6, monocyte chemoattractant protein-1, interleukin-8, and cyclooxygenase-2).

In SALS and mouse mutant SOD1 (mSOD1) cases, motor neuron degeneration has been found to start at the synaptic end and moves toward soma in a retrograde dying-back fashion.[60,61] Dysfunction of neuromuscular transmission has been traced back to the NMJ, where the presynaptic machinery is affected first. During the early symptomatic stages, only a few synapses show compromised function, and presumably healthy synapses are able to compensate for the damaged ones. As the disease progresses and more synapses deteriorate, different rates of degeneration have been observed. Some axon branches sprout and regenerate to compensate for lost synapses and therefore could be manipulated to maintain innervation if taken advantage before the disease progresses to a point where the functional synapses can no longer mediate synaptic transmission.[69] Mouse models[76] have shown that motor neurons controlling the hind limbs are affected long before symptoms appear, while slow motor neurons are more resistant (and compensate for those that are not). This means that ALS potentially involves selective degeneration of NMJs by differentiating between physiological subtypes of axons.

8.5.2 CU,ZNSOD MUTATIONS AND ALS

Cu,ZnSOD accounts for 1% of the total cytosolic protein content in neurons, and mutations in this enzyme are known to cause about 20% of FALS cases. Remarkably, over 150 single-residue mutations at N70 sites within Cu,ZnSOD occur in this protein of 154 amino acids in total.[64–66,79] The underlying molecular mechanism unifying these mutations is still under debate. The majority of inherited SOD1 mutations are dominant, and two copies of a mutation may cause a much earlier onset.[80,81] The exception to this is the relatively common D90A SOD1 mutation, which mostly causes ALS as a recessive trait, in addition to appearing sporadically.[77,78]

Detailed structural and biochemical analyses of Cu,ZnSOD structures and enzymatic mechanisms[62,82] have included comparisons to the bacterial form of Cu,ZnSOD[83] in addition to SOD2, the human mitochondrial MnSOD.[84–86] These studies have provided basis to evaluate the diverse mutations occurring across the Cu,ZnSOD architecture,[12,50] leading to a hypothesis that accounts for all mutations, which is "framework destabilization."[50,56,68] Here, each of the diverse set of mutations can result in local unfolding events that contribute to a globally defective, self-aggregating protein. Cu,ZnSOD can form amyloid-like fibers under certain conditions, and ALS mutations can accelerate this process.[68,147,148] The aggregating forms of Cu,ZnSOD may deleteriously co-aggregate with other key cellular proteins.[68,87–89] These aggregates have been shown to be immunoreactive to Cu,ZnSOD antibodies and moreover are very stable in the presence of either detergents or reducing agents.

In FALS, mutant Cu,ZnSOD was observed to co-precipitate with the copper chaperone for SOD (CCS),[90] nitric oxide synthase (NOS), and phosphorylated neurofilaments.[91] Defects in mitochondrial transport or mitochondrial damage may have deleterious effects in FALS, and Cu,ZnSOD has been observed in purified human mitochondria. The formation of Cu,ZnSOD-mediated protein aggregates could promote mitochondrial defects that drive apoptosis via caspase activation,[92] which would be consistent with motor neuron pathophysiology in FALS.[93–97] Several research groups have attempted to define a correlation between mutant protein stability and the clinical phenotype by conducting studies on mutant forms of Cu,ZnSOD both *in vitro* and in cultured cells.[98–101] However, no clear correlation was observed, and this could be due to potential undefined roles of the metals ions in Cu,ZnSOD structural stability,[102] aberrant oxidative modifications of the free cysteine residues within the protein,[99,103] or undefined interactions of mutant Cu,ZnSOD with distinct cellular components. Such cellular components could include known interactors having important roles in the stress responses, for example, Derlin-1, Rac-1, Hsc70, and CCS involved in protein folding/maturation,[90,104–106] in addition to the vesicular transport–associated proteins, for example, chromogranin and the dynein heavy chain.[107–109]

These toxic effects of mutant Cu,ZnSOD protein may impact multiple cell types (e.g., neurons and glia) and are additive.[110] Mutant Cu,ZnSOD also has pleiotropic effects in cells, as the endoplasmic reticulum,[111] mitochondria, and peroxisomes[112] in motor neurons show characteristic anomalies. It also modifies intracellular signaling,[113] activates apoptosis pathways,[114,115] promotes irregular free radical production,[116] affects the integrity of the neuronal cytoskeleton, and decreases levels of crucial receptors (e.g., Glur2).[117] Mutant Cu,ZnSOD can also be secreted outside cells, which can induce neuroinflammatory responses.[107]

8.5.3 Cu,ZnSOD and Mitochondria

Mitochondria generate ROS that can result in oxidative damage to mitochondrial DNA and proteins. ROS is produced on both sides of the inner membrane, but the accumulation of ROS in the matrix is limited by antioxidant machinery, which includes MnSOD (SOD2). Interestingly, the deacetylase activity of SIRT3 regulates the activity of MnSOD as MnSOD is acetylated on residue Lys68, and this posttranslational modification decreases its activity. Past work has found that a decrease in SIRT3 may be essential for metabolic reprogramming and a shift to glycolysis, characteristics of cancer cells.[125,126] The antioxidant machinery of the inner membrane space (IMS) is limited compared to that of the matrix, and therefore, it relies on Cu,ZnSOD activity to prevent irreversible oxidative damage. The accumulation of ROS in the IMS increases stress on the mitochondria and may result in its collapse, and therefore, any decreases in the expression or activity of SIRT3 result in an increased demand for Cu,ZnSOD.

Past studies have discovered that misfolded Cu,ZnSOD can localize to both the outer[117] and IMSs of the mitochondria. Accumulation on the outer membrane can cause damaging interactions with several critical cellular proteins including Bcl2[118] and VDAC.[119] Alternately, if it is localized to the IMS, it can accumulate, misfold, and theoretically interfere with proteins in the mitochondria.[120–122] Effects on the mitochondria are supported by evidence from cultured motor neuron cells, where mutant Cu,ZnSOD causes anomalies in mitochondrial function, morphology, and axonal transport.[107,123] Also, transgenic mice expressing the SOD1 G93A mutation in the IMS (and not in cytoplasm) were found to develop some symptoms of ALS (e.g., motor defects and mitochondrial irregularities).[124] Additionally, Cu,ZnSOD knockout mice affected by peripheral neuropathy can be rescued by targeting wild-type Cu,ZnSOD to the IMS.

8.6   TREATMENT OF ALS

Current treatments of ALS are largely focused on palliative care, particularly increasing the quality of life for the patient, especially during the ending period of life and helping the patient to adjust to lifestyle limitations. These include but are not limited to consuming food that is easy to swallow, administering medications for muscle spasticity, fatigue, sleep, and depression, as well as installing noninvasive ventilation that can extend survival in nonbulbar patients.[127]

Research efforts are focused on the inhibition of disease progression and treatment of damage present. Biochemical and pharmacological advancements are providing new better understandings of the mechanistic nature of the disease, potentially opening new therapeutic regimes to better treat patients. Along these lines, creating a data bank of ALS protein biomarkers (e.g., TDP-43 and cystatin C) from noninvasive analyses would benefit not only the diagnosis and monitoring progression but also in identifying biological pathways that could be targeted therapeutically. The prognosis of the patient could potentially be improved by discovering better disease markers, which could reduce the duration between initial symptom presentation and diagnosis that now averages 14 months.[37] Thus, at the time of diagnosis, ALS has already progressed significantly, with multiple complications accelerating motor neuron death likely beyond any regenerative mechanisms or capacity.

8.6.1   PHARMACOLOGICAL INTERVENTIONS

Riluzole, an anti-excitotoxicity drug, is the only currently approved treatment for ALS.[10] It preserves motor neuron function by decreasing toxic glutamate levels at glutamatergic nerve terminals by inactivating sodium channels, inhibiting glutamate release, and blocking postsynaptic actions of NMDA receptors.[128] With this treatment, patients typically live for 2–3 additional months and without substantial improvements to muscle strength and limb and bulbar function. To combat this problem, the dual-acting drug dextromethorphan/quinine[129] has been developed to treat the pseudobulbar effects of uncontrolled laughing or crying. Dextromethorphan also inhibits glutamatergic signaling, where excessive release of glutamate is considered to be an underlying cause of neuronal damage, while quinine helps to increase its dextromethorphan bioavailability, and together they have been found to provide moderate benefit to a small group of patients.

Other new drugs are currently in development, such as the antibiotic ceftriaxone that upregulates the GLT-1 glutamate transporter, where chronic loss of glutamate transport can produce a loss of motor neurons, and in ALS GLT-1, loss can be dramatic. Other drugs include a high dose of methylcobalamin vitamin B-12 that reduces homocysteine-mediated toxicity in NSC-34 cells and the antioxidant dexpramipexole that targets the mitochondria and has shown to slow down the function decline and improve survivability.

8.6.2   Cu,ZnSOD-TARGETING THERAPIES[79]

While there are several different mutant transgenic mice models (e.g., TARDBP) available, the use of Cu,ZnSOD models has predominated, likely because SOD1 gene was the first to be identified causing the disease. Studies in mice revealed that the SOD1 gene is expendable but is required under physiologically stressful conditions, such as following neuron loss after axonal injury.[130] Thus, both reducing Cu,ZnSOD expression and disturbing aggregation are favored strategies. Astrocyte cell lines derived from patient progenitor cells were found to be toxic to motor neurons, through a mechanism involving the secretion of uncharacterized factors. Both FALS- and SALS-derived cells, but not non-ALS-derived astrocyte cells, had common pathway changes (NF-κB, MAPK, JNK, and AKT), and knocking down of SOD1 rescued the motor neuron killing phenotype in four of six cell lines examined.[144] This study reaffirms the use of SOD1-targeted therapeutics in the context of SALS, and it also suggests that such cell cultures could prove useful models for therapeutic screening in the absence of an all-encompassing ALS disease model.

Many small molecules and siRNAs are being used to downregulate and reduce SOD levels. siRNA-based strategies have shown to be somewhat successful, but its delivery is a problem as siRNA cannot pass the blood–brain barrier. The development of CSF-infused delivery method for Isis-SOD1RX antisense oligos looks promising. As for small molecule therapies, the hydroxylamine drug arimoclomol (Orphazyme) induces the heat shock response that results in a decrease in ubiquitin-positive aggregates in G93A SOD1 mouse models and is now being tested in SOD1 FALS patients. Edaravone, a free radical scavenger, was recently found to ameliorate ALS symptoms and diminish SOD aggregate deposition in interior horn cells. Also, immunization through vaccination with mutant Cu,ZnSOD or metal-free Cu,ZnSOD may be a promising approach aimed at prevention.

8.6.3 Melatonin, Mitochondria, and ALS[131]

Disruption of mitochondrial function can produce a host of deleterious effects, affecting a number of metabolic pathways in the cell, including the Krebs cycle, cholesterol and lipid synthesis, and fatty acid oxidation, as well as ATP synthesis. Deleterious effects may include the damaging overproduction of ROS, reduced cellular energy, and potentially apoptosis. In generating ATP, the respiratory chain can produce ROS, particularly $O_2{}^{\bullet-}$, and thus, antioxidant defense mechanisms are in play to limit potential cellular damage. These defenses include oxidizable molecules, nonenzymatic proteins, and antioxidant enzymes that include the SODs. An increase in net ROS production may result in one of several diseases and/or rapid aging and which may include the major neurodegenerative diseases, for example, Alzheimer's, Parkinson's and Huttington's disease (HD) in addition to ALS, all of which have been linked to mitochondrial dysfunction.

Since oxidative stress is regarded as a key contributor to ALS progress, the identification and development of antioxidant molecules to combat this disease are of significant interest. Interestingly, one of the oxidizable molecules that helps protect against ROS is melatonin, which is a regulator of mitochondrial bioenergetics function, as unlike other antioxidants, it is selectively taken up by mitochondrial membranes.[131] Melatonin can inhibit nerve degeneration and death of motor neurons, through inhibiting key pathways that include the Rip2/caspase-1 pathway, caspase-3 overexpression and activation and blocking the release of mitochondrial cytochrome c. Melatonin crosses physiological barriers due to its ampipathic nature.[133] Melatonin has direct antioxidant effects and indirect antioxidant activites,[134,135] including supporting SOD, glutathione peroxidase, and glutathione reductase activities.[136–138] This is in addition to stimulating glutathione production,[139] promoting conversion of oxidized glutathione to the reduced form, and activating DNA repair enzymes.[140–143] Support for potential therapeutic roles of melatonin in treating ALS includes that orally administered melatonin delayed ALS progression and extended survival in mice; also, the premature death and hind limb tremor onset were observed to be delayed by 25%. Moreover, in humans, an overall reduction in serum protein carbonyls was observed, which provide a surrogate marker for oxidative stress, and serum protein carbonyls were originally elevated in ALS patients and were then normalized to control values only after four months of melatonin treatment. Overall, this would indicate that combining melatonin to current treatments, such as riluzole, could potentially improve patient conditions.[132]

8.7 NON–Cu,ZnSOD ALS

Proteins involved in RNA metabolism, vesicle trafficking, or proteasomal functions[144] have been observed to cause ALS when mutated, and they show oxidative stress and mitochondrial dysfunction, similar to that found in SOD1 mutations. Oxidative stress appears to play key roles in disease progression as supported by aberrant oxidative stress biomarkers occurring in SALS subjects.[145] The antioxidant glutathione (GSH) is lower in the motor cortex of ALS patients,[146] and in SOD1 mutant mice, levels of GSH are lower, which appears to promote mitochondrial pathology and neuron decline.[147] The increased levels of oxidative stress in neurons can directly damage proteins, resulting in the accumulation of protein inclusions, which are prominent in mSOD1-mediated ALS. A decrease in GSH level was observed to aggravate mSOD1 insolubility, while a significant increase in GSH rescued mSOD1 solubility.[148]

Mutations in the RNA metabolism protein, TDP-43, can cause ALS, and depletion of GSH in neuronal cultures stimulates the formation of TDP-43 inclusions.[149] Oxidative stress appears important for TDP-43-triggered cell death, as the expression of TDP-43 in motor neurons induces oxidative stress, mitochondrial damage, and nuclear accumulation of the nuclear protein E2–related factor 2 Nrf2[75] that is an indicator and modulator of oxidative stress. Additionally, the expression of TDP-43 in yeast has been shown to result in increased oxidative stress, apoptosis, and necrosis.[150] Wild-type and mutant TDP-43 aggregation is driven by improper disulfide bond formation in the TDP-43 RNA recognition motif.[151,152] TDP-43 aggregation occurs with the aid of stress granules that are formed under oxidative stress, and these stress granules contain RNA helicases, RNA nucleases, and RNA-binding proteins,[153] as well as mRNAs that can induce cell dysfunction or act as nucleation sites for larger protein aggregates.[154]

In addition to oxidative stress, the accumulation of aggregated SOD1 in the mitochondria and misfolding of SOD1 with mitochondrial proteins causes damage to mitochondria.[122,155] Such mitochondrial damage occurs in other ALS-related proteins including TDP-43 and the RNA-binding protein FUS/translocated in sarcoma (TLS) both of which can aggregate and associate with mitochondria, in addition to valosin-containing protein (VCP) that can affect the mitochondrial function. FUS accumulation leads to defects in the energy metabolism and protein degradation pathways by directly interacting with their key regulators.[155,156] VCP functions as a ubiquitin segregase and mutant VCP results in mitochondrial uncoupling, and the resulting decrease in mitochondrial membrane potential causes an increase in O_2 consumption and reduced ATP synthesis.[157] VCP mutations account for about 2% of FALS cases, where slow motor neuron loss is accompanied by TDP-43 accumulation and aggregation.[158,159]

The expression of wild-type or mutant TDP-43 in the motor neuron–like NSC34 cell line results in mitochondrial damage.[75,160] Also, TDP-43 was observed to disrupt the interaction between endoplasmic reticulum and mitochondria, which has important functions in the cell, as well as cellular Ca^{2+} homeostasis.[161] Multiple studies in mice have demonstrated the roles of TDP-43 in mitochondrial damage; in one study in transgenic mice, wild-type human TDP-43 expression was observed to occur, equivalent to an ALS-like pathology, that is, mitochondrial aggregation, motor deficits, and early death.[162] Here, the moderate overexpression of hTDP-43 resulted in TDP-43 truncation, an increase in both cytoplasmic and nuclear ubiquitin levels, and intranuclear and cytoplasmic aggregates that were immunopositive for phosphorylated TDP-43.[162] A separate study conducted on hTDP-43 expression noted that the mouse motor neurons have abnormal inclusions that are TDP-43 positive and also contain FUS/TLS. There was also massive accumulation of mitochondria in TDP-43-negative cytoplasmic inclusions in motor neurons and a lack of mitochondria in motor axon terminals, in addition to immature NMJs.[163] Mice expressing mutant TDP-43 have also been noted to have abnormal mitochondrial transport and morphology.[164] While in heterozygous TDP-43 mice, there was a decrease in the expression of Parkin and the fatty acid transporter CD36, along with an increase in fatty acids, HDL cholesterol, and glucose in the blood. This was in addition to abnormal neuronal mitochondrial cristae formation, a reduction in motor neurons by 90% that only caused a slight motoric impairment, suggesting that these heterozygous mice may be useful as a predisease model of ALS.[165]

8.8   CONCLUSIONS

Research efforts have demonstrated the central aspects of ROS in promoting the ALS clinical phenotype. As ALS is currently an incurable disease and significant research efforts are being undertaken to gain further understanding of the molecular genetics behind the disease, the ROS-linked mechanisms lead to aberrant cellular physiology and toxic inclusions. At present, therapeutic strategies aim to slow down the pace of the disease. Ultimately, however, future efforts will work with the aid of development of earlier diagnoses to inhibit disease progression potentially by ROS inhibitors and also to block the initial events leading to neuronal death to prevent damage to the patient's motor ability before it happens.

REFERENCES

1. Shukla, V., Mishra, S.K., and Pant, H.C. 2011. Oxidative stress in neurodegeneration. *Adv. Pharmacol. Sci.* 2011, No. 572634. 1–13.
2. Cui, H., Kong, Y., and Zhang, H. 2012. Oxidative stress, mitochondrial dysfunction, and aging. *J. Signal Transduct.* 2012, No. 646354. 1–13.
3. Ferraiuolo, L., Kirby, J., Greierson, A.J., Sendtner, M., and Shaw, P.J. 2011. Molecular pathways of motor neuron injury in amyotrophic lateral sclerosis. *Nat. Rev. Neurol.* 7:616–630.
4. Contestabile, A. 2011. Amyotrophic lateral sclerosis: From research to therapeutic attempts and therapeutic perspectives. *Curr. Med. Chem.* 18:5655–5665.
5. Kiernan, M.C., Vucic, S., Cheah, B.C., Turner, M.R., Eisen, A., Hardiman, O., Burrell, J.R., and Zoing, M.C. 2011. Amyotrophic lateral sclerosis. *Lancet* 377:942–955.
6. Wijesekera, L.C. and Leigh, P.N. 2009. Amyotrophic lateral sclerosis. *Orphanet J. Rare Dis.* 4:3.
7. Joyce, P., Fratta, P., Fisher, E.M., and Acevedo-Arozena, A. 2011. SODI and TDP-43 animal models of amyotrophic lateral aclerosis: Recent advances in understanding disease toward the development of clinical treatments. *Mamm. Genome* 22:420–448.
8. Swarup, V. and Julien, J.P. 2011. ALS pathogenesis: Recent insights from genetics and mouse models. *Prog. Neuropsychopharmacol. Biol. Psychiatry* 35:363–369.
9. ALS: Amyotrophic Lateral Sclerosis. https://www.mda.org/disease/amyotrophic-lateral-sclerosis/signs-and-symptoms/stages-of-als, accessed January 1, 2016.
10. Bensimon, G., Lacomblez, L., and Meininger, V. 1994. ALS/Riluzole study group. A controlled trial of riluzole in amyotrophic lateral sclerosis. *N. Engl. J. Med.* 330:585–591.
11. Kiernan, M.C. 2005. Riluzole: A glimmer of hope in the treatment of motor neurone disease. *Med. J. Aust.* 182:319–320.
12. Deng, H.X., Hentati, A., Tainer, J.A., Iqbal, Z., Cayabyab, A., Hung, W.Y., Getzoff, E.D. et al. 1993. Amyotrophic lateral sclerosis and structural defects in Cu,Zn superoxide dismutase. *Science* 261:1047–1051.
13. Rosen, D.R., Siddique, T., Patterson, D., Figlewicz, D.A., Sapp, P., Hentati, A., Donaldson, D. et al. 1993. Mutations in Cu/Zn super-oxide dismutase gene are associated with familial amyotrophic lateral sclerosis. *Nature* 362:59–62.
14. Del Bo, R., Tiloca, C., Pensato, V., Corrado, L., Ratti, A., Ticozzi, N., Corti, S. et al. 2011. Novel optineurin mutations in patients with familial and sporadic amyotrophic lateral sclerosis. *J. Neurol. Neurosurg. Psychiatry* 82:1239–1243.
15. Strong, M.J. 2010. The evidence for altered RNA metabolism in amyotrophic lateral sclerosis (ALS). *J. Neurol. Sci.* 288:1–12.
16. Deng, H.X., Chen, W., Hong, S.T., Boycott, K.M., Gorrie, G.H., Siddique, N., Yang, Y. et al. 2011. Mutations in UBQLN2 cause dominant X-linked juvenile and adult-onset ALS and ALS/dementia. *Nature* 477:211–215.
17. Daoud, H. and Rouleau, G.A. 2011. A role for ubiquilin 2 mutations in neurodegeneration. *Nat. Rev. Neurol.* 7:599–600.
18. DeJesus-Hernandez, M., Mackenzie, I.R., Boeve, B.F., Boxer, A.L., Baker, M., Rutherford, N.J., Nicholson, A.M. et al. 2011. Expanded GGGGCC hexanucleotide repeat in noncoding region of C9ORF72 causes chromosome 9p-linked FTD and ALS. *Neuron* 72:245–256.
19. Renton, A.E., Majounie, E., Waite, A., Simón-Sánchez, J., Rollinson, S., Gibbs, J.R., Schymick, J.C. et al. 2011. A hexanucleotide repeat expansion in C9ORF72 is the cause of chromosome 9p21-Linked ALS-FTD. *Neuron* 72:245–268.
20. Wood, H. A hexanucleotide repeat expansion in C9ORF72 links amyotrophic lateral sclerosis and frontotemporal dementia. *Nat. Rev. Neurol.* 7:595.
21. Strong, M.J., Hudson, A.J., and Alvord, W.G. 1991. Familial amyotrophic lateral sclerosis; 1850–1989. *Can. J. Neurol. Sci.* 18:45–58.
22. Cudkowicz, M.E. McKenna-Yasek, D., Sapp, P.E., Chin, W., Geller, B., Hayden, D.L., Schoenfeld, D.A., Hosler, B.A., Horvitz, H.R., and Brown, R.H. 1997. Epidemiology of mutations in superoxide dismutase in amyotrophic lateral aclerosis. *Ann. Neurol.* 41:210–221.

23. Eisen, A., Schulzer, M., MacNeil, M., Pant, B., and Mak, E. 1993. Duration of amyotrophic lateral sclerosis is age dependent. *Muscle Nerve* 16:27–32.
24. Traynor, B.J., Codd, M.B., Corr, B., Forde, C., Frost, E., and Hardiman, O. 2000. Amyotrophic lateral sclerosis mimic syndromes: A population-based study. *Arch. Neurol.* 57:109–113.
25. Byrne, S.C. and Hardiman, O. 2010. Rate of familial amyotrophic lateral sclerosis: A systematic review and meta-analysis. *Neurology* 74:A56.
26. Conwi, R.A. 2006. Preventing familial ALS: A clinical trial may be feasible but is an efficacy trial warranted? *J. Neurol. Sci.* 251:1–2.
27. Haberlandt, W.F. 1959. Genetic aspects of amyotrophic lateral sclerosis and progressive bulbar paralysis. *Acta Genet. Med. Gemellol. (Roma)* 8:369–374.
28. Murros, K. and Fogelholm, R. 1983. Amyotrophic lateral sclerosis in Middle-Finland: An epidemiological study. *Acta Neurol. Scand.* 67:41–47.
29. Williams, D.B., Floate, D.A., and Leicester, J. 1988. Familial motor neuron disease: Differing penetrance in large pedigrees. *J. Neurol. Sci.* 86:215–230.
30. Andersen, P.M. and Al-Chalabi, A. 2011. Clinical genetics of amyotrophic lateral sclerosis: What do we really know? *Nat. Rev. Neurol.* 7:603–615.
31. Hardiman, O., van den Berg, L.H., and Kiernan, M.C. 2011. Clinical diagnosis and management of amyotrophic lateral sclerosis. *Nat. Rev. Neurol.* 7:639–649.
32. Lomen-Hoerth, C. 2008. Amyotrophic lateral sclerosis from bench to bedside. *Semin. Neurol.* 28:205–211.
33. Cristini, J. 2006. Misdiagnosis and missed diagnoses in patients with ALS. *JAAPA* 19:29–35.
34. Dabby, R., Lange, D.J., Trojaborg, W., Hays, A.P., Lovelace, R.E., Brannagan, T.H., and Rowland, L.P. 2001. Inclusion body myositis mimicking motor neuron disease. *Arch. Neurol.* 58:1253–1256.
35. Weihl, C.C. and Lopate, G. 2006. Motor neuron disease associated with copper deficiency. *Muscle Nerve* 34:789–793.
36. Silani, V., Messina, S., Poletti, B., Morelli, C., Doretti, A., Ticozzi, N., and Maderna, L. 2011. The diagnosis of amyotrophic lateral sclerosis in 2010. *Arch. Ital. Biol.* 149:5–27.
37. Chio, A. 1999. ISIS survey: An international study on the diagnostic process and its implications in amyotrophic lateral sclerosis. *J. Neurol.* 246:III1–III5.
38. Davenport, R.J., Swingler, R.J., Chancellor, A.M., and Warlow, C.P. 1996. Avoiding false positive diagnoses of motor neuron disease: Lessons from the Scottish Motor Neuron Disease Register. *J. Neurol. Neurosurg. Psychiatry* 60:147–151.
39. Ludolph, A.C. and Knirsch, U. 1999. Problems and pitfalls in the diagnosis of ALS. *J. Neurol. Sci.* 165:S14–S20.
40. Sathasivam, S. 2010. Motor neurondisease: Clinical features, diagnosis, diagnostic pitfalls and prognostic markers. *Singapore Med. J.* 51:367–373.
41. Okamoto, K., Mizuno, Y., and Fujita, Y. 2008. Bunina bodies in amyotrophic lateral sclerosis. *Neuropathology* 28:109–115.
42. Leigh, P.N., Whitwell, H., Garofalo, O., Buller, J., Swash, M., Martin, J.E., Gallo, J.M., Weller, R.O., and Anderton, B.H. 1991. Ubiquitin-immunoreactive intraneuronal inclusions in amyotrophic lateral sclerosis. Morphology, distribution, and specificity. *Brain* 114: 775–788.
43. Plaitakis, A. and Caroscio, J.T. 1987. Abnormal glutamate metabolism in amyotrophic lateral sclerosis. *Ann. Neurol.* 22:575–579.
44. Barbosa, L.F., Cerqueira, F.M., Macedo, A.F., Garcia, C.C., Angeli, J.P., Schumacher, R.I., Sogayar, M.C. et al. 2010. Increased SOD1 association with chromatin, DNA damage, p53 activation, and apoptosis in a cellular model of SOD1-linked ALS. *Biochim. Biophys. Acta* 1802:462–471.
45. Bilsland, L.G., Sahai, E., Kelly, G., Golding, M., Greensmith, L., and Schiavo, G. 2010. Deficits in axonal transport precede ALS symptoms in vivo. *Proc. Natl. Acad. Sci. U.S.A.* 107:20523–20528.
46. King, A.E., Dickson, T.C., Blizzard, C., Woodhouse, A., Foster, S.S., Chung, R., and Vickers, J.C. 2011. Neuron-glia interactions underlie ALS-like axonal cytoskeletal pathology. *Neurobiol. Aging* 32:459–469.
47. Zhu, H.N., Shi, P., Wei, Y.M., Zhang, J.Y., and Gal, J. 2010. Mitochondrial dysfunction is a converging point of multiple pathological pathways in amyotrophic lateral sclerosis. *J. Alzheimers Dis.* 20:S311–S324.
48. Miana-Mena, F.J., González-Mingot, C., Larrodé, P., Muñoz, M.J., Oliván, S., Fuentes-Broto, L., Martínez-Ballarín, E., Reiter, R.J., Osta, R., and García, J.J. 2011. Monitoring systemic oxidative stress in an animal model of amyotrophic lateral sclerosis. *J. Neurol.* 258:762–769.

49. Wuolikainen, A., Moritz, T., Marklund, S.L., Antti, H., and Andersen, P.M. 2011. Disease-related changes in the cerebrospinal fluid metabolome in amyotrophic lateral sclerosis detected by GC/TOFMS. *PLoS One* 6:e17947.

50. Perry, J.J.P., Shin, D.S., Getzoff, E.D., and Tainer, J.A. 2010. The structural biochemistry of the super-oxide dismutases. *Biochim. Biophys. Acta* 1804:245–262.

51. Getzoff, E.D., Cabelli, D.E., Fisher, C.L., Parge, H.E., Viezzoli, M.S., Banci, L., and Hallewell, R.A. 1992. Faster superoxide dismutase mutants designed by enhancing electrostatic guidance. *Nature* 358:347–351.

52. Tainer, J.A., Getzoff, E.D., Beem, K.M., Richardson, J.S., and Richardson, D.C. 1982. Determination and analysis of the 2 A-structure of copper, zinc superoxide dismutase. *J. Mol. Biol.* 160:181–217.

53. Getzoff, E.D., Tainer, J.A., Stempien, M.M., Bell, G.I., and Hallewell, R.A. 1989. Evolution of CuZn superoxide dismutase and the Greek key beta-barrel structural motif. *Proteins* 5:322–336.

54. Getzoff, E.D., Tainer, J.A., Weiner, P.K., Kollman, P.A., Richardson, J.S., and Richardson, D.C. 1983. Electrostatic recognition between superoxide and copper, zinc superoxide dismutase. *Nature* 306:287–290.

55. Hallewell, R.A., Imlay, K.C., Lee, P., Fong, N.M., Gallegos, C., Getzoff, E.D., Tainer, J.A. et al. 1991. Thermostabilization of recombinant human and bovine CuZn superoxide dismutases by replacement of free cysteines. *Biochem. Biophys. Res. Commun.* 181:474–480.

56. Shin, D.S., Didonato, M., Barondeau, D.P., Hura, G.L., Hitomi, C., Berglund, J.A, Getzoff, E.D., Cary, S.C., and Tainer J.A. 2009. Superoxide dismutase from the eukaryotic thermophile *Alvinella pompejana*: Structures, stability, mechanism, and insights into amyotrophic lateral sclerosis. *J. Mol. Biol.* 385:1534–1555.

57. Cardoso, R.M., Silva, C.H., Ulian de Araujo, A.P., Tanaka, T., Tanaka, M., and Garratt, R.C. 2004. Structure of the cytosolic Cu,Zn superoxide dismutase from Schistosoma mansoni. *Acta Crystallogr. D: Biol. Crystallogr.* 60:1569–1578.

58. Hough, M.A. and Hasnain, S.S. 2003. Structure of fully reduced bovine copper zinc superoxide dismutase at 1.15 A. *Structure* 11:937–946.

59. Strange, R.W., Antonyuk, S.V., Hough, M.A., Doucette, P.A., Valentine, J.S., and Hasnain, S.S. Variable metallation of human superoxide dismutase: Atomic resolution crystal structures of Cu-Zn, Zn-Zn and As-isolated wild-type enzymes. *J. Mol. Biol.* 356:1152–1162.

60. Fischer, L.R., Culver, D.G., Tennant, P., Davis, A.A., Wang, M., Castellano-Sanchez, A., Khan, J., Polak, M.A., and Glass, J.D. 2004. Amyotrophic lateral sclerosis is a distal axonopathy: Evidence in mice and man. *Exp. Neurol.* 185:232–240.

61. Rocha, M.C., Pousinha, P.A., Correia, A.M., Sebastião, A.M., and Ribeiro, J.A. 2013. Early changes of neuromuscular transmission in the SOD1(G93A) mice model of als start long before motor symptoms onset. *PLoS One* 8:e73846.

62. Parge, H.E., Hallewell, R.A., and Tainer, J.A. 1992. Atomic structures of wild-type and thermostable mutant recombinant human Cu,Zn superoxide dismutase, *Proc. Natl. Acad. Sci. U.S.A.* 89:6109–6113.

63. Bordo, D., Djinovic, K., and Bolognesi M. 1994. Conserved patterns in the Cu,Zn superoxide dismutase family, *J. Mol. Biol.* 238:366–386.

64. Dobson, C.M. 2001. Protein folding and its links with human disease. *Biochem. Soc. Symp.* 68: 1–26.

65. Andersen, P.M. 2001. Genetics of sporadic ALS. *Amyotroph. Lateral Scler. Other Motor Neuron Disord.* 2:S37–S41.

66. Gaudette, M., Hirano, M., and Siddique, T. 2000. Current status of SOD1 mutations in familial amyo-trophic lateral sclerosis, *Amyotroph. Lateral Scler. Other Motor Neuron Disord.* 1:83–89.

67. Perry, J.J.P., Fan, L., and Tainer, J.A. 2007. Developing master keys to brain pathology, cancer and aging from the structural biology of proteins controlling reactive oxygen species and DNA repair. *Neuroscience* 145:1280–1299.

68. DiDonato, M., Craig, L., Huff, M.E., Thayer, M.M., Cardoso, R.M., Kassmann, C.J., Lo, T.P. et al. 2003. ALS mutants of human superoxide dismutase form fibrous aggregates via framework destabilization. *J. Mol. Biol.* 332:601–615.

69. Pollari, E. Goldsteins, G., Bart, G., Koistinaho, J., and Giniatullin, R. 2014. The role of oxidative stress in degeneration of the neuromuscular junction in amyotrophic lateral sclerosis. *Front. Cell. Neurosci.* 8:1–8.

70. Naumenko, N., Pollari, E., Kurronen, A., Giniatullina, R., Shakirzyanova, A., Magga, J., Koistinaho, J., and Giniatullin, R. 2011. Gender-specific mechanism of synaptic impairment and its prevention by GCSF in a mouse model of ALS. *Front. Cell. Neurosci.* 5:26.

71. Shaw, P.J., Ince, P.G., Falkous, G., and Mantle, D. 1995. Oxidative damage to protein in sporadic motor neuron disease spinal cord. *Ann. Neurol.* 38:691–695.

72. Ferrante, R., Browne, S.E., Shinobu, L.A., Bowling, A.C., Baik, M.J., MacGarvey, U., Kowall, N.W., Brown Jr., R.H., and Beal, M.F. 1997. Evidence of increased oxidative damage in both sporadic and familial amyotrophic lateral sclerosis. *J. Neurochem.* 69:2064–2074.

73. Smith, R.G., Henry, Y.K., Mattson, M.P., and Appel, S.H. 1998. Presence of 4-hydroxynonenal in cerebrospinal fluid of patients with sporadic amyotrophic lateral sclerosis. *Ann. Neurol.* 44:696–699.

74. Chang, Y., Kong, Q., Shan, X., Tian, G., Ilieva, H., Cleveland, D.W., Rothstein, J.D., Borchelt, D.R., Wong, P.C., and Lin, C.G. 2008. Messenger RNA oxidation occurs early in disease pathogenesis and promotes motor neuron degeneration in ALS. *PLoS One* 3:2849.

75. Duan, W., Li, X., Shi, J., Guo, Y., Li, Z., and Li, C. 2010. Mutant TAR DNA-binding protein-43 induces oxidative injury in motor neuron-like cell. *Neuroscience* 169:1621–1629.

76. Dibaj, P., Steffens, H., Zschüntzsch, J., Nadrigny, F., Schomburg, E.D., Kirchhoff, F., and Neusch, C. 2011. In vivo imaging reveals distinct inflammatory activity of CNS microglia versus PNS macrophages in a mouse model for ALS. *PLoS One* 6:e17910.

77. Jonsson, P.A., Backstrand, A., Andersen, P.M., Jacobsson, J., Parton, M., Shaw, C., Swingler, R. et al. 2002. CuZn-superoxide dismutase in D90A heterozygotes from recessive and dominant ALS pedigrees. *Neurobiol. Dis.* 10:327–333.

78. Mancuso, M., Filosto, M., Naini, A., Rocchi, A., Del Corona, A., Sartucci, F., Siciliano, G., and Murri, L. 2002. A screening for superoxide dismutase-1 D90A mutation in Italian patients with sporadic amyotrophic lateral sclerosis. *Amyotroph. Lateral Scler. Other Motor Neuron Disord.* 3:215–218.

79. Pratt, A.J., Getzoff, E.D., and Perry, J.J.P. 2012. Amyotrophic lateral sclerosis: Update and new developments. *Degener. Neurol. Neuromuscul. Dis.* 2:1–14.

80. Marucci, G., Morandi, L., Bartolomei, E., Salvi, F., Pession, A., Righi, A., Lauria, G., and Foschini, M.P. 2007. Amyotrophic lateral sclerosis with mutation of the Cu/Zn superoxide dismutase gene (SOD1) in a patient with Down syndrome. *Neuromuscul. Disord.* 17:673–676.

81. Hayward, C., Brock, D.J., Minns, R.A., and Swingler, R.J. 1998. Homozygosity for Asn86Ser mutation in the CuZn superoxide dismutase gene produces a severe clinical phenotype in a juvenile onset case of familial amyotrophic lateral sclerosis. *J. Med. Genet.* 35:174.

82. Tainer, J.A., Getzoff, E.D., Richardson, J.S., and Richardson, D.C. 1983. Structure and mechanism of copper, zinc superoxide dismutase. *Nature* 306:284–287.

83. Bourne, Y., Redford, S.M., Steinman, H.M., Lepock, J.R., Tainer, J.A., and Getzoff, E.D. 1996. Novel dimeric interface and electrostatic recognition in bacterial Cu,Zn superoxide dismutase. *Proc. Natl. Acad. Sci. U.S.A.* 93:12774–12779.

84. Borgstahl, G.E., Parge, H.E., Hickey, M.J., Beyer, W.F. Jr., Hallewell, R.A., and Tainer, J.A. 1992. The structure of human mitochondrial manganese superoxide dismutase reveals a novel tetrameric interface of two 4-helix bundles. *Cell* 71:107–118.

85. Guan, Y., Hickey, M.J., Borgstahl, G.E., Hallewell, R.A., Lepock, J.R., O'Connor, D., Hsieh, Y., Nick, H.S., Silverman, D.N., and Tainer, J.A. 1998. Crystal structure of Y3 4 F mutant human mitochondrial manganese superoxide dismutase and the functional role of tyrosine 34. *Biochemistry* 37:4722–4730.

86. Perry, J.J.P., Hearn, A.S., Cabelli, D.E., Nick, H.S., Tainer, J.A., and Silverman, D.N. 2009. Contribution of human manganese superoxide dismutase tyrosine 34 to structure and catalysis. *Biochemistry* 48:3417–3424.

87. Bruijn, L.I., Houseweart, M.K., Kato, S., Anderson, K.L., Anderson, S.D., Ohama, E., Reaume, A.G., Scott, R.W., and Cleveland D.W. 1998. Aggregation and motor neuron toxicity of an ALS-linked SOD1 mutant independent from wild-type SOD1. *Science* 281:1851–1854.

88. Cleveland, D.W. and Rothstein, J.D. 2001. From Charcot to Lou Gehrig: Deciphering selective motor neuron death in ALS. *Nat. Rev. Neurosci.* 2:806–819.

89. Johnston, J.A., Dalton, M.J., Gurney, M.E., and Kopito R.R. 2000. Formation of high molecular weight complexes of mutant Cu,Zn-superoxide dismutase in a mouse model for familial amyotrophic lateral sclerosis. *Proc. Natl. Acad. Sci. U.S.A.* 97:12571–12576.

90. Kato, S. Sumi-Akamaru, H. Fujimura, H., Sakoda, S., Kato, M., Hirano, A., Takikawa, M., and Ohama, E. 2001. Copper chaperone for superoxide dismutase co-aggregates with superoxide dismutase 1 (SOD1) in neuronal Lewy body-like hyaline inclusions: An immunohistochemical study on familial amyotrophic lateral sclerosis with SOD1 gene mutation. *Acta Neuropathol. (Berl.)* 102:233–238.

91. Chou, S.M., Wang, H.S., and Komai, K. 1996. Colocalization of NOS and SOD1 in neurofilament accumulation within motor neurons of amyotrophic lateral sclerosis: An immunohistochemical study. *J. Chem. Neuroanat.* 10:249–258.

92. Ferri, K.F. and Kroemer, G. 2001. Mitochondria-the suicide organelles. *BioEssays* 23:111–115.
93. Durham, H.D., Roy, J., Dong, L., and Figlewicz, D.A. 1997. Aggregation of mutant Cu/Zn superoxide dismutase proteins in a culture model of ALS. *J. Neuropathol. Exp. Neurol.* 56:523–530.
94. Pasinelli, P., Borchelt, D.R., Houseweart, M.K., Cleveland, D.W., and Brown Jr., R.H. 1998. Caspase-1 is activated in neural cells and tissue with amyotrophic lateral sclerosis-associated mutations in copper-zinc superoxide dismutase. *Proc. Natl. Acad. Sci. U.S.A.* 95:15763–15768.
95. Rabizadeh, S., Gralla, E.B., Borchelt, D.R., Gwinn, R., Valentine, J.S., Sisodia, S., Wong, P., Lee, M., Hahn, H., and Bredesen, D.E. 1995. Mutations associated with amyotrophic lateral sclerosis convert superoxide dismutase from an antiapoptotic gene to a proapoptotic gene: Studies in yeast and neural cells. *Proc. Natl. Acad. Sci. U.S.A.* 92:3024–3028.
96. Li, M., Ona, V.O., Guegan, C., Chen, M., Jackson-Lewis, V., Andrews, L.J., Olszewski, A.J. et al. 2000. Functional role of caspase-1 and caspase-3 in an ALS transgenic mouse model. *Science* 288:335–339.
97. Pasinelli, P., Houseweart, M.K., Brown Jr. R.H., and Cleveland, D.W. 2000. Caspase-1 and -3 are sequentially activated in motor neuron death in Cu,Zn superoxide dismutase-mediated familial amyotrophic lateral sclerosis. *Proc. Natl. Acad. Sci. U.S.A.* 97:13901–13906.
98. Wang, Q., Johnson, J.L., Agar, N.Y., and Agar, J.N. 2008. Protein aggregation and protein instability govern familial amyotrophic lateral sclerosis patient survival. *PLoS Biol.* 6:e170.
99. Vassall, K.A., Stubbs, H.R., Primmer, H.A., Tong, M.S., Sullivan, S.M., Sobering, R. Srinivasan, S. et al. 2011. Decreased stability and increased formation of soluble aggregates by immature superoxide dismutase do not account for disease severity in ALS. *Proc. Natl. Acad. Sci. U.S.A.* 108:2210–2215.
100. Bystrom, R., Andersen, P.M., Grobner, G., and Oliveberg, M. 2010. SOD1 mutations targeting surface hydrogen bonds promote amyotrophic lateral sclerosis without reducing apo-state stability. *J. Biol. Chem.* 285:19544–19552.
101. Prudencio, M., Hart, P.J., Borchelt, D.R., and Andersen, P.M. 2009. Variation in aggregation propensities among ALS-associated variants of SOD1: Correlation to human disease. *Hum. Mol. Genet.* 18: 3217–3226.
102. Lindberg, M.J., Tibell, L., and Oliveberg, M. 2002. Common denominator of Cu/Zn superoxide dismutase mutants associated with amyotrophic lateral sclerosis: Decreased stability of the apo state. *Proc. Natl. Acad. Sci. U.S.A.* 99:16607–16612.
103. Proctor, E.A., Ding, F., and Dokholyan, N.V. 2011. Structural and thermodynamic effects of post-translational modifications in mutant and wild type Cu,Zn superoxide dismutase. *J. Mol. Biol.* 408:555–567.
104. Nishitoh, H., Kadowaki, H., Nagai, A, Maruyama, T., Yokota, T., Fukutomi, H., Noguchi, T., Matsuzawa, A., Takeda, K., and Ichijo, H. 2008. ALS-linked mutant SOD1 induces ER stress- and ASK1-dependent motor neuron death by targeting Derlin-1. *Genes Dev.* 22:1451–1464.
105. Harraz, M.M., Marden, J.J., Zhou, W., Zhang, Y., Williams, A., Sharov, V.S., Nelson, K. et al. 2008. SOD1 mutations disrupt redox-sensitive Rac regulation of NADPH oxidase in a familial ALS model. *J. Clin. Invest.* 118:659–670.
106. Wang, J., Farr, G.W., Zeiss, C.J., Rodriguez-Gil, D.J., Wilson, J.H., Furtak, K., Rutkowski, D.T. et al. 2009. Progressive aggregation despite chaperone associations of a mutant SOD1-YFP in transgenic mice that develop ALS. *Proc. Natl. Acad. Sci. U.S.A.* 106:1392–1397.
107. Urushitani, M., Sik, A., Sakurai, T., Nukina, N., Takahashi, R., and Julien, J.P. 2006. Chromogranin-mediated secretion of mutant superoxide dismutase proteins linked to amyotrophic lateral sclerosis. *Nat. Neurosci.* 9:108–118.
108. Kieran, D., Hafezparast, M., Bohnert, S., Dick, J.R., Martin, J., Schiavo, G., Fisher, E.M., and Greensmith, L. 2005. A mutation in dynein rescues axonal transport defects and extends the life span of ALS mice. *J. Cell Biol.* 169:561–567.
109. Zhang, F., Strom, A.L., Fukada, K., Lee, S., Hayward, L.J., and Zhu, H. 2007. Interaction between familial amyotrophic lateral sclerosis (ALS)-linked SOD1 mutants and the dynein complex. *J. Biol. Chem.* 282:16691–16699.
110. Boillee, S., Vande Velde, C., and Cleveland, D.W. 2006. ALS: A disease of motor neurons and their non-neuronal neighbors. *Neuron* 52:39–59.
111. Kikuchi, H., Almer, G., Yamashita, S., Guégan, C., Nagai, M., Xu, Z., Sosunov, A.A., McKhann 2nd, G.M., and Przedborski, S. 2006. Spinal cord endoplasmic reticulum stress associated with a microsomal accumulation of mutant superoxide dismutase-1 in an ALS model. *Proc. Natl. Acad. Sci. U.S.A.* 103:6025–6030.
112. Higgins, C.M., Jung, C., and Xu, Z. 2003. ALS-associated mutant SOD1G93A causes mitochondrial vacuolation by expansion of the intermembrane space and by involvement of SOD1 aggregation and peroxisomes. *BMC Neurosci.* 4:16.

113. Breckenridge, D.G., Germain, M., Mathai, J.P., Nguyen, M., and Shore, G.C. 2003. Regulation of apoptosis by endoplasmic reticulum pathways. *Oncogene* 22:8608–8618.
114. Yoshihara, T., Ishigaki, S., Yamamoto, M., Liang, Y., Niwa, J., Takeuchi, H., Doyu, M., and Sobue, G. 2002. Differential expression of inflammation- and apoptosis-related genes in spinal cords of a mutant SOD1 transgenic mouse model of familial amyotrophic lateral sclerosis. *J. Neurochem.* 80:158–167.
115. Raoul, C., Estévez, A.G., Nishimune, H., Cleveland, D.W., deLapeyrière, O., Henderson, C.E., Haase, G., and Pettmann, B. 2002. Motoneuron death triggered by a specific pathway downstream of Fas. potentiation by ALS-linked SOD1 mutations. *Neuron* 35:1067–1083.
116. Beckman, J.S., Estévez, A.G., Crow, J.P., and Barbeito, L. 2001. Superoxide dismutase and the death of motoneurons in ALS. *Trends Neurosci.* 24:S15–S20.
117. Shaw, P.J. and Eggett, C.J. 2000. Molecular factors underlying selective vulnerability of motor neurons to neurodegeneration in amyotrophic lateral sclerosis. *J. Neurol.* 247(Suppl. 1):I17–I27.
118. Vande Velde, C., Miller, T.M., Cashman, N.R., and Cleveland, D.W. 2008. Selective association of misfolded ALS-linked mutant SOD1 with the cytoplasmic face of mitochondria. *Proc. Natl. Acad. Sci. U.S.A.* 105:4022–4027.
119. Pedrini, S., Sau, D., Guareschi, S., Bogush, M., Brown Jr., R.H, Naniche, N., Kia, A., Trotti, D., and Pasinelli, P. 2010. ALS-linked mutant SOD1 damages mitochondria by promoting conformational changes in Bcl-2. *Hum. Mol. Genet.* 19:2974–2986.
120. Vijayvergiya, C., Beal, M.F., Buck, J., and Manfredi, G. 2005. Mutant superoxide dismutase 1 forms aggregates in the brain mitochondrial matrix of amyotrophic lateral sclerosis mice. *J. Neurosci.* 25:2463–2470.
121. Kawamata, H. and Manfredi, G. 2008. Different regulation of wild- type and mutant Cu,Zn superoxide dismutase localization in mammalian mitochondria. *Hum. Mol. Genet.* 17:3303–3317.
122. Ferri, A., Cozzolino, M., Crosio, C., Nencini, M., Casciati, A., Gralla, E.B., Rotilio, G., Valentine, J.S., and Carrì, M.T. 2006. Familial ALS-superoxide dismutases associate with mitochondria and shift their redox potentials. *Proc. Natl. Acad. Sci. U.S.A.* 103:13860–13865.
123. Magrané, J., Hervias, I., Henning, M.S., Damiano, M., Kawamata, H., and Manfredi, G. 2009. Mutant SOD1 in neuronal mitochondria causes toxicity and mitochondrial dynamics abnormalities. *Hum. Mol. Genet.* 18:4552–4564.
124. Igoudjil, A., Magrané, J., Fischer, L.R., Kim, H.J., Hervias, I., Dumont, M., Cortez, C., Glass, J.D., Starkov, A.A., and Manfredi, G. 2011. In vivo pathogenic role of mutant SOD1 localized in the mitochondrial intermembrane space. *J. Neurosci.* 31:15826–15837.
125. Finley, L.W., Carracedo, A., Lee, J., Souza, A., Egia, A., Zhang, J., Teruya-Feldstein, J. et al. 2011. SIRT3 opposes reprogramming of cancer cell metabolism through HIF1alpha destabilization. *Cancer Cell* 19:416–428.
126. Bell, E.L., Emerling, B.M., Ricoult, S.J., and Guarente, L. 2011. SirT3 suppresses hypoxia inducible factor 1alpha and tumor growth by inhibiting mitochondrial ROS production. *Oncogene* 30:2986–2996.
127. Bourke, S.C., Tomlinson, M., Williams, T.L., Bullock, R.E., Shaw, P.J., and Gibson, G.J. 2006. Effects of non- invasive ventilation on survival and quality of life in patients with amyotrophic lateral sclerosis: A randomised controlled trial. *Lancet Neurol.* 5:140–147.
128. Doble, A. 1996. The pharmacology and mechanism of action of riluzole. *Neurology* 47:S233–S241.
129. Garnock-Jones, K.P. 2011. Dextromethorphan/quinidine: In pseudobulbar affect. *CNS Drugs* 25:435–445.
130. Reaume, A.G., Elliott, J.L., Hoffman, E.K., Kowall, N.W., Siwek, D.F., Wilcox, H.M., Flood, D.G. et al. 1996. Motor neurons in Cu/Zn superoxide dismutase-deficient mice develop normally but exhibit enhanced cell death after axonal injury. *Nat. Genet.* 13:43–47.
131. Gane, S.A., Dar, T., Bhat, A., Dar, K., Anees, S., Masood, A., and Zargar, M.A. 2015. Melatonin: A potential antioxidant therapeutic agent for mitochondrial dysfunctions and related disorders. *Rejuvenation Res.* 2011:1–16.
132. Weishaupt, J.H., Bartels, C., Polking, E., Dietrich, J., Rohde, G., Poeggeler, B., Mertens, N. et al. 2006. Reduced oxidative damage in ALS by high-dose enteral melatonin treatment. *J. Pineal Res.* 41:313–323.
133. Reiter, R.J. and Tan, D.X. 2003. Melatonin: A novel protective agent against oxidative injury of the ischemic/reperfused heart. *Cardiovasc. Res.* 58:10–19.
134. Tan, D.X., Reiter, R.J., and Manchester, L.C. 2002. Chemical and physical properties and potential mechanisms: Melatonin as a broad spectrum antioxidant and free radical scavenger. *Curr. Top. Med. Chem.* 2:181–197.
135. Hardeland, R. 2005. Antioxidative protection by melatonin. Multiplicity of mechanism from radical detoxification or radical avoidance. *Endocrine* 27:119–130.

136. Reiter, R.J., Tan, D.X., Sainz, R.M., Mayo, J.C., and Lopez, B.S. 2002. Melatonin: Reducing the toxicity and increasing the efficacy of drugs. *J. Pharm. Pharmacol.* 54:1299–1321.

137. Rodriguez, C., Mayo, J.C., Sainz, R.M., Antolín, I., Herrera, F., Martín, V., and Reiter, R.J. 2004. Regulation of antioxidant enzymes: A significant role for melatonin. *J. Pineal Res.* 36:1–9.

138. Reiter, R.J., Tan, D.X., Gitto, E., Sainz, R.M., Mayo, J.C., Leon, J., Manchester, L.C., Vijayalaxmi, K.E., and Maldonado, M.D. 2005. Melatonin as an antioxidant: Physiology versus pharmacology. *J. Pineal Res.* 39:215–216.

139. Winiarska, K., Fraczyk, T., Malinska, D., Drosaz, J., and Bryla, J. 2006. Melatonin attenuates diabetes-induced oxidative stress in rabbits. *J. Pineal Res.* 40:168–176.

140. Vijayalaxmi, R.J. and Meltz, M.L. 1995. Melatonin protects human blood lymphocytes from radiation induced chromosome damage. *Mutat. Res.* 34:623–631.

141. Gulcin, I., Buyukokuroglu, M.E., Oktay, M., and Kufrevioglu, O.I. 2002. On the in vitro antioxidative properties of melatonin. *J. Pineal Res.* 33:167–171.

142. Vijayalaxmi, R.J., Reiter, R.J., Herman, T.S., and Meltz, M.L. 1996. Melatonin and radioprotection from genetic damage: In vivo/in vitro studies with human volunteers. *Mutat. Res.* 37:221–228.

143. El-Missiry, M.A., Fayed, T.A., El-Sawy, M.R., and El-Sayed, A.A. 2007. Ameliorative effect of melatonin against gamma-irradiation-induced oxidative stress and tissue injury. *Ecotoxicol. Environ. Saf.* 66:278–286.

144. Carrì, M.T., Valle, C., Bozzo, F., and Cozzolino, M. 2015. Oxidative stress and mitochondrial damage: Importance in non-SOD1 ALS. *Front. Cell. Neurosci.* 9:1–6.

145. D'Amico, E., Factor-Litvak, P., Santella, R.M., and Mitsumoto, H. 2013. Clinical perspective on oxidative stress in sporadic amyotrophic lateral sclerosis. *Free Radic. Biol. Med.* 65:509–527.

146. Weiduschat, N., Mao, X., Hupf, J., Armstrong, N., Kang, G., Lange, D.J., Mitsumoto, H., and Shungu, D.C. 2014. Motor cortex glutathione deficit in ALS measured in vivo with the J-editing technique. *Neurosci. Lett.* 570:102–107.

147. Vargas, M.R., Johnson, D.A., and Johnson, J.A. 2011. Decreased glutathione accelerates neurological deficit and mitochondrial pathology in familial ALS-linked hSOD1(G93A) mice model. *Neurobiol. Dis.* 43:543–551.

148. Ferri, A., Fiorenzo, P., Nencini, M., Cozzolino, M., Pesaresi, M.G., Valle, C., Sepe, S., Moreno, S., and Carrì, M.T. 2010. Glutaredoxin 2 prevents aggregation of mutant SOD1 in mitochondria and abolishes its toxicity. *Hum. Mol. Genet.* 19:4529–4542.

149. Iguchi, Y., Katsuno, M., Takagi, S., Ishigaki, S., Niwa, J., Hasegawa, M., Tanaka, F., and Sobue, G. 2012. Oxidative stress induced by glutathione depletion reproduces pathological modifications of TDP-43 linked to TDP-43 proteinopathies. *Neurobiol. Dis.* 45:862–870.

150. Braun, R.J., Sommer, C., Carmona-Gutierrez, D., Khoury, C.M., Ring, J., Büttner, S., and Madeo, F. 2011. Neurotoxic 43-kDa TAR DNA-binding protein (TDP-43) triggers mitochondrion-dependent programmed cell death in yeast. *J. Biol. Chem.* 286:19958–19972.

151. Cohen, T.J., Hwang, A.W., Unger, T., Trojanowski, J.Q., and Lee, V.M. 2012. Redox signalling directly regulates TDP-43 via cysteine oxidation and disulphide cross-linking. *EMBO J.* 31:1241–1252.

152. Shodai, A., Morimura, T., Ido, A., Uchida, T., Ayaki, T., Takahashi, R., Kitazawa, S. et al. 2013. Aberrant assembly of RNA recognition motif 1 links to pathogenic conversion of TAR DNA-binding protein of 43 kDa (TDP-43). *J. Biol. Chem.* 288:14886–14905.

153. Parker, S.J., Meyerowitz, J., James, J.L., Liddell, J.R., Crouch, P.J., Kanninen, K.M., and White, A.R. 2012. Endogenous TDP-43 localized to stress granules can subsequently form protein aggregates. *Neurochem. Int.* 60:415–424.

154. Colombrita, C., Zennaro, E., Fallini, C., Weber, M., Sommacal, A., Buratti, E., Silani, V., and Ratti, A. 2009. TDP-43 is recruited to stress granules in conditions of oxidative insult. *J. Neurochem.* 111:1051–1061.

155. Sánchez-Ramos, C., Tierrez, A., Fabregat-Andrés, O., Wild, B., Sánchez-Cabo, F., Arduini, A., Dopazo, A., and Monsalve, M. 2011. PGC-1 regulates translocated in liposarcoma activity: Role in oxidative stress gene expression. *Antioxid. Redox Signal.* 15:325–337.

156. Wang, T., Jiang, X., Chen, G., and Xu, J. 2015. Interaction of amyotrophic lateral sclerosis/frontotemporal lobar degeneration-associated fused-in-sarcoma with proteins involved in metabolic and protein degradation pathways. *Neurobiol. Aging* 36:527–535.

157. Bartolome, F., Wu, H.C., Burchell, V.S., Preza, E., Wray, S., Mahoney, C.J., Fox, N.C. et al. 2013. Pathogenic VCP mutations induce mitochondrial uncoupling and reduced ATP levels. *Neuron* 78:57–64.

158. Yin, H.Z., Nalbandian, A., Hsu, C.I., Li, S., Llewellyn, K.J., Mozaffar, T., Kimonis, V.E., and Weiss, J.H. 2012. Slow development of ALS-like spinal cord pathology in mutant valosin-containing protein gene knock-in mice. *Cell Death Dis.* 3:e374.

159. Nalbandian, A., Llewellyn, K.J., Badadani, M., Yin, H.Z., Nguyen, C., Katheria, V., Watts, G. et al. 2013. A progressive translational mouse model of human valosin-containing protein disease: The VCP(R155H/+) mouse. *Muscle Nerve* 47:260–270.

160. Hong, K., Li, Y., Duan, W., Guo, Y., Jiang, H., Li, W., and Li, C. 2012. Full-length TDP-43 and its C-terminal fragments activate mitophagy in NSC34 cell line. *Neurosci. Lett.* 530:144–149.

161. Stoica, R., De Vos, K.J., Paillusson, S., Mueller, S., Sancho, R.M., Lau, K.F., Vizcay-Barrena, G. et al. 2014. ER-mitochondria associations are regulated by the VAPB-PTPIP51 interaction and are disrupted by ALS/FTD-associated TDP-43. *Nat. Commun.* 5:3996.

162. Xu, Y.F., Gendron, T.F., Zhang, Y.J., Lin, W.L., D'Alton, S., Sheng, H., Casey, M.C. et al. 2010. Wild-type human TDP-43 expression causes TDP-43 phosphorylation, mitochondrial aggregation, motor deficits and early mortality in transgenic mice. *J. Neurosci.* 30:10851–10859.

163. Shan, X., Chiang, P.M., Price, D.L., and Wong, P.C. 2010. Altered distributions of Gemini of coiled bodies and mitochondria in motor neurons of TDP-43 transgenic mice. *Proc. Natl. Acad. Sci. U.S.A.* 107:16325–16330.

164. Magrané, J., Cortez, C., Gan, W.B., and Manfredi, G. 2014. Abnormal mitochondrial transport and morphology are common pathological denominators in SOD1 and TDP43 ALS mouse models. *Hum. Mol. Genet.* 23:1413–1424.

165. Stribl, C., Samara, A., Trüembach, D., Peis, R., Neumann, M., Fuchs, H., Gailus-Durner, V. et al. 2014. Mitochondrial dysfunction and decrease in body weight of a transgenic knock-in mouse model for TDP-43. *J. Biol. Chem.* 289:10769–10784.

9 Alzheimer's Disease, Oxidative Stress, and Neuroprotective Approaches

Anami Ahuja, Kapil Dev, and Pankaj K. Tyagi

CONTENTS

ABSTRACT

Alzheimer's disease (AD) is a progressive neurodegenerative disease with a complex pathological phenotype characterized by amyloid plaques and neurofibrillary tangles. Therapeutic strategies targeting the pathological aspect of the disease have achieved success in preclinical models but have failed to achieve success in clinical therapeutic efficacy. The progressive and multifaceted degenerative phenotype of Alzheimer's disease suggests that successful treatment strategies need to be equally multifaceted and disease stage specific. Oxidative stress is a key element in the pathophysiology of Alzheimer's disease; it not only temporally precedes the pathological lesions of the disease but also activates cell signaling pathways, which, in turn, contribute to lesion formation and stimulate cellular responses. Therapeutic strategies targeting oxidative stress have shown promising results in Alzheimer's treatment. This chapter provides insights into the role of oxidative stress and neuroprotective approaches in Alzheimer's disease.

9.1 INTRODUCTION

Alzheimer's disease (AD) is one of the most prevalent chronic neurodegenerative diseases affecting around 44 million people worldwide and is characterized by progressive cognitive decline (Selkoe et al. 2011). The pathological hallmarks of AD are neuronal loss, amyloid plaques, neurofibrillary tangles (NFTs), and gliosis (Parajuli et al. 2013). Till date, several hypotheses such as amyloid cascade hypothesis (Karran et al. 2011), cholinergic hypothesis (Terry and Buccafusco 2003), Tau hypothesis (Maccioni et al. 2010), and oxidative stress hypothesis (Praticò 2008) have been proposed for Alzheimer's pathogenesis. But none of these hypotheses have been able to explain all aspects of disease progression. The most widely accepted hypothesis for AD-related cognitive dysfunction and neuropathogenesis is the "amyloid cascade hypothesis," which posits "all pathological changes in AD brain occur as a downstream of excessive accumulation of β-amyloid in the central nervous system (CNS)" (Hardy and Higgins 1992). Amyloid β (Aβ) exists in both soluble and fibrillar forms. Aβ is produced by the proteolytic cleavage of membrane-associated β-amyloid precursor protein (APP). Processing of APP occurs by two major protease pathways. Cleavage of APP at N-terminus of the Aβ region by β-secretase and at the C-terminus by γ-secretase represents the amyloidogenic pathway for the processing of APP to form Aβ (Chauhan and Chauhan 2006). Alternatively, APP can also be processed by α-secretase, which cleaves within the Aβ sequence and does not produce Aβ (Vardy et al. 2005). The amyloid cascade hypothesis was further supported by the discovery of the autosomal dominant mutations in presenilin (PSEN)-1 and 2, both of which are homologous proteins that can form the catalytic active site of γ-secretase and could result in AD (Sherrington et al. 1995). The apolipoprotein E (APOE) gene represents the major genetic risk factor for AD (Karran et al. 2011). Studies conducted on platelet derived growth factor-β (PDGF) promoter expressing amyloid precursor protein (PDAPP) transgenic mice that harbored the human APOE genes have revealed that these cognate proteins mediate the clearance of Aβ (Kim et al. 2009). The amyloid cascade model has emerged as the organizing element in the studies of both familial and sporadic forms of AD (Demetrius et al. 2015). However, there are frequent conflicts between empirical observations and predictions of this model. Some of the most prominent anomalies are as follows:

1. Neuritic plaques continuously occur in the brain of cognitively unimpaired individuals (Demetrius et al. 2015).
2. There is a weak correlation between the density of plaques and the degree of dementia (Demetrius et al. 2015).
3. Cognitively intact individuals have a significantly large incidence of pre-/postmortem-detected amyloid plaques (Wischik et al. 2014).

Considering these anomalies along with several others and the consistent failure of clinical trials of therapeutic strategies based on this hypothesis have led researchers to look beyond the amyloid cascade hypothesis for alternative therapeutic strategies.

9.2 CHOLINERGIC HYPOTHESIS

Cholinergic hypothesis explains the degeneration of the acetylcholine (ACh)-containing neurons in the basal forebrain and the loss of the cholinergic transmission in the cerebral cortex as a principal cause of cognitive decline observed in AD patients (Francis et al. 1999). Data supporting precise association of basal forebrain cholinergic projections in the etiology of memory dysfunction grew extensively after the discovery that a major loss of cholinergic neurons in the medial septal (MS) nucleus and nucleus basalis magnocellularis (NBM) is observed in AD (Gibbs 2010). Degeneration of cholinergic neurons in the nucleus basalis of Meynert and the loss of cholinergic inputs to the neocortex and hippocampus are the two most consistent neurotransmitter abnormalities observed in AD patients (Schliebs and Arendt 2006). Postmortem analyses of AD suffering brains have revealed a decrease in choline acetyltransferase (ChAT), ACh release, acetylcholinesterase, as well as nicotinic and muscarinic receptors in the cerebral cortex and hippocampus (Auld et al. 2002; Giacobini 2003). These cholinergic deficits have been shown to correlate positively with the cognitive impairment observed in AD (Schliebs and Arendt 2006). Studies on transgenic mice and humans have revealed that the disturbance of axonal transport in cholinergic neurons is one of the earliest signs of AD progression (Stokin et al. 2005). Specific mouse models of AD have been shown to develop deficits in basal forebrain cholinergic function (particularly muscarinic receptor signaling), in addition to deficits in hippocampal function and cognitive performance (Goto et al. 2008; Machova et al. 2008; Wang et al. 2009). Aβ-dependent inactivation of the JAK2/STAT3 axis in the hippocampal neurons causes memory impairment in AD (Chiba et al. 2008). A loss of calcium-binding protein (calbindin) in the basal forebrain cholinergic neurons corresponds with the appearance of NFTs before manifestation of dementia (Geula 1998). Patients of mild dementia have early impairment of presynaptic cholinergic receptors and cholinesterase activity in the cerebral cortex (Herholz et al. 2008). These findings collectively suggest that a progressive decline in basal forebrain cholinergic function, culminating in the loss of cholinergic neurons, contributes significantly to cognitive impairment associated with advanced age and with AD-related dementia. The three commercially available cholinesterase inhibitors to treat symptoms related to memory, thinking, language, judgment, and other thought processes in AD are as follows:

1. Donepezil (Aricept) approved to treat all stages of AD
2. Rivastigmine (Exelon) approved to treat mild to moderate AD
3. Galantamine (Razadyne) approved to treat mild to moderate AD

9.3 TAU HYPOTHESIS

Tau hypothesis was initially proposed to account for the observations that the Tau-tangle pathology was observed prior to Aβ plaque formation (Braak and Braak 1991) and the density of Tau inclusions was more closely correlating with cognitive decline and disease progression than Aβ plaques (Arriagada et al. 1992; Mitchell et al. 2002). The Tau hypothesis states "Tau hyperphosphorylation as an independent and primary cause of AD" (Lansdall 2014). In AD brains, hyperphosphorylated Tau is the major component of both NFTs in pyramidal neurons and neuropil threads in distal dendrites (Finder 2010). NFTs are filamentous inclusions of Tau, which occur both in AD and in other tauopathies (Querfurth and LaFerla 2010). The primary function of the microtubule-associated Tau protein is the stabilization of microtubules (Lansdall 2014). Tau protein is critical for normal neuronal activity in the mammalian brain (Iqbal et al. 2005). In normal physiological conditions, Tau is in a constant dynamic equilibrium, on and off the microtubules (Ballatore et al. 2007). This equilibrium is thought to be controlled primarily by the phosphorylation state of Tau, which in turn is determined by the actions of kinases and phosphatases. Indeed, frequent cycles of binding and detachment of Tau from the microtubules (corresponding to phosphorylation and dephosphorylation, respectively) may be needed to allow effective axonal transport (Ballatore et al. 2007). In Tau-mediated neurodegeneration, aberrant Tau phosphorylation is an important phenomenon as the microtubule-binding ability of Tau is posttranslationally regulated primarily by serine-/threonine-directed phosphorylation. In case of AD, there is an abnormal disengagement of Tau from the microtubules because of the increased rate of phosphorylation or decreased rate of dephosphorylation of Tau. Several other pathological events such as Aβ-mediated toxicity, oxidative stress, and inflammation contribute either independently or in combination for an abnormal detachment of Tau from the microtubules (Moreira et al. 2005; King et al. 2006). Studies have revealed that oxidative stress could be responsible for detrimental covalent modifications of Tau, which include the formation of intermolecular disulphide bridges and tyrosine nitration (Ballatore et al. 2007). Such modifications are likely to cause misfolding, hyperphosphorylation, and aggregation, thereby contributing to abnormal disengagement of Tau from microtubules (Duan et al. 2012). Aggregation of Tau after disengagement from microtubules results in the formation of small nonfibrillary Tau deposits normally referred to as "pretangles." A structural rearrangement is involved in the formation of the characteristic β-pleated sheet during the transition from pretangles to paired helical filaments (PHFs) (Ballatore et al. 2007). Finally, PHFs further self-assemble to form NFTs, disturbing and impairing axonal transport (Finder 2010). Toxic forms of Tau protein eventually *choke* the neuron by preventing the possibility of normal neuronal metabolism and causing progressive neurodegeneration (Iqbal et al. 2005; Wischik et al. 2014).

9.4 OXIDATIVE STRESS

Oxidative stress is an important mediator of the onset, progression, and pathogenesis of neurodegenerative diseases (Ahuja et al. 2015), arteriosclerosis (Perrotta and Aquila 2015), cancer (Thanan et al. 2014), diabetes (Rosales-Corral et al. 2015), and renal diseases (Sureshbabu et al. 2015). The human brain utilizes about 20% of the body's total basal oxygen and subsequently generates relatively high level of reactive oxygen species (ROS) (Shulman et al. 2004). The sources of ROS generation in AD brain are multifaceted, including mitochondria (Wang et al. 2005), redox active metals (Everett et al. 2014; Ahuja et al. 2015), and inflammation via activated microglia (Perry et al. 2010). Under physiological conditions, ROS production is a normal consequence of cellular processes that is tightly controlled by antioxidants like glutathione, α-tocopherol (Vitamin E), carotenoids, and ascorbic acid, as well as by antioxidant enzymes such as catalase and glutathione peroxidases, which detoxify hydrogen peroxide (H_2O_2) by converting it to O_2 and H_2O. However, when ROS levels exceed the antioxidant capacity of a cell under disease condition or by age or metabolic demand, a deleterious condition occurs by oxidative stress causing molecular damage, promoting neuronal adaptation, and leading to critical failure of biological function (Su et al. 2008).

9.4.1 MITOCHONDRIA, OXIDATIVE STRESS, AND AD

According to the mitochondrial cascade hypothesis, mitochondrial dysfunction is the primary event that causes Aβ deposition, synaptic degeneration, and NFTs formation in case of sporadic AD (Swerdlow et al. 2014). Mitochondria plays an important role in ROS production (Moreira et al. 2005). Mitochondria produce ATP at the inner membrane through coupling of oxidative phosphorylation with respiration, which is the major endogenous source of ROS (Fariss 2005). The production of mitochondrial superoxide ($O_2^{\cdot-}$) occurs primarily at discrete points in the electron transport chain (ETC) at complexes 1 and 3 and in components of tricarboxylic acid (TCA) cycle, including α-ketoglutarate dehydrogenase (Reddy 2006). When the ETC is inhibited, electrons accumulate in the early stages of the ETC (complex 1 and coenzyme Q), where they are donated directly to molecular oxygen to give an $O_2^{\cdot-}$. The $O_2^{\cdot-}$ is detoxified by the mitochondrial Mn-superoxide dismutase (MnSOD) to give rise to H_2O_2. Subsequently, H_2O_2 is converted into H_2O either by glutathione peroxidase or catalase. Chronic exposure to ROS results in oxidative damage to the mitochondrial and cellular proteins, lipids, and nucleic acids. Acute exposure to ROS can inactivate the TCA cycle aconitase and iron–sulfur centers of the ETC at complexes 1, 2, and 3, resulting in the shutdown of the mitochondrial energy production (Reddy and Beal 2005). Mitochondrial DNA (mt-DNA) and protein are more susceptible to oxidative damage because of their proximity to ROS, lack of protective histones, and limited repair mechanisms (Wang et al. 2005). Oxidation of mitochondria results in mt-DNA strand breaks, DNA–DNA and DNA–protein cross-linking, and base modifications (Stuart and Brown 2006). All these mitochondrial alterations have been associated with AD and several other neurodegenerative diseases (Wang et al. 2005).

Studies on AD mouse models and postmortem human brains have indicated that the loss of mitochondrial integrity plays an important role in synaptic dysfunction (Calkins et al. 2012). Structural and metabolic changes, including increased fragmentation and decreased fusion, are a prominent feature observed in the AD brain (Silva et al. 2012). The Aβ is also responsible for the loss of mitochondrial potential in astrocytes that induces oxidative stress by the activation of NADPH oxidase in astrocytes leading to Aβ-induced neuronal death (Abramov et al. 2004). Oxidative stress–mediated neuronal apoptosis induced by Aβ operates by eliciting stress-activated protein kinase (SAPK)-dependent multiple regulations of proapoptotic mitochondrial pathways involving both p53 and Bcl (Tamagno et al. 2003). Increased amyloid production in mutant APP–expressed cells has also been reported to result in enhanced nitric oxide (NO) production and mitochondrial dysfunction leading to cell death (Keil et al. 2004). Studies on APOE4 allele have revealed that it can impair mitochondrial function and integrity leading to AD (Harris et al. 2003). The importance of mitochondrial dysfunction in AD pathology has been confirmed by several studies, and therefore, targeting mitochondria for alternative therapies could lead to early therapeutic interventions in AD.

9.4.2   Redox Active Metals, Oxidative Stress, and AD

Transition metal homeostasis is severely perturbed in AD with extracellular pooling of copper in amyloid and intraneuronal accumulation of iron. Copper is an essential trace element present in all tissues and is required for cellular respiration, peptide amidation, neurotransmitter biosynthesis, pigment formation, and connective tissue strengthening (Ahuja et al. 2015). Copper plays a crucial role in normal brain functioning, but its potent redox activity demands tight regulation to maintain the integrity of copper homeostasis (Hung et al. 2012). Copper dyshomeostasis can result in copper displacement, causing inadvertent interactions between copper and cellular components, which can enhance the production of ROS and the formation of neurotoxic copper–protein aggregates, which eventually lead to neuronal cell death (Hung et al. 2012). The neurotoxic interaction between Aβ peptide and copper results in AD progression (Barnham et al. 2003). Aβ peptide binds to the copper with very high affinity (7×10^{-18} M), reducing Cu (II) to Cu (I) with a catalytic generation of H_2O_2 and Aβ aggregation (Ahuja et al. 2015). As a consequence of copper-mediated oxidation, Aβ purified from human plaques was observed to have fewer histidine and tyrosine residues (Atwood et al. 2000). Copper promotes dityrosine cross-linking of Aβ, which enhances the rate of Aβ aggregation (Atwood et al. 2004). *In vitro* studies revealed that certain fragments in the four-repeat microtubule-binding domain of Tau (residues 256–273, 287–304, and 306–336) aggregate in the presence of copper, thus inducing Tau phosphorylation and aggregation (Ma et al. 2006; Zhou et al. 2007).

Apart from the aforementioned domains, microtubule-binding domain residues 287–293 and 310–324 of a 198-amino acid fragment of Tau have also been shown to bind copper in nuclear magnetic resonance (NMR) studies (Soragni et al. 2008). NFTs have also been shown to bind copper in a redox-dependent manner, acting as a source for ROS within the neuron (Sayre et al. 2000). It has been demonstrated that chronic copper exposure induces Tau hyperphosphorylation and promotes the Tau pathology in a mouse model of AD (Kitazawa et al. 2009).

Iron is the most abundant transition metal in the human brain. It plays an important role in oxygen transport, formation, and maintenance of the neuronal network, as well as in numerous aspects of DNA and enzyme processes, including neurotransmitter synthesis (Ali-Rahmani et al. 2014). In AD patients, the overaccumulation of iron is observed in the hippocampus, cerebral cortex, and basal nucleus of Meynert colocalized with AD lesions, senile plaques, and NFT (Su et al. 2008). Redox activity of the iron–Aβ interaction is suggested to facilitate iron binding to His6, His13, and His14 of Aβ (Nakamura et al. 2007; Bousejra-ElGarah et al. 2011). Binding of iron to these sites has been shown to generate H_2O_2 by the Fenton reaction and also promotes Aβ aggregation (Huang et al. 1999). Studies on Aβ-PP transgenic mice have revealed that extensive iron accumulation results in JNK/SAPK signaling pathway activation, which leads to Aβ-induced cell death (Smith et al. 1997). Iron has also been shown to affect the phosphorylation status of Tau (Ayton et al. 2013).

9.4.3   Neuroprotective Strategies, Oxidative Stress, and AD

Oxidative stress plays a crucial role in AD progression by promoting Aβ deposition, Tau hyperphosphorylation, and the subsequent loss of synapses and neurons (Chen and Zhong 2014). In addition, AD is a typical example of complex multifactorial disease, and therapeutic strategies targeting only a particular aspect of the disease progression have not been effective yet. The currently approved drugs such as donepezil, rivastigmine and galantamine are helpful in delaying the disease progression but are not able to stop or reverse disease progression. Several anti-AD strategies such as antiamyloid and anti-Tau, antimitochondrial dysfunction drugs, and neurotrophins are under evaluation in the clinical trials for safety and efficacy (Mecocci and Polidori 2012). Since oxidative stress is a core aspect of both AD onset and progression, several therapeutic strategies are now being designed to prevent or slow down ROS-mediated damages in AD (Mecocci and Polidori 2012). Antioxidants are being evaluated as a potential therapeutics for eliminating

ROS and exerting neuroprotective effects on neurons in AD. MitoQ is one of the best characterized mitochondria-targeted antioxidants that prevent cognitive decline and early neuropathology in transgenic mouse models of AD (McManus et al. 2011). MitoQ is a mitochondria-targeted ubiquinol, produced by covalently attaching a ubiquinone moiety to the lipophilic decyltriphenylphosphonium (dTPP) cation through a 10-carbon aliphatic carbon chain (Ng et al. 2014). The ubiquinone moiety of MitoQ is then inserted into the lipid bilayer of the mitochondrial matrix and rapidly reduced by complex II to ubiquinol, which acts as the active antioxidant. This ubiquinol moiety is continuously recycled to the active antioxidant by the respiratory chain (Magwere et al. 2006). The ubiquinone moiety of MitoQ reacts rapidly with $O_2^{\cdot-}$ and thus may enhance mitochondrial antioxidant defense by complimenting $O_2^{\cdot-}$ degradation by MnSOD (Maroz et al. 2009). MitoQ may also prevent damage downstream of $O_2^{\cdot-}$ by directly detoxifying lipid-derived radicals and reactive nitrogen species or indirectly through effective recycling of α-tocopherol (James et al. 2007). Low nanomolar concentrations of MitoQ prevented the Aβ-induced death of cortical neurons in the cell culture (McManus et al. 2011). MitoQ is reported to block the elevated oxidative stress and it also prevented synaptic loss, astrogliosis, and increased Aβ burden in AD transgenic mouse models (McManus et al. 2011).

Another antioxidant called Szeto–Schiller peptide (SS)-31 possesses a unique aromatic-cationic sequence motif, which alternates between aromatic and basic residues, enabling them to enter cells in an energy-independent and nonsaturable manner without a need for peptide transporters (Zhao et al. 2004). SS-31 contains a dimethyltyrosine (DMT) residue designed to scavenge a variety of ROS and inhibit lipid peroxidation *in vitro* (Zhao et al. 2004; Szeto 2008). The SS-31 peptides are targeted to the inner mitochondrial membrane to prevent apoptosis, necrosis, oxidative stress, and inhibition of the mitochondrial ETC (Szeto 2008). SS-31 protects neuronal cells against tert-butyl-hydroperoxide (tBHP)-induced mitochondrial depolarization and apoptotic cell death by reducing intracellular ROS, decreasing markers of apoptotic cell death and caspase activity (Zhao et al. 2005). It decreases mitochondrial ROS production and inhibits the mitochondrial permeability transition (MPT) and mitochondrial swelling (Zhao et al. 2004). SS-31 also prevents Ca^{2+}-induced cytochrome-c release and inhibits 3-NP-induced activation of the MPT pore and mitochondrial depolarization in isolated mitochondria (Zhao et al. 2004). SS-31 mediates significant neuroprotection against cell death induced by H_2O_2 in SOD1 neuronal cells (Chaturvedi and Beal 2008). Studies on mouse models of AD have revealed that SS-31 mitigates the effects of oligomeric Aβ, such as decreased anterograde mitochondrial movement, increased mitochondrial fission, decreased fusion and structurally damaged mitochondria, abnormal mitochondrial and synaptic proteins, defective mitochondrial function, and apoptotic neuronal death (Calkins et al. 2011). SS-31 was able to restore mitochondrial transport and synaptic viability and decreased the percentage of defective mitochondria, demonstrating its protective effect from Aβ toxicity (García-Escudero et al. 2013). Curcumin is one of the most promising candidates as neuroprotective drugs, because of its ability to inhibit Aβ formation, clearance of existing Aβ, anti-inflammatory, antioxidant, and its copper and iron chelation ability (Walker and Lue 2007). *In vitro* and *in vivo* studies on curcumin have demonstrated that curcumin produced a dose-dependent decrease in the formation of fibrillary β-amyloid$_{1-40}$ and β-amyloid$_{1-42}$ and also destabilized fibrils that had already formed, thus breaking up the β-sheet conformation seen in AD plaques (Huang et al. 2012; Potter 2013). Curcumin also reduced β-amyloid toxicity by decreasing the activity of GSK-3β and stimulating the protective Wnt/β-catenin pathway in APPswe-transfected SY5Y cells (Zhang et al. 2011). In a mouse model, SOD activity was significantly decreased after mice were treated with curcumin, suggesting antioxidant capabilities of curcumin (Huang et al. 2012). When compared to other standard antioxidants, curcumin was found to have free radical scavenging properties similar to α-tocopherol, butylated hydroxyanisole (BHA), and butylated hydroxytoluene (BHT) (Ak and Gülçin 2008). An *in vitro* study in rat cortical neuron cells found that curcumin displayed a strong ability to scavenge $O_2^{\cdot-}$, being four times more effective than a known scavenger trolox (Yao and Xue 2014). Several other promising candidates such as MitoVitE, MitoPBN, acetyl-L-carnitine

(ALCAR), R-alpha lipoic acid (LA), pramipexole, vitamin E (α-tocopherol), quercetin, and resveratrol are being investigated for their potential as a neuroprotective drug against AD (Ansari et al. 2009; Kim et al. 2010; Dumont and Beal 2011).

9.5 CONCLUSIONS

Oxidative stress plays a crucial role in the onset and progression of AD pathogenesis. Brain is one of the most susceptible organs to oxidative damage due to the presence of high concentrations of oxidizable polyunsaturated fatty acids, a higher rate of oxygen consumption per unit mass, along with a relatively modest antioxidant defense system. The generation and accumulation of ROS within cells is detrimental and can exacerbate the disease progression. Therapeutic strategies are now being designed to prevent or slow down ROS-mediated damages in AD. Several antioxidants have shown promising results in reducing the elevated oxidative stress, preventing synaptic loss, astrogliosis, and increased Aβ burden associated with AD. However, there are several challenges that need to be addressed for effective utilization of therapeutic strategies targeting oxidative stress. Further studies are required to account for the observed differences in clinical trials of these therapeutic strategies. An effective delivery mechanism needs to be developed for the delivery of neuroprotective agents to the target cells. Antioxidant therapies have a great potential to be utilized as an alternative therapy for AD.

REFERENCES

Abramov, A., Canevari, L., and Duchen, MR., 2004. Beta-amyloid peptides induce mitochondrial dysfunction and oxidative stress in astrocytes and death of neurons through activation of NADPH oxidase. *Journal of Neuroscience*, 24 (2), 565–575.

Ahuja, A., Dev, K., Tanwar, R., Selwal, K., and Tyagi, P., 2015. Copper mediated neurological disorder: Visions into amyotrophic lateral sclerosis, Alzheimer and Menkes disease. *Journal of Trace Elements in Medicine and Biology*, 29, 11–23.

Ak, T. and Gülçin, İ., 2008. Antioxidant and radical scavenging properties of curcumin. *Chemico-Biological Interactions*, 174 (1), 27–37.

Ali-Rahmani, F., Schengrund, C., and Connor, J., 2014. HFE gene variants, iron, and lipids: A novel connection in Alzheimer's disease. *Frontiers in Pharmacology*, 5 (165), 1–19.

Ansari, M., Abdul, H., Joshi, G., Opii, W., and Butterfield, D., 2009. Protective effect of quercetin in primary neurons against Aβ(1–42): Relevance to Alzheimer's disease. *The Journal of Nutritional Biochemistry*, 20 (4), 269–275.

Arriagada, P., Growdon, J., Hedley-Whyte, E., and Hyman, B., 1992. Neurofibrillary tangles but not senile plaques parallel duration and severity of Alzheimer's disease. *Neurology*, 42 (3), 631–631.

Atwood, C., Huang, X., Khatri, A., Scarpa, R., Kim, Y., Moir, R., Tanzi, R., Roher, A., and Bush, A., 2000. Copper catalyzed oxidation of Alzheimer Abeta. *Cellular and Molecular Biology*, 46 (4), 777–783.

Atwood, C., Perry, G., Zeng, H., Kato, Y., Jones, W., Ling, K., Huang, X. et al., 2004. Copper mediates dityrosine cross-linking of Alzheimer's amyloid-β. *Biochemistry*, 43 (2), 560–568.

Auld, D., Kornecook, T., Bastianetto, S., and Quirion, R., 2002. Alzheimer's disease and the basal forebrain cholinergic system: Relations to β-amyloid peptides, cognition, and treatment strategies. *Progress in Neurobiology*, 68 (3), 209–245.

Ayton, S., Lei, P., and Bush, A., 2013. Metallostasis in Alzheimer's disease. *Free Radical Biology and Medicine*, 62, 76–89.

Ballatore, C., Lee, V., and Trojanowski, J., 2007. Tau-mediated neurodegeneration in Alzheimer's disease and related disorders. *Nature Reviews Neuroscience*, 8 (9), 663–672.

Barnham, K., McKinstry, W., Multhaup, G., Galatis, D., Morton, C., Curtain, C., Williamson, N. et al., 2003. Structure of the Alzheimer's disease amyloid precursor protein copper binding domain. A regulator of neuronal copper homeostasis. *Journal of Biological Chemistry*, 278 (19), 17401–17407.

Bousejra-ElGarah, F., Bijani, C., Coppel, Y., Faller, P., and Hureau, C., 2011. Iron(II) binding to amyloid-β, the Alzheimer's peptide. *Inorganic Chemistry*, 50 (18), 9024–9030.

Braak, H. and Braak, E., 1991. Neuropathological stageing of Alzheimer-related changes. *Acta Neuropathologica*, 82 (4), 239–259.

Calkins, M., Manczak, M., Mao, P., Shirendeb, U., and Reddy, P., 2011. Impaired mitochondrial biogenesis, defective axonal transport of mitochondria, abnormal mitochondrial dynamics and synaptic degeneration in a mouse model of Alzheimer's disease. *Human Molecular Genetics*, 20 (23), 4515–4529.

Calkins, M., Manczak, M., and Reddy, P., 2012. Mitochondria-targeted antioxidant SS31 prevents amyloid beta-induced mitochondrial abnormalities and synaptic degeneration in Alzheimer's disease. *Pharmaceuticals*, 5 (12), 1103–1119.

Chaturvedi, R. and Beal, M., 2008. Mitochondrial approaches for neuroprotection. *Annals of the New York Academy of Sciences*, 1147 (1), 395–412.

Chauhan, V. and Chauhan, A., 2006. Oxidative stress in Alzheimer's disease. *Pathophysiology*, 13 (3), 195–208.

Chen, Z. and Zhong, C., 2014. Oxidative stress in Alzheimer's disease. *Neuroscience Bullentin*, 30 (2), 271–281.

Chiba, T., Yamada, M., Sasabe, J., Terashita, K., Shimoda, M., Matsuoka, M., and Aiso, S., 2008. Amyloid-β causes memory impairment by disturbing the JAK2/STAT3 axis in hippocampal neurons. *Molecular Psychiatry*, 14 (2), 206–222.

Demetrius, L., Magistretti, P., and Pellerin, L., 2015. Alzheimer's disease: The amyloid hypothesis and the inverse Warburg effect. *Frontiers in Physiology*, 5 (522), 1–20.

Duan, Y., Dong, S., Gu, F., Hu, Y., and Zhao, Z., 2012. Advances in the pathogenesis of Alzheimer's disease: Focusing on tau-mediated neurodegeneration. *Translational Neurodegeneration*, 1 (1), 24.

Dumont, M. and Beal, M., 2011. Neuroprotective strategies involving ROS in Alzheimer disease. *Free Radical Biology and Medicine*, 51 (5), 1014–1026.

Everett, J., Cespedes, E., Shelford, L., Exley, C., Collingwood, J., Dobson, J., van der Laan, G., Jenkins, C., Arenholz, E., and Telling, N., 2014. Ferrous iron formation following the co-aggregation of ferric iron and the Alzheimer's disease peptide -amyloid (1–42). *Journal of the Royal Society Interface*, 11 (95), 20140165–20140165.

Fariss, M., 2005. Role of mitochondria in toxic oxidative stress. *Molecular Interventions*, 5 (2), 94–111.

Finder, V., 2010. Alzheimer's disease: A general introduction and pathomechanism. *Journal of Alzheimer's Disease*, 22 (Suppl. 3), 5–19.

Francis, P., Palmer, A., Snape, M., and Wilcock, G., 1999. The cholinergic hypothesis of Alzheimer's disease: A review of progress. *Journal of Neurology, Neurosurgery and Psychiatry*, 66 (2), 137–147.

García-Escudero, V., Martín-Maestro, P., Perry, G., and Avila, J., 2013. Deconstructing mitochondrial dysfunction in Alzheimer disease. *Oxidative Medicine and Cellular Longevity*, 2013 (162152), 1–13.

Geula, C., 1998. Abnormalities of neural circuitry in Alzheimer's disease: Hippocampus and cortical cholinergic innervation. *Neurology*, 51 (1, Suppl. 1), S18–S29.

Giacobini, E., 2003. Cholinergic function and Alzheimer's disease. *International Journal of Geriatric Psychiatry*, 18 (S1), S1–S5.

Gibbs, R., 2010. Estrogen therapy and cognition: A review of the cholinergic hypothesis. *Endocrine Reviews*, 31 (2), 224–253.

Goto, Y., Niidome, T., Hongo, H., Akaike, A., Kihara, T., and Sugimoto, H., 2008. Impaired muscarinic regulation of excitatory synaptic transmission in the APPswe/PS1dE9 mouse model of Alzheimer's disease. *European Journal of Pharmacology*, 583 (1), 84–91.

Hardy, J. and Higgins, G., 1992. Alzheimer's disease: The amyloid cascade hypothesis. *Science*, 256 (5054), 184–185.

Harris, F., Brecht, W., Xu, Q., Tesseur, I., Kekonius, L., Wyss-Coray, T., Fish, J. et al., 2003. Carboxyl-terminal-truncated apolipoprotein E4 causes Alzheimer's disease-like neurodegeneration and behavioral deficits in transgenic mice. *Proceedings of the National Academy of Sciences*, 100 (19), 10966–10971.

Herholz, K., Weisenbach, S., and Kalbe, E., 2008. Deficits of the cholinergic system in early AD. *Neuropsychologia*, 46 (6), 1642–1647.

Huang, H., Chang, P., Dai, X., and Jiang, Z., 2012. Protective effects of curcumin on amyloid-β-induced neuronal oxidative damage. *Neurochemical Research*, 37 (7), 1584–1597.

Huang, X., Atwood, C., Hartshorn, M., Multhaup, G., Goldstein, L., Scarpa, R., Cuajungco, M. et al., 1999. The Aβ peptide of Alzheimer's disease directly produces hydrogen peroxide through metal ion reduction. *Biochemistry*, 38 (24), 7609–7616.

Hung, Y., Bush, A., and Cherny, R., 2012. Copper and Alzheimer disease: The good, the bad and the ugly. Y.V. Li and J.H Zhang (eds.), *Metal Ion in Stroke*, Springer Series in Translational Stroke Research, pp. 609–645.

Iqbal, K., del C. Alonso, A., Chen, S., Chohan, M., El-Akkad, E., Gong, C., Khatoon, S. et al., 2005. Tau pathology in Alzheimer disease and other tauopathies. *Biochimica et Biophysica Acta (BBA)-Molecular Basis of Disease*, 1739 (2–3), 198–210.

James, A., Sharpley, M., Manas, A., Frerman, F., Hirst, J., Smith, R., and Murphy, M., 2007. Interaction of the mitochondria-targeted antioxidant MitoQ with phospholipid bilayers and ubiquinone oxidoreductases. *Journal of Biological Chemistry*, 282 (20), 14708–14718.

Karran, E., Mercken, M., and Strooper, B., 2011. The amyloid cascade hypothesis for Alzheimer's disease: An appraisal for the development of therapeutics. *Nature Reviews Drug Discovery*, 10 (9), 698–712.

Keil, U., Bonert, A., Marques, C., Scherping, I., Weyermann, J., Strosznajder, J., Müller-Spahn, F. et al., 2004. Amyloid β-induced changes in nitric oxide production and mitochondrial activity lead to apoptosis. *Journal of Biological Chemistry*, 279 (48), 50310–50320.

Kim, J., Basak, J., and Holtzman, D., 2009. The role of apolipoprotein E in Alzheimer's disease. *Neuron*, 63 (3), 287–303.

Kim, J., Lee, H., and Lee, K., 2010. Naturally occurring phytochemicals for the prevention of Alzheimer's disease. *Journal of Neurochemistry*, 112 (6), 1415–1430.

King, M., Kan, H., Baas, P., Erisir, A., Glabe, C., and Bloom, G., 2006. Tau-dependent microtubule disassembly initiated by prefibrillar -amyloid. *The Journal of Cell Biology*, 175 (4), 541–546.

Kitazawa, M., Cheng, D., and LaFerla, F., 2009. Chronic copper exposure exacerbates both amyloid and tau pathology and selectively dysregulates cdk5 in a mouse model of AD. *Journal of Neurochemistry*, 108 (6), 1550–1560.

Lansdall, C., 2014. An effective treatment for Alzheimer's disease must consider both amyloid and tau. *Bioscience Horizons*, 7 (0), hzu002–hzu002.

Ma, Q., Li, Y., Du, J., Liu, H., Kanazawa, K., Nemoto, T., Nakanishi, H., and Zhao, Y., 2006. Copper binding properties of a tau peptide associated with Alzheimer's disease studied by CD, NMR, and MALDI-TOF MS. *Peptides*, 27 (4), 841–849.

Maccioni, R., Farías, G., Morales, I., and Navarrete, L., 2010. The revitalized tau hypothesis on Alzheimer's disease. *Archives of Medical Research*, 41 (3), 226–231.

Machová, E., Jakubík, J., Michal, P., Oksman, M., Iivonen, H., Tanila, H., and Doležal, V., 2008. Impairment of muscarinic transmission in transgenic APPswe/PS1dE9 mice. *Neurobiology of Aging*, 29 (3), 368–378.

Magwere, T., West, M., Riyahi, K., Murphy, M., Smith, R., and Partridge, L., 2006. The effects of exogenous antioxidants on lifespan and oxidative stress resistance in Drosophila melanogaster. *Mechanisms of Ageing and Development*, 127 (4), 356–370.

Maroz, A., Anderson, R., Smith, R., and Murphy, M., 2009. Reactivity of ubiquinone and ubiquinol with superoxide and the hydroperoxyl radical: Implications for in vivo antioxidant activity. *Free Radical Biology and Medicine*, 46 (1), 105–109.

McManus, M., Murphy, M., and Franklin, J., 2011. The mitochondria-targeted antioxidant MitoQ prevents loss of spatial memory retention and early neuropathology in a transgenic mouse model of Alzheimer's disease. *Journal of Neuroscience*, 31 (44), 15703–15715.

Mecocci, P. and Polidori, M., 2012. Antioxidant clinical trials in mild cognitive impairment and Alzheimer's disease. *Biochimica et Biophysica Acta (BBA)-Molecular Basis of Disease*, 1822 (5), 631–638.

Mitchell, T., Mufson, E., Schneider, J., Cochran, E., Nissanov, J., Han, L., Bienias, J. et al., 2002. Parahippocampal tau pathology in healthy aging, mild cognitive impairment, and early Alzheimer's disease. *Annals of Neurology*, 51 (2), 182–189.

Moreira, P., Smith, M., Zhu, X., Nunomura, A., Castellani, R., and Perry, G., 2005. Oxidative stress and neurodegeneration. *Annals of the New York Academy of Sciences*, 1043 (1), 545–552.

Nakamura, M., Shishido, N., Nunomura, A., Smith, M., Perry, G., Hayashi, Y., Nakayama, K., and Hayashi, T., 2007. Three histidine residues of Amyloid-β peptide control the redox activity of copper and iron. *Biochemistry*, 46 (44), 12737–12743.

Ng, L., Gruber, J., Cheah, I., Goo, C., Cheong, W., Shui, G., Sit, K., Wenk, M., and Halliwell, B., 2014. The mitochondria-targeted antioxidant MitoQ extends lifespan and improves healthspan of a transgenic caenorhabditis elegans model of Alzheimer disease. *Free Radical Biology and Medicine*, 71, 390–401.

Parajuli, B., Sonobe, Y., Horiuchi, H., Takeuchi, H., Mizuno, T., and Suzumura, A., 2013. Oligomeric amyloid β induces IL-1β processing via production of ROS: Implication in Alzheimer's disease. *Cell Death and Disease*, 4 (12), e975.

Perrotta, I. and Aquila, S., 2015. The role of oxidative stress and Autophagy in Atherosclerosis. *Oxidative Medicine and Cellular Longevity*, 2015, 1–10.

Perry, V., Nicoll, J., and Holmes, C., 2010. Microglia in neurodegenerative disease. *Nature Reviews Neurology*, 6 (4), 193–201.

Potter, P., 2013. Curcumin: A natural substance with potential efficacy in Alzheimer's disease. *Journal of Experimental Pharmacology*, 5, 23–31

Praticò, D., 2008. Oxidative stress hypothesis in Alzheimer's disease: A reappraisal. *Trends in Pharmacological Sciences*, 29 (12), 609–615.

Querfurth, H. and LaFerla, F., 2010. Alzheimer's disease. *New England Journal of Medicine*, 362 (4), 329–344.

Reddy, P., 2006. Amyloid precursor protein-mediated free radicals and oxidative damage: Implications for the development and progression of Alzheimer's disease. *Journal of Neurochemistry*, 96 (1), 1–13.

Reddy, P. and Beal, M., 2005. Are mitochondria critical in the pathogenesis of Alzheimer's disease? *Brain Research Reviews*, 49 (3), 618–632.

Rosales-Corral, S., Tan, D., Manchester, L., and Reiter, R., 2015. Diabetes and Alzheimer disease, two overlapping pathologies with the same background: Oxidative stress. *Oxidative Medicine and Cellular Longevity*, 2015, 1–14.

Sayre, L., Perry, G., Harris, P., Liu, Y., Schubert, K., and Smith, M., 2000. In Situ oxidative catalysis by neurofibrillary tangles and senile plaques in Alzheimer's disease. *Journal of Neurochemistry*, 74 (1), 270–279.

Schliebs, R. and Arendt, T., 2006. The significance of the cholinergic system in the brain during aging and in Alzheimer's disease. *Journal of Neural Transmission*, 113 (11), 1625–1644.

Selkoe, D., Mandelkow, E., and Holtzman, D., 2011. Deciphering Alzheimer disease. *Cold Spring Harbor Perspectives in Medicine*, 2 (1), a011460–a011460.

Sherrington, R., Rogaev, E., Liang, Y., Rogaeva, E., Levesque, G., Ikeda, M., Chi, H. et al., 1995. Cloning of a gene bearing missense mutations in early-onset familial Alzheimer's disease. *Nature*, 375 (6534), 754–760.

Shulman, R., Rothman, D., Behar, K., and Hyder, F., 2004. Energetic basis of brain activity: Implications for neuroimaging. *Trends in Neurosciences*, 27 (8), 489–495.

Silva, D., Selfridge, J., Lu, J., E. L., Cardoso, S., and Swerdlow, R., 2012. Mitochondrial abnormalities in Alzheimer's disease: Possible targets for therapeutic intervention. *Advances in Pharmacology*, 64, 83–126.

Smith, M., Harris, P., Sayre, L., and Perry, G., 1997. Iron accumulation in Alzheimer disease is a source of redox-generated free radicals. *Proceedings of the National Academy of Sciences*, 94 (18), 9866–9868.

Soragni, A., Zambelli, B., Mukrasch, M., Biernat, J., Jeganathan, S., Griesinger, C., Ciurli, S., Mandelkow, E., and Zweckstetter, M., 2008. Structural characterization of binding of Cu(II) to tau protein. *Biochemistry*, 47 (41), 10841–10851.

Stokin, G., Lillo, C., Falzone, T., Brusch, R., Rockenstein, E., Mount, S., Raman, R. et al., 2005. Axonopathy and transport deficits early in the pathogenesis of Alzheimer's disease. *Science*, 307 (5713), 1282–1288.

Stuart, J. and Brown, M., 2006. Mitochondrial DNA maintenance and bioenergetics. *Biochimica et Biophysica Acta (BBA)-Bioenergetics*, 1757 (2), 79–89.

Su, B., Wang, X., Nunomura, A., Moreira, P., Lee, H., Perry, G., Smith, M., and Zhu, X., 2008. Oxidative stress signaling in Alzheimers disease. *Current Alzheimer Research*, 5 (6), 525–532.

Sureshbabu, A., Ryter, S., and Choi, M., 2015. Oxidative stress and autophagy: Crucial modulators of kidney injury. *Redox Biology*, 4, 208–214.

Swerdlow, R., Burns, J., and Khan, S., 2014. The Alzheimer's disease mitochondrial cascade hypothesis: Progress and perspectives. *Biochimica et Biophysica Acta (BBA)-Molecular Basis of Disease*, 1842 (8), 1219–1231.

Szeto, H., 2008. Mitochondria-targeted cytoprotective peptides for ischemia–reperfusion injury. *Antioxidants & Redox Signaling*, 10 (3), 601–620.

Tamagno, E., Parola, M., Guglielmotto, M., Santoro, G., Bardini, P., Marra, L., Tabaton, M., and Danni, O., 2003. Multiple signaling events in amyloid β-induced, oxidative stress-dependent neuronal apoptosis. *Free Radical Biology and Medicine*, 35 (1), 45–58.

Terry, A. and Buccafusco, J., 2003. The cholinergic hypothesis of age and alzheimer's disease-related cognitive deficits: Recent challenges and their implications for novel drug development. *Journal of Pharmacology and Experimental Therapeutics*, 306 (3), 821–827.

Thanan, R., Oikawa, S., Hiraku, Y., Ohnishi, S., Ma, N., Pinlaor, S., Yongvanit, P., Kawanishi, S., and Murata, M., 2014. Oxidative stress and its significant roles in neurodegenerative diseases and cancer. *International Journal of Molecular Sciences*, 16 (1), 193–217.

Vardy, E., Catto, A., and Hooper, N., 2005. Proteolytic mechanisms in amyloid-β metabolism: Therapeutic implications for Alzheimer's disease. *Trends in Molecular Medicine*, 11 (10), 464–472.

Walker, D. and Lue, L., 2007. Anti-inflammatory and immune therapy for Alzheimers disease: Current status and future directions. *Current Neuropharmacology*, 5 (4), 232–243.

Wang, J., Xiong, S., Xie, C., Markesbery, W., and Lovell, M., 2005. Increased oxidative damage in nuclear and mitochondrial DNA in Alzheimer's disease. *Journal of Neurochemistry*, 93 (4), 953–962.

Wang, Y., Greig, N., Yu, Q., and Mattson, M., 2009. Presenilin-1 mutation impairs cholinergic modulation of synaptic plasticity and suppresses NMDA currents in hippocampus slices. *Neurobiology of Aging*, 30 (7), 1061–1068.

Wischik, C., Harrington, C. and Storey, J., 2014. Tau-aggregation inhibitor therapy for Alzheimer's disease. *Biochemical Pharmacology*, 88 (4), 529–539.

Yao, E. and Xue, L., 2014. Therapeutic effects of curcumin on Alzheimer's disease. *Advances in Alzheimer's Disease*, 03 (04), 145–159.

Zhang, X., Yin, W., Shi, X., and Li, Y., 2011. Curcumin activates Wnt/β-catenin signaling pathway through inhibiting the activity of GSK-3β in APPswe transfected SY5Y cells. *European Journal of Pharmaceutical Sciences*, 42 (5), 540–546.

Zhao, K., Luo, G., Giannelli, S., and Szeto, H., 2005. Mitochondria-targeted peptide prevents mitochondrial depolarization and apoptosis induced by tert-butyl hydroperoxide in neuronal cell lines. *Biochemical Pharmacology*, 70 (12), 1796–1806.

Zhao, K., Zhao, G., Wu, D., Soong, Y., Birk, A., Schiller, P., and Szeto, H., 2004. Cell-permeable peptide antioxidants targeted to inner mitochondrial membrane inhibit mitochondrial swelling, oxidative cell death, and reperfusion injury. *Journal of Biological Chemistry*, 279 (33), 34682–34690.

Zhou, L., Du, J., Zeng, Z., Wu, W., Zhao, Y., Kanazawa, K., Ishizuka, Y., Nemoto, T., Nakanishi, H., and Li, Y., 2007. Copper (II) modulates in vitro aggregation of a tau peptide. *Peptides*, 28 (11), 2229–2234.

10 Oxidative Stress and Mitochondrial Dysfunction in Huntington's and Other Neurodegenerative Diseases

Abhishek Chandra and Ashu Johri

CONTENTS

ABSTRACT

Over the last few decades, there has been an overwhelming evidence for a crucial role of mitochondrial dysfunction and oxidative stress in neurodegenerative diseases. Oxidative stress and mitochondrial dysfunction are intertwined, and one leads to the other turning on a cycle of toxic effects, leading ultimately to cellular demise. Mitochondrial dysfunction has catastrophic effects especially for neurons because of their high energy demands, polarized cellular architecture, and limited regenerative capacity. In this chapter, reactive oxygen species production and mitochondrial health will be addressed in the backdrop of major neurodegenerative disorders affecting millions of people worldwide. We also discuss in a greater detail the evidence for and therapeutic approaches targeting oxidative stress and mitochondrial dysfunction in Huntington's disease.

10.1 INTRODUCTION

10.1.1 NEURODEGENERATIVE DISEASES

There exists a large group of disabling disorders of the nervous system, characterized by the relative selective and progressive death of neuronal subtypes. For example, in Parkinson's disease (PD), there is a selective depletion of dopaminergic neurons in the basal ganglia and substantia nigra, and in Huntington's disease (HD), there is a preferential and progressive loss of medium spiny neurons (MSNs) in the striatum. Alzheimer's disease (AD), PD, amyotrophic lateral sclerosis (ALS), and HD are prototypical examples of neurodegenerative diseases, and these and other polyglutamine disease such as prion diseases are often regarded as proteinopathies—diseases in which a particular protein or set of proteins misfolds and aggregates [1].

Mutations have been identified in an increasing number of genes underlying specific forms of neurodegeneration—for example, genes encoding amyloid precursor protein (APP) and presenilins 1 and 2 in AD, parkin and α-synuclein in PD, Cu/Zn superoxide dismutase in ALS, tau in fronto-temporal dementia, prion protein in Creutzfeldt–Jakob and Gerstmann–Straussler diseases, and at least eight different proteins in the polyglutamine diseases.

AD is defined by progressive impairments in memory and cognition and by the presence of extracellular neuritic plaques and intracellular neurofibrillary tangles (NFTs). β-amyloid peptide (Aβ) is the major component of the plaque, while the tangles are composed of hyperphosphorylated tau proteins. Advanced age is the greatest risk factor for AD, and it has early versus the late onset and sporadic and autosomal dominant types.

PD is the second most common neurodegenerative disorder after AD. Clinically, PD is character-ized by the triad of resting tremor, bradykinesia, and rigidity. These symptoms are considered to be a direct consequence of neurodegeneration and loss of dopaminergic neurons. Pathologically, the hallmark feature of PD is loss of pigmented dopaminergic neurons in the substantia nigra and the presence of abnormal protein aggregates called Lewy bodies, which are cytoplasmic eosinophilic inclusions composed of the presynaptic protein α-synuclein.

ALS is a progressive neurodegenerative disease that targets motor neurons in the brain and spinal cord, resulting in muscle weakness, atrophy, and eventual death. Although there have been exten-sive research efforts investigating the pathogenesis of ALS, its etiology is still largely unknown. In about 20% of the familial ALS cases, the disease is associated with one or more mutations in the gene that encodes Cu–Zn superoxide dismutase (SOD1). Dominant mutations in two DNA-/RNA-binding proteins, transactive response DNA binding protein 43 kDa (TDP-43) and fused in sarcoma/translocated in sarcoma (FUS/TLS), are also reported and account for ~5% and 4% of ALS cases, respectively [2]. Mutations in the valosin-containing protein (VCP, also known as tran-sitional endoplasmic reticulum [ER] ATPase) gene were recently reported to be the cause of 1%–2% of familial ALS cases. The ultimate cause of neuronal death and the debilitating phenotypes in ALS is currently unknown; however, several studies have reported mitochondrial damage and dysfunc-tion in human ALS patients and in SOD1 mutant transgenic mice [3].

HD is a dominantly inherited progressive neurodegenerative disease, caused by a variable length CAG repeat expansion in the huntingtin (Htt) gene that translates into an abnormally long poly-glutamine repeat in the mutant huntingtin (mHtt) protein. The disease is characterized by progres-sive motor impairment, personality changes, psychiatric illness, and gradual intellectual decline. Pathologically, there is a preferential and progressive loss of the MSNs in the striatum, as well as cortical atrophy, and degeneration of other brain regions later in the disease.

Other than the aforementioned diseases, examples of neurodegenerative disorders include prion dis-eases (Creutzfeldt–Jakob and Gerstmann–Straussler diseases), spinocerebellar ataxia (SCA), spinal mus-cular atrophy (SMA), Friedreich's ataxia, and Lewy body disease, some of which are discussed in detail in [4]. Chapters 8 and 9 of this book addresses some of the other neurodegenerative diseases; this chapter will mostly be focusing on oxidative stress and mitochondrial dysfunction leading to HD.

10.1.2 OXIDATIVE STRESS

The harmful effects of reactive oxygen species (ROS), causing damage to macromolecules, such as proteins, lipids, polysaccharides, and nucleic acids, are termed "oxidative stress." Free radicals are molecules or molecular fragments containing one or more unpaired electrons in their atomic

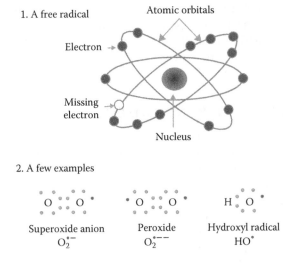

1. A free radical

Atomic orbitals

Electron →

Missing electron

Nucleus

2. A few examples

Superoxide anion
$O_2^{\bullet -}$

Peroxide
$O_2^{\bullet --}$

Hydroxyl radical
$HO^{\bullet}$

FIGURE 10.1 **(See color insert.)** Free radicals are molecules or molecular fragments that have one or more unpaired electrons in their atomic or molecular orbitals.

or molecular orbitals [5] (Figure 10.1). The unpaired electron(s) usually provide a large amount of reactivity to the free radical. Radicals derived from oxygen represent the most important class of radical species generated in living systems [6]. Molecular oxygen (dioxygen) has a unique electronic configuration and is itself a radical. ROS are normally generated during reactions of cellular metabolism. ROS include superoxide ($O_2^{-\bullet}$), hydroxyl, and peroxyl free radicals, as well as nitrogen intermediates (NO and peroxynitrite [ONOO–]). The addition of one electron to dioxygen forms the ($O_2^{-\bullet}$) [6]. Production of this radical occurs mostly within the mitochondrion. Under the stress conditions, an excess of $O_2^{\bullet -}$ releases free iron from iron-containing molecules, which can participate in the Fenton reaction, generating the highly reactive hydroxyl radical ($^{\bullet}OH$). Additional reactive radicals derived from oxygen that can be formed in living systems are peroxyl radicals ($ROO^{\bullet}$). The simplest peroxyl radical is $HOO^{\bullet}$, which is the protonated form of superoxide $O_2^{-\bullet}$, and is usually termed either $^{\bullet}OH$ or perhydroxyl radical (Figure 10.1). It is known that $^{\bullet}OH$ initiates fatty acid peroxidation [7]. $O_2^{-\bullet}$ can also rapidly react with NO in the extracellular space to form $ONOO^-$, which can readily cross cell membranes and damage intracellular components [8].

ROS are produced constantly by aerobic cells through diverse metabolic pathways. They serve as specific signaling molecules in both normal and pathological conditions, and their transient generation (see Table 10.1 for the extremely short half-lives of free radicals) within boundaries is essential

TABLE 10.1

Examples of Free Radicals with Their Respective Half-Lives

Free Radical Half-Life at 37°C		
Radical	**Symbol**	**Half-Life Time (in Seconds)**
Hydroxyl	OH	1×10^{-9}
Singlet oxygen	$^1O_2$	1×10^{-6}
Superoxide	O_2^-	1×10^{-6}
Alkoxyl	OL	1×10^{-6}
Peroxyl	LOO	10×10^{-3}
Nitric oxide	NO	Few seconds

to maintain homeostasis. ROS can inflict oxidative molecular damage to lipids, proteins, and DNA when their production overwhelms the capacity of antioxidant systems. The imbalance is induced by ROS mainly of mitochondrial origin. Biological effects initiated by ROS can elicit a wide range of phenotypic responses that vary from the activation of gene expression, proliferation to growth arrest, and to senescence or cell death [9–11].

10.2   MITOCHONDRIA: SUPERORGANELLES?

Mitochondria (Greek: *Mitos* = thread + *khondrion* = small grain or granule), originally termed as "powerhouses" of the cell, are now well known to house hundreds of biochemical reactions from energy production to amino acid and lipid synthesis, to hormone production. In addition, mitochondria are also involved in apoptosis, regulation of calcium homeostasis, cellular differentiation, signaling, cell death, and control of cell cycle and cell growth. During the last decade alone, there has been a tremendous focus on these superorganelles with the increasing vital understanding of how they move, fuse, and divide, and how each of these processes is altered in the course of neurodegenerative diseases. There is extensive biochemical cross talk between mitochondria and other subcellular organelles, such as the ER, peroxisomes, and endosomes, which is essential for the variety of cellular functions that mitochondria take part in.

Adenosine triphosphate (ATP), the cell's "energy currency," is the end product of a series of pathways involving oxidation of substrates, mainly carbohydrates and fat, in cytosol (glycolysis) and in mitochondria (pyruvate decarboxylation, tricarboxylic acid cycle [TCA or Krebs cycle], and oxidative phosphorylation [OXPHOS/respiratory chain complex]). The respiratory chain complex consists of four distinct multisubunit complexes (I–IV) and two electron carriers, which generate a proton gradient across the mitochondrial inner membrane, which in turn drives ATP synthase (complex V) to generate ATP (Figure 10.2).

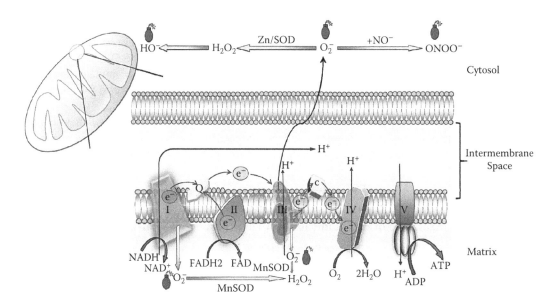

FIGURE 10.2   (See color insert.) The respiratory chain complex resides in the inner mitochondrial membrane and consists of four distinct multisubunit complexes (I–IV) and two electron carriers (CoQ and Cyt c) that generate a proton gradient across the mitochondrial inner membrane, which in turn drives ATP synthase (complex V) to generate ATP. During energy transduction, a small number of electrons *leak* to oxygen prematurely to generate superoxide, instead of contributing to the reduction of oxygen to water.

10.2.1 Mitochondria and Oxidative Stress

Mitochondria are thought to be important sources, targets, and sink of ROS ([12] and references therein). The mitochondrial electron transport chain is the main source of ATP in the mammalian cell and thus is essential for life. During energy transduction, a small number of electrons "leak" to oxygen prematurely to generate $O_2^{-\bullet}$, instead of contributing to the reduction of oxygen to water [13] (Figure 10.2). $O_2^{-\bullet}$ is produced from both complexes I and III of the electron transport chain, and once in its anionic form, it is too strongly charged to readily cross the inner mitochondrial membrane. It has been demonstrated that complex I–dependent $O_2^{-\bullet}$ is exclusively released into the matrix and that no detectable levels escape from intact mitochondria [14]. It has been shown that oxidative stress stimulates mitochondrial fission; the addition of H_2O_2 to cultured cerebellar granule neurons induced mitochondrial fragmentation within 1 h of treatment [15]. It was also shown that NO causes increased mitochondrial fission in neurons, prior to the onset of neuronal loss in a mouse model of stroke [16]. On the other hand, the expression of mitofusin (Mfn) gene to facilitate fusion or a dominant-negative dynamin-related protein (Drp)1 to block fission in cultured neurons was protective against oxidative insults [15,16]. The generation of ROS is increased in damaged mitochondria and in cells with compromised mitochondrial function. Acute exposure to relatively high levels of ROS, especially in the presence of calcium, can induce the mitochondrial permeability transition, uncouple OXPHOS with catastrophic effects on mitochondrial energetics, and contribute to cytotoxicity via necrosis and/or apoptosis. Oxidative stress within mitochondria can lead to a vicious cycle in which ROS production progressively increases especially when the mitochondrial ROS defense system is overwhelmed, leading, in turn, to progressive augmentation of damage. If such damaged mitochondria escape destruction and fuse with normal, healthy mitochondria, it may turn into a chain reaction, resulting in defective energy production with catastrophic consequences for the cell.

10.2.2 Mitochondrial Dysfunction, Oxidative Stress, and Neurodegeneration

The intrinsic properties of neurons make them highly vulnerable to the detrimental effects of ROS: high metabolic rates, a rich composition of fatty acids prone to peroxidation, high intracellular concentrations of transition metals, capable of catalyzing the formation of reactive $^{\bullet}OH$, low levels of antioxidants, inability to divide, and reduced capability to regenerate. Neurons have intense energy demands, which are met by mitochondria.

Nucleic acid oxidation occurs in neurons during disease and is detected as elevated levels of 8-hydroxy-2-deoxyguanosine (8-OHdG) in DNA and 8-hydroxyguanosine in RNA. $^{\bullet}OH$-mediated DNA damage often results in strand breaks, DNA–protein cross-linking, and base modifications. All of these events can lead to neuronal injury. It is known that the generation of ROS results in an attack not only on DNA but also on other cellular components involving polyunsaturated fatty acid residues of phospholipids, which are extremely sensitive to oxidation [17]. Once formed, peroxyl radicals (ROO$^{\bullet}$) can be rearranged via a cyclization reaction to endoperoxides (precursors of malondialdehyde, MDA), with the final product of the peroxidation process being MDA. The major aldehyde product of lipid peroxidation other than MDA is 4-hydroxy-2-nonenal (HNE). Increased production of ROS also results in protein oxidation. The side chains of all amino acid residues of proteins, in particular cysteine and methionine residues of proteins, are susceptible to oxidation by the action of ROS/reactive nitrogen species (RNS) [18]. Oxidation of cysteine residues may lead to the reversible formation of mixed disulfides between protein thiol groups (–SH) and low–molecular weight thiols, in particular glutathione (GSH, S-glutathiolation). The concentration of carbonyl groups, generated by many different mechanisms, is a good measure of ROS-mediated protein oxidation. The generation of isoprostanes has been shown to be a sensitive measure of lipid peroxidation, which is increased in cerebrospinal fluid (CSF) of HD patients [19].

Oxidative stress increases with age in the brain, and the ability of cells to respond to oxidative protein damage also declines, contributing to the buildup of oxidatively damaged proteins. ROS are often present in brain regions affected by neurodegenerative diseases. Increased oxidative alterations to proteins such as α-synuclein in PD, β-amyloid in AD, and SOD1 in ALS may result in increased protein misfolding and impaired degradation, leading to toxic accumulation of insoluble aggregates in the diseased brains and an exacerbation of neurodegeneration. There is extensive evidence of increased oxidative stress and impaired mitochondrial function, and, more recently, evidence has emerged for impaired mitochondrial dynamics (shape, size, fission–fusion, distribution, movement, etc.) in PD, HD, ALS, and AD. Here, we provide a concise discussion of the major findings in recent years and the progress made in the field of HD, highlighting the importance of healthy mitochondria for a healthy neuron.

10.3 OXIDATIVE STRESS AND MITOCHONDRIAL DYSFUNCTION IN HD

In HD, mHtt protein fragments with the expanded polyglutamine domain bind to other proteins more tightly and aggressively, activating a variety of damaging and compensatory molecular pathways, thus rendering the neurons more vulnerable to oxidative stress and other types of stressors. There is evidence for a link between mHtt protein, and mitochondrial abnormalities and subsequent impaired mitochondrial function and dynamics that could ultimately lead to neuronal damage and degeneration in affected brain regions of patients with HD [20–26]. mHtt was shown to interact directly with mitochondria and with mitochondrial fission complex and trip the balance in favor of mitochondrial fission that is harmful for mitochondria and in combination with impaired mitophagy can be deadly for neurons. mHtt also interferes with mitochondrial trafficking along axons and dendrites [27]. The expression of mHtt causes abnormal mitochondrial ultrastructure, impaired calcium buffering, bioenergetic defects, and mitochondrial DNA deletions, all of which may be a consequence of a failure to maintain a balance between mitochondrial fission and fusion [22,23].

Evidence of oxidative DNA damage (increased levels of the oxidative stress biomarker 8-OHdG) and loss of mitochondrial function (reduced levels of the electron transport proteins cytochrome b and cytochrome c oxidase 1) were found in individuals with HD, and these patients also exhibited markedly increased levels of mHtt oligomers in the cerebral cortex. Importantly, immunolabeling of brain samples showed the presence of mHtt oligomers in the nuclei of neurons and mitochondria of patients with HD. HD MSNs show a preferential vulnerability to the toxic effects of mHtt protein probably owing to reduced numbers of mitochondria due to reduced levels of peroxisome proliferator–activated receptor (PPAR)-γ coactivator-1α (PGC-1α) [21]. MSNs also have long projections as compared to interneurons, which could make them preferentially vulnerable to increased mitochondrial fission [21,26,28].

There is overwhelming evidence demonstrating oxidative damage in HD: (1) elevated markers of oxidative damage (hemeoxygenase, 3-nitrotyrosine, MDA) in human HD striatum and cortex by immunohistochemistry, (2) increased levels of MDA and 4-hydroxynonenal in HD patients using biochemical assays, (3) increased lipid peroxidation in plasma that correlates with the degree of severity in patients with HD, (4) increased global oxidative stress, a reduction in antioxidant systems, that correlates with disease stage in patients with HD, (5) increased cytoplasmic lipofuscin and increased DNA fragmentation in HD patients where the latter correlates with CAG repeat length, (6) oxidative modifications of aldolase C, GFAP, tubulin, γ-enolase, and creatine kinase B in both striatum and cortex from HD patients, (7) inability of mitochondria from striatal neurons of postmortem brains of HD patients to handle large Ca^{2+} loads, and, more recently, (8) oxidation of mitochondrial enzymes (pyridoxal kinase and antiquitin 1) resulting in decreased catalytic activity in the striatum samples of HD patients, providing a link to the bioenergetic deficits observed in HD [29–40]. Pyridoxal kinase and antiquitin 1 oxidation could result in decreased pyridoxal 5-phosphate availability, which in turn is necessary as a cofactor in transamination, synthesis of GSH, and synthesis of GABA and dopamine, two neurotransmitters that play a key role in HD pathology. It is well known that the administration of the mitochondrial toxins, 3-nitropropionic acid (3-NP) and malonate, both of which are selective inhibitors of succinate dehydrogenase, to non-human primates and rodents results in central nervous system (CNS) lesions that selectively target medium-sized spiny neurons within the striatum, recapitulating the regional and neuronal specificity of pathologic events in HD. In these toxins and in the excitotoxin models of HD, increased oxidative damage has been consistently observed [41]. All the aforementioned findings most of which are also paralleled in transgenic and neurotoxin mouse models of HD suggest that oxidative stress plays an extremely important role in the pathogenesis of HD.

At the same time, extensive efforts have been dedicated to the study of mitochondrial dysfunction and bioenergetic deficits in HD, and it has been shown that there is (1) increased lactate in the cerebral cortex and basal ganglia by nuclear magnetic resonance spectroscopy, (2) reduced enzymatic activity of OXPHOS complexes II and III and aconitase in the basal ganglia, (3) abnormal mitochondrial membrane depolarization in patient lymphoblast, (4) abnormal ultrastructure of mitochondria in cortical biopsies obtained from patients with both juvenile and adult-onset HD, and (5) pathologic grade–dependent reductions in number and size of mitochondria, along with significantly reduced levels of Tfam—a regulator of mtDNA transcription and replication, and PGC-1α—a key transcriptional regulator of energy metabolism and mitochondrial biogenesis in HD postmortem brain tissue. There is abnormal mitochondrial dynamics: mitochondrial fission proteins Drp1 and Fis1 and mitochondrial matrix protein CypD are significantly increased and mitochondrial fusion proteins Mfn1, Mfn2, Opa1, and Tomm40 are significantly decreased, dysregulation of mitochondrial fission–fusion events, and abrupt mitochondrial trafficking along axons and dendrites in patients with HD [20,21,24–26,42–47]. In transgenic mouse models of HD—there is impaired brain creatine kinase activity and significant alterations in levels of high-energy phosphate intermediates as well as impairment of the PGC-1α pathway, and in striatal cells from mutant Htt knock-in mice, mitochondrial respiration and ATP production are significantly impaired [42–47]. A milestone discovery for mitochondrial involvement in pathogenesis of HD was that the phenotypic and neuropathological features of HD can be modeled in rodents and primates, with the mitochondrial toxin 3-NP [48].

10.4 THERAPEUTIC APPROACHES TARGETING OXIDATIVE STRESS AND MITOCHONDRIAL DYSFUNCTION IN HD

As discussed in the previous section, there is no dearth of data supporting the case for an increased oxidative stress and impaired mitochondrial function in HD [41,47,49]. Consequently, there has been immense focus on therapies targeting oxidative stress and mitochondrial dysfunction in HD:

10.4.1 THERAPIES TARGETING NRF2/ARE PATHWAY

Synthetic triterpenoids (TPs) are analogues of oleanolic acid and are powerful inhibitors of oxidative stress and cellular inflammatory processes. Synthetic TP compounds are potent inducers of the antioxidant response element/Nuclear factor (erythroid-derived 2)-like 2, Kelch-like ECH-associated protein 1 (ARE/Nrf2/Keap1) signaling pathway. Following activation by TP, Nrf2 dissociates from Keap1, translocates to the nucleus, and binds to the ARE promoter sequences, leading to coordinated induction of a battery of cytoprotective genes including antioxidants and anti-inflammatory genes. Recently, we have tested neuroprotective effects of the synthetic TP CDDO-methylamide (CDDO-MA), which is a potent activator of the Nrf2/ARE signaling pathway [50]. CDDO-MA produced marked protection in the 3-NP rat model and both the acute and chronic 1-methyl-4-phenyl-1,2,3,6-tetrahydropyridine (MPTP) mouse models. CDDO-MA exerted significant protection against tertbutylhydroperoxide-induced ROS in vitro. It increased the expression of genes involved in mitochondrial biogenesis, as well as those involved in GSH synthesis and in the expression of antioxidant enzymes [50]. Triterpenoids display neuroprotective effects in transgenic mouse models of ALS, HD and AD.

Several bioenergetic agents have efficacy in improving mitochondrial function including creatine, coenzyme Q10, nicotinamide, riboflavin, lipoic acid, L-carnitine, pyruvate, etc. (Reviewed in [49,51]). Coenzyme Q (CoQ) is an essential biologic factor of electron transport chain, where it accepts electrons from complexes I and II. It also serves as an important antioxidant in mitochondrial lipid membranes. We showed that oral administration of CoQ10 protects against lesions produced by aminooxyacetic acid and the mitochondrial toxins malonate and 3-NP. High-dose CoQ10 significantly extends survival and improves motor performance, grip strength, and brain atrophy in R6/2 HD mice in a dose-dependent manner. Furthermore, we found that the combination of creatine and CoQ10 exerts additive neuroprotective effects in the MPTP model of PD, the 3-NP model of HD, and in a transgenic mouse model of HD [52]. A recent phase III clinical trial in PD was unsuccessful [53]; however, trials in HD and Friedreich's ataxia are continuing.

10.4.2 SMALL MOLECULE/COMPOUND THERAPEUTICS

Several compounds are being developed, which can specifically target mitochondria. These include compounds such as mitoQ, a form of coenzyme Q linked to triphosphonium ions, resulting in selective accumulation within mitochondria; there are also novel peptide antioxidants termed SS31 and SS20 that bind to the inner mitochondrial membrane, which have not been tested in HD but are neuroprotective in transgenic mouse models of ALS as well as in neurotoxin models and in vitro models of AD [54,55]. SS31 and SS20 provide neuroprotection and decrease oxidative stress in the MPTP-induced model of PD [56]. A mitochondrial antioxidant (TEMPOL) coupled to mitochondrial targeting segment of gramicidin S (XJB-5-131) showed excellent neuroprotective effects in a transgenic mouse model of HD, where it preserved mitochondrial function, improved behavior, and enhanced neuronal survival [57]. Dexpramipexole, which is an isomer of the dopamine agonist pramipexole that accumulates in mitochondria, exerts antioxidant effects, and inhibits the activation of the mitochondrial permeability transition, showed efficacy in a phase II clinical trial in ALS, where it produced improvement on the ALS functional rating scale, as well as on mortality [58]. However, a phase III trial showed that dexpramipexole was generally well tolerated but did not differ from placebo on any prespecified efficacy end-point measurement [59]. Other small molecules, such as mitochondrial division inhibitor-1 (mdivi-1), dynasore, and P110, that block mitochondrial fission are also being developed to combat neurodegeneration [24,60–63].

In light of the direct toxic effects of mHtt protein and the direct interaction of mHtt with mitochondria, it also makes sense to *simply* lower the expression levels of mHtt at the level of DNA or RNA, which is then expected to reduce the downstream deleterious effects of the protein that lead to the manifestations of HD. Such therapeutics involve RNA interference using short interfering RNA, translational repression using single-stranded DNA-based antisense oligonucleotides, and transcriptional repression using zinc finger proteins ([64], reviewed in [65]).

10.4.3 THERAPIES TARGETING PGC-1 PATHWAY IN COMBINATION WITH OTHERS

PGC-1α pathway in combination with sirtuins, specifically silent mating type information regulation 2 homolog 1 (Sirt 1) and Sirt 3, PPAR-α, β, and γ, and 5' adenosine monophosphate-activated protein kinase (AMPK) is being extensively explored for use as a therapeutic intervention in HD [66,67]. One particular aspect of PGC-1α, that is, its ability to activate a diverse set of metabolic programs including mitochondrial and antioxidant response, puts it at the center stage. Indeed, overexpression of PGC-1α was shown to enhance the mitochondrial membrane potential and reduce mitochondrial toxicity in in vitro models of HD, and lentiviral delivery of PGC-1α to the striatum of R6/2 HD mice completely prevented striatal atrophy at the site of PGC-1α injection [42,43]. A cross of the mice inducibly overexpressing PGC-1α with a transgenic mouse model of HD showed that increased PGC-1α function could ameliorate neuronal loss and some of the neurological symptoms of HD [68]. PGC-1α overexpression eliminates aggregates of mHtt protein in the brains of the HD mice by switching on the expression of transcription factor EB (TFEB), a master regulatory transcription factor that activates genes in the autophagy–lysosome pathway of protein turnover [68,69]. These findings highlight the important role of PGC-1α in maintaining mitochondrial quality control, accelerating mitochondrial biogenesis, and increasing ATP generation.

Administration of a PPAR-γ agonist, thiazolidinedione, was shown to produce beneficial effects on weight loss, mHtt aggregates, and global ubiquitination profiles in R6/2 mice [70]. In STHdhQ111 cells, PPAR-γ activation by rosiglitazone prevented the mitochondrial dysfunction and oxidative stress that occur when mutant striatal cells are challenged with pathological increases in calcium [71]. Recently, Jin et al. [72] showed that rosiglitazone significantly attenuated mHtt-induced toxicity in striatal cells and significantly improved motor function and attenuated hyperglycemia in N171-82Q HD mice. Rosiglitazone rescued brain-derived neurotrophic factor (BDNF) deficiency in the cerebral cortex, prevented PGC-1α reduction, increased Sirt6 protein levels, and prevented the loss of orexin-A-immunopositive neurons in the brain of N171-82Q HD mice [72]. We previously showed that bezafibrate, which is a pan-PPAR agonist, improved the expression of PGC-1α and downstream target genes, improved behavioral deficits, survival, and striatal atrophy, and reduced oxidative damage in the R6/2 transgenic mouse model of HD [73]. The Sirt 1 activator, resveratrol, increases the activity of PGC-1α and improves mitochondrial activity as a consequence of its deacetylation of PGC-1α, which increases its effects on liver, fat, and muscle metabolism. The metabolic effects of resveratrol were shown to result from competitive inhibition of cyclic adenosine monophosphate (cAMP)-degrading phosphodiesterases, leading to increased cAMP levels, which in turn leads to an increase in intracellular Ca^{+2} levels and activation of CamKKβ–AMPK pathway. This cascade culminates into an increase in NAD^+ and activation of Sirt1 [74]. Resveratrol protects against 3-NP-induced motor and behavioral deficits [75,76]. Resveratrol treatment of the N171-82Q HD transgenic mice produced increased PGC-1α, reduced the apparent vacuolization in brown adipose tissue, and reduced glucose levels [77]. Although there is some controversy surrounding the use of resveratrol, these studies serve as a foundation for the exploration of more sirtuin agonists by itself or in combination with other potential PGC-1α activators.

PGC-1α plays an important role in the suppression of oxidative stress, and it induces mitochondrial uncoupling proteins and antioxidant enzymes, including SOD1, SOD2 (MnSOD), and Gpx-1 [78]. The mitochondrial antioxidant enzymes form the first line of defense against mitochondrial ROS, including SOD2, the enzyme that scavenges $O_2^{-\bullet}$ to produce H_2O_2, and peroxiredoxin III (Prx3), V (Prx5), mitochondrial thioredoxin (Trx2), and mitochondrial thioredoxin reductase (TrxR2). PGC-1α was found to be directly associated with the regulatory promoter sequences of SOD2, uncoupling protein-2 (UCP-2), and Prx5 [79]. PGC-1α controls the expression of Sirt3 in mitochondria, which in turn activates SOD2 by deacetylating it and reduces ROS [80–82]. The ability of Sirt3 activation to enhance fatty acid oxidation may also be beneficial [83]. Therefore, therapeutic approaches targeting PGC-1α may be beneficial both in improving mitochondrial function and biogenesis as well as in restoring the expression of antioxidant enzymes and ameliorating oxidative damage in HD.

10.5 CONCLUSION

HD is a devastating illness, and at present, there is no cure for it; management of the disease is limited to a few treatment options for the control of some of the hyperkinetic movement and psychiatric problems. Several interesting potential targets for drug intervention to fight neurodegeneration in HD are being developed. Some of these are single molecules, whereas others utilize key factors in complex and intricate biochemical pathways. Single molecule approaches include direct targeting of the Htt gene and its protein product mHtt with RNAi, antisense RNA oligonucleotides, and antibodies. The use of siRNA can decrease mHtt expression and ameliorate the phenotype in mouse models of HD, and promising results have been shown with antisense oligonucleotides infused directly into the lateral ventricles of mouse models of HD. Although these approaches have shown promise in transgenic mice, there are obstacles that need to be overcome to ensure success in the clinic, such as delivery to the appropriate sites, and to overcome immunologic defenses. Therapeutic approaches targeting PGC-1α and its regulators have beneficial effects in treating HD in cell culture and mouse models, and it remains to be seen how one or more of these agents fare in clinical trials. Agents that can transcriptionally activate the Nrf2/ARE pathway, leading to the increased expression of antioxidant enzymes, and chaperone proteins, as well as reduce the production of oxidants by iNOS and cyclooxygenase-2, also confer neuroprotection in HD. An advantage of these approaches is that they modulate endogenous neuroprotective pathways and therefore hold great promise for the development of neuroprotective therapies to ameliorate or halt the progression of HD.

REFERENCES

1. Menzies FM, Fleming A, Rubinsztein DC. Compromised autophagy and neurodegenerative diseases. *Nat. Rev. Neurosci.* 2015; 16: 345–357.
2. Da Cruz S, Cleveland DW. Understanding the role of TDP-43 and FUS/TLS in ALS and beyond. *Curr. Opin. Neurobiol.* 2011; 21: 904–919.
3. Beal MF. Mitochondria take center stage in aging and neurodegeneration. *Ann. Neurol.* 2005; 58: 495–505.
4. Ahmad SI, Ed. *Neurodegenerative Diseases*, Springer, New York, 2012.
5. Halliwell B, Gutteridge JMC. Oxidative stress. In: *Free Radicals in Biology and Medicine*, 3rd ed., Halliwell B and Gutteridge JMC, Eds., Oxford University Press, Oxford, U.K., 1999.
6. Miller DM, Buettner GR, Aust SD. Transition metals as catalysts of "autoxidation" reactions. *Free Radic. Biol. Med.* 1990; 8: 95–108.
7. Aikens J, Dix TA. Perhydroxyl radical (HOO.) initiated lipid peroxidation: The role of fatty acid hydroperoxides. *J. Biol. Chem.* 1991; 266: 15091–15098.
8. Beckman JS, Koppenol WH. Nitric oxide, superoxide, and peroxynitrite: The good, the bad, and ugly. *Am. J. Physiol.* 1996; 271: C1424–C1437.
9. Droge W. Free radicals in the physiological control of cell function. *Physiol. Rev.* 2002; 82: 47–95.
10. Hensley K et al. Reactive oxygen species, cell signaling, and cell injury. *Free Radic. Biol. Med.* 2000; 28: 1456–1462.
11. Finkel T. Reactive oxygen species and signal transduction. *IUBMB Life* 2001; 52: 3–6.
12. Starkov AA. The role of mitochondria in reactive oxygen species metabolism and signaling. *Ann. N. Y. Acad. Sci.* 2008; 1147: 37–52.
13. Kovacic P et al. Mechanism of mitochondrial uncouplers, inhibitors, and toxins: Focus on electron transfer, free radicals, and structure-activity relationships. *Curr. Med. Chem.* 2005; 12: 2601–2623.
14. Muller FL, Liu Y, Van Remmen H. Complex III releases superoxide to both sides of the inner mitochondrial membrane. *J. Biol. Chem.* 2004; 279: 49064–49073.
15. Jahani-Asl A et al. Mitofusin 2 protects cerebellar granule neurons against injury-induced cell death. *J. Biol. Chem.* 2007; 282: 23788–23798.
16. Barsoum MJ et al. Nitric oxide-induced mitochondrial fission is regulated by dynamin-related GTPases in neurons. *EMBO J.* 2006; 25: 3900–3911.
17. Siems WG, Grune T, Esterbauer H. 4-Hydroxynonenal formation during ischemia and reperfusion of rat small intestine. *Life Sci.* 1995; 57: 785–789.
18. Stadtman ER. Role of oxidant species in aging. *Curr. Med. Chem.* 2004; 11: 1105–1112.
19. Montine TJ et al. Cerebrospinal fluid F2-isoprostanes are elevated in Huntington's disease. *Neurology* 1999; 52: 1104–1105.
20. Browne SE, Beal MF. The energetics of Huntington's disease. *Neurochem. Res.* 2004; 29: 531–546.
21. Kim J et al. Mitochondrial loss, dysfunction and altered dynamics in Huntington's disease. *Hum. Mol. Genet.* 2010; 19: 3919–3935.
22. Costa V et al. Mitochondrial fission and cristae disruption increase the response of cell models of Huntington's disease to apoptotic stimuli. *EMBO Mol. Med.* 2010; 2: 490–503.
23. Song W et al. Mutant huntingtin binds the mitochondrial fission GTPase dynamin-related protein-1 and increases its enzymatic activity. *Nat. Med.* 2011; 17: 377–382.
24. Johri A, Chaturvedi RK, Beal MF. Hugging tight in Huntington's. *Nat. Med.* 2011; 17: 245–246.
25. Johri A et al. Truncated peroxisome proliferator-activated receptor-γ coactivator 1α splice variant is severely altered in Huntington's disease. *Neurodegener. Dis.* 2011; 8: 496–503.

26. Shirendeb U et al. Abnormal mitochondrial dynamics, mitochondrial loss and mutant huntingtin oligomers in Huntington's disease: Implications for selective neuronal damage. *Hum. Mol. Genet.* 2011; 20: 1438–1455.

27. Reddy PH, Mao P, Manczak M. Mitochondrial structural and functional dynamics in Huntington's disease. *Brain Res. Rev.* 2009; 61: 33–48.

28. Lucas EK et al. PGC-1α provides a transcriptional framework for synchronous neurotransmitter release from parvalbumin-positive interneurons. *J Neurosci.* 2014; 34: 14375–14387.

29. Browne SE, Ferrante RJ, Beal MF. Oxidative stress in Huntington's disease. *Brain Pathol.* 1999; 9: 147–163.

30. Stoy N et al. Tryptophan metabolism and oxidative stress in patients with Huntington's disease. *J. Neurochem.* 2005; 93: 611–623.

31. Chen CM et al. Increased oxidative damage and mitochondrial abnormalities in the peripheral blood of Huntington's disease patients. *Biochem. Biophys. Res. Commun.* 2007; 359: 335–340.

32. Tunez I et al. Important role of oxidative stress biomarkers in Huntington's disease. *J. Med. Chem.* 2011; 54: 5602–5606.

33. Braak H, Braak E. Allocortical involvement in Huntington's disease. *Neuropathol. Appl. Neurobiol.* 1992; 18: 539–547.

34. Tellez-Nagel I, Johnson AB, Terry RD. Studies on brain biopsies of patients with Huntington's chorea. *J. Neuropathol. Exp. Neurol.* 1974; 33: 308–332.

35. Butterworth NJ et al. Trinucleotide (CAG) repeat length is positively correlated with the degree of DNA fragmentation in Huntington's disease striatum. *Neuroscience* 1998; 87: 49–53.

36. Portera-Cailliau C et al. Evidence for apoptotic cell death in Huntington disease and excitotoxic animal models. *J. Neurosci.* 1995; 15: 3775–3787.

37. Sorolla MA et al. Proteomic and oxidative stress analysis in human brain samples of Huntington disease. *Free Radic. Biol. Med.* 2008; 45: 667–678.

38. Tabrizi SJ et al. Biochemical abnormalities and excitotoxicity in Huntington's disease brain. *Ann. Neurol.* 1999; 45: 25–32.

39. Lim D et al. Calcium homeostasis and mitochondrial dysfunction in striatal neurons of Huntington disease. *J. Biol. Chem.* 2008; 283: 5780–5789.

40. Sorolla MA et al. Protein oxidation in Huntington disease affects energy production and vitamin B6 metabolism. *Free Radic. Biol. Med.* 2010; 49: 612–621.

41. Browne SE, Beal MF. Oxidative damage in Huntington's disease pathogenesis. *Antioxid. Redox Signal.* 2006; 8: 2061–2073.

42. Cui L et al. Transcriptional repression of PGC-1alpha by mutant huntingtin leads to mitochondrial dysfunction and neurodegeneration. *Cell* 2006; 127: 59–69.

43. Weydt P et al. Thermoregulatory and metabolic defects in Huntington's disease transgenic mice implicate PGC-1alpha in Huntington's disease neurodegeneration. *Cell Metab.* 2006; 4: 349–362.

44. Bossy-Wetzel E, Petrilli A, Knott AB. Mutant huntingtin and mitochondrial dysfunction. *Trends Neurosci.* 2008; 31: 609–616.

45. Zhang SF et al. Impaired brain creatine kinase activity in Huntington's disease. *Neurodegener. Dis.* 2011; 8: 194–201.

46. Mochel F et al. Early alterations of brain cellular energy homeostasis in huntington disease models. *J. Biol. Chem.* 2012; 287: 1361–1370.

47. Johri A, Beal MF. Mitochondrial dysfunction in neurodegenerative diseases. *J. Pharmacol. Exp. Ther.* 2012; 342: 619–630.

48. Beal MF et al. Neurochemical and histologic characterization of striatal excitotoxic lesions produced by the mitochondrial toxin 3-nitropropionic acid. *J. Neurosci.* 1993; 13: 4181–4192.

49. Johri A, Beal MF. Antioxidants in Huntington's disease. *Biochim. Biophys. Acta* 2012; 1822: 664–674.

50. Yang L et al. Neuroprotective effects of the triterpenoid, CDDO methyl amide, a potent inducer of Nrf2-mediated transcription. *PLoS One* 2009; 4: e5757.

51. Beal MF. Therapeutic approaches to mitochondrial dysfunction in Parkinson's disease. *Parkinsonism Relat. Disord.* 2009; 15(Suppl. 3): S189–S194.

52. Yang L et al. Combination therapy with coenzyme Q10 and creatine produces additive neuroprotective effects in models of Parkinson's and Huntington's diseases. *J. Neurochem.* 2009; 109: 1427–1439.

53. Parkinson Study Group QE3 Investigators et al. A randomized clinical trial of high-dosage coenzyme Q10 in early Parkinson disease: No evidence of benefit. *JAMA Neurol.* 2014; 71: 543–552.

54. Petri S et al. Cell-permeable peptide antioxidants as a novel therapeutic approach in a mouse model of amyotrophic lateral sclerosis. *J. Neurochem.* 2006; 98: 1141–1148.

55. Calkins MJ, Manczak M, Reddy PH. Mitochondria-targeted antioxidant SS31 prevents Amyloid beta-induced mitochondrial abnormalities and synaptic degeneration in Alzheimer's disease. *Pharmaceuticals (Basel)* 2012; 5: 1103–1119.
56. Yang L et al. Mitochondria targeted peptides protect against 1-methyl-4-phenyl-1,2,3,6-tetrahydropyridine neurotoxicity. *Antioxid. Redox Signal.* 2009; 11: 2095–2104.
57. Xun Z et al. Targeting of XJB-5-131 to mitochondria suppresses oxidative DNA damage and motor decline in a mouse model of Huntington's disease. *Cell Rep.* 2012; 2: 1137–1142.
58. Cudkowicz M et al. The effects of dexpramipexole (KNS-760704) in individuals with amyotrophic lateral sclerosis. *Nat. Med.* 2011; 17: 1652–1656.
59. Cudkowicz ME et al. Dexpramipexole versus placebo for patients with amyotrophic lateral sclerosis (EMPOWER): A randomised, double-blind, phase 3 trial. *Lancet Neurol.* 2013; 12: 1059–1067.
60. Macia E et al. Dynasore, a cell-permeable inhibitor of dynamin. *Dev. Cell.* 2006; 10: 839–850.
61. Liu T et al. Cooperative role of RanBP9 and P73 in mitochondria-mediated apoptosis. *Cell Death Dis.* 2013; 4: e476.
62. Xie N et al. Inhibition of mitochondrial fission attenuates Aβ-induced microglia apoptosis. *Neuroscience* 2014; 256: 36–42.
63. Song HL et al. β-Amyloid is transmitted via neuronal connections along axonal membranes. *Ann. Neurol.* 2014; 75: 88–97.
64. Wang N et al. Neuronal targets for reducing mutant huntingtin expression to ameliorate disease in a mouse model of Huntington's disease. *Nat. Med.* 2014; 20: 536–541.
65. Wild EJ, Tabrizi SJ. Targets for future clinical trials in Huntington's disease: What's in the pipeline? *Mov. Disord.* 2014; 29: 1434–1445.
66. Johri A, Chandra A, Beal MF. PGC-1α, mitochondrial dysfunction, and Huntington's disease. *Free Radic. Biol. Med.* 2013; 62: 37–46.
67. Chandra A, Johri A, Beal MF. Prospects for neuroprotective therapies in prodromal Huntington's disease. *Mov. Disord.* 2014; 29: 285–293.
68. Tsunemi T et al. PGC-1alpha rescues Huntington's disease proteotoxicity by preventing oxidative stress and promoting TFEB function. *Sci. Transl. Med.* 2012; 4: 142ra197.
69. Settembre C et al. TFEB links autophagy to lysosomal biogenesis. *Science* 2011; 332: 1429–1433.
70. Chiang MC et al. Modulation of energy deficiency in Huntington's disease via activation of the peroxisome proliferator-activated receptor gamma. *Hum. Mol. Genet.* 2010; 19: 4043–4058.
71. Quintanilla RA et al. Rosiglitazone treatment prevents mitochondrial dysfunction in mutant huntingtin-expressing cells: Possible role of peroxisome proliferator-activated receptor-gamma (PPARgamma) in the pathogenesis of Huntington disease. *J. Biol. Chem.* 2008; 283: 25628–25637.
72. Jin J et al. Neuroprotective effects of PPAR-γ agonist rosiglitazone in N171-82Q mouse model of Huntington's disease. *J. Neurochem.* 2013; 125: 410–419.
73. Johri A et al. Pharmacologic activation of mitochondrial biogenesis exerts widespread beneficial effects in a transgenic mouse model of Huntington's disease. *Hum. Mol. Genet.* 2012; 21: 1124–1137.
74. Park SJ et al. Resveratrol ameliorates aging-related metabolic phenotypes by inhibiting cAMP phosphodiesterases. *Cell* 2012; 148: 421–433.
75. Kumar P et al. Effect of resveratrol on 3-nitropropionic acid-induced biochemical and behavioural changes: Possible neuroprotective mechanisms. *Behav. Pharmacol.* 2006; 17: 485–492.
76. Binienda ZK et al. Assessment of 3-nitropropionic acid-evoked peripheral neuropathy in rats: Neuroprotective effects of acetyl-l-carnitine and resveratrol. *Neurosci. Lett.* 2010; 480: 117–121.
77. Ho DJ et al. Resveratrol protects against peripheral deficits in a mouse model of Huntington's disease. *Exp. Neurol.* 2010; 225: 74–84.
78. St-Pierre J et al. Suppression of reactive oxygen species and neurodegeneration by the PGC-1 transcriptional coactivators. *Cell* 2006; 127: 397–408.
79. Wu Z et al. Mechanisms controlling mitochondrial biogenesis and respiration through the thermogenic coactivator PGC-1. *Cell* 1999; 98: 115–124.
80. Qiu X et al. Calorie restriction reduces oxidative stress by SIRT3-mediated SOD2 activation. *Cell Metab.* 2010; 12: 662–667.
81. Tao R et al. Sirt3-mediated deacetylation of evolutionarily conserved lysine 122 regulates MnSOD activity in response to stress. *Mol. Cell* 2010; 40: 893–904.
82. Chen Y et al. Melatonin protects hepatocytes against bile acid-induced mitochondrial oxidative stress via the AMPK-SIRT3-SOD2 pathway. *Free Radic. Res.* 2015; 29: 1–32.
83. Hirschey MD et al. SIRT3 deficiency and mitochondrial protein hyperacetylation accelerate the development of the metabolic syndrome. *Mol. Cell* 2011; 44: 177–190.

11 Free Radicals, Oxidative Stress, and Epilepsy

Jaroslava Folbergrová

CONTENTS

ABSTRACT

It is evident that oxidative stress plays an important role in epilepsy. Existing findings indicate that reactive oxygen and/or nitrogen species (ROS and RNS) are involved in seizure-induced brain injury and lead to mitochondrial dysfunction, evident particularly as a marked inhibition of respiratory chain complex I, detected both in humans and in several models of epilepsy in adult and immature animals. Other targets sensitive to oxidative damage are DNA, particularly mitochondrial DNA (mtDNA), and phospholipids of biological membranes. Oxidative stress also substantially affects the properties of the neurovascular unit, increases the permeability of the blood–brain barrier, and alters the regulation of regional cerebral blood flow. The findings suggest that the redox status shifts to a more oxidized state and the persisting oxidized environment may favor oxidative posttranslational modifications of sensitive targets. The selected ROS and/or RNS scavengers have been shown to prevent or at least substantially attenuate seizure-induced ROS production, mitochondrial dysfunction, and brain injury associated with status epilepticus.

It can thus be assumed that directly targeting oxidative stress by scavenging free radicals and/or by induction of the intrinsic antioxidant pathways is a promising strategy for the development of novel epilepsy therapies.

11.1 INTRODUCTION

Epilepsy is one of the most common neurological disorders of the brain, characterized by spontaneous recurrent seizures, affecting more than 50 million people worldwide. It is well established, both from clinical experience and from studies on various experimental models of epilepsy, that prolonged seizure activity can lead to irreversible brain damage [1–4]. Epilepsy often occurs in infants and children where convulsions may lead, in addition, to adverse effects on brain maturation and to serious functional consequences later in life, such as various cognitive deficits, especially those affecting learning and memory [5,6].

Epileptogenesis, the process by which a normal brain becomes epileptic, is a gradual process following the initial acute brain insults (such as stroke, trauma, status epilepticus [SE], etc.), and it involves many molecular, cellular, and network changes that result in brain reorganization and the conversion of normal brain to epileptic [7].

Preventing the development of epilepsy or at least attenuation of seizure-induced damage and long-term consequences of SE requires understanding the key mechanisms involved to allow the development of novel efficient therapies.

11.2 OXIDATIVE STRESS IN EPILEPSY

Accumulating evidence indicates that free radicals, oxidative stress, and mitochondrial dysfunction are implicated in the pathogenesis of many neurological diseases, including epilepsy [8–13]. Free radicals, especially reactive oxygen species (ROS), are normal intermediates in aerobic metabolism of all cells, and they can be produced by a number of pathways, but the major source in the CNS is believed to be mitochondria [14–16]. An important contributor to the free radical pool is also nitric oxide and other reactive nitrogen species (RNS) [17]. It is now well established that ROS and RNS have a dual biological role to play: at low or moderate concentrations they play an important role in normal physiological functions influencing several signaling pathways [18,19]. On the other hand, at excessive levels, ROS and/or RNS are harmful to living systems. Various endogenous antioxidant defense mechanisms exist, both enzymatic and nonenzymatic, which within certain limits can counteract the increased production of ROS and/or RNS. However, under pathological conditions of extremely enhanced production of free radicals, the endogenous antioxidant capacity can become overwhelmed, leading to oxidative stress and potential oxidative damage to cellular proteins, lipids, nucleic acids (with consequent accumulation of dysfunctional proteins, lipid peroxidative products, damaged nuclear and mitochondrial DNA), disruption of redox signaling, and ultimately neuronal death [20,21].

11.2.1 Free Radical Formation and Seizure-Induced Brain Damage

The formation of free radicals and their involvement in seizure-induced brain injury have been demonstrated both in vitro and in several in vivo model systems of seizures, by direct determination of ROS, by measuring indirect markers of oxidative stress or from the amelioration of seizure-induced brain damage after treatment with exogenous antioxidants or free radical scavengers [12,22–33,46].

It is well established that seizure-induced brain damage has an excitotoxic character when excessive glutamate (due to an excessive release and potentially insufficient uptake) causes overstimulation of postsynaptic ionotropic glutamate receptors, with subsequent accumulation of intracellular calcium, initiating a cascade of calcium-triggered events, including increased ROS and/or RNS production, mitochondrial dysfunction, and ultimately neuronal injury or death [1,9,11,34].

11.2.1.1 Glutamate Transporters

Importantly, it has been shown that high-affinity astroglial and neuronal glutamate transporters, which are important for maintaining low levels of synaptic glutamate, are extremely sensitive to oxidative stress, resulting in reduced uptake function [35]. Their failure can contribute to the increased excitability seen during the development of spontaneous seizures. Interestingly, a decrease in glial glutamate transporter expression was observed in patients with temporal lobe epilepsy and hippocampal sclerosis [36]. Focal reduction of excitatory amino acid transporters was reported in human neocortical epileptic foci [37]. Genetically altered mice lacking the glial transporter, GLT-1, have been shown to develop spontaneous seizures [38]. On the other hand, increased glial glutamate transporter, EAAT2, expression have been shown recently to reduce epileptogenic process following pilocarpine-induced status epilepticus (SE) [39].

11.3 FREE RADICALS AND MITOCHONDRIAL DYSFUNCTION

Mitochondrial function/dysfunction in relation to epilepsy has been studied in humans [40,41] and in several models of epilepsy in adult and immature animals [10,11,13,42,43]. In patients with temporal lobe epilepsy and hippocampal sclerosis, in addition to the neuropathological abnormalities (a severe loss of neurons in the CA1, CA3, and CA4 hippocampal subfields), functional defects of mitochondria have been reported in the areas of epileptogenesis [40].

11.3.1 RESPIRATORY CHAIN COMPLEX I

Severe impairment of respiratory chain complex I (NADH-oxidoreductase) activity was observed in CA3 hippocampal subfields in these patients [40]. Similar observations concern the vulnerable CA1 and CA3 hippocampal subfields of rats after pilocarpine-induced or kainic acid–induced SE [42,44,45]. In addition, similar signs of mitochondrial dysfunction also occur in immature animals with seizures [31]. The marked decrease in mitochondrial complex I activity was observed in cerebral cortex mitochondria of immature rats during the acute phase of seizures induced by homocysteic acid (HCA), and this decrease persisted during long periods of survival, corresponding to epileptogenesis [46]. Concurrently, with the decreased activities of complex I, significant increases in three markers of mitochondrial oxidative damage (3-nitrotyrosine [3-NT], 4-hydroxynonenal [4-HNE], and protein carbonyls) were detected [46]. The decrease in complex I activity was substantially attenuated by an acute treatment with selected free radical scavengers—SOD mimetics Mn(III) tetrakis (1-methyl-4-pyridyl), porphyrine pentachloride (MnTMPYP), and 4-hydroxy-2,2,6,6-tetramethylpiperidine-1-oxyl (Tempol)—as well as a selective peroxynitrite scavenger and a decomposition catalyst 5,10,15,20-Tetrakis (4-sulfonatophenyl) porphyrinate Iron (III) (FeTPPS) [31,46]. All these findings strongly suggest that oxidative modification (inactivation) of complex I is very likely responsible for the sustained deficiency of complex I activity, in accordance with extreme sensitivity of this enzyme to ROS and RNS.

11.3.1.1 Posttranslational Oxidative Modifications of Complex I

It was postulated that oxidative modification might be localized on some critical subunit of complex I and thus lead to its decreased activity. This possibility has been confirmed by Ryan et al. [45]. Using the kainate model of SE in adult rats, they reported an increased level of protein carbonyls concomitantly with decreased activity of complex I. Mass spectrometry analysis identified specific metal-catalyzed carbonylation to arginine 76 within the 75 kDa subunit of complex I. Computational-based molecular modeling studies predicted that carbonylation at this site can induce substantial structural alterations to the protein complex, leading to the impaired function.

Also, other posttranslational oxidative modifications of complex I can occur, such as S-nitrosation (S-nitrosylation) of some of its protein thiols or nitration of tyrosine (and/or tryptophan) residues within the complex [46 and references therein]. Pearce et al. reported that irreversibly inhibited activity of complex I was accompanied by the formation of 3-NT in 3 out of 46 subunits [47].

An increase in mitochondrial levels of 3-NT was demonstrated concomitantly with decreased complex I activity during and following seizures in immature rats [46] and, recently, also in adult rats in the kainate model of temporal lobe epilepsy [17].

In addition to complex I, other sensitive targets can also be attacked by ROS and/or RNS. Tyrosine nitration has been detected on manganese-superoxide dismutase (SOD2) with concomitant enzyme inactivation or on glutamine synthetase (GS) accompanied by the reduction of enzyme activity [48–50]. Since MnSOD is an important enzyme for removing superoxide anions in mitochondria and GS for reducing glutamate levels, their inhibition can result in increased oxidative stress.

11.3.1.2   Potential Consequences of the Decreased Activity of Complex I

Besides potential impairment of sufficient energy production, there may also be other consequences. It is well established that complex I is not only a target for ROS and RNS, but it is also an important source of ROS and/or RNS production, especially when partially inhibited [51–54]. It has been shown that overproduction of ROS may be responsible for triggering the activation of specific kinase pathways, for the release of cytochrome c and other proapoptotic factors, and ultimate cell death [55,56]. It can be assumed that increased ROS and/or RNS formation may contribute to neuronal injury demonstrated in several models of epilepsy [10,11].

11.4   OXIDATIVE DAMAGE TO DNA

Another target sensitive to oxidative damage is DNA, particularly mtDNA, where repair mechanisms are limited. Mitochondrial DNA damage escaping repair is critical, since resulting mutations can be propagated and can expand to levels affecting the function of enzymes containing mtDNA-encoded subunits, for example, respiratory chain complexes I and IV, and the FOF1-ATP synthase [57].

11.5   OXIDATIVE DAMAGE TO LIPIDS

Polyunsaturated fatty acids present in the phospholipids of biological membranes and cardiolipin localized within mitochondrial inner membrane are highly susceptible to oxidation by ROS, as evident from an increase in several markers of lipid peroxidation, detected following SE induced both in adult and in immature animals [28,31,46,58].

11.6   REDOX STATUS

Changes in cellular and mitochondrial redox status, evaluated by measuring GSH/GSSG (reduced glutathione and its disulfide redox partner) and/or CoASH/CoASSG (reduced CoA and its disulfide with GSH) ratios, were detected following SE. The ratios decreased in the hippocampus of adult rats after SE induced with lithium pilocarpine or KA and the alteration persisted throughout the epileptogenesis and chronic epilepsy [59,60]. Interestingly, a decrease in both GSH levels and glutathione reductase activity in brain regions of patients with epilepsy has been reported [61]. These findings suggest that the redox status shifts to a more oxidized state and the persisting oxidized environment may thus favor oxidative posttranslational modifications of sensitive targets, as described earlier.

11.7   OTHER SENSITIVE TARGETS

It should be noted that oxidative stress also substantially affects the properties of the neurovascular unit, increases the permeability of the blood–brain barrier, and alters the regulation of regional cerebral blood flow [57].

11.8   ANTIOXIDANT DEFENSE OF THE BRAIN

Several endogenous enzymatic and nonenzymatic antioxidant defense mechanisms exist. The superoxide anion ($O_2^{\cdot-}$) is one of the most reactive ROS and it is the precursor of other reactive species that are controlled via multiple enzyme systems, comprising superoxide dismutase (SOD), catalase, glutathione peroxidase (GPX), glutathione reductase (GR), and thioredoxin/peroxiredoxin systems, which, according to some authors, represents the major contribution to the mitochondrial H_2O_2 removal [62–64]. There are many studies concerning antioxidant enzymes under seizures induced by various convulsants [65]. It should be noted that the analyses of antioxidant enzymes in adult animals were performed during different time intervals after the administration of convulsants; different results were reported. Concerning SOD, increased, decreased, and no change in the activity were detected. Whether these differences may reflect different intensities and durations of SE or different extents of free radical formation is not known at present. Interestingly, increased SOD activity in adult animals was always accompanied by an increased activity of catalase and/or GPX, provided that the activities of these enzymes were analyzed. On the contrary, in immature brain, the situation may be different. When seizures were induced in immature rats by homocysteic acid, the activity of total SOD, SOD1 (CuZnSOD), and particularly SOD2 (MnSOD) at selected time intervals significantly increased, and upregulation of SOD2 was also confirmed in mitochondria at the protein level [65]. However, the upregulation of SOD2 was not accompanied by similar changes in the activity of catalase and/or GPX as may be expected as a compensatory mechanism to the increased H_2O_2. The increased levels of H_2O_2 (due to its insufficient removal) can lead, in the presence of reduced transition metals via Fenton reaction, to the formation of highly toxic hydroxyl radicals. It can thus be assumed that insufficient antioxidant defense in immature brain can have potentially serious consequences.

As to the nonenzymatic antioxidant defense mechanisms, several substances such as vitamin C, vitamin E, and DL-α-lipoic acid have been shown to provide a partial effect in various models of seizures in adult animals. However, clinical trials have been so far controversial [10]. The most abundant intracellular antioxidant defense is reduced glutathione (GSH), which is oxidized to its disulfide redox partner GSSG. GSH/GSSG ratio is a commonly used biomarker of redox status and oxidative stress as already mentioned.

11.9   MODULATION OF OXIDATIVE STRESS, MITOCHONDRIAL DYSFUNCTION, AND EPILEPSY-INDUCED BRAIN INJURY

Existing data clearly indicate the association of seizures and SE with oxidative stress. Many efforts have thus been aimed at developing substances capable of detoxifying ROS and RNS and their damaging effects. Synthetic metalloporphyrin catalytic antioxidants (small molecule mimics of SOD and/or catalase and some of them also potent detoxifiers of lipid peroxides and peroxynitrite) appear to be a promise as a novel neuroprotective agents [66]. Some of these compounds have been shown to be effective in animal models of epilepsy, lessening oxidative stress and neuronal damage induced by SE [12,32]. The increased formation of $O_2^{\cdot-}$ associated with SE induced in immature rats by HCA was completely blocked by two SOD mimetics, MnTMPYP and Tempol [26]. These scavengers and also selective peroxynitrite scavenger FeTPPS provided significant attenuation of complex I inhibition [31,46]. In addition, treatment with these antioxidants resulted in partial amelioration of neuronal degeneration associated with SE [26,67]. An improvement in neuroprotective effect can be expected when redox-modulating metalloporphyrin catalytic antioxidants with markedly enhanced lipophilicity become available commercially [68].

It can be assumed that antioxidants interacting with multiple targets can have a better chance for a beneficial effect than single target therapies. A promising substance appears to be resveratrol (RES) (3,5,4'-tri-hydroxy-*trans*-stilbene), a natural polyphenolic compound present in red wine, whose neuroprotective effect has been reported in various models of neurological disorders in adult animals. Resveratrol has an ability to enter the brain after a peripheral administration, and it has no adverse effects. It has been shown that RES, besides its direct antioxidant effect, has multiple cellular effects, interfering with several signaling pathways, including transcription of endogenous antioxidant genes [69–71]. Recently, it was reported that RES activates Nrf2 (nuclear factor erythroid 2-related factor 2), which is an essential transcription factor regulating the expression of numerous antioxidant genes, via binding to the antioxidant response element and plays a crucial role in cellular defence against oxidative stress [72]. A growing body of evidence highlights Nr2f as a promising therapeutic target in different neurodegenerative and neuroinflammatory diseases. A recent study by Mazzuferi et al. [73] showed that overexpression of Nrf2 in adult mice with pilocarpine-induced SE provided an enormous protective effect.

11.10 CONCLUSIONS

It is now well established that oxidative stress plays an important role in epilepsy. The existing findings indicate that reactive oxygen and/or nitrogen species (ROS and RNS) are involved in seizure-induced brain injury. They lead to mitochondrial dysfunction, evident particularly as a marked inhibition of respiratory chain complex I, detected both in humans and in several models of epilepsy in adult and immature animals. Other targets sensitive to oxidative damage are DNA, particularly mtDNA, and phospholipids of biological membranes. Oxidative stress also substantially affects the properties of the neurovascular units, increases the permeability of the blood–brain barrier, and alters the regulation of regional cerebral blood flow. The findings suggest that the redox status shifts to a more oxidized state and the persisting oxidized environment may favor oxidative posttranslational modifications of sensitive targets, as discussed. Experimental data indicate that selected ROS and/or RNS scavengers can prevent or at least substantially attenuate seizure-induced ROS production, mitochondrial dysfunction, and brain injury associated with status epilepticus.

It can thus be assumed that substances with antioxidant properties combined with conventional therapies might provide a beneficial effect in the treatment of epilepsy.

ACKNOWLEDGMENTS

Supports by grants #309/05/2015, #309/08/0292, #P303/10/0999, and #15-08565S from the Czech Science Foundation and by ERC CZ LL 1204 from MEYS CR are gratefully acknowledged. The author expresses thanks to all collaborators who have contributed to experimental work and to Dr. Jakub Otáhal for his valuable comments.

REFERENCES

1. Meldrum, B.S. 1993. Excitotoxity and selective neuronal loss in epilepsy. *Brain Pathol* 3:405–412.
2. Kubová, H., Druga, R., Lukasiuk, K. et al. 2001. Status epilepticus causes necrotic damage in the mediodorsal nucleus of the thalamus in immature rats. *J Neurosci* 21:3593–3599.
3. Wasterlain, C.G., Niquet, J., Thompson, K.W. et al. 2002. Seizure-induced neuronal death in the immature brain. *Prog Brain Res* 135:335–353.
4. Folbergrová, J., Druga, R., Otáhal, J., Haugvicová, R., Mareš, P., Kubová, H. 2005.Seizures induced in immature rats by homocysteic acid and the associated brain damage are prevented by group II metabotropic glutamate receptor agonist (2R,4R)-4-aminopyrrolidine-2,4-dicarboxylate. *Exp Neurol* 192:420–436.
5. Lynch, M., Sayin, U., Bownds, J., Janumpalli, S., Sutula, T. 2000. Long-term consequences of early postnatal seizures on hippocampal learning and plasticity. *Eur J Neurosci* 12:2252–2264.
6. Kubová, H., Mareš, P., Suchomelová, L., Brožek, G., Druga, R., Pitkänen, A. 2004. Status epilepticus in immature rats leads to behavioural and cognitive impairment and epileptogenesis. *Eur J Neurosci* 19:3255–3265.
7. Pitkänen, A., Lukasiuk, K. 2011. Mechanisms of epileptogenesis and potential treatment targets. *Lancet Neurol* 10:173–186.

8. Patel, M. 2004. Mitochondrial dysfunction and oxidative stress: Cause and consequence of epileptic seizures. *Free Radic Biol Med* 37:951–962.

9. Lin, M.T., Beal, M.F. 2006. Mitochondrial dysfunction and oxidative stress in neurodegenerative diseases. *Nature* 443:787–795.

10. Waldbaum, S., Patel, M. 2010. Mitochondria, oxidative stress,and temporal lobe epilepsy. *Epilepsy Res* 88:23–45.

11. Folbergrová, J., Kunz, W.S. 2012. Mitochondrial dysfunction in epilepsy. *Mitochondrion* 12:35–40.

12. Liang, L.P., Waldbaum, S., Rowley, S. et al. 2012. Mitochondrial oxidative stress and epilepsy in SOD2 deficient mice: Attenuation by a lipophilic metalloporphyrin. *Neurobiol Dis* 45:1068–1076.

13. Rowley, S., Patel, M. 2013. Mitochondrial involvement and oxidative stress in temporal lobe epilepsy. *Free Radic Biol Med* 62:121–131.

14. Liu, Y., Fiskum, G., Schubert, D. 2002. Generation of reactive oxygen species by the mitochondrial electron transport chain. *J Neurochem* 80:780–787.

15. Turrens, J.F. 2003. Mitochondrial formation of reactive oxygen species. *J Physiol (London)* 552(Pt 2):335–344.

16. Murphy, M.P. 2009. How mitochondria produce reactive oxygen species. *Biochem J* 417:1–13.

17. Ryan, K., Liang, L.P., Rivard, C., Patel, M. 2014. Temporal and spatial increase of reactive nitrogen species in the kainate model of temporal lobe epilepsy. *Neurobiol Dis* 64:8–15.

18. Drőge, W. 2002. Free radicals in the physiological control of cell function. *Physiol Rev* 82:47–95.

19. Sena, L.A., Chandel, N.S. 2012. Physiological roles of mitochondrial reactive oxygen species. *Mol Cell* 48:158–167.

20. Jones, D.P. 2006. Disruption of mitochondrial redox circuitry in oxidative stress. *Chem Biol Interact* 163:38–53.

21. Sayre, L.M., Perry, G., Smith, M.A. 2008. Oxidative stress and neurotoxicity. *Chem Res Toxicol* 21:172–188.

22. Frantseva, M.V., Perez Velazquez, J.L., Hwang, P.A., Carlen, P.L. 2000. Free radical production correlates with cell death in an in vitro model of epilepsy. *Eur J Neurosci* 12:1431–1439.

23. Kovács, R., Schuchmann, S., Gabriel, S., Kann, O., Kardos, J., Heinemann, U. 2002. Free radical-mediated cell damage after experimental status epilepticus in hippocampal slice cultures. *J Neurophysiol* 88:2909–2918.

24. Rauca, C., Zerbe, R., Jantze, H. 1999. Formation of free hydroxyl radicals after pentylentetrazol-induced seizures and kindling. *Brain Res* 847:347–351.

25. Peterson, S.L., Morrow, D., Liu, S., Liu, K.J. 2002. Hydroethidine detection of superoxide production during the lithium-pilocarpine model of status epilepticus. *Epilepsy Res* 49:226–238.

26. Folbergrová, J., Otáhal, J., Druga, R. 2012. Brain superoxide anion formation in immature rats during seizures: Protection by selected compounds. *Exp Neurol* 233:421–429.

27. Bruce, A.J., Baudry, M. 1995. Oxygen free radicals in rat limbic structures after kainate-induced seizures. *Free Radic Biol Med* 18:993–1002.

28. Dal-Pizzol, F., Klamt, F., Vianna, M.M.R. et al. 2000. Lipid peroxidation in hippocampus early and late after status epilepticus induced by pilocarpine or kainic acid in Wistar rats. *Neurosci Lett* 291:179–182.

29. Liang, L.P., Ho, Y.S., Patel, M. 2000. Mitochondrial superoxide production in kainate-induced hippocampal damage. *Neuroscience* 101:563–570.

30. Ueda,Y., Yokoyama, H., Nakajima, A., Tokumaru, J., Doi,T., Mitsuyama, Y. 2002. Glutamate excess and free radical formation during and following kainic acid-induced status epilepticus. *Exp Brain Res* 147:219–226.

31. Folbergrová, J., Ješina, P., Drahota, Z. et al. 2007. Mitochondrial complex I inhibition in cerebral cortex of immature rats following homocysteic acid-induced seizures. *Exp Neurol* 204:597–609.

32. Rong, Y., Doctrow, S.R., Tocco, G., Baudry, M. 1999. EUK-134, a synthetic superoxide dismutase and catalase mimetic, prevents oxidative stress and attenuates kainate-induced neuropathology. *Proc Natl Acad Sci USA* 96:9897–9902.

33. Folbergrová, J., Druga, R., Otáhal, J., Haugvicová, R., Mareš, P., Kubová, H. 2006. Effect of free radical spin trap *N-tert*-butyl-α-phenylnitrone (PBN) on seizures induced in immature rats by homocysteic acid. *Exp Neurol* 201:105–119.

34. Schinder, A.F., Olson, E.C., Spitzer, N.C., Montal, M. 1996. Mitochondrial dysfunction is a primary event in glutamate neurotoxicity. *J Neurosci* 16:6125–6133.

35. Trotti, D., Danbolt, N.C., Volterra, A. 1998. Glutamate transporters are oxidant-vulnerable: A molecular link between oxidative and excitotoxic neurodegeneration?. *Trends Pharmacol Sci* 19:328–334.

36. Mathern, G.W., Mendoza, D., Lozada, A. et al. 1999. Hippocampal GABA and glutamate transporter immunoreactivity in patients with temporal lobe epilepsy. *Neurology* 52:453–472.
37. Rakhade, S.N., Loeb, J.A. 2008. Focal reduction of neuronal glutamate transporters in human neocortical epilepsy. *Epilepsia* 49:226–236.
38. Tanaka, K., Watase, K., Manabe, T. et al. 1997. Epilepsy and exacerbation of brain injury in mice lacking the glutamate transporter GLT-1. *Science* 276:1699–1702.
39. Kong, Q., Takahashi, K., Schulte, D. et al. 2012. Increased glial glutamate transporter EAAT2 expression reduces epileptogenic process following pilocarpine-induced status epilepticus. *Neurobiol Dis* 47:145–154.
40. Kunz, W.S., Kudin, A.P., Vielhaber, S. et al. 2000. Mitochondrial complex I deficiency in the epileptic focus of patients with temporal lobe epilepsy. *Ann Neurol* 48:766–773.
41. Lee, Y.M., Kang, H.C., Lee, J.S. et al. 2008. Mitochondrial respiratory chain defects: Underlying etiology in various epileptic conditions. *Epilepsia* 49:685–690.
42. Kudin, A.P., Kudina, T.A., Seyfried, J. et al. 2002. Seizure-dependent modulation of mitochondrial oxidative phosphorylation in rat hippocampus. *Eur J Neurosci* 15:1105–1114.
43. Folbergrová, J. 2013. Oxidative stress in immature brain following experimentally-induced seizures. *Physiol Res* 62(Suppl. 1):S39–S48.
44. Chuang, Y.Ch., Chang, A.Y.W., Lin, J.W. et al. 2004. Mitochondrial dysfunction and structural damage in the hippocampus during kainic acid-induced status epilepticus. *Epilepsia* 45:1202–1209.
45. Ryan, K., Backos, D.S., Reigan, P., Patel, M. 2012. Post-translational oxidative modification and inactivation of mitochondrial complex I in epileptogenesis. *J Neurosci* 32:11250–11258.
46. Folbergrová, J., Ješina, P., Haugvicová, R., Lisý, V., Houštěk, J. 2010. Sustained deficiency of mitochondrial complex I activity during long periods of survival after seizures induced in immature rats by homocysteic acid. *Neurochem Int* 56:394–403.
47. Pearce, L.L., Kanai, A.J., Epperly, M.W., Peterson, J. 2005. Nitrosative stress results in irreversible inhibition of purified mitochondrial complex I and III without modification of cofactors. *Nitric Oxide* 13:254–263.
48. MacMillan-Crow, L.A., Crow, J.P., Thompson, J.A. 1998. Peroxynitrite-mediated inactivation of manganese superoxide dismutase involves nitration and oxidation of critical tyrosine residues. *Biochemistry* 37:1613–1622.
49. Yamakura, F., Taka, H., Fujimura, T., Murayama, K. 1998. Inactivation of human manganese-superoxide dismutase by peroxynitrite is caused by exclusive nitration of tyrosine 34 to 3-nitrotyrosine. *J Biol Chem* 273:14085–14089.
50. Bidmon, H.J., Gorg, B., Palomero-Gallagher, N. 2008. Glutamine synthetase becomes nitrated and its activity is reduced during repetitive seizure activity in the pentylentetrazole model of epilepsy. *Epilepsia* 49:1733–1748.
51. Sipos, I., Tretter, L., Adam-Vizi, V. 2003. Quantitative relationship between inhibition of respiratory complexes and formation of reactive oxygen species in isolated nerve terminals. *J Neurochem* 84:112–118.
52. Kudin, A.P., Bimpong-Buta, N.Y.B., Vielhaber, S., Elger, C.E., Kunz, W.S. 2004. Characterization of superoxide-producing sites in isolated brain mitochondria. *J Biol Chem* 279:4127–4135.
53. Kussmaul, L., Hirst, J. 2006. The mechanism of superoxide production by NADH:ubiquinone oxidoreductase (complex I) from bovine heart mitochondria. *Proc Natl Acad Sci USA* 103:7607–7612.
54. Fato, R., Bergamini, C., Leoni, S., Strocchi, P., Lenaz, G. 2008. Generation of reactive oxygen species by mitochondrial complex I: Implications in neurodegeneration. *Neurochem Res* 33:2487–2501.
55. Perier, C., Tieu, K., Guégan, C. et al. 2005. Complex I deficiency primes Bax-dependent neuronal apoptosis through mitochondrial oxidative damage. *Proc Natl Acad Sci USA* 102:19126–19131.
56. Marella, M., Seo, B.B., Matsuno-Yagi, A., Yagi, T. 2007. Mechanism of cell death caused by complex I defects in a rat dopaminergic cell line. *J Biol Chem* 282:24146–24156.
57. Otáhal, J., Folbergrová, J., Kovacs, R., Kunz, W.S., Maggio, N. 2014. Epileptic focus and alteration of metabolism. In *Int Rev Neurobiol; Modern Concepts of Focal Epileptic Networks*, P. Jiruška, M. DeCurtis and J.G.R. Jefferys, Eds., Elsevier, 114, 209–243.
58. Patel, M., Liang, L.P., Hou, H. 2008. Seizure-induced formation of isofurans: Novel products of lipid peroxidation whose formation is positively modulated by oxygen tension. *J Neurochem* 104:264–270.
59. Liang, L.P., Patel, M. 2006. Seizure-induced changes in mitochondrial redox status. *Free Radic Biol Med* 40:316–322.
60. Waldbaum, S., Liang, L.P., Patel, M. 2010. Persistent impairment of mitochondrial and tissue redox status during lithium-pilocarpine-induced epileptogenesis. *J Neurochem* 115:1172–1182.

61. Mueller, S.G., Trabesinger, A.H., Boesiger, P., Wieser, H.G. 2001. Brain glutathione levels in patients with epilepsy measured by in vivo (1)H-MRS. *Neurology* 57:1422–1427.

62. Chen, Y., Cai, J., Jones, D.P. 2006. Mitochondrial thioredoxin in regulation of oxidant-induced cell death. *FEBS Lett* 580:6596–6602.

63. Cox, A.G., Winterbourn, C.C., Hampton, M.B. 2010. Mitochondrial peroxiredoxin involvement in antioxidant defence and redox signaling. *Biochem J* 425:313–325.

64. Drechsel, D.A., Patel, M. 2010. Respiration-dependent H_2O_2 removal in brain mitochondria via the thioredoxin/peroxiredoxin system. *J Biol Chem* 285:27850–27858.

65. Folbergrová, J., Ješina, P., Nůsková, H., Houštěk, J. 2013. Antioxidant enzymes in cerebral cortex of immature rats following experimentally-induced seizures: Upregulation of mitochondrial MnSOD (SOD2). *Int J Dev Neurosci* 31:123–130.

66. Patel, M., Day, B.J. 1999. Metalloporphyrin class of therapeutic catalytic antioxidants. *Trends Pharmacol Sci* 20:359–364.

67. Folbergrová, J., Otáhal, J., Druga, R. 2011. Effect of Tempol on brain superoxide anion production and neuronal injury associated with seizures in immature rats. *Epilepsia* 52(Suppl. 6):51.

68. Sheng, H., Chaparro, R.E., Sasaki, T. et al. 2014. Metalloporphyrins as therapeutic catalytic oxidoreductants in central nervous system disorders. *Antioxid Redox Signal* 20:2437–2465.

69. Kroon, P.A., Iyer, A., Chunduri, P., Chan, V., Brown, L. 2010. The cardiovascular nutrapharmacology of resveratrol: Pharmacokinetics, molecular mechanisms and therapeutic potential. *Curr Med Chem* 17:2442–2455.

70. Sahebkar, A. 2010. Neuroprotective effects of resveratrol: Potential mechanisms. *Neurochem Int* 57:621–622.

71. Shetty, A.K. 2011. Promise of resveratrol for easing status epilepticus and epilepsy. *Pharmacol Ther* 131:269–286.

72. Kesherwani, V., Atif, F., Yousuf, S., Agrawal, S.K. 2013. Resveratrol protects spinal cord dorsal column from hypoxic injury by activating Nrf-2. *Neuroscience* 241:80–88.

73. Mazzuferi, M., Kumar, G., Van Eyll, J. et al. 2013. Nrf2 defense pathway: Experimental evidence for its protective role in epilepsy. *Ann Neurol* 74:560–568.

12 Oxidative Stress and Neuropsychiatric Disorders in the Life Spectrum

Akihiko Nunomura, Toshio Tamaoki, Nobutaka Motohashi, Hyoung-gon Lee, Xiongwei Zhu, and George Perry

CONTENTS

ABSTRACT

In the life spectrum, the brain is particularly vulnerable to oxidative stress (OS) during the neuro-developmental period and brain senescence. Recently accumulating evidence suggests that OS is an upstream event in the pathophysiology of a diverse spectrum of neuropsychiatric disorders, which is typically demonstrated in Alzheimer's disease (AD) and schizophrenia (SZ), the archetypal disorders of neurodegeneration and neurodevelopment, respectively. Indeed, a considerable number of genetic and environmental factors of these disorders are tightly associated with increased oxidative damage or decreased antioxidant capacities, suggesting that OS is located at the convergence point of gene–environment interactions of the disorders.

12.1 INTRODUCTION

The brain is particularly vulnerable to oxidative stress (OS) because of its high rate of oxygen consumption, abundant lipid content, and relative paucity of antioxidant enzymes compared with other organs (Nunomura et al. 2006). A modest level of OS is essential to maintain cellular physiological function through the activation of stress resistance systems. However, when the level of OS increases beyond the compensation of antioxidant defense, cellular macromolecules are oxidatively damaged (Bishop et al. 2010). Many lines of evidence have indicated that OS is involved not only in the process of brain aging (Gemma et al. 2007) but also in the pathogenesis of neurodegenerative disorders, including AD, Parkinson's disease (PD), and amyotrophic lateral sclerosis (ALS) (Jenner 2003; Lin and Beal 2006; Nunomura et al. 2006, 2012a; Ferraiuolo et al. 2011).

Given the notion that the developing brain is particularly vulnerable to oxidative insults because of its immature endogenous free radical scavenging system (Ikonomidou and Kaindl 2011), it is not surprising that recently accumulating evidence suggests an association of OS with psychiatric disorders such as schizophrenia (SZ), bipolar disorders, and depression (Ng et al. 2008; O'Donnell et al. 2014). Therefore, neurodegenerative and neurodevelopmental disorders may have common underlying pathomechanisms tightly connecting to OS and accelerating neuronal aging (Campos et al. 2014). In this chapter, we focus on AD and SZ, which emerge at opposite ends of the life spectrum, and summarize evidence for a role of OS in the pathophysiology of these disorders as an upstream event. Recently, growing evidence on brain networks, including default mode network that are vulnerable to both categories of neurodegenerative and neurodevelopmental disorders (Douaud et al. 2014), has provided novel insights into the OS-related pathomechanisms underlying a diverse spectrum of neuropsychiatric disorders.

12.2 VULNERABILITY TO OS IN NEURONAL DEVELOPMENT AND SENESCENCE

Vulnerable periods during the development of the nervous system extending from the embryonic period through adolescence are sensitive to environmental insults because they are dependent on the temporal and regional emergence of critical developmental processes, that is, proliferation, migration, differentiation, synaptogenesis, myelination, and apoptosis (Rice and Barone 2000). Indeed, half of all lifetime cases with psychiatric disorders start by the age of 14 years and three fourths by the age of 24 years (Kessler et al. 2005), and increased oxidative damage and decreased antioxidant capacities have been described in psychiatric disorders (Ng et al. 2008; O'Donnell et al. 2014).

Given the "last in, first out" theory that age-related brain degeneration mirrors development with the areas of the brain thought to develop later also degenerating earlier, there might be a common brain network linking development, aging, and vulnerability to disorders. Indeed, a recent study on

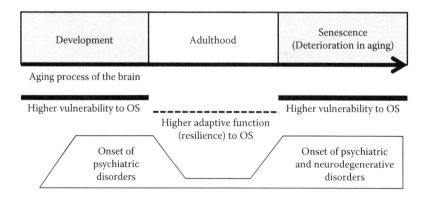

FIGURE 12.1 Brain aging and vulnerability to oxidative stress (OS). The brain is vulnerable to OS particularly in developmental and senescent periods, which is potentially a fundamental mechanism common to psychiatric and neurodegenerative disorders.

healthy individuals reveals the transmodal network whose life span pattern of age-related change intrinsically supports this model of mirroring development and aging (Douaud et al. 2014). Of particular interest, this network of brain regions, which develops relatively late during adolescence and shows accelerated degeneration in old age compared with the rest of the brain, characterizes areas of heightened vulnerability to unhealthy developmental and aging processes, as exemplified by SZ and AD, respectively. Moreover, these regions include core nodes of the default mode network (Braga et al. 2013), which is a strong candidate considering its preferential vulnerability to aging and sensitivity to OS and disease states (Kesler 2014).

Taken together, vulnerability to psychiatric and neurodegenerative disorders in developmental period and in senescence could be explained, at least in part, by sensitivity to OS (Figure 12.1).

12.3   AD AND OS

12.3.1   OS Markers in Postmortem AD Brains

Oxidatively modified products of nucleic acids (e.g., 8-hydroxydeoxyguanosine and 8-hydroxyguanosine) and proteins (e.g., 3-nitrotyrosine and protein carbonyls), as well as products by lipid peroxidation (e.g., 4-hydroxynonenal, 8-isoprostane, acrolein, malondialdehyde, and thiobarbituric acid reactive substances) and glycoxidation (e.g., carboxymethyl-lysine and pentosidine), all known markers of oxidative damage, have been demonstrated in postmortem brains of AD (Nunomura et al. 2006, 2012a; Sonnen et al. 2008; Mangialasche et al. 2009). Moreover, alterations in expression or activities of components of antioxidant defense such as superoxide dismutase (SOD), catalase, glutathione peroxidase (GPX), glutathione (GSH) reductase, and human 8-oxoguanine DNA glycosylase 1 have also been reported in the AD brain.

When we review a number of previous studies, we found no consistent trend of increase or decrease in these components of antioxidant enzymes as well as metal-binding protein (e.g., transferrin and ceruloplasmin) and nonenzymatic antioxidants (e.g., vitamin E and GSH) (Sonnen et al. 2008; Mangialasche et al. 2009). The inconsistency among studies on the levels of the antioxidant components is probably explained by variations in compensatory changes of the components, which is supported by the finding that the Trolox equivalent antioxidant capacity increases with pathological progression in the postmortem brain of AD (Tayler et al. 2010).

12.3.2   Involvement of OS in the Pathophysiology of AD as an Upstream Event

A considerable number of studies suggested the involvement of OS into the pathophysiology of AD, especially at the early stage of the pathological cascade (Table 12.1); actually, oxidative damage to nucleic acids is more prominent in the postmortem brain and cerebrospinal fluids from AD patients with a shorter disease duration (Nunomura et al. 2001; Abe et al. 2002). Moreover, increased oxidative damage at the prodromal stages has been observed in the postmortem brain and cerebrospinal fluid obtained from patients with preclinical AD and mild cognitive impairment (Praticò et al. 2002; Ringman et al. 2008; Bradley et al. 2010; Nunomura et al. 2012b). Co-appearance of oxidative damage and intraneuronal accumulation of amyloid β (Aβ), the initial pathology of AD, is demonstrated in the vulnerable neuronal population of the AD brain (Nunomura et al. 2010). Strikingly, both induced pluripotent stem cell (iPSC) model and transgenic animal model of AD have indicated that elevation of OS markers is simultaneously with or even prior to the intraneuronal Aβ accumulation (McManus et al. 2011; Kondo et al. 2013). Temporal primacy of OS to Aβ deposition is also supported by the findings from the postmortem brain with Down syndrome, where AD-related pathologies are chronologically evident in early adulthood (Nunomura et al. 2000). Indeed, the amount of Aβ deposition in the brain is modified by OS levels in amyloid precursor protein (*APP*) transgenic mice crossed with *manganese (Mn)-SOD* knockout or overexpressing animals (Li et al. 2004; Dumont et al. 2009).

TABLE 12.1

Evidence for a Role of Oxidative Stress (OS) as an Upstream Event in the Pathophysiology of Alzheimer's Disease (AD) and Schizophrenia (SZ)

Materials/Subjects	AD	SZ
iPSC model	iPSC-derived neural cells from AD patients: Increase in ROS and OS markers and upregulation of OS-responsive gene expression in association with Aβ oligomer accumulation (Kondo et al. 2013)	iPSC-derived neural progenitor cells from SZ patients: Increase in ROS (Paulsen Bda et al. 2012) and upregulation of OS-responsive gene expression (Brennand et al. 2015)
Genetically modified animal model	1. Triple transgenic mice with *PSEN1*, *APP*, and *tau* transgenes: OS and mitochondrial dysfunction prior to other pathological changes of AD including intraneuronal Aβ accumulation (McManus et al. 2011) 2. *APP* transgenic mice crossed with *Mn-SOD* knockout animals: Acceleration of the brain Aβ accumulation (Li et al. 2004) 3. *APP* transgenic mice crossed with *Mn-SOD* overexpressing animals: Reduction in the brain Aβ accumulation and memory deficit (Dumont et al. 2009)	1. GSH synthetic enzyme *GCLM* knockout mice: Induction of SZ-like brain structural, electrophysiological, neurochemical, and behavioral alterations in adolescence and early adulthood (Steullet et al. 2010; das Neves Duarte et al. 2012; Kulak et al. 2012). Reversal of the alterations by GSH precursor NAC (Cabungcal et al. 2013; 2014) 2. Known SZ animal model of NMDA receptors conditional knockout mice: Increase in ROS in the cerebral cortex at early life, which is accompanied by the onset of SZ-like behavioral phenotype (Jiang et al. 2013b)
Human subjects	1. Down syndrome: Oxidative damage prior to Aβ plaque deposition in the postmortem brain (Nunomura et al. 2000) 2. Preclinical AD and MCI: Increase in oxidative damage in the postmortem brain and CSF (Praticò et al. 2002; Ringman et al. 2008; Bradley et al. 2010; Nunomura et al. 2012b) 3. AD patients with shorter disease duration: More prominent oxidative damage to nucleic acids in the postmortem brain and CSF (Nunomura et al. 2001; Abe et al. 2002)	1. Untreated SZ patients: Lower plasma catalase and SOD compared with treated SZ patients (Raffa et al. 2009) 2. First-episode SZ patients: Decrease in serum/plasma total antioxidant capacity and red blood cell catalase and SOD (Flatow et al. 2013) 3. Recent-onset SZ patients: Decrease in CSF Cu/Zn-SOD (Coughlin et al. 2013)

Abbreviations: Aβ, amyloid β; APP, amyloid precursor protein; CSF, cerebrospinal fluid; Cu/Zn-SOD, copper/zinc-superoxide dismutase; GCLM, glutamate cysteine ligase modulatory subunit; GSH, glutathione; iPSC, induced pluripotent stem cell; MCI, mild cognitive impairment; Mn-SOD, manganese-superoxide dismutase; NAC, *N*-acetylcysteine; NMDA, *N*-methyl-D-aspartate; *PSEN1*, presenilin 1; ROS, reactive oxygen species; SOD, superoxide dismutase.

12.3.3 GENETIC FACTORS OF AD AND OS

Mutations in three genes—the *APP* gene, the presenilin 1 (*PSEN1*) gene, and the presenilin 2 (*PSEN2*) gene—are inherited in an autosomal dominant fashion and directly lead to early-onset AD. These mutations are generally considered to contribute to the pathogenesis of AD by leading an increase in the production of total Aβ or highly amyloidogenic Aβ42 isoform (Selkoe 2011). Intriguingly, all these mutations are associated with not only altered Aβ metabolism but also increased neuronal oxidative damage or increased neuronal vulnerability to OS, which has been demonstrated by experimental induction of these mutations into cell lines (Guo et al. 1997; Hashimoto et al. 2002;

Marques et al. 2003) or transgenic/knock-in animals (Praticò et al. 2001; Schuessel et al. 2006). Moreover, the ε4 allele of the gene encoding apolipoprotein E (*APOE*), a major genetic risk factor that influences the susceptibility to late-onset AD, is associated with OS. *In vitro*, *APOE* shows allele-specific antioxidant activity, with *APOE* ε2 allele the most effective and *APOE* ε4 allele being the least effective (Miyata and Smith 1996). *In vivo*, an increased level of oxidative damage is found in the brains of *APOE* knockout mice, which can be reversed by the induction of human wild–type gene *APOE* ε3 but not AD susceptibility gene *APOE* ε4 (Yao et al. 2004).

12.3.4 Environmental Factors of AD and OS

A considerable number of environmental and other nongenetic risk factors are identified for AD, including smoking, exposure to pesticides and aluminum in drinking water, and preexisting medical conditions such as traumatic brain injury, cerebrovascular disease, obesity, hypertension, diabetes, and dyslipidemia; these risk factors are associated with an increase in OS, while several protective factors for AD such as intake of vitamin C, vitamin E, and folic acid; fish consumption; moderate drinking of red wine; and regular exercise are associated with elevated capacity of antioxidant defense (Nunomura et al. 2006, 2007).

12.3.5 OS as a Convergence Point for Genetic and Environmental Susceptibility to Neurodegenerative Disorders

Of particular interest, not only in AD but also in other neurodegenerative disorders, we can find a close relationship between OS and disease-specific genetic and environmental factors. Indeed, mutations in the *parkin* gene, the *PTEN-induced putative kinase 1 (PINK1)* gene and the *DJ-1* gene found in familial PD, as well as mutations in the *copper/zinc (Cu/Zn)-SOD* gene and the *TAR DNA-binding protein (TARDBP)* gene found in familial ALS, are associated with OS (Nunomura et al. 2007; Ferraiuolo et al. 2011). Nongenetic risk factors for PD, such as traumatic brain injury and exposure to pesticides and metals, as well as those for ALS, such as smoking and exposure to pesticides, are associated with an increase in OS (Nunomura et al. 2007; Ingre et al. 2015). Therefore, OS is potentially a key pathological event that is located at the convergence point of genetic and environmental susceptibility to major neurodegenerative disorders.

12.4 SZ AND OS

12.4.1 OS Markers in the Postmortem SZ Brain

Markers of oxidative damage to cellular macromolecules, including 8-hydroxydeoxyguanosine and 8-hydroxyguanosine in the hippocampus (Nishioka and Arnold 2004; Che et al. 2010), 3-nitrotyrosine in the prefrontal cortex (Andreazza et al. 2010), and 4-hydroxynonenal in the anterior cingulate cortex (Wang et al. 2009), have been demonstrated in postmortem brains from patients with SZ. Moreover, decreased levels in the components of antioxidant defense system—decrease in GPX and GSH in the prefrontal cortex and caudate nucleus and decrease in GSH reductase in the caudate nucleus—have been observed in postmortem SZ brains (Yao et al. 2006; Gawryluk et al. 2011).

12.4.2 Involvement of OS in the Pathophysiology of SZ as an Upstream Event

A growing body of evidence is accumulating to show the involvement of OS in the pathophysiology of SZ as an upstream event during neurodevelopment (Table 12.1). Decreased antioxidant enzyme activity has been observed in blood samples from patients with the first-episode psychosis (Flatow et al. 2013) or untreated SZ (Raffa et al. 2009) and in cerebrospinal fluid from recent-onset SZ patients

(Coughlin et al. 2013). In concordance, increased levels of OS and OS-responsive gene expression have been demonstrated in iPSC-derived neural progenitor cells from SZ patients (Paulsen Bda et al. 2012; Brennand et al. 2015). An increase in OS in vulnerable neurons is also observed in a well-known animal model of SZ, the conditional gene knockout mouse with the functional elimination of *N*-methyl-D-aspartate (NMDA) receptors specifically from cortical interneurons (Jiang et al. 2013b). Furthermore, a GSH synthetic enzyme *glutamate cysteine ligase modulatory subunit (GCLM)*-knockout mouse shows SZ-like brain structural, electrophysiological, neurochemical, and behavioral alterations in adolescence and early adulthood, which indicates that the congenital defect in the antioxidant defense system induces SZ-like phenotypes. Actually, the SZ-like phenotypes observed in this mouse model are selective loss of the cortical parvalbumin-positive interneurons (Steullet et al. 2010), disturbance in the generation of γ-oscillations (Steullet et al. 2010), alterations in the level of cortical glutamine and glutamate (das Neves Duarte et al. 2012), and alterations in locomotive activity and prepulse inhibition (Kulak et al. 2012), which are prevented by a juvenile and adolescent treatment with a GSH precursor, *N*-acetylcysteine (Cabungcal et al. 2013, 2014).

12.4.3 Genetic Factors of SZ and OS

Genetic polymorphisms and copy number variation in genes encoding antioxidant enzymes—*Mn-SOD* (Akyol et al. 2005), a GSH synthetic enzyme *glutamate cysteine ligase catalytic subunit (GCLC)* (Gysin et al. 2007) and *GCLM* (Tosic et al. 2006), *NADPH quinone oxidoreductase 2 (NQO2)* (Harada et al. 2003), *paraoxonase 1 (PON1)* (Kucukali et al. 2008), and *GSH S-transferase (GST)* (Rodriguez-Santiago et al. 2010)—are associated with SZ. A frameshift mutation in *Glyoxalase I (GLO1)* gene encoding an enzyme required for the cellular detoxification of reactive carbonyl compounds has been found in several patients with SZ (Arai et al. 2010). Interestingly, besides the association with these redox regulatory genes, several known susceptible genes for SZ including *Disrupted-in-Schizophrenia-1 (DISC1)*, *Neuregulin 1 (NRG1)*, *proline dehydrogenase (PRODH)*, and *G72* are associated with increased levels of OS or decreased antioxidant capacities (O'Donnell et al. 2014). Indeed, *DISC1* and *G72* are associated with mitochondrial dysfunction (Park et al. 2010; Otte et al. 2011), and *NRG1* and *PRODH* are associated with disrupted cellular defense against reactive oxygen species (Goldshmit et al. 2001; Krishnan et al. 2008).

12.4.4 Environmental Factors of SZ and OS

Possible environmental factors of SZ such as infection, hypoxia, and malnutrition in the pre- and perinatal period as well as illicit substance use in adolescence are potentially associated with an increase in OS (Brown 2011). Psychosocial stress is also associated with an increase in OS in vulnerable brain regions, which has been demonstrated in an animal model of social isolation stress or restraint stress (Jiang et al. 2013a,b).

12.4.5 OS as a Convergence Point for Genetic and Environmental Susceptibility to Psychiatric Disorders

In the NMDA receptor knockout mouse model of SZ, a marked increase in OS, particularly in the cortical parvalbumin-positive interneurons, is rapidly exacerbated by postweaning social isolation, but treatment with antioxidants abolishes OS and partially alleviates the SZ-like behavioral phenotypes (Cabungcal et al. 2013, 2014). These findings strongly suggest that genetic and environmental factors of SZ synergistically enhance OS that causes disease phenotypes.

There is a possible association of genetic and nongenetic risk factors with OS not only in SZ but also in depression and bipolar disorders. Genetic polymorphisms in *Mn-SOD*, *GPX*, and *GST* are associated with depression, and polymorphism in *PON1* is associated with both depression and bipolar disorders (Maes et al. 2011; Moylan et al. 2014). As for non-genetic risk factors, cardiovascular disorders, stroke, and diabetes are associated with depression, and psychosocial stress and obesity are associated with both depression and bipolar disorders (Simon et al. 2006; Etain et al. 2008; Valkanova and Ebmeier 2013), all of which are associated with an increase in OS (Moylan et al. 2014). Therefore, OS is located at the convergence point of interaction between genetic and environmental factors of the major psychiatric disorders, as we have seen similarly in the major neurodegenerative disorders.

12.5 CONCLUSION

In this chapter, we focus on AD and SZ as archetypal disorders of neurodegeneration and neurodevelopment, respectively, and discuss the involvement of OS in these disorders as an upstream pathological event. In the life spectrum, specific brain regions such as default mode network may show vulnerability to OS in both early neurodevelopmental processes and later age–related neurodegeneration. It is suggested that OS is a convergence point for genetic and environmental susceptibilities to not only neurodegenerative but also psychiatric disorders. In other words, OS potentially plays a central role in the pathomechanisms that integrate gene–environment interactions in a diverse spectrum of neuropsychiatric disorders. Further investigations into the development of useful OS biomarkers and efficacious OS-targeting interventions may shed light on a promising approach for establishing novel prophylactic and preemptive strategies against neuropsychiatric disorders.

REFERENCES

Abe, T., Tohgi, H., Isobe, C. et al. 2002. Remarkable increase in the concentration of 8-hydroxyguanosine in cerebrospinal fluid from patients with Alzheimer's disease. *J Neurosci Res* 70:447–450.

Akyol, O., Yanik, M., Elyas, H. et al. 2005. Association between Ala-9Val polymorphism of Mn-SOD gene and schizophrenia. *Prog Neuropsychopharmacol Biol Psychiatry* 29:123–131.

Andreazza, A.C., Shao, L., Wang, J.F. et al. 2010. Mitochondrial complex I activity and oxidative damage to mitochondrial proteins in the prefrontal cortex of patients with bipolar disorder. *Arch Gen Psychiatry* 67:360–368.

Arai, M., Yuzawa, H., Nohara, I. et al. 2010. Enhanced carbonyl stress in a subpopulation of schizophrenia. *Arch Gen Psychiatry* 67:589–597.

Bishop, N.A., Lu, T., Yankner, B.A. 2010. Neural mechanisms of ageing and cognitive decline. *Nature* 464:529–535.

Bradley, M.A., Markesbery, W.R., Lovell, M.A. 2010. Increased levels of 4-hydroxynonenal and acrolein in the brain in preclinical Alzheimer disease. *Free Radic Biol Med* 48:1570–1576.

Braga, R.M., Sharp, D.J., Leeson, C. et al. 2013. Echoes of the brain within default mode, association, and heteromodal cortices. *J Neurosci* 33:14031–14039.

Brennand, K., Savas, J.N., Kim, Y. et al. 2015. Phenotypic differences in hiPSC NPCs derived from patients with schizophrenia. *Mol Psychiatry* 20:361–368.

Brown, A.S. 2011. The environment and susceptibility to schizophrenia. *Prog Neurobiol* 93:23–58.

Cabungcal, J.H., Counotte, D.S., Lewis, E.M. et al. 2014. Juvenile antioxidant treatment prevents adult deficits in a developmental model of schizophrenia. *Neuron* 83:1073–1084.

Cabungcal, J.H., Steullet, P., Kraftsik, R. et al. 2013. Early-life insults impair parvalbumin interneurons via oxidative stress: Reversal by *N*-acetylcysteine. *Biol Psychiatry* 73:574–582.

Campos, P.B., Paulsen, B.S., Rehen, S.K. 2014. Accelerating neuronal aging in in vitro model brain disorders: A focus on reactive oxygen species. *Front Aging Neurosci* 6:292.

Che, Y., Wang, J.F., Shao, L. et al. 2010. Oxidative damage to RNA but not DNA in the hippocampus of patients with major mental illness. *J Psychiatry Neurosci* 35:296–302.

Coughlin, J.M., Ishizuka, K., Kano, S.I. et al. 2013. Marked reduction of soluble superoxide dismutase-1 (SOD1) in cerebrospinal fluid of patients with recent-onset schizophrenia. *Mol Psychiatry* 18:10–11.

das Neves Duarte, J.M., Kulak, A., Gholam-Razaee, M.M. et al. 2012. *N*-acetylcysteine normalizes neurochemical changes in the glutathione-deficient schizophrenia mouse model during development. *Biol Psychiatry* 71:1006–1014.

Douaud, G., Groves, A.R., Tamnes, C.K. et al. 2014. A common brain network links development, aging, and vulnerability to disease. *Proc Natl Acad Sci USA* 111:17648–17653.

Dumont, M., Wille, E., Stack, C. et al. 2009. Reduction of oxidative stress, amyloid deposition, and memory deficit by manganese superoxide dismutase overexpression in a transgenic mouse model of Alzheimer's disease. *FASEB J* 23:2459–2466.

Etain, B., Henry, C., Bellivier, F. et al. 2008. Beyond genetics: Childhood affective trauma in bipolar disorder. *Bipolar Disord* 10:867–876.

Ferraiuolo, L., Kirby, J., Grierson, A.J. et al. 2011. Molecular pathways of motor neuron injury in amyotrophic lateral sclerosis. *Nat Rev Neurol* 7:616–630.

Flatow, J., Buckley, P., Miller, B.J. 2013. Meta-analysis of oxidative stress in schizophrenia. *Biol Psychiatry* 74:400–409.

Gawryluk, J.W., Wang, J.F., Andreazza, A.C. et al. 2011. Decreased levels of glutathione, the major brain antioxidant, in post-mortem prefrontal cortex from patients with psychiatric disorders. *Int J Neuropsychopharmacol* 14:123–130.

Gemma, C., Vila, J., Bachstetter, A., Bickford, P.C. 2007. Oxidative stress and the aging brain: From theory to prevention. In *Brain Aging: Models, Methods, and Mechanisms*, Riddle, D.R., Ed., pp. 353–374, Boca Raton, FL: CRC Press.

Goldshmit, Y., Erlich, S., Pinkas-Kramarski, R. 2001. Neuregulin rescues PC12-ErbB4 cells from cell death induced by H_2O_2. Regulation of reactive oxygen species levels by phosphatidylinositol 3-kinase. *J Biol Chem* 276:46379–46385.

Guo, Q., Sopher, B.L., Furukawa, K. et al. 1997. Alzheimer's presenilin mutation sensitizes neural cells to apoptosis induced by trophic factor withdrawal and amyloid β-peptide: Involvement of calcium and oxyradicals. *J Neurosci* 17:4212–4222.

Gysin, R., Kraftsik, R., Sandell, J. et al. 2007. Impaired glutathione synthesis in schizophrenia: Convergent genetic and functional evidence. *Proc Natl Acad Sci USA* 104:16621–16626.

Harada, S., Tachikawa, H., Kawanishi, Y. 2003. A possible association between an insertion/deletion polymorphism of the NQO2 gene and schizophrenia. *Psychiatr Genet* 13:205–209.

Hashimoto, Y., Niikura, T., Ito, Y. et al. 2002. Neurotoxic mechanisms by Alzheimer's disease-linked N141I mutant presenilin 2. *J Pharmacol Exp Ther* 300:736–745.

Ikonomidou, C., Kaindl, A.M. 2011. Neuronal death and oxidative stress in the developing brain. *Antioxid Redox Signal* 14:1535–1350.

Ingre, C., Roos, P.M., Piehl, F. et al. 2015. Risk factors for amyotrophic lateral sclerosis. *Clin Epidemiol* 7:181–193.

Jenner, P. 2003. Oxidative stress in Parkinson's disease. *Ann Neurol* 53(Suppl. 3):S26–S36.

Jiang, Z., Cowell, R.M., Nakazawa, K. 2013a. Convergence of genetic and environmental factors on parvalbumin-positive interneurons in schizophrenia. *Front Behav Neurosci* 7:116.

Jiang, Z., Rompala, G.R., Zhang, S. et al. 2013b. Social isolation exacerbates schizophrenia-like phenotypes via oxidative stress in cortical interneurons. *Biol Psychiatry* 73:1024–1034.

Kesler, S.R. 2014. Default mode network as a potential biomarker of chemotherapy-related brain injury. *Neurobiol Aging* (Suppl. 2):S11–S19.

Kessler, R.C., Berglund, P., Demler, O. et al. 2005. Lifetime prevalence and age-of-onset distributions of DSM-IV disorders in the National Comorbidity Survey Replication. *Arch Gen Psychiatry* 62:593–602.

Kondo, T., Asai, M., Tsukita, K. et al. 2013. Modeling Alzheimer's disease with iPSCs reveals stress phenotypes associated with intracellular Aβ and differential drug responsiveness. *Cell Stem Cell* 12:487–496.

Krishnan, N., Dickman, M.B., Becker, D.F. 2008. Proline modulates the intracellular redox environment and protects mammalian cells against oxidative stress. *Free Radic Biol Med* 44:671–681.

Kucukali, C.I., Aydin, M., Ozkok, E. et al. 2008. Paraoxonase-1 55/192 genotypes in schizophrenic patients and their relatives in Turkish population. *Psychiatr Genet* 18:289–294.

Kulak, A., Cuenod, M., Do, K.Q. 2012. Behavioral phenotyping of glutathione-deficient mice: Relevance to schizophrenia and bipolar disorder. *Behav Brain Res* 226:563–570.

Li, F., Calingasan, N.Y., Yu, F. et al. 2004. Increased plaque burden in brains of APP mutant MnSOD heterozygous knockout mice. *J Neurochem* 89:1308–1312.

Lin, M.T., Beal, M.F. 2006. Mitochondrial dysfunction and oxidative stress in neurodegenerative diseases. *Nature* 443:787–795.

Maes, M., Galecki, P., Chang, Y.S. et al. 2011. A review on the oxidative and nitrosative stress (O&NS) pathways in major depression and their possible contribution to the (neuro) degenerative processes in that illness. *Prog Neuropsychopharmacol Biol Psychiatry* 35:676–692.

Mangialasche, F., Polidori, M.C., Monastero, R. et al. 2009. Biomarkers of oxidative and nitrosative damage in Alzheimer's disease and mild cognitive impairment. *Ageing Res Rev* 8:285–305.

Marques, C.A., Keil, U., Bonert, A. et al. 2003. Neurotoxic mechanisms caused by the Alzheimer's disease-linked Swedish amyloid precursor protein mutation: Oxidative stress, caspases, and the JNK pathway. *J Biol Chem* 278:28294–28302.

McManus, M.J., Murphy, M.P., Franklin, J.L. 2011. The mitochondria-targeted antioxidant MitoQ prevents loss of spatial memory retention and early neuropathology in a transgenic mouse model of Alzheimer's disease. *J Neurosci* 31:15703–15715.

Miyata, M., Smith, J.D. 1996. Apolipoprotein E allele-specific antioxidant activity and effects on cytotoxicity by oxidative insults and β-amyloid peptides. *Nat Genet* 14:55–61.

Moylan, S., Berk, M., Dean, O.M. et al. 2014. Oxidative & nitrosative stress in depression: Why so much stress? *Neurosci Biobehav Rev* 45:46–62.

Ng, F., Berk, M., Dean, O. et al. 2008. Oxidative stress in psychiatric disorders: Evidence base and therapeutic implications. *Int J Neuropsychopharmacol* 11:851–876.

Nishioka, N., Arnold, S.E. 2004. Evidence for oxidative DNA damage in the hippocampus of elderly patients with chronic schizophrenia. *Am J Geriatr Psychiatry* 12:167–175.

Nunomura, A., Castellani, R.J., Zhu, X. et al. 2006. Involvement of oxidative stress in Alzheimer disease. *J Neuropathol Exp Neurol* 65:631–641.

Nunomura, A., Moreira, P.I., Castellani, R.J. et al. 2012a. Oxidative damage to RNA in aging and neurodegenerative disorders. *Neurotox Res* 22:231–248.

Nunomura, A., Moreira, P.I., Lee, H.G. et al. 2007. Neuronal death and survival under oxidative stress in Alzheimer and Parkinson diseases. *CNS Neurol Disord Drug Targets* 6:411–423.

Nunomura, A., Perry, G., Aliev, G. et al. 2001. Oxidative damage is the earliest event in Alzheimer disease. *J Neuropathol Exp Neurol* 60:759–767.

Nunomura, A., Perry, G., Pappolla, M.A. et al. 2000. Neuronal oxidative stress precedes amyloid-β deposition in Down syndrome. *J Neuropathol Exp Neurol* 59:1011–1017.

Nunomura, A., Tamaoki, T., Motohashi, N. et al. 2012b. The earliest stage of cognitive impairment in transition from normal aging to Alzheimer disease is marked by prominent RNA oxidation in vulnerable neurons. *J Neuropathol Exp Neurol* 71:233–241.

Nunomura, A., Tamaoki, T., Tanaka, K. et al. 2010. Intraneuronal amyloid β accumulation and oxidative damage to nucleic acids in Alzheimer disease. *Neurobiol Dis* 37:731–737.

O'Donnell, P., Do, K.Q., Arango, C. 2014. Oxidative/nitrosative stress in psychiatric disorders: Are we there yet? *Schizophr Bull* 40:960–962.

Otte, D.M., Sommersberg, B., Kudin, A. et al. 2011. *N*-acetylcysteine treatment rescues cognitive deficits induced by mitochondrial dysfunction in G72/G30 transgenic mice. *Neuropsychopharmacology* 36:2233–2243.

Park, Y.U., Jeong, J., Lee, H. et al. 2010. Disrupted-in-schizophrenia 1 (DISC1) plays essential roles in mitochondria in collaboration with Mitofilin. *Proc Natl Acad Sci USA* 107:17785–17790.

Paulsen Bda, S., de Moraes Maciel, R., Galina, A. et al. 2012. Altered oxygen metabolism associated to neurogenesis of induced pluripotent stem cells derived from a schizophrenic patient. *Cell Transplant* 21:1547–1559.

Praticò, D., Clark, C.M., Liun, F. et al. 2002. Increase of brain oxidative stress in mild cognitive impairment: A possible predictor of Alzheimer disease. *Arch Neurol* 59:972–976.

Praticò, D., Uryu, K., Leight, S. et al. 2001. Increased lipid peroxidation precedes amyloid plaque formation in an animal model of Alzheimer amyloidosis. *J Neurosci* 21:4183–4187.

Raffa, M., Mechri, A., Othman, L.B. et al. 2009. Decreased glutathione levels and antioxidant enzyme activities in untreated and treated schizophrenic patients. *Prog Neuropsychopharmacol Biol Psychiatry* 33:1178–1183.

Rice, D., Barone, Jr. S. 2000. Critical periods of vulnerability for the developing nervous system: Evidence from humans and animal models. *Environ Health Perspect* 108(Suppl. 3):511–533.

Ringman, J.M., Younkin, S.G., Praticò, D. et al. 2008. Biochemical markers in persons with preclinical familial Alzheimer disease. *Neurology* 71:85–92.

Rodriguez-Santiago, B., Brunet, A., Sobrino, B. et al. 2010. Association of common copy number variants at the glutathione *S*-transferase genes and rare novel genomic changes with schizophrenia. *Mol Psychiatry* 15:1023–1033.

Schuessel, K., Frey, C., Jourdan, C. et al. 2006. Aging sensitizes toward ROS formation and lipid peroxidation in PS1M146L transgenic mice. *Free Radic Biol Med* 40:850–862.

Selkoe, D.J. 2011. Resolving controversies on the path to Alzheimer's therapeutics. *Nat Med* 17:1060–1065.

Simon, G.E., Von Korff, M., Saunders, K. et al. 2006. Association between obesity and psychiatric disorders in the US adult population. *Arch Gen Psychiatry* 63:824–830.

Sonnen, J.A., Breitner, J.C., Lovell, M.A. et al. 2008. Free radical-mediated damage to brain in Alzheimer's disease and its transgenic mouse models. *Free Radic Biol Med* 45:219–230.

Steullet, P., Cabungcal, J.H., Kulak, A. et al. 2010. Redox dysregulation affects the ventral but not dorsal hippocampus: Impairment of parvalbumin neurons, gamma oscillations, and related behaviors. *J Neurosci* 30:2547–2558.

Tayler, H., Fraser, T., Miners, J.S. et al. 2010. Oxidative balance in Alzheimer's disease: Relationship to APOE, Braak tangle stage, and the concentrations of soluble and insoluble amyloid-β. *J Alzheimers Dis* 22:1363–1373.

Tosic, M., Ott, J., Barral, S. et al. 2006. Schizophrenia and oxidative stress: Glutamate cysteine ligase modifier as a susceptibility gene. *Am J Hum Genet* 79:586–592.

Valkanova, V., Ebmeier, K.P. 2013. Vascular risk factors and depression in later life: A systematic review and meta-analysis. *Biol Psychiatry* 73:406–413.

Wang, J.F., Shao, L., Sun, X. et al. 2009. Increased oxidative stress in the anterior cingulate cortex of subjects with bipolar disorder and schizophrenia. *Bipolar Disord* 11:523–529.

Yao, J., Petanceska, S.S., Montine, T.J. et al. 2004. Aging, gender and APOE isotype modulate metabolism of Alzheimer's Aβ peptides and F-isoprostanes in the absence of detectable amyloid deposits. *J Neurochem* 90:1011–1018.

Yao, J.K., Leonard, S., Reddy, R. 2006. Altered glutathione redox state in schizophrenia. *Dis Markers* 22:83–93.

13 Reactive Oxygen Species and Antioxidant Therapies for Multiple Sclerosis Treatment

*Elżbieta Miller, Dominika Książek-Winiarek,
Tomasz Włodarczyk, and Andrzej Głąbiński*

CONTENTS

ABSTRACT

Multiple sclerosis (MS) is a complex disease with multifactorial pathogenesis and different clinical courses. Inflammation, oxidative stress, and redox processes are important factors in MS pathophysiology. Additionally, oxidative stress appears to provide an important link between environmental and genetic risk factors. Investigations of effective antioxidative therapy for MS are very reasonable due to the heterogeneous nature of this disease. Over the last decade, various therapies and exogenous compounds were designed to affect immune response in the central nervous system (CNS). Recent studies suggested that neuroprotection is strictly connected with signaling pathways involving special redox-sensitive transcription nuclear factor (erythroid-derived 2)-related factor 2 (Nrf2). Moreover, it creates a promising target for future MS therapies like new oral treatment with fumaric acid ester that has been recently approved for the treatment of early stage of MS (relapsing–remitting (RR)).

In this chapter, the mechanisms of oxidative stress, current pharmacological treatment, main antioxidative therapies, and structures of the CNS involved in MS are presented. Additionally, clinical subtypes with neurological symptoms are shown.

13.1 INTRODUCTION

Multiple sclerosis (MS) is an autoimmune, chronic inflammatory, and neurodegenerative disease of the central nervous system (CNS). Worldwide, approximately one million people suffer from MS, and women outnumber men with a ratio of 1.5:1. The focal demyelination of the CNS is a crucial histopathological feature of MS (Miller, 2012).

The course of this heterogeneous disease is unpredictable and is characterized by recurrent neurological relapses and/or progression (Milijković and Spasojević, 2013). A majority of MS patients at the early phase exhibit immune-mediated response with microglial overactivity and inflammation mainly in white matter patches (plaques) (Milijković and Spasojević, 2013; Miller et al., 2013). There are a variety of clinical symptoms characteristic for early and progressive stages. From the early signs of MS, vision impairments due to optic neuritis and sensation deficits such as burning or prickling are the most frequent. In chronic form of MS, progression leads to paresis/paralysis, mainly of lower extremities, spasticity, ataxia, fatigue, tremor, aphasia, loss of bladder control, constipation, and incontinence. Moreover, cognitive impairment is very characteristic of MS, mainly as a memory and concentration dysfunction. In recent years, a number of clinical studies have reported an association between brain atrophy and cognitive impairment in MS. The occurrence of neurological symptoms is the basis of MS diagnosis.

The periodical flare-up of neurological symptoms is named relapse (Bielekova and Martin, 2004; Miller, 2012). It has to be more than one relapse with dissemination in time and space to diagnose MS. The higher number of relapses at the early stage of MS has worsened the prognosis. Other diagnostic methods such as magnetic resonance imaging (MRI), cerebrospinal fluid (CSF) analyses for immunoglobulin, and oligoclonal banding are needed for the confirmation of the clinical state.

Four subtypes of MS have been described: relapsing–remitting (RRMS), secondary progressive (SPMS), progressive relapsing (PRMS), and primary progressive (PPMS) (Lublin and Reingold, 1996). The most common subtype (about 85%) is RRMS course with periods of exacerbations with new symptoms followed by complete or partial recovery (remission). RRMS is dominated by multifocal CNS inflammation. At about 10–20 years of disease duration, 85% of RRMS converts to SPMS characterized by the continuous progression of neurological symptoms with or without the occurrence of relapses and remissions. SPMS is characterized mainly by neuronal and axonal degeneration and CNS atrophy with steadily increasing reduction in the number of nerve fibers (Ksiazek-Winiarek et al., 2013; Miller et al., 2013). Such disease evolution from RR to progressive forms of MS is related to loss of oligodendrocytes (ODCs) and neurons. The PPMS concerns only about 10% of people with MS. This clinical subtype is characterized by earlier axonal loss and less-pronounced inflammation processes. The progressive forms of MS (PPMS and SPMS) are dominated by neuroaxonal degeneration that correlates with CNS atrophy and a range of neurological alterations (Milijković and Spasojević, 2013). In spite of investigating MS for a long time, many elements of its complex pathogenesis are still unknown. Oxidative stress appears to provide an important link between environmental and genetic risk factors. Various therapies were designed to affect immune response in the CNS. Recently, an important implication of redox alterations in both early and progressive stages of MS was studied successfully (Ortiz et al., 2013).

13.2 CNS CELL TYPES AND THEIR ROLE IN MS

13.2.1 Oligodendrocytes

ODCs are very important cells for MS pathophysiology due to their ability to produce myelin sheaths (Table 13.1 and Figure 13.1). A single ODC can modulate a variety of biological processes of approximately 50 axons. Myelin sheaths are composed of 80% of lipids and 20% of proteins. Myelin proteins are the key biochemical goals in MS immune therapy (Bradl and Lassman, 2010). The CNS can recruit oligodendrocyte progenitor cells (OPCs) that are capable of proliferating, migrating, and differentiating into mature myelinating ODCs (Watzlawik et al., 2010). However, ODCs are unable to completely rebuild the myelin sheath, and repeated attacks successively lead to less-effective remyelination until scar-like plaques (scleroses) are built up around the damaged axons. The capacity for remyelination decreases with age, probably due to a lower ability of OPC to differentiate in the elderly people. OPCs are highly vulnerable to inflammation and oxidative stress as they have a high metabolic rate, high intracellular iron, and low concentrations of the antioxidant glutathione. They also express plenty of molecules, rendering them more susceptible to inflammatory cytokines or high calcium levels than others. ODCs are responsible for remyelination especially in the early phases of MS (Bauman and Pham-Dinh, 2001).

13.2.2 Astrocytes

Neuroanatomically, there are gray (protoplasmic) and white (fibrous) matter astrocytes located in different CNS regions. The human cerebral neocortex contains a variety of astrocyte types. Moreover, there is a long-standing recognition of an extended family of astroglia for cells that are similar but also very different from protoplasmic and fibrous astrocytes, such as Bergmann glia of the cerebellum or Müller glia in the retina, pituicytes in the neurohypophysis, cribrosocytes at the optic nerve head, and others. In addition, these new cell types express various astrocyte-related molecules such as S100β, glutamine synthetase, and others and present functions similar to astrocytes in a specific manner for their locations. Astrocytes have the ability to produce a wide range of pro- or anti-inflammatory molecules (Sofroniew and Vinters, 2010), and as such they can modulate this process. In the neuroinflammatory state, astrocytes release interleukin (IL)-6, tumor necrosis factor (TNF)-α, and IL-1β with their impact on higher blood–brain barrier (BBB) permeability (Abott et al., 2006), resulting in significant influx of immune cells, ions, and toxic molecules into the CNS. They are involved in the crucial processes of neuronal homeostasis, as well as BBB integrity, neurotransmitter regulation, synapsis functioning, and myelination. The astrocytes have connection with CNS blood vessels; therefore, they are very actively engaged in immunity-related actions in all neuroinflammation disorders including MS. Astrocytes support neurons in normal functioning and BBB regulation (Nair et al., 2008).

13.2.3 Microglial Cells

Microglial cells are resident macrophages of the CNS that can eliminate cell debris and/or atypical cells (classical phagocytic cells). They can produce a wide range of inflammatory mediators and effectors (cytokines, prostaglandins, reactive oxygen and nitrogen species [ROS and RNS]) and can deregulate major histocompatibility complex (MHC) molecules. Maintaining the homeostasis of the CNS, especially in pathological conditions such as neurodegenerative processes, is the most important function of microglia (Milijković and Spasojević, 2013). Macrophages are not only able to eliminate microbes and infected cells but are also involved in autoimmunological processes. They have a role in the early stage of brain immunity and adaptation to this process. There are three types of macrophages: meningeal, perivascular, and choroid plexus. The dendritic cells (DCs) are

TABLE 13.1

The Standard Disease-Modifying Treatment for Multiple Sclerosis

Line of DMT	Drug Description	Date of FDA Approval	Disease Type	Mode of Action	Side Effects	Route of Administration	References
First	IFN-β 1a (Avonex, Rebif)—glycoprotein produced in mammalian cells by using the natural human gene sequence; IFN-β 1b (Extavia, Betaseron)—protein produced by DNA recombinant technology using *Escherichia coli*	Betaseron (1993) Avonex (1996) Rebif (2002) Extavia (2009)	RRMS CIS SPMS	Enhancement of suppressor T-cell activity, reduction of proinflammatory cytokine production, downregulation of antigen presentation, inhibition of lymphocyte trafficking into the CNS	Injection site reactions, flu-like symptoms (low-grade fever, myalgias, headache, fatigue, chills), mild liver-enzyme elevation, leukopenia, depression, attempted suicide, insomnia, abdominal pain, seizures, congestive heart failure, hypertonia, hyper- or hypothyroidism	Betaseron—s.c. Avonex—i.m. Rebif—s.c. Extavia—s.c.	Paty and Li (1993); Jacobs et al. (1996); Ebers (1998); Kappos (1998); Walther and Hohlfeld (1999); Reder et al. (2010)
	Glatiramer acetate (Copaxone)—a random polymer composed of four most common amino acids of myelin basic protein (glutamic acid, lysine, alanine, tyrosine)	1996 (20 mg daily) 2014 (40 mg TIW)	RRMS CIS	Wide range affinity for T-lymphocyte receptors, suppression of encephalitogenic response to myelin immunogens, an immunomodulator shifting lymphocyte polarization from Th1 proinflammatory type to Th2 cell profile	Local injection-site reactions, chest tightness, shortness of breath, palpitations, anxiety, flushing	s.c.	Teitelbaum et al. (1971); Yong (2002); Boster et al. (2011)
	Teriflunomide (Aubagio)— an active metabolite of leflunomide, an oral isoxazole	2012	RRMS	Inhibition of *de novo* pyrimidine synthesis by altering dihydroorotate dehydrogenase resulting in decreased proliferation of rapidly dividing cells such as T and B cells	Elevated liver enzymes, trigeminal neuralgia, diarrhea, alopecia, nausea, creatine kinase elevation, hepatotoxicity, teratogenicity, leukopenia, peripheral neuropathy, acute renal failure, headache, hypertension, rash, infections, paresthesia, arthralgia	Oral	Comi et al. (2012); Munier-Lehmann et al. (2013); Novartis Pharmaceuticals Corporation (2014a); Bar-Or et al. (2014)

(Continued)

TABLE 13.1 (*Continued*)
The Standard Disease-Modifying Treatment for Multiple Sclerosis

Line of DMT	Drug Description	Date of FDA Approval	Disease Type	Mode of Action	Side Effects	Route of Administration	References
	Dimethyl fumarate (Tecfidera)—derivative of fumarate, a product of intermediary metabolism in the citric acid cycle	2013	RRMS CIS	Activation of Nrf2 pathway—the cellular response to oxidative stress, DMF metabolite (monomethyl fumarate) is an agonist of nicotinic acid receptor, exerts anti-inflammatory, antioxidant, neuroprotective effects, may alter the function of macrophages, microglia, lymphocytes	Lymphopenia, flushing, gastrointestinal symptoms (abdominal discomfort or pain and diarrhea), headache, transaminitis, nausea, PML development, anaphylaxis, and angioedema	Oral	Litjens et al. (2003); Meili-Butz et al. (2008); Linker et al. (2011); FDA Drug Safety Communication (2013); Tecfidera-FDA (2014)
Second	Natalizumab (Tysabri)—humanized monoclonal antibody against α4-integrin on the surface of mononuclear immune cells	2004	RRMS	Inhibition of migration of all leukocytes (except neutrophils) from the peripheral circulation across the endothelium into inflamed parenchymal tissue	PML development, herpes encephalitis, and meningitis, pneumonia, urinary tract infections, hepatotoxicity, hypersensitivity reactions, gastroenteritis, headache, fatigue, arthralgia, depression, diarrhea	i.v.	Noseworthy and Kirkpatrick (2005); Polman et al. (2006); Rudick et al. (2006); Bezabeh et al. (2010)
	Fingolimod (Gilenya)—sphingosine-1-phosphate-receptor modulator	2010	RRMS	Inhibition of lymphocytes egress from secondary lymphoid tissues and the thymus, reduction of cells' number in peripheral circulation	Fatal infections, bradycardia, atrioventricular block, hypertension, macular edema, skin cancer, elevated liver-enzyme levels, lymphopenia, PML development, headache, diarrhea, posterior reversible encephalopathy syndrome	Oral	Brinkmann (2009); Chun and Hartung (2010); Kappos et al. (2010); Novartis Pharmaceuticals Corporation (2014b); FDA Drug Safety Communication (2013

(*Continued*)

TABLE 13.1 (Continued)
The Standard Disease-Modifying Treatment for Multiple Sclerosis

Line of DMT	Drug Description	Date of FDA Approval	Disease Type	Mode of Action	Side Effects	Route of Administration	References
	Mitoxantrone (Novantrone)—an anthracenedione chemotherapeutic agent, type 2 topoisomerase inhibitor	2000	RRMS PRMS SPMS	Cytotoxic effect on immune system cells, inhibition of B cell, T cell, and macrophage proliferation, impairment of antigen presentation, impairment of cytokine secretion	Dose-dependent cardiac toxicity, menstrual irregularities, ablation of the menstrual cycle, drug-related acute leukemia, teratogenicity	i.v.	Shenkenberg and Von Hoff (1986); Edan et al. (1997); Ghalie et al. (2002); Hartung et al. (2002); Edan et al. (2011)
Third	Alemtuzumab (Lemtrada)—humanized monoclonal antibody against the surface antigen CD52 present in a variety of immune cells including T, B, and NK cells, monocytes, macrophages, and eosinophils	2014	RRMS	Global immunosuppression, rapid depletion of monocytes, B and T lymphocytes through antibody-dependent cell-mediated lysis, complex reorganization of the immune system following depletion	Autoimmune diseases, for example, autoimmune thyroiditis, autoimmune cytopenias, idiopathic thrombocytopenia purpura, Goodpasture's syndrome, nephropathies, opportunistic infections, neoplasms, infusion-associated reactions	i.v.	Coles et al. (2008); Cossburn et al. (2011); Cuker et al. (2011); Costelloe et al. (2012); Perumal et al. (2012); Menge et al. (2014)

Abbreviations: DMT, disease modifying treatment; RRMS, relapsing-remitting multiple sclerosis; PRMS, progressive-relapsing multiple sclerosis; CIS, clinically isolated syndrome; TIW, 3 times weekly; CNS, central nervous system; PML, progressive multifocal leukoencephalopathy; Nrf2, nuclear factor (erythroid-derived 2)-like 2; i.v., intravenous; s.c., subcutaneous; i.m., intramuscular.

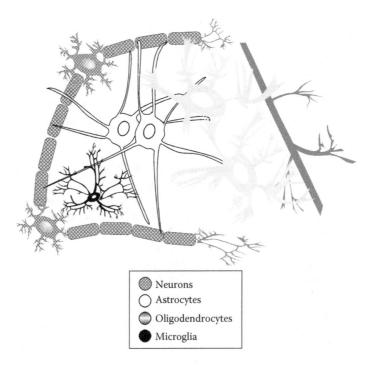

FIGURE 13.1 Central nervous system cell types.

professional antigen-presenting cells (APCs) that provide the initiation of the autoimmunity in MS. Generally, there is no DC and neutrophils in the CNS in physiological state, while T and B cells are present in low numbers (Block and Hong, 2005).

13.2.4 Blood–Brain Barrier

The BBB plays a critical role not only in molecule transport but also as a CNS defense against neurotoxins. This anatomical structure helps to maintain a constant level of [K⁺] at ~2.4–2.9 mM, as well as other major ions including calcium (Ca^{2+}) and magnesium (Mg^{2+}), and also pH levels. Under physiological conditions, only few leukocytes are present in the CNS, but during pathological states, peripheral leukocytes pass through the CSF, the parenchymal perivascular space, and the subarachnoid space. It has been described that aggressive CD4⁺ T lymphocytes cross the BBB and blood–CSF barrier in the experimental autoimmune encephalomyelitis (EAE)—an animal model of MS. The BBB damage is a cardinal feature of active MS, which allows the influx of inflammatory cells. In addition, immunomodulatory treatment with β-interferon or glatiramer acetate leads to decrease in the BBB permeability during relapse (Patel and Frey, 2015).

13.3 NITRIC OXIDE AND NEURAL FUNCTION

Nitric oxide (NO) is an essential factor in controlling and maintaining physiological cerebral blood flow, neurotransmission as well as synaptic plasticity, modulation of neuroendocrine functions, memory formation, and behavioral activity. Moreover, this omnipresent multipotential molecule can modulate blood flow, thrombosis, and neural activity. NO is produced by the isoforms of Ca^{2+}/calmodulin-dependent nitric oxide synthases (NOS) in all CNS cells including neurons, endothelial cells, and macro- and microglial cells. Neuroinflammation results in increased NO production as an effect of the inducible NOS (iNOS) in glial cells. The essential role of NO in both physiology and pathology is the reaction of NO with superoxide and the formation of the powerful oxidant peroxynitrite ONOO⁻ (Pacher et al., 2007). Superoxide is rapidly removed by high concentrations of the main antioxidative enzyme superoxide dismutase (SOD), whose isoenzymes are located in the cytoplasm, mitochondria, and extracellular compartments. *In vivo*, the half-life of NO is less than a second, whereas the concentrations of NO relevant for cellular signaling can persist in phosphate-buffered saline for an hour. There is no enzyme able to catalyze any reaction as fast as it is required to form peroxynitrite from NO and superoxide (Pacher et al., 2007). What is more, even a slightly increased concentration of superoxide and NO (10-fold greater rate) can amplify the peroxynitrite production by 100-fold. Therefore, the activation of CNS microglia stimulates the elevation of superoxide and NO concentration, and 1,000-fold higher production leads to the increased formation of peroxynitrite by a 1,000,000-fold (Denicola et al., 1998). Peroxynitrite is a strong oxidant that can react with various molecules at a relatively slow rate and can diffuse quite far on a cellular scale. The CNS actively synthesized NO using NOS. There are three types of NOS: neuronal NOS (nNOS) or NOS1, iNOS or NOS2, and endothelial NOS (eNOS) or NOS3. The iNOS (NOS2) was isolated from macrophages, and it is readily induced by proinflammatory cytokines in many tissues. NOS2 is not regulated by changes in intracellular calcium and is often called a high-output source of NO (Pacher et al., 2007). There is also eNOS synthase that is found not only in endothelium but also in other cells like neurons. The main role of eNOS is its reaction leading to higher levels of NO (gas), which is a small hydrophobic molecule that is able to cross the cell membranes by diffusion (Calabrese et al., 2009), which in turn facilitates very fast information transition in short distances. In the physiological state, the level of oxidant including peroxynitrite is negligible. The higher than normal persistent flux of peroxynitrite causes the oxidation and potential damage and death of cells. The half-life of peroxynitrite is short (about 10–20 ms), but it is sufficient to cross biological membranes and to diffuse one to two cell diameters. Chronic neuroinflammation is the hallmark of all neurodegenerative diseases including MS. Therefore, peroxynitrite may be an initiating factor in the lipid peroxidation of myelin sheaths. Moreover, peroxynitrite-induced tyrosine nitration of proteins may contribute to the BBB damage. The main inhibitor of tyrosine nitration is uric acid that protects the BBB against the migration of proinflammatory cells into the CNS. Plasma uric acid level is proposed as a biomarker for monitoring MS activity (Guerrero et al., 2011). As the nitrite and free nitrotyrosine are elevated in the CSF, they are suggested as potential markers of neurodegeneration (Onufriev, 2010). Moreover, iNOS expression decreases in demyelinated plaques from MS patients when inflammation is reduced (Liu et al., 2001).

13.4 MS TRADITIONAL TREATMENT

Currently, the reduction in the number of relapses and lower radiological activity in MRI are the main parameters estimated in clinical trials. However, the onset of progression and the level of disability after 5–10 years and later are also very useful in long-term observations of the effectiveness of therapies. The pharmacological treatment of MS is focused mainly on the early phase of MS, especially the RRMS state. The routine treatment in acute relapses is still using the high doses of intravenous corticosteroids due to their strong anti-inflammatory effects. In 2015, the National MS Society updated the list of drugs that are approved by the U.S. Food and Drug Administration (FDA) (National MS Society, n.d.). Currently, there are 12 disease-modifying drugs for treatment-relapsing subtypes of MS. The drug descriptions with side effects are provided in Table 13.2.

13.5 MECHANISMS OF OXIDATIVE DAMAGE IN MS

Oxidative stress may contribute to the disease mechanisms in both the relapsing–remitting and progressive phases of MS through its involvement in inflammation and axonal degeneration,

TABLE 13.2

Central Nervous System (CNS) Cells Involved in Multiple Sclerosis

CNS Cell	Physiological Function	Role in MS
Neuron	Transmits information as an electrical and chemical signal	Gradual decrease in myelin sheath and axons results in slowing down or blockade of impulses transmission
Oligodendrocyte	Support and insulate axons by the production of the myelin sheath	Proteins of myelin sheath are the immune targets (myelin basic protein (MBP), myelin oligodendrocyte glycoprotein (MOG), and proteolipid protein (PLP))
Astrocyte	Control optimal physical and metabolic state for neuronal activities	Neuronal redox homeostasis
Microglial–macrophages	Control homeostasis for neural activities	Upregulate major histocompatibility complex (MHC), produce inflammatory mediators
Microglial–dendritic cell	Professional antigen-presenting cells	Initiation of immune response

respectively. Therefore, redox therapy can be a useful therapeutic strategy in oxidative stress–related disorders such as MS. There are a variety of different molecules with antioxidant properties that are divided into endogenous antioxidants synthesized in the human body and exogenous compounds such as minerals, supplements, vitamins, and others (Miller et al., 2013). Epidemiological studies present that supplementation with antioxidants lowers the risk of heart diseases, some neurodegenerative diseases, and cancers (Miller et al., 2013). The expression of the most of the endogenous antioxidative enzymes is regulated by the signaling pathways involving nuclear factor (erythroid-derived 2)-related factor 2 (Nrf2) and is activated by ROS. In neurons of MS patients, the higher expression of Nrf2/antioxidant response element (ARE) is related to the more pronounced oxidative stress damage in these cells (Motohashi and Yamamoto, 2004). Over 200 Nrf2-driven genes involved in redox reaction were described, such as SODs, glutathione peroxidases (GPxs), catalase (CAT), peroxiredoxins, and NADPH:quinone oxidoreductase. However, the CAT activity in the CNS is very low compared with peripheral tissues. The major redox enzyme in the CNS is GPx. Unlike neurons, astrocytes are much more effective in detoxification and redox potential processes. In active demyelinating MS lesions, SOD1 staining was detected in foamy macrophages and astrocytes. Peroxiredoxins are antioxidant enzymes that are expressed in different CNS cell types (neurons, astrocytes, and endothelial cells) (Van Horssen et al., 2011). Oxidative injury is the main feature of both degenerative processes and normal aging. Moreover, it might be the good explanation for the occurrence of noninflammatory processes in MS pathogenesis (Lee et al., 2012). Therefore, oxidative stress is an attractive therapeutic target. Current approaches to diminish oxidative injury, including supplementation with high doses of exogenous antioxidants, presented rather low therapeutic effectiveness in clinical trials (Miller et al., 2013). In all types of cells, the expression of antioxidative enzymes is controlled by transcriptional regulation, which implicates many groups of antioxidant proteins and transcription factors (TFs) including peroxisome proliferator–activated receptor γ coactivator 1-α (PGC1a) and NrF2. PGC1a and NrF2 are two key TFs in humans that are able to regulate antioxidant activity. PGC1a is a pleiotropic transcriptional coactivator that has the ability to regulate mitochondrial functioning. NrF2 activates the expression of a variety of cytoprotective, detoxifying, and anti-inflammatory genes through the binding to their ARE in regulatory sections. These antioxidant genes have been suggested as the therapeutic target for neurodegenerative diseases or diabetic complications. NrF2 supports neuronal survival in pathological states such as neurodegeneration and nerve injury (Xiong et al., 2015).

Under normal conditions, the level of Nrf2 protein is kept low by the E3 ubiquitin ligase KEAP1 (Kelch-like ECH-associated protein 1), which ubiquitinates Nrf2 in the cytoplasm and targets it for degradation by the 26S proteasome. Therefore, KEAP1 is the key point of cellular oxidative stress. Current studies have shown that the KEAP1-Nrf2 system is an important therapeutic target in neurodegenerative disorders including MS (Canning et al., 2015). High expression of Nrf2 was observed in degenerating cells in both apoptotic and necrotic states. Recent *in vivo* study has analyzed MS lesions by whole-genome microarray and revealed differential expression of various Nrf2-responsive genes in different CNS cell types that might be related to demyelination process (Licht-Mayer et al., 2015). Therefore, the redox strategies might be a very important issue in reducing oxidative tissue damage. However, most of the antioxidants are not able to cross the BBB and have a narrow therapeutic window.

13.6 POSSIBLE ANTIOXIDATIVE THERAPIES IN MS

13.6.1 FUMARIC ACID ESTERS

Recently, dimethyl fumarate (DMF) has been recommended for the treatment of RRMS due to its positive results on clinical and radiological outcomes observed in several clinical trials. Several reports have indicated that DMF exerts both immunomodulatory and neuroprotective properties. It was shown that the main beneficial action of DMF is related to Nrf2. DMF stabilizes the Nrf2, activates its transcriptional activity, and leads to the subsequent expression of phase II detoxifying enzymes (Begleiter et al., 2003; Wierinckx et al., 2005). Moreover, the immediate metabolite of DMF—monomethyl fumarate (MMF)—has the ability to directly modify the inhibitor of Nrf2—KEAP1—at cysteine residue. After such modification, Nrf2 is released from the inhibitory complex and translocates to the nucleus to regulate the expression of the ARE-containing genes, thus protecting affected tissues from oxidative damage (Lee et al., 2005).

In vitro studies revealed that DMF increased the survival of murine neurons and both human and rodent astrocytes against the effects of oxidative stress. Such results were confirmed and extended to *in vivo* studies utilizing a chronic myelin oligodendrocyte glycoprotein (MOG)–EAE disease model, and increased Nrf2 expression and translocation were observed in neurons, ODCs, and astrocytes, with a more prominent effect on oligodendroglial cells. Moreover, one of the Nrf2-regulated detoxifying enzymes—NADP(H) quinoline oxidoreductase-1 (NQO1)—has been upregulated predominantly in glial cells, after the DMF treatment of EAE mice (Linker et al., 2011). EAE mice also show reduced oxidative damage and subsequent reduced nerve fiber demyelination, leading to more pronounced axonal preservation and improved motor function (Schilling et al., 2006; Linker et al., 2011). The importance of Nrf2 upregulation by DMF is further supported by the observations from Nrf2-knockout mice, which present enhanced astrogliosis and myelinopathy in the cerebellum, and also more severe MOG–EAE course with enhanced oxidative damage (Hubbs et al., 2007; Johnson et al., 2009). DMF may also exert its impact on Nrf2 activation in astrocytes via the regulation of histone deacetylase (HDAC) expression, the known silencer of gene transcription (Kalinin et al., 2013).

In vitro studies utilizing primary cultures of CNS cells have shown that DMF and MMF elevate cells' redox potential and increase glutathione (GSH) and ATP levels in a concentration-dependent manner (Albrecht et al., 2012; Scannevin et al., 2012; Huang et al., 2015). Moreover, DMF was also able to enhance the recycling of glutathione (Albrecht et al., 2012). However, the contradictory data regarding the role of DMF on GSH level exist. Schmidt et al. have reported *in vitro* reduced GSH level in astrocytes treated with DMF (Schmidt and Dringen, 2010). One of the possible explanations is that DMF as an unsaturated carboxylic acid ester initially binds and afterward releases glutathione (Ghoreschi et al., 2011); however, additional studies are needed to further clarify this issue. DMF may induce the elevation of other antioxidant molecules, like carnitine and ascorbic acid, the last being one of the major small antioxidant molecules present in the brain (May, 2012; Huang et al., 2015). Ascorbic acid has been shown to promote peripheral nerve myelination and membrane integrity during oxidative stress (Buettner, 1993). Experiments conducted on primary CNS cell cultures indicated that fumaric acid esters (FAEs) may also influence the mitochondrial membrane potential. The protective effect of DMF and MMF on CNS cells during oxidative challenge may be mediated, at least partially, by the improved functioning of mitochondria after such treatment, as it was shown that impaired mitochondria accumulate in axons of MS patients (Lee et al., 2012; Scannevin et al., 2012).

Earlier studies conducted on FAEs used in patients with psoriasis (Fumaderm) indicated that DMF and MMF are modulators of T-cell subsets and APCs (Linker et al., 2008). It was shown that fumarates regulate the expression of molecules engaged in the inflammatory reaction. *In vitro* dermatologic studies reported the effects of fumarates on Th2 shift, decreased the expression of proinflammatory cytokines such as IL-2 and TNF-α, and reduced the expression of adhesion molecules (intercellular adhesion molecule 1, E-selectin, and vascular cell adhesion molecule 1) (de Jong et al., 1996; Vandermeeren et al., 1997; Ockenfels et al., 1998; Loewe et al., 2002; Litjens et al., 2004). It was also observed that at higher doses, FAEs may induce apoptosis (Treumer et al., 2003). However, results obtained for MS are ambiguous. Studies by Schilling et al. (2006) conducted on EAE have shown that DMF results in a significant reduction of macrophage/microglia infiltration, with no meaningful effect observed for T-cell subsets. Other studies regarding the role of MMF reported that its action reduced the migratory activity of T lymphocytes but not macrophages (Dehmel et al., 2014). As no changes in matrix metalloproteinases, chemokine receptors, or adhesion molecules have been observed, it was concluded that MMF has impacted lymphocytes migration by changing their activation state (Dehmel et al., 2014). Moreover, recent studies have also indicated that DMF inhibits DC maturation and function and also suppresses Th1 and Th17 differentiation (Peng et al., 2012). Studies conducted by Albrecht et al. indicated that neuroprotective concentrations of DMF exert immunomodulatory activity leading to altered expression of cytokines in mice splenocytes (Albrecht et al., 2012). It was shown that the effect is dependent on the genetic background of mice strain used in the experiment. In C57BL/6 splenocytes, IL-2, IL-17 but not IL-4 or IL-5 were decreased, whereas IL-2, IL-4, IL-5, IL-6, and IL-17 were reduced in spleen cells obtained from the Swiss Jim Lambert (SJL) mice (Albrecht et al., 2012). DMF exerting long-term effect in neuronal cells may inhibit the nuclear translocation of NFκB, resulting in decreased expression of its target genes implicated in the expression of proinflammatory molecules, such as cytokines, chemokines, and adhesion molecules (Lee et al., 2012).

More recently, the new mechanism of DMF action has been suggested. It was reported that DMF may reduce the amount of neutrophils infiltrating the CNS during EAE in a hydroxycarboxylic acid receptor 2 (HCA$_2$)-dependent fashion. HCA$_2$ is highly expressed in neutrophils and is potentially responsible for the interaction between these cells and endothelium. DMF as an HCA$_2$ agonist may reduce adhesion and chemotaxis of neutrophils by activating HCA$_2$ (Chen et al., 2014).

Few reports are available regarding the Nrf2 and DMF interplay in MS patients. The histopathological study provided by Metz et al. (Metz et al., 2015) reports that 12 months of DMF therapy result in sixfold higher amount of Nrf2-positive nuclei in the biopsy samples of the brain from RRMS patients compared with controls (without DMF therapy). The highest nuclear Nrf2 signal was observed in astrocytes, whereas other cells presented mainly cytoplasmic staining (Ruprecht and Friedemann, 2015). The partial myelin preservation observed in MS patients under DMF treatment may be in part due to a direct action of this drug on ODCs (Huang et al., 2015). However, elevated nuclear Nrf2 presence was observed in ODCs in actively demyelinating lesions in patients with acute, relapsing, and progressive disease, calling into the question the effect of additional Nrf2 upregulation in the late stages of MS. It may suggest that Nrf2 induction in ODCs in initial MS phases is protective, while fumarate-induced Nrf2 activation in active lesions may have no or opposite effect (Licht-Mayer et al., 2015). Upregulation of Nrf2 was also seen in the spinal cord sections from untreated MS patients, suggesting the involvement of the naturally occurring defense reaction against oxidative stress.

13.6.2 MELATONIN

Melatonin (*N*-acetyl-5-methoxytryptamine) is produced mainly by the pineal gland and released into the CSF of the third ventricle of the brain during the night (Galano et al., 2011). This potent antioxidant can react not only with superoxide anion, NO, ONOO⁻, but may also inhibit the activity of NO synthases (iNOS, nNOS). Moreover, it can decrease the levels of proinflammatory cytokines such as IFN, IL-6, TNF-α, and IL-12. Melatonin is able to activate the main antioxidative enzymes. Current studies have shown that the decreased level of melatonin might have a negative impact on fatigue, fluctuation of symptoms of mood disorder (Esposito and Cuzzocrea, 2010), and triglyceride level in MS (Ghorbani et al., 2013). An important feature of melatonin is its ability to easily cross the BBB. Our recent studies showed that 10 mg daily per 30 day's melatonin supplementation caused statistically significant increase of SOD and GPx and decrease of MDA in erythrocytes of SPMS patients (Miller et al., 2013). Sandyk and Awerbuch (1992) suggest the association between nocturnal melatonin secretion and duration of clinical symptoms of MS patients. Therefore, the activity of the pineal gland may decline with the progression of the disease. However, the other studies are not consistent with this statement (Naseem and Parvez, 2014).

13.7 OTHER THERAPIES

13.7.1 CANNABINOIDS

Cannabinoids are a variety of chemical compounds that activate cannabinoid receptors. In Europe, they were used at the end of the nineteenth century to alleviate a wide range of conditions, including pain, spasms, dysentery, depression, sleep disturbance, and loss of appetite (Russo and Guy, 2006). There are endocannabinoids, phytocannabinoids (cannabis and plant compounds), and synthetic cannabinoids. The most psychoactive cannabis is the phytocannabinoid, Δ9-tetrahydrocannabinol (THC). Cannabidol (CBD) has been approved for MS therapy to reduce inflammation, pain, and spasticity. Its anti-inflammatory properties have been experimentally supported in *in vitro* and *in vivo* studies utilizing models of inflammation (De Lago et al., 2012; Zajicek et al., 2013). Two cannabinoid receptors named CB_1 and CB_2 were described. The CB_1 receptors are related to all psychotropic and motor impairment. Synthetic THC (dronabinol) and its derivative nabilone are therapeutic agents with antiemetic and analgesic properties. CB_1 receptors were discovered in the terminals of central and peripheral neurons, and they are responsible for neurotransmitter modulation (Saito et al., 2012). Synthetic cannabinoids such as WIN 55212-2 and HU211 can protect oligodendrocyte progenitor cells (OPCs) from the apoptosis induced by the withdrawal of trophic support (Mecha et al., 2012). The CB_2 receptors are expressed not only in immune cells but also in the CNS, where their presence is related to microglia activation. In the physiological state of the CNS, there is no expression of CB_2. However, CB_2 expression on CNS neurons is still controversial (Zhornitsky and Potvin, 2012). It is suggested that the role of CB_2 cannabinoid receptors is connected not only with lower production of microglia active molecules such as TNF-α, IL-1β, IL-12, IL-23 but also with the regulation of cell migration (macrophages, neutrophils, and others) (Kozela et al., 2011). Endocannabinoids are released in response to a wide range of neuronal insults, and their levels are elevated in the CSF and peripheral lymphocytes of MS patients. The activation of CB_2 receptor resulted in the activation of immune cells, whereas the CB_1 promotes neural cells' survival, which is the hallmark of neuroprotection. Side effects associated with cannabinoid administration are very limited. Moreover, cannabinoids might represent a low-cost therapy (Saito et al., 2012).

13.7.2   Vitamin D

The active form of vitamin D is 1,25-dihydroxyvitamin D (1,25(OH)2D). Hypovitaminosis D is a key environmental risk factor in MS and also may have an immunomodulatory role in the brain (Miller, 2012). Vitamin D can alter the T-cell phenotype into a more anti-inflammatory and regulated state, with the inhibition of Th1 and Th17 cells and the promotion of Th2 and Treg cells. Supplementation with vitamin D has a positive effect on MS patients as a preventive factor after high dose of sterydotherapy (Pedersen et al., 2007). Epidemiological studies have reported lower incidence of MS in the region near the equator. Furthermore, the appropriate level of vitamin D in childhood and adolescence is connected with a lower risk of MS. *In vitro* studies presented that vitamin D prevents IL-12, IL-2, and interferon-gamma production together with the inhibition of B cells' production. Most of the studies reported lower plasma level of vitamin D in MS patients compared to healthy controls. However, in studies conducted in Switzerland and Finland, no differences in vitamin D prevalence were observed between MS patients and healthy controls (Eskandari et al., 2015).

13.7.3   Cryostimulation

The first reports of thermal sensitivity in MS have derived from Charles Prosper Ollivier d'Angers (1824), who described that a hot bath induced numbness in the right leg and reduced feeling and dexterity in the hands in MS patients (Ollivier, 1824). Approximately 60%–80% of the MS patients experience temporary worsening of neurological symptoms as a result of increased body temperature by exposure to high-temperature (41°C–43°C) stimulus such as infrared heating lamps or warm baths (Nelson et al., 1958). Even a slight increase in the body temperature (0.5°C) in MS with demyelinated lesions can cause nerve conduction block (Scott et al., 2014). The brain physiological functioning is related to intense heat production. Moreover, cerebral biochemical processes are very sensitive to even slight temperature changes. Brain hypothermia, with its broader, pleiotropic effects, represents the most potent neuroprotectant in laboratory studies. Temperature changes of 1°C or less can result in functional alterations in various areas of the nervous system, indicating the high thermal sensitivity of the brain. The significance of thermal impact on several principal neurophysiological properties, such as resting potential, action potential, nerve conduction velocity, and synaptic transmission, is well established. For example, the impairment of memory encoding starts at a body temperature of 36.7°C and progresses to the point that 70% of information normally retained is lost at approximately 34°C–35°C (Wang et al., 2014).

Hypothermia is used as neuroprotective therapy in neonate's hypoxic-ischemic encephalopathy, depression, brain injury, MS, and cerebral ischemia (Miller et al., 2010). Even little fluctuation of temperature in the CNS is able to change main biochemical processes such as hemodynamic, excitotoxicity, inflammation, calcium-dependent intercellular signaling, apoptosis, and edema (Miller, 2015). The main effect of hypothermia is cell protection against hypoxia by delaying the pace of cell damage caused by the formation of free radicals, tissue edema, and chemical metabolites. In the late 1970s, in Japan, cryogenic temperatures (–130°C) were used. Whole-body cryotherapy is often named cryostimulation. The basis of this therapy is the exposure of the human body to extremely low temperatures (–130°C) in a cryogenic chamber. One session lasts 3 minutes per day, and the whole standard procedure consists of 10 sessions. Currently, cryostimulation is the additional therapy in sport medicine, inflammatory and rheumatic diseases, as well as psychiatric disorders especially for anxiety-depressive syndrome (Miller et al., 2010). Low temperature is the cause of lower demand for oxygen and lowered cell metabolism. Therefore, it leads to reduction of oxidative stress. Our clinical studies have suggested the decrease in total antioxidant status (TAS) level in depressive MS patients compared with nondepressive MS patients. Additionally, after 10 sessions of cryostimulation (Miller et al., 2011), we have observed the increase in TAS level in the plasma of depressive MS patients similar to the values obtained for healthy controls. It is still unknown how cryostimulation might affect oxidative stress, and further studies are needed in this area. The results of Miller et al.'s study of 10 exposures of MS patients to cryostimulation show increased concentration of uric acid (the main human antioxidant). Moreover, our studies suggest that the decreased TAS level observed in the plasma of MS patients is related to low concentrations of the main endogenous antioxidant, uric acid. Therefore, cryostimulation is a possible additional antioxidative therapy used especially in progressive stages of MS. A long-lasting study with the measurement of uric acid level in the plasma of MS patients presented higher concentration of uric acid even 3 months after completing 10 sessions of cryostimulation (Miller et al., 2013).

13.8 CONCLUSION

The potential of redox therapies is still improving our knowledge about cellular processes in MS. The individually tailored treatment for MS patients according to more precise diagnosis is the essential challenge for scientists. Many therapies are designed to affect redox alterations in MS both in initiation and in progression stages of disease. Recent studies have confirmed the huge success of DMF enhancing the antioxidant responses, which is mainly attributed to its jointed inhibitory effects on proinflammatory immune responses together with its high efficacy and safety (Chiurchiu, 2014).

ABBREVIATIONS

APC	Antigen-presenting cells
ARE	Antioxidant response element
ATP	Adenosine triphosphate
BBB	Blood–brain barrier
CAT	Catalase
CSF	Cerebrospinal fluid
DC	Dendritic cells
DMF	Dimethyl fumarate
EAE	Experimental autoimmune encephalomyelitis
FAEs	Fumaric acid esters
FDA	Food and Drug Administration
GPx	Glutathione peroxidases
GSH	Glutathione
HCA_2	Hydroxycarboxylic acid receptor 2
HDAC	Histone deacetylase
ICAM-1	Intercellular adhesion molecule 1
IL	Interleukin
KEAP1	Kelch ECH–associating protein 1
MHC	Major histocompatibility complex
MMF	Monomethyl fumarate
MMPs	Matrix metalloproteinases
MRI	Magnetic resonance imaging
NO	Nitric oxide
NOS	Nitric oxide synthases
Nrf2	Nuclear factor (erythroid-derived 2)-related factor 2
NQO1	NADP(H) quinoline oxidoreductase-1
ODC	Oligodendrocyte
OPC	Oligodendrocyte progenitor cells
PGC1a	Proliferator-activated receptor γ coactivator 1-α
RNS	Reactive nitrogen species

ROS   Reactive oxygen species
SOD   Superoxide dismutase
TAS   Total antioxidant status
TNF   Tumor necrosis factor
VCAM-1   Vascular cell adhesion molecule 1

REFERENCES

Abott, N.J., Ronnback, L., Hansson E., 2006. Astrocyte-endothelial interactions at the blood-brain barrier. *Nat. Rev. Neurosci.* 7: 41–53.

Albrecht, P., Bouchachia, I., Goebels, N., Henke, N., Hofstetter, H., Issberner, A., Kovacs, Z. et al., 2012. Effects of dimethyl fumarate on neuroprotection and immunomodulation. *J. Neuroinflammation* 9: 163.

Bar-Or, A., Pachner, A., Menguy-Vacheron, F., Kaplan, J., Wiendl, H., 2014. Teriflunomide and its mechanism of action in multiple sclerosis. *Drugs* 74: 659–674.

Bauman, N., Pham-Dinh, D., 2001. Biology of oligodendrocytes and myelin in the mammalian central nervous system. *Physiol. Rev.* 81: 871–927.

Begleiter, A., Sivananthan, K., Curphey, T.J., Bird, R.P., 2003. Induction of NAD(P)H quinone: Oxidoreductase 1 inhibits carcinogen-induced aberrant crypt foci in colons of Sprague-Dawley rats. *Cancer Epidemiol. Biomarkers Prev.* 12: 566–572.

Bezabeh, S., Flowers, C.M., Kortepeter, C., Avigan, M., 2010. Clinically significant liver injury in patients treated with natalizumab. *Aliment. Pharmacol. Ther.* 31: 1028–1035.

Bielekova, B., Martin, R., 2004. Development of biomarkers in multiple sclerosis. *Brain* 127: 1463–1478.

Biogen Idec Inc., 2014. Tecfidera (dimethylfumarate). http://www.accessdata.fda.gov/drugsatfda_docs/label/2014/20406-3s003s008s010lbl.pdf, accessed November 25, 2014.

Block, M.L., Hong, J.S., 2005. Microglia and inflammation-mediated neurodegeneration: Multiple triggers with a common mechanism. *Prog Neurobiol.* 76(2): 77–98.

Boster, A., Bartoszek, M.P., O'Connell, C., Pitt, D., Racke, M., 2011. Efficacy, safety, and cost-effectiveness of glatiramer acetate in the treatment of relapsing–remitting multiple sclerosis. *Ther. Adv. Neurol. Disord.* 4: 319–332.

Bradl, M., Lassman, H., 2010. Oligodendrocytes: Biology and pathology. *Acta Neuropathol.* 119: 37–53.

Brinkmann, V., 2009. FTY720 (fingolimod) in multiple aclerosis: Therapeutic effects in the immune and the central nervous system. *Br. J. Pharmacol.* 158: 1173–1182.

Buettner, G.R., 1993. The pecking order of free radicals and antioxidants: Lipid peroxidation, α-tocopherol, and ascorbate. *Arch. Biochem. Biophys.* 300: 535–543.

Calabrese, V., Cornelius, C., Rizzarelli, E., 2009. Nitric oxide in cell survival: A jams molecule. *Antioxid. Redox Signal.* 11: 2717–2739.

Canning, P., Sorrell, F.J., Bullock, A.N., 2015. Structural basis of Keap1 interactions with Nrf2, free radical biology and medicine. *Acta Neuropathol.* 88(Pt B): 101–107.

Chen, H., Assmann, J.C., Krenz, A., Rahman, M., Grimm, M., Karsten, C.M., Köhl, J.N. et al., 2014. Hydroxycarboxylic acid receptor 2 mediates dimethyl fumarate's protective effect in EAE. *J. Clin. Invest.* 124: 2188–2192.

Chiurchiu, V., 2014. Novel targets in multiple sclerosis: To oxidative stress and beyond. *Curr. Top. Med. Chem.* 14: 2590–2599.

Chun, J., Hartung, H.-P., 2010. Mechanism of action of oral fingolimod (FTY720) in multiple sclerosis. *Clin. Neuropharmacol.* 33: 91–101.

Coles, A.J., Compston, D., Selmaj, K.W., Lake, S.L., Moran, S., Margolin, D.H., Norris, K., Tandon, P., 2008. Alemtuzumab vs. interferon beta-1a in early multiple sclerosis. *N. Engl. J. Med.* 359: 1786–1801.

Comi, G., Jeffery, D., Kappos, L., Montalban, X., Boyko, A., Rocca, M.A., Filippi, M., 2012. Placebo-controlled trial of oral laquinimod for multiple sclerosis. *N. Engl. J. Med.* 366: 1000–1009.

Cossburn, M., Pace, A.A., Jones, J., Ali, R., Ingram, G., Baker, K., Hirst, C. et al., 2011. Autoimmune disease after alemtuzumab treatment for multiple sclerosis in a multicenter cohort. *Neurology* 77: 573–579.

Costelloe, L., Jones, J., Coles, A., 2012. Secondary autoimmune diseases following alemtuzumab therapy for multiple sclerosis. *Expert Rev. Neurother.* 12: 335–341.

Cuker, A., Coles, A.J., Sullivan, H., Fox, E., Goldberg, M., Oyuela, P., Purvis, A., Beardsley, D.S., Margolin, D.H., 2011. A distinctive form of immune thrombocytopenia in a phase 2 study of alemtuzumab for the treatment of relapsing-remitting multiple sclerosis. *Blood* 24: 6299–6305.

de Jong, R., Bezemer, A.C., Zomerdijk, T.P.L., van de Pouw-Kraan, T., Ottenhoff, T.H.M., Nibbering, P.H., 1996. Selective stimulation of T helper 2 cytokine responses by the anti-psoriasis agent monomethylfumarate. *Eur. J. Immunol.* 26: 2067–2074.

De Lago, E., Moreno-Martet, M., Cabranes, A., Ramos, J.A., Fernandez-Ruiz, J., 2012. Cannabinoids ameliorate disease progression in a model of multiple sclerosis in mice, acting preferentially through CB1 receptor-mediated anti-inflammatory effects. *Neuropharmacology* 62: 2299–2308.

Dehmel, T., Döbert, M., Pankratz, S., Leussink, V.I., Hartung, H.P., Wiendl, H., Kieseier, B.C., 2014. Monomethylfumarate reduces in vitro migration of mononuclear cells. *Neurol. Sci.* 35: 1121–1125.

Denicola, A., Souza, J.M., Radi, R., 1998. Diffusion of peroxynitrite across erythrocute membranes. *Proc. Natl. Acad. Sci. USA* 95: 3566–3571.

Ebers, G.C., 1998. Randomised double-blind placebo-controlled study of interferon β-1a in relapsing/remitting multiple sclerosis. *Lancet* 352: 1498–1504.

Edan, G., Comi, G., Le Page, E., Leray, E., Rocca, M.A., Filippi, M., 2011. Mitoxantrone prior to interferon beta-1b in aggressive relapsing multiple sclerosis: A 3-year randomised trial. *J. Neurol. Neurosurg. Psychiatry* 82: 1344–1350.

Edan, G., Miller, D., Clanet, M., Confavreux, C., Lyon-Caen, O., Lubetzki, C., Brochet, B. et al., 1997. Therapeutic effect of mitoxantrone combined with methylprednisolone in multiple sclerosis: A randomised multicentre study of active disease using MRI and clinical criteria. *J. Neurol. Neurosurg. Psychiatry* 62: 112–118.

Eskandari G., Ghajarzadeh, M., Yekaninejad, M.S., Sahraian, M.A., Gorji, R., Rajaei, F., Norouzi-Javidan, A., Faridar, A., Azimi, A., 2015. Comparison of serum vitamin D level in multiple sclerosis patients, their siblings, and healthy controls. *Iran. J. Neurol.* 14: 81–85.

Esposito, E., Cuzzocrea, S., 2010. Antiinflammatory activity of melatonin in central nervous system. *Curr. Neuropharmacol.* 8: 228–242.

FDA Drug Safety Communication, 2013. FDA investigating rare brain infection in patient taking Gilenya (fingolimod). http://www.fda.gov/drugs/drugsafety/ucm366529.htm, accessed August 29, 2013.

FDA Drug Safety Communication, 2014. FDA warns about case of rare brain infection PML with MS drug Tecfidera (dimethylfumarate). http://www.fda.gov/Drugs/Drug Safety/ucm424625.htm, accessed November 25, 2014.

Galano, A., Tan, D.X., Reiter, R.J., 2011. Melatonin as a natural ally against oxidative stress: A physicochemical examination. *J. Pineal. Res.* 51: 1–16.

Ghalie, R.G., Mauch, E., Edan, G., Hartung, H.P., Gonsette, R.E., Eisenmann, S., Le Page, E., Butine, M.D., Goodkin, D.E., 2002. A study of therapy-related acute leukaemia after mitoxantrone therapy for multiple sclerosis. *Mult. Scler.* 8: 441–445.

Ghorbani, A., Salari, M., Shaygannejad, V., Norouzi, R., 2013. The role of melatonin in the pathogenesis of multiple sclerosis: A case-control study. *Int. J. Prev. Med.* 4: 180–184.

Ghoreschi, K., Brück, J., Kellerer, C., Deng, C., Peng, H., Rothfuss, O., Hussain, R.Z. et al., 2011. Fumarates improve psoriasis and multiple sclerosis by inducing type II dendritic cells. *J. Exp. Med.* 208: 2291–2303.

Guerrero, A.L., Gutierrez, F., Iglesias, F., 2011. Serum uric acid levels in multiple sclerosis patients inversely correlate with disability. *Neurol. Sci.* 32: 347–350.

Hartung, H.-P., Gonsette, R., Konig, N., Kwiecinski, H., Guseo, A., Morrissey, S.P., Krapf, H., Zwingers, T., 2002. Mitoxantrone in progressive multiple sclerosis: A placebo-controlled, double-blind, randomised, multicentre trial. *Lancet* 360: 2018–2025.

Huang, H., Taraboletti, A., Shriver, L.P., 2015. Dimethyl fumarate modulates antioxidant and lipid metabolism in oligodendrocytes. *Redox Biol.* 5: 169–175.

Hubbs, A.F., Benkovic, S.A., Miller, D.B., O'Callaghan, J.P., Battelli, L., Schwegler-Berry, D., Ma, Q., 2007. Vacuolar leukoencephalopathy with widespread astrogliosis in mice lacking transcription factor Nrf2. *Am. J. Pathol.* 170: 2068–2076.

Jacobs, L.D., Cookfair, D.L., Rudick, R.A., Herndon, R.M., Richert, J.R., Salazar, A.M., Fischer, J.S. et al., 1996. Intramuscular interferon beta-1a for disease progression in relapsing multiple sclerosis. *Ann. Neurol.* 39: 285–294.

Johnson, D.A., Amirahmadi, S., Ward, C., Fabry, Z., Johnson, J.A., 2009. The absence of the pro-antioxidant transcription factor Nrf2 exacerbates experimental autoimmune encephalomyelitis. *Toxicol. Sci.* 114: 237–246.

Kalinin, S., Polak, P.E., Lin, S.X., Braun, D., Guizzetti, M., Zhang, X., Rubinstein, I., Feinstein, D.L., 2013. Dimethyl fumarate regulates histone deacetylase expression in astrocytes. *J. Neuroimmunol.* 263: 13–19.

Kappos, L., 1998. Placebo-controlled multicentre randomised trial of interferon β-1b in treatment of secondary progressive multiple sclerosis. *Lancet* 352: 1491–1497.

Kappos, L., Radue, E.-W., O'Connor, P., Polman, C., Hohlfeld, R., Calabresi, P., Selmaj, K. et al., 2010. A placebo-controlled trial of oral fingolimod in relapsing multiple sclerosis. *N. Engl. J. Med.* 362: 387–401.

Kozela, E., Nirit, L., Kaushansky, N., Eilam, R., Rimmerman, N., Levy, R., Ben-Nun, A., Juknat, A., Vogel, Z., 2011. Cannabidiol inhibits pathogenic T cells, decreases spinal microglial activation and ameliorates multiple sclerosis-like disease in C57BL/6. *Br. J. Pharmacol.* 163: 1507–1519.

Ksiazek-Winiarek, D.J., Kacperska, M.J., Glabinski, A., 2013. MicroRNAs as novel regulators of neuroinflammation. *Mediators Inflamm.* 2013: 172351.

Lee, D.H., Gold, R., Linker, R.A., 2012. Mechanisms of oxidative damage in multiple sclerosis and neurodegenerative diseases: Therapeutic modulation via fumaric acid esters. *Int. J. Mol. Sci.* 13(9): 11783–11803.

Lee, J.-M., Li, J., Johnson, D.A., Stein, T.D., Kraft, A.D., Calkins, M.J., Jakel, R.J., Johnson, J.A., 2005. Nrf2, a multi-organ protector? *FASEB J.* 19: 1061–1066.

Licht-Mayer, S., Wimmer, I., Traffehn, S., Metz, I., Brück, W., Bauer, J., Bradl, M., Lassmann, H., 2015. Cell type-specific Nrf2 expression in multiple sclerosis lesions. *Acta Neuropathol.* 130(2): 263–277.

Linker, R.A., Lee, D.-H., Ryan, S., van Dam, A.M., Conrad, R., Bista, P., Zeng, W. et al., 2011. Fumaric acid esters exert neuroprotective effects in neuroinflammation via activation of the Nrf2 antioxidant pathway. *Brain* 134: 678–692.

Linker, R.A., Lee, D.-H., Stangel, M., Gold, R., 2008. Fumarates for the treatment of multiple sclerosis: Potential mechanisms of action and clinical studies. *Expert Rev. Neurother.* 8: 1683–1690.

Litjens, N., Nibbering, P., Barrois, A., Zomerdijk, T., Van Den Oudenrijn, A., Noz, K., Rademaker, M. et al., 2003. Beneficial effects of fumarate therapy in psoriasis vulgaris patients coincide with downregulation of type 1 cytokines. *Br. J. Dermatol.* 148: 444–451.

Litjens, N.H.R., Rademaker, M., Ravensbergen, B., Rea, D., van der Plas, M.J.A., Thio, B., Walding, A., van Dissel, J.T., Nibbering, P.H., 2004. Monomethylfumarate affects polarization of monocyte-derived dendritic cells resulting in down-regulated Th1 lymphocyte responses. *Eur. J. Immunol.* 34: 565–575.

Liu, J.S., Zhao, M.L., Brosnan, C.F., Lee, S.C., 2001. Expression of inducible nitric oxide synthetase and nitrityrosine in multiple sclerosis lessions. *Am. J. Pathol.* 158: 2057–2066.

Loewe, R., Holnthoner, W., Gröger, M., Pillinger, M., Gruber, F., Mechtcheriakova, D., Hofer, E., Wolff, K., Petzelbauer, P., 2002. Dimethylfumarate inhibits TNF-induced nuclear entry of NF-κB/p65 in human endothelial cells. *J. Immunol.* 168: 4781–4787.

Lublin, F.D., Reingold, S.C., 1996. Defining the clinical course of multiple sclerosis: Results of an international survey. National Multiple Sclerosis Society (USA) advisory committee on clinical trials of new agents in multiple sclerosis. *Neurology* 46(4): 907–911.

May, J., 2012. Vitamin C transport and its role in the central nervous system, in *Water Soluble Vitamins*, Stanger, O., ed. Springer, Dordrecht, the Netherlands, pp. 85–103.

Mecha, M., Torrao, A.S., Mestre, L., Carrillo-Salinas, F.J., Mechoulam, R., Guaza, C., 2012. Cannabidiol protects oligodendrocyte progenitor cells from inflammation-induced apoptosis by attenuating endoplasmic reticulum stress. *Cell. Death. Dis.* 28(3): e331. doi:10.1038/cddis.2012.71.

Meili-Butz, S., Niermann, T., Fasler-Kan, E., Barbosa, V., Butz, N., John, D., Brink, M., Buser, P.T., Zaugg, C.E., 2008. Dimethyl fumarate, a small molecule drug for psoriasis, inhibits nuclear factor-κB and reduces myocardial infarct size in rats. *Eur. J. Pharmacol.* 586: 251–258.

Menge, T., Stüve, O., Kieseier, B.C., Hartung, H.-P., 2014. Alemtuzumab: The advantages and challenges of a novel therapy in MS. *Neurology* 83: 87–97.

Metz, I., Traffehn, S., Straßburger-Krogias, K., Keyvani, K., Bergmann, M., Nolte, K., Weber, M.S. et al., 2015. Glial cells express nuclear Nrf2 after fumarate treatment for multiple sclerosis and psoriasis. *Neurol. Neuroimmunol. Neuroinflamm.* 2: e99.

Milijković, D., Spasojević, I., 2013. Multiple sclerosis: Molecular mechanisms and therapeutic opportunities. *Antioxid. Redox Signal.* 19: 2286–2334.

Miller, E., 2012. Multiple sclerosis. *Adv. Exp. Med. Biol.* 724: 222–238.

Miller, E., Mrowicka, M., Malinowska, K., Mrowicki, J., Saluk-Juszczak, J., Kędziora, J. et al., 2010. Effects of whole body cryotherapy on oxidative stress in multiple sclerosis patients. *J. Therm. Biol.* 35: 406–410.

Miller, E., Mrowicka, M., Malinowska, K., Mrowicki, J., Saluk-Juszczak, J., Kędziora, J. et al., 2011. Effects of whole body cryotherapy on a total antioxidative status and activity of antioxidant enzymes in blood of depressive multiple sclerosis patients. *World J. Biol. Psychiatry* 12: 223–227.

Miller, E., Saluk, J., Morel, A., Wachowicz, B., 2013a. Long-term effects of whole body cryostimulation on uric acid concentration in plasma of secondary progressive multiple sclerosis patients. *Scand. J. Clin. Lab. Invest.* 73: 635–640.

Miller, E., Wachowicz, B., Majsterek, I., 2013b. Advances in antioxidative therapy of multiple sclerosis. *Curr. Med. Chem.* 20(37): 4720–4730.

Miller, E., Walczak, A., Majsterek, I., Kędziora, J., 2013c. Melatonin reduces oxidative stress in the erythrocytes of multiple sclerosis patients with secondary progressive clinical course. *J. Neuroimmunol.* 257: 97–101.

Motohashi, H., Yamamoto, M., 2004. Nrf2-Keap1 defines a physiologically important stress response mechanism. *Trends Mol. Med.* 10: 549.

Munier-Lehmann, H., Vidalain, P.-O., Tangy, F., Janin, Y.L., 2013. On dihydroorotate dehydrogenases and their inhibitors and uses. *J. Med. Chem.* 56: 3148–3167.

Nair, A., Frederick, T.J., Miller, S.D., 2008. Astrocytes in multiple sclerosis: A product of environment. *Cell. Mol. Life Sci.* 65: 2702–2720.

Naseem, M., Parvez, S., 2014. Role of melatonin in traumatic brain injury and spinal cord injury. *ScientificWorldJournal* 2014: 586270.

National MS Society: http://www.nationalmssociety.org/Treating-MS/Medications.

Nelson, D.A., Jeffreys, W.H., McDowell, D.F., 1958. Effects of induced hyperthermia on some neurological diseases. *AMA Arch. Neurol. Psychiatry* 79: 31–39.

Noseworthy, J.H., Kirkpatrick, P., 2005. Natalizumab. *Nat. Rev. Drug Discov.* 4: 101–102.

Novartis Pharmaceuticals Corporation, 2014a. Aubagio (teriflunomide). http://www.accessdata.fda.gov/drugsatfda_docs/label/2014/202992s001lbl.pdf, accessed October 17, 2014.

Novartis Pharmaceuticals Corporation, 2014b. Gilenya (fingolimod). http://www.accessdata.fda.gov/drugsatfda_docs/label/2014/022527s009lbl.pdf, accessed April 30, 2014.

Ockenfels, H.M., Schultewolter, T., Ockenfels, G., Funk, R., Goos, M., 1998. The antipsoriatic agent dimethylfumarate immunomodulates T-cell cytokine secretion and inhibits cytokines of the psoriatic cytokine network. *Br. J. Dermatol.* 139: 390–395.

Ollivier, C.P., 1824. *De la moelle epiniere et de ses maladies.* Crevot, Paris, France.

Onufriev, M.V., 2010. Nitrosative stress in the brain: Autoantibodies to nitrotyrosine in liquor as a potential marker. *Neurochem. J.* 4: 228–234.

Ortiz, G.G., Pacheco-Moises, F.P., Bitzer-Quintero, O.K., Ramirez-Anguiano, A.C., Flores-Alvarado, L.J., Ramirez-Ramirez, V.R., Macias-Islas, M.A., Torres-Sanchez, E.D., 2013. Immunology and oxidative stress in multiple sclerosis: Clinical basic approach. *Clin. Dev. Immunol.* 2013: 708659.

Pacher, P., Backman, J.S., Liaudet, L., 2007. Nitric oxide and peroxynitrite in heath and disease. *Phys. Rev.* 87: 315–424.

Patel, J.P., Frey, B.N., 2015. Disruption in the blood-brain barrier: The missing link between brain and body inflammation in bipolar disorder? *Neural Plast.* 2015: 708306.

Paty, D.W., Li, D.K.B., 1993. Interferon beta-1b is effective in relapsing-remitting multiple sclerosis: II. MRI analysis results of a multicenter, randomized, double-blind, placebo-controlled trial. UBC MS/MRI Study Group and the IFNB Multiple Sclerosis Study Group. *Neurology* 43: 662.

Pedersen, L.B., Nashold, F.E., Spach, K.M., Hayes, C.E., 2007. 1,25-dihydroxyvitamin D3 reverses experimental autoimmune encephalomyelitis by inhibiting chemokine synthesis and monocyte trafficking. *J Neurosci Res.* 85(11): 2480–2490.

Peng, H., Guerau-de-Arellano, M., Mehta, V.B., Yang, Y., Huss, D.J., Papenfuss, T.L., Lovett-Racke, A.E., Racke, M.K., 2012. Dimethyl fumarate inhibits dendritic cell maturation via nuclear factor κB (NF-κB) and extracellular signal-regulated kinase 1 and 2 (ERK1/2) and mitogen stress-activated kinase 1 (MSK1) signaling. *J. Biol. Chem.* 287: 28017–28026.

Perumal, J.S., Foo, F., Cook, P., Khan, O., 2012. Subcutaneous administration of alemtuzumab in patients with highly active multiple sclerosis. *Mult. Scler. J.* 18: 1197–1199.

Polman, C.H., O'Connor, P.W., Havrdova, E., Hutchinson, M., Kappos, L., Miller, D.H., Phillips, J.T. et al., 2006. A randomized, placebo-controlled trial of natalizumab for relapsing multiple sclerosis. *N. Engl. J. Med.* 354: 899–910.

Reder, A.T., Ebers, G.C., Traboulsee, A., Li, D., Langdon, D., Goodin, D.S., Bogumil, T. et al., 2010. Cross-sectional study assessing long-term safety of interferon-β-1b for relapsing-remitting MS. *Neurology* 74: 1877–1885.

Rudick, R.A., Stuart, W.H., Calabresi, P.A., Confavreux, C., Galetta, S.L., Radue, E.-W., Lublin, F.D. et al., 2006. Natalizumab plus interferon beta-1a for relapsing multiple sclerosis. *N. Engl. J. Med.* 354: 911–923.

Ruprecht, K., Friedemann, P., 2015. Does Nrf2 help nerves to survive? *Neurol. Neuroimmunol. Neuroinflamm.* 2(3): e105. doi:10.1212.

Russo, E., Guy, G.W., 2006. A tale of two cannabinoids: The therapeutic rationale for combining tetrahydro-cannabinol and cannabidiol. *Med. Hypotheses* 66: 234–246.

Saito, V.M., Rezende, R.M., Teixeira, A.L., 2012. Cannabinoid modulation of neuroinflammatory disorders. *Curr. Neuropharmacol.* 10(2): 159–166.

Sandyk, R., Awerbuch, G.I., 1992. Nocturnal plasma melatonin and alpha-melanocyte stimulating hormone levels during exacerbation of multiple sclerosis. *Int. J. Neurosci.* 67: 173–186.

Scannevin, R.H., Chollate, S., Jung, M.-y., Shackett, M., Patel, H., Bista, P., Zeng, W. et al., 2012. Fumarates promote cytoprotection of central nervous system cells against oxidative stress via the nuclear factor (erythroid-derived 2)-like 2 pathway. *J. Pharmacol. Exp. Ther.* 341: 274–284.

Schilling, S., Goelz, S., Linker, R., Luehder, F., Gold, R., 2006. Fumaric acid esters are effective in chronic experimental autoimmune encephalomyelitis and suppress macrophage infiltration. *Clin. Exp. Immunol.* 145: 101–107.

Schmidt, M.M., Dringen, R., 2010. Fumaric acid diesters deprive cultured primary astrocytes rapidly of glutathione. *Neurochem. Int.* 57: 460–467.

Scott, L.D., Wilson, T.E., White, A.T., Frohman, E.M., 2014. Thermoregulation in multiple sclerosis. *Front. Neurosci.* 8: 307. doi:10.3389/fnins.2014.00307.

Shenkenberg, T.D., Von Hoff, D.D., 1986. Mitoxantrone: A new anticancer drug with significant clinical activity. *Ann. Intern. Med.* 105: 67–81.

Sofroniew, M.V., Vinters, H.V., 2010. Astrocytes: Biology and pathology. *Acta Neuropathol.* 119: 7–35.

Teitelbaum, D., Meshorer, A., Hirshfeld, T., Arnon, R., Sela, M., 1971. Suppression of experimental allergic encephalomyelitis by a synthetic polypeptide. *Eur. J. Immunol.* 1: 242–248.

Treumer, F., Zhu, K., Glaser, R., Mrowietz, U., 2003. Dimethylfumarate is a potent inducer of apoptosis in human T cells. *J. Invest. Dermatol.* 121: 1383–1388.

Van Horssen, J., Witte, M.E., Schreibelt, G., de Viries, H.E., 2011. Radical changes in multiple sclerosis pathogenesis. *Biochim. Biophys. Acta* 1812: 141–150.

Vandermeeren, M., Janssens, S., Borgers, M., Geysen, J., 1997. Dimethylfumarate is an inhibitor of cytokine-induced E-selectin, VCAM-1, and ICAM-1 expression in human endothelial cells. *Biochem. Biophys. Res. Commun.* 234: 19–23.

Walther, E.U., Hohlfeld, R., 1999. Multiple sclerosis: Side effects of interferon beta therapy and their management. *Neurology* 53: 1622.

Wang, H., Wang, B., Normoyle, K.P., Jackson, K., Spitler, K., Sharrock, M.F., Miller, C.M., 2010. Brain temperature and its fundamental properties: A review for clinical neuroscientists. *J. Appl. Physiol.* 109: 1531–1537.

Watzlawik, J., Warrington, A.E., Rodriguez, M., 2010. Importance of oligodendrocyte protection BBB. *Expert Rev. Neurother.* 10(3): 441–457.

Wierinckx, A., Brevé, J., Mercier, D., Schultzberg, M., Drukarch, B., Van Dam, A.-M., 2005. Detoxication enzyme inducers modify cytokine production in rat mixed glial cells. *J. Neuroimmunol.* 166: 132–143.

Xiong, W., MacColl Garfinkel, A.E., Li, Y., Benowitz, L.I., Cepko, C.L., 2015. NRF2 promotes neuronal survival in neurodegeneration and acute nerve damage. *J. Clin. Invest.* 125: 1433–1445.

Yong, V.W., 2002. Differential mechanisms of action of interferon-β and glatiramer acetate in MS. *Neurology* 59: 802–808.

Zajicek, J., Ball, S., Wright, D., Nunn, A., Miller, D., Cano, M.G., McManus, D., Mallik, S., Hobart, J., 2013. Effect of dronabinol on progression in progressive multiple sclerosis (CUPID): A randomised, placebo-controlled trial. *Lancet Neurol.* 12: 857–865.

Zhornitsky, S., Potvin, S., 2012. Cannabidiol in humans—The quest for therapeutic targets. *Pharmaceuticals* 5: 529–552.

Section IV

Autoimmune Diseases Caused by ROS

14 Pulmonary Oxidant–Antioxidant Dysregulation in Asthma

Ahmed Nadeem, Ahmed M. El-Sherbeeny,
Mohammed M. Al-Harbi, and Sheikh F. Ahmad

CONTENTS

ABSTRACT

Asthma is a very common inflammatory airway disease affecting millions of people throughout the world. Many studies have been carried out on the role of oxidants and antioxidants on asthmatic patients. This chapter attempts to present comprehensive studies on asthma with an emphasis on correlations between pulmonary oxidants/antioxidants and lung function. The lung has several defense networks employing antioxidants to counteract reactive oxygen species (ROS) produced during normal physiology. During asthma attack, innate/adaptive immune cells such as macrophages, neutrophils, eosinophils, and epithelial cells in lung produce significant amounts of ROS. Excessive ROS generation during asthmatic inflammation overwhelms antioxidant defenses and leads to their depletion/dysfunction. This usually leads to pulmonary injury through the dysfunction of lipid membranes, proteins, and DNA, resulting in increased vascular permeability, mucus secretion, smooth muscle contraction, and epithelial shedding. This chapter describes how pulmonary superoxide dismutase (SOD) activity and total nonenzymatic antioxidant abilities are important predictors of lung function in asthma. Therefore, boosting of endogenous antioxidants either through dietary or pharmacological intervention may be beneficial through the attenuation of ROS-mediated damage in asthmatic lung.

14.1 INTRODUCTION

Asthma is a complex inflammatory airway lung disease characterized by eosinophilic/neutrophilic inflammation and airway hyperreactivity. Eosinophilic inflammation is associated with mild/moderate/severe asthma, whereas neutrophilic inflammation is usually found in severe asthma. Repeated allergen challenge in genetically predisposed individuals causes the release of a variety of inflammatory mediators that collectively produce characteristic features of asthma. Reactive oxygen species (ROS) generated from innate or adaptive immune cells in the lung also contribute toward the pathophysiology of asthma. ROS generation has been shown to be associated with airway hyperresponsiveness, mucus hypersecretion, increased vascular permeability, increased epithelial shedding, and decreased responsiveness to glucocorticoid/beta-2-agonist therapy (Barnes and Drazen 2002; Nadeem et al. 2003, 2008; Nadeem and Mustafa 2006; Reszka et al. 2011; Lee and Yang 2012; Zuo et al. 2013; Stephenson et al. 2015).

ROS production is a normal function of the lung as part of its oxidative metabolism taking place in the mitochondria of different cellular compartments. The lung has multiple protective networks in the form of various enzymatic and nonenzymatic antioxidants (Droge 2002). Both enzymatic and nonenzymatic antioxidants such as catalase, glutathione peroxidase (GPx), superoxide dismutase (SOD), vitamin E, vitamin C, albumin, uric acid, and glutathione (GSH) act in concert to avert any attack of ROS on different pulmonary biomolecules and save them from oxidative attack (Heffner and Repine 1989; Chaudiere and Ferrari-Iliou 1999; Evans and Halliwell 2001; Nadeem et al. 2003; Halliwell 2006).

The asthmatic lung has an ongoing cycle of inflammation, which keeps producing ROS through different innate and adaptive immune cells (Droge 2002; Nadeem et al. 2003; Sackesen et al. 2008; Zuo et al. 2013). Excess ROS production overpowers the normal antioxidative capacity of the lung, thus leading to impairment in the function of different lung components such as epithelium, endothelium, and airways smooth muscle (Boueiz and Hassoun 2009). ROS-mediated impairment in different pulmonary structures results from the modification of lipid membranes, protein structure, and DNA. ROS generation impairs membrane ionic gradients, causes dysfunction in protein function/antioxidant enzyme activity, and alters cellular signaling through DNA damage. These modifications in cellular structures may collectively be responsible for airway smooth muscle (ASM) hypertrophy, subepithelial fibrosis, and airway remodeling, which are common features of asthmatic airways (Nadeem and Mustafa 2006; Sackesen et al. 2008; Höhn et al. 2013).

This chapter attempts to provide an extensive review of the oxidant–antioxidant balance in the asthmatic lung. Furthermore, it will try to address any relationships between lung oxidant–antioxidant markers and lung function in asthmatics.

14.2 SOURCES OF OXIDANTS IN THE LUNG

14.2.1 Primary Oxidants

There are two types of oxidants produced in the lung. Primary oxidants are produced directly by immune cells, whereas secondary oxidants arise due to interaction between two primary oxidants or a primary oxidant and a metal ion/halide ion. Multiple enzymes in the lung have the capacity to produce primary oxidants such as the superoxide radical ($O_2^{-\bullet}$), hydrogen peroxide (H_2O_2), and nitric oxide (NO); however, under normal physiology, their production is highly controlled and quarantined because they are used for specific purposes such as microbial destruction. Phagocytic NADPH oxidase (NOX2) is the principal source of $O_2^{-\bullet}$; however, other isoforms of NOX2 are found in nonphagocytic cells (NOX1, NOX3, NOX4, NOX5, DUOX1, and DUOX2), which can also produce $O_2^{-\bullet}$ or H_2O_2. It is noteworthy to mention here that H_2O_2 can also be directly produced by DUOXs from epithelial cells in the respiratory tract (Fischer 2009). Otherwise, most of the H_2O_2 production results from the dismutation of $O_2^{-\bullet}$. Neutrophils, eosinophils, monocytes, and macrophages have high expression of the NOX2 enzyme, which consists of membrane-bound NOX2 (gp91phox), the transmembrane protein p22phox, cytosolic subunits (p47phox, p40phox, and p67phox), and small GTP-binding protein Rac. Different stimuli have the capacity to activate NOX2, which leads to the assembly of cytosolic subunits with membrane-associated subunits for the production of $O_2^{-\bullet}$ (Babior 1999; Bedard and Krause 2007). Recently, it has been shown that nonphagocytic NOXs usually also present in endothelial cells, cardiac myocytes, fibroblasts, and adipocytes play important roles in different cellular signaling and processes (Bedard and Krause 2007; Jiang et al. 2011; Lee and Yang 2012).

NO is another free radical; however, its oxidizing potential on its own is very limited. The functional role of NO in the lung is dependent on its local concentration and presence of other ROS. NO produces most of its damaging effects on biomolecules when it is converted to other secondary oxidizing species such as peroxynitrite ($ONOO^-$). NO is produced by innate or adaptive immune cells in the respiratory system through nitric oxide synthases (NOSs), which exist in the following three isoforms:

1. Constitutive neural NOS (NOS-I or nNOS)
2. Inducible NOS (NOS-II or iNOS)
3. Constitutive endothelial NOS (NOS-III or eNOS) (Ricciardolo et al. 2004)

All of them are present in the lung, but iNOS plays an important role under inflammatory conditions where it produces NO in much greater quantity from innate/adaptive immune cells and serves as an indicator of oxidative/nitrosative stress (Horvath 1998; Ghosh and Erzurum 2011; Sugiura and Ichinose 2011; Radi 2013).

14.2.2 SECONDARY OXIDANTS AND MARKERS OF OXIDANT-MEDIATED DAMAGE ON BIOMOLECULES

Primary oxidants may lead to the production of secondary oxidants through eosinophil- or neutrophil-derived peroxidases such as eosinophil peroxidase (EPO) or myeloperoxidase (MPO), respectively. EPO and MPO can convert H_2O_2 to form reactive halogen species such as hypochlorous acid (HOCl) and hypobromous acid (HOBr) in the presence of halide ions. NO can react with either $O_2^{-\bullet}$ or H_2O_2 to produce nitrating species, which can be utilized by EPO/MPO/DUOXs to nitrate proteins (Andreadis et al. 2003; Sugiura and Ichinose 2011; Voraphani et al. 2014). These secondary oxidants serve a specific purpose of killing foreign pathogens, but they may become toxic to the biomolecules such as lipids, proteins, and DNA if produced indiscriminately during airway inflammation. Halogenation/nitration/carbonylation of proteins has been reported after an attack from different primary or secondary oxidants such as reactive halogen species, H_2O_2 and $ONOO^-$. Proteins are attacked by ROS by the modification of amino acid side chains (such as methionine, tyrosine, histidine, tryptophan, and cysteine) or the protein backbone itself (protein carbonyls). Proteins are also indirectly modified through lipid peroxidation aldehydes, which cross-link oxidized proteins into protein aggregates. This structural modification of the proteins usually leads to the dysfunction of the protein, increased susceptibility to proteolytic degradation, and inactivation of antioxidant enzymes (Höhn et al. 2013; Nadeem et al. 2014).

The phospholipid component of cellular membranes is a highly vulnerable target due to the susceptibility of polyunsaturated fatty acid (PUFA) side chains to ROS attack. Lipid peroxidation may be initiated by any primary free radical, which has sufficient reactivity to extract a hydrogen atom from a reactive methylene group of an unsaturated fatty acid such as hydroxyl radicals, alkoxyl radicals, peroxyl radicals, and alkyl radicals. Cleavage of the carbon bonds during lipid peroxidation reactions results in the formation of alkanals (malondialdehyde and 4-hydroxynonenal) and hydrocarbon alkane gases, for example, ethane and pentane (Brigelius-Flohe and Traber 1999; Halliwell 2006; Höhn et al. 2013). These end products are measured as markers of lipid peroxidation/oxidative stress in different components of the lung such as exhaled breath condensate (EBC), bronchoalveolar lavage fluid (BALF), and airway epithelial cells (AECs). ROS attack on PUFA not only affects the structural and functional integrity of the membrane but also break downs products such as aldehydes and 4-hydroxynonenal, which can also escape from the membrane and cause disturbances at a distance. They can amplify inflammation by cross-linking oxidatively damaged proteins, which may alter enzymatic activity and transmembrane ionic gradients, ultimately causing altered cellular homeostasis (Halliwell 2006; Höhn et al. 2013).

The genetic material of the cell, DNA, is constantly getting altered due to attack by endogenously generated ROS. However, cells have extremely efficient repair mechanisms to repair the damage and hence avoid any error being copied during replication. Under inflammatory conditions where both primary and secondary oxidants are generated indiscriminately and excessively, DNA damage is much more likely to happen in the form of single-strand breaks, which gives rise to oxidation products, such as 8-oxodeoxyguanine (8-oxodG) and thymine glycol, as well as alkylation products. DNA damage may either lead to altered cellular signaling or apoptosis of the cell, thereby compromising integrity and normal function of the cell (Cadet et al. 2012; Ray et al. 2012).

14.3 PULMONARY OXIDATIVE STRESS IN ASTHMATICS

14.3.1 INCREASED PRODUCTION OF PRIMARY OXIDANTS IN ASTHMATICS

Primary oxidants such as $O_2^{-\bullet}$, H_2O_2, and NO have been shown to be elevated in different components of the lung (BAL cells, ASM, and EBC). Many studies have shown increased ROS generation in BAL macrophages isolated from subjects with asthma as compared to control subjects (Cluzel et al. 1987; Kelly et al. 1988). More importantly, these studies have shown an inverse relation between ROS generation in the lung and FEV1 (Kelly et al. 1988; Jarjour and Calhoun 1994). A recent study has shown not only increased ROS in the ASM of asthmatics but also expression of NOX4. Moreover, NOX4 expression in the ASM of asthmatics in this study was found to be negatively correlated with FEV1 (Sutcliffe et al. 2012). H_2O_2 has been shown to be increased in the EBC of asthmatic patients as compared to normal subjects (Antczak et al. 1997; Horvath et al. 1998; Al Obaidi and Al Samarai 2008; Fitzpatrick et al. 2009). Exhaled H_2O_2 has been shown to have an inverse correlation with FEV1 (Horvath et al. 1998; Emelyanov et al. 2001). It has also been shown that samples obtained from proximal and distal airways have higher percentages of ROS-producing cells as compared to nonasthmatic controls (Anderson et al. 2011). Usually, H_2O_2 is formed due to dismutation of $O_2^{-\bullet}$, but recent studies have shown that a direct source of H_2O_2 may also be epithelial DUOXs (Voraphani et al. 2014). H_2O_2 concentrations in the ELF have also been shown to be elevated in asthmatics (Fitzpatrick et al. 2009). A recent study from us has shown increased epithelial DUOX2 and ROS generation in response to cockroach extract allergens in allergic mice (Nadeem et al. 2015). Other inflammatory mediators found in asthmatic airways such as histamine may also cause the release of H_2O_2 through the DUOX system (Rada et al. 2014). Moreover, DUOX knockout allergic mice have less air inflammation and airway reactivity than their wild-type counterparts (Chang et al. 2013). These studies show us that not only classical NOX2 but also other isoforms such as NOX4/DUOXs play an important role in the generation of oxidative stress in asthmatic airways. In support of the role of NOXs in the production of oxidative stress in asthmatic airways, a recent study has shown that the inhibition of NOX by apocynin leads to the reduction in markers of airway inflammation such as EBC H_2O_2 and NO (Stefanska et al. 2012).

Increased NO as an indicator of airway inflammation has been reported in numerous studies in asthmatic lungs, that is, EBC, BALF, AM, sputum, and AEC (Kanazawa et al. 1997; Horváth et al. 1998; Guo et al. 2000; Dweik et al. 2001; Anderson et al. 2011). Moreover, more often than not, exhaled/sputum NO levels are also inversely correlated with lung function/airway inflammation in asthmatic patients (Kanazawa et al. 1997; Saleh et al. 1998; Khatri et al. 2001). Furthermore, the allergen challenge increases the exhaled NO level, which depicts an increase in airway inflammation, whereas corticosteroid treatment reduces it (Kharitonov et al. 1995; Saleh et al. 1998; Guo et al. 2000; Khatri et al. 2001; Roos et al. 2014). The source of increased NO in exhaled breath may be AEC iNOS from asthmatic airways (Saleh et al. 1998; Guo et al. 2000; Roos et al. 2014). It appears from these studies that NO originating from iNOS in asthmatic airways may be detrimental to the pulmonary structures via the production of other oxidant species.

14.3.2 Indirect Measurements of Oxidative Stress in Asthmatics

Other measurements of ROS-mediated attack involve oxidative markers on proteins (carbonyls, nitro-/chloro-/bromotyrosine), lipids (isoprostanes, aldehydes, and ethane), and DNA (8-hydroxydeoxyguanosine). These biomarkers, according to the authors, are usually measured in different components of the lung (sputum, BAL fluid, BAL cells, and lung tissue) and are important tools in providing critical information about the overall oxidative stress due to their correlations with disease severity/lung function (Figure 14.1).

Reactive halogen species originating from eosinophils and neutrophils cause extensive protein oxidative damage in asthma. For example, bromotyrosine levels are higher in induced sputum and BAL proteins of asthmatics (Wu et al. 2000; Aldridge et al. 2002). Moreover, an increase in airway inflammation often observed in severe asthmatics appears to increase bromotyrosine

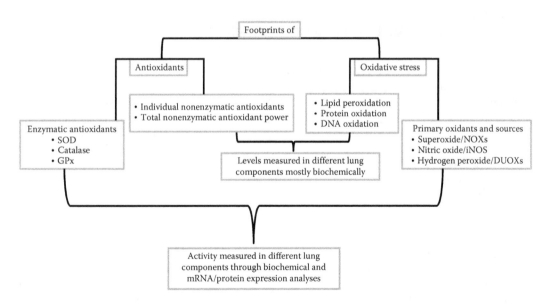

FIGURE 14.1 Footprints of oxidant and antioxidant balance can be measured in various components of the lung, for example, bronchoalveolar lavage inflammatory cells (neutrophils eosinophils, and alveolar macrophages), resident cells (airway epithelial cells and airway smooth muscle cells), or bronchoalveolar lavage fluid. Primary oxidants and enzymatic activities can be measured at both the biochemical and molecular levels, whereas nonenzymatic antioxidants and secondary oxidants are measured mostly biochemically. *Abbreviations*: iNOS, Inducible nitric oxide synthase; GPx, Glutathione peroxidase; SOD, Superoxide dismutase; NOXs, NADPH oxidases; DUOXs, Dual oxidases.

levels in BAL proteins several folds higher as compared to stable or normal subjects (MacPherson et al. 2001). Elevations in chlorotyrosine have also been reported in BAL proteins isolated from allergen-challenged segments of asthmatics (Wu et al. 2000). Peroxynitrite- or H_2O_2-mediated damage in the form of 3-nitrotyrosine has also been shown in the AEC/inflammatory cells/EBC in patients with asthma (Saleh et al. 1998; Guo et al. 2000; Hanazawa et al. 2000; MacPherson et al. 2001; Voraphani et al. 2014). Moreover, 3-nitrotyrosine expression in asthmatic airways correlates inversely with methacholine reactivity/FEV1 (Saleh et al. 1998). Recently, the roles of iNOS and DUOXs have been emphasized in the formation of AEC 3-nitrotyrosine and airway inflammation (Voraphani et al. 2014). Furthermore, bronchial tissues from individuals who die of asthma show that 3-nitrotyrosine immunostaining colocalizes with eosinophils, suggesting a fundamental role for eosinophils in airway inflammation through oxidative stress (MacPherson et al. 2001). One study has also shown the formation of protein carbonyls in BAL after segmental allergen challenge, which correlates strongly with the number of BAL eosinophils (Foreman et al. 1999). Nitration of beta-2-agonist, salbutamol, has also been reported in the presence of MPO/EPO and H_2O_2/nitrite, which may lead to decreased pharmacological activity of salbutamol (Reszka et al. 2011).

Several studies have reported an increase in lipid peroxidation products in the lung of asthmatics (Antczak et al. 1997; Dut et al. 2008; Romieu et al. 2008; Fitzpatrick et al. 2011). Corradi et al. (2003) showed increased aldehydes in EBC after acute exacerbation of asthma in children. Ethane, produced following lipid peroxidation of n-3 fatty acids, has also been reported to be higher in breath condensates of asthmatics (Paredi et al. 2000). A significant negative relation between FEV_1% and EBC MDA has also been shown (Antczak et al. 1997; Romieu et al. 2008). Isoprostanes, produced via free radical–mediated peroxidation of arachidonic acid, have also been found to be

---

TABLE 14.1
Relationship between Markers of Pulmonary Oxidative Stress/Antioxidant Status and Lung Function in Asthma

Marker of Oxidative Stress/ Antioxidant Status	Lung Component Used for Measurement	Correlation with Lung Function (FEV1 or FVC)	Reference(s)
Oxidant generation (hydrogen peroxide/ superoxide generation/ NOX4/nitric oxide)	Airway leukocytes/EBC/ ASM	Negative	Kelly et al. (1988); Jarjour and Calhoun (1994); Antczak et al. (1997); Horvath et al. (1998); Saleh et al. (1998); Emelyanov et al. (2001); Sutcliffe et al. (2012)
Lipid peroxidation (MDA/ Isoprostanes)	EBC	Negative	Battaglia et al. (2005); Barreto et al. (2009); Keskin et al. (2014); Antczak et al. (1997); Paredi et al. (2000); Romieu et al. (2008)
Protein oxidation (nitrotyrosine)	AEC	Negative	Saleh et al. (1998)
DNA oxidation (8-oxodG)	ASM	Negative	Sutcliffe et al. (2012)
Total nonenzymatic antioxidant capacity (PIC)	Sputum	Positive	Kanazawa et al. (2002, 2003)
Enzymatic antioxidant (SOD)	AEC	Positive	Smith et al. (1997); Comhair et al. (2005)
Individual nonenzymatic antioxidant (glutathione)	Sputum	Negative	Wood et al. (2008)

Abbreviations: ASM, Airway smooth muscle; AEC, Airway epithelial cells; EBC, Exhaled breath condensate; 8-oxodG, 8-oxodeoxyguanine; FEV1, Forced expiratory volume in first second; FVC, Forced vital capacity; MDA, Malondialdehyde; PIC, Peroxynitrite inhibitory capacity; SOD, Superoxide dismutase; NOX4, NADPH oxidase 4.

elevated in EBC of asthmatics (Montuschi et al. 1999; Baraldi et al. 2003). Several studies have also shown an inverse relationship between EBC isoprostanes and FEV1 (Battaglia et al. 2005; Barreto et al. 2009; Keskin et al. 2014). Fitzpatrick et al. (2009) have shown increased levels of both MDA and isoprostanes in the BALF of asthmatic children. Moreover, the allergen challenge has been reported to increase EBC isoprostanes in asthmatics (Brussino et al. 2010).

Several studies have measured DNA damage in the lung of asthmatics. A recent study has shown that asthmatic patients have increased 8-oxodG (a biomarker of oxidative damage to DNA) in ASM, which is negatively correlated with FEV1 (Sutcliffe et al. 2012). Another study has shown elevated levels of 8-oxodG in the sputum of asthma patients with a smoking habit (Proklou et al. 2013). Fitzpatrick et al. (2011) have shown increased levels of 8-oxodG in the BALF of asthmatic children. Fortoul et al. (2003) showed that nasal epithelial cells had more DNA strand breaks (another biomarker of oxidative damage to DNA) in asthmatics than in control subjects. These studies show that almost all compartments of asthmatic lung sustain enhanced ROS attack on DNA.

All of these studies support widespread ROS-mediated modification of proteins, lipids, and DNA in the lung of asthmatics. Moreover, several studies show a negative association between primary/ secondary oxidants and other footprints of oxidative stress and lung function (Table 14.1). This suggests that ROS production is associated with a decline in lung function in asthmatic subjects. Finally, ROS not only modify cellular components in the lung and remodel it but may also modify response to the beta-2-agonist therapy.

14.4 ANTIOXIDANTS IN THE LUNG

14.4.1 Nonenzymatic Antioxidants in the Lung

Different pulmonary compartments are enriched with a variety of antioxidants. This is crucial to life as the lung encounters every minute of its life a heavy oxidizing environment both through endogenous and exogenous sources. Complex enzymatic and nonenzymatic antioxidant strategies are available in the lung to prevent oxidant-induced injury. The first line of defense against inhaled environmental/endogenous oxidants is a thin fluid layer covering the respiratory epithelium, known as respiratory tract lining fluid (RTLF). The RTLF is composed of complex antioxidant systems. Nonenzymatic antioxidants comprise vitamin C, urate, reduced GSH, vitamin E, and bilirubin, whereas enzymatic antioxidant comprise extracellular SOD (EC-SOD), catalase, and extracellular GPx. Adjunct antioxidants such as sulfhydryls associated with albumin/other proteins and metal ion–binding proteins such as transferrin/ceruloplasmin also play important roles in the detoxification of ROS through different mechanisms (Heffner and Repine 1989; Halliwell and Gutteridge 1996; Halliwell 2006; Nadeem et al. 2008). Some of these antioxidants are sacrificial in nature as they are consumed by ROS attack, for example, albumin sulfhydryls, while others can be regenerated through cooperation, for example, vitamin E/GSH can be reduced back to their original form. GSH is present in high concentrations in the cells and is the most abundant antioxidant. Vitamin C and GSH are considered to be two major hydrophilic antioxidants, while vitamin E works as a lipophilic antioxidant (Sies 1999; Schafer and Buettner 2001). Among nonenzymatic antioxidants, GSH is the main redox buffer because of its association with other antioxidant networks like NADPH/vitamin E/vitamin C/GSH-Px (Brigelius-Flohe and Traber 1999; Chaudiere and Ferrari Iliou 1999).

Recently, a measurement of total antioxidant capacity (TAC) has gained widespread attention and becomes quite important in providing the overall picture regarding nonenzymatic antioxidant status. This is not unexpected because it is not time-consuming and cumbersome, unlike measurements of individual nonenzymatic antioxidants. It is well established that nonenzymatic antioxidants work cooperatively against ROS attack. Thus, TAC seems to be a much better index of overall nonenzymatic antioxidant potential and may reflect either ROS overload observed during inflammatory states or improvement in antioxidant status due to dietary/pharmacological antioxidant supplementation (Chaudiere and Ferrari-Iliou 1999; Ghiselli et al. 2000; Ginsburg et al. 2011; Nadeem et al. 2014).

14.4.2 ENZYMATIC ANTIOXIDANTS IN THE LUNG

A second arm of defense against ROS-mediated attack is the combined power of enzymatic antioxidants. They are important as they are not easily consumed or destroyed as opposed to nonenzymatic antioxidants. Moreover, they may help in the regeneration of some of the nonenzymatic antioxidants. They work in concert with nonenzymatic antioxidants to form a tightly regulated antioxidant network. Since $O_2^{-\cdot}$ is the primary oxidant produced from a variety of sources, its dismutation by SOD is of primary importance for each cell. There are three forms of SOD:

1. The copper–zinc superoxide dismutase (Cu,Zn-SOD)
2. The manganese superoxide dismutase (Mn-SOD)
3. Extracellular superoxide dismutase (EC-SOD) (Zelko et al. 2002)

All SODs are widely expressed in the human lung. Cu,Zn-SOD, although a cytosolic enzyme, is also found in the nucleus, intermembrane space of mitochondria, and lysosomes. It is expressed in the bronchial epithelium. Mn-SOD is localized in the mitochondria and is important for cell survival. Mn-SOD is mainly expressed in alveolar type II epithelial cells and alveolar macrophages. EC-SOD is primarily localized in the extracellular compartment and has been detected in the bronchial epithelium, alveolar epithelium, and alveolar macrophages. EC-SOD, after being synthesized by vascular smooth muscle cells/fibroblasts, anchors to endothelial surfaces through heparin-binding proteins and provides protection to the extracellular compartment of various tissues. This strategy may also increase the bioavailability of NO, which is released from the endothelium. (Kinnula and Crapo 2003; Fukai and Ushio-Fukai 2011). All of the SOD isoforms cooperatively detoxify $O_2^{-\cdot}$ generated either in the intracellular or in the extracellular compartment.

H_2O_2 coming from the dismutation of either $O_2^{-\cdot}$ or DUOX has high oxidant potential and must be properly detoxified in the cells. For this purpose, two antioxidant systems are at work to tightly regulate the level of H_2O_2. H_2O_2 is reduced to water by either catalase or GPx. Catalase is most effective in the presence of high H_2O_2 concentrations since it cannot be saturated at any H_2O_2 concentration (Nicholls 2012; Sies 2014). Catalase having four ferriprotoporphyrin groups per molecule undergoes alternate divalent oxidation and reduction at its active site in the presence of H_2O_2. Catalase also binds NADPH as a reducing equivalent to prevent the oxidative inactivation of the enzyme by H_2O_2 as it is reduced to water (Kirkman and Gaetani 2007). Furthermore, it also complements GPx as a reducing system for H_2O_2 (Arthur 2000; Sies 2014).

The GPx enzyme system detoxifies a number of toxic organic hydroperoxides to the corresponding nontoxic hydroxyl compounds. Cells have high GSH content in reduced form to ensure active reduction of toxic hydroperoxides. Eight GPxs have been described, which have been divided into two groups. Group 1 consists of GPx-1, GPx-2, GPx-3, GPx-4, and GPx-6 and has selenocysteine within the active site; and group 2 consists of GPx-5, GPx-7, and GPx-8 and has cysteine within the active site. They are expressed in different organs according to their specific requirements (Brigelius-Flohé and Maiorino 2013). GPx-1, also known as cellular GPx, usually detoxifies H_2O_2 and fatty acid peroxides but not esterified peroxyl lipids. GPx-2, also known as gastrointestinal GPx, usually detoxifies dietary peroxides in epithelial cells of gastrointestinal tract. GPx-3, also known as extracellular GPx, detoxifies lipid peroxides in the extracellular compartment of the lung (Arthur 2000; Comhair et al. 2001). GPx-4, also known as phospholipid hydroperoxide GPx, is a membrane-bound enzyme and detoxifies esterified lipids. GPx-5 expressed in epididymis is believed to play a role in male fertility. Other GPxs such as GPx-6, GPx-7, and GPx-8 may be localized in the olfactory epithelium and endoplasmic reticulum, but their function is unknown and currently an active area of research (Brigelius-Flohé and Maiorino 2013).

Basically, enzymatic antioxidants serve two main functions:

1. Scavenging of primary oxidants to prevent them from directly oxidizing proteins/lipids/DNA
2. Inhibition of formation of secondary oxidants such as reactive halogen species, peroxynitrite, and hydroxyl radical (·OH) by converting primary oxidants to nontoxic products (Zelko et al. 2002; Kinnula and Crapo 2003; Nadeem et al. 2014)

Some of the aforementioned enzymatic antioxidants require NADPH as a reducing equivalent. NADPH maintains catalase in the active form and is also used as a cofactor by GSH, which converts oxidized GSH back to reduced GSH. Reduced GSH in turn is used by GPxs to detoxify various lipid peroxides. Intracellular NADPH is generated by the reduction of $NADP^+$ by glucose-6-phosphate dehydrogenase through pentose phosphate pathway. Therefore, NADPH is a crucial factor in the overall redox balance of the cell by buffering the levels of oxidized and reduced GSH and can be considered an essential factor in regulating the antioxidant network (Sies 1999; Deponte 2013).

14.5 PULMONARY ANTIOXIDANT DYSREGULATION IN ASTHMATICS

14.5.1 Alterations in Nonenzymatic Antioxidants in Asthmatics

Evidence for alterations in nonenzymatic antioxidants in asthmatic airways is provided by the findings that show alterations in BALF antioxidants of asthmatics (Kelly et al. 1999; Kanazawa et al. 2002; Fitzpatrick et al. 2011). Many studies have reported deficiencies in individual nonenzymatic antioxidants in asthmatics. Low levels of vitamin C and urate in the RTLF have been observed in adults with mild asthma (Kelly et al. 1999). Vitamin E has also found to be decreased in the bronchial wash of asthmatics (Kelly et al.1999). Disturbed GSH status is reported in asthma, with total (Smith et al. 1993; Beier et al. 2004; Wood et al. 2008) and oxidized GSH being elevated in BALF/alveolar macrophages/sputum (Kelly et al. 1999; Dut et al. 2008; Wood et al. 2008; Fitzpatrick et al. 2009, 2011). Wood et al. (2008) have shown a negative correlation between sputum oxidized GSH and lung function in asthmatics.

Only one study has reported TAC measurement in the lung measured as peroxynitrite inhibitory activity in sputum. Peroxynitrite inhibitory activity was reported to be low in asthmatics as compared to normal controls and was also correlated with the degree of airway obstruction (Kanazawa et al. 2002). Inhaled corticosteroid therapy increased peroxynitrite inhibitory activity in sputum, which was associated with an increase in FEV1 (Kanazawa et al. 2003). This shows that airway inflammation decreases lung antioxidant capacity, which is associated with decline in lung function. In support of this, Yoon et al. (2012) have shown recently that asthma patients with higher TAC levels had better pulmonary function than those with lower TAC levels, and these differences were maintained for the 2-year study period. A recent study has also shown that dietary TAC affects lung function in children with current asthma, that is, asthmatic children with higher TAC had lower likelihood of suffering asthma episodes than children with lower TAC values (Rodríguez et al. 2013). It appears from these studies that if lung TAC is enhanced by dietary/pharmacological agents, there is a likelihood that lung function may improve in asthmatics.

14.5.2 Alterations in Enzymatic Antioxidants in Asthmatics

There are not many studies that have measured different enzymatic antioxidant activities at the molecular level, that is, mRNA/protein expression. Nevertheless, many studies have investigated enzymatic activities in different components of the asthmatic lung biochemically. Although there are several enzymatic antioxidant mechanisms to remove ROS produced in intracellular/extracellular compartments, we are going to limit our discussion only to GPx, catalase, and SOD due to the paucity of data on other enzymatic antioxidants in the lungs of asthmatic subjects.

Several studies show alterations in antioxidant enzymes of asthmatic subjects, which may be reflected either as an increase or as a decrease in different pulmonary compartments. SOD has been found to be decreased in BAL cells and HBEC of asthmatics (De Raeve et al. 1997; Smith et al. 1997; Ghosh et al. 2013). Furthermore, this study showed a direct relationship between HBEC and airway reactivity to methacholine. There is a loss of SOD activity within minutes of antigen challenge in the lungs of asthmatics, which has been attributed to the modification of SOD by ROS (Comhair et al. 2000; Dworski et al. 2011). A recent study has shown that Cu/Zn SOD activity could be due to oxidative inactivation of cysteine (a thiol-containing amino acid similar to GSH), which is required for its functional activity (Ghosh et al. 2013). This is also supported by another study, which shows cysteine oxidation by ROS and its association with decreased glucocorticoid responsiveness due to glucocorticoid receptor modification (Stephenson et al. 2015). Loss of SOD activity in asthma results in an increase of superoxide, which favors peroxynitrite formation, leading to oxidative injury and the nitration of proteins (Andreadis et al. 2003; Nadeem et al. 2014). EC-GPx has been reported to be increased in the AECs isolated from asthmatic lungs (Comhair et al. 2001), whereas it was found to be no different in the BALF of asthmatic children compared to healthy controls (Fitzpatrick et al. 2009). The overexpression of SOD, but not catalase, inhibited the induction of AEC EC-GPx, which shows the importance of $O_2^{-\bullet}$ in the induction of this antioxidant enzyme (Comhair et al. 2001). The increased expression of EC-GPx may help in the detoxification of enhanced lipid peroxides resulting from increased ROS production/decreased SOD activity. BALF catalase activity has also been reported to be 50% lower in asthmatics than in healthy controls, which may be due to the oxidative modification of the cysteine residues in the enzyme (Ghosh et al. 2006). Overall, ROS production either inactivates the enzymatic antioxidants or induces a protective response to counteract oxidative stress. Injury caused to different pulmonary components by ROS may depend on the balance of these factors.

Only pulmonary SOD activity has been consistently shown to be positively correlated with lung function (Smith et al. 1997; Comhair et al. 2005; Table 14.1). It suggests that excess generation of ROS may cause the inactivation of SOD, which may be related to decline in lung function in asthmatics. Excess ROS generation may also cause dysfunction in SOD/catalase activity and the glucocorticoid receptor in asthmatic subjects due to the modification of cysteine residues in the enzyme/glucocorticoid receptor. Finally, it should be noted that not many studies have investigated enzymatic antioxidants at both biochemical and molecular levels. Moreover, the contribution of different enzymatic isoforms and their contribution in different pulmonary components are also required to better understand their significance in disease pathology. This may be confounded by a lack of proper pharmacological tools that help in distinguishing different isoforms as well as access to different compartments of the lung.

14.6 CONCLUSION

Oxidative inflammation observed in the asthmatic lung is partly the result of chronic ROS production in response to repeated allergen/pollutants encounter. This is not well controlled by the antioxidant network present in the lung, which gives rise to dysregulation of enzymatic/nonenzymatic antioxidants (Figure 14.2). This is supported by correlations between parameters of oxidant/antioxidant and lung function in asthmatic subjects. More often than not, oxidative parameters have a negative correlation, whereas nonenzymatic/enzymatic antioxidants have a positive correlation with lung function.

The oxidant–antioxidant imbalance in asthma may also lead to impaired responsiveness to glucocorticoid/beta-2-agonist therapy due to glucocorticoid receptor/beta-2-agonist modification by ROS (Reszka et al. 2011; Stephenson et al. 2015). This emphasizes the importance of the oxidant–antioxidant balance in the lung. Therefore, there is a need to correct this

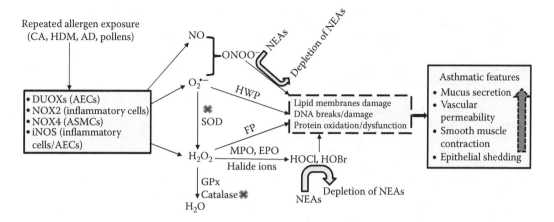

FIGURE 14.2 In response to repeated allergen challenge in allergic individuals, inflammatory cells and resident cells (AEC and ASM) produce primary ROS such as $O_2^{-\bullet}$, H_2O_2, and NO. Primary oxidants give rise to secondary oxidants such as $ONOO^-$, HOCl, and HOBr through various mechanisms. Haber–Weiss/Fenton processes lead to the formation of hydroxyl radical, whereas MPO/EPO in the presence of H_2O_2 forms HOCl and HOBr, which can oxidatively modify proteins, membrane lipids, and DNA. Products of NO metabolism along with H_2O_2 can also produce nitrating species through DUOXs. Primary and secondary oxidants cause depletion of nonenzymatic antioxidants and also inactivation of certain antioxidant enzymes such as SOD/catalase, which leads to oxidative stress due to the formation of lipid peroxides and modification of proteins/DNA structure, thus causing malfunction of the cellular processes such as smooth muscle contractility and vascular permeability. All of these events ultimately amplify the inflammatory process in the asthmatic lung, thus leading to airway remodeling and hyperreactivity. *Abbreviations used*: NO, Nitric oxide; iNOS, Inducible nitric oxide synthase; AEC, Airway epithelial cell; ASMC, Airway smooth muscle cell; ROS, Reactive oxygen species; EPO, Eosinophil peroxidase; MPO, Myeloperoxidase; HOCl, Hypochlorous acid; HOBr, Hypobromous acid; GPx, Glutathione peroxidase; SOD, Superoxide dismutase; NOX, NADPH oxidase; HWP, Haber–Weiss process; FP, Fenton process; $ONOO^-$, Peroxynitrite; H_2O_2, Hydrogen peroxide; $O_2^{-\bullet}$, Superoxide radical; DUOX, Dual oxidase; NEA, Nonenzymatic antioxidant; AD, Animal dander; CA, Cockroach allergens; HDM, House dust mite; × indicates inactivation/dysfunction; ↑ indicates an increase.

oxidant–antioxidant imbalance in the lung of asthmatics. In support of this, recent studies have shown that oral/dietary antioxidant supplementation leads to improvement in lung function and attenuation of airway/systemic inflammation through increase in lung antioxidant potential (Dworski et al. 2011; Hoskins et al. 2012; Wood et al. 2012). However, these studies were conducted on a small scale. Therefore, there is a need to conduct large-scale studies to confirm the role of antioxidants in asthma.

ACKNOWLEDGMENTS

The authors are grateful to the Deanship of Scientific Research and College of Pharmacy Research Center, King Saud University, for their funding and support.

REFERENCES

Aldridge, R., Chan, T., van Dalen, C. et al. 2002. Eosinophil peroxidase produces hypobromous acid in the airways of stable asthmatics. *Free Radic Biol Med* 33:847–856.

Al Obaidi, A.H. and Al Samarai, A.M. 2008. Biochemical markers as a response guide for steroid therapy in asthma. *J Asthma* 45:425–428.

Anderson, J.T., Zeng, M., Li, Q. et al. 2011. Elevated levels of NO are localized to distal airways in asthma. *Free Radic Biol Med* 50:1679–1688.

Andreadis, A.A., Hazen, S.L., Comhair, S.A. et al. 2003. Oxidative and nitrosative events in asthma. *Free Radic Biol Med* 35:213–225.

Antczak, A., Nowak, D., Shariati, B. et al. 1997. Increased hydrogen peroxide and thiobarbituric acid-reactive products in expired breath condensate of asthmatic patients. *Eur Respir J* 10:1235–1241.

Arthur, J. 2000. The glutathione peroxidases. *Cell Mol Life Sci* 57:1825–1835.

Babior, B.M. 1999. NADPH oxidase: An update. *Blood* 93:1464–1476.

Baraldi, E., Carraro, S., Alinovi, R. et al. 2003. Cysteinyl leukotrienes and 8-isoprostanes in exhaled breath condensate of children with asthma exacerbations. *Thorax* 58:505–509.

Barnes, P.J. and J.M. Drazen. 2002. Pathophysiology of asthma: In *Asthma and COPD*, Barnes, P.J. et al., Ed., pp. 342–359. London, U.K.: Academic Press.

Barreto, M., Villa, M.P., Olita, C. et al. 2009. 8-Isoprostane in exhaled breath condensate and exercise-induced bronchoconstriction in asthmatic children and adolescents. *Chest* 135:66–73.

Battaglia, S., den Hertog, H., Timmers, M.C. et al. 2005. Small airways function and molecular markers in exhaled air in mild asthma. *Thorax* 60:639–644.

Bedard, K. and K.H. Krause. 2007. The NOX family of ROS-generating NADPH oxidases: Physiology and pathophysiology. *Physiol Rev* 87:245–313.

Beier, J., Beeh, K.M., Semmler, D. et al. 2004. Increased concentrations of glutathione in induced sputum of patients with mild or moderate allergic asthma. *Ann Allergy Asthma Immunol* 92:459–463.

Boueiz, A. and P.M. Hassoun. 2009. Regulation of endothelial barrier function by reactive oxygen and nitrogen species. *Microvasc Res* 77:26–34.

Brigelius-Flohé, R. and M. Maiorino. 2013. Glutathione peroxidases. *Biochim Biophys Acta* 1830:3289–3303.

Brigelius-Flohé, R. and M.G. Traber. 1999. Vitamin E: Function and metabolism. *FASEB J* 13:1145–1155.

Brussino, L., Badiu, I., Sciascia, S. et al. 2010. Oxidative stress and airway inflammation after allergen challenge evaluated by exhaled breath condensate analysis. *Clin Exp Allergy* 40:1642–1647.

Cadet, J., Loft, S., Olinski, R. et al. 2012. Biologically relevant oxidants and terminology, classification and nomenclature of oxidatively generated damage to nucleobases and 2-deoxyribose in nucleic acids. *Free Radic Res* 46:367–381.

Chang, S., Linderholm, A., Franzi, L. et al. 2013. Dual oxidase regulates neutrophil recruitment in allergic airways. *Free Radic Biol Med* 65:38–46.

Chaudiere, J. and R. Ferrari-Iliou. 1999. Intracellular antioxidants from chemical to biochemical mechanisms. *Food Chem Toxicol* 37:949–962.

Cluzel, M., Damon, M., Chanez, P. et al. 1987. Enhanced alveolar cell luminol-dependent chemiluminescence in asthma. *J Allergy Clin Immunol* 80:195–201.

Comhair, S.A., Bhathena, P.R., Dweik, R.A. et al. 2000. Rapid loss of superoxide dismutase activity during antigen-induced asthmatic response. *Lancet* 355:624.

Comhair, S.A., Bhathena, P.R., Farver, C. et al. 2001.Extracellular glutathione peroxidase induction in asthmatic lungs: Evidence for redox regulation of expression in human airway epithelial cells. *FASEB J* 15:70–78.

Comhair, S.A., Xu, W., Ghosh, S. et al. 2005. Superoxide dismutase inactivation in pathophysiology of asthmatic airway remodeling and reactivity. *Am J Pathol* 166:663–674.

Corradi, M., Folesani, G., Andreoli, R. et al. 2003. Aldehydes and glutathione in exhaled breath condensate of children with asthma exacerbation. *Am J Respir Crit Care Med* 167:395–399.

Deponte, M. 2013. Glutathione catalysis and the reaction mechanisms of glutathione-dependent enzymes. *Biochim Biophys Acta* 183:3217–3266.

De Raeve, H.R., Thunnissen, F.B., Kaneko, F.T. et al. 1997. Decreased Cu, Zn-SOD activity in asthmatic airway epithelium: Correction by inhaled corticosteroid in vivo. *Am J Physiol* 272: L148–L154.

Droge, W. 2002. Free radicals in the physiological control of cell function. *Physiol Rev* 82:47–95.

Dut, R., Dizdar, E.A., Birben, E. et al. 2008. Oxidative stress and its determinants in the airways of children with asthma. *Allergy* 63:1605–1609.

Dweik, R.A., Comhair, S.A., Gaston, B. et al. 2001. NO chemical events in the human airway during the immediate and late antigen-induced asthmatic response. *Proc Natl Acad Sci USA* 98:2622–2627.

Dworski, R., Han, W., Blackwell, T.S. et al. 2011. Vitamin E prevents NRF2 suppression by allergens in asthmatic alveolar macrophages in vivo. *Free Radic Biol Med* 51:516–521.

Emelyanov, A., Fedoseev, G., Abulimity, A. et al. 2001. Elevated concentrations of exhaled hydrogen peroxide in asthmatic patients. *Chest* 120:1136–1139.

Evans, P. and B. Halliwell. 2001. Micronutrients: Oxidant/antioxidant status. *Br J Nutr* 85:S67–S74.

Fischer, H. 2009. Mechanisms and function of DUOX in epithelia of the lung. *Antioxid Redox Signal* 11:2453–2465.

Fitzpatrick, A.M., Teague, W.G., Burwell, L. et al. 2011. Glutathione oxidation is associated with airway macrophage functional impairment in children with severe asthma. *Pediatr Res* 69:154–159.

Fitzpatrick, A.M., Teague, W.G., Holguin, F. et al. 2009. Airway glutathione homeostasis is altered in children with severe asthma: Evidence for oxidant stress. *J Allergy Clin Immunol* 123:146–152.

Foreman, R.C., Mercer, P.F., Kroegel, C. et al. 1999. Role of the eosinophil in protein oxidation in asthma: Possible effects on proteinase/antiproteinase balance. *Int Arch Allergy Immunol* 118:183–186.

Fortoul, T.I., Valverde, M., López Mdel, C. et al. 2003. Single-cell gel electrophoresis assay of nasal epithelium and leukocytes from asthmatic and nonasthmatic subjects in Mexico City. *Arch Environ Health* 58:348–352.

Fukai, T. and M. Ushio-Fukai. 2011. Superoxide dismutases: Role in redox signaling, vascular function, and diseases. *Antioxid Redox Signal* 15:1583–1606.

Ghiselli, A., Serafini, M., Natella, F.S. et al. 2000. Total antioxidant capacity to assess redox status: Critical view and experimental data. *Free Radic Biol Med* 29:1106–1114.

Ghosh, S. and S.C. Erzurum. 2011. Nitric oxide metabolism in asthma pathophysiology. *Biochim Biophys Acta* 1810:1008–1016.

Ghosh, S., Janocha, A.J., Aronica, M.A. et al. 2006. Nitrotyrosine proteome survey in asthma identifies oxidative mechanism of catalase inactivation. *J Immunol* 176:5587–5597.

Ghosh, S., Willard, B., Comhair, S.A . et al. 2013. Disulfide bond as a switch for copper-zinc superoxide dismutase activity in asthma. *Antioxid Redox Signal* 18:412–423.

Ginsburg, I., Kohen, R., Koren, E. et al. 2011. Quantifying oxidant-scavenging ability of blood. *N Engl J Med* 364:883–885.

Guo, F.H., Comhair, S.A., Zheng, S. et al. 2000. Molecular mechanisms of increased nitric oxide (NO) in asthma: Evidence for transcriptional and post-translational regulation of NO synthesis. *J Immunol* 164:5970–5980.

Halliwell, B. 2006. Reactive species and antioxidants. Redox biology is a fundamental theme of aerobic life. *Plant Physiol* 141:312–322.

Halliwell, B. and J.M.C. Gutteridge. 1996. The antioxidants of human extracellular fluids. *Arch Biochem Biophys* 280:1–8.

Hanazawa, T., Kharitonov, S.A., Barnes, P.J. 2000. Increased nitrotyrosine in exhaled breath condensate of patients with asthma. *Am J Respir Crit Care Med* 162:1273–1276.

Heffner, J.A. and J.E. Repine. 1989. State of the art: Pulmonary strategies of antioxidant defense. *Am Rev Respir Dis* 140:531–554.

Höhn, A., König, J., and T. Grune. 2013. Protein oxidation in aging and the removal of oxidized proteins. *J Proteomics* 92:132–159.

Horváth, I., Donnelly, L.E., Kiss, A. et al. 1998. Combined use of exhaled hydrogen peroxide and nitric oxide in monitoring asthma. *Am J Crit Care Med* 158:1042–1046.

Hoskins, A., Roberts, J.L. 2nd, Milne, G. et al. 2012. Natural-source d-α-tocopheryl acetate inhibits oxidant stress and modulates atopic asthma in humans in vivo. *Allergy* 67:676–682.

Jarjour, N.N. and W.J. Calhoun. 1994. Enhanced production of oxygen radicals in asthma. *J Lab Clin Med* 123:131–136.

Jiang, F., Zhang, Y., and G.J. Dusting. 2011. NADPH oxidase-mediated redox signaling: Roles in cellular stress response, stress tolerance, and tissue repair. *Pharmacol Rev* 63:218–242.

Kanazawa, H., Nomura, S., Hirata, K. et al. 2003. Effect of inhaled beclomethasone dipropionate on peroxynitrite inhibitory activity in induced sputum from asthmatic patients. *Chest* 124:1755–1761.

Kanazawa, H., Shiraishi, S., Hirata, K. et al. 2002. Decreased peroxynitrite inhibitory activity in induced sputum in patients with bronchial asthma. *Thorax* 57:509–512.

Kanazawa, H., Shoji, S., Yamada, M. et al. 1997. Increased levels of nitric oxide derivatives in induced sputum in patients with asthma. *J Allergy Clin Immunol* 99:624–629.

Kelly, C., Ward, C., Stenton, C.S. et al. 1988. Number and activity of inflammatory cells in bronchoalveolar lavage fluid in asthma and their relation to airway responsiveness. *Thorax* 2002:684–692.

Kelly, F.J. Mudway, I., Blomberg, A. et al. 1999. Altered lung antioxidant status in patients with mild asthma. *Lancet* 354:482–483.

Keskin, O., Balaban, S., Keskin, M. et al. 2014. Relationship between exhaled leukotriene and 8-isoprostane levels and asthma severity, asthmacontrol level, and asthma control test score. *Allergol Immunopathol* (*Madr*) 42:191–197.

Kharitonov, S.A., O'Connor, B.J., Evans, D.J. et al. 1995. Allergen-induced late asthmatic reactions are associated with elevation of exhaled nitric oxide. *Am J Respir Crit Care Med* 151:1894–1899.

Khatri, S.B., Ozkan, M., McCarthy, K. et al. 2001. Alterations in exhaled gas profile during allergen-induced asthmatic response. *Am J Respir Crit Care Med* 164:1844–1848.

Kinnula, V.L. and J.D. Crapo. 2003. Superoxide dismutases in the lung and human lung diseases. *Am J Respir Crit Care Med* 167:1600–1619.

Kirkman, H.N. and G.F. Gaetani. 2007. Mammalian catalase: A venerable enzyme with new mysteries. *Trends Biochem Sci* 32:44–50.

Lee, I.T. and C.M. Yang. 2012. Role of NADPH oxidase/ROS in pro-inflammatory mediators-induced airway and pulmonary diseases. *Biochem Pharmacol* 84:581–590.

MacPherson, J.C., Comhair, S.A., Erzurum, S.C. et al. 2001. Eosinophils are a major source of nitric oxide-derived oxidants in severe asthma characterization of pathways available to eosinophils for generating reactive nitrogen species. *J Immunol* 166:5763–5772.

Montuschi, P., Corradi, M., Ciabattoni, G. et al. 1999. Increased 8-isoprostane, a marker of oxidative stress, in exhaled condensate of asthma patients. *Am J Respir Crit Care Med* 160:216–220.

Nadeem, A., Alharbi, N.O., Vliagoftis, H. et al. 2015. Protease activated receptor-2 mediated dual oxidase-2 upregulation is involved in enhanced airway reactivity and inflammation in a mouse model of allergic asthma. *Immunology* 145:391–403.

Nadeem, A., Chhabra, S.K., Masood, A. et al. 2003. Increased oxidative stress and altered levels of antioxidants in asthma. *J Allergy Clin Immunol* 111:72–78.

Nadeem, A., Masood, A., and N. Siddiqui. 2008. Oxidant–antioxidant imbalance in asthma: Scientific evidence, epidemiological data and possible therapeutic options. *Ther Adv Respir Dis* 2:215–235.

Nadeem, A. and S.J. Mustafa. 2006. Adenosine receptor antagonists and asthma. *Drug Discov Today: Ther Strategies* 3:269–275.

Nadeem, A., Siddiqui, N., Alharbi, N.O., and M.M. Alharbi. 2014. Airway and systemic oxidant-antioxidant dysregulation in asthma: A possible scenario of oxidants spill over from lung into blood. *Pulm Pharmacol Ther* 29:31–40.

Nicholls, P. 2012. Classical catalase: Ancient and modern. *Arch Biochem Biophys* 525:95–101.

Paredi, P., Kharitonov, S.A., and P.J. Barnes. 2000. Elevation of exhaled ethane concentration in asthma. *Am J Respir Crit Care Med* 162:1450–1454.

Proklou, A., Soulitzis, N., Neofytou, E. et al. 2013. Granule cytotoxic activity and oxidative DNA damage in smoking and nonsmoking patients with asthma. *Chest* 144:1230–1237.

Rada, B., Boudreau, H.E., Park, J.J. et al. 2014. Histamine stimulates hydrogen peroxide production by bronchial epithelial cells via histamine H1 receptor and dual oxidase. *Am J Respir Cell Mol Biol* 50:125–134.

Radi, R. 2013. Peroxynitrite, a stealthy biological oxidant. *J Biol Chem* 288:26464–26472.

Ray, P.D., Huang, B.W., Tsuji, Y. 2012. Reactive oxygen species (ROS) homeostasis and redox regulation in cellular signaling. *Cell Signal* 24:981–990.

Reszka, K.J., Sallans, L., Macha, S. et al. 2011. Airway peroxidases catalyze nitration of the {beta}2-agonist salbutamol and decrease its pharmacological activity. *J Pharmacol Exp Ther* 336:440–449.

Ricciardolo, F.L.M., Sterk, P.J., Gaston, B. et al. 2004. Nitric oxide in health and disease of the respiratory system. *Physiol Rev* 84:731–765.

Rodríguez-Rodríguez, E., Ortega, R.M., González-Rodríguez, L.G. et al. 2014. Dietary total antioxidant capacity and current asthma in Spanish schoolchildren: A case control-control study. *Eur J Pediatr* 173:517–523.

Romieu, I., Barraza-Villarreal, A., Escamilla-Nuñez, C. et al. 2008. Exhaled breath malondialdehyde as a marker of effect of exposure to air pollution in children with asthma. *J Allergy Clin Immunol* 121:903–909.

Roos, A.B., Mori, M., Grönneberg, R. et al. 2014. Elevated exhaled nitric oxide in allergen-provoked asthma is associated with airway epithelial iNOS. *PLoS One* 9(2):e90018.

Sackesen, C., Ercan, H., Dizdar, E. et al. 2008. A comprehensive evaluation of the enzymatic and nonenzymatic antioxidant systems in childhood asthma. *J Allergy Clin Immunol* 122:78–85.

Saleh, D., Ernst, P., Lim, S. et al. 1998. Increased formation of the potent oxidant peroxynitrite in the airways of asthmatic patients is associated with induction of NO synthase: Effect of inhaled glucocorticoid. *FASEB J* 12:929–937.

Schafer, F.Q. and G.R. Buettner. 2001. Redox environment of the cell as viewed through the redox state of the glutathione disulfide/glutathione couple. *Free Radic Biol Med* 30:1191–1212.

Sies, H. 1999. Glutathione and its role in cellular functions. *Free Radic Biol Med* 27:916–921.

Sies, H. 2014. Role of metabolic H2O2 generation: Redox signaling and oxidative stress. *J Biol Chem* 289:8735–8741.

Smith, L.J., Houston, M., Anderson, J. 1993. Increased levels of glutathione in bronchoalveolar lavage fluid from patients with asthma. *Am Rev Respir Dis* 147:1461–1464.

Smith, L.J., Shamsuddin, M., Sporn, P.H. et al. 1997. Reduced superoxide dismutase in lung cells of patients with asthma. *Free Rad Biol Med* 22:1301–1307.

Stefanska, J., Sarniak, A., Wlodarczyk, A. et al. 2012. Apocynin reduces reactive oxygen species concentrations in exhaled breath condensate in asthmatics. *Exp Lung Res* 38:90–99.

Stephenson, S.T., Brown, L.A., Helms, M.N. et al. 2015. Cysteine oxidation impairs systemic glucocorticoid responsiveness in children with difficult-to-treat asthma. *J Allergy Clin Immunol* 136:454–461.

Sugiura, H. and M. Ichinose. 2011. Nitrative stress in inflammatory lung diseases. *Nitric Oxide* 25:138–144.

Sutcliffe, A., Hollins, F., Gomez, E. et al. 2012. Increased nicotinamide adenine dinucleotide phosphate oxidase 4 expression mediates intrinsic airway smooth muscle hypercontractility in asthma. *Am J Respir Crit Care Med* 185:267–274.

Voraphani, N., Gladwin, M.T., Contreras, A.U. et al. 2014. An airway epithelial iNOS-DUOX2-thyroid peroxidase metabolome drives Th1/Th2 nitrative stress in human severe asthma. *Mucosal Immunol* 7:1175–1185.

Wood, L.G., Garg, M.L., Blake, R.J. et al. 2008. Oxidized vitamin E and glutathione as markers of clinical status in asthma. *Clin Nutr* 27:579–586.

Wood, L.G., Garg, M.L., Smart, J.M. et al. 2012. Manipulating antioxidant intake in asthma: A randomized controlled trial. *Am J Clin Nutr* 96:534–543.

Wu, W., Samoszuk, M.K., Comhair, S.A. et al. 2000. Eosinophils generate brominating oxidants in allergen-induced asthma. *J Clin Invest* 105:1455–1463.

Yoon, S.Y., Kim, T.B., Baek, S. et al. 2012. The impact of total antioxidant capacity on pulmonary function in asthma patients. *Int J Tuberc Lung Dis* 16:1544–1550.

Zelko, I.N., Mariani, T.J., and R.J. Folz. 2002. Superoxide dismutase multigene family: A comparison of the CuZn-SOD (SOD1), Mn-SOD (SOD2), and EC-SOD (SOD3) gene structures, evolution, and expression. *Free Radic Biol Med* 33:337–349.

Zuo, L., Otenbaker, N.P., Rose, B.A. et al. 2013. Molecular mechanisms of reactive oxygen species-related pulmonary inflammation and asthma. *Mol Immunol* 56:57–63.

15 Smoking and ROS
Catalyst for Autoimmune Liver Diseases

Tanima Bose

CONTENTS

ABSTRACT

Cigarette smoke contains numerous toxic, carcinogenic, and mutagenic chemicals, stable and unstable free radicals, and reactive oxygen species (ROS), which cause biological oxidative damage. Continuous exposure to these chemicals leads to immense amount of damage to human health either directly or indirectly. In this chapter, I have explained how smoking causes different autoimmune liver disorders, with an introduction to other autoimmune disorders. Among several autoimmune diseases, rheumatoid arthritis (RA), multiple sclerosis (MS), thyroid disease, and primary biliary cirrhosis (PBC) are reported mostly among tobacco-exposed animals. In addition, there is a brief description of how these disorders, especially metabolic diseases, can be treated with different immunosuppressants along with steroids.

15.1 INTRODUCTION

15.1.1 CIGARETTE SMOKE CONTENT

Tobacco use is the leading, prevalent, preventable cause for several diseases around the world. According to the Centers for Disease Control and Prevention (U.S. Department of Health and Human Services 2014), cigarette smoke causes ~480,000 premature deaths in the United States each year and 16 million people suffer from serious diseases caused by smoking. A sad fact is that ~41,000 nonsmokers die in that country, from diseases caused by second-hand smoking.

There are several other ways in which tobacco is consumed other than through cigarette, such as cigars, pipe, snuff, and chewing tobacco. Tobacco smoke is a complex and dynamic mixture of a variety of compounds. Researchers have analyzed both the whole smoke and the individual phases of the smoke which contain either gas or particulate portions.

15.2 SMOKE CONTENT RELATED TO REACTIVE OXYGEN SPECIES (ROS)

Reactive oxygen species (ROS) are highly reactive compounds that mainly contain oxygen in a reactive condition. Along with endogenous ROS formation due to metabolism, they are also produced from pollutants, tobacco, smoke, and ionizing radiation. ROS generation by smoking and consecutive diseases has become one of the alarming causes of unpredictable morbidity and mortality all over the world, causing 4.8 million deaths globally every year (Ezzati and Lopez 2004). It is estimated that smoking-related disease and death will reach >10 million by 2030 (Zavitz et al. 2008). Despite the deadly effects of this habit, >1.1 billion people continue to smoke, representing one-sixth of the world's population (Jha et al. 2002).

Major tobacco products, including cigarette, contain the addictive product nicotine. Upon entering the bloodstream, nicotine can activate the adrenal glands to release adrenaline, which stimulates the central nervous system and increases blood pressure, respiration, and heart rate. There are reports that an additional compound in cigarette smoke, acetaldehyde, can increase nicotine's effect on the brain (Frankel 2011). In addition, cigarette smoke contains ~7357 chemicals of different classes like PAH (polycyclic aromatic hydrocarbon), formaldehyde, benzene, etc., many of which are toxic, carcinogenic, and have different effects on biological systems (Smith and Hansch 2000). Among other potential harmful classes, 1,3-butadiene is specifically and significantly linked to cancer risk, while acrolein and acetaldehyde have the greatest potential to be respiratory irritants, and cyanide, arsenic, and cresols are the primary sources of cardiovascular risk (Fowles and Dybing 2003).

There are two phases of cigarette smoke: a tar or particulate phase and a gaseous phase. Both of these contain debilitating chemicals that cause extensive DNA damage (Bluth et al. 2009; Valavanidis et al. 2009). In addition, cigarette smoking also elevates the apoptotic signal by increasing the amount of CD95, the death receptor, and part of the tumor necrosis family receptor (Fas ligand) in $CD4^+$ and B lymphocytes (Demedts et al. 2006). All these effects conclusively lead to mutations and gene activations responsible for autoimmunity along with apoptotic loads. Smokers demonstrate elevated autoantibody levels (Korpilahde et al. 2004) and decreased antibody titers (Morris et al. 2008) against respiratory antigens, as illustrated in Figure 15.1.

15.3 CIGARETTE SMOKE CAUSING DIFFERENT AUTOIMMUNE DISORDERS (AIDS)

There are a number of relevant studies on smoking and the onset of different kinds of autoimmune diseases, in addition to its direct effects on the immune system,. Among these, the most prominent and confirmed are rheumatoid arthritis (RA), systemic lupus erythematosus (SLE), multiple sclerosis (MS), primary biliary cirrhosis (PBC), and hyperthyroidism. Intriguingly, all such studies on human subjects were done by conducting questionnaire surveys. There are also reports describing how the potent effects on autoimmune diseases in a current, regular smoker differ from those of a former smoker or a nonsmoker. Among the different autoimmune diseases, the risk of developing RA is the highest (Costenbader et al. 2004). Smoking affects the onset of RA at different stages, including (1) age—onset at young age in smokers (Richardson et al. 2000); (2) sex—risk higher in men (Costenbader et al. 2004); (3) abundance—2.4 times more risk in younger smokers (Zandieh et al. 2008); and (4) duration of smoking—risk of developing RA in current, former, and even 20 years after smoking cessation (Malekzadeh et al. 2001). Cigarette smoking is mostly related with rheumatoid factor (RF) and anticyclic citrullinated protein (anti-CCP) antibody seropositivity (Kerkar et al. 2005). Thus, smoking is the most established risk factor for developing RA. Another autoimmune disease at a higher risk of occurring due to smoking is Graves' hyperthyroidism. In contrast to RA, woman and current smokers are more prone to hyperthyroidism (Smith 2003). Smoking is also a high risk factor for Graves' ophthalmopathy and

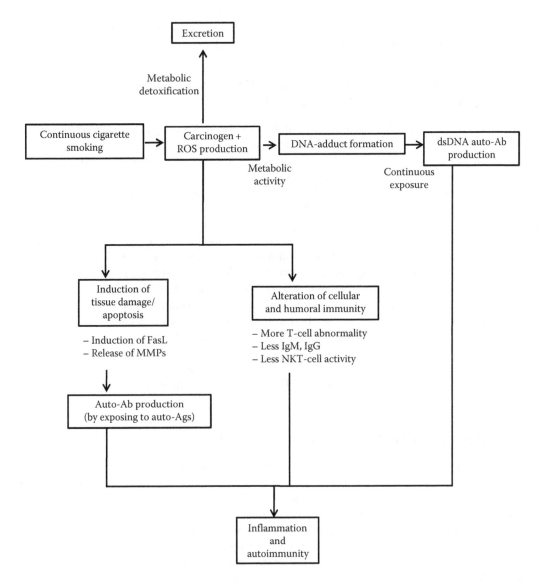

FIGURE 15.1 Effect of smoking on the immune system. The toxic chemicals present in cigarette smoke have an immense effect on the immune system by the production of either ROS or dsDNA autoantibodies (Hecht 1999; Harel-Meir et al. 2007). Here, immune system disorders are only shown as an effect caused by smoking. ROS, Reactive oxygen species, Ab, Antibody; Ag, Antigen; Ig, Immunoglobulin; MMP, Matrix metalloproteinase; NKT cell, Natural killer T cell. (Modified from Bose, T., *Med. Hypotheses,* 84(2), 118, 2014. With permission from Medical Hypotheses.)

the risk declines with smoking cessation (Kleinewietfeld et al. 2013). The reason for this relationship is not properly known, but it is hypothesized to be due to the effects of ROS and autoantibody production (Kerkar et al. 2005).

Another common autoimmune disease occurring due to smoking is SLE. One study has intriguingly made the link between smoking and SLE through inducing double-strand DNA breaks (DSB) and the presence of autoantibodies among smoker subjects (Freemer et al. 2006). In the case of MS, where autoreactive T cells are directed against the components of myelin, cigarette smoking causes worsening of motor functions than in healthy controls (Veldhoen et al. 2008). Nicotine and the free radicals generated by exposure to smoking can cause axonal degeneration or block axonal conduction in already damaged or demyelinated axons (Sollid and Jabri 2013).

15.4 CIGARETTE SMOKE CAUSING AUTOIMMUNE DISORDERS RELATED TO THE METABOLIC SYSTEM

PBC is also an important disease most likely developed due to smoking and should be included in this context (Howel et al. 2000; Parikh-Patel et al. 2001). The exact reason for this association is not known due to lack of cohort studies. There is an established correlation between the severity of liver fibrosis in PBC and continuous exposure to smoking. Interestingly, the number of years exposed to smoking is positively correlated with the adverse histological signs (Zein et al. 2006). It is well documented that smoking is related to the predisposition of this autoimmune disease. It still needs to be investigated whether smoking can pose a cause to a more active and severe form of this disease (Corpechot et al. 2012). There are also studies establishing association between smoking and inflammatory bowel disease (IBD) (Valavanidis et al. 2009) and Crohn's disease (Breuer-Katschinski et al. 1996). Although there are systemic, cellular, and humoral responses in IBD due to smoking, the exact mechanism is not properly investigated. Interestingly, there is a negative relationship between disease prognosis and Crohn's disease, but cigarette smoking can ameliorate the disease features of ulcerative colitis (UC) in some cases. This led to a clinical trial of topical nicotine for improvement in UC patients, but the trial was limited in the actual benefit to the patients (Rubin and Hanauer 2000).

Recent studies have linked a number of other diseases to smoking, including autoimmune hepatitis (Bose 2014). In conclusion, PBC, AIH, and IBD types of autoimmune diseases in the alimentary canal are caused by smoking.

15.5 TREATMENT OF AUTOIMMUNE HEPATITIS

The treatment regime of autoimmune hepatitis (AIH) is carried out employing the same medicines as used for other autoimmune disorders. These are immunosuppressants to suppress the action of overactive immune organs and steroids to reduce the opportunistic infections of different types of pathogens. There are two phases in this treatment: induction of disease remission and maintenance of remission. The target of remission therapy should be normalization of liver enzymes and, ideally, obtaining histological evidence of cessation of active liver inflammation.

Classically, AIH drugs consist of low-dose prednisolone (20–30 mg/kg) with 1 mg/kg azathioprine. The success rate of these drugs is ~80% (Jothimani et al. 2011). Among the rest, 9% are nonresponders and the others are intolerable to the particular treatment (Manns and Vogel 2006). A patient's intolerance to such treatments should be reconsidered, as with standard therapy, relapse occurs in ~90% patients following drug withdrawal (Hegarty et al. 1983), and these patients progress to liver cirrhosis (40%), portal hypertension (25%), and liver failure and death (15%) (Bluth et al. 2009). There are already some alternatives to this standard treatment that have been tested for other autoimmune diseases.

15.5.1 ALTERNATIVE TREATMENT

First, to substitute azathioprine, metabolic mycophenolic acid (MMF) can be used, which helps to ameliorate IBD and RA. MMF was also successfully tested for AIH (Richardson et al. 2000). Due to the 15-fold higher affinity of budesonide to glucocorticoid receptors, it can be a second choice of prednisolone for induction and remission for the patients without the induction of cirrhosis (Hamedani et al. 1997). The use of this type of steroids has been tested and found to have desirable results with a response rate of ~80% in nonresponders to the standard ones (Csepregi et al. 2006;

TABLE 15.1

Alternative Medications for AIH

Immunosuppressant	Azathiorpine, MMF
Antibiotic	Tacrolimus (calcineurin inhibitor), cyclosporin A
Glucocorticoid	Prednisolone, budesonide, ursodiol
CD20 B lymphocyte	Rituximab

Source: Modified from Bose, T., *Med. Hypotheses*, 84(2), 118, 2014. With permission from Medical Hypotheses.

Note: This table lists the number of clinical alternatives to the standard treatment of AIH, other than azathioprine and prednisolone. These alternative treatments can be medicated to patients where conventional medicines do not produce effective improvement in liver conditions.

Zandieh et al. 2008). Another interesting treatment strategy consists of another important immunosuppressant, cyclosporine A. The remission rate in this case was quicker in comparison with the corticosteroids (Malekzadeh et al. 2001), although the long-term side effects are severe, affecting several organs, such as kidney, gum, and blood vessels. The calcineurin inhibitor, tacrolimus, has an emerging role in treating AIH in combination with steroids (Heneghan and McFarlane 2002). Other potential candidates for alternative therapy are rituximab, which induces the apoptosis of B lymphocytes, thereby reducing IgG production (Smith 2003), and rapamycin, which reduces T-cell activation by the inhibition of IL-12 signal transduction (Kerkar et al. 2005). These treatment options are presented in Table 15.1.

15.6 CONCLUSION

Cigarette smoking is a well-known, dangerous cause for a plethora of diseases from mouth, lung, to the alimentary canal. It has direct effects on diseases such as chronic obstructive pulmonary disorders and different forms of cancer. Smoking is also responsible for a variety of autoimmune disorders (e.g., RA and MS). In this chapter, we have concisely discussed the effects of smoking on the autoimmune diseases related to the alimentary canal. Although there are very few studies conducted on the particular topic, they will prove useful for future research. In addition, we discussed the conventional and alternative treatment methods for these kinds of autoimmune disorders. In conclusion, smoking poses a greater risk for autoimmune diseases of the alimentary canal along with a handful of other diseases. Proper investigation in this direction will help create more awareness among the smoker community about the consequences of their actions.

ACKNOWLEDGMENTS

I thank my parents and sister for their continuous support and effort in helping me to achieve my dreams. Also, I acknowledge all my friends who have been helping me to spur my enthusiasm in science and give me the energy to do hard work.

REFERENCES

Bluth, M. H., S. Kohlhoff, K. B. Norowitz, J. I. Silverberg, S. Chice, M. Nowakowski, H. G. Durkin, and T. A. Smith-Norowitz. 2009. Immune responses in autoimmune hepatitis: Effect of prednisone and azathioprine treatment: Case report. *Int J Med Sci* 6(4):177–183.

Bose, T. 2014. Bitter correlationship between autoimmune hepatitis and smoking. *Med Hypotheses* 84(2):118–121.

Breuer-Katschinski, B. D., N. Hollander, and H. Goebell. 1996. Effect of cigarette smoking on the course of Crohn's disease. *Eur J Gastroenterol Hepatol* 8(3):225–228.

Corpechot, C., F. Gaouar, Y. Chretien, C. Johanet, O. Chazouilleres, and R. Poupon. 2012. Smoking as an independent risk factor of liver fibrosis in primary biliary cirrhosis. *J Hepatol* 56(1):218–224.

Costenbader, K. H., D. J. Kim, J. Peerzada, S. Lockman, D. Nobles-Knight, M. Petri, and E. W. Karlson. 2004. Cigarette smoking and the risk of systemic lupus erythematosus: A meta-analysis. *Arthritis Rheum* 50(3):849–857.

Csepregi, A., C. Rocken, G. Treiber, and P. Malfertheiner. 2006. Budesonide induces complete remission in autoimmune hepatitis. *World J Gastroenterol* 12(9):1362–1366.

Demedts, I. K., T. Demoor, K. R. Bracke, G. F. Joos, and G. G. Brusselle. 2006. Role of apoptosis in the pathogenesis of COPD and pulmonary emphysema. *Respir Res* 7:53.

Ezzati, M. and A. D. Lopez. 2004. Regional, disease specific patterns of smoking-attributable mortality in 2000. *Tob Control* 13(4):388–395.

Fowles, J. and E. Dybing. 2003. Application of toxicological risk assessment principles to the chemical constituents of cigarette smoke. *Tob Control* 12(4):424–430.

Frankel, L. K. 2011. The relation of life insurance to public hygiene. 1910. *Am J Public Health* 101(10):1868–1869.

Freemer, M. M., T. E. King, Jr., and L. A. Criswell. 2006. Association of smoking with dsDNA autoantibody production in systemic lupus erythematosus. *Ann Rheum Dis* 65(5):581–584.

Hamedani, R., R. D. Feldman, and B. G. Feagan. 1997. Review article: Drug development in inflammatory bowel disease: Budesonide—A model of targeted therapy. *Aliment Pharmacol Ther* 11(Suppl. 3): 98–107; discussion 107–108.

Harel-Meir, M., Y. Sherer, and Y. Shoenfeld. 2007. Tobacco smoking and autoimmune rheumatic diseases. *Nat Clin Pract Rheumatol* 3(12):707–715.

Hecht, S. S. 1999. Tobacco smoke carcinogens and lung cancer. *J Natl Cancer Inst* 91(14):1194–1210.

Hegarty, J. E., K. T. Nouri Aria, B. Portmann, A. L. Eddleston, and R. Williams. 1983. Relapse following treatment withdrawal in patients with autoimmune chronic active hepatitis. *Hepatology* 3(5):685–689.

Heneghan, M. A. and I. G. McFarlane. 2002. Current and novel immunosuppressive therapy for autoimmune hepatitis. *Hepatology* 35(1):7–13.

Howel, D., C. M. Fischbacher, R. S. Bhopal, J. Gray, J. V. Metcalf, and O. F. James. 2000. An exploratory population-based case-control study of primary biliary cirrhosis. *Hepatology* 31(5):1055–1060.

Jha, P., M. K. Ranson, S. N. Nguyen, and D. Yach. 2002. Estimates of global and regional smoking prevalence in 1995, by age and sex. *Am J Public Health* 92(6):1002–1006.

Jothimani, D., M. E. Cramp, J. D. Mitchell, and T. J. Cross. 2011. Treatment of autoimmune hepatitis: A review of current and evolving therapies. *J Gastroenterol Hepatol* 26(4):619–627.

Kerkar, N., C. Dugan, C. Rumbo, R. A. Morotti, G. Gondolesi, B. L. Shneider, and S. Emre. 2005. Rapamycin successfully treats post-transplant autoimmune hepatitis. *Am J Transplant* 5(5):1085–1089.

Kleinewietfeld, M., A. Manzel, J. Titze, H. Kvakan, N. Yosef, R. A. Linker, D. N. Muller, and D. A. Hafler. 2013. Sodium chloride drives autoimmune disease by the induction of pathogenic TH17 cells. *Nature* 496(7446):518–522.

Korpilahde, T., M. Heliovaara, P. Knekt, J. Marniemi, A. Aromaa, and K. Aho. 2004. Smoking history and serum cotinine and thiocyanate concentrations as determinants of rheumatoid factor in non-rheumatoid subjects. *Rheumatology (Oxford)* 43(11):1424–1428.

Malekzadeh, R., S. Nasseri-Moghaddam, M. J. Kaviani, H. Taheri, N. Kamalian, and M. Sotoudeh. 2001. Cyclosporin A is a promising alternative to corticosteroids in autoimmune hepatitis. *Dig Dis Sci* 46(6):1321–1327.

Manns, M. P. and A. Vogel. 2006. Autoimmune hepatitis, from mechanisms to therapy. *Hepatology* 43(2 Suppl. 1):S132–S144.

Morris, A., M. Netravali, H. M. Kling, T. Shipley, T. Ross, F. C. Sciurba, and K. A. Norris. 2008. Relationship of pneumocystis antibody response to severity of chronic obstructive pulmonary disease. *Clin Infect Dis* 47(7):e64–68.

Parikh-Patel, A., E. B. Gold, H. Worman, K. E. Krivy, and M. E. Gershwin. 2001. Risk factors for primary biliary cirrhosis in a cohort of patients from the united states. *Hepatology* 33(1):16–21.

Richardson, P. D., P. D. James, and S. D. Ryder. 2000. Mycophenolate mofetil for maintenance of remission in autoimmune hepatitis in patients resistant to or intolerant of azathioprine. *J Hepatol* 33(3):371–375.

Rubin, D. T. and S. B. Hanauer. 2000. Smoking and inflammatory bowel disease. *Eur J Gastroenterol Hepatol* 12(8):855–862.

Smith, C. J. and C. Hansch. 2000. The relative toxicity of compounds in mainstream cigarette smoke condensate. *Food Chem Toxicol* 38(7):637–646.

Smith, M. R. 2003. Rituximab (monoclonal anti-CD20 antibody): Mechanisms of action and resistance. *Oncogene* 22(47):7359–7368.

Sollid, L. M. and B. Jabri. 2013. Triggers and drivers of autoimmunity: Lessons from coeliac disease. *Nat Rev Immunol* 13(4):294–302.

U.S. Department of Health and Human Services. 2014. The Health Consequences of Smoking—50 Years of Progress: A Report of the Surgeon General. Atlanta: U.S. Department of Health and Human Services, Centers for Disease Control and Prevention, National Center for Chronic Disease Prevention and Health Promotion, Office on Smoking and Health. http://www.cdc.gov/tobacco/data_statistics/sgr/50th-anniversary/index.htm (accessed December 11, 2015).

Valavanidis, A., T. Vlachogianni, and K. Fiotakis. 2009. Tobacco smoke: Involvement of reactive oxygen species and stable free radicals in mechanisms of oxidative damage, carcinogenesis and synergistic effects with other respirable particles. *Int J Environ Res Public Health* 6(2):445–462.

Veldhoen, M., K. Hirota, A. M. Westendorf, J. Buer, L. Dumoutier, J. C. Renauld, and B. Stockinger. 2008. The aryl hydrocarbon receptor links TH17-cell-mediated autoimmunity to environmental toxins. *Nature* 453(7191):106–109.

Zandieh, I., D. Krygier, V. Wong, J. Howard, L. Worobetz, G. Minuk, H. Witt-Sullivan, and E. M. Yoshida. 2008. The use of budesonide in the treatment of autoimmune hepatitis in Canada. *Can J Gastroenterol* 22(4):388–392.

Zavitz, C. C., G. J. Gaschler, C. S. Robbins, F. M. Botelho, P. G. Cox, and M. R. Stampfli. 2008. Impact of cigarette smoke on T and B cell responsiveness. *Cell Immunol* 253(1–2):38–44.

Zein, C. O., K. Beatty, A. B. Post, L. Logan, S. Debanne, and A. J. McCullough. 2006. Smoking and increased severity of hepatic fibrosis in primary biliary cirrhosis: A cross validated retrospective assessment. *Hepatology* 44(6):1564–1571.

Section V

Cancers Induced by ROS

16 Inflammation and Lung Cancer
Oxidative Stress, ROS, and DNA Damage

*Mónica Gomes, Ana Luísa Teixeira, Ana Coelho,
António Araújo, and Rui Medeiros*

CONTENTS

ABSTRACT

Cancer is the leading cause of death in the world, accounting for more than 25% of all deaths in developed countries, and lung cancer is the most common cause of cancer-related death in the world. The predominant risk factor for this cancer is smoking, accounting for approximately 90% of these lung cancer deaths. Furthermore, lung cancer risk is associated with several indicators of inflammation. The inflammation process is a complex response to stimuli involving the interplay of host cells and signaling molecules, such as angiogenesis factors and chemokines. Inhalation of air pollutants and microorganisms results in lung injury and generation of reactive oxygen species/reactive nitrogen species (ROS/RNS), leading to a cascade of signaling events that trigger the production of proinflammatory cytokines. Inflammation is the primary reaction of a tissue to eliminate pathogenic insult and injured tissue components in order to restore normal physiological functions or replace the irreparable tissue with scar tissue. Cancer and inflammation are closely linked, and many inflammatory conditions increase the risk of cancer development. Matrix metalloproteins are clearly important effectors in inflammation both in physiological situations, such as tissue repair, and in pathological inflammatory conditions and cancer. A better understanding of the role of ROS/RNS in lung inflammation and cancer is probable to inspire new strategies for lung cancer prevention and treatment.

16.1 INTRODUCTION

Cancer is one of the leading causes of death in the world, accounting for more than 25% of all deaths in developed countries, and lung cancer holds the top position in cancer morbidity and mortality among men worldwide (Ferlay et al. 2010). Lung cancer was found to be the most commonly diagnosed cancer and the primary cause of cancer-related mortality for males worldwide and the second leading cause of cancer-related deaths for women (Jemal et al. 2011; Siegel et al. 2012). For the year 2012, it was estimated that lung cancer would account for 26% of all female cancer deaths and 29%

of all male cancer deaths (Siegel et al. 2012). The mortality in this neoplasia, during the first year after diagnosis, is very high, which is about 67% (Grigoryeva et al. 2015). Success of surgical treatment is closely related to the opportunity for early diagnosis of lung cancer. The importance of early diagnosis is confirmed by the 5-year survival rate after radical surgery, so people diagnosed with stages I and II of lung cancer tend to have higher (5-year) survival rate (63.5% and 43.5%, respectively) than people diagnosed with stage III (22.9%) (Grigoryeva et al. 2015).

Lung cancer is rare among young adults, with the average age of occurrence and diagnosis being over 60 years. There are two major classes of lung cancer: primary lung cancer and secondary lung cancer (Azad et al. 2008). Primary lung cancer originates in the lung itself and is further classified into two subtypes: small-cell lung cancer (SCLC) and non-small-cell lung cancer (NSCLC), depending on the morphology of the malignant cells; NSCLC accounts for approximately 85% of all cases of lung cancer (Molina et al. 2008; Chen et al. 2014).

NSCLC is currently defined by pathological characteristics, and the two predominant NSCLC histological phenotypes are adenocarcinoma (ADC ≈ 50%) and squamous cell carcinoma (SCC ≈ 40%) (Chen et al. 2014). Secondary lung cancer is initiated in other organs such as breast or colon and then spreads to the lungs (Azad et al. 2008).

Recently, molecular subtyping of NSCLC has led to the approval of and use of targeted therapies in the frontline setting. Patients with activating mutations in the epidermal growth factor receptor (EGFR) domain and echinoderm microtubule–associated protein-like 4–anaplastic lymphoma kinase (EML4–ALK) translocation benefit from firstline treatment with erlotinib or crizotinib, respectively. These mutations are seen in a relatively small subset of NSCLC patients. They are common in patients with ADC, never smokers, and patients of East Asian origin. Kirsten rat sarcoma viral oncogene homolog (KRAS) is the most common mutation found in NSCLC; however, an effective targeted therapy for this subset of NSCLC does not exist. Despite the addition of new therapies, the median 5-year overall survival of patients in advanced staged disease remains a dismal 1%–2% (Aggarwal 2014).

Several researches indicate that long-term exposure to inhaled carcinogens has the greatest impact on risk of lung cancer. The predominant risk factor for lung cancer is smoking, accounting for approximately 90% of these lung cancer deaths (Azad et al. 2008; Shiels et al. 2013). Mostly, polyaromatic hydrocarbons (PAHs) and nitrosamine 4-(methylnitrosamino)-1-(3-pyridyl)-1-butanone (NNK) are the major components of tobacco that are associated with the etiology of lung cancer (Azad et al. 2008). The lung is vulnerable to a wide range of toxicants and infectious agents with the potential to induce oxidative damage (Azad et al. 2008). An average adult inhales about 10,000 L of air per day, polluted with cigarrete smoke, automobile exhaust, diesel soot, ozone (O_3), sulfur dioxide (SO_2), nitrogen dioxide (NO_2), and varying degrees of other pollutants. Inhalation of such toxic air pollutants and microorganisms results in lung injury and generation of reactive oxygen species/reactive nitrogen species (ROS/RNS), leading to cascades of signaling events that trigger the production of proinflammatory cytokines and chemokines (Azad et al. 2008). Additionally, lung cancer risk is associated with several indicators of inflammation, including pulmonary fibrosis, chronic obstructive pulmonary disease (COPD), and chronic pulmonary infections, even after taking the effects of smoking into consideration (Shiels et al. 2013). Overall, the pathogenesis of lung cancer involves multiple molecular abnormalities accrued over a long period. Although a large number of genetic pathways associated with lung cancer are being discovered, the basic molecular mechanisms involved in lung cancer are still unclear (Azad et al. 2008).

16.2 INFLAMMATION

Inflammation is a physiologic process in response to tissue damage resulting from microbial pathogen infection, chemical irritation, and/or wounding. At the very early stage of inflammation, neutrophils are the first cells to migrate to the inflammatory sites under the regulation of molecules, produced by rapidly responding macrophages and mast cells prestationed in tissues (Lu et al. 2006).

As the inflammation progresses, various types of leukocytes, lymphocytes, and other inflammatory cells are activated and attracted to the inflamed site by a signaling network involving a great number of growth factors, cytokines, and chemokines (Lu et al. 2006).

The most commonly recognized features of cancer-associated inflammation are also those expressed by innate immune system, normally activated in response to stress or infection. The observed chronic inflammation milieu in notable subsets of human cancers is proposed to support tumor growth, plasticity, and resistance to therapy (Grimm et al. 2013). Unluckily, dysregulated persistent inflammation contributes to the chronic phase of many diseases, including maintenance of many cancers. It is accepted that inflammation drives the development of some cancers that adapt to thrive in the oxidant-rich microenvironment as described initially by *co-opting* expression of inflammation mediators (Coussens and Werb 2002; de Visser et al. 2006).

The dynamic role of chronic inflammation in cancer is not novel. In 1863, Rudolf Virchow hypothesized that some irritants associated with tissue injury and resulting cellular inflammation may play a role in cell proliferation and neoplastic development (Balkwill and Mantovani 2001). Based on his observations that normal cellular responses might lead to cancer, he postulated that cancer may develop at sites of chronic inflammation (Balkwill and Mantovani 2001).

Over recent years, several research in this area clearly demonstrated that cell proliferation alone does not produce cancer. However, unlimited proliferation potential of cells is achieved in an environment that is rich in inflammatory cells, producing abundant ROS and RNS promoting unremitting DNA damage, inactivation of apoptosis, upregulation of growth factors and cytokines, and activation of growth-supporting genes (Azad et al. 2008). In recent years, increased understanding of the basic mechanisms involved in inflammation and its physiological systems supported Virchow's hypothesis, establishing an important relationship between cancer and inflammation (Azad et al. 2008).

Inflammation is the primary reaction of a tissue to eliminate pathogenic insult and injured tissue components in order to restore normal physiological functions or replace the irreparable tissue with scar tissue (Azad et al. 2008). The inflammation process is a complex response to stimuli involving the interplay between host cells and signaling molecules, such as proinflammatory and anti-inflammatory cytokines, growth and angiogenesis factors, and chemokines (Shiels et al. 2013).

Inflammation is classified as acute or chronic inflammation, depending on a variety of factors including clinical symptoms and the nature of injury. Acute inflammation is the immediate response, usually of short duration, and results in the release of polymorphonuclear leukocytes (PMNs) so as to eliminate the pathogenic or cytotoxic insult. On the other hand, chronic inflammation is characterized by persistent inflammation, tissue injury, and tissue repair, occurring simultaneously (Azad et al. 2008).

Chronic inflammation induced by various agents including viruses and bacteria is associated with an increased cancer risk due to tissue damage and genetic instability (Linhart et al. 2014). Thereby, repeated tissue damage and regeneration produce increased ROS/RNS from inflammatory cells and then interact with DNA in proliferating epithelium, resulting in permanent genomic alteration such as point mutations, deletions, or rearrangements (Coussens and Werb 2002). Cells respond to DNA damage by activating p53-controlled genes associated with cell cycle and DNA repair, and when the rate of ROS/RNS-mediated DNA damage is extensive, it leads to chronic inflammation (Azad et al. 2008). Chronic inflammation provides a microenvironment rich in inflammatory cells, ROS/RNS, recurring DNA damage, cell-proliferating growth factors, and other growth-supporting stimuli, which increases the frequency of mutations. In pulmonary pathologies such as COPD, fibrosis, and lung carcinogenesis, inflammation is considered as a major precursor or the *hallmark* for cancer development (Azad et al. 2008; Hanahan and Weinberg 2011).

By upregulating key inflammatory molecules, including inducible nitric oxide synthetase (iNOS), cyclooxygenase (COX2), and proinflammatory cytokines and chemokines, tumor cells invoke a chronic inflammatory state that also induces tumor-supporting myeloid cells such as tumor-associated macrophages (TAMs) and myeloid-derived suppressor cells (MDSCs) and drives their infiltration of the tumor microenvironment (TME) (Grimm et al. 2013). While many host cell types including T cells are involved in creating an inflammatory pro-TME, inflammation-directed recruitment of MDSC and macrophage polarization are also important (Grimm et al. 2013).

Actually, tumors are considered as complex tissues with a dynamic and reactive TME. The TME is populated by different nonplastic cells (inflammatory leukocytes, activated fibroblast, and endothelial cells) that actively communicate with cancer cells via chemokines or cytokines (D'Incalci et al. 2014). Recently, it is established that the persistence of inflammatory pathways in the TME is linked with tumor promotion (Diakos et al. 2014; D'Incalci et al. 2014). Among stromal cells, TAMs derived from blood circulating monocytes can functionally be *educated* by tumor cells, through the activity of different cell types of the TME. These cells can promote basically all phases of tumorigenesis and tumor progression, including tumor cell proliferation, invasion, angiogenesis, metastasis formation, and immune suppression (Allavena and Mantovani 2012; Reinartz et al. 2014). Indeed, TAMs play a major role in the production of growth factors (epidermal growth factor, chemokines, interleukins, metalloproteinase, and vascular endothelial growth factor), which promote tumor cell survival and metastatic phenotype (Galmarini et al. 2014).

16.2.1 Inflammation, Oxidative Stress, and DNA Damage

ROS additionally are involved in regulating certain normal cellular processes. When excessive ROS stimulation occurs, it may trigger DNA repair responses in normal cells to remove ROS-mediated DNA damage. For highly active metabolism, cancer cells commonly have higher levels of ROS than normal cells, leading to carcinogenesis by oxidative DNA damage and DNA repair impairment. This nature of high ROS level in cancer cells also provides an opportunity for drug therapy to generate overloading ROS level and induce oxidative stress–induced cell death (Farooqi et al. 2014).

Oxidative stress is an important mechanism in the pathogenesis of many diseases including cancer. The generation of ROS with a consecutive DNA damage is an initial step in carcinogenesis induced by inflammatory processes (Aggarwal 2014). Chronic inflammation induced by various agents including viruses and bacteria is associated with an increased cancer risk due to tissue damage and genetic instability. Oxidative stress with the generation of ROS may occur in chronic infection and inflammation primarily due to the generation of nitric oxide (NO), superoxide anion ($O_2^{-\bullet}$), and other ROS by macrophages and neutrophils that infiltrate the inflamed tissue (Aggarwal 2014). Activated inflammatory cells in various tissues including the liver in turn induce oxidant-generating enzymes such as NADPH oxidase, iNOS, xanthine oxidase (XO), and myeloperoxidase (MPO). In such conditions, ROS and RNS are generated. As a consequence, ROS and RNS can damage DNA, RNA, lipids, and proteins through nitration and oxidation, resulting in an increased mutation load (Aggarwal 2014). Furthermore, cytokines are released in inflammatory tissues, which activate not only the aforementioned enzymes to create ROS and RNS but also NF-κB, a nuclear transcription factor, which among others stimulates COX2, lipoxygenase (LOX), and iNOS, and upregulating these molecules (COX2, LOX, and iNOS) results in an overproduction of ROS and RNS (Aggarwal 2014).

Another process that can influence tumor progression by ROS formation is autophagy (Farooqi et al. 2014). This is a multistep process that maintains cellular homeostasis via the degradation and recycling of long-lived proteins, intracellular aggregates, and damaged organelles (Coussens and Werb 2002). The autophagy may be induced for survival and induction of apoptotic pathways in response to cellular oxidative stress (Farooqi et al. 2014). Recent studies have described a complex role of the autophagy pathway during tumor initiation. On one hand, autophagy protects against the production of ROS in the cells and therefore inhibits their deleterious effect on DNA damage and resulting mutation, which have been extensively described to induce tumorigenesis, defined as the transformation of a normal cell into a cancer cell (de Visser et al. 2006). Autophagy is also described as a tumor suppressor mechanism mainly by preventing ROS accumulation through the elimination of damaged mitochondria that are known to be the major source of ROS production (Coussens and Werb 2002). During in vivo tumor formation, autophagy has been shown to play a major role for the cancer cells to survive under hypoxic stress before the vascularization of the tumor occurs (Coussens and Werb 2002). However, the mechanism is still unclear although several studies suggest a role of autophagy in the regulation of cancer cell metabolism allowing them to meet requirements for rapid proliferation (Coussens and Werb 2002).

High levels of autophagy are indeed observed in hypoxic regions of tumors, and autophagy has been described to be activated by hypoxia and ischemia (glucose deprivation and hypoxia) to promote the survival of cancer cells. Hypoxia induces ROS production leading to the stabilization of hypoxia-inducible factor 1-α (HIF-1α). This factor, a key regulator of oxygen homeostasis, induces mitophagy through the expression of Bcl-2/adenovirus E1B 19-kDa-interacting protein 3 (BNIP3), allowing the cells to survive during prolonged hypoxia by preventing increased levels of ROS production (Balkwill and Mantovani 2001). On the other hand, tumor progression and aggressiveness are characterized by metastasis, epithelial–mesenchymal transition (EMT), and angiogenesis.

Metastasis is a multistep process that allows cancer cells to migrate to distant organ sites (Linhart et al. 2014). EMT is the first step of metastasis and is characterized by the loss of epithelial properties and the acquirement of mesenchymal properties leading to increased cell mobility. Several studies have described a prometastasis role of autophagy. For example, inhibition of autophagy by FIP200 deletion leads to a decrease in metastatic potential associated with an accumulation of damaged mitochondria, which could lead to increased level of ROS. Moreover, increased autophagy in human cancer is associated with metastasis and poor prognosis in patients with melanoma and breast cancer (Coussens and Werb 2002).

16.2.2 ROLE OF INFLAMMATION IN LUNG CANCER

Inflammation is recognized both as a condition that can lead to cancer development and as a condition that can arise due to oncogenic changes in cancer cells (Jafri et al. 2013). The hallmarks of cancer are distinctive and complementary capabilities that enable tumor growth and metastatic dissemination sustaining proliferative signaling, evading growth suppressors, resisting cell death, enabling replicative immortality, inducing angiogenesis, and activating invasion and metastasis (Hanahan and Weinberg 2011). Inflammation has been described as the underlying or enabling characteristic that promotes these hallmarks of cancer (Hanahan and Weinberg 2011).

Cancer and inflammation are closely linked, and many inflammatory conditions increase the risk of cancer development (Jafri et al. 2013). In the NSCLC microenvironment, there is a complex interaction between immune cells and tumor cells as well as other stromal cell types and tissue components. The distribution of these cells and the expression of different inflammatory molecules throughout the TME are to some extents related to tumor progression and survival (Gomes et al. 2014).

The concept of tumor heterogeneity applies not only to tumor epithelial cells but also to the diverse microenvironment with the tumor cells' interaction. Carcinoma cells, in the lung (and others), are closely associated with the extracellular matrix (ECM) and mesenchymal cells such as fibroblasts, infiltrating immune cells, and vasculature (Chen et al. 2014).

Lung tumors develop through a complex process involving many stages such as initiation, promotion, and progression (Hanahan and Weinberg 2011). In lung tumorigenesis, the genesis of new blood and lymphatic vessels supplies necessary nutrients for tumor growth and allows for an influx of immune cells of the myeloid and lymphoid lineages (Chen et al. 2014). In the lung, depending on the type of inflammation, there may be direct effects such as DNA damage, mutation, or an indirect effect or induced effect, induced by activated enzymes such as cytochrome P-450 oxidase or flavin monoxides that produce ROS in the cells, resulting in protein and DNA damage (Azad et al. 2008). The second stage in cancer development is the promotion, which involves clonal expansion of the initiated cells. These initiated cells may undergo promotion under persistent oxidative stress conditions, forming focal lesions from which invasive cancers may originate. Progression is the final stage involving the formation of fully malignant cells from an early neoplastic clone via both genetic and epigenetic mechanisms (Azad et al. 2008). Tumor cells undergo autonomous uncontrolled proliferation with the aid of suitable promoting factors such as EGFR. This factor, a transmembrane receptor with intrinsic tyrosine kinase activity, triggers many transcription factors and is activated in lung tumors (Azad et al. 2008). A very important fact of cancer initiation and progression is genome instability. It was reported that the lack of mismatch in DNA at certain nucleotide level may lead to microsatellite instability in many forms of lung cancers (Massion and Carbone 2003). Despite this endogenous sources of inflammation-induced oxidative stress, exogenous sources such as hyperoxia, radiation, exposure to particulates, and chemical carcinogens are also critical in lung carcinogenesis (Azad et al. 2008). ROS/RNS produced by inflammatory cells also stimulate oncogenes such as c-Jun and c-Fos, and the overexpression of c-Jun was reported to be associated with lung cancer (Azad et al. 2008).

In lung cancers associated with nondestructive agents, such as asbestos and silica, the chronic inflammation in the lung is persistent because of the inability of the immune system to remove these substances. Many of these agents are reported to modulate and activate various transcription factors, producing changes in cell proliferation, differentiation, apoptosis, and inflammation (Azad et al. 2008). Such inflammatory responses increase the incidence of epithelial cancers, including mesothelioma and lung cancer. Cigarrete smoke is a complex preneoplastic agent that may act, in part, by inducing a chronic inflammatory condition by delivering an array of genotoxic carcinogens such as nitrosamines, peroxides, and many potent oxidants into the lungs (Azad et al. 2008). Therefore, inflammatory cells influence the whole organ in tumor development, regulating the growth, migration, and differentiation of all cell types, including neoplastic cells, fibroblasts, and endothelial cells (Azad et al. 2008). Tumors can evade immune surveillance by expressing molecules that maintain tolerance to normal peripheral tissues, including the interaction of the tumor-associated programmed cell death-1 ligand 1 (PDL1) with the immune receptor programmed cell death-1 (PD1, also known as PDCD1). PDL1 is a distal modulator of the immune response whose expression occurs in 40%–50% of NSCLC patients (Guibert et al. 2015). Recently, the use of antibodies targeting the PD1-PDL1 checkpoint has resulted in some marked responses in early-stage clinical trial for a large panel of therapy refractory cancer subtypes, including advanced melanoma, NSCLC, and renal cell cancer, with a proportion of responding patients showing persistent long-term benefit (Chen et al. 2014; Guibert et al. 2015).

16.2.3 MMPs, Lung Cancer, and Inflammation

In inflammation, matrix metalloproteins (MMPs) recruit inflammatory cells during tissue injury, which involves a series of complex morphological changes in cell barrier, cell–cell interaction, and cell–matrix interaction. MMPs also exhibit a wide functional diversity in modulating NSCLC due to their interaction with growth factor receptors, cytokines, chemokines, cell adhesion molecules, apoptotic ligands, and angiogenic factors. MMPs are involved at all junctures of inflammation as well as tumor progression, including proliferation, adhesion, migration, angiogenesis, senescence, apoptosis, cytokine and chemokine bioactivity, and evasion of the immune system (Lopez-Otin and Bond 2008). In lung cancer, the expression of a number of MMPs and their inhibitors is exaggerated and may be causally linked to enhanced tumor progression and metastasis (Sorokin 2010).

MMPs were primarily thought solely to be involved in homeostasis and turnover of the ECM, but recent observations provide evidence suggesting that MMPs act on cytokines, chemokines, and protein mediators to regulate various aspects of inflammation and immunity (Parks et al. 2004). Cancer-associated EMT is known to contribute to tumor progression, increased invasiveness and metastasis, resistance to therapies, and generation of cell populations with stem cell–like characteristics and has been implicated in progression and metastasis of cancer specifically.

EMT is characterized by the loss of cell–cell junctions, polarity, and epithelial markers, and in turn, acquisition of mesenchymal features and motility. Changes associated with this developmental process have been extensively implicated in cancer progression and metastasis. MMP-3 induces EMT associated with malignant transformation via a pathway dependent upon the production of ROS. While the process by which exposure to MMP-3 leads to induction of ROS has been extensively studied, exactly how the MMP-3-induced ROS stimulate EMT remains unknown (Cichon and Radisky 2014).

MMPs have been speculated to play a critical role in various inflammatory diseases, such as acute lung injury, COPD, and cancer. They can regulate the integrity of physical barriers and transmigration of leukocytes from vasculature to tissue. They also regulate the availability and activity of inflammatory mediators, such as cytokines and chemokines. MMPs also generate chemokine gradients in tissue to recruit inflammatory cells to the site of injury or inflammation and can also regulate the survival of inflammatory cells (Parks et al. 2004; Nissinen and Kahari 2014). Immune system plays an important role in cancer cell surveillance by recognizing and attacking cancer cells *in vivo*. However, cancer cells can escape the immune attack in various ways. On the other hand, chronic inflammation is associated with the progression of several types of cancer (Nissinen and Kahari 2014). Inflammation is necessary to promote cancer initiation and progression via vascularization and remodeling of TME, which are important for tumor cell survival. MMPs may also exert immune regulatory function in TME, and they may also help cancer cells escape immunosurveillance (Nissinen and Kahari 2014). Recent findings indicate that MMPs play an important role in the regulation of cytokine and chemokine release and their activation, which are key steps in the immune response (Sorokin 2010). For example, MMP-1, -2, -3, -7, -9, and -12 are able to process pro–tumor necrosis factor-α (TNF-α) into soluble active TNF-α. MMP-2, -3, and -9 also have the ability to cleave IL-1β, generating a more active form. MMP-9 controls the IL-12-dependent proliferation of T lymphocytes. MMP-8, -13, and -14 can cleave IL-8 to generate truncated forms with increased activity. Therefore, inflammatory cytokines and MMPs are interconnected (Sorokin 2010).

The role of MMPs in cancer progression has been extensively studied in various animal models. In addition to tumor cells, stromal cells play an important role in cancer progression, for example, by producing MMPs (Nissinen and Kahari 2014). The association between MMPs and inflammation in cancer progression has also been emphasized. A good example of this association is that the transplantation of wild-type mouse MMP-9 expression bone marrow cells to MMP-9-deficient mice effectively restores the development of cutaneous SCC (Nissinen and Kahari 2014). Cancer progression is regulated by growth factors, chemokines, and cytokines either directly via their angiogenic or angiostatic activity or indirectly by attracting anti- or precancerous inflammatory cells. As discussed earlier, several studies have revealed the proteolytic activation or inhibition of growth factors, cytokines, and chemokines by various MMPs (Nissinen and Kahari 2014).

MMPs are clearly important effectors in inflammation both in physiological situations, such as tissue repair, and in pathological inflammatory conditions and cancer. The association of MMPs with cancer has obviously suggested them as potential therapeutic target (Nissinen and Kahari 2014).

16.3 CONCLUSION

Inflammation can affect every hallmark of tumor development and prognosis as well as the response to therapy. In the NSCLC microenvironment, there is a complex interaction between immune cells and tumor cells as well as other stromal cell types and tissue components. The production of ROS/RNS is critical for normal aerobic metabolism and functioning of several events essential for the organism. Overpowered generation of ROS/RNS is likely to induce chronic inflammatory conditions that may lead to several deleterious effects in the cells. The increased levels of ROS are extensively involved in the mechanisms of chronic lung inflammation and thus contribute to the development of lung cancer. A better understanding of the role of ROS/RNS in lung inflammation and cancer is probable to inspire new strategies for lung cancer prevention and treatment.

REFERENCES

Aggarwal, C. 2014. Targeted therapy for lung cancer: Present and future. *Ann Palliat Med* 3 (3):229–235.

Allavena, P. and A. Mantovani. 2012. Immunology in the clinic review series; focus on cancer: Tumour-associated macrophages: Undisputed stars of the inflammatory tumour microenvironment. *Clin Exp Immunol* 167 (2):195–205.

Azad, N., Y. Rojanasakul, and V. Vallyathan. 2008. Inflammation and lung cancer: Roles of reactive oxygen/nitrogen species. *J Toxicol Environ Health B Crit Rev* 11 (1):1–15.

Balkwill, F. and A. Mantovani. 2001. Inflammation and cancer: Back to Virchow? *Lancet* 357 (9255):539–545.

Chen, Z., C. M. Fillmore, P. S. Hammerman, C. F. Kim, and K. K. Wong. 2014. Non-small-cell lung cancers: A heterogeneous set of diseases. *Nat Rev Cancer* 14 (8):535–546.

Cichon, M. A. and D. C. Radisky. 2014. ROS-induced epithelial-mesenchymal transition in mammary epithelial cells is mediated by NF-kB-dependent activation of Snail. *Oncotarget* 5 (9):2827–2838.

Coussens, L. M. and Z. Werb. 2002. Inflammation and cancer. *Nature* 420 (6917):860–867.

de Visser, K. E., A. Eichten, and L. M. Coussens. 2006. Paradoxical roles of the immune system during cancer development. *Nat Rev Cancer* 6 (1):24–37.

Diakos, C. I., K. A. Charles, D. C. McMillan, and S. J. Clarke. 2014. Cancer-related inflammation and treatment effectiveness. *Lancet Oncol* 15 (11):e493–e503.

D'Incalci, M., N. Badri, C. M. Galmarini, and P. Allavena. 2014. Trabectedin, a drug acting on both cancer cells and the tumour microenvironment. *Br J Cancer* 111 (4):646–650.

Farooqi, A. A., S. Fayyaz, M. F. Hou, K. T. Li, J. Y. Tang, and H. W. Chang. 2014. Reactive oxygen species and autophagy modulation in non-marine drugs and marine drugs. *Mar Drugs* 12 (11):5408–5424.

Ferlay, J., H. R. Shin, F. Bray, D. Forman, C. Mathers, and D. M. Parkin. 2010. Estimates of worldwide burden of cancer in 2008: GLOBOCAN 2008. *Int J Cancer* 127 (12):2893–2917.

Galmarini, C. M., M. D'Incalci, and P. Allavena. 2014. Trabectedin and plitidepsin: Drugs from the sea that strike the tumor microenvironment. *Mar Drugs* 12 (2):719–733.

Gomes, M., A. L. Teixeira, A. Coelho, A. Araujo, and R. Medeiros. 2014. The role of inflammation in lung cancer. *Adv Exp Med Biol* 816:1–23.

Grigoryeva, E. S., D. A. Kokova, A. N. Gratchev, E. S. Cherdyntsev, M. A. Buldakov, J. G. Kzhyshkowska, and N. V. Cherdyntseva. 2015. Smoking-related DNA adducts as potential diagnostic markers of lung cancer: New perspectives. *Exp Oncol* 37 (1):5–12.

Grimm, E. A., A. G. Sikora, and S. Ekmekcioglu. 2013. Molecular pathways: Inflammation-associated nitric-oxide production as a cancer-supporting redox mechanism and a potential therapeutic target. *Clin Cancer Res* 19 (20):5557–5563.

Guibert, N., M. Delaunay, and J. Mazieres. 2015. Targeting the immune system to treat lung cancer: Rationale and clinical experience. *Ther Adv Respir Dis* 9 (3):105–120.

Hanahan, D. and R. A. Weinberg. 2011. Hallmarks of cancer: The next generation. *Cell* 144 (5):646–674.

Jafri, S. H., R. Shi, and G. Mills. 2013. Advance lung cancer inflammation index (ALI) at diagnosis is a prognostic marker in patients with metastatic non-small cell lung cancer (NSCLC): A retrospective review. *BMC Cancer* 13:158.

Jemal, A., F. Bray, M. M. Center, J. Ferlay, E. Ward, and D. Forman. 2011. Global cancer statistics. *CA Cancer J Clin* 61 (2):69–90.

Linhart, K., H. Bartsch, and H. K. Seitz. 2014. The role of reactive oxygen species (ROS) and cytochrome P-450 2E1 in the generation of carcinogenic etheno-DNA adducts. *Redox Biol* 3:56–62.

Lopez-Otin, C. and J. S. Bond. 2008. Proteases: Multifunctional enzymes in life and disease. *J Biol Chem* 283 (45):30433–30437.

Lu, H., W. Ouyang, and C. Huang. 2006. Inflammation, a key event in cancer development. *Mol Cancer Res* 4:221–233.

Massion, P. P. and D. P. Carbone. 2003. The molecular basis of lung cancer: Molecular abnormalities and therapeutic implications. *Respir Res* 4:12.

Molina, J. R., P. Yang, S. D. Cassivi, S. E. Schild, and A. A. Adjei. 2008. Non-small cell lung cancer: Epidemiology, risk factors, treatment, and survivorship. *Mayo Clin Proc* 83 (5):584–594.

Nissinen, L. and V. M. Kahari. 2014. Matrix metalloproteinases in inflammation. *Biochim Biophys Acta* 1840 (8):2571–2580.

Parks, W. C., C. L. Wilson, and Y. S. Lopez-Boado. 2004. Matrix metalloproteinases as modulators of inflammation and innate immunity. *Nat Rev Immunol* 4 (8):617–629.

Reinartz, S., T. Schumann, F. Finkernagel, A. Wortmann, J. M. Jansen, W. Meissner, M. Krause et al. 2014. Mixed-polarization phenotype of ascites-associated macrophages in human ovarian carcinoma: Correlation of CD163 expression, cytokine levels and early relapse. *Int J Cancer* 134 (1):32–42.

Shiels, M. S., R. M. Pfeiffer, A. Hildesheim, E. A. Engels, T. J. Kemp, J. H. Park, H. A. Katki et al. 2013. Circulating inflammation markers and prospective risk for lung cancer. *J Natl Cancer Inst* 105 (24):1871–1880.

Siegel, R., D. Naishadham, and A. Jemal. 2012. Cancer statistics, 2012. *CA Cancer J Clin* 62 (1):10–29.

Sorokin, L. 2010. The impact of the extracellular matrix on inflammation. *Nat Rev Immunol* 10 (10):712–723.

17 ROS and Breast Cancer

Lenora Ann Pluchino and Hwa-Chain Robert Wang

CONTENTS

ABSTRACT

Reactive oxygen species (ROS) play significant roles in induction and progression of breast cell carcinogenesis, leading to breast cancer development. Breast cancer is the most common type of cancer and the leading cause of cancer-related death among women worldwide. Both nonmodifiable (genetic predisposition, age, early menarche, late menopause, etc.) and modifiable (environmental carcinogens, diets, obesity, alcohol consumption, etc.) risk factors contribute to breast cancer development. These risk factors are able to exert their effects on breast cell carcinogenesis through oxidative stress via induction of ROS. The highly reactive and unstable nature of ROS renders them chemically reactive with the potential to damage DNA, amino acids, and lipids, contributing to chromosomal mutagenesis, epigenetic alterations, and cellular carcinogenesis. It is widely recognized that more than 85% of breast cancers are sporadic and attributable to long-term environmental exposure to low quantities of carcinogenic agents; such agents are associated with modifiable risk factors involving oxidative stress. Considering the ever-present threat of exposure to ubiquitous environmental elements capable of generating ROS and the importance of ROS in breast cancer development, it is imperative to identify noncytotoxic agents that possess antioxidant properties capable of effectively intervening in breast cell carcinogenesis. Such agents might be specific fruits, vegetables, spices, and tea, which all contain various bioactive antioxidant compounds capable of inhibiting breast cell carcinogenesis. Thus, a combination of antioxidant dietary agents could be the most optimal course of action against ROS-mediated breast cell carcinogenesis and be routinely used for affordable prevention of breast cancer development.

17.1 INTRODUCTION

Breast cancer is the most common type of cancer affecting women (23% of all cancers diagnosed in women) and the leading cause of cancer-related death in women worldwide [1]. In general, developed countries (in North America, Northern and Western Europe, and Australia) have higher rates of breast cancer than developing countries (in sub-Saharan Africa and East Asia). This difference might be attributable to lifestyles and environmental factors in developed countries and also to low screening rates and incomplete reporting in developing countries [1–3]. Although breast cancer occurs at high rates in developed countries, survival rates exceed 80% due to advanced diagnosis and treatment [1,2]. In contrast, the survival rates in developing countries remain at 50% due to lack of adequate medical care [1].

In North America and Europe, specifically, breast cancer is the most common type of cancer and the second leading cause of cancer-related death among women [2–5]. The National Cancer Institute (NCI) projects >230,000 new cases of invasive breast cancer in American women and >40,000 deaths in 2015 alone [2]. About one in eight women in the United States will likely develop invasive breast cancer during her lifetime. Additionally, 25% of invasive cancer patients suffer terminal disease due to recurrence and metastasis, and an estimated 90% of deaths due to breast cancer are a consequence of metastatic disease [4,5].

It is broadly accepted that over 85% of breast cancers are sporadic and attributable to long-term exposure to sublethal quantities of carcinogenic factors [4,5]. Sporadic cancer develops as a multistep and multipath disease process involving cumulative genetic and epigenetic alterations [6]. These cumulative alterations result in progressively increased cellular carcinogenesis and associated aberrantly regulated signaling modulators and pathways, contributing to breast cancer development from noncancerous to premalignant and malignant stages [4–11].

In this chapter, we intend to introduce the roles reactive oxygen species (ROS) may play in induction and progression of carcinogenesis in breast cells responding to environmental factors, which are also risk factors for breast cancer development.

17.2 CLASSIFICATION OF BREAST CANCER

The different subtypes of breast cancer are diverse and are distinguished by clinical behaviors, histological characteristics, biological markers, gene expression profiles, genetic backgrounds, and responses to clinical treatment [2,3]. Classification of breast cancer subtypes helps optimize medical care and management of individual cases to improve treatment response rates and survival of patients. Breast cancer can be classified based on the presence of estrogen receptor (ER), progesterone receptor (PR), and epidermal growth factor receptor-2 (EGFR-2, also known as human epidermal growth factor receptor 2 or HER2). With information on genetic factors and histology, breast cancer can be further classified into luminal A, luminal B, triple-negative/basal-like, and HER2 subtypes [3] (Table 17.1).

Luminal A and B tumors occur in the inner lining of the mammary duct and account for ~50% of breast cancers. More specifically, luminal A tumors are characterized as ER positive (ER$^+$) and/or PR positive (PR$^+$) and HER2 negative (HER2$^-$) and represent ~40% of all breast cancers. Luminal A tumors are typically slow growing and well differentiated and have the best prognosis of all four subtypes of breast cancer with high survival and low recurrence rates. Luminal B tumors, which account for 10%–20% of all breast cancers, are characterized as ER$^+$ and/or PR$^+$ and HER2$^+$ and may include invasion of cancer cells into nearby lymph nodes. Compared to luminal A cancers, luminal B cancers have a poorer prognosis due to tumors that are larger, fast growing, and moderately differentiated. However, women with luminal B tumors still have considerably high survival rates, although not as high as women with luminal A tumors [3].

Triple-negative breast cancer refers to any breast cancer that is negative for the expression of all three receptors: ER$^-$, PR$^-$, and HER2$^-$. Most triple-negative tumors are classified as basal like because the tumor cells resemble cells of the outer (basal) lining of the mammary duct. Triple-negative tumors account for 10%–20% of all breast cancers and are often aggressive, invasive, and poorly differentiated. Triple-negative tumors do not express any hormone receptors, thereby failing to respond to traditional hormone-based therapies. No targeted therapy is currently available to effectively treat triple-negative breast cancer; thus, the prognosis of triple-negative breast cancer is poorer than that of any other subtype of breast cancer.

HER2 tumors are typically HER2$^+$, ER$^-$, and PR$^-$, and they account for 10%–15% of all breast cancers. Like triple-negative tumors, HER2 tumors grow and spread aggressively and are associated

TABLE 17.1
Subtypes of Breast Cancer

Subtype	Frequency of Occurrence (%)	Hormone Receptor Status	Morphological and Tumor Characteristics	Clinical Outcomes
Luminal A	40	ER^+ and/or PR^+, $HER2^-$	Slow-growing tumors with well-differentiated cells	Good prognosis with high survival and low recurrence rates
Luminal B	10–20	ER^+ and/or PR^+, $HER2^+$	Tumors are larger and faster growing than luminal A with moderately differentiated cells	Poorer prognosis than luminal A tumors but still considerably high survival rates
Triple negative Basal like	10–20	ER^-, PR^-, and $HER2^-$	Aggressive and invasive tumors with poorly differentiated cells resembling cells of the basal lining of the mammary duct	Worst prognosis due to lack of targeted therapies and unresponsiveness to traditional hormone-based chemotherapies
HER2	10–15	ER^-, PR^-, and $HER2^+$	Aggressive and invasive tumors with poorly differentiated cells known to metastasize early and recur often	Poorer prognosis than luminal cancers, but better than triple negative due to the development of targeted therapies to treat $HER2^+$ tumors

with a poorer prognosis than ER^+ (luminal subtypes) breast cancers. HER2 tumors also recur often and have a tendency to metastasize early. However, unlike triple-negative tumors, targeted therapies have been developed to treat $HER2^+$ tumors, resulting in improved prognosis of HER2 subtype breast cancer [3].

17.3 ROS-ASSOCIATED RISK FACTORS FOR BREAST CANCER

A risk factor is defined as any environmental or genetic factor that increases a person's chance of getting a disease. Both nonmodifiable and modifiable risk factors, as listed in Table 17.2, have been

TABLE 17.2
Risk Factors for Breast Cancer

Nonmodifiable Risk Factors	Modifiable Risk Factors
Sex	Environmental factors
Age (>55)	Postmenopausal obesity
Early menarche (<12)/late menopause (>55)	Delay of first pregnancy (>35)
Family history	Oral contraceptives
Inherited BRCA1/2 mutation	Hormone replacement therapies
Personal history of breast cancer	Smoking
Preexisting proliferative breast conditions	Alcohol consumption
High breast density	Lack of breastfeeding
	Lack of physical activity

reported to contribute to breast cancer development. Nonmodifiable risk factors include sex, age, family history, genetic predisposition, early menarche (before age 12), late menopause (after age 55), etc. [1–3]. Modifiable risk factors include environmental and dietary factors, postmenopausal obesity, smoking, alcohol consumption, contraceptive use, hormone replacement therapies (HRTs), etc. [1–3]. Most breast cancers are nongenetic and are caused by chronic exposure to environmental factors, including chemical carcinogens and radiations. Both nonmodifiable and modifiable risk factors have been recognized to involve oxidative stress to exert their effects on carcinogenesis [12].

Oxidative stress plays an important role in the induction and progression of both hereditary and sporadic breast cancer development. Many hereditary factors as well as exposure of cells to environmental factors, including radiation and chemical carcinogens, result in the production and elevation of ROS [13]. ROS elevation contributes to cellular carcinogenesis, tumor growth, and metastatic progression and is considered a distinctive characteristic of cancer cells [12,14]. The short life and unstable nature of ROS renders them chemically reactive with the potential to damage DNA, amino acids, and lipids, thereby altering the structure and hindering the function of cellular components [12]. ROS-mediated oxidation causes various types of DNA damage, leading to chromosomal mutagenesis and contributing to cellular carcinogenesis [13]. For example, the most frequently oxidized base is guanine. Oxidation of guanine results in the 8-oxoguanine (8-Oxo-G) adduct, which is often mispaired with adenine to cause a G–T mutation during DNA replication [13]. Other types of ROS-mediated DNA damage include single- and double-strand breaks, protein–DNA adducts, sister chromatid exchanges, and DNA cross-links [13]. The resulting genomic instability leads to mutations, which silence tumor suppressor genes and activate proto-oncogenes, thereby amplifying malignant potential [14–20]. Growing evidence suggests that ROS also serve as intracellular signaling modulators to mediate various signaling pathways, including the Ras-extracellular signal-regulated kinase (Ras-ERK) pathway. ROS can act as secondary messenger molecules, which are short-lived, produced in response to a stimulus, highly diffusible, and ubiquitous in various types of cells [21]. Accordingly, ROS play important roles in modulating signaling pathways at various stages of cellular carcinogenesis, from initiation through tumor promotion and metastatic progression.

Breast cancer cells are undergoing constant oxidative stress from ROS elevation that is significantly higher than that found in noncancerous cells [16]. Increased oxidative stress in cancer cells may result from alterations in metabolic pathways, aberrant regulation of nicotinamide adenine dinucleotide phosphate (NADPH) oxidase (Nox) activity, oxidation of estrogenic hormones by lactoperoxidase, deprivation of glucose, hypoxia, or infiltration with tumor-associated macrophages [16,20]. ROS can also induce mutations in the *p53* tumor suppressor gene during breast carcinogenesis, permitting DNA damage to be inherited through replication with continued chromosomal rearrangements in offspring cells [17]. In clinical studies, elevated lipid peroxidation and DNA oxidation are frequently detected in the sera and tissues of breast cancer patients. In addition, elevation of antioxidant enzymes, such as superoxide dismutase (SOD) and glutathione peroxidase (GPx) in serum, is also indicative of ongoing oxidative stress [15,17].

17.3.1 ROS and Nonmodifiable Risk Factors

Nonmodifiable risk factors include sex, family history, genetic predisposition, age, early menarche (before age 12), late menopause (after age 55), etc. [2,3]. Being a woman is the single key risk factor for developing breast cancer, as men account for less than 1% of all new breast cancer cases. Women with a family history of breast cancer in a first-degree relative (mother, daughter, sister) have two-fold higher risk over the general population; this risk further increases with the number of first-degree relatives having had the disease [2,3]. Approximately 5%–10% of breast cancers are associated with inherited genetic mutations, especially mutations occurring in the *BRCA1* and *BRCA2* tumor suppressor genes, which are important in repairing damaged DNA for protecting the integrity of cells [1–3]. Loss of BRCA1/2 results in an inability to repair DNA damages, leading to cumulative mutations in other genes. Between 44% and 78% of women with BRCA1 mutations and between 31% and 56% of women with BRCA2 mutations may develop breast cancer by the age of 70 [2,3]. It has been reported that the *BRCA1* gene is absent or mutated in 40%–50% of hereditary breast cancers and 30%–40% of sporadic breast cancers [12,13]. The BRCA1 gene product plays an important role in protecting cells from oxidative damage of DNA [12,13]. BRCA1 induces the gene expression of two antioxidant response transcription factors: nuclear respiratory factors 1 and 2 (Nrf1 and 2). Nrf1/2 induces detoxifying enzymes, such as glutathione-*S*-transferase (GST) and oxidoreductases, to block ROS elevation induced by stress [12,13]. Loss of functional BRCA1 results in failure to control ROS elevation, thereby contributing to induction and maintenance of a neoplastic phenotype in breast tissue.

Sex and age, including age at menarche and menopause, affect lifetime exposure to the reproductive hormone estrogen [3]. The risk for developing breast cancer increases with the length of time breast tissue is exposed to estrogen; about two-thirds of invasive breast cancers occur in women older than 55. ROS elevation itself is an important intrisic risk factor in association with the involvement of estrogen for breast cancer development. Breast tissue is a site for oxidative metabolism of estrogen, which results in ROS production and ROS are reportedly important in the development of estrogen-induced breast cancers [13,19,21–24]. 17β-estradiol is the most abundant and potent natural estrogen in all vertebrates [22]. The oxidative metabolism of 17β-estradiol, via cytochrome P450 1B1 (CYP1B1) in breast tissue, produces the catechol estrogen 4-hydroxyestradiol [13,21–23]. Catechol estrogens can undergo further oxidative metabolism to form quinone metabolites in a reaction that produces hydrogen peroxide (H_2O_2) and superoxide radicals (O_2^-) as by-products [22]. These oxidative metabolites of estrogen are directly genotoxic, via the formation of stable depurinating DNA adducts, and induce genetic mutations or silencing that can predispose an individual to the development of breast cancer [13,21,22]. In *in vitro* studies, estrogen has been shown to increase cell proliferation, activate mitogenic signaling, activate oncogenes, inactivate tumor suppressor genes, and cause chromosomal aberrations and epigenetic alterations [19]. Furthermore, in nude mice, ROS induced by chronic exposure to 4-hydroxyestradiol caused neoplastic transformation of noncancerous human breast epithelial MCF-10A cells with increased malignancy [24]. In clinical studies, a higher level of DNA oxidation was observed in ER$^+$ than in ER$^-$ breast cancer tissues [21]. Thus, the estrogen metabolite 4-hydroxyestradiol mediates the carcinogenic activity of estrogen via its cytogenetic toxicity, its quinone DNA adduct intermediates, and its ability to generate ROS in breast cancer development.

17.3.2 ROS and Modifiable Risk Factors: Estrogen, Obesity, and Alcohol

Oxidative stress, due to increased estrogen exposure, is also associated with modifiable breast cancer risk factors. It was reported that the use of high-dose oral contraceptives increases breast cancer risk by 10%–30% for up to 10 years after ceasing use, though the risk associated with low-dose formulations is unclear [3]. The use of HRTs for >5 years may also increase breast cancer incidence by up to 25% and double the risk of breast cancer mortality in postmenopausal women [3,25,26]. The relative risk of invasive breast cancer increases with extended HRT uses and with HRT use that begins soon after menopause. However, the risk of breast cancer varies with the formulation of HRT and the method of delivery. Combined use of estrogen and progesterone in HRT may result in a four-fold increase in breast cancer incidence over HRT with estrogen alone [25,26]. The use of HRT with combined estrogen and progesterone for >10 years is estimated to result in 15–19 more cancer incidences for every 1000 women in the United Kingdom [25]. With respect to the method of delivery for HRT, the oral administration of HRT is associated with a higher breast cancer risk than transdermal delivery [25,26]. Possibly, oral delivery causes an initial high-dose peak in blood circulation followed by clearance through the liver; in contrast, transdermal delivery results in a low dose of hormones in circulation due to immediate clearance by the liver [26]. The difference in exposure of breast tissue to high versus low doses of hormones may play an important role in the delivery-related increases of breast cancer risk associated with long-term HRT use [26].

Estrogen-induced oxidative stress also applies to the increased breast cancer risk associated with postmenopausal obesity. Fat tissue is the major source of estrogen once the ovaries cease hormone production, and the risk of developing postmenopausal breast cancer is twice as high in obese women than in lean women [3]. Obesity is also associated with chronic inflammation. Chronic inflammation is procarcinogenic to adajcent breast cells through ROS, which are released by inflammatory cells (neutrophils, macrophages, monocytes, etc.), to advance cancer development [12,14,27]. ROS, in turn, are able to contribute to chronic inflammation by inducing the major inflammatory enzyme cyclooxygenase-2 (COX-2), along with inflammatory cytokines such as tumor necrosis factor-alpha (TNF-α) and interleukins-1 and -6 (IL-1, IL-6), chemokines like IL-8, and proinflammatory transcription factors such as nuclear factor kappa-light-chain-enhancer of activated B cells (NF-κB), to promote invasion and metastasis of breast tumors [27].

Lifetime consumption of one alcoholic drink per day or two to five drinks per day can increase the risk of developing breast cancer before the age of 75 by 7%–12% and 41%, respectively, in both pre- and postmenopausal women [3,28,29]. The risk of alcohol-related breast cancer development is especially high in women who consume low levels of antioxidant-rich foods [28,29]. Alcohol consumption may interfere with the metabolism of estrogen that is heavily involved in breast carcinogenesis and absorption of folate antioxidants that reduces the ability to regulate ROS. Alcohol consumption may also promote DNA damage and subsequent mutagenesis induced by oxidative acetaldehyde metabolites. Exposure to alcohol results in the accumulation of acetaldehyde metabolites in mammary tissues, which are able to induce DNA damage and reduce GSH and related antioxidant enzymes, thereby exposing breast cells to mutagens and oxidative stresses leading to breast cell carcinogenesis [28–31].

17.3.3   ROS AND MODIFIABLE RISK FACTORS: RADIATION

The ability of ultraviolet (UV) light and ionizing radiation to generate ROS elevation, causing DNA damage in cells, accounts for the mechanism of their carcinogenic effects [13]. Sunlight is the main source of UV radiation, but UV rays are also given off by tanning beds and lamps. UVB rays, having higher energy than UVA, can directly damage cellular DNA. Direct absorption of UVB radiation by DNA causes the formation of pyrimidine dimers, which cannot be copied by replicative enzymes, resulting in mutations in exposed cells and their progeny [32]. Additionally, cellular exposure to UVB radiation stimulates the production of ROS. UVB-induced intracellular ROS production is mediated by catalase, which converts hydrogen peroxide into less energetic, but still reactive and genotoxic, oxidant species. Through the actions of catalase, high-energy UVB rays are absorbed by the enzyme and converted to reactive chemical intermediates that can be degraded by other antioxidant enzymes. However, the accumulation of excessive UVB-induced ROS, generated through the action of catalase, may lead to oxidative stress and DNA damage, contributing to the development of breast cancer [33].

Ionizing radiations, including gamma rays and X-rays, are high-frequency radiations with high energy sufficient to remove electrons from an atom, causing the atom to become charged or ionized. Ionizing radiation stimulates excitation and radiolysis of water molecules to generate hydroxyl radicals, hydrogen peroxide, and other forms of ROS in biological systems [12,13]; and the deposition of energy from radiation also generates molecular hydrogen and peroxynitrite [12]. Additionally, exposure of cells to ionizing radiation results in upregulated electron transport chain function to increase mitochondrial ROS production, leading to oxidative stress [34].

17.3.4   ROS AND MODIFIABLE RISK FACTORS: CHEMICAL CARCINOGENS

Most breast cancers occur sporadically due to chronic exposure of mammary tissues to low doses of environmental carcinogens. The International Agency for Research on Cancer (IARC) has identified >476 environmental agents as carcinogenic (group 1), probably carcinogenic (group 2), and possibly carcinogenic (group 3) to humans [35]. Of these, 216 have been identified

TABLE 17.3
ROS-Inducing Breast Carcinogens

Agent	Examples	Sources
Radiation	Ultraviolet (UV) radiation	Sunlight, tanning beds/lamps
	Ionizing radiation	X-rays, gamma rays
Organochlorides	Polychlorinated biphenyls (PCBs)	Plastics and electrical products made before 1979; poorly maintained hazardous waste sites
	2,3,7,8-Tetrachlorodibenzo-*p*-dioxin (TCDD)	Waste incineration; metal production; fossil fuel and wood combustion; food sources (meat, dairy, fish)
Metals	Cadmium	Pigments, batteries, metal plating, and plastics; smelting of metal ores; fossil fuel combustion; tobacco smoke; waste incineration; food from farm fields applied with phosphate fertilizers or sewage sludge
Phorbol esters	12-O-tetradecanoylphorbol-13-acetate (TPA)	Croton oil, derived from *Croton tiglium*, present in exfoliating chemical face peels
Antimicrobials	3,4,4′-Trichlorocarbanilide (TCC)	Disinfectants; detergents; soaps; lotions; deodorants
Polycyclic aromatic hydrocarbons (PAHs)	7,12-Dimethylbenz(α)anthracene (DMBA) Benzo[α]pyrene (B[α]P)	Automobile exhaust; tobacco smoke; coal smog; charcoal barbequed and wood fire grilled foods; wood smoke
Heterocyclic amines (HCAs)	2-Amino-1-methyl-6-phenylimidazo[4,5-b]pyridine (PhIP)	Well-done meats cooked by high-temperature methods such as grilling or barbequing

as mammary gland carcinogens tested in animal model studies [35,36]. These agents are detectable in air pollution, foods, consumer products, industrial products, pharmaceuticals, and agricultural products [36]. Among these agents, polychlorinated compounds, metal ions, phorbol esters, antimicrobial agents, polycyclic aromatic hydrocarbons (PAHs), and heterocyclic amines (HCAs) (Table 17.3) have been identified as inducers of oxidative stress in *in vitro* and *in vivo* studies [6,8,36,37,40,42–45,48,51–57,62,69–72].

Polychlorinated biphenyls (PCBs) are polychlorinated compounds that were widely used in electrical and plastic products until they were banned in 1979. However, PCBs are still released into the environment from poorly managed hazardous waste sites, illegal dumping of hazardous waste, leaks from electrical transformers containing PCBs, and burning of some waste in municipal and industrial incinerators [38]. In 2013, the IARC upgraded PCBs to a group 1 carcinogenic agent classification based on their carcinogenic activity as deterimined by animal studies and consistent association of PCB exposure with increased risk of melanoma and suspected risk of non-Hodgkin lymphoma and breast cancer in humans [39]. PCBs have been detected to bioaccumulate in fatty tissues of the breast, where they are metabolized by CYP450 enzymes to PCB dihydroxy compounds. Lactoperoxidase, which is rich in the mammary gland, catalyzes the oxidation of PCB dihydroxy compounds to the corresponding quinones through a semiquinone intermediate in a reaction that generates superoxide as a by-product [40]. PCB-derived quinones are reactive metabolites capable of forming DNA adducts and inducing oxidative DNA damage, contributing to breast cell carcinogenesis [40].

The polychlorinated dibenzodioxin 2,3,7,8-tetrachlorodibenzo-*p*-dioxin (TCDD) is an environmental contaminant generated from incomplete combustion of organic material. It is produced during waste incineration, metal production, fossil fuel and wood combustion, as well as manufacturing of certain chemicals. It is resistant to degradation, resulting in accumulation in human tissues and environmental substances. Low levels of TCDD are ubiquitous in the environment, including air, soil, and agricultural products, which represent the main threat to the public [41]. The IARC has classified TCDD as a group 1 carcinogen since 1997 [35]. TCDD is an aryl hydrocarbon receptor

(AhR) agonist, capable of inducing the AhR signaling pathway, gene expression, oxidative stress, and tumor promotion [42]. Treatment of human breast carcinoma cells with TCDD induces DNA strand breaks, chromosomal abnormalities, and oxidized adducts, via increased intracellular ROS production [42]. TCDD treatment also induces activation of CYP1B1, which mediates the oxidation of 17β-estradiol to produce catechol estrogens, contributing to ROS elevation and oxidative damage in breast cells [42,43].

Cadmium is a heavy metal and a toxic agent to human health, and it is classified as a group 1 carcinogen by the IARC [35,44,45]. Cadmium is formed as a by-product from the smelting of zinc, lead, or copper ores, and it is still currently used in the manufacture of pigments, batteries, metal plating, and plastics [46]. The major sources of airborne cadmium are the burning of fossil fuels and the incineration of municipal waste. Agricultural products are also reportedly contaminated with cadmium [46]. Cadmium is also a major component of cigarette smoke and smokers are detected to have twice as much cadmium in their tissues than nonsmokers [45,46]. Cadmium exposure reportedly induces ROS elevation in *in vitro* and *in vivo* studies and induces ROS-mediated DNA damage in the form of strand breaks, base modifications, and chromosomal aberrations. Cadmium induces ROS elevation possibly through the inactivation of antioxidant enzymes including catalase, SOD, GPx, and glutathione reductase [44]. Accordingly, cadmium-induced genotoxicity, oxidative stress, and interference with cellular redox regulation contribute to its carcinogenicity. Cadmium exposure has been shown to induce transcription factors, such as NF-kB and activator protein-1 (AP-1), and signaling pathways, such as Hedgehog and Wnt pathways, leading to cell growth, cell survival, and carcinogenesis [44]. Cadmium-induced ROS elevation also activates γ-secretase to cleave the transmembrane adhesion protein E-cadherin, resulting in disassembly of adherens junctions and disruption of cell–cell adhesion, contributing to an invasive breast cancer cell phenotype [45]. Cadmium is an established lung cancer carcinogen, however its role in breast cancer development is increasingly recognized.

Phorbol esters are plant-derived organic compounds isolated from croton oil that is derived from the seeds of the flowering herbaceous plant *Croton tiglium*, which is native to Southeast Asia [47]. In ancient Chinese medicine, croton oil was used to cure severe constipation and to treat joint stiffness and discomfort along with other liniment oils. Croton oil is widely used in the production of chemical face peels, but *Croton tiglium* is also used in the production of biodiesel fuel. The phorbol diester 12-O-tetradecanoylphorbol-13-acetate (TPA), also known as phorbol-12-myristate-13-acetate (PMA), is a potent tumor promoter commonly used in experimental animal models to study carcinogenesis [47]. Despite strong evidence of carcinogenicity in animal studies, TPA has not been evaluated by the IARC for its carcinogenic potential in humans [35]. TPA is structurally analogous to diacylglycerol, which is a secondary messenger signaling lipid that activates protein kinase C (PKC) in the regulation of various signaling pathways and cellular metabolic activities [47]. Exposure of human breast epithelial MCF-10A cells to TPA induces ROS elevation, leading to the activaton of the ERK pathway, and prolonged exposure to TPA results in transformation of cells to tumorigenic phenotypes [48].

Triclocarban (3,4,4′-trichlorocarbanilide (TCC)) is a lipophilic, antibacterial compound commonly used in household and personal care products such as disinfectants, detergents, soaps, lotions, and deodorants [49]. TCC is resistant to both chemical and biological wastewater treatments and is therefore among the most commonly detected organic wastewater compounds, in both frequency and concentration, in surface water throughout the world [49]. Additionally, the agricultural use of TCC-containing biosolids allows for its entry into the food chain, thus increasing human exposure to TCC. Although other antibacterial agents such as triclosan have been classified as possibly carcinogenic in humans (group 2B), the carcinogenic potential of TCC in humans has not yet been evaluated by the IARC [35]. TCC can be found in approximately 45% of commercial liquid and bar soaps available for sale in the United States. [50]. Direct dermal exposure to TCC is known to result in transdermal absorption and bioaccumulation in underlying tissues, including mammary tissues [50]. TCC exhibits endocrine-disrupting activity to induce estradiol-dependent activation of

ER-responsive gene expression, a known causative factor for breast cancer development [50]. Despite its bioavailability in breast tissues and its function as an estrogen mimic, the role of TCC exposure in increasing breast cancer risk has not been adequately addressed. In a cellular model of chronically induced breast cell carcinogenesis, long-term exposure to TCC has been shown to induce a precancerous phenotype [51]. Chronic TCC exposure led to constitutive activation of the ERK pathway resulting in the increased expression of Nox-1, which is responsible for nonmitochondrial production of ROS. TCC is able to induce both Nox-dependent and Nox-independent ROS elevation, which induces increased DNA damage in both noncancerous and cancerous breast cells [51]. Therefore, the role of TCC in ROS-mediated breast cancer development warrants further examination.

PAHs are ubiquitous environmental contaminants that are produced from incomplete combustion of organic materials. PAHs are found in automobile exhaust (especially from diesel engines), tobacco smoke, coal smog, charcoal barbequed and wood fire grilled foods, and wood smoke [37]. The PAH 7,12-dimethylbenz(α)anthracene (DMBA) is a potent carcinogen that induces mammary tumors in animal models [36]. The metabolism of DMBA is mediated through AhR, which upregulates the expression of CYP450 enzymes responsible for the enzymatic activation of PAHs. The activation of DMBA leads to the formation of reactive epoxides and quinines capable of generating peroxides and superoxide radicals to induce oxidative stress in the form of lipid peroxidation. Exposure of animals to DMBA induces mammary tumors, increases serum lipid peroxidation levels, and decreases the activities of SOD, catalase, GPx, and nonenzymic antioxidants such as glutathione, vitamin C, and vitamin E [52]. However, the IARC has not yet classified the carcinogenic potential of DMBA in humans [35].

The most well-studied environmental PAH benzo[α]pyrene (B[α]P), classified as a group 1 carcinogen, is capable of inducing ROS elevation and breast cell carcinogenesis [35]. B[α]P can accumulate in human mammary fatty tissue, and epidemiological studies have reported higher levels of PAH-DNA adducts in breast tumor tissues compared with normal breast tissues [53]. B[α]P exerts its biological effects by directly binding to DNA and indirectly producing ROS-mediated damages [54]. Exposure of human mammary epithelial cells to B[α]P induces ROS elevation and oxidative DNA damage [54,55]. Like in DMBA-induced mammary carcinogenesis, the oxidative metabolism of B[α]P is required for its carcinogenic activity. B[α]P binds to AhR to upregulate CYP1A1. B[α]P is then metabolized by CYP1A1 to epoxides, converted to diols by epoxide hydrolases, and then to the active metabolite 7,8-dihydroxy-9,10-epoxy-7,8,9,10-tetrahydro-benzo(a)pyrene (BPDE), which forms stable adducts with DNA [55,56]. The diol intermediates of B[α]P metabolites undergo oxidation by aldo–keto reductase 1A1 (AKR1A1) to form reactive quinones (BPQs). BPQs form DNA adducts and enter a redox cycling pathway to produce ROS, namely, superoxide anions, hydrogen peroxide, and hydroxyl radicals [55,56]. BPQ-induced ROS elevation leads to the activation of the AKT (also known as protein kinase B) and ERK pathways for increased cell proliferation [56]. The activation of the ERK pathway leads to the induction of matrix metalloproteinase-9 (MMP-9) which enhances breast cancer cell migration *in vitro* and mammary tumor metastasis *in vivo*, suggesting that cumulative B[α]P exposure is an important contributor to breast cancer metastasis [57].

HCAs are generated during the cooking of meats from the breakdown of creatine, amino acids, and sugars [58]. The most abundant HCA found in cooked meats is 2-amino-1-methyl-6-phenylimidazo[4,5-*b*]pyridine (PhIP). PhIP abundance increases with increased cooking time and temperature, with particularly high amounts found in well-done meats cooked by high-temperature methods such as grilling or barbequing [58]. PhIP is a dietary carcinogen and is classified as possibly carcinogenic in humans (group 2B) [35]. PhIP requires metabolic activation through CYP1A2-mediated N-oxidation or N-acetyltransferase 2 (NAT2)-mediated O-acetylation [58]. The mutagenic potency of the resulting PhIP metabolites is 100-fold higher than other HCAs due to the stability of its reactive intermediates to form DNA adducts [59]. Detection of PhIP as high as 59 pg/mL in the milk of healthy women indicates that ductal mammary epithelial cells are directly exposed to PhIP [60]. PhIP is known to induce activating mutations in the proto-oncogene

H-Ras at low nanomolar levels, indicating that PhIP can stimulate the activation of oncogenic signaling pathways at levels close to human exposure [61]. Exposure of breast cells to PhIP at nanomolar levels induces transient activation of the Ras-ERK pathway and expression of the downstream ROS-generating enzyme Nox-1 [62]. Nox enzymes are a family of membrane-bound flavoenzymes that produce ROS by transporting electrons across membranes to reduce oxygen to superoxide. The Nox-1 isoform is expressed in breast epithelial cells and has been implicated in breast cancer [63]. Nox-1 is found to be highly expressed in breast tumors, where upregulation of Nox-1 redox signaling reportedly stimulates cell growth and proliferation and contributes to tumorigenesis [63]. Expression of Nox-1 is dependent on Ras activity. Ras upregulates Nox-1 expression through the ERK pathway [64,65]. Increased Nox-derived ROS is required for Ras-transformed phenotypes, including morphological alteration, decreased cell adhesion, increased production of vascular endothelial growth factor (VEGF), tumorigenesis, invasiveness, and angiogenesis [66]. Nox-derived ROS also cause increased DNA damage, which is essential for the initiation of carcinogenesis; thus, transient activation of the Ras-ERK-Nox pathway and ROS elevation are crucial for the initiation of breast cell carcinogenesis induced by exposure to PhIP [62]. Long-term, cumulative exposures of breast cells to PhIP result in constitutive activation of the Ras-ERK-Nox pathway and ROS elevation, as well as cellular acquisition of cancer-associated properties, such as anchorage-independent cell growth, reduced dependence on growth factors for cell growth, increased cell mobility and invasion, increased populations of cancer stem–like cells, and tumorigenicity [62]. Both the constitutively activated Ras-ERK-Nox pathway and ROS elevation are required for maintaining cancer-associated properties acquired by carcinogenic breast cells. Well-done meat consumption was also found to be positively correlated with PhIP-DNA adducts in the breast tissue of women with newly diagnosed breast cancer [67]. Epidemiological studies have indicated a strong correlation between consumption of well-done meats and increased risk of breast cancer, especially in postmenopausal women [58,67,68].

17.4 LONG-TERM EXPOSURE TO MULTIPLE ROS-INDUCING CARCINOGENS

Increasing scientific evidence suggests that chronic exposure to common environmental chemicals and radiation, alone and in combination, is responsible for the currently high rates of breast cancer incidence observed throughout the world [1,8]. Carcinogenic materials in air, water, soil, smoke, and food, as well as occupational exposure and lifestyle factors, all contribute to the complex exposure situation most women consistently experience throughout their lives [8]. Numerous environmental and dietary carcinogens, implicated in the causation of breast cancer, can bioaccumulate in the breast tissue where they may persist together for years [8]. Thus, it is conceivable that long-term direct exposure of breast cells to multiple carcinogenic agents may induce constant oxidative stress and signaling pathway activation involved in breast cancer development. However, the complex role of multicarcinogens in various stages of breast cancer development still remains largely unclear.

In a series of studies to reveal the potency of multiple carcinogens in breast cell carcinogenesis, Wang's group has developed a breast cell model to demonstrate that exposure of noncancerous breast cells to multiple carcinogens results in additive effects on the induction of cellular acquisition of various cancer-associated properties [69–72]. The breast cell model mimics sporadic breast cancer development associated with long-term exposure to low doses of carcinogens. Three carcinogens, the nicotine-derived nitrosamine ketone NNK (4-(methylnitrosamino)-1-(3-pyridyl)-1-butanone), B[α]P, and PhIP, were used in their studies. Tobacco smoking is considered a risk factor for the development of human breast cancer [1–4,8,37], and recent studies indicate that active smoking and second-hand smoke exposure can increase breast cancer risk, especially in postmenopausal women [73–75]. NNK is considered one of the most potent lung carcinogens in tobacco products, but its role in breast carcinogenesis is not fully defined [76].

B[α]P is considered a weak mammary carcinogen, and PhIP is a recognized mammary carcinogen [58,67,68,77,78]. In Wang's model, immortalized, noncancerous human breast epithelial cells are repeatedly exposed to physiologically achievable doses of NNK, B[α]P, and/or PhIP to progressively induce cellular acquisition of various cancer-associated properties, which are hallmarks of mammary epithelial cancer cells, including reduced dependence on growth factors, anchorage-independent growth, cell migration, and acinar-conformational disruption [69–72]. Long-term, cumulative exposures of cells to combined NNK and B[α]P result in additively enhanced acquisition of these cancer-associated properties [69,70]. Short-term exposure of cells to NNK and B[α]P results in transient elevation of ROS, leading to Ras-/Raf-independent ERK pathway activation and subsequent induction of cell proliferation and DNA damage [70]. ROS elevation is induced prior to ERK pathway activation, indicating that ROS may play a role in activating the ERK pathway in NNK/B[α]P-exposed cells [70]. Blockage of ROS elevation by the general antioxidant N-acetyl-L-cysteine during each exposure to NNK and B[α]P results in significant suppression of ERK pathway activation and cellular acquisition of cancer-associated properties, indicating an essential role of ROS-induced ERK pathway activation in breast cell carcinogenesis induced by NNK and B[α]P [70].

A combination of NNK and B[α]P is able to enhance PhIP-induced initiation and progression of breast cell carcinogenesis [71,72]. Initiation of carcinogenesis can be measured by transient end points, including ROS elevation, DNA damage, Ras-ERK-Nox pathway activation, and increased cell proliferation, induced in a single exposure to combined NNK, B[α]P and PhIP (NBP). The cross-talk between transiently induced ROS elevation and Ras-ERK-Nox pathway activation plays an essential role in inducing increased cell proliferation, DNA oxidation, and DNA damage, contributing to enhanced initiation of carcinogenesis induced by combined NBP [72]. Progression of carcinogenesis can be measured by acquisition of various cancer-associated properties: increased cancer stem–like cell population, constitutive ROS elevation, and Ras-ERK-Nox pathway activation, as well as activation of the epithelial-to-mesenchymal transition (EMT) program induced by cumulative exposures to NBP. The Ras-ERK-Nox pathway plays an important role in maintaining increased ROS production, and ROS elevation plays an essential role in maintaining the activated Ras-ERK-Nox pathway for NBP-induced cellular transformation. The cross-talk between the constitutively activated Ras-ERK-Nox pathway and ROS elevation is necessary for maintaining enhanced acquired cancer-associated properties [71,72]. Accordingly, the interaction between Ras-ERK-Nox pathway activation and ROS elevation plays a crucial role in initiating and maintaining an enhanced carcinogenic phenotype in cells exposed to combined NBP.

Nox-produced ROS also play a role in mediating Ras-induced NF-κB signaling [13,18,66]. NF-κB is a family of transcription factors that play critical roles in cell survival, proliferation, inflammation, and immunity. Deregulation of NF-κB activity has been implicated in the etiology of breast cancer, and ROS-induced activation of NF-κB promotes breast malignancy [13]. In unstimulated cells, NF-κB dimers are sequestered by the inhibitory protein IκB in the cytoplasm. Nox-derived ROS stimulate the degradation of IκBα through the activation of IκB kinase α (IKKα), leading to liberation and subsequent nuclear localization of NF-κB proteins (p50, p52, p65, cRel, and RelB), resulting in the activation of various genes involved in cell growth and proliferation, anti-apoptosis, angiogenesis, and metastasis [13]. ROS also activate transcription factors including AP-1, STAT3, Nrf1/2, and hypoxia-inducible factor 1-alpha (HIF1-α), which are involved in the regulation of inflammation, cellular transformation, proliferation, cell cycle progression, apoptosis, tumorigenesis, angiogenesis, invasion/metastasis, and breast cancer stem–like cell generation [13,21]. In addition, ROS act as secondary messengers in various signaling pathways, including p38 MAPK, JNK, PI3K/AKT, phospholipase C, EGFR, and PKC pathways, which are involved in the modulation of various cellular activities [12,15,16]. Accordingly, it is conceivable that the mutual interplay between elevation of ROS and aberrant regulation of various signaling pathways, in addition to ROS-induced oxidative damage and genotoxicity, contributes to breast cancer development [13,21].

17.5   PERSPECTIVE: TARGETING ROS FOR THE INTERVENTION OF BREAST CANCER

Given the ever-present threat of exposure to ubiquitous environmental and dietary elements capable of generating ROS, and the multifaceted role of ROS in promoting breast cancer development and progression, it is imperative to identify noncytotoxic substances that possess antioxidant properties for effective and affordable control of sporadic breast cancers. Numerous studies have indicated that high intake of fruits and vegetables is associated with decreased risk of breast cancer [79–81]. Fruits, vegetables, spices, and tea contain various types of bioactive compounds, such as primary metabolites (vitamins, minerals, etc.) and secondary metabolites (carotenoids, flavonoids, poly-phenols, terpenoids, alkaloids, etc.), that exert antioxidant effects to suppress oxidative stress and inhibit cellular carcinogenesis at various stages of breast cancer development [80,81].

Carotenoids, such as β-carotene (vitamin A) and lycopene, are found in orange- or red-pigmented vegetables and fruits including sweet potatoes, carrots, pumpkin, tomatoes, red peppers, papaya, mango, and cantaloupe, while other carotenoids, such as lutein, are found in dark leafy greens such as spinach and kale [80]. The antioxidant benefits of carotenoids are mediated by their unique physiological function as scavengers of singlet oxygen, especially singlet oxygen generated from UV-induced lipid oxidation or ionizing radiation [80]. Organosulfur compounds, such as indole-3-carbinol and phenethyl isothiocyanate (PEITC), found in cruciferous vegetables like broccoli, cabbage, and cauliflower, and diallyl sulfides and allyl methyl trisulfides found in onions, garlic, and chives can help alleviate oxidative stress by increasing the activity of antioxidant enzymes, such as GST, involved in the detoxification of carcinogen-produced ROS [80,82]. The antioxidants vitamin C, found in citrus fruits, papaya, strawberries, potatoes, kale, and broccoli; vitamin D, found in mushrooms, dairy products, and cereal grains; and vitamin E, found in nuts, fatty fish, plant oils, and dark leafy greens, are effective free radical scavengers capable of protecting cell membranes and DNA from oxidative damage [80,83]. Vitamin C may further help scavenge and reduce nitrite, thereby reducing substrate for nitrosamines such as NNK [83]. Curcumin, a naturally occurring pigment and major component of the spice turmeric, possesses potent antioxidant activity to scavenge ROS including superoxide anions, hydroxyl radicals, and nitrogen dioxide radicals [84]. Physiologically achievable levels of curcumin have been shown to reduce TCC-induced breast cell carcinogenesis by suppressing ROS elevation and subsequent DNA damage [51].

Flavonoids, such as genistein and catechins, are polyphenolic compounds most commonly found in fruits, tea, soybeans, and wine [81]. In addition to their free radical scavenging activity, flavonoids also protect against carcinogenesis by increasing the expression of antioxidant enzymes and decreasing the expression of carcinogen-metabolizing enzymes [85]. Soybeans, containing the isoflavonoid and phytoestrogen genistein, have been shown to reduce the risk of breast cancer by decreasing the CYP450-mediated activation of carcinogens to reduce the formation of reactive metabolites and ROS in breast tissues [81]. Catechins are polyphenolic compounds found in particularly high amounts in green tea [86,87]. Green tea is made from the same tea leaves as black tea but has little to no oxidation during processing to preserve high concentrations of antioxidant components. Green tea catechins (GTCs) have been shown to quench free radicals and free metal ions to prevent ROS formation [88]. GTCs are also able to inhibit CYP450 enzymes, which metabolize carcinogens, and activate antioxidant enzymes to increase free radical scavenger activity [81].

Various dietary substances contain various types of bioactive compounds with distinctive anti-oxidant activities. Therefore, a balanced diet has been recommended for cancer prevention [89]. It is suggested that prevention strategies using combined dietary agents are advantageous over those using individual agents due to higher efficacy and lower toxicity [90]. For instance, consuming a variety of chemopreventive agents from fruits, vegetables, and soy is associated with decreased risk of breast cancer among postmenopausal women in Singapore [91]. Another study showed that 1-acetoxychavicol acetate (ACA) combined with SOD synergistically suppresses ROS genera-tion in inflammatory leukocytes compared to individual agents [92]. ACA is a phenylpropanoid compound found in the root of *Alpinia galanga*, which is a plant in the ginger family used in Indonesian and Thai cuisines. SOD is an ROS-scavenging enzyme present in cereal grains and cruciferous vegetables. The high efficacy of this combination is due to the complementary antioxidant effects of ACA, which prevents ROS generation by inhibiting Nox family enzymes, and SOD, which scavenges ROS. Similar results were seen with ACA/caffeic acid, epigallocatechin-3-gallate (EGCG)/SOD, EGCG/genistein, EGCG/benzyl isothiocyanate, and benzyl isothiocyanate/genistein combinations [92].

Using their chronically induced breast cell carcinogenesis model as a target, Wang's group has demonstrated the increased efficacy of combined dietary agents in the intervention of breast cell carcinogenesis induced by multiple carcinogens [71,72]. Using transient end points and constitutive cancer-associated properties as targets, they demonstrated that co-exposure to physiologically achievable, noncytotoxic levels of combined GTCs (–)-EGCG with (–)-epicatechin-3-gallate (ECG) or combined fungal sterol ergosterol with legume amino acid mimosine is more effective than individual agents in the intervention of ROS-mediated breast cell carcinogenesis induced by cumulative exposures to combined carcinogens including NBP [71,72]. Combined ergosterol and mimosine is more effective than individual agents in blocking carcinogen-induced ROS-mediated DNA oxidation, accounting for their preventive ability to suppress carcinogen-induced cellular carcinogenesis [72]. Thus, it is important to consider the use of combined dietary agents to effectively intervene with breast cell carcinogenesis induced by long-term exposure to multiple carcinogens. A combination of dietary agents could be the most optimal course of action against ROS-mediated breast cell carcinogenesis and be routinely used for affordable prevention of sporadic breast cancer associated with long-term exposure to environmental and dietary carcinogens.

17.6 CONCLUSION

ROS play an important, multifaceted role in various stages of breast cell carcinogenesis leading toward breast cancer development. The ability of ROS to directly bind DNA, amino acids, and lipids enables them to cause genomic mutations and epigenetic alterations that drive the induction and progression of breast cell carcinogenesis. ROS are also able to serve as secondary messengers to modulate various signaling pathways involved in the regulation of inflammation, cell proliferation, apoptosis, angiogenesis, and metastasis. Accordingly, ROS elevation plays an important role in causing cell structure damage and modulating alterations in cell signaling, leading to breast cell malignancy. Evidence suggests that chronic exposure to environmental factors, such as carcinogenic chemicals and radiation, are mainly responsible for the high rates of global breast cancer incidence. Many carcinogenic factors are recognized to exert their effects via induction of oxidative stresses through ROS elevation. However, the complex role of multiple carcinogenic factors in various stages of ROS-mediated breast cancer development still remains largely unclear. Thus, it is important to further evaluate the potency of low quantities of combined environmental and dietary carcinogenic agents, in induction of breast cell carcinogenesis, leading to breast cancer development. In the meantime, it is also imperative to identify dietary agents with antioxidant and ROS-scavenging abilities that can be consumed daily for an accessible and affordable solution to the problem of ROS-mediated breast cancer development.

REFERENCES

1. World Health Organization (2008). *World Cancer Report.* Lyon, France: International Agency for Research on Cancer.
2. National Institutes of Health (2008). National Cancer Institute breast cancer home page. http://www.cancer.gov/cancertopics/types/breast (accessed May 27, 2015).
3. American Cancer Society (2013). *Breast Cancer Facts & Figures 2013–2014.* Atlanta, GA: American Cancer Society.

4. American Cancer Society (2015). What are the risk factors for breast cancer? http://www.cancer.org/Cancer/BreastCancer/DetailedGuide/breast-cancer-risk-factors (accessed May 27, 2015).

5. National Institutes of Health (2015). What you need to know about breast cancer, National Cancer Institute Breast Cancer home page. http://www.nci.nih.gov/publications/patient-education/wyntk-breast-cancer (accessed May 27, 2015).

6. DeBruin LS, Josephy PD (2002). Perspectives on the chemical etiology of breast cancer. *Environmental Health Perspectives* 110(S1): 119–128.

7. Simpson PT, Reis-Filho JS, Gale T, Lakhani SR (2005). Molecular evolution of breast cancer. *Journal of Pathology* 205: 248–254.

8. Gray J, Evans N, Taylor B, Rizzo J, Walker M (2009). State of the evidence: The connection between breast cancer and the environment. *International Journal of Occupational and Environmental Health* 15: 43–78.

9. Geyer FC, Lopez-Garcia MA, Lambros MB, Reis-Filho JS (2009). Genetic characterization of breast cancer and implications for clinical management. *Journal of Cellular and Molecular Medicine* 13: 4090–4103.

10. Eroles P, Bosch A, Pérez-Fidalgo JA, Lluch A (2012). Molecular biology in breast cancer: Intrinsic subtypes and signaling pathways. *Cancer Treatment Reviews* 38: 698–707.

11. Arribas J, Baselga J, Pedersen K, Parra-Palau JL (2011). p95HER2 and breast cancer. *Cancer Research* 71: 1515–1519.

12. Rios-Arrabal S, Artacho-Cordon F, Leon J, Roman-Marinetto E, Del Mar Salinas-Asensio M, Calvente I, Nunez MI (2013). Involvement of free radicals in breast cancer. *SpringerPlus* 2: 404.

13. Acharya A, Das I, Chandhok D, Saha T (2010). Redox regulation in cancer: A double-edged sword with therapeutic potential. *Oxidative Medicine and Cellular Longevity* 3: 23–34.

14. Hanahan D, Weinberg RA (2011). Hallmarks of cancer: The next generation. *Cell* 144: 646–674.

15. Nourazarian AR, Kangari P, Salmaninejad A (2014). Roles of oxidative stress in the development and progression of breast cancer. *Asian Pacific Journal of Cancer Prevention* 15: 4745–4751.

16. Brown NS, Bicknell R (2001). Hypoxia and oxidative stress in breast cancer. Oxidative stress: Its effects on the growth, metastatic potential and response to therapy of breast cancer. *Breast Cancer Research* 3: 323–327.

17. Kang DH (2002). Oxidative stress, DNA damage, and breast cancer. *AACN Clinical Issues* 13: 540–549.

18. Cichon MA, Radisky DC (2014). ROS-induced epithelial-mesenchymal transition in mammary epithelial cells is mediated by NF-kB-dependent activation of Snail. *Oncotarget* 5: 2827–2838.

19. Mahalingaiah PK, Singh KP (2014). Chronic oxidative stress increases growth and tumorigenic potential of MCF-7 breast cancer cells. *PloS One* 9: e87371.

20. Jezierska-Drutel A, Rosenzweig SA, Neumann CA (2013). Role of oxidative stress and the microenvironment in breast cancer development and progression. *Advances in Cancer Research* 119: 107–125.

21. Okoh V, Deoraj A, Roy D (2011). Estrogen-induced reactive oxygen species-mediated signalings contribute to breast cancer. *Biochimica et Biophysica Acta* 1815: 115–133.

22. Jefcoate CR, Liehr JG, Santen RJ, Sutter TR, Yager JD et al. (2000). Tissue-specific synthesis and oxidative metabolism of estrogens. *Journal of the National Cancer Institute Monographs* 2000: 95–112.

23. Okobia MN, Bunker CH (2006). Estrogen metabolism and breast cancer risk—A review. *African Journal of Reproductive Health* 10: 13–25.

24. Okoh VO, Felty Q, Parkash J, Poppiti R, Roy D (2013). Reactive oxygen species via redox signaling to PI3K/AKT pathway contribute to the malignant growth of 4-hydroxy estradiol-transformed mammary epithelial cells. *PloS One* 8: e54206.

25. Beral V, Million Women Study Collaborators (2003). Breast cancer and hormone-replacement therapy in the Million Women Study. *Lancet* 362: 419–427.

26. Opatrny L, Dell'Aniello S, Assouline S, Suissa S (2008). Hormone replacement therapy use and variations in the risk of breast cancer. *British Journal of Obstetrics and Gynaecology* 115: 169–175.

27. Grivennikov SI, Greten FR, Karin M (2010). Immunity, inflammation, and cancer. *Cell* 140: 883–899.

28. Castro GD, Castro JA (2014). Alcohol drinking and mammary cancer: Pathogenesis and potential dietary preventive alternatives. *World Journal of Clinical Oncology* 5: 713–729.

29. Dumitrescu RG, Shields PG (2005). The etiology of alcohol-induced breast cancer. *Alcohol* 35: 213–225.

30. Coronado GD, Beasley J, Livaudais J (2011). Alcohol consumption and the risk of breast cancer. *Salud Publica de Mexico* 53: 440–447.

31. Castro GD, de Castro CR, Maciel ME, Fanelli SL, de Ferreyra EC, Gomez MI, Castro JA (2006). Ethanol-induced oxidative stress and acetaldehyde formation in rat mammary tissue: Potential factors involved in alcohol drinking promotion of breast cancer. *Toxicology* 219: 208–219.

32. Cadet J, Grand A, Douki T (2015). Solar UV radiation-induced DNA Bipyrimidine photoproducts: Formation and mechanistic insights. *Topics in Current Chemistry* 356: 249–275.

33. Heck DE, Vetrano AM, Mariano TM, Laskin JD (2003). UVB light stimulates production of reactive oxygen species: Unexpected role for catalase. *Journal of Biological Chemistry* 278: 22432–22436.

34. Yamamori T, Yasui H, Yamazumi M, Wada Y, Nakamura Y, Nakamura H, Inanami O (2012). Ionizing radiation induces mitochondrial reactive oxygen species production accompanied by upregulation of mitochondrial electron transport chain function and mitochondrial content under control of the cell cycle checkpoint. *Free Radical Biology and Medicine* 53: 260–270.

35. International Agency for Research on Cancer (2006). *IARC Monographs on the Evaluation of Carcinogenic Risks to Humans*. Lyon, France: International Agency for Research on Cancer.

36. Rudel RA, Attfield KR, Schifano JN, Brody JG (2007). Chemicals causing mammary gland tumors in animals signal new directions for epidemiology, chemicals testing, and risk assessment for breast cancer prevention. *Cancer* 109: 2635–2666.

37. Wogan GN, Hecht SS, Felton JS, Conney AH, Loeb LA (2004). Environmental and chemical carcinogenesis. *Seminars in Cancer Biology* 14: 473–486.

38. Environmental Protection Agency (2013). Polychlorinated biphenyls (PCBs): Basic information. http://www.epa.gov/epawaste/hazard/tsd/pcbs/about.htm (accessed May 27, 2015).

39. Lauby-Secretan B, Loomis D, Grosse Y, El Ghissassi F, Bouvard V, Benbrahim-Tallaa L, Guha N, Baan R, Mattock H, Straif K, on behalf of the International Agency for Research on Cancer Monograph Working Group (IARC, Lyon, France) (2013). Carcinogenicity of polychlorinated biphenyls and polybrominated biphenyls. *The Lancet Oncology* 14: 287–288.

40. Oakley GG, Devanaboyina U, Robertson LW, Gupta RC (1996). Oxidative DNA damage induced by activation of polychlorinated biphenyls (PCBs): Implications for PCB-induced oxidative stress in breast cancer. *Chemical Research in Toxicology* 9: 1285–1292.

41. Agency for Toxic Substances and Disease Registry (ATSDR) (1998). *Toxicological Profile for Chlorinated Dibenzo-p-Dioxins*. Atlanta, GA: Public Health Service, U.S. Department of Health and Human Services.

42. Lin PH, Lin CH, Huang CC, Chuang MC, Lin P (2007). 2,3,7,8-Tetrachlorodibenzo-p-dioxin (TCDD) induces oxidative stress, DNA strand breaks, and poly(ADP-ribose) polymerase-1 activation in human breast carcinoma cell lines. *Toxicology Letters* 172: 146–158.

43. Chen ZH, Hurh YJ, Na HK, Kim JH, Chun YJ, Kim DH, Kang KS, Cho MH, Surh YJ (2004). Resveratrol inhibits TCDD-induced expression of CYP1A1 and CYP1B1 and catechol estrogen-mediated oxidative DNA damage in cultured human mammary epithelial cells. *Carcinogenesis* 25: 2005–2013.

44. Hartwig A (2013). Cadmium and cancer. *Metal Ions in Life Sciences* 11: 491–507.

45. Park CS, Kim OS, Yun SM, Jo SA, Jo I, Koh YH (2008). Presenilin 1/gamma-secretase is associated with cadmium-induced E-cadherin cleavage and COX-2 gene expression in T47D breast cancer cells. *Toxicological Sciences* 106: 413–422.

46. Agency for Toxic Substances and Disease Registry (ATSDR) (1997). *Toxicological Profile for Cadmium*. Atlanta, GA: Public Health Service, U.S. Department of Health and Human Services.

47. Goel G, Makkar HP, Francis G, Becker K (2007). Phorbol esters: Structure, biological activity, and toxicity in animals. *International Journal of Toxicology* 26: 279–288.

48. Rakib MA, Kim YS, Jang WJ, Choi BD, Kim JO, Kong IK, Ha YL (2010). Attenuation of 12-O-tetradecanoylphorbol-13-acetate (TPA)-induced gap junctional intercellular communication (GJIC) inhibition in MCF-10A cells by c9,t11-conjugated linoleic acid. *Journal of Agricultural and Food Chemistry* 58: 12022–12030.

49. Brausch JM, Rand GM (2011). A review of personal care products in the aquatic environment: Environmental concentrations and toxicity. *Chemosphere* 82: 1518–1532.

50. Witorsch RJ, Thomas JA (2010). Personal care products and endocrine disruption: A critical review of the literature. *Critical Reviews in Toxicology* 40(S3): 1–30.

51. Sood S, Choudhary S, Wang HC (2013). Induction of human breast cell carcinogenesis by triclocarban and intervention by curcumin. *Biochemical and Biophysical Research Communications* 438: 600–606.

52. Anbuselvam C, Vijayavel K, Balasubramanian MP (2007). Protective effect of Operculina turpethum against 7,12-dimethyl benz(a)anthracene induced oxidative stress with reference to breast cancer in experimental rats. *Chemico-Biological Interactions* 168: 229–236.

53. Yilmaz B, Ssempebwa J, Mackerer CR, Arcaro KF, Carpenter DO (2007). Effects of polycyclic aromatic hydrocarbon-containing oil mixtures on generation of reactive oxygen species and cell viability in MCF-7 breast cancer cells. *Journal of Toxicology and Environmental Health Part A* 70: 1108–1115.

54. Leadon SA, Stampfer MR, Bartley J (1988). Production of oxidative DNA damage during the metabolic activation of benzo[a]pyrene in human mammary epithelial cells correlates with cell killing. *Proceedings of the National Academy of Sciences of the United States of America* 85: 4365–4368.
55. Sigounas G, Hairr JW, Cooke CD, Owen JR, Asch AS, Weidner DA, Wiley JE (2010). Role of benzo[alpha]pyrene in generation of clustered DNA damage in human breast tissue. *Free Radical Biology and Medicine* 49: 77–87.
56. Burdick AD, Davis JW, 2nd, Liu KJ, Hudson LG, Shi H, Monske ML, Burchiel SW (2003). Benzo(a)pyrene quinones increase cell proliferation, generate reactive oxygen species, and transactivate the epidermal growth factor receptor in breast epithelial cells. *Cancer Research* 63: 7825–7833.
57. Guo J, Xu Y, Ji W, Song L, Dai C, Zhan L (2015). Effects of exposure to benzo[a]pyrene on metastasis of breast cancer are mediated through ROS-ERK-MMP9 axis signaling. *Toxicology Letters* 234: 201–210.
58. Zheng W, Lee SA (2009). Well-done meat intake, heterocyclic amine exposure, and cancer risk. *Nutrition and Cancer* 61: 437–446.
59. Felton JS, Knize MG, Wu RW, Colvin ME, Hatch FT, Malfatti MA (2007). Mutagenic potency of food-derived heterocyclic amines. *Mutation Research* 616: 90–94.
60. DeBruin LS, Martos PA, Josephy PD (2001). Detection of PhIP (2-amino-1-methyl-6-phenylimidazo[4,5-b]pyridine) in the milk of healthy women. *Chemical Research in Toxicology* 14: 1523–1528.
61. Yu M, Snyderwine EG (2002). H-ras oncogene mutations during development of 2-amino-1-methyl-6-phenylimidazo[4,5-b]pyridine (PhIP)-induced rat mammary gland cancer. *Carcinogenesis* 23: 2123–2128.
62. Choudhary S, Sood S, Donnell RL, Wang HC (2012). Intervention of human breast cell carcinogenesis chronically induced by 2-amino-1-methyl-6-phenylimidazo[4,5-b]pyridine. *Carcinogenesis* 33: 876–885.
63. Desouki MM, Kulawiec M, Bansal S, Das GM, Singh KK (2005). Cross talk between mitochondria and superoxide generating NADPH oxidase in breast and ovarian tumors. *Cancer Biology and Therapy* 4: 1367–1373.
64. Ferro E, Goitre L, Retta SF, Trabalzini L (2012). The interplay between ROS and Ras GTPases: Physiological and pathological implications. *Journal of Signal Transduction* 2012: 365769.
65. Adachi Y, Shibai Y, Mitsushita J, Shang WH, Hirose K, Kamata T (2008). Oncogenic Ras upregulates NADPH oxidase 1 gene expression through MEK-ERK-dependent phosphorylation of GATA-6. *Oncogene* 27: 4921–4932.
66. Shinohara M, Adachi Y, Mitsushita J, Kuwabara M, Nagasawa A et al. (2010). Reactive oxygen generated by NADPH oxidase 1 (Nox1) contributes to cell invasion by regulating matrix metalloprotease-9 production and cell migration. *Journal of Biological Chemistry* 285: 4481–4488.
67. Zhu J, Chang P, Bondy ML, Sahin AA, Singletary SE, Takahashi S, Shirai T, Li D (2003). Detection of 2- amino-1-methyl-6-phenylimidazo [4,5-b]-pyridine-DNA adducts in normal breast tissues and risk of breast cancer. *Cancer Epidemiology, Biomarkers and Prevention* 12: 830–837.
68. Zheng W, Gustafson DR, Moore D, Hong C-P, Anderson KE, Kushi LH, Sellers TA, Folsom AR (1998). Well-done meat intake and the risk of breast cancer. *Journal of the National Cancer Institute* 90: 1724–1729.
69. Rathore K, Wang HC (2012). Green tea catechin extract in intervention of chronic breast cell carcinogenesis induced by environmental carcinogens. *Molecular Carcinogenesis* 51: 280–289.
70. Rathore K, Choudhary S, Odoi A, Wang HC (2012). Green tea catechin intervention of reactive oxygen species-mediated ERK pathway activation and chronically induced breast cell carcinogenesis. *Carcinogenesis* 33: 174–183.
71. Pluchino LA, Wang HC (2014). Chronic exposure to combined carcinogens enhances breast cell carcinogenesis with mesenchymal and stem-like cell properties. *PLoS One* 9: e108698.
72. Pluchino LA, Liu AK, Wang HC (2015). Reactive oxygen species-mediated breast cell carcinogenesis enhanced by multiple carcinogens and intervened by dietary ergosterol and mimosine. *Free Radical Biology and Medicine* 80: 12–26.
73. Johnson KC, Miller AB, Collishaw NE, Palmer JR, Hammond SK et al. (2010). Active smoking and secondhand smoke increase breast cancer risk: The report of the Canadian Expert Panel on Tobacco Smoke and Breast Cancer Risk. *Tobacco Control* 20: e2.
74. Luo J, Margolis KL, Wactawski-Wende J, Horn K, Messina C, Stefanick ML, Tindle HA, Tong E, Rohan TE (2011). Association of active and passive smoking with risk of breast cancer among postmenopausal women: A prospective cohort study. *BMJ* 342: d1016.
75. Hartz AJ, He T (2013). Cohort study of risk factors for breast cancer in post-menopausal women. *Epidemiology and Health* 35: e2013003.

76. Hecht SS (1999). Tobacco smoke carcinogens and lung cancer. *Journal of the National Cancer Institute* 91: 1194–1210.

77. Rundle A, Tang D, Hibshoosh H, Estabrook A, Schnabel F et al. (2000) The relationship between genetic damage from polycyclic aromatic hydrocarbons in breast tissue and breast cancer. *Carcinogenesis* 21: 1281–1289.

78. Gammon MD, Sagiv SK, Eng SM, Shantakumar S, Gaudet MM et al. (2004). Polycyclic aromatic hydrocarbon-DNA adducts and breast cancer: A pooled analysis. *Archives of Environmental Health* 59: 640–649.

79. Block G, Patterson B, Subar A (1992). Fruit, vegetables, and cancer prevention: A review of the epidemiological evidence. *Nutrition and Cancer* 18: 1–29.

80. Liu RH (2013). Health-promoting components of fruits and vegetables in the diet. *Advances in Nutrition* 4(3): 384–392.

81. Vadodkar AS, Suman S, Lakshmanaswamy R, Damodaran C (2012). Chemoprevention of breast cancer by dietary compounds. *Anticancer Agents in Medicinal Chemistry* 12: 1185–1202.

82. Sparnins VL, Mott AW, Barany G, Wattenberg LW (1986). Effects of allyl methyl trisulfide on glutathione S-transferase activity and BP-induced neoplasia in the mouse. *Nutrition and Cancer* 8: 211–215.

83. Mamede AC, Tavares SD, Abrantes AM, Trindade J, Maia JM, Botelho MF (2011). The role of vitamins in cancer: A review. *Nutrition and Cancer* 63: 479–494.

84. Khan N, Afaq F, Mukhtar H (2008). Cancer chemoprevention through dietary antioxidants: Progress and promise. *Antioxidants and Redox Signaling* 10: 476–510.

85. Dragsted LO, Strube M, Larsen JC (1993). Cancer-protective factors in fruits and vegetables: Biochemical and biological background. *Pharmacology and Toxicology* 72: 116–135.

86. Mukhtar H, Ahmad N (2000). Tea polyphenols: Prevention of cancer and optimizing health. *American Journal of Clinical Nutrition* 71(6): 1698–1702.

87. Chacko SM, Thambi PT, Kuttan R, Nishigaki I (2010). Beneficial effects of green tea: A literature review. *Chinese Medical Journal* 5: 13.

88. Yang CS, Wang X, Lu G, Picinich S (2009). Cancer prevention by tea: Animal studies, molecular mechanisms and human relevance. *Nature Reviews Cancer* 9: 429–439.

89. Steinmetz KA, Potter JD (1996). Vegetables, fruit, and cancer prevention: A review. *Journal of the American Dietetic Association* 96: 1027–1039.

90. Ohigashi H, Murakami A (2004). Cancer prevention with food factors: Alone and in combination. *Biofactors* 22: 49–55.

91. Butler LM, Wu AH, Wang R, Koh WP, Yuan JM, Yu MC (2010). A vegetable-fruit-soy dietary pattern protects against breast cancer among postmenopausal Singapore Chinese women. *American Journal of Clinical Nutrition* 91: 1013–1019.

92. Murakami A, Takahashi D, Koshimizu K, Ohigashi H (2003). Synergistic suppression of superoxide and nitric oxide generation from inflammatory leukocytes by combined food factors. *Mutation Research* 523–524: 151–161.

18 Reactive Oxygen Species and Redox Signaling in Breast Cancer

*Carolina Panis, Bruno Ricardo Barreto Pires,
Ana Carolina de Andrade, Mateus Batista Silva,
and Vanessa Jacob Victorino*

CONTENTS

ABSTRACT

Breast cancer progression depends on host innate and adaptive immune responses. Despite the effect of immune system in tumor destruction, cancer cells may induce the immune cells to support tumor progression. During tumor–host interactions, both tumor and immune cells produce significant amounts of reactive species (RS). The imbalance resulting from RS production and its neutralization is called oxidative stress, and it modulates several cellular processes impacting breast cancer outcome. Two major sensors of these processes are the nuclear factor erythroid–derived 2-related factor 2 (NRF2) and NF-κB-related pathways. The transcriptional factor NF-κB has pro-oxidant capacity, while Nrf2 has neutralizing action. NF-κB is a regulator of innate immunity, and it induces RS production and inflammatory cytokines. Furthermore, high level of NF-κB is found for breast cancer, and its downstream signaling has been implicated in aggressive tumor features. The redox sensor NRF2 is activated under oxidative stress conditions and induces human antioxidant response element in order to control cellular homeostasis; however, it protects tumors against RS caused by chemotherapy leading to treatment resistance. This chapter raises the discussion regarding the redox mechanisms involved in breast cancer development driven by the activation of the Nrf2-NF-κB axis.

18.1 INTRODUCTION

The interactions that occur between the host's immune system and tumor cells result in a wide range of signaling cascades. Beyond genes and proteins affected in this process, a number of free radicals are produced. These reactive species (RS) can modulate a variety of cellular processes in an autocrine or paracrine ways. RS are produced by immune cells; also tumors can yield free RS under

243

appropriate conditions. The imbalance resulting from RS production and its neutralization, known as oxidative stress, can provoke severe modifications in host and tumor cells, varying from survival regulatory effects to cell death. Oxidative stress is frequently found in chronic inflammatory diseases such as breast cancer. Sustained inflammation is a hallmark of cancer that involves endogenous and exogenous components, including RS. Moderate levels of RS are implicated in breast cancer outcome, and infiltrated immune cells are expected to combat cancer through inflammation and also frequently conditioned to promote breast cancer development by generating specific mediators of inflammation. Considering this complex network, two major axes are in constant on–off process, the nuclear factor erythroid–derived 2-related factor 2 (Nrf2) and NF-κB-related pathways. NF-κB represents the pro-oxidant component of redox signaling, while Nrf2 is its neutralizing arm. The imbalance between these signaling networks can result in redox stress, which plays a crucial role in breast cancer development and progression. In this chapter, we discuss the redox mechanisms involved in breast cancer biology driven by the activation of the Nrf2-NF-κB axis.

18.2   IMMUNE COMPONENTS AS SOURCES OF INFLAMMATION IN BREAST CANCER

Breast cancer is most common malignant neoplasia in women worldwide. It results from diverse cellular events sustained by chronic inflammation, cell death resistance, and genomic instability. The development of breast cancer in women is mostly dependent on the capacity of host cells to evade cancer attack using immune responses.

In the early phases of tumor development, cancer cells have to multiply, ensuring the development of sufficient tumoral mass and its protection against immune recognition. After this, the primary tumor mass activates its metastasizing mechanisms and prepares the distant metastatic niche for systemic spreading (Samadi et al. 2015). Therefore, it is possible that established cancer cells may initially deceive immune response by hiding its antigens, with their posterior exposition during spreading. This evasive process seems to depend on the cytokine status in breast cancer patients. During early disease phase, breast cancer patients carry capsulated tumor developing cells that are delimited inside the mammary gland. Such patients present systemically anti-inflammatory status, marked by the predominance of reduced levels of tumor necrosis factor-alpha (TNF-α). On the other hand, women presenting advanced level tumor have a proinflammatory profile in blood, characterized by high TNF-α and IL-1β (Panis et al. 2012). These findings suggest that the systemic cytokine profiling found in breast cancer patients may result from tumor–host immunity interactions and is crucial for disease outcomes.

Breast tumors are frequently infiltrated by immune cells, and the interactions, resulting from infiltrating components and cancer cells, determine breast cancer progression (Ham and Moon 2013). Hence, the first line of defense against cancer is the innate immune response, in which soluble and cellular components trigger inflammation resulting adaptive immunity.

The role of immune systems is tumor destruction; however, evidence points that once infiltrated, the immune cells can easily work favoring breast cancer progression. The pathological epithelial-to-mesenchymal transition (EMT) that occurs in breast stem cells (SCs) has been associated with signals provided by infiltrating tumor cells (Payne and Manjili 2012), suggesting that breast cancer microenvironment drives immune cells to the tolerogenic phenotype.

Macrophages seem to constitute the first line of action for the infiltrating cells in breast tumors. These cells have paradoxical roles in relation to tumors, depending on Th1 or Th2 cell's polarization, which can favor or inhibit tumor growth and angiogenesis (Chawla et al. 2013). In breast cancer, tumor-associated macrophages (TAMs) are largely infiltrated and exhibit M2 macrophage characteristics, which are presented as a tumor-promoting phenotype (Rego et al. 2014). A significant amount of TAMs is reported in breast tumors (Leek et al. 2002), which secretes pro-angiogenic factors and enzymes that modulate angiogenesis and inflammation including cyclooxygenase 2 (COX-2), inducible nitric oxide synthase (iNOS), and matrix metalloproteinases

(MMPs), facilitating the extracellular matrix invasion by cancer cells. Furthermore, TAMs down-regulate the proinflammatory and antitumor responses, prompting breast cancer cells to evade and subsequently metastasize (Obeid et al. 2013). Hence, these mechanisms drive breast cancer outcome, leading to possible poor prognosis in patients (Mukhtar et al. 2012; Tang 2013).

Tumor-associated neutrophils (TAN) categorized as N1 or N2 cells are frequently found in solid tumors and play dual roles in cancer progression. N1 are tumoricidal neutrophils, while N2 are tumor-promoting cells (Houghton 2010). Although less studied, TANs, in association with TAMs, account for about 50% of tumor mass in invasive breast cancers (Queen et al. 2005). In breast tumors, TAN expresses elastase and modulates the killing of cancer cells (Mittendorf et al. 2012). Furthermore, tumor-induced neutrophils suppress the cytotoxic activity of TCD8 lymphocytes and enhance breast cancer metastatization in an IL-17-dependent manner (Coffelt et al. 2015).

Dendritic cells (DCs) have been implicated in the antitumor responses against breast cancer. These are antigen-presenting cells (APCs) that efficiently activate lymphocytes possessing cytotoxic properties. As occurs in other immune cell subtypes, tumor microenvironment polarizes DCs to immunosuppressive regulatory cells that are tolerogenic and supports breast cancer progression (da Cunha et al. 2014).

Myeloid-derived suppressor cells (MDSCs) can also participate as both tumor suppressor and tumor promotor of breast cancer. MDSCs are immature myeloid cells that under normal conditions play a negative regulatory role on other immune cells. In breast cancer, immature MDSC levels in blood increase tumor burden (Diaz-Montero et al. 2009), suggesting a positive link to disease prognosis. On the other hand, tumor-infiltrating MDSCs mediate tumor cells resistance to cytotoxic effect of T lymphocytes, which is a negative link for breast cancer elimination (Lu et al. 2011).

Tumor-infiltrating lymphocytes paradoxically affect breast cancer. Although T cells have to kill tumor cells as their primary function, they seem ineffective in the elimination of breast cancer depending on its subtype. Significant TCD4 infiltration occurs in breast cancer, presenting lymph node metastasis and suggesting a role for these cells in disease progression (Chin et al. 1992; Macchetti et al. 2006). On the other hand, TCD8-infiltrating cells have been correlated with a better prognosis in breast cancer patients (Mahmoud et al. 2011; Park et al. 2012). Furthermore, B and T lymphocytes are co-expressed in breast cancer (Helal et al. 2013), and B cells are significantly increased and phenotypically most aggressive in breast tumors, as those with high grade and basal phenotype (Mahmoud et al. 2012). Therefore, strong evidence exists that infiltrating cells can exert significant effects on breast cancer outcome.

Significant amounts of RS are produced during tumor–host interactions, and both tumor and immune cells are major producers of these species in cancer. Immune cells have a significant role of RS before and after tumor infiltration. Macrophages and neutrophils are important sources of RS and are activated to produce them during antigenic challenges. RS production in phagocytic cells initiates when NADPH oxidase system (NOX) is activated (Bogdan et al. 2000). Multiple roles of RS in immune response include antimicrobial activity, induction of transcription factors, neutrophil extracellular net formation, amplification of autophagy, and macrophages' polarization (Tal et al. 2009; Yost et. al. 2009; Gonzalez-Dosal et al. 2011; Zhang et al. 2013; Paiva and Bozza 2014).

In spite of these physiological roles of RS in controlling unwanted invasions, evidence exists that the free radicals produced during immune responses against tumors play a significant role in cancer progression. The main nontumoral source of RS in cancer is immune-infiltrating cells, such as cancer-associated macrophages (CAMs), cancer-associated fibroblasts (CAFs), and cancer-associated neutrophils (CANs). CAMs contribute to tumor spreading as they secrete proinflammatory cytokines that modulate tumor cells and stromal surround tissue inflammatory response, while CAFs contribute to RS arising in tumors by releasing MMPs and cytokines (Coussens and Werb 2002; Fiaschi and Chiarugi 2012). In this context, it is important to highlight that infiltrating leukocytes can be immunosuppressed by breast cancer, which may compromise its capability to generate RS. Therefore, another source of RS is necessary to explain the sustained oxidative stress found in breast cancer.

Beyond immunity-related components, cancer cells are significant sources of RS (Brown and Bicknell 2001). Under concentrations compatible with cell survival, RS are growth promoters of breast cancer cells (Madeddu et al. 2014), which helps explain why patients with breast cancer are under constant systemic oxidative stress, suggesting that the systemic redox status found in breast cancer results from the tumor–host interactions (Panis et al. 2012; Vera-Ramirez et al. 2012; Victorino et al. 2014; Lemos et al. 2015).

18.3   Nrf2/NF-κB STATUS IN BREAST CANCER

Recent advances allowed understanding the biological events in breast cancer at molecular level. The classical definition of oxidative stress enrolls the imbalance between RS production and its neutralization. Employing our current knowledge, a definition could be improved as the "result from the activation of major controllers of redox genes transcription, the Nrf2 and NF-κB-related pathways." In this context, NF-κB represents the pro-oxidative axis, while Nrf2 contributes with antioxidant neutralizing components. Here, we discuss how these signaling pathways may contribute to redox signaling in breast cancer.

18.3.1   NF-κB Axis

Breast cancer is a heterogeneous disease composed of multiple subtypes with distinct etiology, course of progression, outcome, and molecular features. Based on gene expression profiles, human breast cancer is classified into five subtypes: luminal A, luminal B, luminal human epidermal growth factor receptor 2 (HER2), HER enriched, basal like and claudin low. The basal and claudin-low groups are also known as a single subtype called as "Triple Negative" (TN), due to the low expression of estrogen receptor (ER), progesterone receptor (PR), and HER2/neu markers (Sorlie et al. 2001). The TN group is of great interest because that is not a specific target for developing treatment, and it is associated with a poor prognosis (Polyak and Metzger-Filho 2012). Despite breast cancer heterogeneity, there are common molecular signatures among its subtypes. For a better comprehension of the breast cancer molecular signatures, we should understand the mammary biology since its development.

The major periods of mammary gland development are the embryonic, prepubertal, pubertal, pregnancy, lactation, and involution (Cao and Karin 2003). The differentiation process involves ductal elongation, branching, alveolar proliferation, differentiation of the mammary epithelium, and ultimately, the involution stage with extensive programmed cell death (Geymayer and Doppler 2000). Steroid and polypeptide hormones play pivotal role during this whole process. Estrogen has a crucial role for ductal elongation during the puberty. In pregnancy period, estrogen contributes to ductal elongation, progesterone is important for ductal branching, and prolactin is important for differentiation process (Tiede and Kang 2011). The main downstream targets of these three hormones have been identified for controlling biological processes, such as proliferation and mammary SC differentiation. Prolactin activates Stat5A, which is responsible for forming luminal progenitors in virgin mice, and regulates Elf5 and Gata-3 that are important for the differentiation of luminal progenitors during pregnancy (Tiede and Kang 2011).

The PR, together with parathyroid hormone–related protein (PTHrP), is important for mammary SC expansion during pregnancy through the activation of RANK/NF-κB pathway (Tiede and Kang 2011). PR upregulates the expression of receptor activator of NF-κB ligand (RANKL) in mammary epithelial cells (MECs) and releases the ligand that binds to the receptor RANK expressed in other MECs, activating the RANK/NF-κB/Cyclin D1 pathway that is crucial for lobo alveolar development during pregnancy (Oakes et al. 2006). RANKL/RANK may be paracrine factors responsible for cell proliferation in steroid receptor negative cells (Oakes et al. 2006). The knockout mice for RANK or RANKL result in defective lobo alveolar development and milk secretion during pregnancy. This effect is due to the downregulation of NF-κB activity, the main activator of the mitosis-inducing factor Cyclin D1. RANK level is elevated in mammary SC during the pregnancy, while RANKL is expressed in the surface of luminal cells (Tiede and Kang 2011). The inhibition of pregnancy hormones decreases the RANK signaling and the number of functional mammary SCs (Tiede and Kang 2011).

The NF-κB family consists of five conserved proteins, RelA (p65), RelB, c-Rel, p50, and p52. All of them share the conserved Rel homology domain (RHD), which is responsible for DNA binding, dimerization, and association with the repressor protein IκB. The NF-κB pathway comprises the classical or canonical or alternatively of noncanonical pathway. The classical pathway consists of downstream proinflammatory cytokine receptors, such as TNF-α, IL-1β; toll-like receptor (TLR) family (TLR3, TLR4, TLR7); T-cell receptor (TCR); and B-cell receptor (BCR) that ultimately leads to ubiquitin-dependent degradation of the repressor IkBα through the phosphorylation of the IKK complex (IKKα, IKKβ, and IKKγ). The IkBα degradation releases the p50/RelA dimer to translocate into the nucleus and activates the transcription target genes (Gupta et al. 2010; Vandenabeele et al. 2010). The alternative is downstream of CD40L, RANK, and B cell–activating factor receptor (BAFF-R) pathways that lead to NF-κB-inducing kinase (NIK) activation, which phosphorylates the homodimer IKKα, which in turn phosphorylates the p100 subunit to be processed into p52. It ultimately leads to nuclear translocation of p52/RelB dimer (Gupta et al. 2010).

Since cancer-related inflammation (CRI) was recognized as a hallmark of cancer, NF-κB received a new importance in cancer studies. NF-κB is a regulator of innate immunity, and it is responsible for inducing the expression of COX-2 and nitric oxide synthase (NOS), inflammatory cytokines such as IL-1, IL-6, IL-8, and TNF-α, and chemokines such as CCL2 and CXCL8 (Colotta et al. 2009). Michael Karin's group evidenced that NF-κB activation is sufficient for promoting a chronic inflammation that triggers the gastric cancer in models infected with *Helicobacter pylori*. The mechanism of this activation is based on the response of LPS (lipopolysaccharide) comprised in the membrane of *H. pylori*, a gram-negative bacteria, through TLR-MyD88 signaling (Karin 2006). In addition, NF-κB is a key factor in both radio- and chemoresistance because it is the main regulator of anti-apoptotic genes, such as Bcl-2, Bcl-XL, and Birc5 (Survivin) (Gupta et al. 2010).

In normal mammary physiology, NF-κB is required for ductal development and regulation of the mammary epithelial branching and proliferation (Cao and Karin 2003). Studies in mice showed that NF-κB is activated during pregnancy, regulating positively the lobo alveolar development. Its levels decrease in lactation, when NF-κB acts as a negative regulator of the prolactin/JAK2/STAT5 signaling, which is necessary for alveolar differentiation and milk secretion (Geymayer and Doppler 2000). Finally, NF-κB expression becomes almost undetectable during the involution, which is marked by extensive apoptosis, when it induces the anti-apoptotic signaling in order to achieve a proper tissue remodeling (Cao and Karin 2003).

Josef Penninger's group intensively studied the relevance of the RANKL/RANK pathway in progestin-driven breast cancer; they reported that progesterone or derivatives as progestin transformed the mammary gland epithelial cells into malignant phenotype via IKK/NF-κB activation (Schramek et al. 2010; Sigl and Penninger 2014). The deletion of IKKα in mammary gland cells reduced the progestin-driven murine breast cancer and the number of mammospheres (Schramek et al. 2010; Sigl and Penninger 2014), reinforcing the relevance of the RANK/IKKα/NF-κB axis for the self-revival of breast cancer stem cells (BCSCs).

Regarding breast cancer, constitutive activation of NF-κB contributes to cellular proliferation, angiogenesis, and evasion of apoptosis, and it is mostly described in Her2/neu and TN tumors. Liu et al. (2010) reported that NF-κB is required for the initiation of Her2-positive murine mammary tumor growth. This transcription factor governs the initiation of Her2 tumors, and its inhibition was sufficient to decrease the CD44 (marker of human BCSCs)-positive cell population and reduced the tumor microvessel density in models. When the Her2 murine cells expressed IkBα mutant that represses constitutively the NF-κB pathway, a dramatic reduction was observed in mammosphere number and downregulation of embryonic SC factors Sox2 and Nanog. These findings together provide the evidence that NF-κB pathway contributes to BCSC phenotype.

Pratt et al. (2009) reported that NF-κB is activated during the differentiation of luminal progenitor cells, and it is required for the early steps of tumorigenesis of breast cancer. Microarray data from a cohort of human breast cancer revealed the expression of the canonical NF-κB pathway in Her2-positive samples. The authors also reported that the luminal progenitor cells express NF-κB, whereas the mammary SC population does not, contrary to studies mentioned earlier (Liu et al. 2010; Schramek et al. 2010). However, Merkhofer et al. (2010) proved, using specific inhibitors, that Her2 signal transduction to NF-κB activation was via IKK complex in a phosphatidylinositol 3-kinase (PI3K)/AKT-independent route. Interestingly, they reported that IKKα has a more significant role than IKKβ, the most critical catalytic subunit.

The high levels of NF-κB found in basal-like tumors are due to epidermal growth factor receptor (EGFR) overexpression, which is a positive action of this group (Voduc et al. 2010) and ultimately activates NF-κB (reviewed by Biswas and Iglehart 2006). In a recent report, Kendellen et al. (2014) showed consistently that the canonical and noncanonical NF-κB signaling is required for self-renewal and to form xenograft tumors for both claudine-low and basal-like cells.

Interestingly, Nakshatri et al. (1997) reported that the progression of rat mammary carcinoma from an ER-positive, nonmalignant phenotype to an ER-negative, malignant phenotype was accompanied with a constitutive activation of NF-κB signaling. These findings were in concordance with Wang and collaborators (2007) who described the repression of RelB in ER-positive breast cancer cells. Both studies reinforce the significance of NF-κB signaling to the hormone-independent breast cancer. For many years, the efforts to treat cancer focused on the destruction of tumor cells. However, the modulation of CSC and host microenvironment signaling had no significant advances. The challenge now is to determine how and when some important physiological pathways could be inhibited in order to attack the malignant cells. Hence, NF-κB pathway is a promising target for cancer therapy.

18.3.2 Nrf2 Axis

NRF2 is a transcription factor that regulates the expression of several genes for the cellular antioxidant response program. The antioxidant function of NRF2 was first reported about 20 years ago, and in this research, the overexpression of NRF2 was responsible for the increased expression of human antioxidant response element (ARE)-mediated chloramphenicol acetyltransferase gene expression (Venugopal and Jaiswal 1996). In subsequent reports, it was shown that NRF2 linked with Jun proteins, as c-Jun, Jun-B, and Jun-D, regulates ARE-mediated expression and coordinated induction of antioxidant as NAD(P)H:quinone oxidoreductase 1 (NQO1) and glutathione S-transferase Ya (GST) subunit genes (Venugopal and Jaiswal 1998).

Subsequent research showed that NRF2 activity was under the control of a regulatory protein. A protein named Kelch-like erythroid cell-derived protein with CNC homology (ECH)-associated protein 1 (Keap1) was responsible for suppressing the transcriptional activity of NRF2. Keap1 specifically binds to the amino-terminal regulatory domain of NRF2, known as nei like DNA glycosylase 2 (Neh2), and ensures that the levels of free NRF2 in the cell remain at low levels. The inhibition of Keap1 by conformational changes caused by oxidative stress allows NRF2 to move from the cytoplasm to the nucleus where it can control ARE response (Itoh et al. 1999).

Therefore, NRF2 along with Keap1 are essential mediator of cellular redox signaling for the maintenance of cellular homeostasis. The release of NRF2 from Keap1 allows NRF2 to leave cytoplasm and enter into the nucleus, where it forms a heterodimer with small Maf protein to induce antioxidant defense via AREs in response to oxidative stress. Upon unstressed conditions, Keap1 dimer binds NRF2 via DLG (Asp–Leu–Gly) and ETGE (GluThr–Gly–Glu) motifs at Neh2 domain to undergo proteasomal degradation by controlling its levels (Itoh et al. 1999; Richardson et al. 2015). The ubiquitination of NRF2 is mediated by the association of Keap1 with Cullin-3 (Cul3), whereas Keap1 functions as a substrate adaptor protein for a Cul3/Rbx1 E3 ubiquitin ligase complex. Keap1-dependent proteasomal degradation of Nrf2 is inhibited upon oxidative stress situation (Zhang et al. 2004).

Once cellular redox homeostasis is recovered, Keap1 translocates into the nucleus to dissociate NRF2 from the ARE DNA regulatory sequence. Then, Keap1-NRF2 complex is transported back to cytoplasm by a nuclear export sequence (NES). In the cytoplasm, Keap1 mediates the postinduction repression of NRF2 (Sun et al. 2007). The underlying functions of NRF2 have been extensively investigated, and recent studies confirm that this transcription factor regulates the expression of hundreds of genes that confer proteins with antioxidant properties, such as heme oxygenase-1 (HO-1), NQO1, superoxide dismutase (SOD), GST, and γ-glutamyl cysteine ligase (γ-GCL), in order to control oxidative stress (Keum and Choi 2014).

There are several sources of RS production as endogenous and exogenous factors. The antioxidant response is induced in order to counteract the overproduction of RS to avoid the induction of oxidative stress that leads to damage to lipids, proteins, and DNA of the cell. It is well known that oxidative stress is linked to several diseases, including cancer. In cancer, the extensive oxidant status determines the cellular fate, as mild oxidative stress status favors cellular proliferation (Halliwell 2000, 2007). However, it is important to note that RS are physiologically generated by the organism. In physiological conditions, low to moderate concentrations of this species are necessary to mediate cellular signaling (Halliwell 2007). Imbalance in redox signaling is attributed to the amount of RS generated, the cell compartment where they are produced, and the avaibility of antioxidant defense. When RS are not produced at an appropriate region, time, or levels, damages to cellular structures occur (Nathan and Cunningham-Bussel 2013). Hence, it can be speculated that the deregulated expression of redox sensor as NRF2 may be linked to cancer progression, where it has a tumor suppressive or oncogene role (Moon and Giaccia 2015).

One way the RS induce cancer is via DNA damage and resulting misrepair. The resulting cumulative mutations induce cellular transformation into malignant cells in the breast environment, impacting the development of the disease in early and advanced stages. The constant production of RS also influences the disease aggressiveness (Mencalha et al. 2014). In breast cancer patients, the increase in oxidative status contributes to cancer progression into advanced stages of the disease as a result from the host and tumor inflammatory mediators (Panis et al. 2012). The oxidative stress status in breast cancer can affect the disease outcome, as it can predict the long-term response of patients for disease recurrence in early breast cancer stages (Herrera et al. 2014). In this context, the understanding of NRF2 as a redox sensor and its impact on oxidative signaling during breast cancer development and disease progression is important. In breast cancer cells, it was shown that the treatment with an oxidant influences the NRF2 activity. In this study, the oxidant treatment induced an increased NRF2 nuclear translocation as a mechanism to protect breast cancer cells from the oxidative stress (Habib et al. 2015).

A mammosphere-based culture of breast cancer cells can contribute to tumorigenicity and chemoresistance. Using a model of MCF-7 (luminal phenotype) mammospheres, Ryoo and collaborators (2015) showed that the activation of NFR2 is involved in the enhanced expression of detoxifying/antioxidant genes, such as γ-glutamate cysteine ligase (GCLC), HO-1, aldo–keto reductase 1c1 (AKR1c1), and NQO-1. In addition, NRF2 facilitates mammosphere formation and growth conferring resistance to chemotherapy. The partial knockdown of NRF2 in MCF-7 cells resulted in low expression of genes regarding antioxidant system and consequently increased intracellular RS, enhancing cell death and delaying mammospheres' formation (Ryoo et al. 2015). Moreover, MCF-7 and MDA-MB-231 (TN phenotype) mammosphere culture displays higher tumorigenicity and chemoresistance accompanied by low intracellular levels of RS than their adherent counterparts. In those cells, NRF2 levels are increased and enhanced resistance to chemotherapy and anchorage-independent growth. As expected, this phenomenon is reversed by the inhibition of NRF2 (Wu et al. 2014a).

Mutation at N-terminal domain of Keap1 may occur in breast cancer patients, and this mutation impairs the ability of Keap1 to repress NRF2 activity as it is unable to catalyze the ubiquitination of NRF2 (Nioi and Nguyen 2007), imposing an advantage to the tumor propagation. The importance of NRF2 has been reported to be associated with specific subtypes of breast cancer; for instance, the presence of constitutively active HER2 induces the expression of NRF2-target proteins through the activation of NRF2 transcription activity by signal cross talk or by direct interaction. In addition, this interaction is linked to the transcription of genes related to chemotherapy resistance (Kang et al. 2014). Interestingly, in breast cancer patients bearing HER2-positive tumors, there is an attenuation of oxidative stress (Victorino et al. 2014). Targeting NRF2 in HER2 overexpressed cells (BT-474 cell line) suppresses HER2 levels and sensitizes cells to chemotherapy (Manandhar et al. 2012). In ER-positive breast cancer cells (MCF-7), estrogen stimulates the activity of NRF2 by PI3K and glycogen synthase kinase 3 beta (GSK3β) pathway leading to increased ARE activity (Wu et al. 2014b).

The presence of the tumor suppressor gene BRCA1 seems to be linked with NRF2 signaling, in that in breast cancer, a positive association was found between NRF2 and BRCA1. Under oxidative stress conditions, NRF2 and BRCA1 are induced in MCF-7 cells. NRF2 interacts with CBP and p300 to generate an active transcription complex on the BRCA1 promoter (Wang et al. 2013). In MECs, the BRCA1 loss of function leads to increased ROS formation due to reduced NRF2-mediated antioxidant response. Thereby, it seems that BRCA1 and NRF2 are necessary for an antioxidant response.

BRCA1 is capable of physically interacting with NRF2 in order to promote its stabilization. At oxidative stress, BRCA1 controls Keap1-mediated NRF2 ubiquitination by interacting with ETGE site, leading to the accumulation of NRF2 protein. In the absence of BRCA1, the estrogen treatment restores NRF2 and induces NRF2-dependent antioxidant response (Gorrini et al. 2013).

Understanding the implication of NRF2 signaling during breast cancer development is extremely important. NRF2 modulation is a promisor field considering that its inhibition should impact breast cancer outcome and the response to chemotherapy treatment.

18.4   REDOX SIGNALING AND BREAST CANCER PROGNOSIS

One of the main reasons of cancers is its genomic instability, yielded by DNA damage, mismatch repair, and cumulative mutations. DNA is a target of a wide range of damaging agents, which can react with its components and cause alterations in the genetic content. RS is one of the commonest damaging agents that can react with DNA and participate in every steps of carcinogenesis by triggering aberrant cell signaling (Acharya et al. 2010). RS-driven mutations are especially relevant when affecting tumor suppressor genes in the early phases of breast carcinogenesis (Kang et al. 2014), since such genes have antioxidant roles in cells (Vurusaner et al. 2012).

The transcription factor NF-κB constitutes a proinflammatory axis in cancer, since the downstream components of this pathway are mainly Th1 cytokines. NF-κB is constitutively expressed during breast cancer progression (Nakshatri et al. 1997) and is associated with the EMT (Huber et al. 2004). EMT has been implicated in breast cancer spreading, induced by MMP-3, in an RS-dependent manner (Cichon and Radisky 2014). Th1 cytokines in breast cancer are linked with tumor-promoting effects, and most well-characterized members are TNF-α, IL-6, and IL-1β. Th1 cytokines and RS act as a vicious circle that enhances oxidative/nitrosative stress in breast cancer (Singh et al. 2014).

Evidence suggests a relationship between TNF-α and redox signaling in breast cancer. TNF-α is able to induce intracellular accumulation of RS in breast cells (Kim et al. 2010). Regarding the tumor–host interactions, a study conducted by Herrera and cols (2014) revealed that breast cancer patients carrying the primary breast tumor mass exhibit concomitant higher levels of TNF-α and systemic oxidative stress when compared with women after their tumor removal. Concomitant sustained oxidative stress and high levels of TNF-α in plasma have been demonstrated in patients bearing advanced breast cancer (Panis et al. 2012) and further correlate with disease subtype, especially luminal A and HER-amplified tumors (Herrera et al. 2012). The RS-TNF-α axis in breast cancer may enroll a variety of components. In this context, a proteomic study revealed that patients with advanced breast cancer presenting the upregulation of TNF-α also exhibited augmented expression of the DNA mismatch repair protein PMS2 in both plasma and tumor samples (Panis et al. 2013). Since TNF-α receptor superfamily is associated with cell death by redox-dependent mechanisms (Shen and Pervaiz 2006), it is appropriate to hypothesize that this mechanism may regulate cell fate in breast cancer.

The Nrf2 is a counteracting mechanism to NF-κB effects on cells. Nrf2 is expressed in the nucleus of almost 50% of breast tumors and has correlated with poor disease prognosis and recurrence (Onodera et al. 2014). The aberrant activation of this pathway is associated with breast cancer risk (Hartikainen et al. 2012), and its persistent activation is critical for chemoresistance (Kang et al. 2014). Nrf2 can induce the expression of breast cancer–resistant protein and other drug efflux transporters in normal liver cells (Wu et al. 2012). It suggests that breast carcinomas, which overexpress Nrf2, may also have enhanced levels of multidrug resistance–related proteins. Nrf2 has been implicated in breast cancer chemoresistance by enrolling its inhibitory molecule, the Keap1 adaptor protein. Downregulation of Keap1 increases Nrf2 expression in breast cancer and induces drug resistance (Katri et al. 2015). Additionally, Keap1 aberrant methylation is enrolled in breast carcinogenesis and disease progression (Barbano et al. 2013). Somatic mutations in this adaptor protein are found in 2% of breast carcinomas (Yoo et al. 2012), indicating that breast cancer cells bearing such mutations are protected from RS effects.

In breast cancer, NRF2 status positively associates with pathological parameters such as histological grade, Ki-67 LI, p62 status, and NQO1 status. Furthermore, it is marginally associated with pathological parameters such as lymph node metastasis, HER2 status, and Ki-67 status. NRF2 activity also correlates with clinical outcome, as NRF2 is related to increased incidence of disease recurrence, adverse clinical outcome from patients, and worse prognosis (Onodera et al. 2014). Despite the fact that a marked improvement in the current treatments for all breast cancer subtypes has been observed, resistance to therapy is still noticed for most subtypes. The involvement of NRF2 in acquiring resistance to chemotherapy has been widely investigated in this context. MCF-7 cells that are resistant to chemotherapy treatment with doxorubicin display increased levels of NRF2 and its target genes, such as HO-1 and NQO, in comparison with its counteract (Zhong et al. 2013).

Breast cancer patients bearing luminal tumor may be treated with aromatase inhibitors; however, several patients developed resistance to this therapy. In breast cancer cells with acquired resistance to aromatase inhibitors treatment, it was shown due to an enhancement in NRF2 level as a consequence of lower ubiquitination or degradation of NRF2. Mechanistically, NRF2 contributes to aromatase inhibitor resistance as it increases the levels of biotransformation enzymes, drug transporters, and anti-apoptotic proteins (Khatri et al. 2015).

Resistance to tamoxifen treatment is also observed during the treatment of breast cancer patients. Increased expression of antioxidant protein such as γ-glutamylcysteine ligase heavy chain (γ-GCLh), HO-1, thioredoxin, and peroxiredoxin 1 in tamoxifen-resistant breast cancer cell mediated by NRF2 was found. The upregulation of NRF2 in resistant breast cancer cells may be due to the elevated activities of extracellular signal–regulated kinase (ERK) and p38 kinase pathways (Kim et al. 2008). The silencing of Cul3 in breast cancer cells also increases resistance to oxidative stress and to chemotherapy treatment with paclitaxel and doxorubicin as it promotes increase in NRF2 activity (Loignon et al. 2009). Therefore, recent studies have been focusing on the modulation of NRF2 activity showing promising results. For instance, it was demonstrated that in TN breast cancer cells, the decrease in NRF2 activity leads to improved chemotherapy efficacy (Sabizichi et al. 2014).

The downstream elements that appear after NRF2 activation are the AREs, composed by several proteins bearing antioxidant activity. The involvement of AREs in breast cancer has been extensively investigated. Although breast cancer research advances, the role of AREs is still controversial.

A recent study demonstrated that cancer initiation needs antioxidant synthesis, which is not totally necessary in the late stages of carcinogenesis (Harris et al. 2015). The main proteins that compose the AREs are catalase, SOD, thioredoxin, and the glutathione system.

Catalase and SOD are among the most studied antioxidant defenses in breast cancer. SOD catalyzes the dismutation of superoxide anion into hydrogen peroxide (H_2O_2) that is further decomposed by catalase. These enzymes constitute the first line of defense against pro-oxidants; therefore, their involvement in breast cancer survival is expected. The expression of manganese SOD (MnSOD) genes changes significantly through breast cancer progression from early to advanced stages (Becuwe et al. 2014). This enzyme seems to exert a dual role in breast cancer, since it has been linked with both poor (Ennen et al. 2012) and good prognosis (Kim et al. 2015). This evidence points out to the existence of a regulation of SOD expression in cancer, linked with disease aspects and tumor biological properties (Dhar and St Clair 2012). Enhanced SOD activity in blood has been associated with attenuated oxidative stress in patients bearing tumors with the amplification of the HER2, suggesting a protective role for this antioxidant against systemic redox stress (Victorino et al. 2014).

Breast cancer cells are producers of H_2O_2 (Xiao et al. 2015). Catalase is a very efficient system that eliminates H_2O_2 and is expressed in breast cancer cells under the control of PI3K/Akt/mTor signaling pathway (Glorieux et al. 2014). Its overexpression has been associated with reduced migration and invasiveness in breast cancer cell lineages (Glorieux et al. 2011). In patients, catalase activity is augmented in erythrocytes from women bearing advanced breast cancer (Panis et al. 2012), suggesting a protective role for this enzyme under the pro-oxidant environment found in metastatic breast disease.

Thioredoxin system is a pivotal component for oxidative stress counteract, controlling H_2O_2 levels inside cells. This system is frequently overexpressed in breast tumors (Cha et al. 2009) and interferes in growth and spreading of breast cancer cells (Qu et al. 2011). Further, thioredoxin system possesses a prognostic value associated with worse metastasis-free intervals (Cadenas et al. 2010) and chemoresistance in primary breast tumors (Kim et al. 2005). Thioredoxin levels are augmented in the blood of women with breast cancer (Kilic et al. 2014) and have been suggested as a putative marker for the diagnosis of early disease (Park et al. 2014).

Indeed, AREs are mostly implicated in tumor resistance against RS. The use of antineoplastic drugs that generate RS as their mechanism of action is a clear example. Possibly, RS yielded by chemotherapy are able to kill cancer cells from tumor periphery in the early phases of treatment, reducing antioxidants. This is observed in women with breast cancer undergoing chemotherapy, which presents acute impairment of antioxidant capacity after early doxorubicin treatment (Panis et al. 2012). Meanwhile, the tumoral core can adapt against oxidative stress and upregulate the Nrf2/ARE system, enhancing its antioxidant capacity and becoming the remaining breast cells chemoresistant.

18.5 CONCLUSION

Breast cancer is a solid tumor that is under endogenous and exogenous oxidative stress attack. Autocrine effects of RS seem to be beneficial to breast cancer progression, while paracrine RS (especially when derived from immune cells) may be detrimental. The immune system infiltrates breast tumors, but after immunosuppression, these cells work in favor of tumor cells. This event possibly limits the capacity of these cells to produce RS. The increment in exogenous RS caused by chemotherapy can trigger the overexpression of the antioxidant Nrf2/AREs axis, which protects tumor cells against RS-mediated death. The understanding of the precise mechanisms enrolling RS in breast cancer progression is necessary, since most evidences are based on the variations of pro- and antioxidant components investigated in patients, which present limited interpretation for therapeutic development.

REFERENCES

Acharya, A., Das, I., Chandhok, D., Saha, T. 2010. Redox regulation in cancer: A double-edged sword with therapeutic potential. *Oxidative Medicine and Cell Longevity* 3(1):23–34.

Barbano, R., Muscarella, L.A., Pasculli, B., Valori, V.M., Fontana, A., Coco, M., la Torre, A. et al. 2013. Aberrant Keap1 methylation in breast cancer and association with clinicopathological features. *Epigenetics* 8(1):105–112.

Becuwe, P., Ennen, M., Klotz, R., Barbieux, C., Grandemange, S. 2014. Manganese superoxide dismutase in breast cancer: From molecular mechanisms of gene regulation to biological and clinical significance. *Free Radicals in Biology and Medicine* 77:139–151.

Biswas, D.K., Iglehart, J.D. 2006. Linkage between EGFR family receptors and nuclear factor kappaB (NF-kappaB) signaling in breast cancer. *Journal of Cellular Physiology* 209(3):645–652.

Bogdan, C., Röllinghoff, M., Diefenbach, A. 2000. Reactive oxygen and reactive nitrogen intermediates in innate and specific immunity. *Current Opinion in Immunology* 12(1):64–76.

Brown, N.S., Bicknell, R. 2001. Hypoxia and oxidative stress in breast cancer. Oxidative stress: Its effects on the growth, metastatic potential and response to therapy of breast cancer. *Breast Cancer Research* 3(5):323–327.

Cadenas, C., Franckenstein, D., Schmidt, M., Gehrmann, M., Hermes, M., Geppert, B., Schormann, W. et al. 2010. Role of thioredoxin reductase 1 and thioredoxin interacting protein in prognosis of breast cancer. *Breast Cancer Research* 12(3):R44.

Cao, Y., Karin, M. 2003. NF-kappaB in mammary gland development and breast cancer. *Journal of Mammary Gland Biology and Neoplasia* 8(2):215–223.

Cha, M.K., Suh, K.H., Kim, I.H. 2009. Overexpression of peroxiredoxin I and thioredoxin1 in human breast carcinoma. *Journal of Experimental and Clinical Cancer Research* 28:93.

Chawla, A., Alatrash, G., Wu, Y., Mittendorf, E.A. 2013. Immune aspects of the breast tumor microenvironment. *Breast Cancer Management* 2(3):231–244.

Chin, Y., Janseens, J., Vandepitte, J., Vandenbrande, J., Opdebeek, L., Raus, J. 1992. Phenotypic analysis of tumor-infiltrating lymphocytes from human breast cancer. *Anticancer Research* 12(5):1463–1466.

Cichon, M.A., Radisky, D.C. 2014. ROS-induced epithelial-mesenchymal transition in mammary epithelial cells is mediated by NF-κB-dependent activation of Snail. *Oncotargets* 5(9):2827–2838.

Coffelt, S.B., Kersten, K., Doornebal, C.W., Weiden, J., Vrijland, K., Hau, C.S., Verstegen, N.J. et al. 2015. IL-17-producing γδ T cells and neutrophils conspire to promote breast cancer metastasis. *Nature* 522(7556):345–348.

Colotta, F., Allavena, P., Sica, A., Garlanda, C., Mantovani, A. 2009. Cancer-related inflammation, the seventh hallmark of cancer: Links to genetic instability. *Carcinogenesis* 30(7):1073–1081.

Coussens, L.M., Werb, Z. 2002. Inflammation and cancer. *Nature* 420(6917):860–867.

da Cunha, A., Michelin, M.A., Murta, E.F. 2014. Pattern response of dendritic cells in the tumor microenvironment and breast cancer. *World Journal of Clinical Oncology* 5(3):495–502.

Dhar, S.K., St Clair, D.K. 2012. Manganese superoxide dismutase regulation and cancer. *Free Radicals in Biology and Medicine* 52(11–12):2209–2222.

Diaz-Montero, C.M., Salem, M.L., Nishimura, M.I., Garrett-Mayer, E., Cole, D.J., Montero, A.J. 2009. Increased circulating myeloid-derived suppressor cells correlate with clinical cancer stage, metastatic tumor burden, and doxorubicin-cyclophosphamide chemotherapy. *Cancer Immunology Immunotherapy* 58(1):49–59.

Ennen, M., Minig, V., Grandemange, S., Touche, N., Merlin, J.L., Besancenot, V., Brunner, E., Domenjoud, L., Becuwe, P. 2012. Regulation of the high basal expression of the manganese superoxide dismutase gene in aggressive breast cancer cells. *Free Radicals in Biology and Medicine* 50(12):1771–1779.

Fiaschi, T., Chiarugi, P. 2012. Oxidative stress, tumor microenvironment, and metabolic reprogramming: A diabolic liaison. *International Journal of Cell Biology* 2012:762825.

Geymayer, S., Doppler, W. 2000. Activation of NF-kappaB p50/p65 is regulated in the developing mammary gland and inhibits STAT5-mediated beta-casein gene expression. *FASEB Journal* 14(9):1159–1170.

Glorieux, C., Auquier, J., Dejeans, N., Sid, B., Demoulin, J.B., Bertrand, L., Verrax, J., Calderon, P.B. 2014. Catalase expression in MCF-7 breast cancer cells is mainly controlled by PI3K/Akt/mTor signaling pathway. *Biochemical Pharmacology* 89(2):217–223.

Glorieux, C., Dejeans, N., Sid, B., Beck, R., Calderon, P.B., Verrax, J. 2011. Catalase overexpression in mammary cancer cells leads to a less aggressive phenotype and an altered response to chemotherapy. *Biochemical Pharmacology* 82(10):1384–1390.

Gonzalez-Dosal, R., Horan, K.A., Rahbek, S.H., Ichijo, H., Chen, Z.J., Mieyal, J.J., Hartmann, R., Paludan, S.R. 2011. HSV infection induces production of ROS, which potentiate signaling from pattern recognition receptors: Role for S-glutathionylation of TRAF3 and 6. *PLoS Pathogens* 7(9):e1002250.

Gorrini, C., Baniasadi, P.S., Harris, I.S., Silvester, J., Inoue, S., Snow, B., Joshi, P.A. et al. 2013. BRCA1 interacts with Nrf2 to regulate antioxidant signaling and cell survival. *Journal of Experimental Medicine* 210(8):1529–1544.

Gupta, S.C., Sundaram, C., Reuter, S., Aggarwal, BB. 2010. Inhibiting NF-κB activation by small molecules as a therapeutic strategy. *Biochimica Biophysica Acta* 1799(10):775–787.

Habib, E., Linher-Melville, K., Lin, H.X., Singh, G. 2015. Expression of xCT and activity of system xc- are regulated by NRF2 in human breast cancer cells in response to oxidative stress. *Redox Biology* 5:33–42.

Halliwell, B. 2000. Lipid peroxidation, antioxidants and cardiovascular disease: How should we move forward? *Cardiovascular Research* 47(3):410–418.

Halliwell, B. 2007. Oxidative stress and cancer: Have we moved forward? *Biochemical Journal* 401(1):1–11.

Ham, M., Moon, A. 2013. Inflammatory and microenvironmental factors involved in breast cancer progression. *Archives of Pharmacal Research* 36(12):1419–1431.

Harris, I.S., Treloar, A.E., Inoue, S., Sasaki, M., Gorrini, C., Lee, K.C., Yung, K.Y. et al. 2015. Glutathione and thioredoxin antioxidant pathways synergize to drive cancer initiation and progression. *Cancer Cell* 27(2):211–222.

Hartikainen, J.M., Tengström, M., Kosma, V.M., Kinnula, V.L., Mannermaa, A., Soini, Y. 2012. Genetic polymorphisms and protein expression of NRF2 and Sulfiredoxin predict survival outcomes in breast cancer. *Cancer Research* 72(21):5537–5546.

Helal, T.E., Ibrahim, E.A., Alloub, A.I. 2013. Immunohistochemical analysis of tumor-infiltrating lymphocytes in breast carcinoma: Relation to prognostic variables. *Indian Journal of Pathology and Microbiology* 56(2):89–93.

Herrera, A.C., Panis, C., Victorino, V.J., Campos, F.C., Colado-Simão, A.N., Cecchini, A.L., Cecchini, R. 2012. Molecular subtype is determinant on inflammatory status and immunological profile from invasive breast cancer patients. *Cancer Immunology and Immunotherapy* 61(11):2193–2201.

Herrera, A.C., Victorino, V.J., Campos, F.C., Verenitach, B.D., Lemos, L.T., Aranome, A.M., Oliveira, S.R. et al. 2014. Impact of tumor removal on the systemic oxidative profile of patients with breast cancer discloses lipidperoxidation at diagnosis as a putative marker of disease recurrence. *Clinical Breast Cancer* 14(6):451–459.

Houghton, A.M. 2010. The paradox of tumor-associated neutrophils: Fueling tumor growth with cytotoxic substances. *Cell Cycle* 9(9):1732–1737.

Huber, M.A., Azoitei, N., Baumann, B., Grünert, S., Sommer, A., Pehamberger, H., Kraut, N., Beug, H., Wirth, T. 2004. NF-kappaB is essential for epithelial-mesenchymal transition and metastasis in a model of breast cancer progression. *Journal of Clinical Investigation* 114(4):569–581.

Itoh, K., Wakabayashi, N., Katoh, Y., Ishii, T., Igarashi, K., Engel, J.D., Yamamoto, M. 1999. Keap1 represses nuclear activation of antioxidant responsive elements by Nrf2 through binding to the amino-terminal Neh2 domain. *Genes & Development* 13(1):76–86.

Kang, H.J., Yi, Y.W., Hong, Y.B., Kim, H.J., Jang, Y.J., Seong, Y.S., Bae, I. 2014. HER2 confers drug resistance of human breast cancer cells through activation of NRF2 by direct interaction. *Scientific Reports* 4:7201.

Karin, M. 2006. Nuclear factor-kappaB in cancer development and progression. *Nature* 441(7092):431–436.

Kendellen, M.F., Bradford, J.W., Lawrence, C.L., Clark, K.S., Baldwin, A.S. March 6, 2014. Canonical and noncanonical NF-κB signaling promotes breast cancer tumor-initiating cells. *Oncogene* 33(10):1297–1305.

Keum, Y.S., Choi, B.Y. 2014. Molecular and chemical regulation of the Keap1-Nrf2 signaling pathway. *Molecules* 19(7):10074–10089.

Khatri, R., Shah, P., Guha, R., Rassool, F.V., Tomkinson, A.E., Brodie, A., Jaiswal, A.K. 2015. Aromatase inhibitor-mediated down regulation of INrf2 (Keap1) leads to increased Nrf2 and resistance in breast cancer. *Molecular Cancer Therapeutics* 14(7):1728–1737.

Kilic, N., Yavuz Taslipinar, M., Guney, Y., Tekin, E., Onuk, E. 2014. An investigation into the serum thioredoxin, superoxide dismutase, malondialdehyde, and advanced oxidation protein products in patients with breast cancer. *Annals of Surgical Oncology* 21(13):4139–4143.

Kim, D., Koo, J.S., Lee, S. 2015. Overexpression of reactive oxygen species scavenger enzymes is associated with a good prognosis in triple-negative breast cancer. *Oncology* 88(1):9–17.

Kim, M.J., Kim, D.H., Na, H.K., Surh, Y.J. 2010. TNF-α induces expression of urokinase-type plasminogen activator and β-catenin activation through generation of ROS in human breast epithelial cells. *Biochemical Pharmacology* 80(12):2092–2100.

Kim, S.J., Miyoshi, Y., Taguchi, T., Tamaki, Y., Nakamura, H., Yodoi, J., Kato, K., Noguchi, S. 2005. High thioredoxin expression is associated with resistance to docetaxel in primary breast cancer. *Clinical Cancer Research* 11(23):8425–8430.

Kim, S.K., Yang, J.W., Kim, M.R., Roh, S.H., Kim, H.G., Lee, K.Y., Jeong, H.G., Kang, K.W. 2008. Increased expression of Nrf2/ARE-dependent anti-oxidant proteins in tamoxifen-resistant breast cancer cells. *Free Radicals in Biology and Medicine* 45(4):537–546.

Leek, R.D., Harris, A.L. 2002. Tumor-associated macrophages in breast cancer. *Journal of Mammary Gland Biology and Neoplasia* 7(2):177–189.

Lemos, L.G., Victorino, V.J., Herrera, A.C., Aranome, A.M., Cecchini, A.L., Simão, A.N., Panis, C., Cecchini, R. 2015. Trastuzumab-based chemotherapy modulates systemic redox homeostasis in women with HER2-positive breast cancer. *International Immunopharmacology* 27(1):8–14.

Liu, M., Sakamaki, T., Casimiro, M.C., Willmarth, N.E., Quong, A.A., Ju, X., Ojeifo, J., Jiao, X. et al. 2010. The canonical NF-kappaB pathway governs mammary tumorigenesis in transgenic mice and tumor stem cell expansion. *Cancer Research* 70(24):10464–10473.

Loignon, M., Miao, W., Hu, L., Bier, A., Bismar, T.A., Scrivens, P.J., Mann, K. et al. 2009. Cul3 overexpression depletes Nrf2 in breast cancer and is associated with sensitivity to carcinogens, to oxidative stress, and to chemotherapy. *Molecular Cancer Therapeutics* 8(8):2432–2440.

Lu, T., Ramakrishnan, R., Altiok, S., Youn, J.I., Cheng, P., Celis, E., Pisarev, V., Sherman, S., Sporn, M.B., Gabrilovich, D. 2011. Tumor-infiltrating myeloid cells induce tumor cell resistance to cytotoxic T cells in mice. *Journal of Clinical Investigation* 121(10):4015–4029.

Macchetti, A.H., Marana, H.R., Silva, J.S., de Andrade, J.M., Ribeiro-Silva, A., Bighetti, S. 2006. Tumor-infiltrating CD4+ T lymphocytes in early breast cancer reflect lymph node involvement. *Clinics* 61(3):203–208.

Madeddu, C., Gramignano, G., Floris, C., Murenu, G., Sollai, G., Macciò, A. 2014. Role of inflammation and oxidative stress in post-menopausal oestrogen-dependent breast cancer. *Journal of Cellular and Molecular Medicine* 18(12):2519–2529.

Mahmoud, S.M., Lee, A.H., Paish, E.C., Macmillan, R.D., Ellis, I.O., Green, A.R. 2012. The prognostic significance of B lymphocytes in invasive carcinoma of the breast. *Breast Cancer Research and Treatment* 132(2):545–553.

Mahmoud, S.M., Paish, E.C., Powe, D.G., Macmillan, R.D., Grainge, M.J., Lee, A.H., Ellis, I.O., Green, A.R. 2011. Tumor-infiltrating CD8+ lymphocytes predict clinical outcome in breast cancer. *Journal of Clinical Oncology* 29(15):1949–1955.

Manandhar, S., Choi, B.H., Jung, K.A., Ryoo, I.G., Song, M., Kang, S.J., Choi, H.G., Kim, J.A., Park, P.H., Kwak, M.K. 2012. NRF2 inhibition represses ErbB2 signaling in ovarian carcinoma cells: Implications for tumor growth retardation and docetaxel sensitivity. *Free Radicals in Biology and Medicine* 52(9):1773–1785.

Mencalha, A., Victorino, V.J., Cecchini, R., Panis, C. 2014. Mapping oxidative changes in breast cancer: Understanding the basic to reach the clinics. *Anticancer Research* 34(3):1127–1140.

Merkhofer, E.C., Cogswell, P., Baldwin, A.S. February 25, 2010. Her2 activates NF-kappaB and induces invasion through the canonical pathway involving IKKalpha. *Oncogene* 29(8):1238–1248.

Mittendorf, E.A., Alatrash, G., Qiao, N., Wu, Y., Sukhumalchandra, P., St John, L.S., Philips, A.V. et al. 2012. Breast cancer cell uptake of the inflammatory mediator neutrophil elastase triggers an anticancer adaptive immune response. *Cancer Research* 72(13):3153–3162.

Moon, E.J., Giaccia, A. 2015. Dual roles of NRF2 in tumor prevention and progression: Possible implications in cancer treatment. *Free Radicals in Biology and Medicine* 79:292–299.

Mukhtar, R.A., Moore, A.P., Tandon, V.J., Nseyo, O., Twomey, P., Adisa, C.A., Eleweke, N. et al. 2012. *Annals of Surgical Oncology* 19(12):3979–3986.

Nakshatri, H., Bhat-Nakshatri, P., Martin, D.A., Goulet, R.J. Jr, Sledge, G.W. Jr. 1997. Constitutive activation of NF-kappaB during progression of breast cancer to hormone-independent growth. *Molecular and Cellular Biology* 17(7):3629–3639.

Nathan, C., Cunningham-Bussel, A. 2013. Beyond oxidative stress: An immunologist's guide to reactive oxygen species. *Nature Reviews Immunology* 13(5):349–361.

Nioi, P., Nguyen, T. 2007. A mutation of Keap1 found in breast cancer impairs its ability to repress Nrf2 activity. *Biochemical and Biophysical Research Communications* 362(4):816–821.

Oakes, S.R., Hilton, H.N., Ormandy, C.J. 2006. The alveolar switch: Coordinating the proliferative cues and cell fate decisions that drive the formation of lobule alveoli from ductal epithelium. *Breast Cancer Research* 8(2):207.

Obeid, E., Nanda, R., Fu, Y.X., Olopade, O.I. 2013. The role of tumor-associated macrophages in breast cancer progression (review). *International Journal of Oncology* 43(1):5–12.

Onodera, Y., Motohashi, H., Takagi, K., Miki, Y., Shibahara, Y., Watanabe, M., Ishida, T. et al. 2014. NRF2 immunolocalization in human breast cancer patients as a prognostic factor. *Endocrine-Related Cancer* 21(2):241–252.

Paiva, C.N., Bozza, M.T. 2014. Are reactive oxygen species always detrimental to pathogens? *Antioxidants and Redox Signaling* 20(6):1000–1037.

Panis, C., Herrera, A.C., Victorino, V.J., Campos, F.C., Freitas, L.F., De Rossi, T., Colado Simão, A.N., Cecchini, A.L., Cecchini, R. 2012. Oxidative stress and hematological profiles of advanced breast cancer patients subjected to paclitaxel or doxorubicin chemotherapy. *Breast Cancer Research and Treatment* 133(1):89–97.

Panis, C., Pizzatti, L., Herrera, A.C., Cecchini, R., Abdelhay, E. 2013. Putative circulating markers of the early and advanced stages of breast cancer identified by high-resolution label-free proteomics. *Cancer Letters* 330(1):57–66.

Panis, C., Victorino, V.J., Herrera, A.C., Freitas, L.F., De Rossi, T., Campos, F.C., Simão, A.N. et al. 2012. Differential oxidative status and immune characterization of the early and advanced stages of human breast cancer. *Breast Cancer Research and Treatment* 133(3):881–888.

Park, B.J., Cha, M.K., Kim, I.H. 2014. Thioredoxin 1 as a serum marker for breast cancer and its use in combination with CEA or CA15–3 for improving the sensitivity of breast cancer diagnoses. *BMC Research Notes* 7:7.

Park, M.H., Lee, J.S., Yoon, J.H. 2012. High expression of CX3CL1 by tumor cells correlates with a good prognosis and increased tumor-infiltrating CD8+ T cells, natural killer cells, and dendritic cells in breast carcinoma. *Journal of Surgical Oncology* 106(4):386–392.

Payne, K.K., Manjili, M.H. 2012. Adaptive immune responses associated with breast cancer relapse. *Archivum Immunologiae et Therapia Experimentalis* 60(5):345–350.

Polyak, K., Metzger Filho, O. 2012. SnapShot: Breast cancer. *Cancer Cell* 22(4):562–562.

Pratt, M.A., Tibbo, E., Robertson, S.J., Jansson, D., Hurst, K., Perez-Iratxeta, C., Lau, R., Niu, M.Y. 2009. The canonical NF-kappaB pathway is required for formation of luminal mammary neoplasias and is activated in the mammary progenitor population. *Oncogene* 28(30):2710–2722.

Qu, Y., Wang, J., Ray, P.S., Guo, H., Huang, J., Shin-Sim, M., Bukoye, B.A. et al. 2011. Thioredoxin-like 2 regulates human cancer cell growth and metastasis via redox homeostasis and NF-κB signaling. *Journal of Clinical Investigation* 121(1):212–225.

Queen, M.M., Ryan, RE.., Holzer, R.G., Keller-Peck, C.R., Jorcyk, C.L. 2005. Breast cancer cells stimulate neutrophils to produce oncostatin M: Potential implications for tumor progression. *Cancer Research* 65(19):8896–8904.

Rego, S.L., Helms, R.S., Dréau, D. 2014. Tumor necrosis factor-alpha-converting enzyme activities and tumor-associated macrophages in breast cancer. *Immunology Research* 58(1):87–100.

Richardson, B.G., Jain, A.D., Speltz, T.E., Moore, T.W. 2015. Non-electrophilic modulators of the canonical Keap1/Nrf2 pathway. *Bioorganic & Medicinal Chemistry Letters* 25(11):2261–2268.

Ryoo, I.G., Choi, B.H., Kwak, M.K. 2015. Activation of NRF2 by p62 and proteasome reduction in sphere-forming breast carcinoma cells. *Oncotarget* 6(10):8167–8184.

Sabzichi, M., Hamishehkar, H., Ramezani, F., Sharifi, S., Tabasinezhad, M., Pirouzpanah, M., Ghanbari, P., Samadi, N. 2014. Luteolin-loaded phytosomes sensitize human breast carcinoma MDA-MB 231 cells to doxorubicin by suppressing Nrf2 mediated signalling. *Asian Pacific Journal of Cancer Prevention* 15(13):5311–5316.

Samadi, A.K., Bilsland, A., Georgakilas, A.G., Amedei, A., Amin, A, Bishayee, A., Azmi, A.S. et al. 2015. A multi-targeted approach to suppress tumor-promoting inflammation. *Seminars in Cancer Biology* 35 Suppl:S151–S184.

Schramek, D., Leibbrandt, A., Sigl, V., Kenner, L., Pospisilik, J.A., Lee, H.J., Hanada, R. et al. 2010. Osteoclast differentiation factor RANKL controls development of progestin-driven mammary cancer. *Nature Letters* 468(7320):98–102.

Shen, H.M., Pervaiz, S. 2006. TNF receptor superfamily-induced cell death: Redox-dependent execution. *FASEB Journal* 20(10):1589–1598.

Sigl, V., Penninger, J.M. 2014. RANKL/RANK—From bone physiology to breast cancer. *Cytokine & Growth Factor Reviews* 25(2):205–214.

Singh, R., Shankar, B.S., Sainis, K.B. 2014. TGF-β1-ROS-ATM-CREB signaling axis in macrophage mediated migration of human breast cancer MCF7 cells. *Cell Signaling* 26(7):1604–1615.

Sørlie, T., Perou, C.M., Tibshirani, R., Aas, T., Geisler, S., Johnsen, H., Hastie, T. et al. 2001. Gene expression patterns of breast carcinomas distinguish tumor subclasses with clinical implications. *Proceedings of the National Academy of Sciences U S A* 98(19):10869–10874.

Sun, Z., Zhang, S., Chan, J.Y., Zhang, D.D. 2007. Keap1 controls postinduction repression of the Nrf2-mediated antioxidant response by escorting nuclear export of Nrf2. *Molecular and Cell Biology* 27(18):6334–6349.

Tal, M.C., Sasai, M., Lee, H.K., Yordy, B., Shadel, G.S., Iwasaki, A. 2009. Absence of autophagy results in reactive oxygen species-dependent amplification of RLR signaling. *Proceedings of the National Academy of Sciences U S A* 106(8):2770–2775.

Tang, X. 2013. Tumor-associated macrophages as potential diagnostic and prognostic biomarkers in breast cancer. *Cancer Letters* 332(1):3–10.

Tiede, B., Kang, Y. 2011. From milk to malignancy: The role of mammary stem cells in development, pregnancy and breast cancer. *Cell Research* 21(2):245–257.

Vandenabeele, P., Galluzzi, L., Vanden Berghe, T., Kroemer, G. 2010. Molecular mechanisms of necroptosis: An ordered cellular explosion. *Nature Reviews Molecular Cell Biology* 11(10):700–714.

Venugopal, R., Jaiswal, A.K. 1996. Nrf1 and Nrf2 positively and c-Fos and Fra1 negatively regulate the human antioxidant response element-mediated expression of NAD(P)H:Quinone oxidoreductase1 gene. *Proceedings of the National Academy of Sciences U S A* 93(25):14960–14965.

Venugopal, R., Jaiswal, A.K. 1998. Nrf2 and Nrf1 in association with Jun proteins regulate antioxidant response element-mediated expression and coordinated induction of genes encoding detoxifying enzymes. *Oncogene* 17(24):3145–3156.

Vera-Ramirez, L., Sanchez-Rovira, P., Ramirez-Tortosa, M.C., Ramirez-Tortosa, C.L., Granados-Principal, S., Lorente, J.A., Quiles, J.L. 2012. Oxidative stress status in metastatic breast cancer patients receiving palliative chemotherapy and its impact on survival rates. *Free Radical Research* 46(1):2–10.

Victorino, V.J., Campos, F.C., Herrera, A.C., Colado-Simão, A.N., Cecchini, A.L., Panis, C., Cecchini, R. 2014. Overexpression of HER-2/neu protein attenuates the oxidative systemic profile in women diagnosed with breast cancer. *Tumour Biology* 35(4):3025–3034.

Voduc, K.D., Cheang, M.C., Tyldesley, S., Gelmon, K., Nielsen, T.O., Kennecke, H. 2010. Breast cancer subtypes and the risk of local and regional relapse. *Journal of Clinical Oncology* 28(10):1684–1691.

Vurusaner, B., Poli, G., Basaga, H. 2012. Tumor suppressor genes and ROS: Complex networks of interactions. *Free Radicals in Biology and Medicine* 52(1):7–18.

Wang, Q., Li, J., Yang, X., Sun, H., Gao, S., Zhu, H., Wu, J., Jin, W. 2013. Nrf2 is associated with the regulation of basal transcription activity of the BRCA1 gene. *Acta Biochimica et Biophysica Sinica* 45(3):179–187.

Wang, X., Belguise, K., Kersual, N., Kirsch, K.H., Mineva, N.D., Galtier, F., Chalbos, D., Sonenshein, G.E. 2007. Oestrogen signalling inhibits invasive phenotype by repressing RelB and its target BCL2. *Nature Cell Biology* 9(4):470–478.

Wu, J., Williams, D., Walter, G.A., Thompson, W.E., Sidell, N. 2014b. Estrogen increases Nrf2 activity through activation of the PI3K pathway in MCF-7 breast cancer cells. *Experimental Cell Research* 328(2):351–60.

Wu, K.C., Cui, J.Y., Klaassen, C.D. 2012. Effect of graded Nrf2 activation on phase-I and -II drug metabolizing enzymes and transporters in mouse liver. *PLOS ONE* 7(7):e39006.

Wu, T., Harder, B.G., Wong, P.K., Lang, J.E., Zhang, D.D. 2014a. Oxidative stress, mammospheres and Nrf2-new implication for breast cancer therapy? *Molecular Carcinogenesis* 54(11):1494–1502.

Xiao, C., Liu, Y.L., Xu, J.Q., Lv, S.W., Guo, S., Huang, W.H. 2015. Real-time monitoring of H2O2 release from single cells using nanoporous gold microelectrodes decorated with platinum nanoparticles. *Analyst* 140(11):3753–3758.

Yoo, N.J., Kim, H.R., Kim, Y.R., An, C.H., Lee, S.H. 2012. Somatic mutations of the KEAP1 gene in common solid cancers. *Histopathology* 60(6):943–952.

Yost, C.C., Cody, M.J., Harris, E.S., Thornton, N.L., McInturff, A.M., Martinez, M.L., Chandler, N.B. et al. 2009. Impaired neutrophil extracellular trap (NET) formation: A novel innate immune deficiency of human neonates. *Blood* 113(25):6419–6427.

Zhang, D.D., Lo, S.C., Cross, J.V., Templeton, D.J., Hannink, M. 2004. Keap1 is a redox-regulated substrate adaptor protein for a Cul3-dependent ubiquitin ligase complex. *Molecular and Cellular Biology* 24(24):10941–10953.

Zhang, Y., Choksi, S., Chen, K., Pobezinskaya, Y., Linnoila, I., Liu, Z.G. 2013. ROS play a critical role in the differentiation of alternatively activated macrophages and the occurrence of tumor-associated macrophages. *Cell Research* 23(7):898–914.

Zhong, Y., Zhang, F., Sun, Z., Zhou, W., Li, Z.Y., You, Q.D., Guo, Q.L., Hu, R. 2013. Drug resistance associates with activation of Nrf2 in MCF-7/DOX cells, and wogonin reverses it by down-regulating Nrf2-mediated cellular defense response. *Molecular Carcinogenesis* 52(10):824–834.

19 Reactive Oxygen Species in Melanoma Etiology

Feng Liu-Smith

CONTENTS

ABSTRACT

Melanoma is the deadliest skin cancer that exhibits high genome instability. Reactive oxygen species (ROS) play an essential role in melanomagenesis. The most important environmental risk factor ultraviolet radiation induces ROS generation, through the cellular enzymatic reactions (mainly NADPH oxidases and nitric oxide synthase), nonenzymatic reactions (melanin types, intermediates, and others), and organelle-associated ROS (endoplasmic reticulum, melanosomes, and mitochondria). ROS from these different sources interact with each other and may be maintained at a high level in melanoma cells. Meanwhile, melanoma cells also express high levels of antioxidant regulators including NRF2 and apurinic/apyrimidinic endonuclease/redox-repair protein 1 (APE/REF-1), which maintain a high level of antioxidants including glutathione and thioredoxin. This chapter summarizes the source of ROS in melanoma development and presents evidence that cancer cells achieve a new balance between ROS and antioxidants. A continuous and persistent disruption of the ROS/antioxidant balance by genetic and/or environmental factors may be responsible for genome instability and melanomagenesis.

19.1 INTRODUCTION

Cutaneous melanomas arise from melanin-producing skin pigment cells termed melanocytes, which belong to the neuron crest cell lineage. Human melanocytes play a protective role against the environmental insults such as solar ultraviolet radiation (UVR) or man-made UV lamps, the single most important environmental risk factor for melanoma [1–5]. While melanin biosynthesis is stimulated by UV radiation for preventing further skin damage, the intermediates for melanin biosynthesis produce oxidative stress [6] and are potential hazards for skin damage. There is evidence that melanin-producing organelles (melanosomes) can be damaged by UV radiation, leading

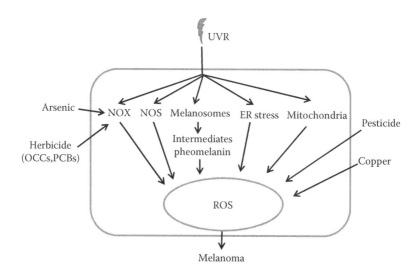

FIGURE 19.1 The source of ROS in melanoma development. The major risk factor UV radiation induces cellular ROS pools via NOX, NOS, ER stress, melanosomes, and mitochondria, which exhibit cross interaction with each other. Other risk factors arsenic and herbicides are shown to induce ROS via at least NOX enzymes; pesticides and copper may also function via an ROS mechanism with unspecified pathways.

to leakage of the toxic intermediates and exacerbating the deleterious effect of UVR via reactive oxygen species (ROS) pathway [7]. Pheomelanins (the type of yellow and red melanin) exhibit pro-oxidant property independent of UV radiation [8]. Furthermore, UVA-induced melanoma in mice required the presence of melanin [9]. This may account at least partly for higher ROS levels observed in melanocytes and melanoma cells compared to fibroblasts [10]. Consequently, mutations in genes crucial for determining melanin types and melanin synthesis are associated with melanoma risk, including melanocortin receptor 1 (Mc1R), microphthalmia transcription factor (MiTF), Agouti signaling protein (ASIP), and tyrosinase [11–15]. In addition, other enzymatically generated ROS, via NADPH oxidase (NOX) and nitric oxide synthase, have also emerged to play an important role in melanoma genesis along with ROS generated in mitochondria and endoplasmic reticulum (ER) [16,17]. UV radiation induces ROS via these cellular organelles and/ or enzymes (Figure 19.1). Other melanoma risk factors such as herbicides, pesticides, and heavy metals may also function via a similar pathway (Figure 19.1). In this chapter, the most recent melanocyte-specific source of ROS and its impact on cellular transformation and tumor progression will be summarized and discussed.

19.2 SOURCE OF ROS IN MELANOCYTES AND MELANOMA

19.2.1 UV Radiation and ROS in Melanoma Development

Epidemiology, genetics, and molecular studies have all confirmed that solar and artificial UV radiations play a pivotal role in melanoma development. UV radiation is classified into three groups by their wavelengths: UVA (320–400 nm), UVB (280–320 nm), and UVC (100–280 nm) [18]. Essentially, no UVC can penetrate the stratosphere and reach the earth surface; while most UVA and a small portion of UVB compose the ground solar UV radiation, with 90%–95% of energy from UVA and ~5%–10% from UVB. UVB can only reach the epidermis layer of human skin, while UVA can reach the dermal/epidermal junction where melanocytes reside. Therefore, there have been debates on which UV wavelength is more important for human melanoma genesis.

Most researchers agree that UVB generates UV-signature mutations mainly via direct interaction with DNA, and UVA causes DNA damage mainly via ROS-mediated oxidative stress, with 8-oxo-7,8-dihydroxyguanine (8-oxoG) and single-strand break (SSB) as the major products. Unique and shared redox signaling pathways in melanocytes and melanomas are involved in the cellular responses to UVA and UVB radiations.

It has long been recognized that UVA radiation causes an increase in ROS in a range of cell types [19]. The increased ROS triggers a chain of signal transduction events leading to DNA repair, cell death, and/or transformation. However, how ROS was generated by UVA has never been explained well. Early research suggested that UVA induces ROS via cellular photosensitizers, such as porphirines or NADH, or mitochondrial dysfunction, neither of which was well characterized [20–23]. More recently, it has become evident that specific ROS-generating enzymes are involved [24,25], including NOXs and nitric oxide synthases (NOSs) [16,26]. The involvement of UV-induced ER stress has also emerged to play a role.

19.2.2 NOX FAMILY

The NOX family enzymes comprise seven members: NOX1, NOX2, NOX3, NOX4, NOX5, DUOX1, and DUOX2, each having different tissue specificity and response to different stimuli [27]. NOX enzymes catalyze NAD(P)H oxidation by oxygen molecules and generate superoxide ($O_2^{\cdot-}$) or hydrogen peroxide (H_2O_2) [27]. In keratinocytes, NOX1 is the major source of ROS induced by UVA [24]. In cultured normal human melanocytes, NOX1 is the only NOX family members that can be detected at RT-PCR level [28]. NOX1 was found in melanoma cell lines and samples from all stages, while NOX4 and NOX5 were found in the subsets of metastatic melanoma samples [28–30].

Total NOX activity was measured after UVA or UVB irradiation in melanoma cell line SK-Mel28. UVA (2.5 J/cm²) induced NOX activity to about sixfold of the baseline 6 h post radiation, which lasted to 24 h (Figure 19.2). Western blot revealed the induction of NOX1 protein accumulation in both SK-Mel28 cell line and normal human melanocytes after UVA radiation [31]. UVB (25 mJ/cm²) also induced NOX activity to 2.5-fold of baseline and lasted to 24 h (Figure 19.2). Both UVA and UVB induced ROS (measured by 2′,7″-dichlorodihydrofluorescein [DCFDA]) in normal melanocytes with a peak 6 h postradiation, consistent with NOX activity kinetics [31]. These results

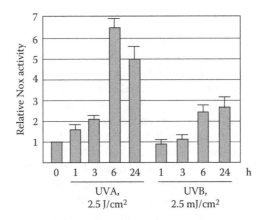

FIGURE 19.2 UVA and UVB induced total NOX activity. SK-Mel28 melanoma cells were exposed to UVA (2.5 J/cm²) and UVB (25 mJ/cm²) and collected at the indicated time points. Total NOX activity was measured as described earlier [28] and was normalized to total protein level first and then again to the unirradiated control cells. Data for each time points represent the average activities for at least three replicates.

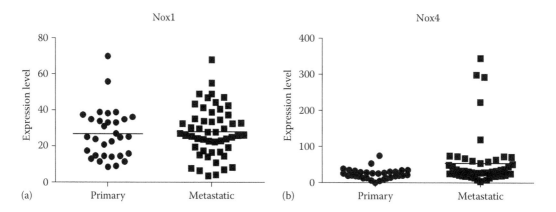

FIGURE 19.3 Expression levels of (a) NOX1 and (b) NOX4 in primary and metastatic melanoma samples. NOX1 and NOX4 mRNA expression levels were retrieved from NCBI GEO database (accession number GDS3966) for 31 primary and 52 metastasis melanoma samples. The level bars represent average expression levels, and each dot represents expression level from one sample.

suggest a role of NOX1 in UV-induced ROS generation in melanocytes, but further experiments are needed to determine whether NOX and/or which NOX (in addition to NOX1) is responsible for the UV-induced ROS generation.

Using gene microarray expression data obtained from the Gene Expression Omnibus (GEO) of National Center of Biotechnology Information [32], we discovered that the average NOX1 mRNA expression level was comparable between 31 primary melanomas and 52 melanoma metastasis (Figure 19.3). In contrast, the average NOX4 expression level was much higher in metastatic melanoma than in primary tumors (Figure 19.3), with a few samples exhibiting extremely high level of NOX4 expression. These data may suggest different roles of these two functionally similar enzymes in melanoma etiology and progression [16]. NOX1 may play a prominent role in melanoma initiation (as suggested by its response to UV radiation), while NOX4 may drive metastasis, which is consistent with the role of NOX4 downstream of AKT pathway driving radial growth phase melanoma into vertical growth phase [29].

19.2.3 NOS Family

NOS family contains three members: NOS1 (nNOS, neuronal NOS), NOS2 (iNOS, inducible NOS), and NOS3 (eNOS, endothelial NOS). All three enzymes catalyze the biosynthesis of nitric oxide (NO), which plays a crucial role in many aspects of physiology and pathology. UVR-mediated NO release in skin induces a serial of reactions including melanogenesis, inflammation, erythema, and immunosuppression [33], as well as protecting keratinocytes from cell damage and cell death [26,34], perhaps dependent on the NO levels. A major reason that NOS enzymes are considered to be a source of ROS is that when NOS is uncoupled, the end product of NOS-catalyzed reaction will be superoxide instead of NO [35]. $O_2^{\cdot-}$ interacts with NO to form extremely reactive peroxynitrite (ONOO$^-$) and exacerbates the UVR effect on cell cycle arrest and cell death [36]. UVB induced protein accumulation of all three NOS enzymes in human keratinocytes, accompanied by a transient elevation of NO [37]. UVB also induced uncoupled NOS3 [37]. Whether similar events occur in human melanocytes remain to be further investigated, where all three NOS isozymes are expressed [38–44]. NOS1 was shown to be induced by UVA up to 72 h after radiation in melanoma cells but was only transiently induced (4–6 h postradiation) by UVB [45]. No upregulation of NOS1 was observed 24 h postradiation by UVB in either primary melanocytes or melanoma cells [45]. Nevertheless, it is clear that NO plays a role in UV-mediated melanogenesis, as a synthetic compound MHY966 was able to inhibit melanin synthesis via inhibiting NO production [46]. NOS1 also plays an important role in melanomagenesis. Evidence from an epidemiology study showed that a single nucleotide polymorphism (SNP) in NOS1 gene (rs2682826, 276C>T) conferred higher risk to melanoma (p = 0.01) [47]. A different hospital-based case–control study also revealed significant associations of melanoma risk with an additional NOS1 SNP (−84G>A, or −84AA+AG/276CT+TT haplotype) but not with the examined NOS2 SNPs [48].

19.2.4 ER Stress and ROS in Melanoma Etiology

In addition to the aforementioned enzymatically catalyzed ROS production, melanocytes also make their ROS pools via subcellular organelles, including mitochondria, ER, and melanosomes [17,49–51]. The mitochondrial pathway was not very well specified in melanocytes and melanoma; most results were suggested from other cell lineage [52]. The melanin and melanosome pathway was summarized elsewhere [17]. The ER pathway is a rapidly developing field, and its impact in melanomagenesis has achieved much progression recently [53–56]. ER processes protein post-translational modifications, which often involve disulfide bond formation, a typical redox process catalyzed by protein disulfide isomerases (PDIs) and ER oxidoreductase 1 (including Ero-1α, Ero-1β, and Ero-1p) [57]. In favor with the formation of disulfide bonds, ER is a relatively oxidizing cellular compartment as compared to nucleus, mitochondria, and cytoplasm [58]. Ero-1 enzymes utilize oxygen molecules for electron acceptors to oxidize PDIs and generate H_2O_2 as by-products [59]. It is estimated that about 25% of total ROS generated during protein synthesis and modification is via Ero-1 enzymes [60]. However, no Ero-1 regulation has been reported so far in melanocytes and melanoma. Most of the current studies regarding ER stress in melanoma are focused on unfolded protein response (UPR) and PDIs. UPR is triggered in ER when newly synthesized proteins are not processed properly and timely due to pathological or physiological stimuli; and ROS are integrated parts of this response due to the crucial roles of Ero-1s and PDIs [61]. UV radiation induced PDIA1 aggregation and inactivation in a cell-free *in vitro* test [62]. UVB triggered an immediate ROS release, which led to ER stress and activation of PERK-peIF2α-CHOP pathway in skin fibroblasts and immortalized human keratinocytes (HaCaT cells) [63]. Blocking ROS generation by antioxidant reduced this UVB-induced ER stress and UPR [63], suggesting that UV-induced ER stress may require ROS.

Proteomic analysis indicated that PDI precursor protein was overexpressed in A375 melanoma cell line as compared to normal melanocytes [64], indicating a chronic ER stress in these tumor cells. Melanoma cells exhibit high ROS levels, which may help to maintain a chronic ER stress. Consistently, the major melanoma oncogene BRAF (mutations found in ~60% of melanomas) induced a chronic ER stress and leads to reduced autophagy and resistance to apoptosis [56], which may help to explain the BRAF-mediated oncogenic pathway in melanomas from a different view point other than the canonical mitogen-activated protein kinases (MAPK) pathway.

Melanomas also express high levels of antioxidant system including master redox regulators apurinic/apyrimidinic endonuclease/redox-repair protein 1 (APE/REF-1) and NRF2 (further discussions later); and PDIA3/ERp57 has a secondary subcellular location in the nucleus and is capable of binding DNA fragments in melanoma cells [65]. Some of these DNA fragments are associated with the APE/REF-1 in an *in vitro* assay [65]. Further investigations are needed to characterize the complex relationships among melanoma oncogenes, ER stress, ROS, and their interactions induced by UV radiation.

It should be noted that ROS derived from these different sources exhibit feedback effect and hence coordinate with each other within a cell. Mitochondrial ROS is known to stimulate NOX-generated ROS [66]. Some NOX enzymes, especially NOX4, are functional units located in ER [67]. NOX1 is regulated by PDIs in vascular smooth cells and promotes cell migration [68,69].

Furthermore, emerging evidence shows that mitochondria and ER are in physical contact in mammalian cells [70]. Since both organelles are important for ROS generation and signaling, the biological function of these physical contacts may be interesting for cellular ROS balance.

19.2.5 OTHER ENVIRONMENTAL RISK FACTORS AND ROS IN MELANOMA DEVELOPMENT

Epidemiology studies suggested that in addition to UV radiation, heavy metals (including chromium, cobalt, and arsenic), herbicides, and pesticides are associated with melanoma risk [71–74]. In the Agricultural Health Study, melanoma incidence showed a borderline association with the use of acetochlor (a herbicide) (RR = 1.61; 95% confidence interval [CI]: 0.98–2.66) [73] and a significant association with maneb/mancozeb, parathion, and carbaryl [72], all of which are linked to ROS and oxidative stress [75–77]. In addition, acetochlor caused DNA SSB associated with excessive ROS production in liver cells of Bufo raddei tadpoles [78]. In a small case–control study, melanoma incidence was strongly associated with nondioxin-like polychlorinated biphenyls (PCBs) and other organochlorine compounds (OCCs) (odds ratio [OR]: 7.02; 95% CI: 2.30–21.43 for the highest quartile) after adjusting the sun sensitivity and sun exposure [79]. These organochlorine compounds are well documented to function via ROS mechanism, perhaps NOX enzymes [80].

Most heavy metals exhibit redox potentials. Much research has pointed to a link of heavy metals with melanoma risk. An early study revealed a significant increase in melanoma incidence in metal-on-metal hip arthroplasty patients [81], with accompanied increase in plasma chromium and cobalt concentrations [71]. However, more recent epidemiology studies with extremely large sample sizes from Finnish Arthroplastry Registry/Finnish Cancer Registry and National Joint Registry of England and Wales showed that there was no significant difference of melanoma incidence between the metal-on-metal replacement patients and patients with alternative bearings, or between the hip replacement patients and general population, for at least 7 years of follow-up time [82,83]. Longer follow-up is needed to draw further conclusions. These results are consistent with a study of toenail metal levels and melanoma [84]. Toenail metal levels of cadmium, chromium, lead, and selenium were not significantly associated with melanoma risk [84]; rather, the copper level is significantly associated with adjusted OR of 9.8 (95% CI: 2.6–36.2) [84]. This is a very interesting discovery indeed, as recent studies showed that copper played a crucial role in the MAPK pathway and melanoma proliferation [85]. Specifically, copper binds to MEK1 and stimulates its kinase activity and signals for cell proliferation and transformation; consequently, copper chelators effectively prevented BRAF^{V600E}-mediated cell transformation [85]. Copper is an essential element exhibiting redox potential between Cu^+ and Cu^{++}, which is utilized by several ROS metabolizing enzymes including SOD1 and SOD3. Copper is also essential for tyrosinase activity and tyrosinase trafficking via ATP7A, which itself requires Cu^{++} for activity [86]. Additional copper chaperones include ATP7B, antioxidant protein 1 (ATOX1), copper transporter 1 (CTR1), SLC31A1, copper transporter 2 (CTR2), and SLC31A2). Most of these copper chaperones are not very well studied in melanocytes and melanoma.

Data mining on the microarray analysis of various stages of melanoma (two to three biopsy samples for each stage) revealed an interesting trend of these CTRs/chaperones in melanoma progression (Figure 19.4) (data from NCBI Omnibus Gene Expression accession # GDS1989 [87]). Compared to benign nevi samples, atypical nevi, melanoma in situ, vertical growth phase, lymph node metastasis, and distal metastatic melanomas mostly exhibited higher expression levels of ATOX1, CTR1, and CTR2 but not ATP7A (Figure 19.4). Since copper is required for tyrosinase activity and melanomas express high level of tyrosinase, whether these associations of high levels of copper chaperones are coincident with melanin contents warrants further investigation. However, CTR1 is also highly expressed in prostate cancer [88], suggesting a role other than melanogenesis of these copper chaperones in melanoma development. The high copper levels may reflect high MAPK pathway activities as a recent study showed that CTR1 is required for MAPK

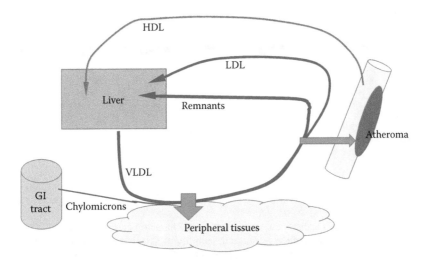

FIGURE 6.1 Basic lipid transport in plasma. The two major classes of lipoproteins are the A-lipoproteins (good cholesterol), represented by high-density lipoproteins or HDL, and the B-lipoproteins (bad cholesterol), represented by chylomicrons, very low-density lipoproteins or VLDL, remnants, and low-density lipoproteins or LDL.

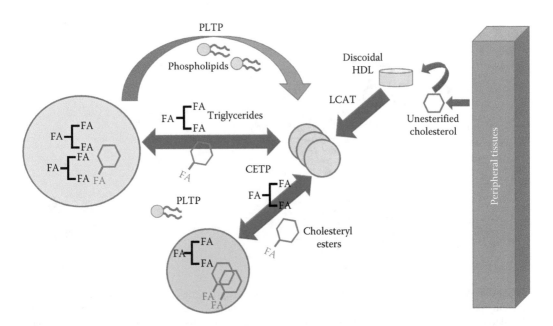

FIGURE 6.2 Key lipid transfer enzymes responsible for the remodeling of plasma lipoproteins. PLTP, phospholipid transfer protein; CETP, cholesteryl ester transfer protein; LCAT, lecithin cholesterol acyl transferase.

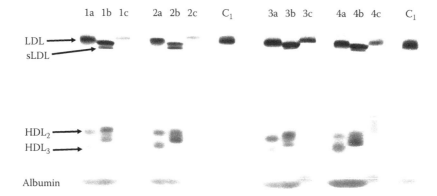

FIGURE 6.3 Effects of lipid transfer enzymes of physical characteristics of plasma lipoproteins: Changes in lipoprotein particle diameters and electrophoretic mobility as assessed by nondenaturing polyacrylamide gradient gel electrophoresis. Whole plasma was incubated at 37°C for 0 h (designated as "a"), 2 h (designated as "b"), and 6 h (designated as "c") for four different individuals (designated as 1–4). C1 designates the control plasma.

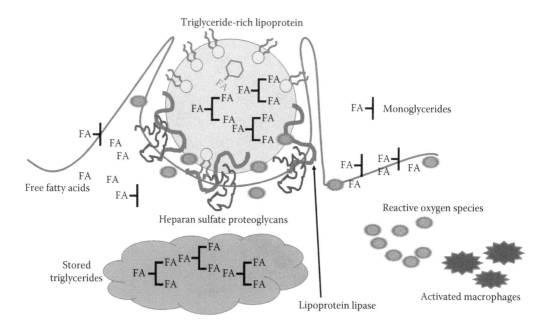

FIGURE 6.4 Schematic representation of the interactions of triglyceride-rich lipoproteins (TRL) with lipoprotein lipase (LPL) anchored to the arterial wall via heparan sulfate proteoglycans. Triglycerides are hydrolyzed to free fatty acids (FA) and monoglycerides that can move across the endothelium and are reconstituted as triglycerides for storage. Reactive oxygen species (ROS) that are generated from activated macrophages can diffuse through the endothelium and seed the fatty acids on plasma lipoproteins.

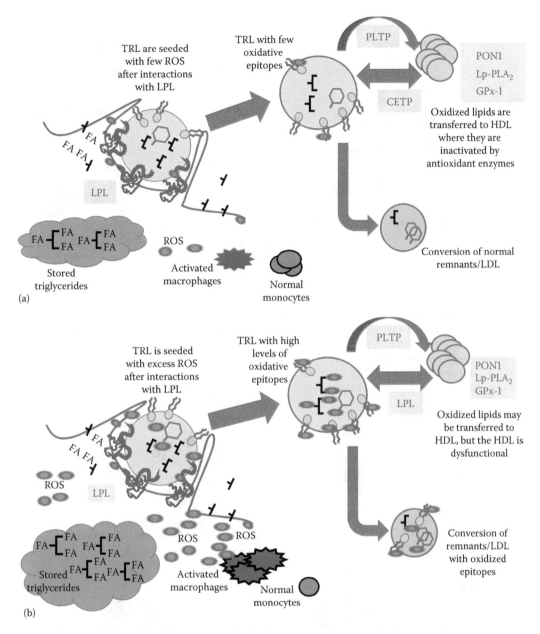

FIGURE 6.5 Schematic representation of the interactions between the arterial wall, triglyceride-rich lipoproteins (TRL), and high-density lipoproteins (HDL) under normal and diseased conditions. (a) Under normal conditions, only a few of the monocytes trapped in the subendothelium are activated with the generation of low quantity of ROS. During interactions of TRL with LPL, there are few ROS to seed the plasma lipoproteins. These ROS are efficiently transferred to circulating HDL and are rapidly inactivated by the antioxidant enzymes on HDL, including paraoxonase-1 (PON-1), lipoprotein-associated phospholipase A_2 (Lp-PLA$_2$), and glutathione peroxidase (GPx-1). The final product is a normal remnant/LDL particle. (b) Under diseased conditions, the arterial wall is inflamed, and most of the monocytes have been transformed to activated macrophages generating excessive quantities of ROS. Interactions of TRL with LPL along the arterial wall result in the seeding of plasma lipoproteins with excess ROS. The excess ROS overwhelm the capability of the HDL-associated antioxidant enzymes and proceed to propagate the oxidative modification process, especially if the highly oxidizable polyunsaturated fatty acids (PUFAs) are present. The net result is the generation of remnant/LDL particles with oxidatively modified epitopes.

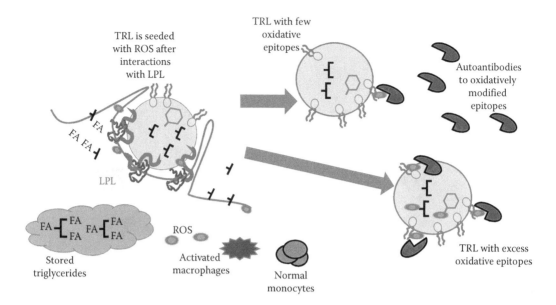

FIGURE 6.6 Interactions of TRL with autoantibodies against oxidized epitopes: During postprandial lipemia, interactions of TRL with LPL along the arterial wall will generate partially hydrolyzed particles seeded by either few ROS or excess ROS depending on the levels of ROS produced by the macrophages. If there are only few oxidized epitopes and there are adequate circulating autoantibodies, the impact of postprandial lipemia on autoantibody titers is negligible. If, however, there are excess oxidized epitopes, the pool of circulating autoantibodies can be acutely depleted resulting in a transient reduction in autoantibody levels.

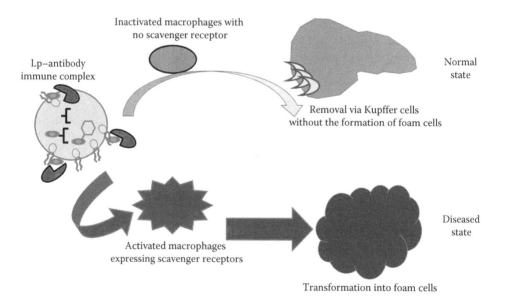

FIGURE 6.7 Metabolic fates of lipoprotein–autoantibody immune complexes. The association of oxidized lipoproteins and autoantibodies is an ongoing process, and the immune complexes are normally taken up by Kupffer cells in the liver and degraded. However, in diseased states with a large number of activated macrophages scattered through the arterial tree and expressing scavenger receptors, there would be higher probability that the immune complexes are taken up by macrophages prior to reaching the liver for normal degradation. In this instance, excess autoantibodies and immune complexes might be expected to accelerate disease progression.

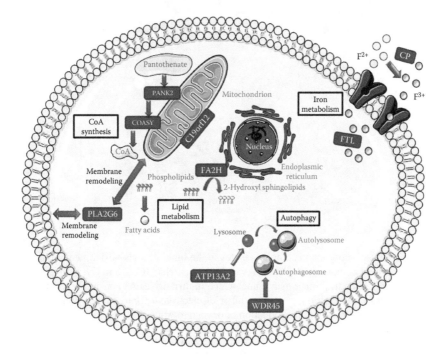

FIGURE 7.2 Localization of NBIA genes in the context of autophagy, lipid, and iron metabolism.

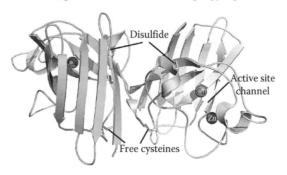

FIGURE 8.1 The human Cu,ZnSOD crystal structure (PDB ID 1PU0). hCu,ZnSOD is a homodimer (green and cyan), with each subunit consisting of a Greek key motif that is a barrel-like domain flanked by extended loops that form the active site cleft containing the Cu and Zn ions (large spheres). A conserved disulfide bond in each subunit provides structural rigidity, while the free cysteine residues are implicated in irreversible unfolding.

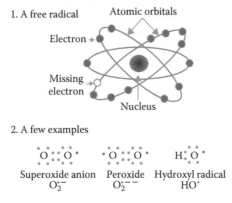

FIGURE 10.1 Free radicals are molecules or molecular fragments that have one or more unpaired electrons in their atomic or molecular orbitals.

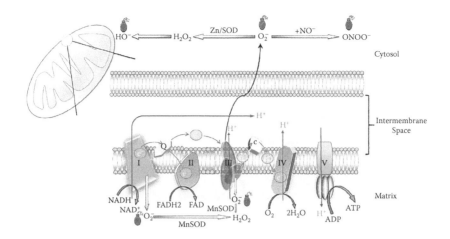

FIGURE 10.2 The respiratory chain complex resides in the inner mitochondrial membrane and consists of four distinct multisubunit complexes (I–IV) and two electron carriers (CoQ and Cyt c) that generate a proton gradient across the mitochondrial inner membrane, which in turn drives ATP synthase (complex V) to generate ATP. During energy transduction, a small number of electrons *leak* to oxygen prematurely to generate superoxide, instead of contributing to the reduction of oxygen to water.

Oxidative stress

Cellular homeostasis	Adaptation to changing circumstances	Reversibly impaired cellular function	Irreversibly impaired cellular function or pathological gain of function	Malignant transformation or death

Pathways to suppress ROS production, catabolize ROS, repair ROS-mediated damage, degrade what cannot be repaired or sequester what cannot be degraded

ROS production at an inappropriate place or time, for too long, at too high a level or of inappropriate forms

FIGURE 20.1 The broad range of reactive oxygen species (ROS) signaling is influenced by ROS production and catabolism and by cellular adaptation. Restriction of ROS production to appropriate subcellular locations, times, levels, molecular species, and for appropriate durations allows ROS to contribute to homeostasis and physiological cell activation. For example, brief pulses of H_2O_2 production at the plasma membrane or at the endosomal membrane mediate signaling in response to the engagement of receptors with cytokines, microbial products, or antigens (left-hand side). When ROS production escapes these restrictions—for example, when there are high levels or sustained production of hydroxyl radicals—macromolecules are damaged ("oxidative stress"). ROS-mediated damage can often be reversed by repair, replacement, degradation, or sequestration of the damaged macromolecules (middle). However, damage that exceeds the capacity of the cell for these responses can lead to cell death (right-hand side). When damage to DNA results in mutagenesis without irreparable double-strand breakage, and when damage to other macromolecules is repaired, the consequence can be malignant transformation rather than the death of the cell (right-hand side). (Reprinted by permission from Macmillan Publishers Ltd: *Nat. Rev. Immunol.*, Nathan, C., Cunningham-Bussel, A. Beyond oxidative stress: An immunologist's guide to reactive oxygen species. 13(5):349–361. © 2013.)

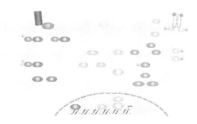

FIGURE 20.2 Examples of transcriptional regulation by ROS acting at the plasma membrane or in the cytosol. (Reprinted by permission from Macmillan Publishers Ltd: *Nat. Rev. Immunol.*, Nathan, C., Cunningham-Bussel, A. Beyond oxidative stress: An immunologist's guide to reactive oxygen species. 13(5):349–361. © 2013.)

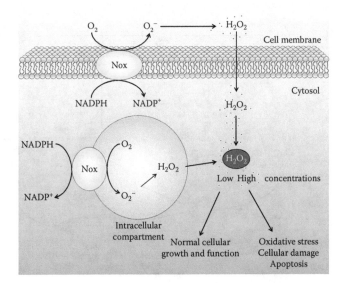

FIGURE 21.1 Hydrogen peroxide (H_2O_2) as a second messenger. NADPH oxidases (Nox) dynamically produce $O_2^{-\bullet}$ and H_2O_2. O_2^- is very rapidly dismutated to H_2O_2 by superoxide dismutase (SOD) and is restricted to certain cellular compartments. H_2O_2 is stable and can penetrate the membrane and reach its targets rapidly, which enables it to act as a second messenger. Thus, intracellular H_2O_2 levels are lowered by passive and active diffusion into the extracellular space. The remaining H_2O_2 induces numerous biomolecular modifications, especially the oxidation of cysteine residues. At low physiological levels, H_2O_2 can promote cell growth, while at high pathological levels, it can create oxidative stress and apoptosis. NADPH, nicotinamide adenine dinucleotide phosphate; Nox, NADPH oxidases.

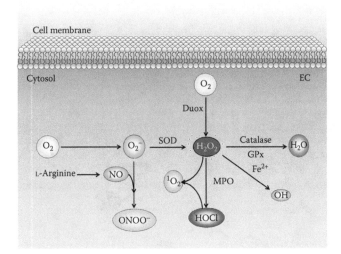

FIGURE 21.2 Biochemical pathways of hydrogen peroxide (H_2O_2) generation and metabolism. Several enzymes can produce intracellular superoxide anions (O_2^-), such as NADPH oxidases (Nox), xanthine oxidase, aldehyde oxidase, cytochrome P450s, the electron transport chain, and urate oxidase. O_2^- is either rapidly dismutated to H_2O_2 spontaneously or catalyzed by superoxide dismutase (SOD). H_2O_2 is also formed by the catalyzed two-electron reduction of molecular oxygen (O_2) or can diffuse in from the extracellular matrix. O_2^- can also react with NO to produce peroxynitrite ($ONOO^-$). H_2O_2 is converted to other reactive oxygen species such as hydroxyl radicals ($^\bullet OH$, with Fe^{2+}) and hypochlorous acid (HOCl, with chloride, catalyzed by myeloperoxidase). H_2O_2 is also neutralized by catalase localized in peroxisomes, myeloperoxidase (MPO), and glutathione peroxidase (GPx). SOD, superoxide dismutase; O_2^-, superoxide anions; $ONOO^-$, peroxynitrite; $^\bullet OH$, hydroxyl radicals; HOCl, hypochlorous acid; MPO, myeloperoxidase; GPx, glutathione peroxidase.

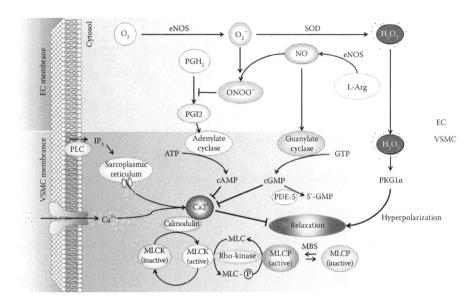

FIGURE 21.3 Interaction between endothelial cells (ECs) and vascular smooth muscle cells (VSMC) during vascular relaxation. The process involves multiple reactive oxygen species (ROS) and their targets as well as several proteins in the RhoA/Rho-kinase signaling pathway. EC, endothelial cells; PGI$_2$, prostacyclin; NO, nitric oxide; EDHF, endothelium-derived hyperpolarizing factors; NOS, nitric oxide synthase; PKG, protein kinase G; MLC, myosin light chain; MLCK, MLC kinase; MLCP, MLC phosphatase; MBS, myosin-binding subunit.

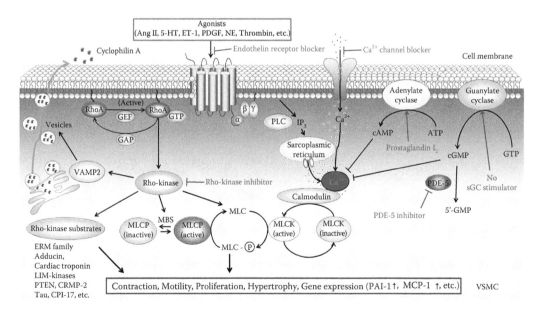

FIGURE 21.4 Extracellular cyclophilin A augments ROS production. Reactive oxygen species (ROS) inducer, such as angiotensin II (AngII), mechanical stress, and environmental factors, promotes cyclophilin A (CyPA) secretion from vascular smooth muscle cells (VSMC). Secreted CyPA activates ERK1/2 and promotes ROS production, contributing to the augmentation of ROS production. ERK1/2, extracellular signal–regulated kinase 1/2.

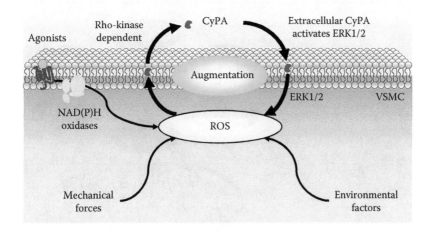

FIGURE 21.5 The RhoA/Rho-kinase signaling pathway and therapeutic targets. Rho GTPases are small GTP-binding proteins that act as molecular switches and regulate many intracellular signaling pathways. RhoA cycles between an inactive GDP-bound and an active GTP-bound conformation, interacting with down-stream targets, including Rho-kinase. The activity of RhoA is controlled by the guanine nucleotide exchange factors (GEFs) that catalyze the exchange of GDP for GTP. GTPase-activating proteins (GAPs) stimulate the intrinsic GTPase activity and inactivate RhoA. Guanine nucleotide dissociation inhibitors (GDIs) block spontaneous RhoA activation. Various substrates of Rho-kinase have been identified, including MYPT-1, MLC, ERM (ezrin, radixin, and moesin) family, adducin, phosphatase and tensin homolog on chromosome ten (PTEN), vesicle-associated membrane protein 2 (VAMP2), and LIM-kinases. Rho-kinase activation promotes cyclophilin A (CyPA) secretion. VSMCs, vascular smooth muscles cells; MLC, myosin light chain; MLCK, MLC kinase; MLCP, MLC phosphatase. (Courtesy of Satoh, K.)

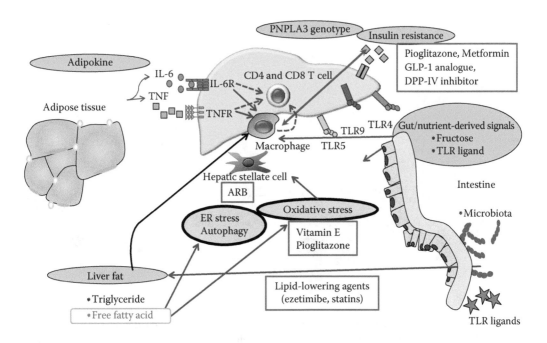

FIGURE 25.1 Nonalcoholic steatohepatitis (NASH) pathogenesis, treatment, and targets.

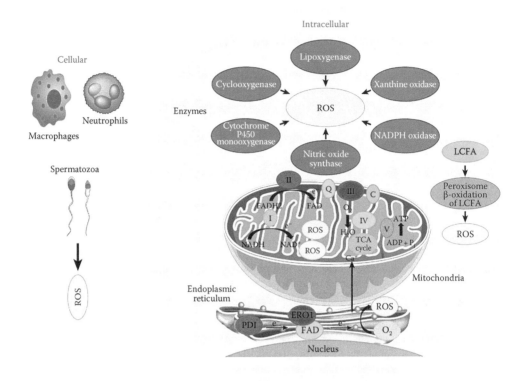

FIGURE 30.1 Major cellular and subcellular components responsible for ROS production in semen. (Reprinted with permission, Cleveland Clinic Center for Medical Art & Photography © 2015. All Rights Reserved.)

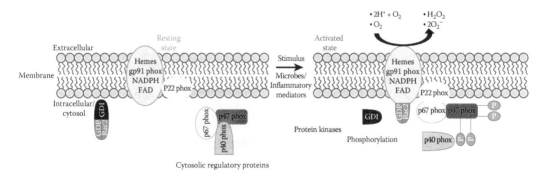

FIGURE 30.2 Mechanistic activation of NADPH oxidase complex and production of superoxide anions. (Reprinted with permission, Cleveland Clinic Center for Medical Art & Photography © 2015. All Rights Reserved.)

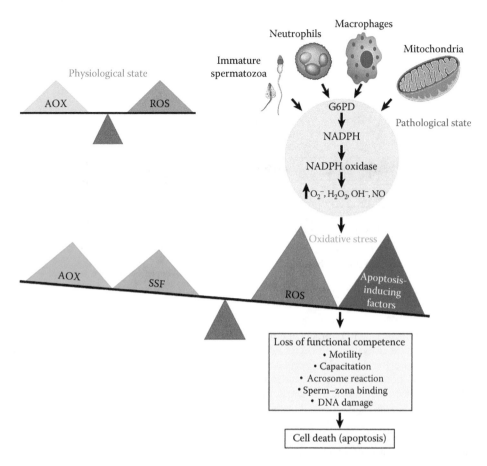

FIGURE 30.3 Key players involved in oxidative stress and sperm apoptosis. (Reprinted with permission, Cleveland Clinic Center for Medical Art & Photography © 2015. All Rights Reserved.)

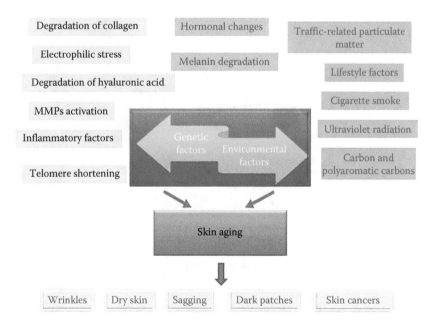

FIGURE 33.1 Factors affecting skin aging.

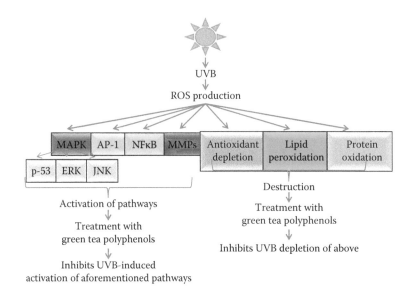

FIGURE 33.3 Endogenous ROS production due to UVB has a wide range of harmful effects on human health. It results in the activation of various pathways like mitogen-activated protein kinases (MAPK) and its family members extracellular signal–regulated kinase (ERK) 1 and 2, c-jun N-terminal kinase (JNK), and p38; activating protein-1 (AP-1); nuclear factor kappa B (NFκB); and matrix metalloproteins (MMPs); all of these ultimately lead to oxidative stress, aging, and disease. They also reduce antioxidants, enhance the release of pro-oxidants, and cause cell membrane and ECM breakdown by lipid peroxidation and protein oxidation. GTPs have been shown to inhibit all these pathways when applied to epidermal cells before or after UVB exposure. They enhance the production of antioxidant enzymes and deplete the production of pro-oxidant enzymes. They also prevent lipid peroxidation and protein oxidation.

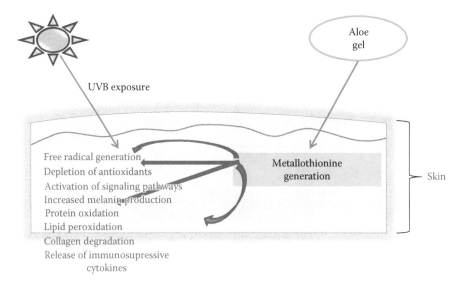

FIGURE 33.4 The exact mechanism by which aloe vera protects skin from sun damage is unknown. But following the application of aloe vera gel, an antioxidant protein, metallothionein, is generated in the skin, which scavenges hydroxyl radicals and regulates SOD and glutathione peroxidase in the skin. It reduces the production and liberation of immunosuppressive cytokines (IL-10) by skin keratinocyte and, hence, prevents UV-induced suppression of delayed type hypersensitivity (DTH). (From Tanuja, Y. et al., *Environ. Toxicol. Pharmacol.*, 39(1), 384, 2015. With permission.)

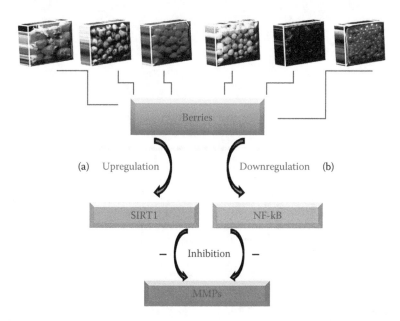

FIGURE 33.5 Berries show antiaging property by inhibiting the activation of MMPs, which are known to degrade collagen by two main pathways: (a) by upregulating SIRT1 gene, which inhibits MMPs, and (b) by downregulating the NF-kB transcription factor, which inhibits MMPs production.

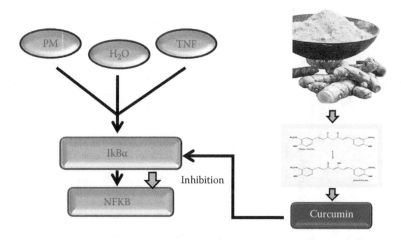

FIGURE 33.6 Mechanism of action of curcumin: Curcumin inhibits the phosphorylation and degradation of IkBα and subsequent translocation in the nucleus.

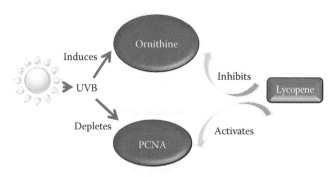

FIGURE 33.7 UVB induces ornithine decarboxylase and MMPs, and, on the other hand, it depletes PCNA. Lycopene reverses the induction and depletion of ornithine decarboxylase and PCNA, respectively.

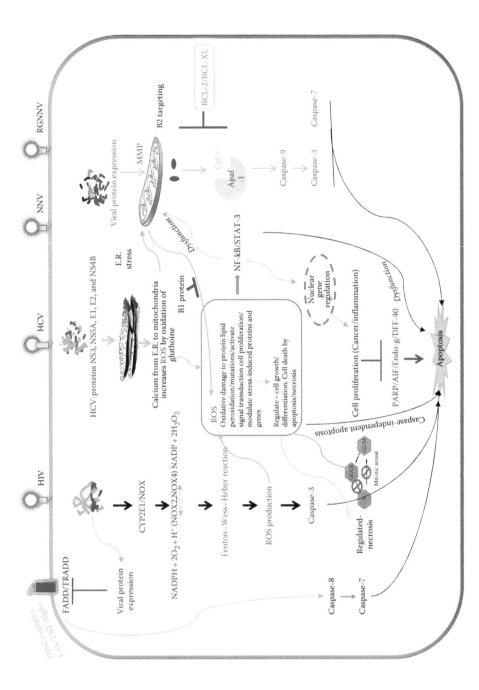

FIGURE 35.1 Schematic diagram of apoptotic cascade and the sites of action of general and specific caspases. The figure also illustrates how ROS produced during viral infection can affect apoptotic cascades. *Abbreviations*: FAS/TNF-α, tumor necrosis factor alpha (death receptors); FADD, Fas-associated death domain; TRADD, tumor necrosis factor receptor–associated death domain; Cas, caspase; AIF, apoptosis-inducing factor; PARP, poly(ADP-ribosyl) polymerase; EndoG, endonuclease G; DFF-40, DNA fragmentation factor 40 kDa; tBID, truncated BID; DISC, death-inducing signaling complex.

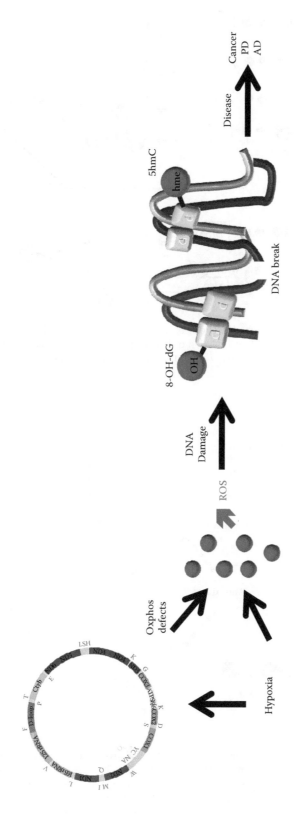

FIGURE 39.1 Effect of oxidative stress on DNA damage and modifications and consequences on disease development. Conditions that induce oxidative stress (e.g., hypoxia, H₂O₂ treatment, mitochondria dysfunction) increase ROS levels, which in turn can affect mitochondrial function and lead to DNA damage as well as DNA modifications. These alterations will alter gene expression and lead to the development and progression of diseases. AD, Alzheimer's disease; PD, Parkinson's disease; 8-OH-dG, 8-hydroxy-2'-deoxyguanosine; 5hmC, 5-hydroxymethylcytosine.

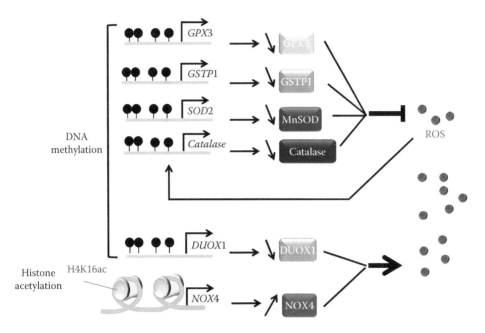

FIGURE 39.2 Effect of epigenetic modifications on the expression of ROS-inhibiting or ROS-producing genes. Epigenetic modifications (DNA methylation or histone post-translational modifications) can alter the expression of numerous genes among which we can find antioxidant (*Catalase*, *SOD*...) or ROS-producing genes (*NOX4*, *DUOX1*...).

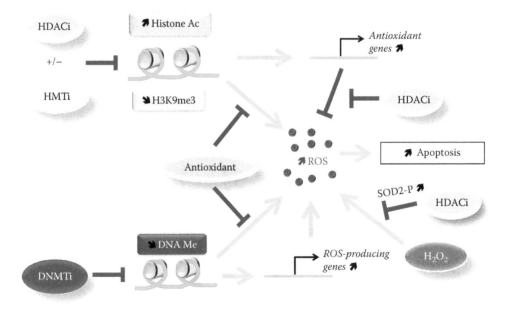

FIGURE 39.3 Interplay between epigenetic-targeting drugs and oxidative stress and its effect on therapeutic efficiency. Inhibitors of epigenetic enzymes (HDACi, HMTi, DNMTi...) used in therapeutics can directly or indirectly regulate the expression of genes involved in the production or degradation of ROS. Depending on the pathology, it might be useful in the future to combine these chemicals with antioxidants or ROS-inducing drugs in order to potentiate their cytotoxic therapeutic effect or inhibit the side effect of ROS production.

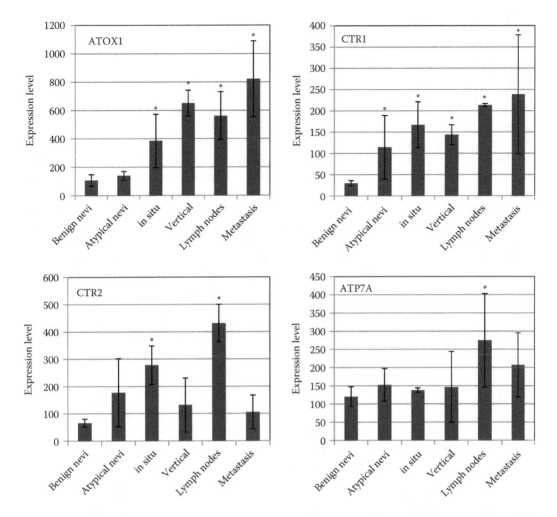

FIGURE 19.4 Expression levels of copper chaperones in nevi and melanomas of various stages. Gene expression data were retrieved from microarray dataset (NCBI GEO accession number GDS1989). Average mRNA levels were calculated from benign nevi, atypical nevi, melanoma in situ, vertical growth phase melanoma, melanoma from lymph nodes, and distal metastatic melanoma, an order that is consistent with the nature course of melanoma development. The stars designate significant difference as compared to benign nevi ($p < 0.05$ by two-side t-test). Further investigation is needed for confirmation of these data as the biopsy samples are limited to two to three in each group.

activation by all three major receptor tyrosine kinases (RTKs), fibroblast growth factor (FGF), platelet-derived growth factor (PDGF), and epidermal growth factor (EGF); these RTKs failed to activate extracellular signal regulated kinase 1/2 (ERK1/2) in CTR1$^{-/-}$ cells [89].

19.3 UV-MEDIATED TANNING RESPONSE AND ROS: THE ROLE OF MiTF AND Mc1R

Early studies on UVA redox signaling were focused on NFκB, AP-1 (activator protein 1) family proteins, and their relationships to the MAPKs [90]. Dose- and time-dependent activation of different MAPKs (p38-MAPK, JNKs, and ERK1/2) and their cellular effects were reviewed comprehensively before [91]; this chapter will focus more on the recent development in the field and how melanocyte-specific factors (especially MiTF and Mc1R) are involved in the redox signal transduction.

19.3.1 Role of MiTF

Skin color is a major intrinsic factor for melanoma development, as evident by one magnitude higher incidence rates in light-skinned Caucasian population [92]. Skin color is determined by more than 200 genes, among which many genes are well characterized and are involved in UV signaling and ROS reaction, including MiTF, Mc1R, and ASIP. MiTF is the master transcriptional factor for melanocytes differentiation, regulating the expression of many genes responsible for melanin biosynthesis and cell survival [93–95]. Mice deficient in *Mitf* exhibited small eye and white coat color [96], demonstrating the crucial role of *Mitf* in melanocytes survival and pigment biosynthesis. UVA is efficient in inducing skin tanning effect, that is, pigment biosynthesis and translocation; hence, it is nature to assume a role of MiTF in UV-induced tanning responses [97,98]. Experiment on human subjects (by Vincent Hearing's group) using 1 MEM (minimal erythema dose) of UV (solar simulated UV, ~95% UVA and ~5% UVB) treatment revealed a nonsignificant increase in MiTF and tyrosinase (by immunohistochemistry staining) for all skin types [99]. The most significant change observed in this study was the relocation of melanin from lower skin layers toward the skin surface. No significant change in MiTF levels was further confirmed in a different experiment after repetitive UV radiation [100]. These results are consistent with our study, which showed no increase in MiTF levels in normal human melanocytes after UVA or UVB radiation up to 48 h [101]. In contrast, there was actually a transient degradation of MiTF protein in UVA-treated melanocytes 4–6 h after irradiation, which is further discussed in the following paragraph. These observations suggest that MiTF may play a crucial role in pigmentation and melanocytes survival during development but may not be crucial for the tanning response, opposite to what other studies have suggested, where UVR-conditioned keratinocytes media stimulated MiTF accumulation at both mRNA and protein levels [102,103].

The crucial role of MiTF in melanocytes survival determined its importance in melanoma development. MiTF E318K mutation was associated with higher melanoma risk as compared to wild-type individuals in an Italian cohort, as well as higher risk for multiple melanomas [11,12,104]. However, these results were not confirmed in a large Polish melanoma population [105]. Wild-type homozygous *Mitf* mouse melanocytes survived better than heterozygous *Mitf* melanocytes (with one nonfunctional allele and one wild-type allele), although there was no difference in the expression of melanin synthesized enzymes tyrosinase, dopachrome tautomerase (DCT), and tyrosinase-related protein 1 (TYRP-1) [106]. *Mitf* was downregulated in all three types of mouse melanocytes after UVR in this same study [106], consistent with our observation in human melanocytes [101]. This raised a question as why UVR downregulated MiTF, as it is the master transcriptional factor for melanogenesis? Taking consideration of UVR-induced ROS, we postulate that this is a response to the UV-induced cellular ROS changes. Data from our laboratory and other laboratories have shown that UV-induced ROS increase is a transient early event occurring within 24 h. Blocking ROS generation by NOX inhibitor diphenyleneiodonium (DPI) led to the increase in MiTF accumulation in B16 melanoma cells [107], suggesting that ROS is a negative regulator for MiTF. Interestingly, MiTF stimulated NOX4 expression downstream of α-melanocyte-stimulating hormone (MSH) [108], which may suggest a feedback loop between ROS and MiTF. This proposed feedback loop can explain why the persistent high-level ROS kill melanocytes—because it diminishes the key melanocytes survival gene MiTF. However, it seems that MiTF is able to bring the redox balance back because MiTF also stimulated the expression of APE/REF-1 [109].

19.3.2 Mc1R in UV-Induced ROS Response

Mc1R is a transmembrane protein that binds to MSH, adrenocorticotropic hormone, and ASIP and plays a crucial role in melanin synthesis and skin pigmentation [110]. Loss-of-function Mc1R mutations are associated with human red hair phenotype and melanoma risk [111,112]. Mutations in ASIP are also associated with melanoma risk [14,15]. As illustrated in Figure 19.5, when Mc1R binds to

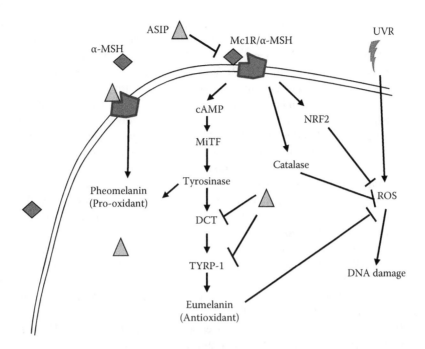

FIGURE 19.5 Function of Mc1R in melanomagenesis and the ROS connection. α-MSH is an Mc1R ligand, which upregulates cAMP signal and stimulates MiTF expression, as well as melanogenesis enzymes tyrosinase, DCT and TYRP-1, leading to eumelanin synthesis. Eumelanin exhibits antioxidant property. ASIP is an Mc1R antagonist, which inhibits α-MSH biding and the downstream melanogenesis enzymes. Recent studies show that ASIP may inhibit both pheomelanin and eumelanin synthesis; however, because pheomelanin synthesis is less dependent on the enzymatic catalysis, it is inhibited to a lesser degree by ASIP, resulting in an apparent phenotype of increased pheomelanin as compared to eumelanin. Pheomelanin exhibits pro-oxidant property.

α-MSH, it activates cAMP pathway, which in turn stimulates the expression of MiTF and biosynthesis of eumelanin. Eumelanin has an intrinsic antioxidant nature [113–115]. When ASIP binds to Mc1R, it prevents α-MSH binding and blocks the downstream events [116]; meanwhile, ASIP inhibits the expression of TYRP-1 and DCT (to a lesser extent tyrosinase), both of which are downstream of tyrosinase in the eumelanin synthesis pathway [116]. Hence, ASIP or loss-of-function Mc1R (unable to bind α-MSH) both lead to decreased eumelanin contents, perhaps pheomelanin as well [117]. It was suggested that pigmentation was a major pathway for Mc1R function in protecting human skin from UV radiation due to its critical role in melanin synthesis; emerging evidence has revealed that the α-MSH/Mc1R may execute its protective function via upregulating cellular antioxidant genes including catalase and NRF2 [118–121], thereby reducing the UV-induced oxidative DNA damage and enhance melanocytes survival [120]. These responses were diminished in melanocytes carrying loss-of-function red hair Mc1R alleles [122], which may account for the higher melanoma incidence rates in red hair population.

19.4 REDOX BALANCE IN NORMAL AND CANCER CELLS: COEVOLUTION OF ROS AND ANTIOXIDANTS

Numerous studies have shown that cancer cells exhibit elevated levels of ROS [123], as well as ROS-metabolizing enzymes including superoxide dismutases (SODs), catalase, glutathione peroxidases (GPXs), thioredoxin (Trx), and thioredoxin reductase (TrxR) [124–126]. Thus, it seems that high ROS levels and a more robust detoxifying antioxidant system form a new balance in cancer cells in order to maintain ROS levels in the nontoxic ranges. One consequence of this new balance is that a more robust antioxidant enzyme system renders cancer cells even more adaptive to new challenges like ROS-inducing chemotherapy drugs [127]. An excellent well-studied example is that with rat pheochromocytoma cells (line PC12). When they are preconditioned by low-level hydrogen peroxide (10–20 µM), they became more resistant to higher toxic levels of hydrogen peroxide [128]. In this way, cancer cells can be viewed as redox adapted cells, which exhibit strong adaptive capacity to many different agents including chemotherapeutic compounds [129]. As a result, the two major antioxidant systems, glutathione (GSH) and Trx, are upregulated in cancer cells as compared to normal cells [130,131]. As GSH exhibits redox cycles between reduced and oxidized forms, the ratio of GSH:GSSG serves as an excellent marker for overall cellular redox status. Trx also cycles between a reduced form carrying free sulfhydryl group and an oxidized internal disulfide form. Upon oxidative stress, a portion of Trx is released into extracellular spaces; hence, the plasma Trx levels can also be used as an index for oxidative stress in diseases including cancer [132–134].

The precise mechanisms of redox adaptation are not completely understood. A number of redox-sensitive transcription factors, including NFκB, Hif-1α, NRF (NFE2L2, nuclear factor [erythroid-derived 2]-like 2), AP-1, and APE/REF-1, may play key roles in the regulation of the melanocyte lineage. Among these transcription factors, NRF2 is a master regulator for redox switching [135]. NRF2 binds to specific DNA elements (antioxidant response element [ARE]) on promoter regions to regulate gene expression [136]. Key NRF2 transcriptional targets include both GSH and Trx systems [137] that are maintained between reduced and oxidized forms by GSH reductase, GPXs, glutaredoxins (Grxs), PRXs, and TrxRs. The reduced forms of GSH and Trx are able to reduce target protein substrates and play major roles in signal transduction. NRF2 upregulates a number of enzymes responsible for synthesis and the maintenance of reduced forms of GSH and Trx [138,139], thus maintaining the redox balance upon oxidative stress.

Emerging evidence shows that a physiologically elevated ROS level stimulates NRF2 activity and provokes an adaptive response to increase cellular antioxidant levels for proliferation and survival [140]. Interestingly, the elevated level of ROS is likely generated by NOX enzymes [141,142]. In fact, NOX1 played an essential role in NRF2 induction by intermittent hypoxia in adenocarcinoma cells [142]. In a mouse experiment, NOX4 expression and activity was upregulated as a result of pathophysiological insult to the heart and was associated with an NRF2-mediated increase of GSH levels in cardiac tissue [141]. The increased antioxidants regulated by NRF2 serve as protective pathway for drug-treated cells, resulting in drug resistance [143,144].

The emerging NOX-ROS-NRF2 axis strongly suggests a coevolution of the ROS-generating system and antioxidant system. We have observed higher NOX1, higher NOS1, higher NRF2, and higher APE/REF-1 levels in human melanoma cell lines as compared to normal human melanocytes [28,44,109] (NRF2 data not shown). Hence, the transformation process may be a process of continuous rebalance of ROS and antioxidants. Only when the excessive ROS triggered by genetic and/or environmental factors are balanced by a coevolving robust antioxidant system, will the cells be able to survive and proliferate. The problem of this process is that the DNA damage may not be repaired, and thus, cells are allowed to proliferate with these damages. When crucial damages (i.e., mutations, such as those observed in oncogenes and tumor suppressors) accumulate, cellular transformation occurs. This process has been perfectly demonstrated by a recent study by DeNicola et al. [145], in which NRF2 actually detoxified oncogene-induced ROS and promoted tumorigenesis.

Taken together, normal cells rely on certain levels of ROS to proliferate, which are balanced by normally operated antioxidant systems. When the ROS level is raised by carcinogens, the antioxidant transcription factor and regulators including NRF2 and APE/REF-1 will be induced concomitantly, resulting in high levels of GSH and Trx antioxidants (Figure 19.6). The antioxidant defense system is upregulated to a level that allows proliferation and cell survival but may not be sufficient to balance off the persistent carcinogen-produced ROS; hence, the end result is that cancer cells appear to exhibit higher ROS levels than normal cells, albeit they also have a higher antioxidant system (Figure 19.6).

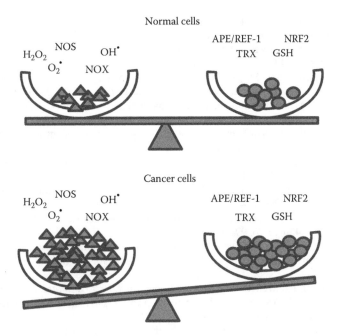

FIGURE 19.6 ROS and antioxidant rebalance in cancer cells. In normal cells, both ROS (represented by $O_2^{\cdot}$, H_2O_2, OH$^{\cdot}$, NOX, and NOS) and antioxidants (represented by NRF2, APE/REF-1, glutathione, and thioredoxin) are maintained and balanced at a relative low level. Cancer development may reflect a process where genetic and/or environmental factors continuously and persistently increase ROS levels, which are continuously counteracted by the cellular antioxidant levels, enabling cells to survive and proliferate. The end result is that the cancer cells exhibit higher levels of ROS and higher levels of antioxidants than normal cells. However, the antioxidant system may not be able to catch up with ROS; hence, the ROS levels will not be completely balanced by the antioxidants in the cancer cells.

19.5 CONCLUSIONS

Several risk factors for melanoma may all function through an ROS induction mechanism; hence, it is reasonable to hypothesize that melanoma is an ROS-driven cancer. Early studies revealed that melanoma exhibited an aberrantly higher level of ROS [10], which may be caused by several possible reasons: (1) the special function of this cell type (synthesize melanins) is associated with ROS production; (2) environmental factors including UVR, heavy metals, herbicides, and pesticide drive ROS production in melanocytes; (3) melanocytes have a compromised antioxidant defense system. The aberrantly high level of ROS may be responsible for the observed high level of genome instability in melanoma as compared to other cancer types [146]. Tumorigenesis (including melanomagenesis) can be viewed as a process of loss of balance between ROS and antioxidant system. Future investigations on the redox balance mechanisms are needed to understand the molecular events that are the foundations for prevention.

REFERENCES

1. Gilchrest, B.A. and M.S. Eller, DNA photodamage stimulates melanogenesis and other photoprotective responses. *J Investig Dermatol Symp Proc*, 1999. **4**(1): 35–40.
2. Hemminki, K. et al., Estimation of genetic and environmental components in colorectal and lung cancer and melanoma. *Genet Epidemiol*, 2001. **20**(1): 107–116.
3. Fears, T.R. et al., Average midrange ultraviolet radiation flux and time outdoors predict melanoma risk. *Cancer Res*, 2002. **62**(14): 3992–3996.

4. Zaidi, M.R., C.P. Day, and G. Merlino, From UVs to metastases: Modeling melanoma initiation and progression in the mouse. *J Invest Dermatol*, 2008. **128**(10): 2381–2391.

5. Lea, C.S. et al., Ambient UVB and melanoma risk in the United States: A case-control analysis. *Ann Epidemiol*, 2007. **17**(6): 447–453.

6. Miranda, M., D. Botti, and M. Di Cola, Possible genotoxity of melanin synthesis intermediates: Tyrosinase reaction products interact with DNA in vitro. *Mol Gen Genet*, 1984. **193**(3): 395–399.

7. Gidanian, S. et al., Melanosomal damage in normal human melanocytes induced by UVB and metal uptake—A basis for the pro-oxidant state of melanoma. *Photochem Photobiol*, 2008. **84**(3): 556–564.

8. Penzella, L. et al., Red human hair pheomelanin is a potent pro-oxidant mediating UV-independent contributory mechanisms of melanomagenesis. *Pigment Cell Melanoma Res*, 2014. **27**(2): 244–252.

9. Noonan, F.P. et al., Melanoma induction by ultraviolet A but not ultraviolet B radiation requires melanin pigment. *Nat Commun*, 2012. **3**: 884.

10. Meyskens, F.L., Jr., P. Farmer, and J.P. Fruehauf, Redox regulation in human melanocytes and melanoma. *Pigment Cell Res*, 2001. **14**(3): 148–154.

11. Yokoyama, S. et al., A novel recurrent mutation in MITF predisposes to familial and sporadic melanoma. *Nature*, 2011. **480**(7375): 99–103.

12. Ghiorzo, P. et al., Prevalence of the E318K MITF germline mutation in Italian melanoma patients: Associations with histological subtypes and family cancer history. *Pigment Cell Melanoma Res*, 2013. **26**(2): 259–262.

13. Sturm, R.A., GSTP1 and MC1R in melanoma susceptibility. *Br J Dermatol*, 2012. **166**(6): 1155–1156.

14. Gudbjartsson, D.F. et al., ASIP and TYR pigmentation variants associate with cutaneous melanoma and basal cell carcinoma. *Nat Genet*, 2008. **40**(7): 886–891.

15. Macgregor, S. et al., Genome-wide association study identifies a new melanoma susceptibility locus at 1q21.3. *Nat Genet*, 2011. **43**(11): 1114–1118.

16. Liu-Smith, F., R. Dellinger, and F.L. Meyskens, Jr., Updates of reactive oxygen species in melanoma etiology and progression. *Arch Biochem Biophys*, 2014. **563**: 51–55.

17. Liu-Smith, F. et al., Amyloids, melanins and oxidative stress in melanomagenesis. *Exp Dermatol*, 2015. **24**(3): 171–174.

18. von Thaler, A.K., Y. Kamenisch, and M. Berneburg, The role of ultraviolet radiation in melanomagenesis. *Exp Dermatol*, 2010. **19**(2): 81–88.

19. de Gruijl, F.R., Photocarcinogenesis: UVA vs UVB. *Methods Enzymol*, 2000. **319**: 359–366.

20. Tyrrell, R.M., Solar ultraviolet A radiation: An oxidizing skin carcinogen that activates heme oxygenase-1. *Antioxid Redox Signal*, 2004. **6**(5): 835–840.

21. Carraro, C. and M.A. Pathak, Studies on the nature of in vitro and in vivo photosensitization reactions by psoralens and porphyrins. *J Invest Dermatol*, 1988. **90**(3): 267–275.

22. Tanaka, M., K. Ohkubo, and S. Fukuzumi, DNA cleavage by UVA irradiation of NADH with dioxygen via radical chain processes. *J Phys Chem A*, 2006. **110**(38): 11214–11218.

23. Anderson, A. et al., A role for human mitochondrial complex II in the production of reactive oxygen species in human skin. *Redox Biol*, 2014. **2C**: 1016–1022.

24. Valencia, A. and I.E. Kochevar, Nox1-based NADPH oxidase is the major source of UVA-induced reactive oxygen species in human keratinocytes. *J Invest Dermatol*, 2008. **128**(1): 214–222.

25. Henri, P. et al., MC1R expression in HaCaT keratinocytes inhibits UVA-induced ROS production via NADPH oxidase- and cAMP-dependent mechanisms. *J Cell Physiol*, 2012. **227**(6): 2578–2585.

26. Suschek, C.V. et al., Even after UVA-exposure will nitric oxide protect cells from reactive oxygen intermediate-mediated apoptosis and necrosis. *Cell Death Differ*, 2001. **8**(5): 515–527.

27. Lambeth, J.D., T. Kawahara, and B. Diebold, Regulation of Nox and Duox enzymatic activity and expression. *Free Radic Biol Med*, 2007. **43**(3): 319–331.

28. Liu, F., A.M. Gomez Garcia, and F.L. Meyskens, Jr., NADPH oxidase 1 overexpression enhances invasion via matrix metalloproteinase-2 and epithelial-mesenchymal transition in melanoma cells. *J Invest Dermatol*, 2012. **132**(8): 2033–2041.

29. Govindarajan, B. et al., Overexpression of Akt converts radial growth melanoma to vertical growth melanoma. *J Clin Invest*, 2007. **117**(3): 719–729.

30. Antony, S. et al., Characterization of NADPH oxidase 5 expression in human tumors and tumor cell lines with a novel mouse monoclonal antibody. *Free Radic Biol Med*, 2013. **65**: 497–508.

31. Liu-Smith, F., F.L. Meyskens, Abstract # CS60 for IPCC 2014, The NRF2 and NOX1 mediated redox balance in UV-induced melanoma development. *Pigment Cell Melanoma Res*, 2014. **27**(5): 849–1001.

32. Xu, L. et al., Gene expression changes in an animal melanoma model correlate with aggressiveness of human melanoma metastases. *Mol Cancer Res*, 2008. **6**(5): 760–769.

33. Cals-Grierson, M.M. and A.D. Ormerod, Nitric oxide function in the skin. *Nitric Oxide*, 2004. **10**(4): 179–193.
34. Suschek, C.V. et al., Nitric oxide fully protects against UVA-induced apoptosis in tight correlation with Bcl-2 up-regulation. *J Biol Chem*, 1999. **274**(10): 6130–6137.
35. Sullivan, J.C. and J.S. Pollock, Coupled and uncoupled NOS: Separate but equal? Uncoupled NOS in endothelial cells is a critical pathway for intracellular signaling. *Circ Res*, 2006. **98**(6): 717–719.
36. Wang, L., Y. Liu, and S. Wu, The roles of nitric oxide synthase and eIF2alpha kinases in regulation of cell cycle upon UVB-irradiation. *Cell Cycle*, 2010. **9**(1): 38–42.
37. Liu, W. and S. Wu, Differential roles of nitric oxide synthases in regulation of ultraviolet B light-induced apoptosis. *Nitric Oxide*, 2010. **23**(3): 199–205.
38. Yang, Z. et al., Nitric oxide initiates progression of human melanoma via a feedback loop mediated by apurinic/apyrimidinic endonuclease-1/redox factor-1, which is inhibited by resveratrol. *Mol Cancer Ther*, 2008. **7**(12): 3751–3760.
39. Yang, J. et al., Nitric oxide activated by p38 and NF-kappaB facilitates apoptosis and cell cycle arrest under oxidative stress in evodiamine-treated human melanoma A375-S2 cells. *Free Radic Res*, 2008. **42**(1): 1–11.
40. Sikora, A.G. et al., Targeted inhibition of inducible nitric oxide synthase inhibits growth of human melanoma in vivo and synergizes with chemotherapy. *Clin Cancer Res*, 2010. **16**(6): 1834–1844.
41. Rouaud, F. et al., Regulation of NADPH-dependent Nitric Oxide and reactive oxygen species signalling in endothelial and melanoma cells by a photoactive NADPH analogue. *Oncotarget*, 2014. **5**(21): 10650–10664.
42. Melo, F.H. et al., Endothelial nitric oxide synthase uncoupling as a key mediator of melanocyte malignant transformation associated with sustained stress conditions. *Free Radic Biol Med*, 2011. **50**(10): 1263–1273.
43. Fecker, L.F. et al., Inducible nitric oxide synthase is expressed in normal human melanocytes but not in melanoma cells in response to tumor necrosis factor-alpha, interferon-gamma, and lipopolysaccharide. *J Invest Dermatol*, 2002. **118**(6): 1019–1025.
44. Ahmed, B. and J.J. Van Den Oord, Expression of the neuronal isoform of nitric oxide synthase (nNOS) and its inhibitor, protein inhibitor of nNOS, in pigment cell lesions of the skin. *Br J Dermatol*, 1999. **141**(1): 12–19.
45. Yang, Z. et al., Targeting nitric oxide signaling with nNOS inhibitors as a novel strategy for the therapy and prevention of human melanoma. *Antioxid Redox Signal*, 2013. **19**(5): 433–447.
46. Choi, Y.J. et al., Suppression of melanogenesis by a newly synthesized compound, MHY966 via the nitric oxide/protein kinase G signaling pathway in murine skin. *J Dermatol Sci*, 2012. **68**(3): 164–171.
47. Ibarrola-Villava, M. et al., Genetic polymorphisms in DNA repair and oxidative stress pathways associated with malignant melanoma susceptibility. *Eur J Cancer*, 2011. **47**(17): 2618–2625.
48. Li, C. et al., Polymorphisms of the neuronal and inducible nitric oxide synthase genes and the risk of cutaneous melanoma: A case-control study. *Cancer*, 2007. **109**(8): 1570–1578.
49. Soengas, M.S., Mitophagy or how to control the Jekyll and Hyde embedded in mitochondrial metabolism: Implications for melanoma progression and drug resistance. *Pigment Cell Melanoma Res*, 2012. **25**(6): 721–731.
50. Bakhshi, J. et al., Coupling endoplasmic reticulum stress to the cell death program in mouse melanoma cells: Effect of curcumin. *Apoptosis*, 2008. **13**(7): 904–914.
51. Fruehauf, J.P. and V. Trapp, Reactive oxygen species: An Achilles' heel of melanoma? *Expert Rev Anticancer Ther*, 2008. **8**(11): 1751–1757.
52. Wittgen, H.G. and L.C. van Kempen, Reactive oxygen species in melanoma and its therapeutic implications. *Melanoma Res*, 2007. **17**(6): 400–409.
53. Cao, S.S. and R.J. Kaufman, Endoplasmic reticulum stress and oxidative stress in cell fate decision and human disease. *Antioxid Redox Signal*, 2014. **21**(3): 396–413.
54. Higa, A. and E. Chevet, Redox signaling loops in the unfolded protein response. *Cell Signal*, 2012. **24**(8): 1548–1555.
55. Syed, D.N. et al., Involvement of ER stress and activation of apoptotic pathways in fisetin induced cytotoxicity in human melanoma. *Arch Biochem Biophys*, 2014. **563**: 108–117.
56. Corazzari, M. et al., Oncogenic BRAF induces chronic ER stress condition resulting in increased basal autophagy and apoptotic resistance of cutaneous melanoma. *Cell Death Differ*, 2015. **22**(6): 946–958.
57. Chaudhari, N. et al., A molecular web: Endoplasmic reticulum stress, inflammation, and oxidative stress. *Front Cell Neurosci*, 2014. **8**: 213.

58. Go, Y.M. and D.P. Jones, Redox compartmentalization in eukaryotic cells. *Biochim Biophys Acta*, 2008. **1780**(11): 1273–1290.

59. Sevier, C.S. and C.A. Kaiser, Ero1 and redox homeostasis in the endoplasmic reticulum. *Biochim Biophys Acta*, 2008. **1783**(4): 549–556.

60. Tu, B.P. and J.S. Weissman, Oxidative protein folding in eukaryotes: Mechanisms and consequences. *J Cell Biol*, 2004. **164**(3): 341–336.

61. Santos, C.X. et al., Mechanisms and implications of reactive oxygen species generation during the unfolded protein response: Roles of endoplasmic reticulum oxidoreductases, mitochondrial electron transport, and NADPH oxidase. *Antioxid Redox Signal*, 2009. **11**(10): 2409–2427.

62. Iqbal, A. et al., Oxidation, inactivation and aggregation of protein disulfide isomerase promoted by the bicarbonate-dependent peroxidase activity of human superoxide dismutase. *Arch Biochem Biophys*, 2014. **557**: 72–81.

63. Farrukh, M.R. et al., Oxidative stress mediated Ca(2+) release manifests endoplasmic reticulum stress leading to unfolded protein response in UV-B irradiated human skin cells. *J Dermatol Sci*, 2014. **75**(1): 24–35.

64. Caputo, E. et al., Characterization of human melanoma cell lines and melanocytes by proteome analysis. *Cell Cycle*, 2011. **10**(17): 2924–2936.

65. Aureli, C. et al., ERp57/PDIA3 binds specific DNA fragments in a melanoma cell line. *Gene*, 2013. **524**(2): 390–395.

66. Daiber, A., Redox signaling (cross-talk) from and to mitochondria involves mitochondrial pores and reactive oxygen species. *Biochim Biophys Acta*, 2010. **1797**(6–7): 897–906.

67. Laurindo, F.R., T.L. Araujo, and T.B. Abrahao, Nox NADPH oxidases and the endoplasmic reticulum. *Antioxid Redox Signal*, 2014. **20**(17): 2755–2775.

68. Janiszewski, M. et al., Regulation of NAD(P)H oxidase by associated protein disulfide isomerase in vascular smooth muscle cells. *J Biol Chem*, 2005. **280**(49): 40813–40819.

69. Pescatore, L.A. et al., Protein disulfide isomerase is required for platelet-derived growth factor-induced vascular smooth muscle cell migration, Nox1 NADPH oxidase expression, and RhoGTPase activation. *J Biol Chem*, 2012. **287**(35): 29290–29300.

70. Elgass, K.D. et al., Analysis of ER-mitochondria contacts by correlative fluorescence microscopy and soft X-ray tomography of mammalian cells. *J Cell Sci*, 2015. **128**(15): 2795–2804.

71. Meyskens, F.L. and S. Yang, Thinking about the role (largely ignored) of heavy metals in cancer prevention: Hexavalent chromium and melanoma as a case in point. *Recent Results Cancer Res*, 2011. **188**: 65–74.

72. Dennis, L.K. et al., Pesticide use and cutaneous melanoma in pesticide applicators in the agricultural heath study. *Environ Health Perspect*, 2010. **118**(6): 812–817.

73. Lerro, C.C. et al., Use of acetochlor and cancer incidence in the Agricultural Health Study. *Int J Cancer*, 2015. **137**(5): 1167–1175.

74. Beane Freeman, L.E. et al., Toenail arsenic content and cutaneous melanoma in Iowa. *Am J Epidemiol*, 2004. **160**(7): 679–687.

75. Domico, L.M. et al., Reactive oxygen species generation by the ethylene-bis-dithiocarbamate (EBDC) fungicide mancozeb and its contribution to neuronal toxicity in mesencephalic cells. *Neurotoxicology*, 2007. **28**(6): 1079–1091.

76. Canales-Aguirre, A.A. et al., Curcumin protects against the oxidative damage induced by the pesticide parathion in the hippocampus of the rat brain. *Nutr Neurosci*, 2012. **15**(2): 62–69.

77. Leomanni, A. et al., Antioxidant and oxidative stress related responses in the Mediterranean land snail Cantareus apertus exposed to the carbamate pesticide Carbaryl. *Comp Biochem Physiol C Toxicol Pharmacol*, 2015. **168**: 20–27.

78. Liu, Y. et al., The role of reactive oxygen species in the herbicide acetochlor-induced DNA damage on Bufo raddei tadpole liver. *Aquat Toxicol*, 2006. **78**(1): 21–26.

79. Gallagher, R.P. et al., Plasma levels of polychlorinated biphenyls and risk of cutaneous malignant melanoma: A preliminary study. *Int J Cancer*, 2011. **128**(8): 1872–1880.

80. Fonnum, F., E. Mariussen, and T. Reistad, Molecular mechanisms involved in the toxic effects of polychlorinated biphenyls (PCBs) and brominated flame retardants (BFRs). *J Toxicol Environ Health A*, 2006. **69**(1–2): 21–35.

81. Visuri, T.I. et al., Cancer incidence and causes of death among total hip replacement patients: A review based on Nordic cohorts with a special emphasis on metal-on-metal bearings. *Proc Inst Mech Eng H*, 2006. **220**(2): 399–407.

82. Makela, K.T. et al., Risk of cancer with metal-on-metal hip replacements: Population based study. *BMJ*, 2012. **345**: e4646.

83. Smith, A.J. et al., Risk of cancer in first seven years after metal-on-metal hip replacement compared with other bearings and general population: Linkage study between the National Joint Registry of England and Wales and hospital episode statistics. *BMJ*, 2012. **344**: e2383.

84. Vinceti, M. et al., Environmental exposure to trace elements and risk of cutaneous melanoma. *J Expo Anal Environ Epidemiol*, 2005. **15**(5): 458–462.

85. Brady, D.C. et al., Copper is required for oncogenic BRAF signalling and tumorigenesis. *Nature*, 2014. **509**(7501): 492–496.

86. Setty, S.R. et al., Cell-specific ATP7A transport sustains copper-dependent tyrosinase activity in melanosomes. *Nature*, 2008. **454**(7208): 1142–1146.

87. Smith, A.P., K. Hoek, and D. Becker, Whole-genome expression profiling of the melanoma progression pathway reveals marked molecular differences between nevi/melanoma in situ and advanced-stage melanomas. *Cancer Biol Ther*, 2005. **4**(9): 1018–1029.

88. Safi, R. et al., Copper signaling axis as a target for prostate cancer therapeutics. *Cancer Res*, 2014. **74**(20): 5819–5831.

89. Tsai, C.Y. et al., Copper influx transporter 1 is required for FGF, PDGF and EGF-induced MAPK signaling. *Biochem Pharmacol*, 2012. **84**(8): 1007–1013.

90. Bachelor, M.A. and G.T. Bowden, UVA-mediated activation of signaling pathways involved in skin tumor promotion and progression. *Semin Cancer Biol*, 2004. **14**(2): 131–138.

91. Bode, A.M. and Z. Dong, Mitogen-activated protein kinase activation in UV-induced signal transduction. *Sci STKE*, 2003. **2003**(167): RE2.

92. Wu, X.C. et al., Racial and ethnic variations in incidence and survival of cutaneous melanoma in the United States, 1999–2006. *J Am Acad Dermatol*, 2011. **65**(5 Suppl 1): S26–S37.

93. Sturm, R.A., Molecular genetics of human pigmentation diversity. *Hum Mol Genet*, 2009. **18**(R1): R9–R17.

94. Cheli, Y. et al., Fifteen-year quest for microphthalmia-associated transcription factor target genes. *Pigment Cell Melanoma Res*, 2010. **23**(1): 27–40.

95. McGill, G.G. et al., Bcl2 regulation by the melanocyte master regulator Mitf modulates lineage survival and melanoma cell viability. *Cell*, 2002. **109**(6): 707–718.

96. Steingrimsson, E., Interpretation of complex phenotypes: Lessons from the Mitf gene. *Pigment Cell Melanoma Res*, 2010. **23**(6): 736–740.

97. Chen, H., Q.Y. Weng, and D.E. Fisher, UV signaling pathways within the skin. *J Invest Dermatol*, 2014. **134**(8): 2080–2085.

98. Liu, J.J. and D.E. Fisher, Lighting a path to pigmentation: Mechanisms of MITF induction by UV. *Pigment Cell Melanoma Res*, 2010. **23**(6): 741–745.

99. Tadokoro, T. et al., Mechanisms of skin tanning in different racial/ethnic groups in response to ultraviolet radiation. *J Invest Dermatol*, 2005. **124**(6): 1326–1332.

100. Brenner, M. et al., Long-lasting molecular changes in human skin after repetitive in situ UV irradiation. *J Invest Dermatol*, 2009. **129**(4): 1002–1011.

101. Liu, F. et al., MiTF links Erk1/2 kinase and p21 CIP1/WAF1 activation after UVC radiation in normal human melanocytes and melanoma cells. *Mol Cancer*, 2010. **9**: 214.

102. Yang, G. et al., Inhibition of PAX3 by TGF-beta modulates melanocyte viability. *Mol Cell*, 2008. **32**(4): 554–563.

103. Shoag, J. et al., PGC-1 coactivators regulate MITF and the tanning response. *Mol Cell*, 2013. **49**(1): 145–157.

104. Sturm, R.A. et al., Phenotypic characterization of nevus and tumor patterns in MITF E318K mutation carrier melanoma patients. *J Invest Dermatol*, 2014. **134**(1): 141–149.

105. Gromowski, T. et al., Prevalence of the E318K and V320I MITF germline mutations in Polish cancer patients and multiorgan cancer risk-a population-based study. *Cancer Genet*, 2014. **207**(4): 128–132.

106. Hornyak, T.J. et al., Mitf dosage as a primary determinant of melanocyte survival after ultraviolet irradiation. *Pigment Cell Melanoma Res*, 2009. **22**(3): 307–318.

107. Zhao, Y., J. Liu, and K.E. McMartin, Inhibition of NADPH oxidase activity promotes differentiation of B16 melanoma cells. *Oncol Rep*, 2008. **19**(5): 1225–1230.

108. Liu, G.S. et al., Microphthalmia-associated transcription factor modulates expression of NADPH oxidase type 4: A negative regulator of melanogenesis. *Free Radic Biol Med*, 2012. **52**(9): 1835–1843.

109. Liu, F., Y. Fu, and F.L. Meyskens, Jr., MiTF regulates cellular response to reactive oxygen species through transcriptional regulation of APE-1/Ref-1. *J Invest Dermatol*, 2009. **129**(2): 422–431.

110. Scherer, D. and R. Kumar, Genetics of pigmentation in skin cancer—A review. *Mutat Res*, 2010. **705**(2): 141–153.

111. Sturm, R.A. et al., Genetic association and cellular function of MC1R variant alleles in human pigmentation. *Ann N Y Acad Sci*, 2003. **994**: 348–358.

112. Bishop, D.T. et al., Genome-wide association study identifies three loci associated with melanoma risk. *Nat Genet*, 2009. **41**(8): 920–925.

113. Wang, Z., J. Dillon, and E.R. Gaillard, Antioxidant properties of melanin in retinal pigment epithelial cells. *Photochem Photobiol*, 2006. **82**(2): 474–479.

114. Meyskens, F.L., Jr., P.J. Farmer, and H. Anton-Culver, Etiologic pathogenesis of melanoma: A unifying hypothesis for the missing attributable risk. *Clin Cancer Res*, 2004. **10**(8): 2581–2583.

115. Brenner, M. and V.J. Hearing, The protective role of melanin against UV damage in human skin. *Photochem Photobiol*, 2008. **84**(3): 539–549.

116. Suzuki, I. et al., Agouti signaling protein inhibits melanogenesis and the response of human melanocytes to alpha-melanotropin. *J Invest Dermatol*, 1997. **108**(6): 838–842.

117. Le Pape, E. et al., Regulation of eumelanin/pheomelanin synthesis and visible pigmentation in melanocytes by ligands of the melanocortin 1 receptor. *Pigment Cell Melanoma Res*, 2008. **21**(4): 477–486.

118. Swope, V.B. et al., Defining MC1R regulation in human melanocytes by its agonist alpha-melanocortin and antagonists agouti signaling protein and beta-defensin 3. *J Invest Dermatol*, 2012. **132**(9): 2255–2262.

119. Garcia-Borron, J.C., Z. Abdel-Malek, and C. Jimenez-Cervantes, MC1R, the cAMP pathway, and the response to solar UV: Extending the horizon beyond pigmentation. *Pigment Cell Melanoma Res*, 2014. **27**(5): 699–720.

120. Song, X. et al., Alpha-MSH activates immediate defense responses to UV-induced oxidative stress in human melanocytes. *Pigment Cell Melanoma Res*, 2009. **22**(6): 809–818.

121. Kokot, A. et al., Alpha-melanocyte-stimulating hormone counteracts the suppressive effect of UVB on Nrf2 and Nrf-dependent gene expression in human skin. *Endocrinology*, 2009. **150**(7): 3197–3206.

122. Abdel-Malek, Z.A. et al., Alpha-MSH tripeptide analogs activate the melanocortin 1 receptor and reduce UV-induced DNA damage in human melanocytes. *Pigment Cell Melanoma Res*, 2009. **22**(5): 635–644.

123. Fruehauf, J.P. and F.L. Meyskens, Jr., Reactive oxygen species: A breath of life or death? *Clin Cancer Res*, 2007. **13**(3): 789–794.

124. Lubos, E., J. Loscalzo, and D.E. Handy, Glutathione peroxidase-1 in health and disease: From molecular mechanisms to therapeutic opportunities. *Antioxid Redox Signal*, 2011. **15**(7): 1957–1997.

125. Kinnula, V.L. and J.D. Crapo, Superoxide dismutases in malignant cells and human tumors. *Free Radic Biol Med*, 2004. **36**(6): 718–744.

126. Arner, E.S. and A. Holmgren, The thioredoxin system in cancer. *Semin Cancer Biol*, 2006. **16**(6): 420–426.

127. Trachootham, D., J. Alexandre, and P. Huang, Targeting cancer cells by ROS-mediated mechanisms: A radical therapeutic approach? *Nat Rev Drug Discov*, 2009. **8**(7): 579–591.

128. Tang, X.Q. et al., Protection of oxidative preconditioning against apoptosis induced by H_2O_2 in PC12 cells: Mechanisms via MMP, ROS, and Bcl-2. *Brain Res*, 2005. **1057**(1–2): 57–64.

129. Maiti, A.K., Genetic determinants of oxidative stress-mediated sensitization of drug-resistant cancer cells. *Int J Cancer*, 2011. **130**(1): 1–9.

130. Estrela, J.M., A. Ortega, and E. Obrador, Glutathione in cancer biology and therapy. *Crit Rev Clin Lab Sci*, 2006. **43**(2): 143–181.

131. Powis, G. and D.L. Kirkpatrick, Thioredoxin signaling as a target for cancer therapy. *Curr Opin Pharmacol*, 2007. **7**(4): 392–397.

132. Nakamura, H., H. Masutani, and J. Yodoi, Extracellular thioredoxin and thioredoxin-binding protein 2 in control of cancer. *Semin Cancer Biol*, 2006. **16**(6): 444–451.

133. Miyazaki, K. et al., Elevated serum level of thioredoxin in patients with hepatocellular carcinoma. *Biotherapy*, 1998. **11**(4): 277–288.

134. Nakamura, H. et al., Expression of thioredoxin and glutaredoxin, redox-regulating proteins, in pancreatic cancer. *Cancer Detect Prev*, 2000. **24**(1): 53–60.

135. Surh, Y.J., J.K. Kundu, and H.K. Na, Nrf2 as a master redox switch in turning on the cellular signaling involved in the induction of cytoprotective genes by some chemopreventive phytochemicals. *Planta Med*, 2008. **74**(13): 1526–1539.

136. Mathers, J. et al., Antioxidant and cytoprotective responses to redox stress. *Biochem Soc Symp*, 2004. **71**: 157–176.

137. Surh, Y.J. et al., Role of Nrf2-mediated heme oxygenase-1 upregulation in adaptive survival response to nitrosative stress. *Arch Pharm Res*, 2009. **32**(8): 1163–1176.

138. Chanas, S.A. et al., Loss of the Nrf2 transcription factor causes a marked reduction in constitutive and inducible expression of the glutathione S-transferase Gsta1, Gsta2, Gstm1, Gstm2, Gstm3 and Gstm4 genes in the livers of male and female mice. *Biochem J*, 2002. **365**(Pt 2): 405–416.

139. Brigelius-Flohe, R. and A. Banning, Part of the series: From dietary antioxidants to regulators in cellular signaling and gene regulation. Sulforaphane and selenium, partners in adaptive response and prevention of cancer. *Free Radic Res*, 2006. **40**(8): 775–787.

140. Khan, N.M. et al., Pro-oxidants ameliorate radiation-induced apoptosis through activation of the calcium-ERK1/2-Nrf2 pathway. *Free Radic Biol Med*, 2011. **51**(1): 115–128.

141. Brewer, A.C. et al., Nox4 regulates Nrf2 and glutathione redox in cardiomyocytes in vivo. *Free Radic Biol Med*, 2011. **51**(1): 205–215.

142. Malec, V. et al., HIF-1 alpha signaling is augmented during intermittent hypoxia by induction of the Nrf2 pathway in NOX1-expressing adenocarcinoma A549 cells. *Free Radic Biol Med*, 2010. **48**(12): 1626–1635.

143. Giudice, A., C. Arra, and M.C. Turco, Review of molecular mechanisms involved in the activation of the Nrf2-ARE signaling pathway by chemopreventive agents. *Methods Mol Biol*, 2010. **647**: 37–74.

144. Acharya, A. et al., Redox regulation in cancer: A double-edged sword with therapeutic potential. *Oxid Med Cell Longev*, 2010. **3**(1): 23–34.

145. DeNicola, G.M. et al., Oncogen-induced Nrf2 transcription promotes ROS detoxification and tumorigenesis. *Nature*, 2011. **475**(7354): 106–109.

146. Greenman, C. et al., Patterns of somatic mutation in human cancer genomes. *Nature*, 2007. **446**(7132): 153–158.

Section VI

*Cardiovascular Diseases
Induced by ROS*

20 Reactive Oxygen Species, Oxidative Stress, and Cardiovascular Disease

Eric L. Johnson

CONTENTS

ABSTRACT

Cardiovascular disease (CVD), including coronary heart disease, stroke, and peripheral arterial disease, is the number one cause of death globally, and according to the World Health Organization, CVD caused 17.5 million deaths in 2012, which is about 31% of the global total. More than 75% of these deaths occurred in low- and middle-income countries [1]. Organizations such as the European Society of Cardiology [2] and the American College of Cardiology [3] promote prevention strategies focused on the well-established manageable risk factors of tobacco use, obesity, physical inactivity, poor dietary habits, hypertension, diabetes, and dyslipidemia.

Over the last three decades, understanding has developed regarding reactive oxygen species (ROS) as contributors to inflammation in the pathophysiology of CVD along with the related conditions of obesity and type 2 diabetes mellitus (T2DM). Also, ROS and inflammation are related to disorders of the neurologic system and certain cancers [4]. ROS play physiologic roles in the maintenance of vascular diameter, oxygen sensing, immune responses, regulation of glucose uptake by skeletal muscle, and gene stability [5,12–14]. In recent years, research shows that ROS and oxidative stress may be a common pathway for the pathobiology of these various disease states [13] (Figure 20.1). In this chapter, the clinically relevant role of ROS and oxidative stress in the development of CVD is examined.

20.1 INTRODUCTION

Cardiovascular disease (CVD) has an etiology rooted in multiple factors. Chronic systemic inflammation characterized by high levels of reactive oxygen species (ROS) occurs early in the disease process. A variety of sources and effects of chronic inflammation are described in this chapter. Immunologic, metabolic, and endothelial functions are reliant on redox balance between physiologic and pathologic processes.

Interleukins (ILs), including IL-1β and IL-6, along with tumor necrosis factor-α (TNF-α) are underlying pathological processes for CVD. C-reactive protein (CRP) is an acute-phase reactive product that accompanies this process. Low-density lipoprotein (LDL), a known risk factor for CVD, is oxidized in the intima, leading to an inflammatory state that creates a difficult environment for normal processes at the cellular level to achieve oxidative homeostasis via the ROS systems [15,16].

ROS are defined as molecules or ions formed by the incomplete one-electron reduction of oxygen. These reactive oxygen intermediates include singlet oxygen, superoxides, peroxides, hydroxyl radical, and hypochlorous acid. These molecules are highly reactive. ROS arise from molecular oxygen conversion to superoxide anion ($O_2^{\cdot-}$); this is a univalent reduction occurring in almost all cells. Phagocyte respiratory bursts and ionizing radiation can also be sources of ROS. ROS contribute to the microbicidal activity of phagocytes; regulation of signal transduction and gene expression; and the oxidative damage to nucleic acids, proteins, and lipids [17]. ROS have a role to play in many physiologic systems, including oxygen sensing, the immune system, skeletal muscle, genomic stability, certain cancers, and vascular function [13].

Mitochondrial function and the endoplasmic reticulum (ER) are highly involved in the metabolism of ROS at the cellular level. The ER is involved in ROS balance through its mechanism of proper protein folding [18]. Protein kinase R–like ER kinase (PERK) pathway–induced antioxidant protection occurs through factors activating transcription factor-4 (ATF4) and Nrf2 [19]. Mitochondria are dynamic organelles, with their function governed by uncoupling proteins 1, 2, and 3 [20]. Abnormal function is tied to obesity and type 2 diabetes mellitus (T2DM) [20,21]. Protective mechanisms include antioxidant enzymes such as superoxide dismutase (SOD), catalase, and glutathione peroxidase (GPx)/reductase [22].

Thus, ROS play a key role in the development of inflammatory states leading to insulin resistance, T2DM, and CVD. As ROS may play a role in regulating and maintaining appropriate physiologic responses, not just as damaging agents, the term "redox regulation" has been proposed as a better descriptor than oxidative stress [13].

20.2 ROS REGULATION BY MITOCHONDRIA AND ER

Usual functions of these intracellular organelles are known to include cellular thermogenesis, ion storage, energy production, intermediary metabolism, and proper protein loading/folding balance [23].

The ER senses signals in a cellular stress environment with the unfolded protein response (UPR) or ER stress response. Protein folding is mediated by enzymes including Grp78 (BiP), Grp 94, protein disulfide isomerase (PDI), calnexin, and calreticulin. UPR function seeks to reestablish homeostasis and alleviate ER stress through increasing folding capacity via the expression of protein folding and downregulation of ER protein client load with resultant inhibition of general protein translation and promoting the degradation of abnormally folded proteins. UPR-mediated cell death may contribute to the pathogenesis of many diseases including cancer, T2DM, neurodegenerative disorders, and atherosclerosis/CVD. Chronic activation of UPR appears to contribute to a number of inflammatory conditions [24–28].

Mitochondria are intimately involved in energy production and metabolism and are characterized by mitochondrial DNA (mtDNA). A major source for cellular ROS production is the mitochondrial electron transport chain (ETC). The $O_2^{\cdot-}$ is formed by semiubiquinone and then further enzymatically converted by SOD to hydrogen peroxide (H_2O_2); additional H_2O_2 can be formed nonenzymatically along with singlet oxygen ($^1O_2$). H_2O_2 may be enzymatically converted to a reactive hydroxyl radical ($^\cdot OH$) in the absence of reduced transition metals and can also be converted to water by catalase and/or GPx [9]. Mitochondrial ETC ROS production is under tight control by uncoupling proteins 1, 2, and 3 as part of its normal function [4]. Excessive ROS in mitochondria or breakdown of the physiologic regulatory system leads to dysfunction. In both ER and mitochondria, this can lead to a vicious cycle of oxidative stress [29].

20.3 BIOMARKERS OF OXIDATIVE STRESS

Recently, Lubrano et al. [30], consistent with increased cardiovascular risk, have described elevated levels of adipokines and oxidative stress evidenced by reduced levels of glutathione, downregulated glutathione-*S*-transferase, increased 4-hydroxynonenal-protein adducts, ROS, and membrane-bound monounsaturated fatty acids, with elevated GPx activity, TNF-α, and other proinflammatory cytokines/chemokines/growth factor with a combination of high adipokine plasminogen activator inhibitor-1 (PAI-1) in obese patients with metabolic syndrome. Also, monoamine oxidases (MAOs) have been identified as another important mitochondrial source of oxidative stress in the cardiovascular system. MAOs are flavoproteins located in the outer mitochondrial membrane, serving as catalysts for the oxidative breakdown of endogenous monoamines and dietary amines, with the constant generation of H_2O_2, aldehydes, and ammonia as by-products [31,32].

20.4 ROS IN VASCULAR FUNCTION, CVD, AND RELATED METABOLIC DISORDERS

With respect to cardiovascular function and vascular diameter regulation in normal vascular function, well-described mitochondrial ROS, $O_2{}^{\bullet-}$, and H_2O_2 are of importance and have been studied in detail [5]. Atherosclerotic pathology includes many factors, including immune system responses from leukocytes, monocytes, macrophages, platelets, and the endothelium. Inflammatory states affect these functions. Diabetes is a common comorbid condition with CVD and is a disease of inflammation and endothelial dysfunction.

However, when exceeding physiologic limitations, ROS are generated by an imbalance between the activity of endogenous pro-oxidative enzymes (such as NADPH oxidase, xanthine oxidase, and mitochondrial respiratory chain) and antioxidative enzymes (such as SOD, GPx, heme oxygenase, and catalase [32]. Nitric oxide activity is decreased via inactivation by ROS; in turn, the normal functions of nitric oxide to physiologically inhibit platelet function and leukocyte adhesion are impaired [32]. Oxidized proteins from the plasma membrane and cytoskeleton increase mitochondrial response and signals for increased cellular inflammation [6]. The imbalance between the production of ROS and the ability to eliminate them at the cellular level is the definition of oxidative stress and has been described to be central in the pathogenesis of CVD as a major contributor to endothelial and smooth muscle in vascular disease [33]. Thus, mitochondrial dysfunction is thought to play a key role in the early development of CVD, and ER protective functions with regard to folding are also disrupted, worsening the state of chronic inflammation.

Glucolipotoxicity is the term used to describe the abnormal insulin-resistant metabolic states underlying metabolic syndrome, T2DM, and CVD. Free fatty acids and elevated blood glucose levels are prime contributors to tissue damage through known biochemical mechanisms including glycation, glyceraldehyde autooxidation, hexosamine pathway, protein C kinase activation, sorbitol production, and oxidative phosphorylation. This process may additionally contribute to T2DM β-cell dysfunction. Organ damage and failure in the liver and heart can also result from lipid accumulation. Overall, lipotoxicity has a direct connection with oxidative and ER stress [34]. Oxidative stress has a role in hyperlipidemia and lipid peroxidation, contributing to CVD [33].

In those with established T2DM, insulin resistance and glycemic variability are well established as a cause of oxidative stress in CVD, and there is considerable overlap between these two conditions [35,36]. Other dysmetabolic factors including hypertension and lipid disorders are associated with these adverse proinflammatory states [37]. Endothelial dysfunction is characterized by impaired endothelium-dependent vasodilation due to the unavailability of vasodilators such as NO and may be mediated by TNF-α [38]. In diabetes, endothelium activation is a proinflammatory,

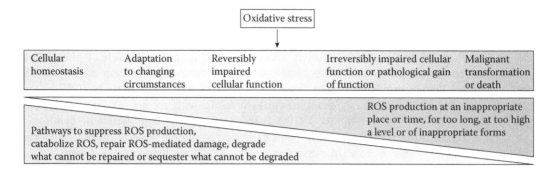

FIGURE 20.1 **(See color insert.)** The broad range of reactive oxygen species (ROS) signaling is influenced by ROS production and catabolism and by cellular adaptation. Restriction of ROS production to appropriate subcellular locations, times, levels, molecular species, and for appropriate durations allows ROS to contribute to homeostasis and physiological cell activation. For example, brief pulses of H_2O_2 production at the plasma membrane or at the endosomal membrane mediate signaling in response to the engagement of receptors with cytokines, microbial products, or antigens (left-hand side). When ROS production escapes these restrictions—for example, when there are high levels or sustained production of hydroxyl radicals—macromolecules are damaged ("oxidative stress"). ROS-mediated damage can often be reversed by repair, replacement, degradation, or sequestration of the damaged macromolecules (middle). However, damage that exceeds the capacity of the cell for these responses can lead to cell death (right-hand side). When damage to DNA results in mutagenesis without irreparable double-strand breakage, and when damage to other macromolecules is repaired, the consequence can be malignant transformation rather than the death of the cell (right-hand side). (Reprinted by permission from Macmillan Publishers Ltd: *Nat. Rev. Immunol.*, Nathan, C., Cunningham-Bussel, A. Beyond oxidative stress: An immunologist's guide to reactive oxygen species. 13(5):349–361. © 2013.)

proliferative, and procoagulatory setting, ultimately leading to arterial narrowing and susceptibility to atheroma deposition in CVD [36] (Figure 20.1). The multiplex protein, NADPH oxidase, expressed in endothelial cells and phagocytes may be the common pathway for these processes and has been demonstrated in animal models [9,39].

Brownlee [40] proposed what is known as the unifying hypothesis of diabetes and vascular injury, where ROS produced within the mitochondria initiate the development of diabetic complications. The hypothesis maintains that ROS formed in the mitochondria migrate to the nucleus, where they cause DNA damage, resulting in the activation of PARP-1. Glyceraldehyde-3-phosphate dehydrogenase (GAPDH) is transported into the nucleus, where it undergoes ADP ribosylation, a reaction that inactivates the glycolytic enzyme. Glycolysis is disrupted by a decline in GAPDH activity, wherein glycolytic intermediates are diverted from glycolysis into pathways of hyperglycemic injury. Later in 2010, this was updated by Brownlee and Giacco [40].

More recent reviews have suggested that there may be elements missing from the unifying hypothesis, including steps in the transfer of superoxide from mitochondria to the DNA in the cell nucleus, uncoupling of the nitric oxide system, and oxidative species generated from the hyperlipidemia present in diabetes, which are closely related to cardiovascular complications [41].

Giacco and Brownlee's [39] review considered metabolic abnormalities of diabetes, which caused mitochondrial $O_2^{\bullet-}$ overproduction in endothelial cells of both large and small vessels and in the myocardium, with the following pathways activated:

- Polyol pathway flux
- Increased formation of advanced glycation end products (AGEs)
- Increased expression of the receptor for AGEs and its activating ligands
- Activation of protein kinase C isoforms
- Overactivity of the hexosamine pathway

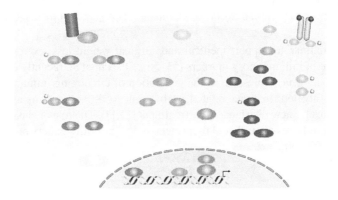

FIGURE 20.2 (See color insert.) Examples of transcriptional regulation by ROS acting at the plasma membrane or in the cytosol. (Reprinted by permission from Macmillan Publishers Ltd: *Nat. Rev. Immunol.*, Nathan, C., Cunningham-Bussel, A. Beyond oxidative stress: An immunologist's guide to reactive oxygen species. 13(5):349–361. © 2013.)

These are thought to be important factors in the genesis of diabetes vascular complications. Also, this process directly inactivates the following two critical anti-atherosclerotic enzymes:

1. Endothelial nitric oxide synthase
2. Prostacyclin

Resultant ROS activate a number of proinflammatory pathways. Pathway-selective insulin resistance increases mitochondrial ROS production from free fatty acids and by the inactivation of anti-atherosclerosis enzymes by ROS.

Adipokines, first described in the 1990s, control numerous physiological processes, including appetite and satiety, energy metabolism, hemostasis, blood pressure, insulin secretion and sensitivity, adipogenesis, endothelial function, adipocyte functions, and fat distribution. There are many proinflammatory mediators among adipokines, including ILs, TNF-α, chemokines, and growth factors [42]. Leptin, PAI-1, chemerin, MCP-1, and retinol-binding-protein-4 are a specific adipokine profile for body weight. Some adipokines, including TNF-α, ILs, chemokines, and growth factors, are produced by circulating blood leukocytes, tissue macrophages, and keratinocytes. Adipose tissue is thought to be the only source of adiponectin, leptin, chemerin, visfatin, and PAI-1 [43]. Excess adipose tissue in obesity is well known to have active mediators including cytokines and adipokines accompanying the increase in monocytes and macrophages leading to a global state of pathophysiologic inflammation and resultant production of ROS stressing the redox system [43,44] (Figure 20.2).

In summary, ROS play a key role in normal redox balance. A chronic inflammatory milieu leading to a disturbance in these systems is mechanistically tied to CVD and accompanying clinical disorders of diabetes, hypertension, and lipid disorders.

20.5 CLINICAL AND THERAPEUTIC CONSIDERATIONS

Managing risk factors commonly associated with CVD, such as hypertension, lifestyle modification, diabetes, dyslipidemia, and tobacco use, are cornerstones for comprehensive prevention and treatment. Aspirin, with its well-described antiplatelet and anti-inflammatory effects, is a widely used preventive measure in these clinical settings [2,3,45]. In addition to hypertension control via the renin–angiotensin system, anti-inflammatory effects of angiotensin-converting enzyme inhibitors (ACEIs) and angiotensin receptor blockers (ARBs) also exist [46]. Statin drugs exert an anti-inflammatory effect in addition to lipid lowering through the inhibition of the HMG-CoA reductase pathway [47,48].

Oxidative stress from hyperglycemia and glucose variability is well described in the literature. Some diabetes treatments available or in development specifically address postprandial hyperglycemia [49–52].

Ongoing data continue to support medical and surgical weight loss, exercise, and healthy diet on the improvement of inflammatory markers [53–56]. Ellsorth et al. recently reviewed weight loss as a means of altering gene expression in the generation of circulating immune cells and vascular endothelium [57]. Targeted therapies related to different ROS mechanisms and the cellular level provide opportunities for new pharmacologic treatment [13,14]. The role of classic antioxidants such as vitamin C, vitamin E, resveratrol, and β-carotene is still not fully understood in the reduction of cardiovascular risk [58–60].

20.6 CONCLUSION

Imbalance in the reactive oxidative system from inflammation plays a role in many disease processes, including CVD. Redox regulation and inflammation in tissue and cell environments with alteration of cellular function are also involved. Changes occur early in the timeline of development of overt clinical disease. The treatment of traditional cardiovascular risk factors through healthy lifestyle approaches, medications, and medical or surgical weight loss appears to have some impact on this negative inflammatory environment at many levels. Potential mechanisms for future interventions may also exist at the cellular level with respect to central signaling in the generation of ROS and redox balance.

REFERENCES

1. World Health Organization (WHO). Cardiovascular diseases. http://www.who.int/cardiovascular_diseases/en/. Accessed July 3, 2015.
2. European Society of Cardiology (ESC). ESC clinical practice guidelines. http://www.escardio.org/Guidelines-&-Education/Clinical-Practice-Guidelines/ESC-Clinical-Practice-Guidelines-list/listing. Accessed July 3, 2015.
3. American College of Cardiology (ACC). Guidelines. http://www.acc.org/guidelines. Accessed July 3, 2015.
4. Kumar V, Abbas A, Aster, J. *Robbins and Cotran Pathologic Basis of Disease*, 9th edn. Philadelphia, PA: Saunders, An Imprint of Elsevier, 2015, pp. 69–111.
5. Liu Y, Zhao H, Li H, Kalyanaraman C, Nicolosi AC, Gutterman DD. Mitochondrial sources of H_2O_2 generation play a key role in flow-mediated dilation in human coronary resistance arteries. *Circ Res.* 2003;93(6):573–580.
6. Go Y-M, Park H, Koval M et al. A key role for mitochondria in endothelial signaling by plasma cysteine/cystine redox potential. *Free Radic Biol Med.* 2010;48(2):275–283.
7. Guzy RD, Schumacker PT. Oxygen sensing by mitochondria at complex III: The paradox of increased reactive oxygen species during hypoxia. *Exp Physiol.* 2006;91(5):807–819.
8. Semenza GL. Hypoxia-inducible factor 1 and cancer pathogenesis. *IUBMB Life.* 2008;60(9):591–597.
9. Dröge W. Free radicals in the physiological control of cell function. *Physiol Rev.* 2002;82(1):47–95.
10. Yang Y, Bazhin AV, Werner J, Karakhanova S. Reactive oxygen species in the immune system. *Int Rev Immunol.* 2013;32(3):249–270.
11. Merry TL, McConell GK. Do reactive oxygen species regulate skeletal muscle glucose uptake during contraction? *Exerc Sport Sci Rev.* 2012;40(2):102–105.
12. Sandström ME, Zhang S-J, Bruton J et al. Role of reactive oxygen species in contraction-mediated glucose transport in mouse skeletal muscle. *J Physiol.* 2006;575(1):251–262.
13. Alfadda A, Sallam R. Reactive oxygen species in health and disease. *J Biomed Biotechnol.* 2012;2012:936486.
14. Rajendran R, Garva R, Krstic-Demonacos M, Demonacos C. Sirtuins: Molecular traffic lights in the crossroad of oxidative stress, chromatin remodeling, and transcription. *J Biomed Biotechnol.* 2011;2011:368276. doi:10.1155/2011/368276.

15. Kaptoge S, Seshasai SR, Gao P et al. Inflammatory cytokines and risk of coronary heart disease: New prospective study and updated meta-analysis. *Eur Heart J.* 2014;35:578–589.

16. Heinrich PC, Castell JV, Andus T. Interleukin-6 and the acute phase response. *Biochem J.* 1990;265:621–636.

17. *IUPAC Glossary of Terms Used in Toxicology*, 2nd edn. National Library of Medicine. http://www.nlm.nih.gov/. Accessed July 3, 2015.

18. Rutkowski DT, Kaufman RJ. A trip to the ER: Coping with stress. *Trends Cell Biol.* 2004;14:20–28.

19. Cullinan SB, Zhang D, Hannink M, Arvisais E, Kaufman RJ, Diehl JA. Nrf2 is a direct PERK substrate and effector of PERK-dependent cell survival. *Mol Cell Biol.* 2003;23(20):7198–7209.

20. Patti ME, Corvera S. The role of mitochondria in the pathogenesis of type 2 diabetes. *Endocr Rev.* 2010;31(3):364–395.

21. Zhang CY, Baffy G, Perret P et al. Uncoupling protein-2 negatively regulates insulin secretion and is a major link between obesity, β cell dysfunction, and type 2 diabetes. *Cell.* 2001;105(6):745–755.

22. Devasagayam TP, Tilak JC, Boloor KK, Sane KS, Ghaskadbi SS, Lele RD. Free radicals and antioxidants in human health: Current status and future prospects. *J Assoc Physicians India.* 2004;52:794–804.

23. Hall JE. *Guyton and Hall Textbook of Medical Physiology*, 13th edn. Elsevier, Inc., 2016.

24. Rutkowski DT, Kaufman RJ. A trip to the ER: Coping with stress. *Trends Cell Biol.* 2004;14:20–28.

25. Ron D, Walter P. Signal integration in the endoplasmic reticulum unfolded protein response. *Nat Rev Mol Cell Biol.* 2007;8:519–529.

26. Tabas I, Ron D. Integrating the mechanisms of apoptosis induced by endoplasmic reticulum stress. *Nat Cell Biol.* 2011;13:184–190.

27. Marciniak SJ, Ron D. Endoplasmic reticulum stress signaling in disease. *Physiol Rev.* 2006;86:1133–1149.

28. Wu J, Kaufman RJ. From acute ER stress to physiological roles of the unfolded protein response. *Cell Death Differ.* 2006;13:374–384.

29. Mathers J, Fraser JA, McMahon M, Saunders RDC, Hayes JD, McLellan LI. Antioxidant and cytoprotective responses to redox stress. *Biochem Soc Symp.* 2004;71:157–176.

30. Lubrano C, Valacchi G, Specchia P, Gnessi L, Rubanenko EP, Shuginina EA, Trukhanov AI, Korkina LG, De Luca C. Integrated haematological profiles of redox status, lipid, and inflammatory protein biomarkers in benign obesity and unhealthy obesity with metabolic syndrome. *Oxid Med Cell Longev.* 2015;2015:490613. doi:10.1155/2015/490613.

31. Duicu OM, Lighezan R, Sturza A et al. Monoamine oxidases as potential contributors to oxidative stress in diabetes: Time for a study in patients undergoing heart surgery. *BioMed Res Int.* 2015;2015:515437.

32. Pitocco D, Zaccardi F, Di Stasio E et al. Oxidative stress, nitric oxide, and diabetes. *Rev Diabet Stud.* 2010;7(1):15–25. doi:10.1900/RDS.2010.7.15.

33. Libby P, Ridker PM, Maseri A. Inflammation and atherosclerosis. *Circulation.* 2002;105(9):1135–1143.

34. Poitout V, Robertson RP. Glucolipotoxicity: Fuel excess and β-cell dysfunction. *Endocr Rev.* 2008;29(3):351–366. doi:10.1210/er.2007-0023.

35. Johnson EL. Glycemic variability in type 2 diabetes mellitus: Oxidative stress and macrovascular complications. *Adv Exp Med Biol.* 2012;771:139–154.

36. Pitocco D, Tesauro M, Alessandro R, Ghirlanda G, Cardillo C. Oxidative stress in diabetes: Implications for vascular and other complications. *Int J Mol Sci.* 2013;14(11):21525–21550. doi:10.3390/ijms141121525.

37. Mann DL, Zipes DP, Libby P, Bonow RO, and Braunwald, E. *Braunwald's Heart Disease: A Textbook of Cardiovascular Medicine*, 10th edn. Philadelphia, PA: Saunders, An Imprint of Elsevier Inc., 2015.

38. Rask-Madsen C, Dominguez H, Ihlemann N, Hermann T, Kober L, Torp-Pedersen C. Tumor necrosis factor-alpha inhibits insulin's stimulating effect on glucose uptake and endothelium-dependent vasodilation in humans. *Circulation.* 2003;108:1815–1821.

39. Giacco F, Brownlee M. Oxidative stress and diabetic complications. *Circ Res.* 2010;107:1058–1070.

40. Brownlee M. Biochemistry and molecular cell biology of diabetic complications. *Nature.* 2001; 414:813–820.

41. Schaffer SW, Jong CJ, Mozaffari M. Role of oxidative stress in diabetes-mediated vascular dysfunction: Unifying hypothesis of diabetes revisited. *Vascul Pharmacol.* 2012;57(5–6):139–149. doi:10.1016/j.vph.2012.03.005.

42. Tesauro M, Canale MP, Rodia G, di Daniele N, Lauro D, Scuteri A, Cardillo C. Metabolic syndrome, chronic kidney, and cardiovascular diseases: Role of adipokines. *Cardiol Res Pract.* 2011;2011:653182.

43. Kershaw EE, Flier JS. Adipose tissue as an endocrine organ. *J Clin Endocrinol Metab.* 2004;89:2548–2556.

44. Li ZY, Wang P, Miao CY. Adipokines in inflammation, insulin resistance and cardiovascular disease. *Clin Exp Pharmacol Physiol.* 2011;38(12):888–896.

45. Gaglia MA Jr., Clavijo L. Cardiovascular pharmacology core reviews: Aspirin. *J Cardiovasc Pharmacol Ther.* 2013;18(6):505–513.

46. Rosei EA, Rizzoni D, Muiesan ML, Sleiman I, Salvetti M, Monteduro C, Porteri E. CENTRO (CandEsartaN on aTherosclerotic Risk factors) study investigators. Effects of candesartan cilexetil and enalapril on inflammatory markers of atherosclerosis in hypertensive patients with non-insulin-dependent diabetes mellitus. *J Hypertens.* 2005;23(2):435–444.

47. Musial J, Undas A, Gajewski P, Jankowski M, Sydor W, Szczeklik A. Anti-inflammatory effects of simvastatin in subjects with hypercholesterolemia. *Int J Cardiol.* 2001;77(2–3):247–253.

48. Moreira DM, da Silva RL, Vieira JL, Fattah T, Lueneberg ME, Gottschall CA. Role of vascular inflammation in coronary artery disease: Potential of anti-inflammatory drugs in the prevention of athero-thrombosis. Inflammation and anti-inflammatory drugs in coronary artery disease. *Am J Cardiovasc Drugs.* 2015;15(1):1–11.

49. Johnson EL. Glycemic variability in type 2 diabetes mellitus: Oxidative stress and macrovascular complications. *Adv Exp Med Biol.* 2012;771:139–154.

50. Pitocco D, Tesauro M, Alessandro R, Ghirlanda G, Cardillo C. Oxidative stress in diabetes: Implications for vascular and other complications. *Int J Mol Sci.* 2013;14(11):21525–21550. doi:10.3390/ijms141121525.

51. American Diabetes Association. Standards of medical care in diabetes. *Diabetes Care.* 2015;38:S1–S2.

52. American Association of Clinical Endocrinologists. AACE/ACE Comprehensive Diabetes Management Algorithm 2015. *Endocr Pract.* 2015;21:438–447.

53. Strasser B, Berger K, Fuchs D. Effects of a caloric restriction weight loss diet on tryptophan metabolism and inflammatory biomarkers in overweight adults. *Eur J Nutr.* 2015;54(1):101–107.

54. Strasser B, Arvandi M, Siebert U. Resistance training, visceral obesity and inflammatory response: A review of the evidence. *Obes Rev.* 2012;13(7):578–591.

55. Engström G, Site-Flondell D, Lindblad B, Janzon L, Lindgärde F. Risk of treatment of peripheral arterial disease is related to inflammation-sensitive plasma proteins: A prospective cohort study. *J Vasc Surg.* 2004;40(6):1101–1105.

56. Lasselin J, Magne E, Beau C, Ledaguenel P, Dexpert S, Aubert A, Layé S, Capuron L. Adipose inflammation in obesity: Relationship with circulating levels of inflammatory markers and association with surgery-induced weight loss. *J Clin Endocrinol Metab.* 2014;99(1):E53–E61.

57. Ellsworth DL Mamula KA, Blackburn HL et al. Importance of substantial weight loss for altering gene expression during cardiovascular lifestyle modification. *Obesity (SilverSpring).* 2015;23(6):1312–1319.

58. Gostner J, Ciardi C, Becker K, Fuchs D, Sucher R. Immunoregulatory impact of food antioxidants. *Curr Pharm Des.* 2014;20(6):840–849.

59. Mangge H, Becker K, Fuchs D, Gostner JM. Antioxidants, inflammation and cardiovascular disease. *World J Cardiol.* 2014;6(6):462–477.

60. Murr C, Winklhofer-Roob BM, Schroecksnadel K, Maritschnegg M, Mangge H, Böhm BO, Winkelmann BR, März W, Fuchs D. Inverse association between serum concentrations of neopterin and antioxidants in patients with and without angiographic coronary artery disease. *Atherosclerosis.* 2009;202(2):543–549.

61. Nathan C, Cunningham-Bussel A. Beyond oxidative stress: An immunologist's guide to reactive oxygen species. *Nat Rev Immunol.* 2013;13(5):349–361.

21 Reactive Oxygen Species in the Cardiovascular System

Hiroaki Shimokawa and Kimio Satoh

CONTENTS

ABSTRACT

Reactive oxygen species (ROS) play a crucial role in the pathogenesis of cardiovascular diseases. However, vascular-derived hydrogen peroxide (H_2O_2) also serves as an important signaling molecule in the cardiovascular system at its low physiological concentrations. At low concentrations, H_2O_2 can act as a second messenger, transducing the oxidative signal into biological responses through posttranslational protein modification. These structural changes ultimately lead to altered cellular function. The intracellular redox status is closely regulated by the balance between oxidant and antioxidant systems, and their imbalance can cause oxidative or reductive stress, leading to cellular damage and dysregulation. For example, excessive H_2O_2 deteriorates vascular functions and promotes vascular disease through multiple pathways. The RhoA/Rho-kinase pathway plays an important role in various fundamental cellular functions, including the production of excessive reactive oxygen species (ROS), leading to the development of cardiovascular diseases. Rho-kinase (ROCK1 and ROCK2) belongs to the family of serine/threonine kinases and is an important downstream effector of the small GTP-binding protein RhoA. Rho-kinase plays a crucial role in the pathogenesis of vasospasm, arteriosclerosis, ischemia/reperfusion injury, hypertension, pulmonary hypertension (PH), stroke, and heart failure. Thus, Rho-kinase inhibitors may be useful for the treatment of cardiovascular diseases in humans. Furthermore, cyclophilin A (CyPA) is secreted from vascular smooth muscle cells, which augments the destructive effects of ROS. Thus, it is important to understand the H_2O_2 signaling and the roles of downstream effectors such as CyPA in the vascular system in order to develop new therapeutic strategies for cardiovascular diseases. In this chapter, we will discuss the dual roles of vascular-derived H_2O_2 in mediating vascular functions (physiological roles) and promoting vascular diseases (pathological roles).

21.1 INTRODUCTION

Endothelial cells (ECs) and vascular smooth muscle cells (VSMCs) secrete a variety of vasoactive substances that contribute to vascular protection and vascular remodeling [1,2]. Furthermore, the growth factors secreted from VSMCs play important roles in the remodeling process as they mediate various cellular responses [3]. Oxidative stress is one of the important stimuli that modulate VSMC function and promote VSMC growth by inducing autocrine/paracrine growth mechanisms [4]. Oxidative stress is generated by high levels of ROS, including superoxide anions ($O_2^{-\bullet}$), hydrogen peroxide (H_2O_2), and hydroxyl radical ($^\bullet OH$) [5]. ROS have been shown to promote cell proliferation and hypertrophy in a concentration-dependent manner [6]. Moreover, excessive ROS production can cause DNA damage and harmful protein oxidation, ultimately promoting vascular diseases. Therefore, excessive ROS promote the development of cardiovascular diseases [7–9], although their specific molecular targets are unclear.

Although high levels of ROS contribute to vascular disease development, strictly controlled ROS formation also mediates important physiological functions in vascular walls. For example, H_2O_2 plays a crucial role as a signaling molecule at very low concentrations [10] as we have previously demonstrated that it is one of the endothelium-derived hyperpolarizing factors (EDHFs) that modulate vascular tone especially in microvessels [11–13] and human coronary arteries [14]. In contrast, H_2O_2 induces constrictions in a concentration-dependent manner [15,16]. A plausible explanation for why or how ROS contribute simultaneously to both vascular protection and vascular diseases remains to be elucidated [17].

21.2 REGULATION OF ROS GENERATION

Intracellular redox status equilibrium is maintained by the balance between oxidants (e.g., ROS) and antioxidants that can scavenge ROS in the cell [18]. Excessive ROS damage mitochondrial proteins and further increase ROS levels, thus forming a vicious cycle of oxidative damage. In addition to the generation of ROS in mitochondria, several enzymes can also generate intracellular ROS [18], for example, nicotinamide adenine dinucleotide phosphate (NADPH) oxidases (Nox), which dynamically produce $O_2^{-\bullet}$ and H_2O_2 (Figure 21.1). Importantly, the production of endothelial H_2O_2 for EDHF responses appears to depend in part on endothelial NOS (eNOS) [19]. eNOS also produces NO with a resultant production of cyclic guanosine monophosphate (cGMP), and NO can react with $O_2^{-\bullet}$ to produce peroxynitrite ($ONOO^-$) (Figure 21.2) [20]. Other enzymes, such as cytochrome P450 and xanthine oxidase (XO), also produce intracellular ROS under pathological conditions [18].

The amount of ROS produced by mitochondria and oxidizing enzymes is offset by the presence of antioxidants [21]. For example, $O_2^{-\bullet}$ is rapidly dismutated to H_2O_2 by superoxide dismutase (SOD) and is therefore highly restricted to certain cellular compartments (Figure 21.1). Among the ROS, H_2O_2 is most stable and can penetrate the membrane and rapidly reach its cellular targets, acting as a second messenger. However, H_2O_2 is neutralized by catalase localized in the peroxisome, which catalyzes the decomposition of H_2O_2 to oxygen and water (Figure 21.2). In addition, peroxiredoxins also reduce H_2O_2 levels, whereby the peroxidatic thiol reacts with the oxidant to form sulfenic acid. Peroxiredoxin is then regenerated by the antioxidant protein thioredoxin 1 (Trx1), thus balancing the intracellular redox state [22]. Trx1 also works as a signaling intermediate that can sense redox state imbalances and correct these asymmetries and transduces signals to other effectors, including transcription factors and kinases [18]. The dual roles of ROS, particularly H_2O_2 as both a protective and pathological agent, appear to be particularly important in vascular health.

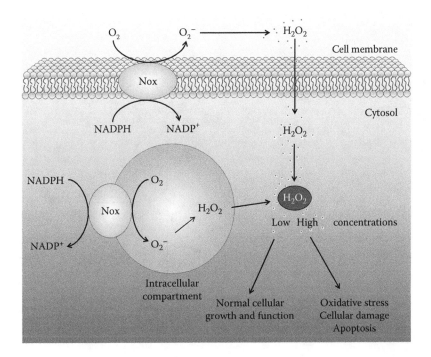

FIGURE 21.1 **(See color insert.)** Hydrogen peroxide (H_2O_2) as a second messenger. NADPH oxidases (Nox) dynamically produce $O_2^{-\bullet}$ and H_2O_2. O_2^- is very rapidly dismutated to H_2O_2 by superoxide dismutase (SOD) and is restricted to certain cellular compartments. H_2O_2 is stable and can penetrate the membrane and reach its targets rapidly, which enables it to act as a second messenger. Thus, intracellular H_2O_2 levels are lowered by passive and active diffusion into the extracellular space. The remaining H_2O_2 induces numerous biomolecular modifications, especially the oxidation of cysteine residues. At low physiological levels, H_2O_2 can promote cell growth, while at high pathological levels, it can create oxidative stress and apoptosis. NADPH, nicotinamide adenine dinucleotide phosphate; Nox, NADPH oxidases.

21.3 ROS IN THE VASCULAR SYSTEM

In the vascular wall, ROS are generated by several enzymes, including NADPH oxidases, XO, enzymes in the mitochondrial respiratory chain, and lipoxygenases [23]. Physiological levels of ROS regulate cell function, proliferation, and normal levels of cell death. For example, at low concentrations, H_2O_2 plays an important role in endothelial function and vascular relaxation [11–13]. Endothelium-dependent relaxation is mediated primarily by prostacyclin, NO, and EDHF (Figure 21.3) [24,25]. Feletou and Vanhoutte and Chen et al. were the first to describe, independently in 1988, the existence of EDHFs, and they are primarily found in resistance vessels, which are the principal regulators of vascular resistance and thus blood pressure [26,27]. Thus, the regulation of EDHF is the predominant mechanism for controlling vasodilation [28].

The contribution of H_2O_2 to EDHF-dependent vasodilation of resistance vessels can primarily be attributed to the oxidation of protein kinase G, subunit 1α (PKG1α) [11–13,29]. Recent work from the Eaton Laboratory has demonstrated that PKG can be activated by an oxidation mechanism, where the homodimer complex forms an interprotein disulfide [30]. This oxidation to the disulfide state is important for the catalytic activity of PKG. In ECs, PKG activity is also regulated by intracellular cGMP levels, which can be modified by NO produced by shear stress and agonists, such as bradykinin, acetylcholine (ACh), and adenosine (Figure 21.2) [31]. Furthermore, vascular smooth muscle expressing PKG resistant to H_2O_2 oxidation (i.e., Cys42 to Ser mutant)

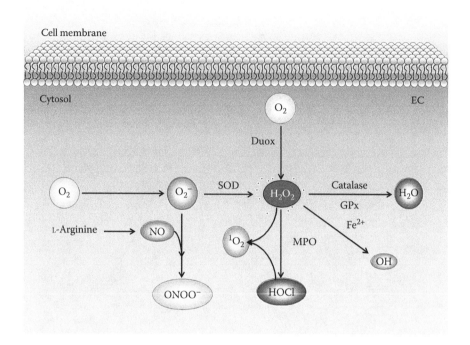

FIGURE 21.2 (See color insert.) Biochemical pathways of hydrogen peroxide (H_2O_2) generation and metabolism. Several enzymes can produce intracellular superoxide anions (O_2^-), such as NADPH oxidases (Nox), xanthine oxidase, aldehyde oxidase, cytochrome P450s, the electron transport chain, and urate oxidase. O_2^- is either rapidly dismutated to H_2O_2 spontaneously or catalyzed by superoxide dismutase (SOD). H_2O_2 is also formed by the catalyzed two-electron reduction of molecular oxygen (O_2) or can diffuse in from the extracellular matrix. O_2^- can also react with NO to produce peroxynitrite ($ONOO^-$). H_2O_2 is converted to other reactive oxygen species such as hydroxyl radicals ($^\cdot OH$, with Fe^{2+}) and hypochlorous acid (HOCl, with chloride, catalyzed by myeloperoxidase). H_2O_2 is also neutralized by catalase localized in peroxisomes, myeloperoxidase (MPO), and glutathione peroxidase (GPx). SOD, superoxide dismutase; O_2^-, superoxide anions; $ONOO^-$, peroxynitrite; $^\cdot OH$, hydroxyl radicals; HOCl, hypochlorous acid; MPO, myeloperoxidase; GPx, glutathione peroxidase.

displays altered relaxation properties that may be due in part to changes in protein interactions with substrates such as RhoA, myosin phosphatase target subunit (MYPT-1), and BKCa channels [31] (Figure 21.3). These effects were demonstrated using knock-in mice overexpressing redox-dead Cys42Ser PKG1α that fail to form a PKG disulfide and importantly those mice are hypertensive as compared with control mice [29,31].

The mechanism of H_2O_2-induced hyperpolarization seems to be complex and varies depending on the type of blood vessels tested. For example, Ca^{2+}/calmodulin-dependent protein kinase kinase β (CaMKKβ), caveolin-1, and PKG1α in murine microvessels have all been shown to play a substantial role in the enhanced EDHF-mediated response mechanism utilizing small amounts of $O_2^{-\cdot}$ and H_2O_2 in vascular ECs [32]. Bone marrow (BM) also appears to play an important role in modulating microvascular EDHF [33]. To address this notion, we transplanted BM from wild-type (WT) mice or eNOS$^{-/-}$ mice into male eNOS$^{-/-}$ mice and found that the reduced endothelium-dependent relaxations and hyperpolarizations of mesenteric arteries to ACh in eNOS$^{-/-}$ mice were markedly improved when transplanted with WT-BM but not with eNOS$^{-/-}$-BM [33]. Furthermore, the enhanced endothelium-dependent relaxations by WT-BM transplantation were abolished by catalase, indicating that the improved responses were mediated by H_2O_2.

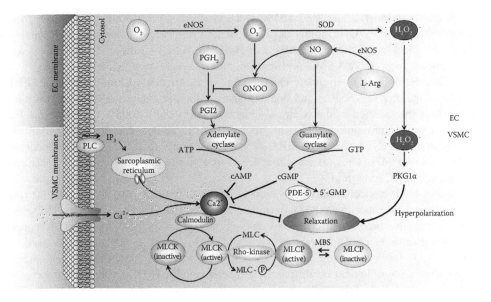

FIGURE 21.3 **(See color insert.)** Interaction between endothelial cells (ECs) and vascular smooth muscle cells (VSMC) during vascular relaxation. The process involves multiple reactive oxygen species (ROS) and their targets as well as several proteins in the RhoA/Rho-kinase signaling pathway. EC, endothelial cells; PGI_2, prostacyclin; NO, nitric oxide; EDHF, endothelium-derived hyperpolarizing factors; NOS, nitric oxide synthase; PKG, protein kinase G; MLC, myosin light chain; MLCK, MLC kinase; MLCP, MLC phosphatase; MBS, myosin-binding subunit.

The role of H_2O_2 as an EDHF has led to extensive research on the importance and complexity of endothelium-derived relaxing factors. Although our understanding of this vascular-derived oxidant is continually expanding, further studies are needed to clarify the physiological role of H_2O_2 in vascular homeostasis.

21.3.1 EXCESSIVE ROS PROMOTE VASCULAR DISEASE

To date, most research on H_2O_2 has focused on its pathological roles. Diabetic vascular dysfunction is associated with an increase in ROS [34]. Cardiovascular diseases often result from imbalances in the levels of oxidative species in the cell. The $O_2^{-\bullet}$-producing oxidases in the vascular system, including eNOS, cyclooxygenase, lipoxygenase, P450 monooxygenase, and NADPH oxidases, can be stimulated to produce excess ROS by external stimuli such as mechanical stretch, pressure, shear stress, hypoxia and by humoral factors such as angiotensin II (AngII) [23,35,36]. Excessive ROS target multiple biomolecules, causing numerous cellular complications, including lipid peroxidation, protein oxidation/inactivation, and DNA damage/mutations [23]. Furthermore, increased $O_2^{-\bullet}$ levels attenuate endothelium-dependent relaxation, promote contraction in VSMCs through the formation of $\bullet$OH [37,38], and become H_2O_2 spontaneously or through SOD-dependent dismutation (Figure 21.2). Although H_2O_2 is important for endothelial function and vascular relaxation at physiological low concentrations, pathological higher concentrations of ROS are hazardous to the cells, leading to endothelial dysfunction and VSMC proliferation [11,13]. Furthermore, H_2O_2 is converted by endogenous peroxidases into either H_2O and O_2 or $\bullet$OH, which are known to cause endothelium-dependent contractions through the production of vasoconstrictor prostanoids in VSMCs, leading to additional cellular damage [39].

Twenty years ago, very little was known about the proliferative response induced in VSMCs after arterial injury. ROS generated during arterial injury were considered, at least in part, to be responsible for this cellular response. Using xanthine/XO to generate ROS, Rao and Berk demonstrated that ROS stimulate VSMC proliferation in vitro and that H_2O_2 is the primary molecule responsible for xanthine/XO-induced VSMC DNA synthesis [40]. Both $O_2^{-\bullet}$ and H_2O_2 stimulate VSMC growth, but only $O_2^{-\bullet}$ rapidly activates MAP kinase, suggesting that additional signal events are involved in the mitogenic effects of H_2O_2 [9]. Based on these reports, excess amount of ROS appears to promote VSMC proliferation and increase the potential to develop vascular diseases. Interestingly, ROS stimulate extracellular signal–regulated kinases 1 and 2 (ERK1/2) in a biphasic manner in VSMCs, and one explanation for the delayed ERK1/2 activation is an involvement of the secreted oxidative stress–induced *F*actors (SOXFs) [41]. Cyclophilin A (CyPA) is one of the major proteins released into the medium in response to ROS [42]. Importantly, human recombinant CyPA stimulates ERK1/2 and DNA synthesis in VSMCs in a concentration-dependent manner [42]. Thus, extracellular CyPA is a novel growth factor that contributes to ROS-induced VSMC growth [43].

Exogenous H_2O_2 has commonly been used to elucidate the role of H_2O_2 in vascular function [14,44]. In the recent paper from the Gutterman laboratory, the EC50 of exogenous H_2O_2-induced dilation in human coronary arteries was approximately 1–3×10^{-5} mol/L. This concentration is similar to those used in the other studies [44]. However, the concentrations used for the exogenous H_2O_2 application were higher than those reported for endogenously generated H_2O_2 ($<10^{-6}$ mol/L) in physiological conditions. The discrepancy of H_2O_2 concentrations between the exogenous stimulation and the endogenous generation would be explained in part by the fact that the limited amount of exogenous H_2O_2 (1%–15%) diffuses and reaches into the intracellular target components [14,44].

21.3.2 Oxidative Stress and the Vascular Rho-Kinase System

Among the Rho family of small G proteins, RhoA is the best-characterized protein that acts as a molecular switch to elicit a variety of cellular responses (Figure 21.4) [45]. The activity of RhoA is controlled by the guanine nucleotide exchange factors (GEFs) that catalyze the exchange of GDP

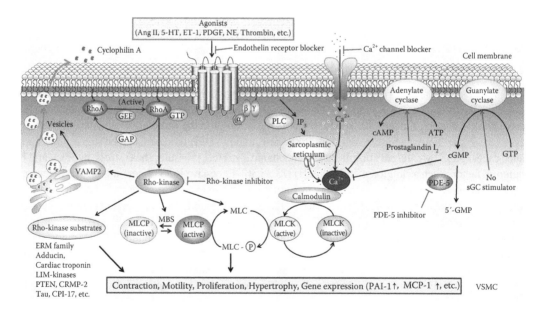

FIGURE 21.4 (See color insert.) Extracellular cyclophilin A augments ROS production. Reactive oxygen species (ROS) inducer, such as angiotensin II (AngII), mechanical stress, and environmental factors, promotes cyclophilin A (CyPA) secretion from vascular smooth muscle cells (VSMC). Secreted CyPA activates ERK1/2 and promotes ROS production, contributing to the augmentation of ROS production. ERK1/2, extracellular signal–regulated kinase 1/2.

for GTP [46]. In contrast, GTPase-activating proteins (GAPs) stimulate the intrinsic GTPase activity and inactivate RhoA [47]. Additionally, guanine nucleotide dissociation inhibitors (GDIs) block spontaneous RhoA activation (Figure 21.4) [48].

Agonists bind to G protein–coupled receptors and induce contraction by increasing both cytosolic Ca^{2+} concentration and Rho-kinase activity through mediating GEF (Figure 21.4). Rho-kinase (ROCK1 and ROCK2) enhances myosin light chain (MLC) phosphorylation through the inhibition of myosin-binding subunit (MBS) of myosin phosphatase and mediates agonists-induced VSMC contraction (Figure 21.4) [49,50]. Phosphorylation of MLC is a key event in the regulation of VSMC contraction. MLC is phosphorylated by Ca^{2+}/calmodulin-activated MLC kinase (MLCK) and dephosphorylated by MLC phosphatase (MLCP) (Figure 21.4). Many substrates of Rho-kinase have been identified, including MLC, MBS or MYPT-1, ERM (ezrin, radixin, and moesin) family, adducin, phosphatase and tensin homolog on chromosome ten (PTEN), and LIM-kinases (Figure 21.4).

The interaction between ECs and VSMCs plays an important role in regulating vascular integrity and vascular homeostasis (Figure 21.3). It has been demonstrated that both endothelial NO production and NO-mediated signaling in VSMCs are targets and effectors of the RhoA/Rho-kinase pathway (Figure 21.4). In ECs, the RhoA/Rho-kinase pathway negatively regulates NO production. In VSMCs, the RhoA/Rho-kinase pathway activates gene expression and secretion of growth factors, which promote VSMC proliferation and vascular remodeling [3,51,52]. Recent evidence suggests that many other stimuli that modulate VSMC functions, including ROS, promote VSMC growth by inducing auto/paracrine growth mechanisms [4]. Among the auto/paracrine factors, CyPA has been identified as an ROS-responsive protein that is secreted from VSMCs upon activation of the RhoA/Rho-kinase system [53,54]. The extracellular CyPA decreases eNOS expression in aortic ECs [55], suggesting the role of RhoA/Rho-kinase for the negative regulation of endothelial NO production. The initial investigations in our laboratory on the therapeutic importance of Rho-kinase were previously summarized [56].

Rho-kinase is substantially involved in the vascular effects of various vasoactive factors, including AngII, thrombin, platelet-derived growth factor, extracellular nucleotides, and urotensin [57–65] (Figure 21.4). Changes in vascular redox state are a common pathway involved in the pathogenesis of atherosclerosis, aortic aneurysms, and vascular stenosis. Vascular ROS formation can be stimulated by mechanical stretch, pressure, shear stress, hypoxia, and growth factors all of which activate the RhoA/Rho-kinase system. Interestingly, the pleiotropic effects of HMG-CoA reductase inhibitors (statins) are mediated in part by reduced synthesis of the isoprenoids that are responsible for the posttranslational modulation of intracellular proteins [36]. Since membrane localization and GTPase activity of small GTP-binding proteins are dependent on isoprenylation, the pleiotropic effects of statins have been considered to be mediated by the inhibition of those small GTP-binding proteins [66]. However, we have previously demonstrated that this is not the case and that clinical doses of statins mainly inhibit Rac1 in mice and humans [67]. Subsequently, we demonstrated that small GTP-binding protein dissociation stimulator (SmgGDS) plays a crucial role in the pleiotropic effects of statins in mice and humans [68]. Importantly, statins and selective Rho-kinase inhibitors also completely block the secretion of CyPA from VSMCs [53,69]. Thus, Rho-kinase plays an important role in mediating various cellular functions, not only VSMC contraction but also actin cytoskeleton organization, cytokinesis, and ROS augmentation [56,70–74]. Taken together, the RhoA/Rho-kinase system plays a crucial role in the development of cardiovascular disease through endothelial dysfunction, VSMC contraction/proliferation, and inflammation.

21.3.3 RHO-KINASE AND INFLAMMATION

VSMCs secrete several growth factors in response to oxidative stress, including CyPA [42,53]. Extracellular CyPA stimulates ERK1/2, Akt, and JAK in VSMCs that contribute to ROS production [75,76]. The secretion of CyPA from VSMCs is a highly regulated pathway that involves vesicle transport and plasma membrane binding [53]. Rho GTPases including RhoA are key regulators in signaling pathways linked to actin cytoskeletal rearrangement [77]. RhoA plays a central role in vesicular trafficking pathways by controlling the organization of actin cytoskeleton. Active participation of Rho GTPases is required for secretion. Dominant-negative mutants of RhoA inhibit ROS-induced CyPA secretion [53]. Myosin II is involved in secretory mechanisms as a motor for vesicle transport [78]. Rho-kinase (ROCK1 and ROCK2), downstream effectors of RhoA, mediates myosin II activation via phosphorylation and inactivation of myosin II light chain phosphatase [49]. Consistently, Rho-kinase inhibitor completely blocks ROS-induced CyPA secretion [53,54]. Altogether, myosin II-mediated vesicle transport is required for CyPA secretion from VSMCs. Extracellular CyPA stimulates adhesion molecule expression in ECs, directly recruits inflammatory cells, and promotes the activation of matrix metalloproteinases (MMPs) through its extracellular receptor basigin [79–83]. All of these roles of extracellular CyPA derive from the activation of Rho-kinase in the cardiovascular system. Thus, CyPA is one of the key mediators of Rho-kinase that augments ROS formation, affecting ECs, VSMCs, and inflammatory cells [74–76].

21.3.4 RHO-KINASE AND CARDIOVASCULAR DISEASES

Rho-kinase plays a crucial role in the ROS augmentation and vascular inflammation. Accumulating evidence indicates that Rho-kinase inhibitors have broad pharmacological properties [56,71,84]. The beneficial effects of long-term inhibition of Rho-kinase for the treatment of cardiovascular diseases have been demonstrated in various animal models, such as coronary vasospasm, arteriosclerosis, PH, and heart failure. Gene transfer of dominant-negative Rho-kinase reduced neointimal formation in pigs and pressure-overload-induced cardiac hypertrophy and failure in mice [85,86]. Long-term treatment with a Rho-kinase inhibitor suppressed neo-intimal formation, constrictive remodeling, in-stent restenosis, and the development of cardiac allograft vasculopathy [87–92].

Among the Rho-kinase-mediated vascular diseases, coronary artery spasm plays an important role in variant angina, myocardial infarction, and sudden death [93]. Cortisol, one of the important stress hormones, causes coronary hyperreactivity through the activation of Rho-kinase in pigs in vivo [94]. Intracoronary administration of fasudil and hydroxyfasudil inhibited coronary spasm in pigs in vivo [89,95,96]. Fasudil is effective in preventing coronary spasm in patients with vaso-spastic angina and those with microvascular angina [97,98]. The clinical trials of the effects of fasudil in Japanese patients with stable effort angina demonstrated that the long-term oral treatment with the Rho-kinase inhibitor is effective in ameliorating exercise tolerance [99]. We have also recently demonstrated that Rho-kinase activity in circulating neutrophils is a useful biomarker for the diagnosis and disease activity assessment in patients with vasospastic angina [100–103].

Rho-kinase is also substantially involved in the pathogenesis of PH as it is associated with hypoxic exposure, endothelial dysfunction, VSMC proliferation, enhanced ROS production, and inflammatory cell migration. Chronic hypoxic exposure of mice induces vascular remodeling [104]. We have demonstrated that pulmonary vascular dysfunction plays a crucial role in the development of hypoxia-induced PH, for which Rho-kinase plays a crucial role [54,86,105–107]. Rho-kinase promotes the secretion of CyPA from VSMCs, and extracellular CyPA stimulates VSMC proliferation in vivo and in vitro [41,42,69]. Extracellular CyPA induces EC adhesion molecule expression, induces apoptosis [55], and is a chemoattractant for inflammatory cells [69,79,108]. Therefore, the extracellular CyPA may contribute to hypoxia-induced PH. Long-term treatment with fasudil suppressed the development of monocrotaline-induced PH in rats and hypoxia-induced PH in mice [109,110]. On the other hand, statins and Rho-kinase inhibitor reduce the secretion of CyPA from VSMCs, and pravastatin ameliorates hypoxia-induced PH in mice [53,54]. Thus, the inhibition of CyPA secretion by statins or Rho-kinase inhibitors may be involved in the therapeutic effects of these medications on PH. Furthermore, we have recently demonstrated the crucial role of ROCK2 in the development of hypoxia-induced PH using VSMC-specific ROCK2-deficient mice [106]. Consistently, we found Rho-kinase activation in patients with PH [111]. Furthermore, a selective Rho-kinase inhibitor fasudil significantly reduced pulmonary vascular resistance in patients with PH [112,113].

21.3.5 ROLE OF CYPA IN ROS-INDUCED VASCULAR DISEASE

In 1984, intracellular CyPA was identified as the main target for the immunosuppressive drug, cyclosporine [114]. Cyclophilins are a family of highly conserved and ubiquitous proteins termed immunophilins [115]. The most abundant cyclophilin is CyPA, which is widely distributed in almost all tissues [116]. Owing to its enzymatic properties, cellular localization, and role in protein folding, CyPA is classified into a diverse set of proteins termed "foldases" [117]. It catalyzes the cis–trans isomerization of peptidyl–prolyl bonds of certain proteins (PPIase activity) and acts as an accelerant during protein folding and assembly. In addition to its role in protein folding, CyPA has recently been demonstrated to have a variety of functions, including intracellular trafficking, signal transduction, and transcriptional regulation [118,119]. Importantly, CyPA plays a crucial role in the translocation of Nox enzymes such as p47phox, which are known to contribute to VSMC proliferation and development of vascular diseases [23,120]. Since ROS production by Nox enzymes activates other oxidase systems, CyPA and Nox enzymes amplify ROS formation in a synergistic manner, leading to increased oxidative stress.

It is now known that CyPA is secreted from VSMCs via a highly regulated pathway, which involves vesicle transport and plasma membrane binding in response to oxidative stress [53]. The expression of RhoA and Cdc42 (cell division control protein 42) dominant-negative mutants and a Rho-kinase inhibitor blocked ROS-induced CyPA secretion [53,121]. These results suggest that CyPA is secreted from VSMCs through a process that requires ROS production, RhoA/Rho-kinase activation, and vesicle formation [74]. In VSMCs, extracellular CyPA stimulates ERK1/2, Akt, and JAK (Janus protein tyrosine kinase) that contribute to ROS production [75,76] (Figure 21.5). In ECs, extracellular CyPA augments proinflammatory pathways, including enhanced expression of adhesion molecules, and promotes atherosclerosis [55,79]. In inflammatory cells, extracellular CyPA also works as a chemoattractant in cooperation with other cytokines and chemokines. Although the protein basigin has been proposed to serve as an extracellular receptor for CyPA in inflammatory cells [122], the identity of CyPA receptors in ECs and VSMCs remains unknown. Further knowledge of the extracellular CyPA receptors on vascular cells will contribute to the development of novel therapies for cardiovascular diseases.

Changes in vascular redox state and extracellular CyPA are commonly involved in the pathogenesis of vascular restenosis, aortic aneurysms, atherosclerosis, and cardiac hypertrophy [55,69,121,123]. We demonstrated that CyPA (both intracellular and extracellular) contributes to atherosclerosis by promoting EC apoptosis and EC expression of leukocyte adhesion molecules. These actions stimulate inflammatory cell migration, enhance ROS production, increase proliferation of macrophages and VSMCs, and increase proinflammatory signal transduction in VSMC [55]. In the context of atherosclerosis, CyPA is regarded as a proinflammatory and pro-atherogenic molecule. The role of CyPA in inflammation was discovered using mice lacking both apolipoprotein E (ApoE$^{-/-}$) and CyPA (CyPA$^{-/-}$) that appeared to be protected from atherosclerosis development. The atheroprotection observed in ApoE$^{-/-}$CyPA$^{-/-}$ mice was due to the decreased levels of inflammation mediated by the absence of CyPA [124]. The vascular endothelium expresses a large array of vital proteins that function in normal cellular processes, the loss of which lead to the initiation of atherosclerosis [125]. For example, eNOS function is critical for vascular homeostasis via generation of NO and its loss is pro-atherogenic. Furthermore, the progression of atherosclerosis is associated with decreases in both eNOS expression and NO production. In the ApoE$^{-/-}$CyPA$^{-/-}$ mice, aortic staining revealed significantly higher eNOS expression as compared with ApoE$^{-/-}$ mice, indicating that CyPA plays a role in regulating the eNOS/NO levels [55]. Moreover, shear stress–induced eNOS expression was significantly increased when CyPA siRNA was used to silence CyPA in human umbilical vein endothelial cells (HUVECs) [55]. In addition, CyPA knockdown in HUVEC increased eNOS

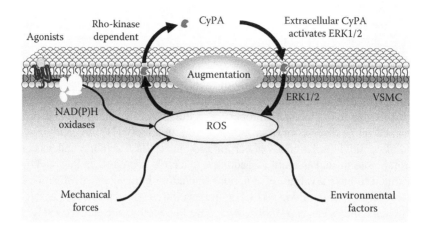

FIGURE 21.5 **(See color insert.)** The RhoA/Rho-kinase signaling pathway and therapeutic targets. Rho GTPases are small GTP-binding proteins that act as molecular switches and regulate many intracellular signaling pathways. RhoA cycles between an inactive GDP-bound and an active GTP-bound conformation, interacting with downstream targets, including Rho-kinase. The activity of RhoA is controlled by the guanine nucleotide exchange factors (GEFs) that catalyze the exchange of GDP for GTP. GTPase-activating proteins (GAPs) stimulate the intrinsic GTPase activity and inactivate RhoA. Guanine nucleotide dissociation inhibitors (GDIs) block spontaneous RhoA activation. Various substrates of Rho-kinase have been identified, including MYPT-1, MLC, ERM (ezrin, radixin, and moesin) family, adducin, phosphatase and tensin homolog on chromosome ten (PTEN), vesicle-associated membrane protein 2 (VAMP2), and LIM-kinases. Rho-kinase activation promotes cyclophilin A (CyPA) secretion. VSMCs, vascular smooth muscles cells; MLC, myosin light chain; MLCK, MLC kinase; MLCP, MLC phosphatase. (Courtesy of Satoh, K.)

promoter activity and eNOS mRNA levels, whereas overexpression of CyPA reduced eNOS protein and mRNA levels. Both the antioxidants, *N*-acetyl cysteine (NAC) and Tiron, reversed this CyPA-mediated inhibition of eNOS promoter activity [55]. These findings suggest a novel mechanism by which CyPA promotes atherosclerosis through the suppression of eNOS transcription.

Furthermore, overall ROS production is significantly higher in HUVEC overexpressing CyPA than in cells transfected with the vector control [55]. This suggests that CyPA plays a critical role in ROS generation in ECs as in VSMC and that CyPA likely induces inflammation through ROS-dependent mechanisms in these vascular cells [54,76]. Based on these results, CyPA is likely the primary mediator that augments ROS production, contributing to vascular inflammation and atherogenesis [76].

21.4 CLINICAL IMPLICATIONS OF OXIDATIVE STRESS RESEARCH

The identification of CyPA as a mediator of tissue damage associated with inflammation and oxidative stress provides new insight into the mechanisms of several therapies. We have recently demonstrated that plasma levels of CyPA are significantly elevated in patients with angiographically proven coronary artery disease (CAD) [126]. Importantly, CyPA levels were also elevated in patients with hypertension, diabetes mellitus, smoking, dyslipidemia, and advanced age, all of which are atherosclerotic risk factors and ROS inducers [126]. We also demonstrated that CyPA is a prognostic marker for cardiovascular intervention, such as percutaneous coronary intervention (PCI) and coronary artery bypass graft (CABG) [126]. Furthermore, after the treatment in several individuals, the plasma obtained at the baseline and follow-up revealed a significant reduction after the treatment [126]. Medical treatments that control atherosclerotic risk factors decreased plasma

CyPA levels in CAD patients, suggesting that plasma CyPA is useful for the evaluation of systemic oxidative stress and the therapeutic effect by medication [126]. Taken together, these results suggest that circulating CyPA is a novel biomarker for CAD and plays a crucial and synergistic role in ROS augmentation, contributing to the progression of atherosclerosis [76]. Importantly, CyPA is highly expressed at sites with unstable atherosclerotic plaques, especially those associated with macrophages and foam cells [126]. However, CyPA expression during plaque destabilization in humans and its regulatory mechanism remains elusive. Further research regarding the role of CyPA in the progression of atherosclerosis is necessary to identify potential CyPA-related therapeutic targets.

21.5 FUTURE PERSPECTIVES

Numerous basic and clinical studies have demonstrated that ROS play a major role in the pathogenesis of endothelial dysfunction and atherosclerosis. However, no therapeutic strategies are yet available for the clinical use of antioxidants. We consider that one of the reasons for this dilemma is that low concentrations of ROS, particularly H_2O_2, play an important role in intracellular signaling pathways that are crucial for numerous vascular cell functions. The source/location of ROS production may also be important in determining their physiological or pathological roles. For example, the roles of ROS in ECs, VSMC, and migrating perivascular inflammatory cells would be different between physiological and pathological conditions. In addition, the dual roles of ROS may be somewhat analogous to the beneficial/deleterious actions of NO in cell signaling [20]. Furthermore, the production of ROS in inflammatory cells plays a crucial role in the cellular responses in immune response and infection [127]. Although many strategies to control oxidative stress have previously been tested in various diseases, we need to pay close attention to the existence of a complex network of molecules that exogenously or endogenously contribute to the balance between oxidant and antioxidant systems. Thus, it may be an important clinical strategy to use antioxidants and/or drugs that can prevent the oxidation of selected redox-sensitive targets under certain disease conditions, while allowing the ROS to continue to function in normal processes.

Furthermore, the identification of CyPA as a mediator of oxidative stress–induced tissue damage has provided some additional insight into the mechanisms of several therapies. For example, Rho-kinase inhibitor and simvastatin significantly reduced CyPA secretion from VSMC [53,54]. Indeed, Rho-kinase is an important therapeutic target in cardiovascular diseases, and Rho-kinase inhibition has been reported to reduce AngII-induced abdominal aortic aneurysm (AAA) formation, atherosclerosis, PH, and cardiac hypertrophy [56,57,128,129]. Moreover, angiotensin II type 1 (AT1) receptor blockers and angiotensin-converting enzyme (ACE) inhibitors have been shown to reduce cardiovascular diseases, for which reduced CyPA secretion may be involved as shown in AAA, atherosclerosis, and PH [54,130–132].

Based on the role of extracellular CyPA, it is logical to consider that agents that prevent CyPA receptor binding and reduce circulating CyPA may have therapeutic potentials. Blocking this vicious cycle that augments ROS production through CyPA autocrine/paracrine signaling pathway could be a novel therapeutic tool for controlling cardiovascular diseases. However, the regulation of CyPA expression and the identity of its extracellular receptors remain largely unknown. Thus, further basic and clinical studies are needed to identify CyPA-related therapeutic targets in the future [133].

21.6 CONCLUSION

The identification of the Rho-kinase/CyPA system as a regulator of oxidative stress level has provided some insight into the mechanisms of several therapies. Based on the role of extracellular CyPA, it is logical to consider that agents that prevent CyPA receptor binding and reduce circulating CyPA may have therapeutic potentials. Blocking this vicious cycle that augments ROS production through Rho-kinase-mediated CyPA autocrine/paracrine signaling pathway could be a novel therapeutic tool for controlling cardiovascular diseases.

REFERENCES

1. Shimokawa, H., Tomoike, H., Nabeyama, S., Yamamoto, H., Araki, H., Nakamura, M., Ishii, Y., and Tanaka, K. 1983. Coronary artery spasm induced in atherosclerotic miniature swine. *Science* 221:560–562.
2. Shimokawa, H. 1999. Primary endothelial dysfunction: Atherosclerosis. *J Mol Cell Cardiol* 31:23–37.
3. Berk, B.C. 2001. Vascular smooth muscle growth: Autocrine growth mechanisms. *Physiol Rev* 81:999–1030.
4. Taniyama, Y. and Griendling, K.K. 2003. Reactive oxygen species in the vasculature: Molecular and cellular mechanisms. *Hypertension* 42:1075–1081.
5. Burgoyne, J.R., Oka, S., Ale-Agha, N., and Eaton, P. 2013. Hydrogen peroxide sensing and signaling by protein kinases in the cardiovascular system. *Antioxid Redox Signal* 18:1042–1052.
6. Griendling, K.K. and Ushio-Fukai, M. 1998. Redox control of vascular smooth muscle proliferation. *J Lab Clin Med* 132:9–15.
7. Alexander, R.W. 1995. Theodore cooper memorial lecture. Hypertension and the pathogenesis of atherosclerosis. Oxidative stress and the mediation of arterial inflammatory response: A new perspective. *Hypertension* 25:155–161.
8. Omar, H.A., Cherry, P.D., Mortelliti, M.P., Burke-Wolin, T., and Wolin, M.S. 1991. Inhibition of coronary artery superoxide dismutase attenuates endothelium-dependent and -independent nitrovasodilator relaxation. *Circ Res* 69:601–608.
9. Baas, A.S. and Berk, B.C. 1995. Differential activation of mitogen-activated protein kinases by H_2O_2 and O_2^- in vascular smooth muscle cells. *Circ Res* 77:29–36.
10. Vanhoutte, P.M. 2001. Endothelium-derived free radicals: For worse and for better. *J Clin Invest* 107:23–25.
11. Matoba, T., Shimokawa, H., Nakashima, M., Hirakawa, Y., Mukai, Y., Hirano, K., Kanaide, H., and Takeshita, A. 2000. Hydrogen peroxide is an endothelium-derived hyperpolarizing factor in mice. *J Clin Invest* 106:1521–1530.
12. Morikawa, K., Shimokawa, H., Matoba, T., Kubota, H., Akaike, T., Talukder, M.A., Hatanaka, M. et al. 2003. Pivotal role of Cu,Zn-superoxide dismutase in endothelium-dependent hyperpolarization. *J Clin Invest* 112:1871–1879.
13. Takaki, A., Morikawa, K., Tsutsui, M., Murayama, Y., Tekes, E., Yamagishi, H., Ohashi, J., Yada, T., Yanagihara, N., and Shimokawa, H. 2008. Crucial role of nitric oxide synthases system in endothelium-dependent hyperpolarization in mice. *J Exp Med* 205:2053–2063.
14. Zhang, D.X., Borbouse, L., Gebremedhin, D., Mendoza, S.A., Zinkevich, N.S., Li, R., and Gutterman, D.D. 2012. H_2O_2-induced dilation in human coronary arterioles: Role of protein kinase G dimerization and large-conductance Ca^{2+}-activated K^+ channel activation. *Circ Res* 110:471–480.
15. Katusic, Z.S., Schugel, J., Cosentino, F., and Vanhoutte, P.M. 1993. Endothelium-dependent contractions to oxygen-derived free radicals in the canine basilar artery. *Am J Physiol* 264:H859–H864.
16. Iida, Y. and Katusic, Z.S. 2000. Mechanisms of cerebral arterial relaxations to hydrogen peroxide. *Stroke* 31:2224–2230.
17. Sandow, S.L. 2004. Factors, fiction and endothelium-derived hyperpolarizing factor. *Clin Exp Pharmacol Physiol* 31:563–570.
18. Shao, D., Oka, S., Brady, C.D., Haendeler, J., Eaton, P., and Sadoshima, J. 2012. Redox modification of cell signaling in the cardiovascular system. *J Mol Cell Cardiol* 52:550–558.
19. Takaki, A., Morikawa, K., Murayama, Y., Yamagishi, H., Hosoya, M., Ohashi, J., and Shimokawa, H. 2008. Roles of endothelial oxidases in endothelium-derived hyperpolarizing factor responses in mice. *J Cardiovasc Pharmacol* 52:510–517.
20. Cohen, R.A. and Adachi, T. 2006. Nitric-oxide-induced vasodilatation: Regulation by physiologic s-glutathiolation and pathologic oxidation of the sarcoplasmic endoplasmic reticulum calcium ATPase. *Trends Cardiovasc Med* 16:109–114.
21. Sumimoto, H. 2008. Structure, regulation and evolution of Nox-family NADPH oxidases that produce reactive oxygen species. *FEBS J* 275:3249–3277.
22. Wu, C., Parrott, A.M., Fu, C., Liu, T., Marino, S.M., Gladyshev, V.N., Jain, M.R. et al. 2011. Thioredoxin 1-mediated post-translational modifications: Reduction, transnitrosylation, denitrosylation, and related proteomics methodologies. *Antioxid Redox Signal* 15:2565–2604.
23. Lassegue, B., San, M.A., and Griendling, K. 2012. Biochemistry, physiology, and pathophysiology of NADPH oxidases in the cardiovascular system. *Circ Res* 110:1364–1390.

24. Vanhoutte, P.M., Shimokawa, H., Tang, E.H., and Feletou, M. 2009. Endothelial dysfunction and vascular disease. *Acta Physiol (Oxf)* 196:193–222.

25. Shimokawa, H. 2010. Hydrogen peroxide as an endothelium-derived hyperpolarizing factor. *Pflugers Arch* 459:915–922.

26. Feletou, M. and Vanhoutte, P.M. 1988. Endothelium-dependent hyperpolarization of canine coronary smooth muscle. *Br J Pharmacol* 93:515–524.

27. Chen, G., Suzuki, H., and Weston, A.H. 1988. Acetylcholine releases endothelium-derived hyperpolarizing factor and EDRF from rat blood vessels. *Br J Pharmacol* 95:1165–1174.

28. Feletou, M. and Vanhoutte, P.M. 2007. Endothelium-dependent hyperpolarizations: Past beliefs and present facts. *Ann Med* 39:495–516.

29. Prysyazhna, O., Rudyk, O., and Eaton, P. 2012. Single atom substitution in mouse protein kinase G eliminates oxidant sensing to cause hypertension. *Nat Med* 18:286–290.

30. Burgoyne, J.R., Madhani, M., Cuello, F., Charles, R.L., Brennan, J.P., Schroder, E., Browning, D.D., and Eaton, P. 2007. Cysteine redox sensor in PKGIa enables oxidant-induced activation. *Science* 317:1393–1397.

31. Burgoyne, J.R., Prysyazhna, O., Rudyk, O., and Eaton, P. 2012. cGMP-dependent activation of protein kinase G precludes disulfide activation: Implications for blood pressure control. *Hypertension* 60:1301–1308.

32. Ohashi, J., Sawada, A., Nakajima, S., Noda, K., Takaki, A., and Shimokawa, H. 2012. Mechanisms for enhanced endothelium-derived hyperpolarizing factor-mediated responses in microvessels in mice. *Circ J* 76:1768–1779.

33. Nakajima, S., Ohashi, J., Sawada, A., Noda, K., Fukumoto, Y., and Shimokawa, H. 2012. Essential role of bone marrow for microvascular endothelial and metabolic functions in mice. *Circ Res* 111:87–96.

34. Lu, T., He, T., Katusic, Z.S., and Lee, H.C. 2006. Molecular mechanisms mediating inhibition of human large conductance Ca^{2+}-activated K^+ channels by high glucose. *Circ Res* 99:607–616.

35. Fleming, I., Michaelis, U.R., Bredenkotter, D., Fisslthaler, B., Dehghani, F., Brandes, R.P., and Busse, R. 2001. Endothelium-derived hyperpolarizing factor synthase (Cytochrome P450 2C9) is a functionally significant source of reactive oxygen species in coronary arteries. *Circ Res* 88:44–51.

36. Griendling, K.K., Minieri, C.A., Ollerenshaw, J.D., and Alexander, R.W. 1994. Angiotensin II stimulates NADH and NADPH oxidase activity in cultured vascular smooth muscle cells. *Circ Res* 74:1141–1148.

37. Vanhoutte, P.M. 2000. Say NO to ET. *J Auton Nerv Syst* 81:271–277.

38. Yang, D., Feletou, M., Boulanger, C.M., Wu, H.F., Levens, N., Zhang, J.N., and Vanhoutte, P.M. 2002. Oxygen-derived free radicals mediate endothelium-dependent contractions to acetylcholine in aortas from spontaneously hypertensive rats. *Br J Pharmacol* 136:104–110.

39. Auch-Schwelk, W., Katusic, Z.S., and Vanhoutte, P.M. 1989. Contractions to oxygen-derived free radicals are augmented in aorta of the spontaneously hypertensive rat. *Hypertension* 13:859–864.

40. Rao, G.N. and Berk, B.C. 1992. Active oxygen species stimulate vascular smooth muscle cell growth and proto-oncogene expression. *Circ Res* 70:593–599.

41. Liao, D.F., Jin, Z.G., Baas, A.S., Daum, G., Gygi, S.P., Aebersold, R., and Berk, B.C. 2000. Purification and identification of secreted oxidative stress-induced factors from vascular smooth muscle cells. *J Biol Chem* 275:189–196.

42. Jin, Z.G., Melaragno, M.G., Liao, D.F., Yan, C., Haendeler, J., Suh, Y.A., Lambeth, J.D., and Berk, B.C. 2000. Cyclophilin A is a secreted growth factor induced by oxidative stress. *Circ Res* 87:789–796.

43. Shimokawa, H. 2013. Reactive oxygen species promote vascular smooth muscle cell proliferation. *Circ Res* 113:1040–1042.

44. Schroder, E. and Eaton, P. 2008. Hydrogen peroxide as an endogenous mediator and exogenous tool in cardiovascular research: Issues and considerations. *Curr Opin Pharmacol* 8:153–159.

45. Etienne-Manneville, S. and Hall, A. 2002. Rho GTPases in cell biology. *Nature* 420:629–635.

46. Schmidt, A. and Hall, A. 2002. Guanine nucleotide exchange factors for Rho GTPases: Turning on the switch. *Genes Dev* 16:1587–1609.

47. Bernards, A. 2003. GAPs galore! A survey of putative Ras superfamily GTPase activating proteins in man and *Drosophila*. *Biochim Biophys Acta* 1603:47–82.

48. Olofsson, B. 1999. Rho guanine dissociation inhibitors: Pivotal molecules in cellular signalling. *Cell Signal* 11:545–554.

49. Kimura, K., Ito, M., Amano, M., Chihara, K., Fukata, Y., Nakafuku, M., Yamamori, B. et al. 1996. Regulation of myosin phosphatase by Rho and Rho-associated kinase (Rho-kinase). *Science* 273:245–248.

50. Amano, M., Ito, M., Kimura, K., Fukata, Y., Chihara, K., Nakano, T., Matsuura, Y., and Kaibuchi, K. 1996. Phosphorylation and activation of myosin by Rho-associated kinase (Rho-kinase). *J Biol Chem* 271:20246–20249.

51. Berk, B.C., Alexander, R.W., Brock, T.A., Gimbrone, M.A., Jr., and Webb, R.C. 1986. Vasoconstriction: A new activity for platelet-derived growth factor. *Science* 232:87–90.

52. Griendling, K.K., Berk, B.C., Ganz, P., Gimbrone, M.A., Jr., and Alexander, R.W. 1987. Angiotensin II stimulation of vascular smooth muscle phosphoinositide metabolism. State of the art lecture. *Hypertension* 9:III181–III185.

53. Suzuki, J., Jin, Z.G., Meoli, D.F., Matoba, T., and Berk, B.C. 2006. Cyclophilin A is secreted by a vesicular pathway in vascular smooth muscle cells. *Circ Res* 98:811–817.

54. Satoh, K., Fukumoto, Y., Nakano, M., Sugimura, K., Nawata, J., Demachi, J., Karibe, A. et al. 2009. Statin ameliorates hypoxia-induced pulmonary hypertension associated with down-regulated stromal cell-derived factor-1. *Cardiovasc Res* 81:226–234.

55. Nigro, P., Satoh, K., O'Dell, M.R., Soe, N.N., Cui, Z., Mohan, A., Abe, J., Alexis, J.D., Sparks, J.D., and Berk, B.C. 2011. Cyclophilin A is an inflammatory mediator that promotes atherosclerosis in apolipoprotein E-deficient mice. *J Exp Med* 208:53–66.

56. Shimokawa, H. and Takeshita, A. 2005. Rho-kinase is an important therapeutic target in cardiovascular medicine. *Arterioscler Thromb Vasc Biol* 25:1767–1775.

57. Higashi, M., Shimokawa, H., Hattori, T., Hiroki, J., Mukai, Y., Morikawa, K., Ichiki, T., Takahashi, S., and Takeshita, A. 2003. Long-term inhibition of Rho-kinase suppresses angiotensin II-induced cardiovascular hypertrophy in rats in vivo: Effect on endothelial NAD(P)H oxidase system. *Circ Res* 93:767–775.

58. Funakoshi, Y., Ichiki, T., Shimokawa, H., Egashira, K., Takeda, K., Kaibuchi, K., Takeya, M., Yoshimura, T., and Takeshita, A. 2001. Rho-kinase mediates angiotensin II-induced monocyte chemoattractant protein-1 expression in rat vascular smooth muscle cells. *Hypertension* 38:100–104.

59. Takeda, K., Ichiki, T., Tokunou, T., Iino, N., Fujii, S., Kitabatake, A., Shimokawa, H., and Takeshita, A. 2001. Critical role of Rho-kinase and MEK/ERK pathways for angiotensin II-induced plasminogen activator inhibitor type-1 gene expression. *Arterioscler Thromb Vasc Biol* 21:868–873.

60. Guilluy, C., Bregeon, J., Toumaniantz, G., Rolli-Derkinderen, M., Retailleau, K., Loufrani, L., Henrion, D. et al. 2010. The Rho exchange factor Arhgef1 mediates the effects of angiotensin II on vascular tone and blood pressure. *Nat Med* 16:183–190.

61. van Nieuw Amerongen, G.P., van Delft, S., Vermeer, M.A., Collard, J.G., and van Hinsbergh, V.W. 2000. Activation of RhoA by thrombin in endothelial hyperpermeability: Role of Rho-kinase and protein tyrosine kinases. *Circ Res* 87:335–340.

62. Seasholtz, T.M., Majumdar, M., Kaplan, D.D., and Brown, J.H. 1999. Rho and Rho kinase mediate thrombin-stimulated vascular smooth muscle cell DNA synthesis and migration. *Circ Res* 84:1186–1193.

63. Kishi, H., Bao, J., and Kohama, K. 2000. Inhibitory effects of ML-9, wortmannin, and Y-27632 on the chemotaxis of vascular smooth muscle cells in response to platelet-derived growth factor-BB. *J Biochem* 128:719–722.

64. Sauzeau, V., Le Jeune, H., Cario-Toumaniantz, C., Vaillant, N., Gadeau, A.P., Desgranges, C., Scalbert, E., Chardin, P., Pacaud, P., and Loirand, G. 2000. P2Y(1), P2Y(2), P2Y(4), and P2Y(6) receptors are coupled to Rho and Rho kinase activation in vascular myocytes. *Am J Physiol Heart Circ Physiol* 278:H1751–H1761.

65. Sauzeau, V., Le Mellionnec, E., Bertoglio, J., Scalbert, E., Pacaud, P., and Loirand, G. 2001. Human urotensin II-induced contraction and arterial smooth muscle cell proliferation are mediated by RhoA and Rho-kinase. *Circ Res* 88:1102–1104.

66. Takemoto, M. and Liao, J.K. 2001. Pleiotropic effects of 3-hydroxy-3-methylglutaryl coenzyme a reductase inhibitors. *Arterioscler Thromb Vasc Biol* 21:1712–1719.

67. Rashid, M., Tawara, S., Fukumoto, Y., Seto, M., Yano, K., and Shimokawa, H. 2009. Importance of Rac1 signaling pathway inhibition in the pleiotropic effects of HMG-CoA reductase inhibitors. *Circ J* 73:361–370.

68. Tanaka, S., Fukumoto, Y., Nochioka, K., Minami, T., Kudo, S., Shiba, N., Takai, Y., Williams, C.L., Liao, J.K., and Shimokawa, H. 2013. Statins exert the pleiotropic effects through small GTP-binding protein dissociation stimulator upregulation with a resultant Rac1 degradation. *Arterioscler Thromb Vasc Biol* 33:1591–1600.

69. Satoh, K., Matoba, T., Suzuki, J., O'Dell, M.R., Nigro, P., Cui, Z., Mohan, A. et al. 2008. Cyclophilin A mediates vascular remodeling by promoting inflammation and vascular smooth muscle cell proliferation. *Circulation* 117:3088–3098.

70. Shimokawa, H. 2000. Cellular and molecular mechanisms of coronary artery spasm: Lessons from animal models. *Jpn Circ J* 64:1–12.

71. Shimokawa, H. 2002. Rho-kinase as a novel therapeutic target in treatment of cardiovascular diseases. *J Cardiovasc Pharmacol* 39:319–327.

72. Amano, M., Chihara, K., Kimura, K., Fukata, Y., Nakamura, N., Matsuura, Y., and Kaibuchi, K. 1997. Formation of actin stress fibers and focal adhesions enhanced by Rho-kinase. *Science* 275:1308–1311.

73. Hall, A. 1998. Rho GTPases and the actin cytoskeleton. *Science* 279:509–514.

74. Satoh, K., Fukumoto, Y., and Shimokawa, H. 2011. Rho-kinase: Important new therapeutic target in cardiovascular diseases. *Am J Physiol Heart Circ Physiol* 301:H287–H296.

75. Satoh, K., Nigro, P., and Berk, B.C. 2010. Oxidative stress and vascular smooth muscle cell growth: A mechanistic linkage by cyclophilin A. *Antioxid Redox Signal* 12:675–682.

76. Satoh, K., Shimokawa, H., and Berk, B.C. 2010. Cyclophilin A: Promising new target in cardiovascular therapy. *Circ J* 74:2249–2256.

77. Mackay, D.J. and Hall, A. 1998. Rho GTPases. *J Biol Chem* 273:20685–20688.

78. Neco, P., Giner, D., Viniegra, S., Borges, R., Villarroel, A., and Gutierrez, L.M. 2004. New roles of myosin II during vesicle transport and fusion in chromaffin cells. *J Biol Chem* 279:27450–27457.

79. Jin, Z.G., Lungu, A.O., Xie, L., Wang, M., Wong, C., and Berk, B.C. 2004. Cyclophilin A is a proinflammatory cytokine that activates endothelial cells. *Arterioscler Thromb Vasc Biol* 24:1186–1191.

80. Kim, H., Kim, W.J., Jeon, S.T., Koh, E.M., Cha, H.S., Ahn, K.S., and Lee, W.H. 2005. Cyclophilin A may contribute to the inflammatory processes in rheumatoid arthritis through induction of matrix degrading enzymes and inflammatory cytokines from macrophages. *Clin Immunol* 116:217–224.

81. Damsker, J.M., Bukrinsky, M.I., and Constant, S.L. 2007. Preferential chemotaxis of activated human CD4+ T cells by extracellular cyclophilin A. *J Leukoc Biol* 82:613–618.

82. Yang, Y., Lu, N., Zhou, J., Chen, Z.N., and Zhu, P. 2008. Cyclophilin A up-regulates MMP-9 expression and adhesion of monocytes/macrophages via CD147 signalling pathway in rheumatoid arthritis. *Rheumatology (Oxford)* 47:1299–1310.

83. Wang, L., Wang, C.H., Jia, J.F., Ma, X.K., Li, Y., Zhu, H.B., Tang, H., Chen, Z.N., and Zhu, P. Contribution of cyclophilin A to the regulation of inflammatory processes in rheumatoid arthritis. *J Clin Immunol* 30:24–33.

84. Shimokawa, H. and Rashid, M. 2007. Development of Rho-kinase inhibitors for cardiovascular medicine. *Trends Pharmacol Sci* 28:296–302.

85. Eto, Y., Shimokawa, H., Hiroki, J., Morishige, K., Kandabashi, T., Matsumoto, Y., Amano, M., Hoshijima, M., Kaibuchi, K., and Takeshita, A. 2000. Gene transfer of dominant negative Rho-kinase suppresses neointimal formation after balloon injury in pigs. *Am J Physiol Heart Circ Physiol* 278:H1744–H1750.

86. Ikeda, S., Satoh, K., Kikuchi, N., Miyata, S., Suzuki, K., Omura, J., Shimizu, T. et al. 2014. Crucial role of rho-kinase in pressure overload-induced right ventricular hypertrophy and dysfunction in mice. *Arterioscler Thromb Vasc Biol* 34:1260–1271.

87. Sawada, N., Itoh, H., Ueyama, K., Yamashita, J., Doi, K., Chun, T.H., Inoue, M. et al. 2000. Inhibition of rho-associated kinase results in suppression of neointimal formation of balloon-injured arteries. *Circulation* 101:2030–2033.

88. Shibata, R., Kai, H., Seki, Y., Kato, S., Morimatsu, M., Kaibuchi, K., and Imaizumi, T. 2001. Role of Rho-associated kinase in neointima formation after vascular injury. *Circulation* 103:284–289.

89. Shimokawa, H., Ito, A., Fukumoto, Y., Kadokami, T., Nakaike, R., Sakata, M., Takayanagi, T., Egashira, K., and Takeshita, A. 1996. Chronic treatment with interleukin-1 beta induces coronary intimal lesions and vasospastic responses in pigs in vivo. The role of platelet-derived growth factor. *J Clin Invest* 97:769–776.

90. Shimokawa, H., Morishige, K., Miyata, K., Kandabashi, T., Eto, Y., Ikegaki, I., Asano, T., Kaibuchi, K., and Takeshita, A. 2001. Long-term inhibition of Rho-kinase induces a regression of arteriosclerotic coronary lesions in a porcine model in vivo. *Cardiovasc Res* 51:169–177.

91. Matsumoto, Y., Uwatoku, T., Oi, K., Abe, K., Hattori, T., Morishige, K., Eto, Y. et al. 2004. Long-term inhibition of Rho-kinase suppresses neointimal formation after stent implantation in porcine coronary arteries: Involvement of multiple mechanisms. *Arterioscler Thromb Vasc Biol* 24:181–186.

92. Hattori, T., Shimokawa, H., Higashi, M., Hiroki, J., Mukai, Y., Tsutsui, H., Kaibuchi, K., and Takeshita, A. 2004. Long-term inhibition of Rho-kinase suppresses left ventricular remodeling after myocardial infarction in mice. *Circulation* 109:2234–2239.

93. Takagi, Y., Yasuda, S., Takahashi, J., Takeda, M., Nakayama, M., Ito, K., Hirose, M., Wakayama, Y., Fukuda, K., and Shimokawa, H. 2009. Importance of dual induction tests for coronary vasospasm and ventricular fibrillation in patients surviving out-of-hospital cardiac arrest. *Circ J* 73:767–769.

94. Hizume, T., Morikawa, K., Takaki, A., Abe, K., Sunagawa, K., Amano, M., Kaibuchi, K., Kubo, C., and Shimokawa, H. 2006. Sustained elevation of serum cortisol level causes sensitization of coronary vasoconstricting responses in pigs in vivo: A possible link between stress and coronary vasospasm. *Circ Res* 99:767–775.

95. Katsumata, N., Shimokawa, H., Seto, M., Kozai, T., Yamawaki, T., Kuwata, K., Egashira, K. et al. 1997. Enhanced myosin light chain phosphorylations as a central mechanism for coronary artery spasm in a swine model with interleukin-1beta. *Circulation* 96:4357–4363.

96. Shimokawa, H., Seto, M., Katsumata, N., Amano, M., Kozai, T., Yamawaki, T., Kuwata, K. et al. 1999. Rho-kinase-mediated pathway induces enhanced myosin light chain phosphorylations in a swine model of coronary artery spasm. *Cardiovasc Res* 43:1029–1039.

97. Masumoto, A., Mohri, M., Shimokawa, H., Urakami, L., Usui, M., and Takeshita, A. 2002. Suppression of coronary artery spasm by the Rho-kinase inhibitor fasudil in patients with vasospastic angina. *Circulation* 105:1545–1547.

98. Mohri, M., Shimokawa, H., Hirakawa, Y., Masumoto, A., and Takeshita, A. 2003. Rho-kinase inhibition with intracoronary fasudil prevents myocardial ischemia in patients with coronary microvascular spasm. *J Am Coll Cardiol* 41:15–19.

99. Shimokawa, H., Hiramori, K., Iinuma, H., Hosoda, S., Kishida, H., Osada, H., Katagiri, T. et al. 2002. Anti-anginal effect of fasudil, a Rho-kinase inhibitor, in patients with stable effort angina: A multicenter study. *J Cardiovasc Pharmacol* 40:751–761.

100. Kikuchi, Y., Yasuda, S., Aizawa, K., Tsuburaya, R., Ito, Y., Takeda, M., Nakayama, M., Ito, K., Takahashi, J., and Shimokawa, H. 2011. Enhanced Rho-kinase activity in circulating neutrophils of patients with vasospastic angina: A possible biomarker for diagnosis and disease activity assessment. *J Am Coll Cardiol* 58:1231–1237.

101. Nihei, T., Takahashi, J., Kikuchi, Y., Takagi, Y., Hao, K., Tsuburaya, R., Shiroto, T. et al. 2012. Enhanced Rho-kinase activity in patients with vasospastic angina after the Great East Japan Earthquake. *Circ J* 76:2892–2894.

102. Tsuburaya, R., Yasuda, S., Shiroto, T., Ito, Y., Gao, J.Y., Aizawa, K., Kikuchi, Y. et al. 2012. Long-term treatment with nifedipine suppresses coronary hyperconstricting responses and inflammatory changes induced by paclitaxel-eluting stent in pigs in vivo: Possible involvement of Rho-kinase pathway. *Eur Heart J* 33:791–799.

103. Nihei, T., Takahashi, J., Tsuburaya, R., Ito, Y., Shiroto, T., Hao, K., Takagi, Y. et al. 2014. Circadian variation of Rho-kinase activity in circulating leukocytes of patients with vasospastic angina. *Circ J* 78:1183–1190.

104. Stenmark, K.R., Fagan, K.A., and Frid, M.G. 2006. Hypoxia-induced pulmonary vascular remodeling: Cellular and molecular mechanisms. *Circ Res* 99:675–691.

105. Satoh, K., Kagaya, Y., Nakano, M., Ito, Y., Ohta, J., Tada, H., Karibe, A. et al. 2006. Important role of endogenous erythropoietin system in recruitment of endothelial progenitor cells in hypoxia-induced pulmonary hypertension in mice. *Circulation* 113:1442–1450.

106. Shimizu, T., Fukumoto, Y., Tanaka, S., Satoh, K., Ikeda, S., and Shimokawa, H. 2013. Crucial role of ROCK2 in vascular smooth muscle cells for hypoxia-induced pulmonary hypertension in mice. *Arterioscler Thromb Vasc Biol* 33:2780–2791.

107. Elias-Al-Mamun, M., Satoh, K., Tanaka, S., Shimizu, T., Nergui, S., Miyata, S., Fukumoto, Y., and Shimokawa, H. 2014. Combination therapy with fasudil and sildenafil ameliorates monocrotaline-induced pulmonary hypertension and survival in rats. *Circ J* 78:967–976.

108. Khromykh, L.M., Kulikova, N.L., Anfalova, T.V., Muranova, T.A., Abramov, V.M., Vasiliev, A.M., Khlebnikov, V.S., and Kazansky, D.B. 2007. Cyclophilin A produced by thymocytes regulates the migration of murine bone marrow cells. *Cell Immunol* 249:46–53.

109. Abe, K., Shimokawa, H., Morikawa, K., Uwatoku, T., Oi, K., Matsumoto, Y., Hattori, T. et al. 2004. Long-term treatment with a Rho-kinase inhibitor improves monocrotaline-induced fatal pulmonary hypertension in rats. *Circ Res* 94:385–393.

110. Abe, K., Tawara, S., Oi, K., Hizume, T., Uwatoku, T., Fukumoto, Y., Kaibuchi, K., and Shimokawa, H. 2006. Long-term inhibition of Rho-kinase ameliorates hypoxia-induced pulmonary hypertension in mice. *J Cardiovasc Pharmacol* 48:280–285.

111. Do e, Z., Fukumoto, Y., Takaki, A., Tawara, S., Ohashi, J., Nakano, M., Tada, T. et al. 2009. Evidence for Rho-kinase activation in patients with pulmonary arterial hypertension. *Circ J* 73:1731–1739.

112. Fukumoto, Y., Matoba, T., Ito, A., Tanaka, H., Kishi, T., Hayashidani, S., Abe, K., Takeshita, A., and Shimokawa, H. 2005. Acute vasodilator effects of a Rho-kinase inhibitor, fasudil, in patients with severe pulmonary hypertension. *Heart* 91:391–392.

113. Fukumoto, Y., Yamada, N., Matsubara, H., Mizoguchi, M., Uchino, K., Yao, A., Kihara, Y. et al. 2013. Double-blind, placebo-controlled clinical trial with a rho-kinase inhibitor in pulmonary arterial hypertension. *Circ J* 77:2619–2625.

114. Handschumacher, R.E., Harding, M.W., Rice, J., Drugge, R.J., and Speicher, D.W. 1984. Cyclophilin: A specific cytosolic binding protein for cyclosporin A. *Science* 226:544–547.

115. Marks, A.R. 1996. Cellular functions of immunophilins. *Physiol Rev* 76:631–649.

116. Galat, A. and Metcalfe, S.M. 1995. Peptidylproline cis/trans isomerases. *Prog Biophys Mol Biol* 63:67–118.

117. Theuerkorn, M., Fischer, G., and Schiene-Fischer, C. 2011. Prolyl cis/trans isomerase signalling pathways in cancer. *Curr Opin Pharmacol* 11:281–287.

118. Zhu, C., Wang, X., Deinum, J., Huang, Z., Gao, J., Modjtahedi, N., Neagu, M.R. et al. 2007. Cyclophilin A participates in the nuclear translocation of apoptosis-inducing factor in neurons after cerebral hypoxia-ischemia. *J Exp Med* 204:1741–1748.

119. Krummrei, U., Bang, R., Schmidtchen, R., Brune, K., and Bang, H. 1995. Cyclophilin-A is a zinc-dependent DNA binding protein in macrophages. *FEBS Lett* 371:47–51.

120. Soe, N., Sowden, M., Baskaran, P., Smolock, E., Kim, Y., Nigro, P., and Berk, B.C. 2013. Cyclophilin A is required for angiotensin II-induced p47phox translocation to caveolae in vascular smooth muscle cells. *Arterioscler Thromb Vasc Biol* 33(9):2147–2153.

121. Satoh, K., Nigro, P., Matoba, T., O'Dell, M.R., Cui, Z., Shi, X., Mohan, A. et al. 2009. Cyclophilin A enhances vascular oxidative stress and the development of angiotensin II-induced aortic aneurysms. *Nat Med* 15:649–656.

122. Pushkarsky, T., Zybarth, G., Dubrovsky, L., Yurchenko, V., Tang, H., Guo, H., Toole, B., Sherry, B., and Bukrinsky, M. 2001. CD147 facilitates HIV-1 infection by interacting with virus-associated cyclophilin A. *Proc Natl Acad Sci USA* 98:6360–6365.

123. Satoh, K., Nigro, P., Zeidan, A., Soe, N.N., Jaffre, F., Oikawa, M., O'Dell, M.R. et al. 2011. Cyclophilin A promotes cardiac hypertrophy in apolipoprotein E-deficient mice. *Arterioscler Thromb Vasc Biol* 31:1116–1123.

124. Bell, R.D., Winkler, E.A., Singh, I., Sagare, A.P., Deane, R., Wu, Z., Holtzman, D.M. et al. 2012. Apolipoprotein E controls cerebrovascular integrity via cyclophilin A. *Nature* 485:512–516.

125. Montezano, A.C. and Touyz, R.M. 2012. Molecular mechanisms of hypertension—Reactive oxygen species and antioxidants: A basic science update for the clinician. *Can J Cardiol* 28:288–295.

126. Satoh, K., Fukumoto, Y., Sugimura, K., Miura, Y., Aoki, T., Nochioka, K., Tatebe, S. et al. 2013. Plasma Cyclophilin A is a novel biomarker for coronary artery disease. *Circ J* 77:447–455.

127. Kesarwani, P., Murali, A.K., Al-Khami, A.A., and Mehrotra, S. 2013. Redox regulation of T-cell function: From molecular mechanisms to significance in human health and disease. *Antioxid Redox Signal* 18:1497–1534.

128. Wang, Y.X., Martin-McNulty, B., da Cunha, V., Vincelette, J., Lu, X., Feng, Q., Halks-Miller, M. et al. 2005. Fasudil, a Rho-kinase inhibitor, attenuates angiotensin II-induced abdominal aortic aneurysm in apolipoprotein E-deficient mice by inhibiting apoptosis and proteolysis. *Circulation* 111:2219–2226.

129. Shimizu, T., Fukumoto, Y., Tanaka, S.I., Satoh, K., Ikeda, S., and Shimokawa, H. 2013. Crucial role of ROCK2 in vascular smooth muscle cells for hypoxia-induced pulmonary hypertension in mice. *Arterioscler Thromb Vasc Biol* 33(12):2780–2791.

130. Ejiri, J., Inoue, N., Tsukube, T., Munezane, T., Hino, Y., Kobayashi, S., Hirata, K. et al. 2003. Oxidative stress in the pathogenesis of thoracic aortic aneurysm: Protective role of statin and angiotensin II type 1 receptor blocker. *Cardiovasc Res* 59:988–996.

131. Habashi, J.P., Judge, D.P., Holm, T.M., Cohn, R.D., Loeys, B.L., Cooper, T.K., Myers, L. et al. 2006. Losartan, an AT1 antagonist, prevents aortic aneurysm in a mouse model of Marfan syndrome. *Science* 312:117–121.

132. Cassis, L.A., Rateri, D.L., Lu, H., and Daugherty, A. 2007. Bone marrow transplantation reveals that recipient AT1a receptors are required to initiate angiotensin II-induced atherosclerosis and aneurysms. *Arterioscler Thromb Vasc Biol* 27:380–386.

133. Weintraub, N.L. 2009. Understanding abdominal aortic aneurysm. *N Engl J Med* 361:1114–1116.

22 Reactive Oxygen Species, Antioxidants, Inflammation, and Cardiovascular Disease

Harald Mangge and Johanna M. Gostner

CONTENTS

ABSTRACT

Cardiovascular disease (CVD) has a multifaceted etiology. Dyslipidemia, metabolic syndrome, systemic and local vascular immune-mediated inflammation, oxidative "stress," clotting mechanisms, smooth muscle proliferation, microbiome, and yet unknown biochemical "fingerprints," together with genetic/epigenetic factors, are discussed to be involved. Pathological changes occur long before symptoms become apparent and diagnosis is made. Diminished levels of antioxidants are most likely caused by an overwhelming reactive oxygen species production by activated immune effector cells like macrophages due to a chronic low-grade systemic inflammation. Although antioxidants are considered to act beneficially in this context, the interplay of endogenous and exogenous antioxidants with the overall redox system is not fully understood. Especially, molecular mechanisms of oxidative stress in CVD remain to be clarified.

During cellular immune response, interferon-γ (IFN-γ)-dependent pathways are activated such as tryptophan breakdown by the enzyme indoleamine 2,3-dioxygenase (IDO) in monocyte-derived macrophages, fibroblasts, and endothelial cells. Neopterin, a marker of oxidative stress and immune activation, is produced by guanosine triphosphate-cyclohydrolase I (GTP-CH-I) in macrophages and dendritic cells. Nitric oxide synthase (NOS) is induced in several cell types to generate nitric oxide (NO). NO, despite its low reactivity, is a potent antioxidant involved in the regulation of the vasomotor tone and of several signaling pathways during immune response. NO inhibits the expression and function of IDO. Function of NOS requires the cofactor tetrahydrobiopterin (BH_4), which is produced in humans primarily by fibroblasts and endothelial cells. Neopterin and kynurenine to tryptophan ratio (Kyn/Trp), as an estimate of IDO enzyme activity, are robust markers of immune activation *in vitro* and *in vivo*. Both serve as significant diagnostic parameters to predict cardiovascular and overall mortality. Neopterin concentrations and Kyn/Trp ratio correlate significantly with the lowering of plasma levels of vitamins C, E, and B. Additional analysis of NO metabolites, BH_4, and plasma antioxidants in patients with CVD may improve the understanding of redox regulation and might provide a rationale for potential antioxidant therapies in CVD.

22.1 INTRODUCTION

Cardiovascular diseases (CVDs) remain the world's number one killer, and the prevalence is still increasing [1,2]. CVD comprises the heart and systemic vascular system [3]. In coronary heart disease, including cerebrovascular disease or peripheral arterial disease, the impaired blood vessel function leads to an inadequate blood supply of organs.

Avoiding risk factors, such as smoking, obesogenic lifestyle, physical inactivity, high blood pressure, diabetes, and dyslipidemia, is strongly recommended to prevent CVD. Nevertheless, beside lifestyle, genetic, epigenetic, and environmental factors essentially influence the risk of CVD.

The multifactorial background makes it difficult to unravel initial pathological events, where symptoms are still subclinical. Immune-mediated inflammation plays a key role in both disease initiation and progression [4]. Chronic inflammatory conditions attenuate endogenous antioxidant capacities due to continuous production of high levels of reactive oxygen species (ROS). CVD patients often represent with low antioxidant capacity [5] and enhanced oxidative stress markers [6]. This is usually caused by an increased demand in situations of overwhelming ROS production by activated immune effector cells like macrophages. Nevertheless, an insufficient nutritional uptake may also be involved. Although the intake of exogenous antioxidants is suggested to beneficially interfere with disease-related oxidative stress, the interplay of endogenous and exogenous antioxidants with the overall redox system remains to be elucidated. A better understanding of the redox regulation of the immune responses in CVD will also enable new preventive and therapeutic strategies.

22.2 PATHOLOGY OF ATHEROGENESIS

Atherosclerosis leads to CVD including myocardial infarction, heart failure, stroke, and claudication. A central event is the development of so-called atherosclerotic plaques in the inner lining of arteries. Inflammation, hypertension, hyperglycemia, and dyslipidemia cause endothelial cell "stress" leading to an increased expression of adhesion molecules and consecutive recruitment of leukocytes [7]. Atherosclerotic plaques consist of a necrotic core of different sizes, occasionally calcified regions, accumulated modified and oxidized lipids, increased smooth muscle cells, leukocytes, and foam cells [8]. The "oxidative modification hypothesis" of atherogenesis implies that low-density lipoprotein (LDL) oxidation is an early event in atherosclerosis [9]. LDL particles are retained in the artery wall. Biochemically modified components of these particles induce leukocyte adhesion and intracellular cholesterol accumulation in invaded macrophages [10]. Chronic inflammatory conditions are maintained due to the production of proinflammatory mediators through immune competent cells, mainly T helper cells of type 1, in the lesions [11]. Activated macrophages are involved in atherosclerotic plaque formation, fibrous cap disruption, and thrombus formation.

While in the past atherosclerosis was viewed primarily as the passive process of cholesterol accumulation, recent data support the view that it is a dynamic and complex process involving components of the vascular, immune, metabolic, and endocrine system. Hence, pathological changes occur in a variety of cell types long before symptoms become clinically apparent [12–14].

Atherosclerotic plaque composition and endothelial erosion, rather than the percent stenosis, appear to be critical predictors for the risk of both plaque rupture and subsequent thrombotic exacerbation [2,15]. Disruption of a vulnerable or unstable plaque may lead to a complete occlusion, plaque progression, or result in an acute coronary syndrome (ACS), that is, acute myocardial infarction (AMI), unstable angina (UA), and sudden cardiac death or ischemic stroke in case of carotid plaque destabilization.

22.3 OXIDATIVE STRESS

Although substantial efforts have been made to elucidate molecular details of atherogenesis, the underlying pathological cascades remain not fully understood. Nevertheless, evidence exists that the activation of immune competent cells, leading to local and finally systemic inflammation, and associated heightened oxidative stress are essential events [4]. Of note, also in a large sample of CVD-free adults, Chrysohoou et al. observed an association of pre-hypertension with oxidative stress markers linking to the atherosclerotic process [16]. Both systolic and diastolic blood pressures were inversely correlated with total antioxidant capacity and positively correlated with oxidized LDL (oxLDL) (ATTICA study, 16).

Immune reactions in atherosclerotic lesions are mainly Th1-type responses, as indicated by the dominance of proinflammatory and macrostimulating cytokines found in advanced plaques [11,17,18]. During Th1-type response, IFN-γ is probably the most important trigger for high ROS production in macrophages [19] by phagocytic NADPH oxidase (NOX) [20]. The main reactive species are not only hydrogen peroxide (H_2O_2) and superoxide anion ($O_2^{\cdot-}$) but also reactive nitrogen species (RNS) such as peroxynitrite ($ONOO^-$) and nitrogen dioxide and trioxide [21]. IFN-γ signaling initiates a variety of cellular defense mechanisms such as proinflammatory cytokine production via nuclear factor kappa B (NF-κB) signaling, enhanced antigen presentation [22], and other important pathways, for example, neopterin formation via guanosine triphosphate (GTP)-cyclohydrolase I (GTP-GCH-I) and indoleamine 2,3-dioxygenase (IDO)-mediated tryptophan breakdown [23] (Figure 22.1).

Low levels of ROS are mainly by-products from mitochondrial electron transport chain reactions [24]. They are important regulators of several redox-sensitive pathways [25] involved in the maintenance of cellular homeostasis and act by modifying molecules, enzymes, and transcription

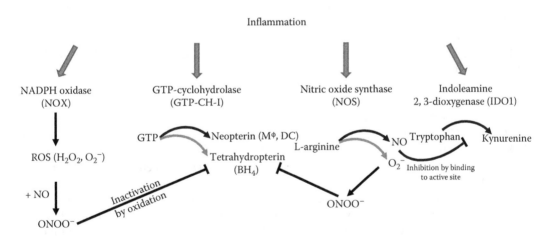

FIGURE 22.1 Regulatory circuits in inflammation and endothelial dysfunction. During inflammation, NADPH oxidase (NOX) produces high levels of reactive oxygen species. T cells and natural killer cells produce interferon-γ, which activates enzymes such as GTP-cyclohydrolase 1 (GTP-CH-I), indoleamine 2,3-dioxygenase (IDO), and inducible nitric oxide synthase (iNOS) in monocyte-derived macrophages (M) and dendritic cells (DC). In endothelial cells, endothelial NOS (eNOS) is constitutively expressed and GTP-CH-I produces tetrahydrobiopterin (BH4), which is a NOS cofactor. BH4 deficiency leads to NOS uncoupling and O_2^- formation, which reacts with NO to form peroxynitrite ($ONOO^-$). In a vicious cycle, $ONOO^-$ oxidizes BH4. In M/DC, GTP-CH-I synthesizes neopterin at the expense of BH4, which contributes to the low activity of iNOS in human M/DC. Furthermore, NO is a reversible inhibitor of the immunoregulatory enzyme IDO. IDO degrades the essential amino acid tryptophan to kynurenine.

factors as well by depleting the endogenous antioxidant pool [21,25,26]. Negative interference with the endogenous redox system in conditions with overwhelming oxidative stress is critical not only due to triggering of immune responses but also through leading to endothelial and smooth muscle dysfunction and thus to the progression of atherosclerosis [27,28].

Proatherogenic oxLDL accumulates in the vascular wall and contributes to the pathogenesis of vascular dysfunction early in the development of atherosclerosis.

After incorporation via scavenger receptors of macrophages, oxLDL leads to transformation into foam cells [29]. Foam cells accumulate lipid droplets in the cytoplasm and secrete extracellular matrix proteins that further support the retention of lipoproteins and attraction of immune cells and hence to the enlargement of the lesion [10].

Oxidation of LDL is considered to occur primarily in the vascular wall [30]. Nevertheless, circulating oxLDL was also detected in patients with type 2 diabetes and metabolic syndrome. Enhanced serum levels of oxLDL, as well as antibodies against its epitopes, are predictive for endothelial dysfunction and coronary heart disease [31,32], although the amount of circulating oxLDL is small compared to total LDL [33]. Elevated oxLDL was found to be predictive for future coronary heart disease events in apparently healthy men [34], suggesting oxLDL as a relevant proatherogenic marker.

Another potential biomarker is the anti-atherogenic high-density lipoprotein (HDL) due to its function in the reverse cholesterol transport and in decreasing lipoprotein oxidation [35]. HDL is involved in several biological processes that counteract inflammation and oxidative stress, by beneficially influencing, for example, pancreatic β-cell function, endothelial vasoreactivity, endothelial apoptosis, epithelial restorative processes, and monocyte activation as well as adhesion molecules expression, thus being highly vasculoprotective [30,36]. Paraoxonase-1 (PON1), a calcium-dependent enzyme, is located at the surface of HDL particles and contributes to the antioxidant and anti-inflammatory role of HDL [37]. In particular, HDL-associated paraoxonase was shown to inhibit the formation of "minimally oxidized" LDL by hydrolyzing biologically active oxidized phospholipids [38]. Nevertheless, the antioxidant effects of HDL have also been demonstrated under calcium-free conditions, arguing that this enzyme may not be the only mechanism by which HDL inhibits LDL oxidation [39].

Plasma HDL cholesterol (HDL-C) levels are inversely associated with CVD risk in preclinical and large epidemiologic studies. Of note, low HDL-C level was identified as a robust predictor of lipid peroxidation irrespective of gender, age, obesity, and inflammatory or metabolic biomarkers in the Styrian Juvenile Obesity (STYJOBS)/Early DEteCTion of Atherosclerosis (EDECTA) study employing 797 participants aged 5–50 years [30].

However, HDL is highly heterogeneous, and the atheroprotective functions of the different HDL subpopulations are not completely understood. Furthermore, current data indicate that therapeutically increased HDL-C levels *per se* do not always correlate with enhanced HDL functions *in vivo* [40,41].

A variety of oxidation events contribute to a pro-oxidant environment in atherosclerotic lesions [28]. Oxidation of LDL is a potent stimulus of vascular ROS formation, as oxLDL-induced O_2^- formation stimulates, for example, the activation of NOX and uncoupling of endothelial nitric oxide synthase (NOS) that promotes endothelial dysfunction [30].

22.4   NEOPTERIN FORMATION

Neopterin, a marker of immune system activation, is produced by GTP-CH-I in macrophages and dendritic cells (DCs) [42,43] and has emerged as an important predictive marker in cardiovascular risk assessment [6]. IFN-γ is the major stimulus for neopterin formation, while other cytokines have only limited stimulatory potential *in vitro*, but some, for example, tumor necrosis factor α (TNF-α), can indirectly enhance IFN-γ-induced neopterin formation [44]. Of note, also proinflammatory compounds like lipopolysaccharide (LPS) can induce neopterin production [44]. The amount of neopterin secreted by human macrophages correlates with their ROS-generating capacity *in vitro* [45], and neopterin concentration in body fluids is considered as an indicator for immune activation–associated oxidative stress [46].

Activation of GTP-CH-I leads to the production of neopterin and 6,7,8-tetrahydrobiopterin (BH4). Human monocyte–derived macrophages and DCs are the most important source of neopterin and its partially reduced derivative 7,8-dihydroneopterin, both present in relative constant ratio in human serum [46], but not of BH4, due to the relative deficiency of pyruvoyl-tetrahydropterin synthase (PTPS) [47] (Figure 22.1). In contrast, cells from other animal species and other human cell types such as endothelial cells or fibroblasts preferentially produce BH4, which is needed as a cofactor of several monooxygenases including NOS, phenylalanine 4-hydroxylase, or tyrosine 3-hydroxylase [48].

Elevated neopterin concentrations were shown in several diseases associated with chronic immune actions such as viral, bacterial, and parasite infections; autoimmune or malignant tumour diseases; and during rejection episodes in allograft recipients [49–52].

Also, patients with CVD may present with increased neopterin concentrations, supporting the involvement of chronic immune activation, in particular of macrophages, in atherogenesis. Urinary neopterin was reported to be increased in AMI [53], and elevated plasma neopterin levels were found after acute cerebral ischemia [54]. In patients with atherosclerosis, plasma neopterin levels were found to be elevated in about 50% of hospitalized patients undergoing conservative or surgical therapy [55]. In this study, neopterin levels could be associated with different clinical stages of atherosclerosis, and results indicated that neopterin might be a valuable parameter in activity staging and follow-up. In another study enrolling 561 individuals (Ischemic Heart Disease and Stroke Prevention Study, Bruneck, Italy), the extent of carotid atherosclerosis was shown to correlate with the increase in neopterin concentrations, while soluble interleukin-2 receptor, another immune activation marker, did not [56]. In chronic and ACSs, neopterin levels were significantly increased in patients and more pronounced in patients with AMI shortly after the onset of symptoms [57].

In patients admitted with UA pectoris, a correlation was found between the degree of immune cell activation measured by neopterin concentration and the extent of angiographically determined atherosclerosis and the degree of myocardial ischemia [58]. A number of recent studies show that neopterin can be used as an independent marker for CVD and as a predictor of future cardiovascular events in patients with coronary artery disease (CAD) [6]. Rapid coronary artery disease (CAD progression) in 124 patients with stable angina pectoris was found to be associated with increased concentrations of neopterin and high-sensitivity C-reactive protein (CRP) and raised concentrations of biochemical indicators of atheromatous plaque disruption and endothelial activation such as soluble intercellular adhesion molecule 1 (sICAM-1) and matrix metalloproteinase-9 (MMP-9) [59]. In follow-up prospective studies in patients with chronic stable angina pectoris [60] and in Mediterranean patients with non-ST elevation (NSTE) ACSs, that is, UA and NSTE myocardial infarction (MI) [61], neopterin levels were found to be elevated in those patients with adverse coronary events. Johnston et al. found a correlation of serum neopterin with thrombolysis in MI risk scores in ACSs [62]. In critical limb ischemia, cardiac rhythm disturbances and ischemic electrocardiogram changes were found to be related to inflammatory mediators including neopterin levels and predicted 1-year mortality rate [63]. Furthermore, neopterin was considered as a significant factor in the process of plaque inflammation and destabilization in human coronary atherosclerotic lesions [64]. The PRavastatin Or atorVastatin Evaluation Infection Therapy–Thrombolysis In Myocardial Infarction (PROVE IT–TIMI 22) trial assessed the relationship between plasma neopterin levels and the risk of death or acute coronary events (nonfatal MI or UA) over 2 years in over 3000 patients and further confirmed the long-term prognostic value of neopterin [65].

Of note, in patients with CAD, higher neopterin concentrations were associated with a decline in levels of several antioxidant compounds and vitamins such as ascorbic acid, γ-tocopherol, lycopene, lutein, zeaxanthin, and α- and β-carotene [5], indicating again the close relation of immune activation and oxidative stress.

Neopterin concentrations not only correlate with IFN-γ-induced ROS production [45] but also have pro-oxidant properties itself and can trigger the oxidizing potential of ROS as well as of reactive chlorine and nitrogen species in a dose-dependent fashion [66]. Of note, Herpfer et al. showed that neopterin is able to enhance $ONOO^-$ and Cu(II)-mediated LDL oxidation, whereas 7,8-dihydroneopterin mainly protects LDL from oxidation [67]. Additionally, neopterin may enhance the effects of $ONOO^-$ in the processes of protein nitration [68]. The pro-oxidant nature of neopterin suggests a role for the molecule in the antimicrobial and antitumoral action of macrophages [69].

The property of neopterin to interfere with and enhance the effects of various ROS might be of central relevance also in atherogenesis. Furthermore, neopterin was shown to induce an atherothrombotic phenotype in human coronary endothelial cells by promoting cellular adhesion molecules (ICAM 1 and VCAM 1) and tissue factor (TF) expression mediated by the activation of NF-κB [70].

22.5 TRYPTOPHAN BREAKDOWN

In parallel to neopterin formation, during cellular immune response, other IFN-γ-dependent pathways are activated, such as tryptophan (Trp) breakdown by IDO not only in monocyte-derived macrophages but also in fibroblast, endothelial, and epithelial cells [71,72] (Figure 22.1). IDO catalyzes the rate-limiting step in the conversion of Trp and other indole derivates to kynurenine (Kyn) [73]. Both expression and activity of the heme-containing enzyme IDO are sensible to redox regulation, and IDO enzyme itself can exert antioxidant activity by the scavenging of O_2^- [74,75]. The estimation of Kyn to Trp ratio (Kyn/Trp), expressed as μmol Kyn per mmol Trp, can be used as a measure of IDO enzyme activity both *in vitro* and *in vivo* [49,76]. Simultaneous measurement of immune activation markers such as neopterin, IFN-γ, or soluble interleukin receptors allows to relate circulating Trp levels with inflammation-induced IDO activity, as hepatic tryptophan 2,3-dioxygenase (TDO) could also degrade Trp. TDO, however, is regulated by tryptophan content and steroid hormones such as glucocorticoids [77,78], while IDO is strongly induced in response to several proinflammatory stimuli such as IFN-γ, TNF-α, or LPS [44,72].

Depletion of the essential amino acid Trp contributes to the development of an antiproliferative environment and represents an effective antimicrobial and antitumoral strategy of the immune system [79]. Also, T-cell proliferation depends on Trp availability, and thus, IDO activation is a metabolic checkpoint of immune regulation and is crucially involved in the control of T-cell proliferation and in the generation of T regulatory cells (Tregs) and thus in the suppression of autoimmune responses and promotion of tolerance [80].

Metabolic control by the reduction of Trp levels may slow down hematopoiesis in addition to other proinflammatory stimuli that may affect the growth and differentiation of erythroid progenitor cells. Hence, in patients with inflammation-induced anemia, the Kyn/Trp was found to inversely correlate with hemoglobin levels [80,81].

Accelerated Trp breakdown was reported in patients with coronary heart disease [82], and IDO activity correlated significantly with several risk factors for atherosclerosis in the Cardiovascular Risk in Young Finns Study [83]. Niinsalo et al. showed that IDO activity positively correlated with carotid artery intima/media thickness, an early marker of atherosclerosis, although this association did not remain significant after adjustment with classical risk factors of atherosclerosis in this patient group [84].

In inflammatory diseases, including CVD, a concurrent increase in neopterin production and tryptophan degradation is usually observed, and the prognostic ability of neopterin is likely to relate to its association with IFN-γ in the atherogenic process, while IDO activity accounts for immunobiochemical changes induced during Th1-type immune response and may account for several additional aspects observed during disease progression [23]. For example, depressive symptoms have been associated with the increased risk of CAD and poor prognosis among patients with existing CAD [85]. As tryptophan is a precursor for the biosynthesis of serotonin by tryptophan-5-hydroxylase, the lowered Trp availability in inflammatory conditions may limit the biosynthesis of the neurotransmitter serotonin and thus enhance the increased susceptibility for lowered mood and depression [86].

22.6 NITRIC OXIDE

NOS converts L-arginine into citrulline, thereby synthesizing NO. Free NO can migrate through cell membranes by diffusion up to micrometer distances, and although its reactivity is relatively low, NO is a potent antioxidant molecule that can protect from ROS damage [87]. Nevertheless, NO is a free radical and can undergo oxidation to nitrite and nitrate, react with $O_2^{\cdot-}$ to form $ONOO^-$, or bind to transition metals [88]. NO signaling is strongly concentration dependent, and although endogenous NO is essentially involved in many physiological processes and beneficial in a variety of circumstances, its reaction products may mediate nitrosative and oxidative stress. Otherwise, NO products also own protective facets. Plasma NO circulates primarily complexed in S-nitrosothiol species [89] that are suggested to be a transport and buffer system that controls intercellular NO exchange. S-nitrosylation of the proteome is a unique form of posttranslational modification that can have significant consequences for protein function and cell phenotype. In particular, in the cardiovascular system, S-nitrosothiols were shown to exert many actions, including promoting vasodilation, inhibiting platelet aggregation, and regulating Ca $(^{2+})$ channel function of myocytes [90]. The impact of S-nitroso and also N-nitroso protein formation on the reduction of free NO under inflammatory conditions *in vivo* has still to be investigated [91,92].

Endothelial and neuronal NOS are constitutively expressed and produce NO at low concentrations, while inducible NOS is activated, for example, in macrophages in response to proinflammatory stimuli giving rise to higher NO output [93]. Endothelial dysfunction, for example, vasodilation and/or platelet inhibition, a key feature of early atherosclerosis, is associated with the reduced availability of endothelium-derived NO [94]. Defects in NO production, metabolism, and response have been described to be involved in mechanisms. For example, in the presence of $O_2^{\cdot-}$, $ONOO^-$ formation may be a factor that limits NO bioavailability. Beside being strongly vasoconstrictory, $ONOO^-$ has been shown to oxidize the NOS cofactor BH4, thereby leading to eNOS uncoupling and O_2^- production [95], thus starting a vicious cycle (Figure 22.1). Reduced vascular BH4 levels were found in rat and mice models of atherosclerosis and diabetes [96].

High NO output and generation of RNS via iNOS contribute to cellular defense strategies in inflammation. However, although this has been reported for several species, including mice, until now, large output of NO could not be equally demonstrated in human macrophages, and the contribution of human iNOS is still controversially discussed [97,98]. Human macrophages produce neopterin at the expense of BH4, and low BH4 leads to NOS enzyme uncoupling. Furthermore, neopterin pro-oxidant properties may compensate for deficient NO and $ONOO^-$ production [99].

Of note, NO inhibits IDO expression and function by reversibly binding to the active site heme [100]. Induction of IDO and NOS in IFN-γ-mediated inflammatory response is suggested to be functionally cross-regulated [101]. The absence of NO-mediated immune regulation may be involved in increased IDO activity at the site of inflammation. Of note, elevated tryptophan degradation was reported in patients with coronary heart disease [23].

22.7 POTENTIAL ROLE OF TH2 RESPONSES IN CVD

Th1 responses are in general proinflammatory and known to be proatherogenic, while Th2 cells are usually involved in helminthic and allergic responses. The role of Th2 cells in atherosclerosis seems to be very complex and even contradictory. A potential protective role of Th2 response is discussed in few studies [102,103], while Ait-Oufella et al. assume a potential proatherogenic function of Th2 cells within the complex interaction scenario of CD4+ T-cell subsets in atherosclerosis. Thus, the exact role of the Th2 response remains to be elucidated based on an improved understanding of the complex interplay between Th1, Th2, and other T-cell populations such as Th17 and Tregs within the atherosclerotic scenario [18]. Overall, Th cell subset polarization in atherosclerosis is less distinct in humans than in mice [104].

A high cholesterol diet of ApoE(–/–) mice with different T-cell subset polarizations resulted in the increased development of atherosclerosis in the aortic root and abdominal aorta with predominantly Th1-like immune responses (ApoE(–/–) BL/6 mice) in comparison to animals with Th2 predominance (ApoE(–/–) BALB/c) [105]. A potential of IL-4 to limit Th1 cell responses and reduce lesion size was observed in several murine atherosclerotic models [106,107].

Only recently, Engelbersten et al. reported an association between Th2 immunity and reduced risk of MI, as a high number of Th2 cells were associated with a decreased mean common carotid intima-media thickness and reduced risk of AMI in women, and IL-4 was independently associated with a reduced risk of CVD [108]. Although some limitations, for example, differences in lymphocyte number between healthy men and women or the use of long-term cryoconserved cells, this study provides first hints for the clinical importance of an improved understanding of Th2-type responses in CVD. However, again in contrast to these positive, protective attributes, Shimizu et al. suggested a role for Th2 cells and cytokines in the promotion of arterial aneurysm formation [109].

22.8   ANTIOXIDANT IN CVD THERAPY

Oxidative stress triggers inflammation and endothelial disruption in atherogenesis. A number of studies showed that exogenous antioxidants can modulate endothelium-dependent vasodilation responses, endothelium–leukocyte interactions, and balance between pro- and antithrombotic properties [110]. Accordingly, antioxidant therapy was suggested to beneficially interfere with the development and progression of atherosclerosis.

Th1/Th2 balance is crucially dependent on redox events; while Th1 responses prevail at oxidative conditions, Th2 responses were shown to be supported by "antioxidative stress" [111]. Thus, imbalances of Th1/Th2 cytokines may be involved in CVD as a mechanism of immunotoxicity. As Th1 and Th2 responses cross-regulate each other to balance immune responses [112], the suppression of the Th1-type response by antioxidants would favor Th2-type reactions. Of note, several types of antioxidants were shown to reduce the production of IFN-γ and neopterin in peripheral blood mononuclear cells *in vitro.*

A number of studies reported an inverse relationship between plasma antioxidants, or total antioxidant capacity, and CVDs [5,16]. Low intake of antioxidant, in particular of vitamins, was suggested to be associated with an increased risk of CVD [113,114]. Thus, the findings of an inverse correlation between concentrations of antioxidants compounds and vitamins and disease risk could relate to an increased requirement for antioxidant molecules during inflammatory diseases, and insufficient supply with these compounds may further accelerate disease process.

This assumption, however, has not been conclusively proven in clinical trials and is still controversially discussed in the literature [115–118]. A number of studies were performed to assess the impact of antioxidant supplementation, such as vitamin E, β-carotene, α-lipoic acid, coenzyme Q10, alone or in combination, on cardiovascular health [119], leading to different conclusions. For example, several studies were made with vitamin E, and due to its fat solubility, it is part of cell membranes and lipoprotein particles, where it counteracts oxidation events. Vitamin E–mediated protection from oxidative stress and atherosclerotic plaque formation has been shown *in vitro* and in mouse models [119]. Although several studies were performed with vitamin E supplementation, some reported reduction of risk of fatal and nonfatal AMI, while others found a slight increase in mortality with high-dose vitamin E treatment [119,120]. No final suggestion can be made about the impact of vitamin E supplementation, and even a meta-analysis including a large trial number leads to inconsistent results [121].

So far, although a general association of low vitamin levels with oxidative stress–related conditions is established, no clear evidence for a beneficial effect of vitamin supplementation exists. The association between vitamin deficiency in patients and disease symptoms is suggested to result mainly from the inflammation-associated consumption of oxidation-sensitive vitamins [23,116,122], which may lead to a variety of secondary effects.

Besides being part of the antioxidant defense system, vitamins may act as enzyme cofactors. For example, low B vitamin (folic acid and B12) availability leads to impaired remethylation and thus accumulation of homocysteine, as they are essential cofactors in homocysteine–methionine metabolism. Homocysteine is considered as an independent risk marker for CVD [123]; however, data indicate that it accumulates secondarily to immune activation [124,125]. Thus, the impact of such markers has to be critically evaluated when assessing the effect of vitamin supplementation.

A major aim in the treatment of atherosclerosis is the prevention of LDL oxidation. Lipid-lowering drugs such as statins and niacin are in use for a long time, alone or together, for cardiovascular protection in patients with coronary disease and low plasma levels of HDL [127]. However, combination therapies with antioxidant vitamins seemed even to counteract the beneficial effect of statin/niacin therapy [126,127].

A broader understanding of antioxidant action is clearly warranted. Beside their direct effects in the prevention of biomolecule oxidation by being oxidized themselves, several antioxidants mediate a variety of effects that are of longer duration, as they may induce signaling changes in the biological system [127] or modulate enzymatic function by acting as cofactors [75]. However, a variety of drugs may also act as antioxidants; thus, antioxidant vitamins could interfere with pharmaco-relevant signaling pathways. The following example should highlight potential interferences.

Statins are inhibitors of 3-hydroxy-3-methylglutaryl-co-enzyme A (HMG-CoA) reductase, and their lipid-lowering effects are suggested to reduce the risk of coronary heart disease [128], although therapeutic efficacy is limited [129].

The primary mechanism of statin action is suggested to be the reduction of LDL cholesterol, but several clinical trials have suggested that statins exert pleiotropic effect that contributes to therapeutic efficacy. Statins act as effective antioxidants not only by inhibiting the generation of ROS but also by interfering with NOX and NOS, antioxidant enzymes, lipid peroxidation and LDL cholesterol oxidation [130]. In *in vitro* studies with vascular smooth muscle and mononuclear cells, treatment with atorvastatin could reduce NF-κB activation and expression of proinflammatory cytokines and chemokines [131]. In human peripheral blood mononuclear cells and in monocytic cell lines, atorvastatin was shown to suppress stimulation-induced neopterin formation and tryptophan degradation, suggesting that immunoreactivity of both T cells and monocyte-derived macrophages is downregulated by this statin [132].

Treatment with several statins could promote Th2 polarization of CD4+ T cells primed *in vitro* with anti-CD3 antibody and splenic antigen-presenting cells [133]. These findings strongly suggest that statins contribute to the regulation of Th1/Th2 cell balance also *in vivo*. In endothelial cells, statins were shown to be involved in restorative processes by increasing NO bioavailability and promoting reendothelialization [134]. Of note, lovastatin was able to prevent neopterin-induced activation of human coronary artery endothelial cells (HCAECs) *in vitro* by interfering with NF-κB activation and decreasing the expression of CAMs and TF [70]. Furthermore, lovastantin reduced CRP-induced NF-κB activation expression in human umbilical vein endothelial cells (HUVECs) [135]. Beside NF-κB, the activation of the inflammatory transcription factors activator protein 1 (AP-1) and hypoxia-inducible factor 1 alpha (Hif-1α) was shown to be downregulated in human endothelial and vascular smooth muscle cells upon treatment with HMG-CoA reductase inhibitors [136]. In line with the reported antioxidant and anti-inflammatory properties, statin use has been associated with lower neopterin levels in patients [137,138].

The influence on redox balance and Th1-type signaling pathways such as neopterin formation and tryptophan breakdown has been described for a variety of (potentially) cardioprotective antioxidant drugs and vitamins, for example, aspirin [139], atorvastatin [137], vitamin C, and vitamin E [140] and seems to be a common mechanism. Furthermore, circulating vitamin E was shown to increase upon statin therapy [141,142]. Thus, due to inferences with common pathways, therapeutic efficacy might change when combining several antioxidant drugs and supplements. Furthermore, antioxidant composition may differ between patients, and estimation of antioxidant profiles before therapy could be useful to select candidate patients who would profit from antioxidant therapies [143,144]. Excessive antioxidant consumption may lead to adverse reactions ranging from favoring of allergy and asthma to an increase in mortality [145–147]. So far, for patients who respond well to statin/niacin therapy, additional supplementation might only be advantageous when nutritional deficiencies are still detectable. Nevertheless, this hypothesis has to be investigated in more detail.

22.9 NUTRITION, ANTIOXIDANTS, AND CVD

The strong relationship between redox status, immune response, and metabolism is supported by the close association of metabolic diseases such as diabetes, obesity, and metabolic syndrome with CVD [148]. Tissues that are important in metabolism are suggested to have an evolutionary potential to mediate inflammatory responses [149]. Metabolic and immune response pathways are closely cross-regulated to respond to the energetic demands necessary during immune activation. Several metabolic and immune cells show similarities on a genetic and functional level, for example, pre-adipocytes can transdifferentiate into macrophages [150] and activate similar transcriptional responses [151].

In contrast to classical activation of the immune system, for example, by infection or tissue injury, inflammation may also be induced by metabolic triggers. The so-called metaflammation [149] or para-inflammation [152] is crucially involved in the development of chronic diseases such as diabetes, fatty liver disease, and CVD.

A variety of dietary factors are able to produce cardiometabolic imprints that predispose to disease development. For example, increased consumption of trans-fatty acids (TFA) is supposed to activate pathways that are linked to insulin resistance syndrome. High TFA intake was found to be associated with harmful changes in serum lipids, systemic inflammation, and endothelial function, and prospective observational studies demonstrate strong positive associations with the risk of MI, coronary heart disease death, and sudden death [153]. Changes of traditional nutrition patterns, as it is the case, for example, in India, where "westernization" led to an increase in the uptake of sugar, salt, high-fat diary products, and TFA-rich food, are at least partially responsible for an about threefold increase in the prevalence of CVD and diabetes in the latter part of the twentieth century [154].

Excessive intake of antioxidants is a burden of modern life due to the omnipresence of preservatives, food colorants, and vitamin supplements in the "western diet." Although nutritional deficiencies may still exist for some specific ones, overall antioxidant stress may favor a Th2 environment by suppressing Th1 responses (Figure 22.2). In combination with a high caloric diet and low physical activity, this may contribute to the development of obesity [117]. Food additives such as sodium benzoate, propionic acid, sodium sulfite, ascorbic acid, and curcumin were shown to suppress Th1-type immune response *in vitro* [155,156]. Antioxidant food additives also interfere with hedonic saturation circuits, as they have been shown to inhibit leptin release in cultured LPS-stimulated murine adipocytes in a dose- and time-dependent manner [157]. Lowering the amount of circulating leptin is suggested to contribute to an obesogenic environment, as the reduced satiety effect in turn could lead to compensatory antioxidant craving and thus even more food intake [117]. Leptin is considered a proinflammatory cytokine with proatherogenic features, as it increases monocyte chemoattractant protein-1 (MCP-1) and endothelin-1 secretion by endothelial cells, enhances oxidative stress, promotes migration and proliferation of smooth muscle cells, and increases platelet aggregation, thus facilitating thrombosis [158]. In the initial phase of obesity-related inflammation, leptin is predictively associated with interleukin-6 plasma levels in juveniles [159]. However, leptin resistance, which later develops during obesity, also favors atherogenesis.

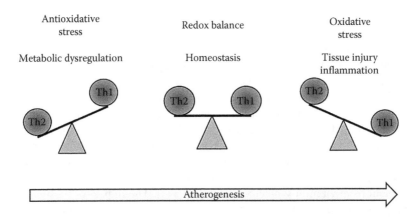

FIGURE 22.2 Dysregulation of redox balance and Th1/Th2 balance in the course of atherogenesis. Excessive antioxidant intake in combination with other risk factors such as a high-caloric diet and low physical exercise leads to suppression of Th1-type immunity, thereby favoring Th2-associated development of allergies and asthma and promoting juvenile obesity. Factors such as high blood pressure and hyperlipidemia lead to shear stress and tissue injury. Inflammatory reactions are associated with high ROS generation, which results in immunotoxicity due to the oxidation of biomolecules (lipids, proteins, etc.).

Obesity-related immune-mediated systemic inflammation was found to be associated with the development of the metabolic syndrome and altered Trp metabolism. However, across life span from juvenility to adulthood, differences in the Trp breakdown rate were observed. While juvenile overweight/obese individuals showed a decreased unaltered Kyn/Trp ratio in comparison to control, overweight/obese adults had significantly elevated Kyn serum levels and an increased Kyn/Trp ratio [160]. Thus, while in younger patients Th2-type responses might be favored, potentially due to the high antioxidant intake, overwhelming inflammation with Th1-type cytokines may predispose to the development of atherosclerosis in adult age.

Epidemiological observations suggest that consumption of certain foods rich in bioactive compounds, for example, vitamins E and C, polyphenols and carotenoids such as lycopene and β-carotene, and coenzyme Q10, leads to a decrease in atherosclerotic risk, and such an antioxidant-rich diet is supposed to be particularly effective in the early stages of atherosclerosis by preventing LDL oxidation and the oxidative lesion of endothelium [161,162].

22.10 CONCLUSION

It is well established that oxidative modifications contribute to important clinical manifestations of CVD such as endothelial dysfunction and plaque disruption. However, due to the poor performance of antioxidant strategies in limiting atherosclerosis and cardiovascular events, it remains to be answered if oxidative modification is a causal factor for the initiation or is an injurious response to atherogenesis [95]. The strong interconnection of metabolic and inflammatory pathways suggests that metabolically induced inflammatory processes should be considered as early or even primary events [149]. Many data support that there is a large time span between initial pathological changes and the onset of clinical manifestations. This time frame could be used for preventive strategies; however, a better understanding of disease development and more sensitive detection methods would be a prerequisite.

A detailed knowledge on inflammatory and redox-regulated processes would also allow a better adaption of treatment regimes. Stable biochemical markers are necessary to control disease courses and treatment efficacy. In this context, for example, neopterin is a useful indicator of the immune activation status and oxidative stress [6], and Kyn/Trp ratio accounts for the aspects of immune regulation via IDO and represents an important metabolic checkpoint. Combined measurements of multiple markers, such as additional determination of lipoprotein, NO metabolites, BH4, and plasma antioxidants, will be helpful to understand redox regulation in health and disease and may allow to best discriminate between different clinical diagnostic categories and to evaluate treatment strategies.

In summary, a general evaluation of the effect of an "antioxidant therapy" is not possible at the moment. While vitamin supplementation might be beneficial under certain circumstances, a variety of studies indicate no or even adverse effects when administered alone and even more when used in combination with lipid-lowering agents. However, also for statin and niacin treatment, a panel of adverse effects has been described [163,164]. Although antioxidant supplementation may have some benefit to counteract secondary symptoms, their role in CAD seems to be of moderate importance [116]. Surveillance of the antioxidant status before and during therapy would allow seek out patients who could benefit from vitamin supplementation [143,144]. The impact of lifestyle factors such as nutrition and physical exercise, however, has turned out as a major factor in CVD prevention and also in influencing treatment efficacy [165].

REFERENCES

1. World Health Organization. Global status report on noncommunicable diseases 2010. Geneva, Switzerland: World Health Organization, 2011.
2. Hansson GK. Inflammation, atherosclerosis, and coronary artery disease. *N. Engl. J. Med.* 2005; 352, 1685–1695.
3. World Health Organization. Cardiovascular diseases (CVDs). Fact sheet No. 317, Updated January 2015. http://www.who.int/mediacentre/factsheets/fs317/en/index.html (accessed March 31, 2016).
4. Libby P, Ridker PM, Maseri A. Inflammation and atherosclerosis. *Circulation* 2002; 105, 1135–1143.
5. Murr C, Winklhofer-Roob BM, Schroecksnadel K, Maritschnegg M, Mangge H, Bohm BO, Winkelmann BR, Marz W, Fuchs D. Inverse association between serum concentrations of neopterin and antioxidants in patients with and without angiographic coronary artery disease. *Atherosclerosis* 2009; 202, 543–549.
6. Fuchs D, Avanzas P, Arroyo-Espliguero R, Jenny M, Consuegra-Sanchez L, Kaski JC. The role of neopterin in atherogenesis and cardiovascular risk assessment. *Curr. Med. Chem.* 2009; 16, 4644–4653.
7. Libby P, Ridker PM, Hansson GK. Progress and challenges in translating the biology of atherosclerosis. *Nature* 2011; 473, 317–325.
8. Galkina E, Ley K. Immune and inflammatory mechanisms of atherosclerosis (*). *Annu. Rev. Immunol.* 2009; 27, 165–197.
9. Chisolm GM, Steinberg D. The oxidative modification hypothesis of atherogenesis: An overview. *Free Radic. Biol. Med.* 2000; 28, 1815–1826.
10. Tabas I. Macrophage death and defective inflammation resolution in atherosclerosis. *Nat. Rev. Immunol.* 2010; 10, 36–46.
11. Frostegard J. Immunity, atherosclerosis and cardiovascular disease. *BMC Med.* 2013; 11, 117.
12. Ghazalpour A, Doss S, Yang X, Aten J, Toomey EM, van Nas A, Wang S, Drake TA, Lusis AJ. Thematic review series: The pathogenesis of atherosclerosis. Toward a biological network for atherosclerosis. *J. Lipid Res.* 2004; 45, 1793–1805.
13. Ramsey SA, Gold ES, Aderem A. A systems biology approach to understanding atherosclerosis. *EMBO Mol. Med.* 2010; 2, 79–89.
14. Libby P. Inflammation in atherosclerosis. *Arterioscler. Thromb. Vasc. Biol.* 2012; 32, 2045–2051.
15. Corti R, Hutter R, Badimon JJ, Fuster V. Evolving concepts in the triad of atherosclerosis, inflammation and thrombosis. *J. Thromb. Thrombolysis* 2004; 17, 35–44.
16. Chrysohoou C, Panagiotakos DB, Pitsavos C, Skoumas J, Economou M, Papadimitriou L, Stefanadis C. The association between pre-hypertension status and oxidative stress markers related to atherosclerotic disease: The ATTICA study. *Atherosclerosis* 2007; 192, 169–176.
17. Uyemura K, Demer LL, Castle SC, Jullien D, Berliner JA, Gately MK, Warrier RR, Pham N, Fogelman AM, Modlin RL. Cross-regulatory roles of interleukin (IL)-12 and IL-10 in atherosclerosis. *J. Clin. Invest.* 1996; 97, 2130–2138.
18. Ait-Oufella H, Taleb S, Mallat Z, Tedgui A. Cytokine network and T cell immunity in atherosclerosis. *Semi. Immunopathol.* 2009; 31:23–33.

19. Nathan CF, Murray HW, Wiebe ME, Rubin BY. Identification of interferon-gamma as the lymphokine that activates human macrophage oxidative metabolism and antimicrobial activity. *J. Exp. Med.* 1983; 158, 670–689.
20. Rossi F, Zatti M. Biochemical aspects of phagocytosis in polymorphonuclear leucocytes. NADH and NADPH oxidation by the granules of resting and phagocytizing cells. *Experientia* 1964; 20, 21–23.
21. Wink DA, Hines HB, Cheng RY, Switzer CH, Flores-Santana W, Vitek MP, Ridnour LA, Colton CA. Nitric oxide and redox mechanisms in the immune response. *J. Leukoc. Biol.* 2011; 89, 873–891.
22. Szabo SJ, Sullivan BM, Peng SL, Glimcher LH. Molecular mechanisms regulating Th1 immune responses. *Annu. Rev. Immunol.* 2003; 21, 713–758.
23. Schroecksnadel K, Frick B, Winkler C, Fuchs D. Crucial role of interferon-gamma and stimulated macrophages in cardiovascular disease. *Curr. Vasc. Pharmacol.* 2006; 4, 205–213.
24. Le Bras M, Clement MV, Pervaiz S, Brenner C. Reactive oxygen species and the mitochondrial signaling pathway of cell death. *Histol. Histopathol.* 2005; 20, 205–219.
25. Valko M, Leibfritz D, Moncol J, Cronin MT, Mazur M, Telser J. Free radicals and antioxidants in normal physiological functions and human disease. *Int. J. Biochem. Cell Biol.* 2007; 39, 44–84.
26. Gostner JM, Becker K, Fuchs D, Sucher R. Redox regulation of the immune response. *Redox Rep.* 2013; 18, 88–94.
27. Griendling KK, FitzGerald GA. Oxidative stress and cardiovascular injury: Part II: Animal and human studies. *Circulation* 2003; 108, 2034–2040.
28. Griendling KK, FitzGerald GA. Oxidative stress and cardiovascular injury: Part I: Basic mechanisms and in vivo monitoring of ROS. *Circulation* 2003; 108, 1912–1916.
29. Peiser L, Mukhopadhyay S, Gordon S. Scavenger receptors in innate immunity. *Curr. Opin. Immunol.* 2002; 14, 123–128.
30. Zelzer S, Fuchs D, Almer G, Raggam RB, Prüller F, Truschnig-Wilders M, Schnedl W et al. High density lipoprotein cholesterol level is a robust predictor of lipid peroxidation irrespective of gender, age, obesity, and inflammatory or metabolic biomarkers. *Clin. Chim. Acta* 2011; 412(15–16), 1345–1349.
31. Fraley AE, Tsimikas S. Clinical applications of circulating oxidized low-density lipoprotein biomarkers in cardiovascular disease. *Curr. Opin. Lipidol.* 2006; 17, 502–509.
32. Holvoet P, Vanhaecke J, Janssens S, Van de WF, Collen D. Oxidized LDL and malondialdehyde-modified LDL in patients with acute coronary syndromes and stable coronary artery disease. *Circulation* 1998; 98, 1487–1494.
33. Shoji T, Nishizawa Y, Fukumoto M, Shimamura K, Kimura J, Kanda H, Emoto M, Kawagishi T, Morii H. Inverse relationship between circulating oxidized low density lipoprotein (oxLDL) and anti-oxLDL antibody levels in healthy subjects. *Atherosclerosis* 2000; 148, 171–177.
34. Meisinger C, Baumert J, Khuseyinova N, Loewel H, Koenig W. Plasma oxidized low-density lipoprotein, a strong predictor for acute coronary heart disease events in apparently healthy, middle-aged men from the general population. *Circulation* 2005; 112, 651–657.
35. Khera AV, Cuchel M, Llera-Moya M, Rodrigues A, Burke MF, Jafri K, French BC et al. Cholesterol efflux capacity, high-density lipoprotein function, and atherosclerosis. *N. Engl. J. Med.* 2011; 364, 127–135.
36. Chapman MJ, Ginsberg HN, Amarenco P, Andreotti F, Boren J, Catapano AL, Descamps OS et al. Triglyceride-rich lipoproteins and high-density lipoprotein cholesterol in patients at high risk of cardiovascular disease: Evidence and guidance for management. *Eur. Heart J.* 2011; 32, 1345–1361.
37. Ferretti G, Bacchetti T, Masciangelo S, Bicchiega V. HDL-paraoxonase and membrane lipid peroxidation: A comparison between healthy and obese subjects. *Obesity (Silver Spring)* 2010; 18, 1079–1084.
38. Watson AD, Berliner JA, Hama SY, La Du BN, Faull KF, Fogelman AM, Navab M. Protective effect of high density lipoprotein associated paraoxonase. Inhibition of the biological activity of minimally oxidized low density lipoprotein. *J. Clin. Invest.* 1995; 96, 2882–2891.
39. Graham A, Hassall DG, Rafique S, Owen JS. Evidence for a paraoxonase-independent inhibition of low-density lipoprotein oxidation by high-density lipoprotein. *Atherosclerosis* 1997; 135, 193–204.
40. Escola-Gil JC, Cedo L, Blanco-Vaca F. High-density lipoprotein cholesterol targeting for novel drug discovery: Where have we gone wrong? *Expert Opin. Drug Discov.* 2014; 9(2), 119–124.
41. Singh IM, Shishehbor MH, Ansell BJ. High-density lipoprotein as a therapeutic target: A systematic review. *JAMA* 2007; 298, 786–798.
42. Huber C, Batchelor JR, Fuchs D, Hausen A, Lang A, Niederwieser D, Reibnegger G, Swetly P, Troppmair J, Wachter H. Immune response-associated production of neopterin. Release from macrophages primarily under control of interferon-gamma. *J. Exp. Med.* 1984; 160, 310–316.

43. Wirleitner B, Reider D, Ebner S, Bock G, Widner B, Jaeger M, Schennach H et al. Monocyte-derived dendritic cells release neopterin. *J. Leukoc. Biol.* 2002; 72, 1148–1153.

44. Werner-Felmayer G, Werner ER, Fuchs D, Hausen A, Reibnegger G, Wachter H. Tumour necrosis factor-alpha and lipopolysaccharide enhance interferon-induced tryptophan degradation and pteridine synthesis in human cells. *Biol. Chem. Hoppe Seyler* 1989; 370, 1063–1069.

45. Nathan CF. Peroxide and pteridine: A hypothesis on the regulation of macrophage antimicrobial activity by interferon gamma. *Interferon* 1986; 7, 125–143.

46. Murr C, Widner B, Wirleitner B, Fuchs D. Neopterin as a marker for immune system activation. *Curr. Drug Metab.* 2002; 3, 175–187.

47. Werner ER, Werner-Felmayer G, Fuchs D, Hausen A, Reibnegger G, Yim JJ, Pfleiderer W, Wachter H. Tetrahydrobiopterin biosynthetic activities in human macrophages, fibroblasts, THP-1, and T 24 cells. GTP-cyclohydrolase I is stimulated by interferon-gamma, and 6-pyruvoyl tetrahydropterin synthase and sepiapterin reductase are constitutively present. *J. Biol. Chem.* 1990; 265, 3189–3192.

48. Werner-Felmayer G, Golderer G, Werner ER. Tetrahydrobiopterin biosynthesis, utilization and pharmacological effects. *Curr. Drug Metab.* 2002; 3, 159–173.

49. Fuchs D, Forsman A, Hagberg L, Larsson M, Norkrans G, Reibnegger G, Werner ER, Wachter H. Immune activation and decreased tryptophan in patients with HIV-1 infection. *J. Interf. Res.* 1990; 10, 599–603.

50. Murr C, Fuith LC, Widner B, Wirleitner B, Baier-Bitterlich G, Fuchs D. Increased neopterin concentrations in patients with cancer: Indicator of oxidative stress? *Anticancer Res.* 1999; 19, 1721–1728.

51. Sucher R, Schroecksnadel K, Weiss G, Margreiter R, Fuchs D, Brandacher G. Neopterin, a prognostic marker in human malignancies. *Cancer Lett.* 2010; 287, 13–22.

52. Reibnegger G, Aichberger C, Fuchs D, Hausen A, Spielberger M, Werner ER, Margreiter R, Wachtehr H. Posttransplant neopterin excretion in renal allograft recipients—A reliable diagnostic aid for acute rejection and a predictive marker of long-term graft survival. *Transplantation* 1991; 52, 58–63.

53. Melichar B, Gregor J, Solichova D, Lukes J, Tichy M, Pidrman V. Increased urinary neopterin in acute myocardial infarction. *Clin. Chem.* 1994; 40, 338–339.

54. Anwaar I, Gottsater A, Lindgarde F, Mattiasson I. Increasing plasma neopterin and persistent plasma endothelin during follow-up after acute cerebral ischemia. *Angiology* 1999; 50, 1–8.

55. Tatzber F, Rabl H, Koriska K, Erhart U, Puhl H, Waeg G, Krebs A et al. Elevated serum neopterin levels in atherosclerosis. *Atherosclerosis* 1991; 89, 203–208.

56. Weiss G, Willeit J, Kiechl S, Fuchs D, Jarosch E, Oberhollenzer F, Reibnegger G, Tilz GP, Gerstenbrand F, Wachter H. Increased concentrations of neopterin in carotid atherosclerosis. *Atherosclerosis* 1994; 106, 263–271.

57. Schumacher M, Halwachs G, Tatzber F, Fruhwald FM, Zweiker R, Watzinger N, Eber B, Wilders-Truschnig M, Esterbauer H, Klein W. Increased neopterin in patients with chronic and acute coronary syndromes. *J. Am. Coll. Cardiol.* 1997; 30, 703–707.

58. Gurfinkel EP, Scirica BM, Bozovich G, Macchia A, Manos E, Mautner B. Serum neopterin levels and the angiographic extent of coronary arterial narrowing in unstable angina pectoris and in non-Q-wave acute myocardial infarction. *Am. J. Cardiol.* 1999; 83, 515–518.

59. Zouridakis E, Avanzas P, Arroyo-Espliguero R, Fredericks S, Kaski JC. Markers of inflammation and rapid coronary artery disease progression in patients with stable angina pectoris. *Circulation* 2004; 110, 1747–1753.

60. Avanzas P, Arroyo-Espliguero R, Quiles J, Roy D, Kaski JC. Elevated serum neopterin predicts future adverse cardiac events in patients with chronic stable angina pectoris. *Eur. Heart J.* 2005; 26, 457–463.

61. Kaski JC, Consuegra-Sanchez L, Fernandez-Berges DJ, Cruz-Fernandez JM, Garcia-Moll X, Marrugat J, Mostaza J, Toro-Cebada R, Gonzalez-Juanatey JR, Guzman-Martinez G. Elevated serum neopterin levels and adverse cardiac events at 6 months follow-up in Mediterranean patients with non-ST-segment elevation acute coronary syndrome. *Atherosclerosis* 2008; 201, 176–183.

62. Johnston DT, Gagos M, Raio N, Ragolia L, Shenouda D, Davis-Lorton MA, De Leon JR. Alterations in serum neopterin correlate with thrombolysis in myocardial infarction risk scores in acute coronary syndromes. *Coron. Artery Dis.* 2006; 17, 511–516.

63. Barani J, Mattiasson I, Lindblad B, Gottsater A. Cardiac function, inflammatory mediators and mortality in critical limb ischemia. *Angiology* 2006; 57, 437–444.

64. Adachi T, Naruko T, Itoh A, Komatsu R, Abe Y, Shirai N, Yamashita H et al. Neopterin is associated with plaque inflammation and destabilisation in human coronary atherosclerotic lesions. *Heart* 2007; 93, 1537–1541.

65. Ray KK, Morrow DA, Sabatine MS, Shui A, Rifai N, Cannon CP, Braunwald E. Long-term prognostic value of neopterin: A novel marker of monocyte activation in patients with acute coronary syndrome. *Circulation* 2007; 115, 3071–3078.

66. Weiss G, Fuchs D, Hausen A, Reibnegger G, Werner ER, Werner-Felmayer G, Semenitz E, Dierich MP, Wachter H. Neopterin modulates toxicity mediated by reactive oxygen and chloride species. *FEBS Lett.* 1993; 321, 89–92.

67. Herpfer I, Greilberger J, Ledinski G, Widner B, Fuchs D, Jurgens G. Neopterin and 7,8-dihydroneopterin interfere with low density lipoprotein oxidation mediated by peroxynitrite and/or copper. *Free Radic. Res.* 2002; 36, 509–520.

68. Widner B, Baier-Bitterlich G, Wede I, Wirleitner B, Fuchs D. Neopterin derivatives modulate the nitration of tyrosine by peroxynitrite. *Biochem. Biophys. Res. Commun.* 1998; 248, 341–346.

69. Hoffmann G, Wirleitner B, Fuchs D. Potential role of immune system activation-associated production of neopterin derivatives in humans. *Inflamm. Res.* 2003; 52, 313–321.

70. Cirillo P, Pacileo M, De Rosa S, Calabro P, Gargiulo A, Angri V, Granato-Corigliano F et al. Neopterin induces pro-atherothrombotic phenotype in human coronary endothelial cells. *J. Thromb. Haemost.* 2006; 4, 2248–2255.

71. Byrne GI, Lehmann LK, Kirschbaum JG, Borden EC, Lee CM, Brown RR. Induction of tryptophan degradation in vitro and in vivo: A gamma-interferon-stimulated activity. *J. Interferon Res.* 1986; 6, 389–396.

72. Werner ER, Bitterlich G, Fuchs D, Hausen A, Reibnegger G, Szabo G, Dierich MP, Wachter H. Human macrophages degrade tryptophan upon induction by interferon-gamma. *Life Sci.* 1987; 41, 273–280.

73. Taylor MW, Feng GS. Relationship between interferon-gamma, indoleamine 2,3-dioxygenase, and tryptophan catabolism. *FASEB J.* 1991; 5, 2516–2522.

74. Thomas SR, Stocker R. Redox reactions related to indoleamine 2,3-dioxygenase and tryptophan metabolism along the kynurenine pathway. *Redox Rep.* 1999; 4, 199–220.

75. Schroecksnadel K, Fischer B, Schennach H, Weiss G, Fuchs D. Antioxidants suppress Th1-type immune response in vitro. *Drug Metab. Lett.* 2007; 1, 166–171.

76. Widner B, Werner ER, Schennach H, Wachter H, Fuchs D. Simultaneous measurement of serum tryptophan and kynurenine by HPLC. *Clin. Chem.* 1997; 43, 2424–2426.

77. Knox WE. The regulation of tryptophan pyrrolase activity by tryptophan. *Adv. Enzyme Regul.* 1966; 4, 287–297.

78. Mangge H, Stelzer I, Reininghaus EZ, Weghuber D, Postolache TT, Fuchs D. Disturbed tryptophan metabolism in cardiovascular disease. *Curr. Med. Chem.* 2014; 21(17), 1931–1937.

79. Brandacher G, Winkler C, Schroecksnadel K, Margreiter R, Fuchs D. Antitumoral activity of interferon-gamma involved in impaired immune function in cancer patients. *Curr. Drug Metab.* 2006; 7, 599–612.

80. Munn DH, Mellor AL. Indoleamine 2,3 dioxygenase and metabolic control of immune responses. *Trends Immunol.* 2013; 34, 137–143.

81. Fuchs D, Hausen A, Reibnegger G, Werner ER, Werner-Felmayer G, Dierich MP, Wachter H. Immune activation and the anaemia associated with chronic inflammatory disorders. *Eur. J. Haematol.* 1991; 46, 65–70.

82. Weiss G, Schroecksnadel K, Mattle V, Winkler C, Konwalinka G, Fuchs D. Possible role of cytokine-induced tryptophan degradation in anaemia of inflammation. *Eur. J. Haematol.* 2004; 72, 130–134.

83. Pertovaara M, Raitala A, Juonala M, Lehtimaki T, Huhtala H, Oja SS, Jokinen E, Viikari JS, Raitakari OT, Hurme M. Indoleamine 2,3-dioxygenase enzyme activity correlates with risk factors for atherosclerosis: The cardiovascular risk in young finns study. *Clin. Exp. Immunol.* 2007; 148, 106–111; Sucher R, Fischler K, Oberhuber R, Kronberger I, Margreiter C, Ollinger R, Schneeberger S et al. IDO and regulatory T cell support are critical for cytotoxic T lymphocyte-associated Ag-4 Ig-mediated long-term solid organ allograft survival. *J. Immunol.* 2012; 188, 37–46.

84. Niinisalo P, Raitala A, Pertovaara M, Oja SS, Lehtimäki T, Kähönen M, Reunanen A et al. Indoleamine 2,3-dioxygenase activity associates with cardiovascular risk factors: The Health 2000 study. *Scand. J. Clin. Lab. Invest.* 2008; 68, 767–770.

85. Wellenius GA, Mukamal KJ, Kulshreshtha A, Asonganyi S, Mittleman MA. Depressive symptoms and the risk of atherosclerotic progression among patients with coronary artery bypass grafts. *Circulation* 2008; 117, 2313–2319.

86. Widner B, Laich A, Sperner-Unterweger B, Ledochowski M, Fuchs D. Neopterin production, tryptophan degradation, and mental depression—What is the link? *Brain Behav. Immun.* 2002; 16, 590–595.

87. Wink DA, Vodovotz Y, Grisham MB, DeGraff W, Cook JC, Pacelli R, Krishna M, Mitchell JB. Antioxidant effects of nitric oxide. *Methods Enzymol.* 1999; 301, 413–424.

88. Stamler JS, Singel DJ, Loscalzo J. Biochemistry of nitric oxide and its redox-activated forms. *Science* 1992; 258, 1898–1902.

89. Stamler JS, Jaraki O, Osborne J, Simon DI, Keaney J, Vita J, Singel D, Valeri CR, Loscalzo J. Nitric oxide circulates in mammalian plasma primarily as an S-nitroso adduct of serum albumin. *Proc. Natl. Acad. Sci. USA* 1992; 89, 7674–7677.

90. Maron BA, Tang SS, Loscalzo J. S-nitrosothiols and the S-nitrosoproteome of the cardiovascular system. *Antioxid. Redox Signal.* 2013; 18, 270–287.

91. Rassaf T, Bryan NS, Kelm M, Feelisch M. Concomitant presence of N-nitroso and S-nitroso proteins in human plasma. *Free Radic. Biol. Med.* 2002; 33, 1590–1596.

92. Feelisch M, Rassaf T, Mnaimneh S, Singh N, Bryan NS, Jourd'Heuil D, Kelm M. Concomitant S-, N-, and heme-nitros(yl)ation in biological tissues and fluids: Implications for the fate of NO in vivo. *FASEB J.* 2002; 16, 1775–1785.

93. Thomas DD, Ridnour LA, Isenberg JS, Flores-Santana W, Switzer CH, Donzelli S, Hussain P et al. The chemical biology of nitric oxide: Implications in cellular signaling. *Free Radic. Biol. Med.* 2008; 45, 18–31.

94. Stocker R, Keaney JF, Jr. Role of oxidative modifications in atherosclerosis. *Physiol. Rev.* 2004; 84, 1381–1478.

95. Laursen JB, Somers M, Kurz S, McCann L, Warnholtz A, Freeman BA, Tarpey M, Fukai T, Harrison DG. Endothelial regulation of vasomotion in apoE-deficient mice: Implications for interactions between peroxynitrite and tetrahydrobiopterin. *Circulation* 2001; 103, 1282–1288.

96. Bendall JK, Douglas G, McNeill E, Channon K, Crabtree MJ. Tetrahydrobiopterin in cardiovascular health and disease. *Antioxid. Redox Signal.* 2014; 20(18), 3040–3077.

97. Fang FC, Nathan CF. Man is not a mouse: Reply. *J. Leukoc. Biol.* 2007; 81, 580.

98. Schneemann M, Schoeden G. Macrophage biology and immunology: Man is not a mouse. *J. Leukoc. Biol.* 2007; 81, 579.

99. Fuchs D, Murr C, Reibnegger G, Weiss G, Werner ER, Werner-Felmayer G, Wachter H. Nitric oxide synthase and antimicrobial armature of human macrophages. *J. Infect. Dis.* 1994; 169, 224–225.

100. Thomas SR, Terentis AC, Cai H, Takikawa O, Levina A, Lay PA, Freewan M, Stocker R. Post-translational regulation of human indoleamine 2,3-dioxygenase activity by nitric oxide. *J. Biol. Chem.* 2007; 282, 23778–23787.

101. Alberati-Giani D, Malherbe P, Ricciardi-Castagnoli P, Kohler C, Denis-Donini S, Cesura AM. Differential regulation of indoleamine 2,3-dioxygenase expression by nitric oxide and inflammatory mediators in IFN-gamma-activated murine macrophages and microglial cells. *J. Immunol.* 1997; 159, 419–426.

102. Olson NC, Sallam R, Doyle MF, Tracy RP, Huber SA. T helper cell polarization in healthy people: Implications for cardiovascular disease. *J. Cardiovasc. Transl. Res.* 2013; 6, 772–786.

103. Magen E, Borkow G, Bentwich Z, Mishal J, Scharf S. Can worms defend our hearts? Chronic helminthic infections may attenuate the development of cardiovascular diseases. *Med. Hypotheses* 2005; 64, 904–909.

104. Libby P, Lichtman AH, Hansson GK. Immune effector mechanisms implicated in atherosclerosis: From mice to humans. *Immunity* 2013; 38, 1092–1104.

105. Schulte S, Sukhova GK, Libby P. Genetically programmed biases in Th1 and Th2 immune responses modulate atherogenesis. *Am. J. Pathol.* 2008; 172, 1500–1508.

106. George J, Shoenfeld Y, Gilburd B, Afek A, Shaish A, Harats D. Requisite role for interleukin-4 in the acceleration of fatty streaks induced by heat shock protein 65 or Mycobacterium tuberculosis. *Circ. Res.* 2000; 86, 1203–1210.

107. Huber SA, Sakkinen P, David C, Newell MK, Tracy RP. T helper-cell phenotype regulates atherosclerosis in mice under conditions of mild hypercholesterolemia. *Circulation* 2001; 103, 2610–2616.

108. Engelbertsen D, Andersson L, Ljungcrantz I, Wigren M, Hedblad B, Nilsson J, Bjorkbacka H. T-helper 2 immunity is associated with reduced risk of myocardial infarction and stroke. *Arterioscler. Thromb. Vasc. Biol.* 2013; 33, 637–644.

109. Shimizu K, Shichiri M, Libby P, Lee RT, Mitchell RN. Th2-predominant inflammation and blockade of IFN-gamma signaling induce aneurysms in allografted aortas. *J. Clin. Invest.* 2004; 114, 300–308.

110. Pratico D. Antioxidants and endothelium protection. *Atherosclerosis* 2005; 181, 215–224.

111. Poljsak B, Milisav I. The neglected significance of "antioxidative stress". *Oxid. Med. Cell Longev.* 2012; 2012, 480895.

112. Romagnani S. Type 1 T helper and type 2 T helper cells: Functions, regulation and role in protection and disease. *Int. J. Clin. Lab. Res.* 1991; 21, 152–158.

113. Rimm EB, Stampfer MJ, Ascherio A, Giovannucci E, Colditz GA, Willett WC. Vitamin E consumption and the risk of coronary heart disease in men. *N. Engl. J. Med.* 1993; 328, 1450–1456.

114. Stampfer MJ, Hennekens CH, Manson JE, Colditz GA, Rosner B, Willett WC. Vitamin E consumption and the risk of coronary disease in women. *N. Engl. J. Med.* 1993; 328, 1444–1449.

115. Katsiki N, Manes C. Is there a role for supplemented antioxidants in the prevention of atherosclerosis? *Clin. Nutr.* 2009; 28, 3–9.

116. Fuchs D, Sperner-Unterweger B. Can intake of extra antioxidants delay the development and progression of atherosclerosis? *Atherosclerosis* 2013; 226, 43–44.

117. Mangge H, Summers K, Almer G, Prassl R, Weghuber D, Schnedl W, Fuchs D. Antioxidant food supplements and obesity-related inflammation. *Curr. Med. Chem.* 2013; 20, 2330–2337.

118. Riccioni G, Bucciarelli T, Mancini B, Corradi F, Di Ilio C, Mattei PA, D'Orazio N. Antioxidant vitamin supplementation in cardiovascular diseases. *Ann. Clin. Lab. Sci.* 2007; 37, 89–95.

119. Sachidanandam K, Fagan SC, Ergul A. Oxidative stress and cardiovascular disease: Antioxidants and unresolved issues. *Cardiovasc. Drug Rev.* 2005; 23, 115–132.

120. Saremi A, Arora R. Vitamin E and cardiovascular disease. *Am. J. Ther.* 2010; 17, e56–e65.

121. Gerss J, Köpcke W. The questionable association of vitamin E supplementation and mortality—Inconsistent results of different meta-analytic approaches. *Cell Mol. Biol. (Noisy-le-grand)* 2009; 55, OL1111–OL1120.

122. Mangge H, Weghuber D, Prassl R, Haara A, Schnedl W, Postolache TT, Fuchs D. The role of vitamin D in atherosclerosis inflammation revisited: More a bystander than a player? *Curr. Vasc. Pharmacol.* 2015; 13(3), 392–398.

123. McCully KS. Homocysteine and vascular disease. *Nat. Med.* 1996; 2, 386–389.

124. Schroecksnadel K, Grammer TB, Boehm BO, März W, Fuchs D. Total homocysteine in patients with angiographic coronary artery disease correlates with inflammation markers. *Thromb. Haemost.* 2010; 103, 926–935.

125. Schroecksnadel K, Walter RB, Weiss G, Mark M, Reinhart WH, Fuchs D. Association between plasma thiols and immune activation marker neopterin in stable coronary heart disease. *Clin. Chem. Lab. Med.* 2008; 46, 648–654.

126. Brown BG, Zhao XQ, Chait A, Fisher LD, Cheung MC, Morse JS, Dowdy AA et al. Simvastatin and niacin, antioxidant vitamins, or the combination for the prevention of coronary disease. *N. Engl. J. Med.* 2001; 345, 1583–1592.

127. Virmani A, Pinto L, Binienda Z, Ali S. Food, nutrigenomics, and neurodegeneration—Neuroprotection by what you eat! *Mol. Neurobiol.* 2013; 48, 353–362; Cheung MC, Zhao XQ, Chait A, Albers JJ, Brown BG. Antioxidant supplements block the response of HDL to simvastatin-niacin therapy in patients with coronary artery disease and low HDL. *Arterioscler. Thromb. Vasc. Biol.* 2001; 21, 1320–1326.

128. Hausenloy DJ, Yellon DM. Targeting residual cardiovascular risk: Raising high-density lipoprotein cholesterol levels. *Heart* 2008; 94, 706–714.

129. Baginsky P. Should we treat all patients with coronary heart disease or the equivalent with statins? *Curr. Atheroscler. Rep.* 2009; 11, 28–35.

130. Stoll LL, McCormick ML, Denning GM, Weintraub NL. Antioxidant effects of statins. *Drugs Today (Barc.)* 2004; 40, 975–990.

131. Ortego M, Bustos C, Hernandez-Presa MA, Tunon J, Diaz C, Hernandez G, Egido J. Atorvastatin reduces NF-kappaB activation and chemokine expression in vascular smooth muscle cells and mononuclear cells. *Atherosclerosis* 1999; 147, 253–261.

132. Schroecksnadel K, Frick B, Winkler C, Wirleitner B, Weiss G, Fuchs D. Atorvastatin suppresses homocysteine formation in stimulated human peripheral blood mononuclear cells. *Clin. Chem. Lab. Med.* 2005; 43(12), 1373–1376.

133. Hakamada-Taguchi R, Uehara Y, Kuribayashi K, Numabe A, Saito K, Negoro H, Fujita T, Toyo-oka T, Kato T. Inhibition of hydroxymethylglutaryl-coenzyme a reductase reduces Th1 development and promotes Th2 development. *Circ. Res.* 2003; 93, 948–956.

134. Wolfrum S, Jensen KS, Liao JK. Endothelium-dependent effects of statins. *Arterioscler. Thromb. Vasc. Biol.* 2003; 23, 729–736.

135. Lin R, Liu J, Peng N, Yang G, Gan W, Wang W. Lovastatin reduces nuclear factor kappaB activation induced by C-reactive protein in human vascular endothelial cells. *Biol. Pharm. Bull.* 2005; 28, 1630–1634.

136. Dichtl W, Dulak J, Frick M, Alber HF, Schwarzacher SP, Ares MP, Nilsson J, Pachinger O, Weidinger F. HMG-CoA reductase inhibitors regulate inflammatory transcription factors in human endothelial and vascular smooth muscle cells. *Arterioscler. Thromb. Vasc. Biol.* 2003; 23, 58–63.

137. Neurauter G, Wirleitner B, Laich A, Schennach H, Weiss G, Fuchs D. Atorvastatin suppresses interferon-gamma-induced neopterin formation and tryptophan degradation in human peripheral blood mononuclear cells and in monocytic cell lines. *Clin. Exp. Immunol.* 2003; 131, 264–267; van Haelst PL, Liem A, van Boven AJ, Veeger NJ, van Veldhuisen DJ, Tervaert JW, Gans RO, Zijlstra F. Usefulness of elevated neopterin and C-reactive protein levels in predicting cardiovascular events in patients with non-Q-wave myocardial infarction. *Am. J. Cardiol.* 2003; 92, 1201–1203.

138. Walter RB, Fuchs D, Weiss G, Walter TR, Reinhart WH. HMG-CoA reductase inhibitors are associated with decreased serum neopterin levels in stable coronary artery disease. *Clin. Chem. Lab. Med.* 2003; 41, 1314–1319.

139. Schroecksnadel K, Winkler C, Wirleitner B, Schennach H, Fuchs D. Aspirin down-regulates tryptophan degradation in stimulated human peripheral blood mononuclear cells in vitro. *Clin. Exp. Immunol.* 2005; 140, 41–45.

140. Winkler C, Frick B, Schroecksnadel K, Schennach H, Fuchs D. Food preservatives sodium sulfite and sorbic acid suppress mitogen-stimulated peripheral blood mononuclear cells. *Food Chem. Toxicol.* 2006; 44, 2003–2007.

141. Cangemi R, Loffredo L, Carnevale R, Pignatelli P, Violi F. Statins enhance circulating vitamin E. *Int. J. Cardiol.* 2008; 123, 172–174.

142. Cangemi R, Loffredo L, Carnevale R, Perri L, Patrizi MP, Sanguigni V, Pignatelli P, Violi F. Early decrease of oxidative stress by atorvastatin in hypercholesterolaemic patients: Effect on circulating vitamin E. *Eur. Heart J.* 2008; 29, 54–62.

143. Violi F, Loffredo L, Musella L, Marcoccia A. Should antioxidant status be considered in interventional trials with antioxidants? *Heart* 2004; 90, 598–602.

144. Vardi M, Levy NS, Levy AP. Vitamin E in the prevention of cardiovascular disease: The importance of proper patient selection. *J. Lipid Res.* 2013; 54, 2307–2314.

145. Gostner J, Ciardi C, Becker K, Fuchs D, Sucher R. Immunoregulatory impact of food antioxidants. *Curr. Pharm. Des.* 2014; 20(6), 840–849.

146. Zaknun D, Schroecksnadel S, Kurz K, Fuchs D. Potential role of antioxidant food supplements, preservatives and colorants in the pathogenesis of allergy and asthma. *Int. Arch. Allergy. Immunol.* 2012; 157, 113–124.

147. Bjelakovic G, Nikolova D, Gluud C. Antioxidant supplements and mortality. *Curr. Opin. Clin. Nutr. Metab. Care* 2014; 17, 40–44.

148. Katagiri H, Yamada T, Oka Y. Adiposity and cardiovascular disorders: Disturbance of the regulatory system consisting of humoral and neuronal signals. *Circ. Res.* 2007; 101, 27–39.

149. Hotamisligil GS. Inflammation and metabolic disorders. *Nature* 2006; 444, 860–867.

150. Charriere G, Cousin B, Arnaud E, Andre M, Bacou F, Penicaud L, Casteilla L. Preadipocyte conversion to macrophage. Evidence of plasticity. *J. Biol. Chem.* 2003; 278, 9850–9855.

151. Hotamisligil GS, Erbay E. Nutrient sensing and inflammation in metabolic diseases. *Nat. Rev. Immunol.* 2008; 8, 923–934.

152. Medzhitov R. Origin and physiological roles of inflammation. *Nature* 2008; 454, 428–435.

153. Mozaffarian D, Willett WC. Trans fatty acids and cardiovascular risk: A unique cardiometabolic imprint? *Curr. Atheroscler. Rep.* 2007; 9, 486–493.

154. Sivasankaran S. The cardio-protective diet. *Ind. J. Med. Res.* 2010; 132, 608–616.

155. Maier E, Kurz K, Jenny M, Schennach H, Ueberall F, Fuchs D. Food preservatives sodium benzoate and propionic acid and colorant curcumin suppress Th1-type immune response in vitro. *Food Chem. Toxicol.* 2010; 48, 1950–1956.

156. Winkler C, Schroecksnadel K, Schennach H, Fuchs D. Vitamin C and E suppress mitogen-stimulated peripheral blood mononuclear cells in vitro. *Int. Arch. Allergy Immunol.* 2007; 142, 127–132.

157. Ciardi C, Jenny M, Tschoner A, Ueberall F, Patsch J, Pedrini M, Ebenbichler C, Fuchs D. Food additives such as sodium sulphite, sodium benzoate and curcumin inhibit leptin release in lipopolysaccharide-treated murine adipocytes in vitro. *Br. J. Nutr.* 2012; 107, 826–833.

158. Yang R, Barouch LA. Leptin signaling and obesity: Cardiovascular consequences. *Circ. Res.* 2007; 101, 545–559.

159. Stelzer I, Zelzer S, Raggam RB, Prüller F, Truschnig-Wilders M, Meinitzer A, Schnedl WJ et al. Link between leptin and interleukin-6 levels in the initial phase of obesity related inflammation. *Transl. Res.* 2012; 159, 118–124.

160. Mangge H, Summers KL, Meinitzer A, Zelzer S, Almer G, Prassl R, Schnedl WJ et al. Obesity-related dysregulation of the Tryptophan-Kynurenine metabolism: Role of age and parameters of the metabolic syndrome. *Obesity (Silver Spring)* 2014; 22(1), 195–201.
161. Kaliora AC, Dedoussis GV, Schmidt H. Dietary antioxidants in preventing atherogenesis. *Atherosclerosis* 2006; 187, 1–17.
162. Kaliora AC, Dedoussis GV. Natural antioxidant compounds in risk factors for CVD. *Pharmacol. Res.* 2007; 56, 99–109.
163. Sakamoto K, Kimura J. Mechanism of statin-induced rhabdomyolysis. *J. Pharmacol. Sci.* 2013; 123, 289–294.
164. Rhodes T, Norquist JM, Sisk CM, McQuarrie K, Trovato A, Liao J, Miller T, Maccubbin D, Watson DJ. The association of flushing bother, impact, treatment satisfaction and discontinuation of niacin therapy. *Int. J. Clin. Pract.* 2013; 67, 1238–1246.
165. Mangge H, Becker K, Fuchs D, Gostner JM. Antioxidants, inflammation and cardiovascular disease. *World J. Cardiol.* 2014; 6, 462–477.

23 Carotenoids, ROS, and Cardiovascular Health

Maria Alessandra Gammone

CONTENTS

ABSTRACT

Carotenoids are a class of natural pigments found principally in plants. They have potential antioxidant biological properties because of their chemical structure and interaction with biological membranes, which could be exploited as an inexpensive means of cardiovascular prevention. In fact, the resistance of low-density lipopolysaccharides to oxidation is increased by high dietary antioxidant intake. Further properties of carotenoids leading to a potential reduction of cardiovascular risk are represented by lowering of blood pressure, reduction of proinflammatory cytokines and markers of inflammation (such as C-reactive protein), and improvement of insulin sensitivity in muscle, liver, and adipose tissues. In addition, recent nutrigenomic studies have focused on the exceptional ability of carotenoids in modulating the expression of specific genes involved in cell metabolism. The aim of this chapter is to describe how carotenoids can improve cardiovascular health.

23.1 INTRODUCTION

Lipid peroxidation is a process usually generated physiologically in the body, mainly by the effect of several reactive oxygen species (ROS), and is crucial in the pathogenesis of human diseases. Even if the enzymatic and nonenzymatic natural antioxidant defenses are present, they may be overcome, and oxidative stress occurs through a self-propagating chain reaction, which can result in significant tissue damage. Both oxidative stress and chronic inflammation are the main pathophysiological contributors to diabetes and cardiovascular disease (CVD) development. In fact, CVDs start from a local redox unbalance to endothelial dysfunction, phlogoses, and inappropriate vascular remodeling, which slowly advances to atherosclerosis and CVD. Many antioxidant phytochemicals have been discovered, which show anti-inflammatory effects and hence can prevent chronic disease. They include carotenoids, ubiquitous in the vegetable kingdom, and are classified, according to their chemical structure, into carotenes (β-carotene and lycopene) and xanthophylls (lutein, fucoxanthin, zeaxanthin, β-cryptoxanthin, and astaxanthin) [1]. Their antioxidant properties represent the main mechanism for beneficial health effects. In this respect, they could be helpful in preventing a number of diseases induced by ROS.

23.2 ASTAXANTHIN

Astaxanthin is a red soluble xanthophyll, which is abundant in the marine biological species, including microalgae, plankton, krill, fish, and other seafood. The red coloration of salmon and crustaceans is due to the presence of astaxanthin [2]. Astaxanthin has considerable promising applications against oxidative stress–related diseases, such as cancers, chronic inflammatory diseases, metabolic syndrome, diabetes, liver diseases, gastrointestinal disorders, neurodegenerative diseases, and even CVD. Other metabolic syndromes such as atherosclerotic vascular damage can be averted by the use of one or more of these antioxidants. Due to its unique chemical structure, astaxanthin has a strong antioxidant activity and makes it a potent quencher of ROS and nitrogen oxygen species (NOS) [3]. This compound can capture intracellular free radicals and transfer them to the extracellular side, where they can be neutralized by water-soluble antioxidants such as vitamins C and E through a close synergy between hydrosoluble and liposoluble antioxidants [4].

Astaxanthin improves blood lipid profile by reducing low-density lipopolysaccharides (LDL) cholesterol and triglycerides (TG), increasing high-density lipoprotein (HDL) cholesterol, and decreasing markers of lipid peroxidation, inflammation, and thrombosis [5]. Consumption of astaxanthin from 1.8 to 21.6 mg/day for at least 2 weeks has been shown to ameliorate TG and HDL together with increased adiponectin in human, with a significant inhibition of LDL oxidation and a significant reduction of high-sensitive C-reactive proteins (hs-CRPs), which is considered an important indicator of CVD [6]. A decrease in DNA damage, measured using plasma 8-hydroxy-2′-deoxyguanosine (8-OH-dG), was also reported 4 weeks after the treatment [7]. Another potential cardiovascular benefit is its modulatory effect on nitric oxide (NO) and thus reduction in hypertension. Also, there might be a potential therapeutic role in the management of myocardial injury, oxidized LDL, rethrombosis after thrombolysis, and other cardiac diseases, such as atrial fibrillation [3].

23.3 FUCOXANTHIN

Fucoxanthin is an orange xanthophyll that is present in edible brown seaweeds, such as *Undaria pinnatifida*, *Laminaria japonica*, and *Sargassum fulvellum*. Its structure includes an unusual allenic bond, an epoxide group, and a conjugated carbonyl group in polyene chain with antioxidant properties [8]. Dietary fucoxanthin is hydrolyzed to fucoxanthinol in the gastrointestinal tract by digestive enzymes such as lipase and cholesterol esterase and then converted to amarouciaxanthin A in the liver [9]. Fucoxanthin acts on the reduction of major cardiovascular risk factors, such as obesity, diabetes, high blood pressure, chronic inflammation, and lowering of TG and LDL concentrations [10]. Adaptive thermogenesis by uncoupling protein-1 (UCP-1) could be a physiological defense against obesity. UCP-1 expression is the key of whole-body energy expenditure, and its dysfunction contributes to the development of obesity [11]. This protein, situated in the mitochondrial inner membrane, dissipates the pH gradient generated by oxidative phosphorylation, releasing chemical energy as heat. Fucoxanthin was found to induce both protein and mRNA expression of UCP-1 in white adipose tissue (WAT), leading to oxidation of fatty acids and heat production, with a significant reduction in body weight, fat, systolic/diastolic blood pressure, TG, CRP, glutamic pyruvic transaminase (GPT), glutamic oxaloacetic transaminase (GOT), γ-glutamyl transpeptidase (γ-GT) levels and a significant increase in resting energy expenditure (REE) measured by indirect calorimetry [12]. These findings give a clue for new dietary antiobesity therapies. A reduction in body weight and fat in obese individuals results in the downregulation of inflammatory markers and prevents metabolic syndrome. The potential antidiabetic effects of fucoxanthin are attributable to the ability of this compound to induce weight loss and WAT reduction.

The adipocyte has recently been recognized as an endocrine organ for its role in the secretion of biologically active mediators, such as chemokines, leptin, adiponectin, resistin, tumor necrosis factor alpha (TNF-α), and monocyte chemoattractant protein-1 (MCP-1). Certain adipokines are reported to alter insulin sensitivity and glucose and lipid metabolism in muscle, liver, and adipose tissues [13]. The chronic low-grade inflammation elicited by proinflammatory mediators in the WAT leads to insulin resistance and increased cardiovascular risk [14]. Fucoxanthinol also prevents inflammation and insulin resistance by inhibiting the production of NO and prostaglandin-2 (PGE-2) through the downregulation of inducible nitric oxide synthase (iNOS), which is NO-producing enzyme related to the pathogenesis of inflammation and overexpression of WAT in obese mice, and COX-2 mRNA expression and adipocytokine production in WAT. An interesting extrametabolic benefit of fucoxanthin administration in rodents was the induction of the synthesis of docosahexaenoic acid (DHA) in the liver, resulting in improvements in lipid profile [15]. Experiments on stroke-prone spontaneously hypertensive rats (SHRSP) show the possible protective role of fucoxanthin in CVD. Also, *U. pinnatifida* powder delayed the incidence of stroke signs and increased the life span of SHRSP rats [16]. Thus, fucoxanthin may have a potential role in modulation and prevention of human diseases, particularly in reducing the incidence of CVD.

23.4 LYCOPENE

Lycopene is the red pigment, which can be found in high concentrations in some fruits such as red grapefruits and watermelons and vegetables such as tomato. It is an efficient antioxidant, which was demonstrated to scavenge ROS, inhibit lipid peroxidation, and reinforce the immune system. In fact, it is a lipophilic molecule transported in blood by lipoproteins, which accumulates in human tissues including the vasculature. Low plasmatic levels of antioxidant vitamins A, E, β-carotene, and lycopene are associated with removing early carotid atherosclerotic lesions, in particular reduced carotid intima-media thickness and lower incidences of cardiovascular accidents, such as coronary heart disease and stroke [17]. The anti-atherogenic effect of lycopene is associated with anti-inflammatory activities, improved lipid homeostasis, antioxidation leading to the inhibition of LDL peroxidation, and vascular endothelium protection. In fact, lycopene decreases vascular oxidative stress and inflammation, decreases blood lipid biomarkers of oxidative stress in vivo, and attenuates adhesion molecule expression and interactions between monocytes and endothelial cells [18]. This anti-inflammatory effect was recognized by inhibiting interleukine-1 (IL-1) secretion, which is a key factor in inflammatory processes inducing the synthesis of other proinflammatory cytokines, adhesion molecules, chemotactic factors, and acute-phase proteins [19]. The production of phlogistic mediators, with the subsequent recruitment of leukocytes to the intima, is involved in the early formation of atherosclerotic lesions, leading to the chronic inflammatory process of atherosclerosis [20]. In addition, lycopene displayed positive effects on NO levels, contributing to vasodilatation, thus resulting more effective than by fluvastatin in reducing the cardiovascular risk [21]. For this reason, dietary intake of lycopene, especially if diet is also rich in extra-virgin olive oil, can reduce CVD risk [22]. However, many factors can affect lycopene bioavailability and absorption including season, dietary sources, food composition, and processing such as cooking that can transform all-*trans*-lycopene to *cis*-lycopene, which is better absorbed [23]. So, higher levels of lycopene were found when tomatoes were consumed cooked rather than raw. On the contrary, too much heat treatment (more than 2 h at 100°C) reduces its content and its beneficial effects [24].

23.5 LUTEIN

Lutein is a dietary xanthophyll found in the human retina in high concentration. It is an isomer of the carotenoid zeaxanthin, with identical chemical formulas. It can be found in several foods, such as egg yolk, orange juice, honeydew melon, and especially occurring in dark green vegetables. Lutein, which has been shown to prevent lipid peroxidation, is well known to be protective against age-related macular degeneration and senile cataract, whose major risk factor is oxidative stress [25]. In fact, lutein has a strong ROS-scavenging capacity, and it blocks the activation of the ubiquitous nuclear transcription factor NF-kB (nuclear factor kappa-light-chain-enhancer of activated B cells) and the degradation of the inhibitor kB (I-kB), thus decreasing inducible gene transcription and synthesis of inflammatory markers such as cytokines, iNOS, PGE-2, monocyte chemotactic protein-1 (MCP-1), and macrophage inflammatory protein-2 (MIP-2) and also reducing oxidative stress [26,27]. This antioxidant and anti-inflammatory capacity also promotes cardiovascular health and reduces CVD risk. Plasmatic lutein and oxidized LDL are inversely correlated (lutein has a strong ROS-scavenging ability so that its higher levels determine a reduction in LDL oxidation), suggesting its potent antioxidant and anti-inflammatory effects on aortic tissue, which may protect against the development of atherosclerosis [28]. In fact, serum levels of lutein were significantly lower in atherosclerosis than in controls, and it was inversely associated with carotid stiffness [29]. A beneficial cardiovascular effect of lutein was also linked to prevention of hypertension, and a higher concentration was generally inversely associated with an increased systolic blood pressure and incidental hypertension. Subjects with higher lutein levels show lower baseline blood pressure, generally with lower risk of future hypertension, independent of smoking status [30]. ROS are known to be mediators of myocardial ischemia/reperfusion damage (the restoration of blood flow to ischemic regions, with increased generation of ROS). Lutein can protect myocardium from ischemia/reperfusion injury by decreasing oxidative stress and myocyte apoptosis. Limiting myocardial injury may prevent contractile dysfunction, reducing cardiovascular morbidity and mortality [31].

23.6 ZEAXANTHIN

Zeaxanthin is an oxygenated non–provitamin A xanthophyll, whose major dietary sources include corn, eggs, orange juice, honeydew melon, and dark green leafy vegetables, especially kale, turnip greens, spinach, and broccoli. Increased intake of it from both food sources and supplements is positively correlated with increased macular pigment density, which is suggested to lower the risk of macular degeneration [32]. In addition to scavenge ROS, zeaxanthin may prevent oxidative damage to protein, lipid, and DNA by regulating other cellular antioxidant systems. The protective effects of zeaxanthin against protein oxidation, lipid peroxidation, and DNA damage are comparable to α-tocopherol. Supplementation with zeaxanthin or α-tocopherol decreases oxidized glutathione (GSH) and increases the intracellular reduced GSH levels and GSH/GSSG (glutathione dysulfide) ratio, especially in response to oxidative stress. Thus, zeaxanthin, one of the major intracellular antioxidants, acts directly or indirectly by regulating GSH synthesis and therefore increasing GSH levels. As a consequence, intracellular redox status on oxidative stress improves and the susceptibility to H_2O_2-induced cell death reduces. Moreover, it is inversely correlated with the right common carotid artery stiffness, pulse wave velocity, and elastic modulus. The Beijing atherosclerosis and the Los Angeles atherosclerosis studies also found the inverse association between plasma lutein and early atherosclerosis, and further studies confirmed that higher levels of zeaxanthin may be protectively beneficial to arterial health [33].

23.7 β-CRYPTOXANTHIN

β-Cryptoxanthin is a xanthophyll with provitamin A activity, whose main dietary sources are oranges, peach, tangerines, and tropical fruits such as papaya. An increase in the intake of β-cryptoxanthin results in reduced risk of developing inflammatory disorders such as polyarthritis and rheumatoid arthritis [34]. Circulating antioxidants, such as scavengers of ROS and inhibitors of oxidative damage and inflammation, might also have a role in the prevention of CVD. Epidemiologic studies displayed that CRP and oxidized LDL levels, which have been linked to the development of CVD, are inversely related to serum concentrations of circulating antioxidants, including β-cryptoxanthin [35]. A recent report found that obesity is negatively related to serum concentrations of β-cryptoxanthin and positively related to CRP resulting in reduced CVD risk [36].

23.8 β-CAROTENE

β-Carotene is the most widely studied carotenoids for its provitamin A activity and its abundance in fruits and vegetables, such as carrot, orange, kale, spinach, turnip greens, apricot, and tomato. It increases immunological functions by enhancing lymphocyte proliferation and possesses antioxidant capacity. The enrichment of LDL with β-carotene in vitro has been shown to reduce the susceptibility of LDL to oxidative modification [37]. Another interesting mechanism of its role in preventing CVD is the modulation of vascular NO bioavailability by employing its reducing activity. In fact, the earliest pathogenic events in atherosclerosis are represented by the overexpression of cell surface adhesion molecules, which causes the binding of normally nonthrombogenic circulating cells, such as monocytes, to the endothelium. The activation of NF-kB pathway subsequently triggers the upregulation of the expression of the vascular cell adhesion molecule-1 (VCAM-1), intercellular cell adhesion molecule-1 (ICAM-1), and E-selectin in response to various inflammatory cytokines [38]. NO, constitutively generated by endothelial cells, plays an important role in the maintenance of vascular homeostasis and in the proinflammatory response that characterizes the early stages of atherosclerosis by inhibiting the vascular inflammatory response by blocking NF-kB nuclear transfer. β-carotene affects NF-kB-dependent expression of adhesion molecule and protects NO bioavailability, thereby reducing TNF-α-induced nitro-oxidative stress. In vascular inflammation, the presence of high concentrations of β-carotene is associated with a significant increase in NO level and bioavailability, which lead to a downregulation of NF-kB-dependent adhesion molecules in endothelial cells [39]. The 9-*cis*-β-carotene isomer, present in the highest levels in the alga *Dunaliella bardawil*, showed positive results too in that a recent study demonstrated that combined treatment with the drug bezafibrate and *Dunaliella* powder enhanced the effect of the drug on HDL elevation in human apolipoprotein [40]. A 9-*cis*-β-carotene-rich diet can inhibit atherosclerosis by reducing non-HDL plasma cholesterol concentrations and by inhibiting fatty liver development and inflammation in a mouse model of atherosclerosis [41]. Similar to rexinoids, the 9-*cis* β-carotene-rich diet significantly reduced mRNA levels of CYP7a, the rate-limiting enzyme of bile acid synthesis, and, consequently, it may reduce cholesterol absorption in the intestine [42]. The expression of other genes, ABCG-1, 5, and 8, involved in cholesterol metabolism, was reduced, and these liver transporters play a role in excreting cholesterol and hence in atherogenesis.

β-Carotene can also control the body fat reserves [43]. In mature adipocytes, it is metabolized to retinoic acid (RA), which decreases the expression of peroxisome proliferator-activated receptor alpha and CCAAT/enhancer-binding protein. This protein is a key lipogenic transcription factor, thus reducing the lipid content of mature adipocytes. Hence, a diet rich in β-carotene and fat directs toward energy expenditure, but in the absence of β-carotene, adipocytes store energy as fat. In fact, the circulating β-carotene levels are inversely correlated with the risk of type 2 diabetes and obesity, which are important cardiovascular risk factors. However, these benefits are associated with dietary consumption and seem to disappear when β-carotene is administered as a pharmacological supplement, otherwise have harmful effects in some subpopulations. Administration of synthetic all-*trans*-β-carotene to smokers seems to decrease both lung cancer and CVD incidence [44]. More research will certainly provide valuable information on the trial effects and potential subsequent effects of intervention with these antioxidant vitamins.

23.9   CONCLUSIONS

The pathophysiology of many chronic and acute conditions, especially of CVD, is explained by inflammation and oxidative stress. Apart from sex, age, and genetic factors that cannot be modified, lifestyle and dietary intervention can be considered as important means of prevention and treatment of cardiovascular risk factors and disease. For this, supplementation of diet with high antioxidant compounds, such as polyphenols, vitamins, and carotenoids, should be beneficial. Evidence exists that carotenoids possess antioxidant properties due to their chemical structure and interaction with biological components, through their antioxidant and anti-inflammatory activities helping against cardiovascular risk factors such as inflammation, hyperlipidemia, hypertension, insulin resistance, and obesity. Consequent improvements in blood pressure baseline levels, reduction of inflammation, and correction of dyslipidemias can lead to an improvement of cardiovascular health. Further research in these directions could find preventive and therapeutic strategies in order to reduce the risk of developing CVD, with promising applications and little or no side effects.

REFERENCES

1. D'Orazio N, Gammone MA, Gemello E, DeGirolamo M, Cusenza S, Riccioni G. Marine bioactives: Pharmacological properties and potential applications against inflammatory diseases. *Mar. Drugs* 2012;10:812–833.
2. Schweigert F. *Metabolism of Carotenoids in Mammals*. Basel, Switzerland: Birkhauser Verlag; 1998.
3. Pashkow FJ, Watumull DG, Campbell CL. Astaxanthin: A novel potential treatment for oxidative stress and inflammation in cardiovascular disease. *Am. J. Cardiol.* 2008;101:58–68.
4. Nakano M, Onodera A, Saito E, Tanabe M, Yajima K, Takahashi J. Effect of astaxanthin in combination with α-tocopherol or ascorbic acid against oxidative damage in diabetic ODS rats. *J. Nutr. Sci. Vitaminol.* 2008;54:329–334.
5. Khan SK, Malinski T, Mason RP, Kubant R, Jacob RF, Fujioka K. Novel astaxanthin prodrug CDX-085 attenuates thrombosis in mouse model. *Thromb. Res.* 2010;126:299–305.
6. Yoshida H, Yanai H, Ito K, Tomono Y, Koikeda T, Tsukahara H. Administration of natural astaxanthin increases serum HDL-c and adiponectin in subjects with mild hyperlipidemia. *Atherosclerosis* 2010;209:520–523.
7. Park JS, Chyun JH, Kim YK, Line LL, Chew BP. Astaxanthin decreased oxidative stress and inflammation and enhanced immune response in humans. *Nutr. Metab.* 2010;7:18.
8. Hu T, Liu D, Chen Y, Wu J, Wang S. Antioxidant activity of sulfated polysaccharide fractions extracted from *Undaria pinnitafida in vitro. Int. J. Biol. Macromol.* 2010;46:193–198.
9. Asai A, Sugawara T, Ono H, Nagao A. Biotransformation of fucoxanthinol in amarouciaxanthin A in mice and Hep-G2 cells: Formation and cytotoxicity of fucoxanthin metabolites. *Drug Metab. Dispos.* 2004;32:205–211.
10. Asai A, Yonekura L, Nagao A. Low bioavailability of dietary epoxy-xanthophylls in humans. *Br. J. Nutr.* 2008;100:273–277.
11. Gammone MA, Gemello E, Riccioni G, D'Orazio N. Marine bioactives and potential applications in sport. *Mar. Drugs* 2014;12:2357–2382.
12. D'Orazio N, Gemello E, Gammone MA, DeGirolamo M, Ficoneri C, Riccioni G. Fucoxantin: A treasure from sea. *Mar. Drugs* 2012;10:604–616.
13. Shoelson SE, Lee J, Goldfine AB. Inflammation and insulin resistance. *J. Clin. Invest.* 2006;116:1793–1801.
14. Matsuzawa Y, Shimomura I, Kihara S, Funahashi T. Importance of adipokines in obesity-related diseases. *Horm. Res.* 2003;60:56–59.
15. Park HJ, Lee MK, Park YB, Shin YC, Choi MS. Beneficial effects of *Undaria pinnatifida* ethanol extract on diet-induced-insulin resistance in C57BL/6J mice. *Food Chem. Toxicol.* 2010;13:357–363.
16. Ikeda K, Kitamura A, Machida H, Watanabe M, Negishi H, Hiraoka J. Effect of *Undaria pinnatifida* on the development of cerebrovascular diseases in stroke prone spontaneously hypertensive rats. *Clin. Exp. Pharmacol. Physiol.* 2003;30:44–48.
17. Riccioni G, Bucciarelli T, D'Orazio N, Palumbo N, DiIlio E, Corradi F et al. Plasma antioxidants and asymptomatic carotid atherosclerotic disease. *Ann. Nutr. Metab.* 2008;53:86–90.
18. Martin KR, Wu D, Meydani M. The effect of carotenoids on the expression of cell surface adhesion molecules and binding of monocytes to human aortic endothelial cells. *Atherosclerosis* 2000;150:265–274.
19. Chi H, Messas E, Levine RA, Graves DT, Amar S. IL-1 receptor signaling mediates atherosclerosis associated with bacterial exposure and/or a high-fat diet in a murine apolipoprotein E heterozygote model: Pharmacotherapeutic implications. *Circulation* 2004;110:1678–1685.
20. Libby P. Inflammation in ATS. *Nature* 2002;420:868–874.

21. Hu MY, Li YL, Jiang CH, Liu ZQ, Qu SL, Huang YM. Comparison of lycopene and fluvastatin effects on atherosclerosis induced by a high-fat diet in rabbits. *Nutrition* 2008;24:1030–1038.

22. Kiran DK, Ahuja M, Pittaway JKB, Ball MJ. Effects of olive oil and tomato lycopene combination on serum lycopene, lipid profile, and lipid oxidation. *Nutrition* 2006;22:259–265.

23. Gutierrez RJM, deCastro LMD. Lycopene: The need for better methods for characterization and determination. *Trends Anal. Chem.* 2007;26:163–170.

24. Graziani G, Pernice R, Lanzuise S, Vitaglione P, Anese M, Fogliano V. Effect of peeling and heating on carotenoid content and antioxidant activity of tomato and tomato-virgin olive oil systems. *Eur. Food Res. Technol.* 2003;216:116–121.

25. Beebe DC, Holekamp NM, Shui YB. Oxidative damage and the prevention of age-related cataracts. *Ophthalmic Res.* 2010;44:155–165.

26. Jin XH, Ohgami K, Shiratori K, Suzuki Y, Hirano T, Koyama Y. Inhibitory effects of lutein on endo toxin-induced uveitis in Lewis rats. *Invest. Ophthalmol. Vis. Sci.* 2006;47:2562–2568.

27. Ashino T, Yamanaka R, Yamamoto M, Shimokawa H, Sekikawa K, Iwakura Y. Negative feedback regulation of LPS-induced inducible NOS gene expression by HMOX-1 induction in macrophages. *Mol. Immunol.* 2008;45:2106–2115.

28. Kim JE, Leite JO, DeOgburn R, Smyth JA, Clark RM, Fernandez ML. A lutein-enriched diet prevents cholesterol accumulation and decreases ox-LDL and inflammatory cytokines in the aorta of guinea pigs. *J. Nutr.* 2011;141:1458–1463.

29. Zou Z, Xu X, Huang Y, Xiao X, Ma L, Sun T. High serum level of lutein may be protective against early ATS: The Beijing atherosclerosis study. *Atherosclerosis* 2011;219:789–793.

30. Hozawa A, Jacobs JDR, Steffes MW, Gross MD, Steffen LM, Lee DH. Circulating carotenoid concentrations and incident hypertension: The coronary artery risk development in young adults (CARDIA) study. *J. Hypertens.* 2009;27:237–242.

31. Adluri RS, Thirunavukkarasu M, Zhan L, Maulik N, Svennevig K, Bagchi M et al. Cardio-protective efficacy of a novel antioxidant mix vitaepro against *ex vivo* myocardial ischemia-reperfusion injury. *Cell Biochem. Biophys.* 2011;30:9300–9307.

32. Mares-Perlman JA, Millen AE, Ficek TL, Hankinson SE. The body of evidence to support a protective role for lutein and zeaxanthin in delaying chronic disease. *J. Nutr.* 2002;132:518S–524S.

33. Zweier JL, Talukder MA. The role of oxidants and free radicals in reperfusion injury. *Cardiovasc. Res.* 2006;70:181–190.

34. Pattison DJ, Symmons DPM, Lunt M, Welch A, Bingham SA, Day NE. Dietary β-cryptoxanthin and inflammatory polyarthritis: Results from a population-based prospective study. *Am. J. Clin. Nutr.* 2005;82:451–455.

35. Kritchevsky SB, Bush AJ, Pahor M, Gross MD. Serum carotenoids and markers of inflammation in non-smokers. *Am. J. Epidemiol.* 2000;152:1065–1071.

36. Suzuki K, Ito Y, Ochiai J. Relationship between obesity and serum markers of oxidative stress and inflammation in Japanese. *Asian Pac. J. Cancer Prev.* 2003;4:259–266.

37. Jialal I, Norkus EP, Cristol L, Grundy SM. Beta-carotene inhibits the oxidative modification of low-density lipoprotein. *Biochem. Biophys. Acta* 1991;1086:134–138.

38. Robbins M, Topol EJ. Inflammation in acute coronary syndromes. *Cleve. Clin. J. Med.* 2002;69:130–142.

39. DiTomo P, Canali R, Ciavardelli D, DiSilvestre S, DeMarco A, Giardinelli A. β-carotene and lycopene affect endothelial response to TNF-a reducing nitro-oxidative stress and interaction with monocytes. *Mol. Nutr. Food Res.* 2011;55:1–11.

40. Shaish A, Harari A, Hananshvili L, Cohen H, Bitzur R, Luvish T. 9-cis-β-Carotene-rich powder of the alga *Dunaliella bardawil* increases plasma HDL in fibrate treated patients. *Atherosclerosis* 2006;189:215–221.

41. Harari A, Harats D, Marko D, Cohen H, Barshack I, Kamari Y. A 9-cis-β-Carotene–enriched diet inhibits atherogenesis and fatty liver formation in LDL-R knockout mice. *J. Nutr. Dis.* 2008;138:1923–1930.

42. Hubacek JA, Bobkova D. Role of cholesterol 7α-hydroxylase (CYP7A1) in nutrigenetics and pharmacogenetics of cholesterol lowering. *Mol. Diagn. Ther.* 2006;10:93–100.

43. Lobo GP, Amengual J, Li HNM, Golczak M, Bonet ML, Palczewski K. β-Carotene decreases PPR–α activity and reduces lipid storage capacity of adipocytes in a β-carotene oxygenase 1-dependent manner. *J. Biol. Chem.* 2010;285:27891–2799.

44. Omenn GS, Goodman G, Thornquist M, Cherniack MG, Cullen MR, Glass A. The β-carotene and retinol efficacy trial (CARET) for chemoprevention of lung cancer in high risk populations: Smokers and asbestos-exposed workers. *Cancer Res.* 1994;54(7), 2038s–2043s.

Section VII

Other Uncommon Diseases Induced by Oxidative Stress

24 Oxidative Stress and Sickle Cell Disease

Danilo Grünig Humberto da Silva,
Edis Belini Junior, Claudia Regina Bonini-Domingos,
and Eduardo Alves de Almeida

CONTENTS

ABSTRACT

Sickle cell disease (SCD) embraces a group of genetic hemolytic disorders associated with high morbidity and mortality. SCD is characterized by a complex pathophysiology initiated by hemoglobin S (HbS) polymerization that triggers a cascade of pathological events, including vaso-occlusion episodes, hemolysis, endothelial dysfunction, inflammation, hypercoagulability, reperfusion injury, and hypoxemia, leading to devastating clinical manifestations. Although SCD is one of the first disorders to be clearly defined at the molecular level, the genetic understanding of the basis for the disease expression variability is still not fully explained. In this intriguing scenario, oxidative stress plays a major role because it acts as both causing and being caused by SCD complications.

24.1 MORE THAN A HUNDRED-YEAR-OLD DISORDER

Sickle cell disease (SCD) is a multisystem disease; clinically, it is one of the most important hemoglobinopathies, associated with episodes of acute illness and progressive organ damage, and is one of the most common severe monogenic disorders worldwide (Weatherall et al. 2005). Knowledge of a disease heralded by painful episodes and leading to early death has existed in Africa for much more than a century (Stuart and Nagel 2004). However, in 1910, Herrick first described the characteristic sickle-shaped erythrocytes (Herrick [1910] 2001). Later, Pauling et al. (1949) showed the abnormal electrophoretic mobility of hemoglobin in an affected individual, identifying hemoglobin S (HbS) and defining SCD as the first "molecular disease." HbS and normal hemoglobin (HbA) were among the first proteins sequenced, and this analysis showed that the charge difference detected by Pauling and his collaborators was due to a mutation from glutamate to valine at the sixth site of the two β chains of hemoglobin (HBB[glu6val]) (Ingram 1957). Subsequent discoveries deciphered the structure of HbS, elucidating the molecular basis of its function (Perutz et al. 1960). These studies placed SCD at the leading edge of investigations to elucidate the molecular basis of human diseases (Frenette and Atweh 2007).

The term SCD embraces a group of genetic conditions in which pathology results from the inheritance of a single nucleotide substitution in *beta-globin* gene (*HBB*; c.20A>T; rs334) either homozygously or as a double heterozygote with another interacting gene (Weatherall et al. 2005, Steinberg and Sebastiani 2012, Serjeant 2013). Although more than 15 different genotypes have been identified as causing SCD, homozygosity for the sickle mutation (HbSS) is the most common and severe form of the disease and is often referred to as sickle cell anemia (SCA) (Rees et al. 2010). The other main types of SCD are hemoglobin SC (HbSC) disease and the various forms of HbS/β-thalassemia, being the spectrum of resulting conditions, therefore, influenced by the geography of individual hemoglobin genes and further by a multitude of genes other than the one directly involved (*HBB*S*) (Nagel 1991, Rees et al. 2010, Serjeant 2013). In this way, SCD is a strikingly variable condition. The incidence of most clinical complications varies markedly both with time in the same individual and between different individuals (Rees and Gibson 2012).

The pathophysiological hallmark of SCD is the intracellular polymerization of HbS upon deoxygenation (Bunn 1997). The elongated polymers damage red blood cells (RBCs) by increasing the cytoplasmic viscosity, inducing cation leaks through the membrane, which causes dehydration, and increasing the expression of adhesion molecules, allowing sickle cells to form homotypic aggregations and heterotypic adhesion to endothelial cells and white blood cells (WBCs) (Kaul et al. 1989, Frenette 2002). These changes lead to RBCs occluding and damaging blood vessels. This vaso-occlusion causes tissue ischemia, resulting in a further cascade of pathological events, including hemolysis, endothelial dysfunction, inflammation, hypercoagulability, reperfusion injury, and hypoxemia (Stuart and Nagel 2004, Rees et al. 2010). These complications have a cyclic nature in which a chronic and systemic oxidative stress acts as both causing and being caused by them. Thus, oxidative stress biomarkers are potentially useful both to identify patients at high risk of oxidative damage and to evaluate antioxidant therapies (Nur et al. 2011).

Although many general principles of molecular genetics and cell biology have been established in SCD, its phenotypic heterogeneity and variable clinical severity have not been fully explained (Sangokoya et al. 2010). Several genetic association studies have been done trying to link single nucleotide polymorphisms (SNPs) with particular complications of SCD (Fertrin and Costa 2010, Menzel et al. 2010, Sebastiani et al. 2010, Flanagan et al. 2011, Bhatnagar et al. 2013, Galarneau et al. 2013). However, studies investigating genetic markers directly involved in regulating oxidative stress and vice versa are scarce, taking into consideration the fact that SCD is characterized by a lifelong continuous oxidative stress (Jeney et al. 2002, Kato et al. 2009), which might be involved in transcriptional, translational, or posttranslational regulation of many physiologic pathways. Moreover, a long-term goal of association studies of interrelated genetic and oxidative markers is to identify genes and pathways that might be therapeutically manipulated in novel treatment approaches.

24.2   FROM AN AMINO ACID SUBSTITUTION TO CHRONIC AND SYSTEMIC OXIDATIVE CONSEQUENCES

24.2.1   SICKLE ERYTHROCYTES: AN IMMEASURABLE SOURCE OF REACTIVE OXYGEN SPECIES

The primary event responsible for all the complications of SCD is the polymerization of deoxygenated HbS due to the crucial role of the mutation to valine. On deoxygenation within the microcirculation, HbS molecules alter their configuration, and hydrophobic contacts are formed between valine of one HbS molecule and alanine, phenylalanine, and leucine from an adjacent HbS molecule (Wishner et al. 1975, Fronticelli and Gold 1976, Dykes et al. 1979, Carragher et al. 1988). This crystallization produces a polymer nucleus, which grows and fills the erythrocyte, disrupting its architecture and flexibility and promoting cellular dehydration, with physical and oxidative cellular stress (Brittenham et al. 1985). Nevertheless, HbS polymerization is reversible; fibers *melt* as oxygen is taken up by the HbS and the normal discoid shape returns (Stuart and Nagel 2004). This way, one of the important pro-oxidant sources in SCD is sickle erythrocytes. The unstable auto-oxidative HbS and increased metabolic turnover due to recurrent HbS polymerization and depolymerization cause increased reactive oxygen species (ROS) generation (Hebbel et al. 1988, Banerjee and Kuypers 2004) and increased lipid oxidation products when compared with HbA-containing erythrocytes (Rice-Evans et al. 1986, Hebbel et al. 1988, Sheng et al. 1998, Aslan et al. 2000). This increased and unremitting ROS generation results in excessive antioxidant consumption and thus antioxidant deficiency in sickle erythrocytes (Banerjee and Kuypers 2004, Reid et al. 2006, Chaves et al. 2008). Thus, the ability to deal with oxidant stress is compromised and challenged in SCD, resulting in an unbalanced redox state, an altered thiol redox metabolism, and protein and lipid damage, leading to premature aging of these cells.

Direct measures of ROS, such as superoxide ($O_2^{\cdot-}$), hydrogen peroxide (H_2O_2), and hydroxyl radicals ($^{\cdot}OH$), are too unstable to be clinically useful as biomarkers (Rees and Gibson 2012). However, indirect markers of oxidative damage to tissues and proteins are abundant. RBC glutathione (GSH) concentrations and glutamine:glutamate are perhaps the best-established markers (Rees and Gibson 2012). For instance, glutamine:glutamate ratios correlated inversely with tricuspid regurgitant jet (TRJ) velocities, a marker of pulmonary hypertension risk (Morris et al. 2008). Many other markers of oxidative stress have been studied, all broadly correlating with other indicators of severity without adding specific new information (Rees and Gibson 2012) and with contradictory results among some studies (Silva et al. 2013). Thus, it is worthy to seek new, more specific and sensitive oxidative stress markers related to RBC redox metabolism in SCD patients. In this way, Kuypers (2014) suggested the investigation of ergothioneine (ergo), another RBC antioxidant thiol, that is the second most abundant one. However, the reasons for this abundance in hematopoietic tissues have not been investigated. Interestingly, HbS, which increases Hb pro-oxidant activity, selectively depletes ergo over GSH, suggesting that ergo has a specialized GSH-independent function in protecting RBCs against pro-oxidative hemoglobin. The implications of the presence of HbS in erythrocyte redox metabolism and the related markers studied have been the subject of recent reviews (Silva et al. 2013, Kuypers 2014).

24.2.2 VASO-OCCLUSIVE EVENTS AND HEMOLYSIS: OVERWHELMING AND CHRONIC OXIDATIVE STRESS SOURCES

The formation of HbS polymers negatively affects the RBC ability to maintain its normal morphology, and it has long been considered that the radical shape change in sickle erythrocytes under low oxygen leads to their inability to properly deform and pass through the microvasculature (Knee et al. 2007, Vekilov 2007). Moreover, HbS polymers cause damage to the RBC membrane. In addition, the mutated globin can undergo autoxidation and precipitate on the inner surface of the RBC membrane, causing membrane damage via iron-mediated generation of ROS (Browne et al. 1998). Among many changes that result from the damage to the sickled RBC membrane is their propensity to adhere (Hebbel 2008, Kaul et al. 2009), leading to vaso-occlusive events (VOE) with ischemia–reperfusion injury. The vaso-occlusive process is now believed to comprise a multistep process involving interactions among sickled RBC, activated leukocytes, endothelial cells, platelets, and plasma proteins. Recurrent VOE, processes of ischemia–reperfusion and consequent vascular endothelial cell activation and injury (Chiang and Frenette 2005), induces a continuous inflammatory response in the SCD individuals that is propagated by elevated levels of inflammatory cytokines, a decreased bioavailability of nitric oxide (NO), and oxidative stress (Conran et al. 2009). However, research has revealed that the VOE in SCD patients is far more complex, and the exact mechanisms remain incompletely elucidated (Manwani and Frenette 2013).

The VOE and hemolysis are the two major SCD pathophysiological processes (Rees et al. 2010), and the latter may be considered the major generating source of ROS. In the RBC, oxidative reactions are inhibited by an extensive antioxidant system. For extracellular Hb, the oxidative reactions are not completely neutralized by the available antioxidant system (Rifkind et al. 2014). Under normal conditions, this potential source of oxidative stress is minimized by haptoglobin (Hp) (Kato 2009). Hp avidly binds to αβ dimers of Hb and forms a highly stable Hb–Hp complex that quickly escorts it to the CD163 protein, in which the Hp–Hb complex is bound and cleared from plasma, depleting plasma Hp in the process (Schaer et al. 2007). However, in SCD patients, plasma Hp levels are low due to chronic hemolysis (Muller-Eberhard et al. 1968) that overwhelms endogenous plasma Hp and other scavenging mechanisms. Thus, the impairment of these adaptive antioxidant and anti-inflammatory mechanisms causes pathological effects involving oxidative reactions of extracellular Hb.

Hemolysis increases the concentration of free plasma Hb, which in the ferrous (Fe^{2+}) valence state is readily available to participate in Fenton-based redox reactions (Belcher et al. 2010). In addition, the increase in plasma Hb contributes to the increased ability of both deoxy- and oxy-HbS to participate in scavenging reactions with NO (Belcher et al. 2010). The binding of NO to deoxy-Hb results in the formation of a stable Fe^{2+}Hb–NO complex, while the reaction between NO and oxy-Hb can form methemoglobin (metHb) and nitrate (Aslan and Freeman 2007). The high propensity for these reactions leads to the decrease in NO bioavailability, becoming problematic in SCD because it has been reported that these patients have high levels of cell-free Hb and metHb (Morris et al. 2000, Reiter and Gladwin 2003, Jeffers et al. 2006). NO is normally produced by the endothelium, regulates basal vasodilator tone, and inhibits platelet and hemostatic activation and transcriptional expression of nuclear factor κB (NF-κB)-dependent adhesion molecules, such as vascular cell adhesion molecule-1, intercellular cell adhesion molecule-1, and the selectins (Palmer et al. 1988, De Caterina et al. 1995). Thus, the effect of NO scavenging via cell-free Hb has a role in increasing vasoconstriction and decreased oxygen delivery through the abatement of NO-dependent vasodilation (Chen et al. 2008) and also contributes to the hypercoagulable state of SCD (Ataga et al. 2008).

Moreover, free plasma Hb generates ROS, such as the $HO^\bullet$ and $O_2^{\bullet-}$ (Repka and Hebbel 1991), which is a potent scavenger of NO (Reiter et al. 2002). The interaction between $O_2^{\bullet-}$ and NO is more detrimental than their individual actions. While under normal physiological conditions $O_2^{\bullet-}$ can easily be dismutated by superoxide dismutase (SOD), the overproduction of $O_2^{\bullet-}$ seen in SCD overwhelms the defenses of body and reacts at diffusion-limited rates with NO. The reaction of NO and $O_2^{\bullet-}$ results in the generation of peroxynitrite ($ONOO^-$): a powerful and highly reactive oxidant. The generation of $ONOO^-$ is favored over spontaneous $O_2^{\bullet-}$ dismutation and NO autoxidation; the interaction is faster than both reaction of NO with Hb and reaction of superoxide with SOD (Aslan et al. 2000). $ONOO^-$ can form $HO^\bullet$ and nitrogen dioxide (NO_2), two other potent oxidants (Aslan et al. 2000). Therefore, the reaction between NO and $O_2^{\bullet-}$ has two main negative consequences because it not only decreases the concentration of NO but also produces more reactive free radicals (Chirico and Pialoux 2012).

The potential oxidative stress resulting from extracellular Hb is further exacerbated by a dramatic increase in rates of autoxidation for partially oxygenated Hb formed in the microcirculation and by an increase in the rate of autoxidation of Hb dimers formed when the Hb tetramer dissociates into dimers at the reduced Hb concentration in plasma (Zhang et al. 1991). The lower molecular weight of Hb dimers also facilitates the translocation of Hb from the circulation to the vasculature and other tissues sensitive to Hb oxidative reactions (Kato 2009). Moreover, with extracellular Hb, the body not only needs to deal with the autoxidation reaction and the resultant formation of $O_2{}^{\cdot-}$ and H_2O_2 but also needs to consider the secondary oxidative reactions involving reactions of H_2O_2 with Hb, as Rifkind et al. (2014) have brightly revised. In summary, it has been shown that a direct reaction of H_2O_2 with the iron of Fe(II)-Hb can undergo a Fenton reaction producing the highly reactive hydroxyl radical (Sadrzadeh et al. 1984), which was previously evidenced in sickle erythrocytes (Hebbel 1985). In addition, two-electron oxidation of Fe(II)-Hb by H_2O_2 produces the Fe(IV)-ferrylHb that may be a source for oxidative damage and toxic effects, but such deleterious effects have not been demonstrated (Rifkind et al. 2014). The increased formation of $O_2{}^{\cdot-}$ and H_2O_2 due to autoxidation of extracellular Hb also results in the oxidation of the functional ferrous Hb to Fe(III)-metHb. Although high concentrations of metHb are associated with renal dysfunction, according to high Hb levels in the urine (hemoglobinuria) (Tracz et al. 2007), direct pathological effects of metHb have not been established (Rifkind et al. 2014).

Regardless of metHb direct deleterious effects, along the hemolytic process, the heme dissociates from metHb due to its lower affinity for metHb than ferrous Hb (Bunn and Jandl 1968). Heme is a low–molecular weight hydrophobic molecule that can be taken up by cell membranes, plasma proteins, and lipids. In the same manner as Hp–Hb-scavenging system, plasma hemopexin (Hpx) sequesters heme in an inert, nontoxic form and transports it to the liver for catabolism and excretion, preventing heme's pro-oxidant and proinflammatory effects (Kato 2009, Schaer et al. 2014). However, once more the characteristic hemolytic condition of SCD patients overwhelms endogenous scavenging capacity of plasma Hpx (Muller-Eberhard et al. 1968), enabling further deleterious effects of heme and by-products. For instance, heme reaction with low-density lipoproteins (LDL) has been reported to produce the more toxic oxidized LDL (Balla et al. 1991). Furthermore, when heme associates with plasma proteins like albumin, it is also translocated to various tissues (Schaer et al. 2013), where it can further generate toxic effects. Heme can also bind to certain receptors, transcription factors, and enzymes. These interactions can alter cellular function, metabolism, and gene transcription. The altered gene transcription is the basis for a heme-induced proinflammatory effect (Wagener et al. 2001).

In addition, hemolysis further impairs NO bioavailability through the release of erythrocyte arginase-1 into plasma, which competes with endothelial NO synthase (eNOS) for the substrate L-arginine, metabolizing it into ornithine (Krajewski et al. 2008). eNOS is an enzyme consisting of a reductase domain and an oxygenase domain that produces NO. Normally, through the catalytic action of eNOS, tetrahydrobiopterin (BH_4) transfers an electron in the oxygenase domain, converting L-arginine into NO and L-citrulline. However, under certain conditions, eNOS can produce $O_2{}^{\cdot-}$ rather than NO. The two main cofactors of this mechanism, L-arginine and BH_4, are reduced due to previous repercussions of SCD, therefore, compounding an additional generating source of $O_2{}^{\cdot-}$ (Morris et al. 2005, Wood et al. 2006) and thus increasing the hyperoxidative status of SCD patients.

Hemolysis along with VOE are also associated with the release of microparticles (MPs) into the bloodstream (Allan et al. 1982). MPs are submicrometer membrane vesicles or "membrane dust" shed by compromised cells including RBCs (Piccin et al. 2007). The MPs display cell surface proteins that indicate their cellular origin. Moreover, MPs may also present other markers on their surface, for example, markers of cellular activation (Elsayh et al. 2013). Most MPs are highly procoagulant, expressing annexin V binding sites, and are capable of interacting with other cells to participate in various physiologic and pathologic processes, especially thrombosis and inflammation (Freyssinet 2003, Horstman et al. 2004). Oxidative stress is one of the risk factors involved in increasing the MPs' formation (Wang et al. 2007). At the same time, oxidative stress is generated by MPs (Helal et al. 2011). In conclusion, it is the eminent cyclic and interrelated nature of the SCD pathophysiological processes, where oxidative stress should be looked as a key element (Figure 24.1).

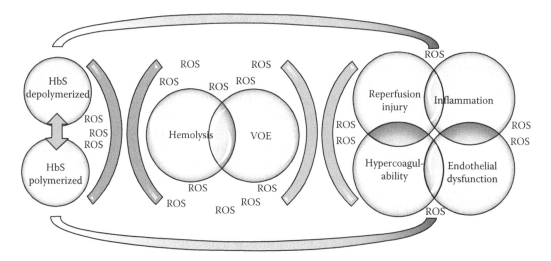

FIGURE 24.1 The cyclic and interrelated nature of pathological events in SCD: From the pathophysiological hallmark of SCD, intracellular polymerization of HbS that can result in hemolysis and VOE, until further cascade of complications, it is striking the number of ROS-generating processes. These processes are responsible for a lifelong continuous oxidative stress that should be looked as a major element in the SCD pathophysiology. HbS, hemoglobin S; ROS, reactive oxygen species; VOE, vaso-occlusive events.

24.3 ONE SINGLE-POINT MUTATION BUT DIFFERENT GENETIC PATHWAYS INVOLVED IN OXIDATIVE STRESS REGULATION

24.3.1 KEAP1-CUL3/NRF2 PATHWAY: A PROMISING TARGET FOR SCD PHENOTYPIC MODULATION

The molecular basis for HbS formation is known, but only the mutation is not sufficient to explain the heterogeneous phenotype and variable clinical severity found in SCD patients (Belini et al. 2015). Thus, several studies have been proposed for identifying genetic modifiers of SCD phenotypes that might lead to more precise prognostic tests and to the development of more specific and effective therapies. In this way, the discovery and characterization of genetic variation by emblematic projects such as the HapMap Project and the 1000 Genomes Project led to new insights into the selective and recombination forces that shape our genome (Abecasis et al. 2010, Altshuler et al. 2010). Together, advances in the development of high-throughput DNA genotyping and sequencing technologies and the knowledge of the genetic variation in our genome allowed the discovery of several DNA sequence variants associated with common human diseases and complex traits by genome-wide association studies (GWAS) (Manolio et al. 2009, Lettre 2012).

For SCD, fetal hemoglobin (HbF) and the alpha thalassemia co-inheritance are the two most thoroughly studied genetic modifiers of SCD pathophysiology (Steinberg and Adewoye 2006, Steinberg 2009). Among SCD patients, HbF concentrations vary from 0.1% to 30%, and high HbF concentrations, in general, reduce the severity of disease because HbF dilutes the amount of HbS, but importantly, by failing to be incorporated into the HbS polymer, HbF disrupts HbS polymerization (Steinberg 2005). In addition, some studies have shown that increased HbF decreases oxidative stress markers (e.g., lipid peroxidation levels) and increases the antioxidant capacity in both transgenic knockout sickle mice and SCD patients (Dasgupta et al. 2010, Silva et al. 2011, Belini et al. 2012, Torres et al. 2012). About 30% of SCD patients have coincidental alpha thalassemia, and these carriers have less hemolysis, higher hematocrit, lower mean corpuscular volume (MCV), and lower reticulocyte counts (Embury et al. 1982, Higgs et al. 1982, Higgs 2013). By reducing the mean cellular HbS concentration and erythrocyte density, alpha thalassemia increases the sickle erythrocyte life span and, consequently, the hematocrit (Steinberg et al. 1984, Thomas et al. 1997).

However, genetic modifiers directly involved in regulating oxidative stress and vice versa are scarce, as well as association studies of interrelated genetic and oxidative markers, as mentioned previously. In this sense, the nuclear factor-erythroid 2-related factor 2 (NRF2) has been shown to be an important master regulator of genes related with detoxification and antioxidant responses, anti-inflammatory functions, and cytoprotective functions in several diseases such as multiple sclerosis (Bomprezzi 2015), cancer (Moon and Giaccia 2015), epilepsy (Lim et al. 2014), cardiovascular disease (Cominacini et al. 2015), diabetes (Jiménez-Osorio et al. 2015), and other diseases.

Under conditions of oxidative stress, NRF2 is not degraded by their oppressors and their levels increase in the cytoplasm, and, thus, it translocates into and binds to the antioxidant response element (ARE) present on the promoter of several genes (Ma 2013). The binding of NRF2 to ARE leads to transcription of about 23 target genes involved in cellular defense against oxidative damage, for example, genes responsible for the synthesis of SOD, catalase (CAT), glutathione peroxidase 1 (Gpx1), phase II detoxification enzymes such as heme oxygenase 1 (HO-1), glutamate cysteine ligase catalytic and modifier subunits (GCLC and GCLM, respectively), glutathione S-transferase (GST), and NADPH quinone oxidoreductase 1 (NQO1) (Owusu-Ansah et al. 2015, Pall and Levine 2015). On the other hand, under unstressed conditions, NRF2 forms a complex with Kelch-like ECH-associated protein 1 (Keap1), leading to polyubiquitination and subsequent degradation by 26S proteasome in the cytoplasm and blocking the expression of antioxidant genes (Magesh et al. 2012). Keap1 serves as a negative regulator of NFR2, and it has five functional domains. The bric-a-brac–tramtrack–broad complex (BTB) domain is responsible for NRF2 ubiquitination through binding to cullin E3 ubiquitin ligase (Cul3)-based ubiquitin E3 ligase (Andérica-Romero et al. 2013, No et al. 2014). Therefore, Keap1-Cul3/NRF2 is a major cellular pathway that protects normal cells against oxidative and xenobiotic damage (Kansanen et al. 2013) (Figure 24.2).

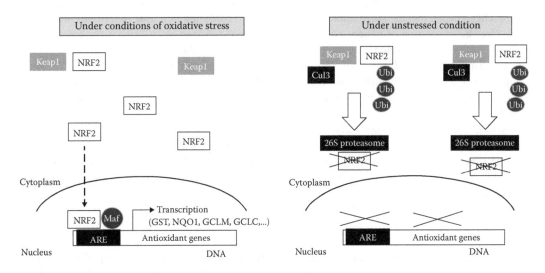

FIGURE 24.2 The Keap1-Cul3/NRF2 cellular pathway: Under conditions of oxidative stress, NRF2 dissociates from Keap1 and is protected from degradation. NRF2 translocates into the nucleus, where it will heterodimerize with MAF proteins and then binds to the antioxidant response element (ARE), which is located on the promoter of several antioxidant genes. Under unstressed condition, NRF2 forms a complex with Keap1, Cul3, and other proteins. The NRF2 suffers the polyubiquitination and subsequent degradation by the 26S proteasome. NRF2, nuclear factor-erythroid 2-related factor 2; Keap1, Kelch-like ECH-associated protein 1; ARE, antioxidant response element; Maf, musculoaponeurotic fibrosarcoma; Cul3, cullin E3 ubiquitin ligase; Ubi, ubiquitin.

The relationship between the NRF2 pathway and cellular defense mechanism of the erythrocyte came from studies involving NRF2-deficient mice, and these mice developed hemolytic anemia, increased sensitivity to oxidative stress, decreased GSH levels, and decreased expression levels of NRF2-dependent genes (Neumann et al. 2003, Lee et al. 2004). In SCD, the balance between oxidative stress and antioxidant capacity of the erythrocyte is significantly altered and the basis for such reduced capacity to defend against oxidative stress is currently unknown, and this reduced antioxidant capacity makes SCD erythrocytes susceptible to oxidative insult and hemolysis (Tatum and Chow 1996, Amer et al. 2006, Silva et al. 2013). In this way, the study of NRF2 pathway can bring useful information for understanding the role of various antioxidant genes in SCD.

The importance of NRF2 pathway in SCD was highlighted by the results from a study in which the reduced NRF2 levels were associated with increased miR-144 in HbSS reticulocytes and with decreased GSH regeneration and attenuated antioxidant capacity in HbSS erythrocytes, thereby providing a possible mechanism for the reduced oxidative stress tolerance and increased anemia severity in SCD patients (Sangokoya et al. 2010). Through this observation, other study examined whether inhibiting the 26S proteasome using the proteasome inhibitory drugs could promote the nuclear translocation of NRF2 and thereby could induce an antioxidant response. The results showed that proteasome inhibitors induced the ROS production and thereby increased NRF2-dependent antioxidant enzyme transcripts, elevated cellular GSH levels in the K562 cell line and in erythroid burst forming units (BFU-E) generated from peripheral blood mononuclear cells of SCD patients (Pullarkat et al. 2014).

Given the studies on NRF-2, it has been suggested that the use of therapies targeting the NRF2 pathway in SCD and the use of triterpenoids would have beneficial effects against oxidative stress (Owusu-Ansah et al. 2015). In many disease states in which oxidative and/or inflammatory stress has a crucial role in pathogenesis, the NRF2 is able to suppress stress and thus prevent disease complications, besides the ability of pharmacological agents to activate NRF2, which is an essential component of their desirable actions for oxidative and/or inflammatory states (Sporn and Liby 2012). These drugs include sulforaphane (Kensler et al. 2012) and curcumin (Shureiqi and Baron 2011) (for the prevention of cancer), dimethyl fumarate (Linker et al. 2011) (for the treatment of multiple sclerosis), bardoxolone methyl (Pergola et al. 2011) (for the treatment of diabetic nephropathy), and resveratrol (Palsamy and Subramanian 2011) (for multiple indications). In addition to synthetic drugs, there are health-promoting factors that have been shown to act, at least in part, by increasing the NRF2 such as phenolics (tocopherols/tocotrienols), carotenoids, fish oil, and exercise (Pall and Levine 2015).

Possible therapeutic strategies that can be used for SCD are the Keap1-Cul3/NRF2 pathway, but it is not yet known how these elements and their target genes are expressed and regulated in SCD that is considered a chronic oxidative stress condition. However, additional studies are needed to better understand this pathway in order to develop more efficient therapies against oxidative stress in SCD.

24.4 CONCLUSION

It is clear that SCD presents a clinical course highly variable due to its complex pathophysiology that can be affected by a number of modifying factors including genetic and biochemical ones. In this way, the understanding of SCD pathophysiology is gradually rising. Among the new evidences, oxidative stress processes have been increasingly related to this complexity. However, studies investigating molecular strategies directly involved in regulating oxidative stress remain scarce, as well as studies seeking more specific and sensitive oxidative stress markers related to sickle erythrocyte redox metabolism. According to the overview summary, oxidative stress should be looked at as a key element in further investigations of SCD.

REFERENCES

Abecasis, G. R., D. Altshuler, A. Auton et al. 2010. A map of human genome variation from population-scale sequencing. *Nature* 467(7319):1061–1073.

Allan, D., A. R. Limbrick, P. Thomas et al. 1982. Release of spectrin-free spicules on reoxygenation of sickled erythrocytes. *Nature* 295(5850):612–613.

Altshuler, D. M., R. A. Gibbs, L. Peltonen et al. 2010. Integrating common and rare genetic variation in diverse human populations. *Nature* 467(7311):52–58.

Amer, J., H. Ghoti, E. Rachmilewitz et al. 2006. Red blood cells, platelets and polymorphonuclear neutrophils of patients with sickle cell disease exhibit oxidative stress that can be ameliorated by antioxidants. *Br J Haematol* 132(1):108–113.

Andérica-Romero, A. C., I. G. González-Herrera, A. Santamaria et al. 2013. Cullin 3 as a novel target in diverse pathologies. *Redox Biol* 1(1):366–372.

Aslan, M. and B. A. Freeman. 2007. Redox-dependent impairment of vascular function in sickle cell disease. *Free Radic Biol Med* 43(11):1469–1483.

Aslan, M., D. Thornley-Brown, and B. A. Freeman. 2000. Reactive species in sickle cell disease. *Ann NY Acad Sci* 899:375–391.

Ataga, K. I., C. G. Moore, C. A. Hillery et al. 2008. Coagulation activation and inflammation in sickle cell disease-associated pulmonary hypertension. *Haematologica* 93(1):20–26.

Balla, G., H. S. Jacob, J. W. Eaton et al. 1991. Hemin: A possible physiological mediator of low density lipoprotein oxidation and endothelial injury. *Arterioscler Thromb* 11(6):1700–1711.

Banerjee, T. and F. A. Kuypers. 2004. Reactive oxygen species and phosphatidylserine externalization in murine sickle red cells. *Br J Haematol* 124(3):391–402.

Belcher, J. D., J. D. Beckman, G. Balla et al. 2010. Heme degradation and vascular injury. *Antioxid Redox Signal* 12(2):233–248.

Belini Jr., E., D. G. H. Da Silva, L. De Souza Torres et al. 2012. Oxidative stress and antioxidant capacity in sickle cell anaemia patients receiving different treatments and medications for different periods of time. *Ann Hematol* 91(4):479–489.

Belini Jr., E., D. G. Silva, S. Torres Lde et al. 2015. Severity of Brazilian sickle cell disease patients: Severity scores and feasibility of the Bayesian network model use. *Blood Cells Mol Dis* 54(4):321–327.

Bhatnagar, P., E. Barron-Casella, C. J. Bean et al. 2013. Genome-wide meta-analysis of systolic blood pressure in children with sickle cell disease. *PLoS One* 8(9):e74193.

Bomprezzi, R. 2015. Dimethyl fumarate in the treatment of relapsing-remitting multiple sclerosis: An overview. *Ther Adv Neurol Disord* 8(1):20–30.

Brittenham, G. M., A. N. Schechter, and C. T. Noguchi. 1985. Hemoglobin S polymerization: Primary determinant of the hemolytic and clinical severity of the sickling syndromes. *Blood* 65(1):183–189.

Browne, P., O. Shalev, and R. P. Hebbel. 1998. The molecular pathobiology of cell membrane iron: The sickle red cell as a model. *Free Radic Biol Med* 24(6):1040–1048.

Bunn, H. F. 1997. Pathogenesis and treatment of sickle cell disease. *N Engl J Med* 337(11):762–729.

Bunn, H. F. and J. H. Jandl. 1968. Exchange of heme among hemoglobins and between hemoglobin and albumin. *J Biol Chem* 243(3):465–475.

Carragher, B., D. A. Bluemke, B. Gabriel et al. 1988. Structural analysis of polymers of sickle cell hemoglobin. I. Sickle hemoglobin fibers. *J Mol Biol* 199(2):315–331.

Chaves, M. A., M. S. Leonart, and A. J. do Nascimento. 2008. Oxidative process in erythrocytes of individuals with hemoglobin S. *Hematology* 13(3):187–192.

Chen, K., B. Piknova, R. N. Pittman et al. 2008. Nitric oxide from nitrite reduction by hemoglobin in the plasma and erythrocytes. *Nitric Oxide* 18(1):47–60.

Chiang, E. Y. and P. S. Frenette. 2005. Sickle cell vaso-occlusion. *Hematol Oncol Clin North Am* 19(5):771–784.

Chirico, E. N. and V. Pialoux. 2012. Role of oxidative stress in the pathogenesis of sickle cell disease. *IUBMB Life* 64(1):72–80.

Comincini, L., C. Mozzini, U. Garbin et al. 2015. Endoplasmic reticulum stress and NRF2 signaling in cardiovascular diseases. *Free Radic Biol Med.* 88(Pt B):233–242. doi: 10.1016/j.freeradbiomed.2015.05.027.

Conran, N., C. F. Franco-Penteado, and F. F. Costa. 2009. Newer aspects of the pathophysiology of sickle cell disease vaso-occlusion. *Hemoglobin* 33(1):1–16.

Dasgupta, T., M. Fabry, and D. Kaul. 2010. Antisickling property of fetal hemoglobin enhances nitric oxide bioavailability and ameliorates organ oxidative stress in transgenic-knockout sickle mice. *Am J Physiol Regul Integr Comp Physiol* 298(2):402.

De Caterina, R., P. Libby, H. B. Peng et al. 1995. Nitric oxide decreases cytokine-induced endothelial activation. Nitric oxide selectively reduces endothelial expression of adhesion molecules and proinflammatory cytokines. *J Clin Invest* 96(1):60–68.

Dykes, G. W., R. H. Crepeau, and S. J. Edelstein. 1979. Three-dimensional reconstruction of the 14-filament fibers of hemoglobin S. *J Mol Biol* 130(4):451–472.

Elsayh, K. I., A. M. Zahran, T. B. El-Abaseri et al. 2013. Hypoxia biomarker, oxidative stress, and circulating microparticles in pediatric patients with thalassemia in upper Egypt. *Clin Appl Thromb Hemost* 20(5):536–545. doi: 10.1177/1076029612472552.

Embury, S. H., A. M. Dozy, J. Miller et al 1982. Concurrent sickle-cell anemia and alpha-thalassemia: Effect on severity of anemia. *N Engl J Med* 306(5):270–274.

Fertrin, K. Y. and F. F. Costa. 2010. Genomic polymorphisms in sickle cell disease: Implications for clinical diversity and treatment. *Expert Rev Hematol* 3(4):443–458.

Flanagan, J. M., D. M. Frohlich, T. A. Howard et al. 2011. Genetic predictors for stroke in children with sickle cell anemia. *Blood* 117(24):6681–6684.

Frenette, P. S. 2002. Sickle cell vaso-occlusion: Multistep and multicellular paradigm. *Curr Opin Hematol* 9(2):101–106.

Frenette, P. S. and G. F. Atweh. 2007. Sickle cell disease: Old discoveries, new concepts, and future promise. *J Clin Invest* 117(4):850–858.

Freyssinet, J. M. 2003. Cellular microparticles: What are they bad or good for? *J Thromb Haemost* 1(7):1655–1662.

Fronticelli, C. and R. Gold. 1976. Conformational relevance of the beta-6-Glu replaced by Val mutation in the beta subunits and in the beta(1-55) and beta(1-30) peptides of hemoglobin S. *J Biol Chem* 251(16):4968–4972.

Galarneau, G., S. Coady, M. E. Garrett et al. 2013. Gene-centric association study of acute chest syndrome and painful crisis in sickle cell disease patients. *Blood* 122(3):434–442.

Hebbel, R. P. 1985. Auto-oxidation and a membrane-associated "Fenton reagent": A possible explanation for development of membrane lesions in sickle erythrocytes. *Clin Haematol* 14(1):129–140.

Hebbel, R. P. 2008. Adhesion of sickle red cells to endothelium: Myths and future directions. *Transfus Clin Biol* 15(1–2):14–18.

Hebbel, R. P., W. T. Morgan, J. W. Eaton, and B. E. Hedlund. 1988. Accelerated autoxidation and heme loss due to instability of sickle hemoglobin. *Proc Natl Acad Sci USA* 85(1):237–241.

Helal, O., C. Defoort, S. Robert et al. 2011. Increased levels of microparticles originating from endothelial cells, platelets and erythrocytes in subjects with metabolic syndrome: Relationship with oxidative stress. *Nutr Metab Cardiovasc Dis* 21(9):665–671.

Herrick, J. B. 2001. Peculiar elongated and sickle-shaped red blood corpuscles in a case of severe anemia. *Yale J Biol Med* 74:179–184. First published in 1910.

Higgs, D. R. 2013. The molecular basis of alpha-thalassemia. *Cold Spring Harb Perspect Med* 3(1):a011718.

Higgs, D. R., B. E. Aldridge, J. Lamb et al. 1982. The interaction of alpha-thalassemia and homozygous sickle-cell disease. *N Engl J Med* 306(24):1441–1446.

Horstman, L. L., W. Jy, J. J. Jimenez, and Y. S. Ahn. 2004. Endothelial microparticles as markers of endothelial dysfunction. *Front Biosci* 9:1118–1135.

Ingram, V. M. 1957. Gene mutations in human haemoglobin: The chemical difference between normal and sickle cell haemoglobin. *Nature* 180(4581):326–328.

Jeffers, A., M. T. Gladwin, and D. B. Kim-Shapiro. 2006. Computation of plasma hemoglobin nitric oxide scavenging in hemolytic anemias. *Free Radic Biol Med* 41(10):1557–1565.

Jeney, V., J. Balla, A. Yachie et al. 2002. Pro-oxidant and cytotoxic effects of circulating heme. *Blood* 100(3):879–887.

Jiménez-Osorio, A. S., S. González-Reyes, and J. Pedraza-Chaverri. 2015. Natural NRF2 activators in diabetes. *Clin Chim Acta* 448:182–192.

Kansanen, E., S. M. Kuosmanen, H. Leinonen et al. 2013. The Keap1-NRF2 pathway: Mechanisms of activation and dysregulation in cancer. *Redox Biol* 1:45–49.

Kato, G. J. 2009. Haptoglobin halts hemoglobin's havoc. *J Clin Invest* 119(8):2140–2142. doi: 10.1172/JCI40258.

Kato, G. J., R. P. Hebbel, M. H. Steinberg et al. 2009. Vasculopathy in sickle cell disease: Biology, pathophysiology, genetics, translational medicine, and new research directions. *Am J Hematol* 84(9):618–625.

Kaul, D. K., M. E. Fabry, and R. L. Nagel. 1989. Microvascular sites and characteristics of sickle cell adhesion to vascular endothelium in shear flow conditions: Pathophysiological implications. *Proc Natl Acad Sci USA* 86(9):3356–3360.

Kaul, D. K., E. Finnegan, and G. A. Barabino. 2009. Sickle red cell-endothelium interactions. *Microcirculation* 16(1):97–111.

Kensler, T. W., D. Ng, S. G. Carmella et al. 2012. Modulation of the metabolism of airborne pollutants by glucoraphanin-rich and sulforaphane-rich broccoli sprout beverages in Qidong, China. *Carcinogenesis* 33(1):101–107.

Knee, K. M., C. K. Roden, M. R. Flory et al. 2007. The role of beta93 Cys in the inhibition of Hb S fiber formation. *Biophys Chem* 127(3):181–193.

Krajewski, M. L., L. L. Hsu, and M. T. Gladwin. 2008. The proverbial chicken or the egg? Dissection of the role of cell-free hemoglobin versus reactive oxygen species in sickle cell pathophysiology. *Am J Physiol Heart Circ Physiol* 295(1):H4–H7.

Kuypers, F. A. 2014. Hemoglobin s polymerization and red cell membrane changes. *Hematol Oncol Clin North Am* 28(2):155–179.

Lee, J. M., K. Chan, Y. W. Kan et al. 2004. Targeted disruption of NRF2 causes regenerative immune-mediated hemolytic anemia. *Proc Natl Acad Sci USA* 101(26):9751–9756.

Lettre, G. 2012. The search for genetic modifiers of disease severity in the beta-hemoglobinopathies. *Cold Spring Harb Perspect Med* 2(10):015032. Doi: 10.1101/cshperspect.a015032a015032.

Lim, J. L., M. M. Wilhelmus, H. E. de Vries et al. 2014. Antioxidative defense mechanisms controlled by NRF2: State-of-the-art and clinical perspectives in neurodegenerative diseases. *Arch Toxicol* 88(10):1773–1786.

Linker, R. A., D. H. Lee, S. Ryan et al. 2011. Fumaric acid esters exert neuroprotective effects in neuroinflammation via activation of the NRF2 antioxidant pathway. *Brain* 134(Pt 3):678–692.

Ma, Q. 2013. Role of NFR2 in oxidative stress and toxicity. *Annu Rev Pharmacol Toxicol* 53:401–426.

Magesh, S., Y. Chen, and L. Hu. 2012. Small molecule modulators of Keap1-NRF2-ARE pathway as potential preventive and therapeutic agents. *Med Res Rev* 32(4):687–726.

Manolio, T. A., F. S. Collins, N. J. Cox et al. 2009. Finding the missing heritability of complex diseases. *Nature* 461(7265):747–753.

Manwani, D. and P. S. Frenette. 2013. Vaso-occlusion in sickle cell disease: Pathophysiology and novel targeted therapies. *Blood* 122(24):3892–3898.

Menzel, S., J. Qin, N. Vasavda et al. 2010. Experimental generation of SNP haplotype signatures in patients with sickle cell anaemia. *PLoS One* 5(9):e13004.

Moon, E. J. and A. Giaccia. 2015. Dual roles of NRF2 in tumor prevention and progression: Possible implications in cancer treatment. *Free Radic Biol Med* 79:292–299.

Morris, C. R., G. J. Kato, M. Poljakovic et al. 2005. Dysregulated arginine metabolism, hemolysis-associated pulmonary hypertension, and mortality in sickle cell disease. *JAMA* 294(1):81–90.

Morris, C. R., F. A. Kuypers, S. Larkin et al. 2000. Arginine therapy: A novel strategy to induce nitric oxide production in sickle cell disease. *Br J Haematol* 111(2):498–500.

Morris, C. R., J. H. Suh, W. Hagar et al. 2008. Erythrocyte glutamine depletion, altered redox environment, and pulmonary hypertension in sickle cell disease. *Blood* 111(1):402–410.

Muller-Eberhard, U., J. Javid, H. H. Liem et al. 1968. Plasma concentrations of hemopexin, haptoglobin and heme in patients with various hemolytic diseases. *Blood* 32(5):811–815.

Nagel, R. L. 1991. Severity, pathobiology, epistatic effects, and genetic markers in sickle cell anemia. *Semin Hematol* 28(3):180–201.

Neumann, C. A., D. S. Krause, C. V. Carman et al. 2003. Essential role for the peroxiredoxin Prdx1 in erythrocyte antioxidant defence and tumour suppression. *Nature* 424(6948):561–565.

No, J. H., Y. B. Kim, and Y. S. Song. 2014. Targeting NRF2 signaling to combat chemoresistance. *J Cancer Prev* 19(2):111–117.

Nur, E., B. J. Biemond, H. M. Otten et al. 2011. Oxidative stress in sickle cell disease: Pathophysiology and potential implications for disease management. *Am J Hematol* 86(6):484–489.

Owusu-Ansah, A., S. H. Choi, A. Petrosiute et al. 2015. Triterpenoid inducers of NRF2 signaling as potential therapeutic agents in sickle cell disease: A review. *Front Med* 9(1):46–56.

Pall, M. L. and S. Levine. 2015. NRF2, a master regulator of detoxification and also antioxidant, anti-inflammatory and other cytoprotective mechanisms, is raised by health promoting factors. *Sheng Li Xue Bao* 67(1):1–18.

Palmer, R. M., D. S. Ashton, and S. Moncada. 1988. Vascular endothelial cells synthesize nitric oxide from L-arginine. *Nature* 333(6174):664–666.

Palsamy, P. and S. Subramanian. 2011. Resveratrol protects diabetic kidney by attenuating hyperglycemia-mediated oxidative stress and renal inflammatory cytokines via NRF2-Keap1 signaling. *Biochim Biophys Acta* 1812(7):719–731.

Pauling, L., H. A. Itano, S. J. Singer et al. 1949. Sickle cell anemia a molecular disease. *Science* 110(2865):543–548.

Pergola, P. E., P. Raskin, R. D. Toto et al. 2011. Bardoxolone methyl and kidney function in CKD with type 2 diabetes. *N Engl J Med* 365(4):327–336.

Perutz, M. F., M. G. Rossmann, A. F. Cullis et al. 1960. Structure of haemoglobin: A three-dimensional Fourier synthesis at 5.5-A. resolution, obtained by X-ray analysis. *Nature* 185(4711):416–422.

Piccin, A., W. G. Murphy, and O. P. Smith. 2007. Circulating microparticles: Pathophysiology and clinical implications. *Blood Rev* 21(3):157–171.

Pullarkat, V., Z. Meng, S. M. Tahara et al. 2014. Proteasome inhibition induces both antioxidant and Hb F responses in sickle cell disease via the NRF2 pathway. *Hemoglobin* 38(3):188–195.

Rees, D. C. and J. S. Gibson. 2012. Biomarkers in sickle cell disease. *Br J Haematol* 156(4):433–445.

Rees, D. C., T. N. Williams, and M. T. Gladwin. 2010. Sickle-cell disease. *Lancet* 376(9757):2018–2031.

Reid, M., A. Badaloo, T. Forrester et al. 2006. In vivo rates of erythrocyte glutathione synthesis in adults with sickle cell disease. *Am J Physiol Endocrinol Metab* 291(1):E73–E79.

Reiter, C. D. and M. T. Gladwin. 2003. An emerging role for nitric oxide in sickle cell disease vascular homeostasis and therapy. *Curr Opin Hematol* 10(2):99–107.

Reiter, C. D., X. Wang, J. E. Tanus-Santos et al. 2002. Cell-free hemoglobin limits nitric oxide bioavailability in sickle-cell disease. *Nat Med* 8(12):1383–1389.

Repka, T. and R. P. Hebbel. 1991. Hydroxyl radical formation by sickle erythrocyte membranes: Role of pathologic iron deposits and cytoplasmic reducing agents. *Blood* 78(10):2753–2758.

Rice-Evans, C., S. C. Omorphos, and E. Baysal. 1986. Sickle cell membranes and oxidative damage. *Biochem J* 237(1):265–269.

Rifkind, J. M., J. G. Mohanty, and E. Nagababu. 2014. The pathophysiology of extracellular hemoglobin associated with enhanced oxidative reactions. *Front Physiol* 5:500.

Sadrzadeh, S. M., E. Graf, S. S. Panter, P. E. Hallaway, and J. W. Eaton. 1984. Hemoglobin. A biologic fenton reagent. *J Biol Chem* 259(23):14354–14356.

Sangokoya, C., M. J. Telen, and J. T. Chi. 2010. microRNA miR-144 modulates oxidative stress tolerance and associates with anemia severity in sickle cell disease. *Blood* 116(20):4338–4348.

Schaer, D. J., A. I. Alayash, and P. W. Buehler. 2007. Gating the radical hemoglobin to macrophages: The anti-inflammatory role of CD163, a scavenger receptor. *Antioxid Redox Signal* 9(7):991–999.

Schaer, D. J., P. W. Buehler, A. I. Alayash et al. 2013. Hemolysis and free hemoglobin revisited: Exploring hemoglobin and hemin scavengers as a novel class of therapeutic proteins. *Blood* 121(8):1276–1284.

Schaer, D. J., F. Vinchi, G. Ingoglia et al. 2014. Haptoglobin, hemopexin, and related defense pathways-basic science, clinical perspectives, and drug development. *Front Physiol* 5:415.

Sebastiani, P., N. Solovieff, S. W. Hartley et al. 2010. Genetic modifiers of the severity of sickle cell anemia identified through a genome-wide association study. *Am J Hematol* 85(1):29–35.

Serjeant, G. R. 2013. The natural history of sickle cell disease. *Cold Spring Harb Perspect Med* 3(10):a011783.

Sheng, K., M. Shariff, and R. P. Hebbel. 1998. Comparative oxidation of hemoglobins A and S. *Blood* 91(9):3467–3470.

Shureiqi, I. and J. A. Baron. 2011. Curcumin chemoprevention: The long road to clinical translation. *Cancer Prev Res (Phila)* 4(3):296–298.

Silva, D. G., E. Belini Jr., E. A. de Almeida et al. 2013. Oxidative stress in sickle cell disease: An overview of erythrocyte redox metabolism and current antioxidant therapeutic strategies. *Free Radic Biol Med* 65:1101–1109.

Silva, D. G., E. Belini Jr., L. D. Torres et al. 2011. Relationship between oxidative stress, glutathione S-transferase polymorphisms and hydroxyurea treatment in sickle cell anemia. *Blood Cells Mol Dis* 47(1):23–28.

Sporn, M. B. and K. T. Liby. 2012. NRF2 and cancer: The good, the bad and the importance of context. *Nat Rev Cancer* 12(8):564–571.

Steinberg, M. 2005. Predicting clinical severity in sickle cell anaemia. *Br J Haematol* 129(4):465–481.

Steinberg, M. H. 2009. Genetic etiologies for phenotypic diversity in sickle cell anemia. *Sci World J* 9:46–67.

Steinberg, M. H. and A. H. Adewoye. 2006. Modifier genes and sickle cell anemia. *Curr Opin Hematol* 13(3):131–136.

Steinberg, M. H., W. Rosenstock, M. B. Coleman et al. 1984. Effects of thalassemia and microcytosis on the hematologic and vasoocclusive severity of sickle cell anemia. *Blood* 63(6):1353–1360.

Steinberg, M. H. and P. Sebastiani. 2012. Genetic modifiers of sickle cell disease. *Am J Hematol* 87(8):795–803.

Stuart, M. J. and R. L. Nagel. 2004. Sickle-cell disease. *Lancet* 364(9442):1343–1360.

Tatum, V. L. and C. K. Chow. 1996. Antioxidant status and susceptibility of sickle erythrocytes to oxidative and osmotic stress. *Free Radic Res* 25(2):133–139.

Thomas, P. W., D. R. Higgs, and G. R. Serjeant. 1997. Benign clinical course in homozygous sickle cell disease: A search for predictors. *J Clin Epidemiol* 50(2):121–126.

Torres, L. S., D. G. H. da Silva, E. Belini Jr. et al. 2012. The influence of hydroxyurea on oxidative stress in sickle cell anemia. *Rev Bras Hematol Hemoter* 34(6):421–425.

Tracz, M. J., J. Alam, and K. A. Nath. 2007. Physiology and pathophysiology of heme: Implications for kidney disease. *J Am Soc Nephrol* 18:414–420.

Vekilov, P. G. 2007. Sickle-cell haemoglobin polymerization: Is it the primary pathogenic event of sickle-cell anaemia? *Br J Haematol* 139(2):173–184.

Wagener, F. A., A. Eggert, O. C. Boerman et al. 2001. Heme is a potent inducer of inflammation in mice and is counteracted by heme oxygenase. *Blood* 98(6):1802–1811.

Wang, J. M., Y. J. Huang, Y. Wang et al. 2007. Increased circulating CD31+/CD42− microparticles are associated with impaired systemic artery elasticity in healthy subjects. *Am J Hypertens* 20(9):957–964.

Weatherall, D., K. Hofman, G. Rodgers, J. Ruffin, and S. Hrynkow. 2005. A case for developing North-South partnerships for research in sickle cell disease. *Blood* 105(3):921–923.

Wishner, B. C., K. B. Ward, E. E. Lattman, and W. E. Love. 1975. Crystal structure of sickle-cell deoxyhemoglobin at 5 A resolution. *J Mol Biol* 98(1):179–194.

Wood, K. C., R. P. Hebbel, D. J. Lefer et al. 2006. Critical role of endothelial cell-derived nitric oxide synthase in sickle cell disease-induced microvascular dysfunction. *Free Radic Biol Med* 40(8):1443–1453.

Zhang, L., A. Levy, and J. M. Rifkind. 1991. Autoxidation of hemoglobin enhanced by dissociation into dimers. *J Biol Chem* 266(36):24698–24701.

25 Oxidative Stress in Nonalcoholic Steatohepatitis

Akinobu Takaki, Daisuke Uchida, and Kazuhide Yamamoto

CONTENTS

ABSTRACT

Nonalcoholic fatty liver disease (NAFLD) is a disease with increasing prevalence that is potentially lethal after decades of persistent fat toxicity on the liver and accompanying organs. The pathogenesis for the progression of benign nonalcoholic fatty liver (NAFL) to progressive nonalcoholic steatohepatitis (NASH) is acknowledged as a multiple parallel hit process, including fat deposition, genetic factors, insulin resistance, and oxidative stress. Oxidative stress is one of the main pathogenic factors driving the progression of NAFL to NASH. Excessive accumulation of long chain fatty acids results in activation of the mitochondrial b-oxidation pathway and consequent reactive oxygen species (ROS) production. ROS can mediate oxidative stress responses in hepatocytes, Kupffer cells and hepatic stellate cells (HSCs), resulting in hepatocyte damage as well as proinflammatory and profibrogenic responses. ROS detoxification pathway signaling is often damaged in NAFLD following ROS accumulation. Inactivation of the prooxidant production pathway is presently the primary treatment strategy. However, as oxidative stress is an essential response in living organisms, a new treatment strategy that activates toxic oxidative stress detoxifying pathways while maintaining essential oxidative stress responses is preferred.

25.1 INTRODUCTION

Nonalcoholic fatty liver (NAFL) had long been believed to be a nonprogressive disease called simple steatosis. However, several researchers have more recently reported progressive NAFL in patients with diabetes, called diabetic cirrhosis (Itoh et al. 1979). Ludwig et al. (1980) have named

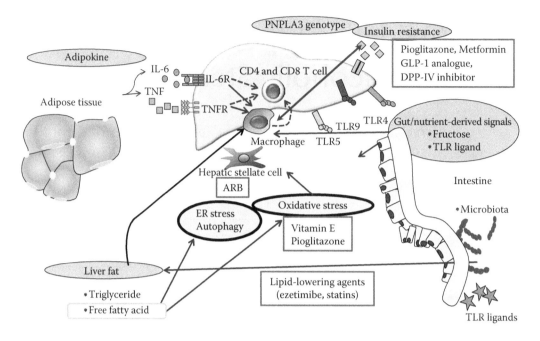

FIGURE 25.1 **(See color insert.)** Nonalcoholic steatohepatitis (NASH) pathogenesis, treatment, and targets.

nonalcoholic steatohepatitis (NASH) from their experience with 20 cases. At first, the disease entity was not widely accepted; however, lifestyle changes have forced clinicians and patients to increasingly accept NAFL as a life-threatening disease from Western to Eastern countries.

Recently, the presence of liver fat deposition has been referred to as nonalcoholic fatty liver disease (NAFLD), which is a common and increasingly prevalent chronic liver disease, representing a hepatic manifestation of the metabolic syndrome first observed in Western countries (Pacifico et al. 2011). Even in Eastern countries, the prevalence of NAFLD is increasing with economic development associated with a Western diet and lifestyle. Most patients with NAFLD exhibit nonprogressive simple fatty liver, that is, NAFL, while some proportion of patients present with a progressive disease, called NASH. NASH is broadly defined by the presence of steatosis with inflammation, hepatocyte degeneration, and liver fibrosis (Matteoni et al. 1999, Brunt et al. 2009) leading to cirrhosis and hepatocellular carcinoma (HCC) (Fassio et al. 2004, Ono et al. 2006, Yatsuji et al. 2009). The mechanisms through which a subset of patients with NAFLD develops NASH are poorly understood. The development of NASH is generally thought of as a "two-hit" process (Day et al. 1998). The first hit is the development of hepatic steatosis due to overeating or lack of exercise, while the second hit includes cellular stresses such as oxidative stress, apoptosis, and gut-derived signals. However, recently uncovered liver fat deposition mechanisms have led us to define NASH pathogenesis as involving a "multiple-parallel-hits" process (Tilg et al. 2010). Reactive oxygen species (ROS) have been accepted as the main feature of NAFLD progression. Here, we review the current understanding of ROS in NAFLD molecular pathogenesis and emerging ROS-controlling treatments to be considered in future therapeutic paradigms (Figure 25.1).

25.2 PATHOGENESIS OF NAFLD

25.2.1 DISEASE-SPECIFIC GENE POLYMORPHISM

A recent genome-wide association study (GWAS) identified the patatin-like phospholipase 3 (PNPLA3) gene polymorphism as a key in the development of NASH (Romeo et al. 2008).

Such genetic characteristics precede the fat deposition component of the "multiple-parallel-hits" process. Several GWAS results of various races have confirmed the importance of PNPLA3 gene polymorphisms in NAFLD (Romeo et al. 2008). This genetic polymorphism differentiates between simple steatosis with or without minimal inflammation and fibrosis that progresses to NASH (Valenti et al. 2010, Kawaguchi et al. 2012). Patients with the NASH-sensitive single nucleotide polymorphism (SNP) rs738409 G/G genotype might progress not only to simple steatosis but also to NASH, probably under the same types of metabolic stimulation. The function of PNPLA3 is not well known, as mice deficient in PNPLA3 develop neither fatty liver nor liver injury. However, the overexpression of sterol regulatory element-binding protein 1c (SREBP-1c) results in its binding to the transcription start site of the mouse PNPLA3 gene, while PNPLA3 knockdown can decrease the intracellular triglyceride content in primary hepatocytes (Qiao et al. 2011). Thus, PNPLA3 might function as a downstream target gene of SREBP-1c to mediate SREBP-1c stimulation of lipid accumulation. A meta-analysis revealed that this variant is associated with increased liver fat content when compared to weight-matched individuals not harboring the PNPLA3 polymorphism as well as increased risk of severe fibrosis, even in the presence of other etiologies of chronic liver disease (Singal et al. 2014).

Not all patients with progressive NASH harbor the PNPLA3 risk allele. Thus, differences in NAFLD patients bearing the characteristics of the PNPLA3 risk allele versus patients without the allele have been demonstrated (Lallukka et al. 2013). PNPLA3-related NAFLD is not characterized by features typical of metabolic syndrome such as hyperinsulinemia, hypertriglyceridemia, and low HDL cholesterol levels (Sookoian et al. 2011). Patients with obesity-related NAFLD exhibit the same distribution of the PNPLA3 genotype as nonobese patients, whereas inflammation-related genes are upregulated in the adipose tissue of these patients.

25.2.2 Obesity and Insulin Resistance

Obesity and insulin resistance are known risk factors for NAFLD and NASH progression. Visceral fat induces the production of several fat-associated cytokines and induces inflammation, even in patients harboring the nonrisk allele of PNPLA3. Hepatic fat induces cellular stresses and inflammation directly to hepatocytes and the surrounding nonparenchymal cells. One of the NASH diagnostic markers, the NAFIC score created for Japanese NASH patients, included the following characteristics: high ferritin, high fasting insulin, and type IV collagen 7S (Sumida et al. 2011). As the fasting serum insulin level was significantly correlated with NASH prevalence, a modified NAFIC score that included the serum fasting insulin level resulted in a stepwise refinement that was more effective in diagnosing NASH (Nakamura et al. 2013).

Insulin stimulates glucose uptake by the skeletal muscle and adipose tissue. Insulin also stimulates the liver to convert excess glucose into glycogen and triglycerides for storage. Insulin binds the insulin receptor and stimulates receptor autophosphorylation and internalization, which in turn recruits and activates insulin receptor substrate proteins 1 and 2 (IRS1/2). IRS1/2 can activate phosphatidylinositol 3-kinase, which converts PI bisphosphonate (PIP2) to PIP3. Cell proliferation transcription factor Akt binds PIP3 and is subsequently phosphorylated and activated. In the muscle and adipose tissue, Akt stimulates the translocation of glucose transporters to the membrane to allow glucose uptake. In the liver, insulin binding promotes fatty acid synthesis through the activation of SREBP-1c. Akt kinase pathways have roles in cell growth, cell proliferation, fibrogenesis, and hepatocarcinogenesis (Larter et al. 2010). Insulin resistance is concordant with NASH (Larter et al. 2006). In the obese, insulin-resistant model *ob/ob* mouse, liver insulin receptor knockout was shown to improve hepatic lipogenesis (Haas et al. 2012). This result indicates that insulin receptor signaling is required for hepatic steatosis. Insulin resistance could be induced by Kupffer cells, as depletion of Kupffer cells could attenuate systemic insulin resistance and improve liver autophagy in mice fed a high-fat diet (Zeng et al. 2015).

25.2.3 Oxidative Stress

Many sources of cellular stress, including oxidative stress, apoptosis, and gut-derived lipopolysaccharide (LPS), trigger an inflammatory response and progressive liver damage (Csak et al. 2011). Oxidative stress appears to be responsible for initiating necroinflammation. Notably, ROS generated during free fatty acid metabolism in microsomes, peroxisomes, and mitochondria comprise an established source of oxidative stress (Pessayre 2007). Mitochondria are the most important cellular source of ROS, and mitochondrial dysfunction might, therefore, play a central role in the pathological mechanisms of NASH. Although the mechanisms of mitochondrial dysfunction are not clearly understood, emerging data suggest that ROS, lipid peroxidation products, and tumor necrosis factor α (TNF-α) are involved in the multiple hits, inducing the progression from NAFL to NASH (Takaki et al. 2014). Many cytokines and inflammatory cells are involved in the pathogenesis of NAFL and even more so in NASH. ROS can educate adaptive inflammatory cells and induce directional migration of resident hepatic profibrogenic cells, resulting in liver inflammation and fibrosis (Novo et al. 2011, Sutti et al. 2014). Even in the setting of NAFL, several inflammatory cell types are present in the liver, and these are believed to affect the progression of NAFL to NASH. Gut microbiota–related stimulants and hepatic lipids also induce immune reactions.

25.2.4 Endoplasmic Reticulum Stress and Autophagy

One epicenter for NAFLD progression-related stress response is the endoplasmic reticulum (ER), a membranous network that functions in the synthesis and assembly of secretory and membrane proteins to achieve their proper conformation. Lipid metabolism–related proteins are enriched, whereas protein synthesis and transport functions are downregulated in obese hepatic ER (Fu et al. 2011). The maintenance of ER function requires high concentrations of intra-ER Ca^{2+}, which is actively controlled by sarco(endo)plasmic reticulum Ca^{2+}-ATPase (SERCA). Free cholesterol accumulation triggers ER stress by altering the critical free cholesterol-to-phospholipid ratio of the ER membrane necessary to maintain its fluidity. Among the ER enzymes, SERCA is particularly sensitive to ER membrane cholesterol content, which can inhibit SERCA conformational changes and activity. Such changes induce a decrease in physiologically high intra-ER Ca^{2+} concentrations, resulting in impaired ER function known as ER stress. Such stress is one of the most important factors for disease progression in NASH, along with hepatocyte apoptosis and hepatic stellate cell (HSC) or Kupffer cell activation.

Autophagy is a cellular self-digestion process that targets damaged organelles and intracellular pathogens. Autophagy suppression has been shown to cause ER stress (Yang et al. 2010). This implies that the onset of hepatic steatosis due to defective autophagy could reflect a combination of both decreased fat degradation by autophagy and increased fat synthesis by ER stress. The chaperone response is blunted in obese mice as a result of dysfunctional X-box binding protein 1 (XBP1), which is a master regulator of ER folding capacity (Park et al. 2010). The serine/threonine kinase mammalian target of rapamycin (mTOR), a component of the mTORC1 complex, is the main inhibitor of autophagy. Hyperinsulinemia induces the activation of Akt/PKB and downstream mTOR resulting in a defect in autophagy (Liu et al. 2009).

25.2.5 Inflammatory Reactions

Inflammation could also be an inducer of hepatic steatosis, and anti-inflammatory cytokine treatment could induce steatosis reduction (Zein et al. 2012). Adipose tissue–derived cytokines promote metabolic disease progression. In obesity, excessive numbers of proinflammatory, M1-like macrophages accumulate in the adipose tissue and liver (Lanthier et al. 2010). Even in simple fatty liver, macrophage infiltration and the expression of the macrophage attractant chemokine monocyte chemotactic protein 1 (MCP1) are significantly increased (Gadd et al. 2014).

Macrophages are important mediators of inflammation and insulin resistance, which are the common processes associated with NAFLD. In advanced NASH, CD4[+] and CD8[+] T-cell infiltration increases, and the levels of inflammatory cytokines, such as interleukin (IL)-6 or IL-8, are also increased (Gadd et al. 2014).

Immune reaction could be induced by oxidative stress and could be an inducer of hepatic steatosis (Navarro et al. 2015). Anti-TNF-α antibody was reported to improve steatosis in *ob/ob* mice, suggesting that primary inflammation can induce secondary steatosis (Li et al. 2003).

25.3 OXIDATIVE STRESS AS MAIN NASH PROGRESSION-RELATED FACTOR

Fatty liver is the basic feature of NAFLD and NASH. Lipid droplets were previously believed to function simply as energy storage structures. However, they are now considered to be complicated organelles that exhibit many processes, such as metabolic, inflammatory, and immunological responses. Lipid toxicity induces multiple effects, such as oxidative stress, ER stress, and immune reactions (Takaki et al. 2014). Triglycerides are the main type of lipid stored in the livers of patients with NAFLD. The toxic oxidative stress–producing lipids present in NASH and the nontoxic lipids in simple steatosis could differ (Yamaguchi et al. 2007). Diacylglycerol acyltransferase 2 (DGAT2) catalyzes the final step in hepatocyte triglyceride biosynthesis. Hepatic steatosis and dietary triglyceride contents induced in a model of obese mice with simple fatty liver are reduced by DGAT2 antisense oligonucleotides in a manner that does not correlate with changes in body weight, adiposity, or insulin sensitivity (Yu et al. 2005). However, DGAT2 antisense oligonucleotide has been shown to increase the levels of hepatic free fatty acids, lipid oxidant stress, lobular necroinflammation, and fibrosis in mice fed a diet deficient in methionine choline (MCD), which generated inflammation and fibrosis with hepatic steatosis, whereas hepatic triglyceride content decreased (Yamaguchi et al. 2007). These results suggest that the pathogenesis and treatment of steatosis in simple fatty liver and in NASH are different. Human genetic variability analysis of lifestyle intervention has shown that the DGAT2 gene polymorphism is related to a decrease in liver fat, while changes in insulin resistance are not correlated (Kantartzis et al. 2009). Since insulin resistance is the key marker for NASH, DGAT2 gene polymorphism might only be associated with nonprogressive fatty liver.

Oxidative stress is involved in the mechanisms of aging, carcinogenesis, and atherosclerotic progression (Takaki et al. 2013). Excessive oxidative stress induced by mitochondrial, peroxisomal, and microsomal ROS in NASH results in apoptosis as well as nuclear and mitochondrial DNA damage. Limited antioxidant defenses contribute to the processes of both NASH and hepatocarcinogenesis (Buganesi 2007, Kawai et al. 2012). Physiologically low levels of ROS are involved in vital cellular processes, indicating that the balance of oxidative and antioxidative responses is important (Mittler et al. 2011). Mitochondrial dysfunction not only impairs fat homeostasis in the liver but also leads to an overproduction of oxidative stress, resulting in the generation of ROS that trigger lipid peroxidation, cytokine overproduction, and cell death. Indeed, ultrastructural alterations, impairment of adenosine triphosphate (ATP) synthesis, and increased ROS production have been reported in liver mitochondria from NASH patients and in a rodent NASH model (Cortez-Pinto et al. 1999, Serviddio et al. 2008). Mitochondria are a principal source of cellular ROS arising from inefficiencies in electron flow along the electron transport chain (ETC). Under physiological conditions, the majority of incompletely reduced ROS, such as superoxide, are detoxified into water, thereby maintaining steady-state oxidant concentrations at relatively low levels (less than 1% of total oxygen consumed by mitochondria) by a variety of antioxidant defenses and repair enzymes (Brand 2010). The mitochondrial capacity to control oxidative balance collapses under continuous oxidative stress. Excess superoxide is generated within injured mitochondria through electron leakage, and the resulting excess of superoxide is converted to hydrogen peroxide (H_2O_2) by superoxide dismutase (SOD). Glutathione peroxidase (GPx) or catalase can metabolize H_2O_2 to nontoxic H_2O; however, the Fenton and/or Haber–Weiss reactions generate the highly reactive and toxic hydroxyl radical.

Iron is the key mineral that induces oxidative stress produced via the Fenton's reaction. Although its role in NASH is not fully understood, iron levels are elevated in NASH, which is an inducer of oxidative stress, and lowering iron levels have resulted in fair outcomes for patients with chronic liver disease (Nelson et al. 2011). However, one-third of early-stage NAFLD patients have iron deficiency associated with the female sex, obesity, and type 2 diabetes (Siddique et al. 2013). We must wait for long-term follow-up studies to confirm whether iron-deficient obese patients progress to NASH and to clarify the role of iron in NAFLD progression.

The mitochondrial proliferation and differentiation program could be impaired in NASH. One of the most important regulators of mitochondrial biogenesis is the transcription coactivator peroxisome proliferator-activated receptor-γ (PPAR-γ)-coactivator-1α (PGC-1α) (Scarpulla 2011), which coordinates the large number of genes required for mitochondrial biogenesis. The activity of PGC-1α is impaired in the fatty liver, which results in decreased mitochondrial biogenesis (Aharoni-Simon et al. 2011).

In NASH-related HCC models, PGC-1α was downregulated in comparison with nontumorous tissues, thus signifying its importance in the normal hepatocyte phenotype (Wang et al. 2012). Decreases in mitochondrial DNA and mitochondrial DNA–encoded polypeptides are representative findings in NASH, whereas mitochondrial DNA content is increased in simple fatty liver (Chiappini et al. 2006). The complementary activation of mitochondrial DNA in simple fatty liver may help to protect the liver from inflammation and fibrosis, whereas decreases in NASH induce progressive inflammation and fibrosis with normal hepatocyte function disturbance.

Nonparenchymal liver cells, such as Kupffer and HSCs, play significant roles in the progression of chronic liver inflammation and fibrosis (Rolo et al. 2012). Excess fatty acid accumulation in hepatocytes induces oxidative stress from mitochondria as well as peroxisomes or microsomes. These cytotoxic ROS and lipid peroxidation products are able to diffuse into the extracellular space, affecting Kupffer cells and HSCs. Cellular oxidative stress from hepatocytes and the direct uptake of free fatty acids or free cholesterol in Kupffer cells activate nuclear factor-κB, which induces the synthesis of TNF-α and several proinflammatory cytokines, such as IL-6 and IL-8 (Hui et al. 2004). Kupffer cells in patients with NASH produce TGF-β, resulting in HSCs acquiring a fibrogenic myofibroblast–like phenotype.

Exposure of primary HSCs or HSC cell lines to H_2O_2 leads to increased gene expression of ER chaperone BiP–binding transmembrane proteins such as inositol-requiring enzyme 1 (IRE1a) or activating transcription factor 4 (ATF4). ER stress in HSCs results in increased autophagy and HSC activation to the fibrogenic status (Hernandez-Gea et al. 2013). The most characteristic features of HSC activation is the loss of cytoplasmic lipid droplets, which are composed of retinyl esters and triglycerides (Blaner et al. 2009).

Autophagy is present in all cell types and is upregulated as an adaptive response during cellular stress to generate intracellular nutrients and energy. Autophagy is upregulated in activated HSC of mice with liver damage. However, cytoplasmic lipid droplets are maintained and remain quiescent in autophagy-defective HSCs, indicating that oxidative stress–induced ER stress and autophagy are a key event in HSC activation (Hernandez-Gea et al. 2012).

25.4 PATHOGENESIS-TARGETED TREATMENT OF NAFLD

Since NAFLD and NASH have emerged as lifestyle-associated diseases, lifestyle intervention is an important and first approach to their treatment. Body weight reduction using a low-calorie, low-fat diet could improve serum transaminase levels or radiologically defined fatty liver. Exercise is also regarded as useful therapy for NAFLD (Watanabe et al. 2015). Pharmacological therapy is adopted for patients who are refractory to lifestyle intervention or for those with an advanced active state of steatohepatitis. A recent GWAS identified the PNPLA3 gene polymorphism as key in the development of NASH. To reduce ROS is the present standard approach to NASH; however, ROS are also essential for the processes such as the prevention of cell infection among others. In addition,

ROS are required for intact autophagosome maturation and intact autophagy, which are required for the process of degradation of aged and defective cellular organelles (Qiao et al. 2015). These results suggest that NASH could be treated with antioxidants; however, physiological ROS should be maintained.

25.4.1 Standard Oxidative Stress–Targeted Treatment

Since vitamin E supplementation is a representative antioxidant drug treatment that has become the standard treatment for NASH (Chalasani et al. 2012), the American Association for the Study of Liver Disease (AASLD) recommends the use of vitamin E at a daily dose of 800 IU, which is a higher dose than usual. Administration of vitamin E improves NAFLD activity scores (NASs) for clinical and histological activity within 2 years but increases insulin resistance and plasma triglyceride levels (Sanyal et al. 2010). However, the recovery of fibrosis progression has not been demonstrated (Hoofnagle et al. 2013).

Controversy surrounds antioxidant therapies because ROS have essential functions in living organisms. There are, at minimum, three problems for adaptation for NAFLD treatment. The first is the difficulty in the evaluation of the effect in living organs. Antioxidants exhibit effective chemical activity *in vitro*; however, there have been many failures in demonstrating *in vivo* effects (Bast et al. 2013). The second problem relates to the already known negative effects of ROS on cerebrovascular disease. Many cerebrovascular studies have investigated the effects of vitamin E. A meta-analysis of the effects of vitamin E on stroke revealed a 10% reduction in ischemic stroke accompanied by a 22% increase in hemorrhagic stroke. Antioxidants are likely to contribute to the progression of cancer (Watson 2013). Antioxidants such as vitamin E administration should be carefully designed for patients at the risk of cerebrovascular disease. As NAFLD is often also involved in metabolic syndrome, cerebrovascular disease risk assessment is required before vitamin E administration. The third problem is the possibility of progression to the development of cancer. Stem cell–like cancer cells have powerful antioxidative properties that protect them from oxidative stress and thus prevent apoptosis (Yae et al. 2012). Oxidative stress on normal cells may induce a transition to a cancer cell phenotype that is highly resistant to further oxidative stress. Inducing oxidative stress under these conditions is an approach that is being investigated as a cancer treatment in several clinical trials (Trachootham et al. 2009). However, this approach is likely to be toxic to normal cells and may lead to the induction of further carcinogenesis. Thus, oxidative stress must be controlled according to clinical circumstances.

25.4.2 Antidiabetic Insulin Resistance Treatment

Since the general characteristics of NASH include obesity and insulin resistance, the use of anti-insulin resistance treatment has become prevalent. Pioglitazone, a PPAR agonist, and metformin, have been clinically shown to improve NASH. Pioglitazone is especially recommended even for patients with nondiabetic NASH as outlined in the AASLD and the European Association for the Study of Liver (EASL) guidelines (Nascimbeni et al. 2013). However, pioglitazone has the frequent undesirable side effect of body weight gain. Metformin is not strongly recommended but regarded as permissible. Metformin has brought about histological improvements in the nondiabetic NASH mouse model (Musso et al. 2012). Metformin induces weight loss that is preferable for NAFLD. However, recent meta-analysis did not show a beneficial effect on serum biochemical responses (Rakoski et al. 2010, Shyangdan et al. 2011). These antidiabetic insulin-sensitizing drugs also have antioxidative effects, making them good candidates for NAFLD therapy.

Both glucagon-like peptide-1 (GLP-1) and gastric inhibitory polypeptide (GIP) are incretins, which are a group of gastrointestinal hormones that cause increased insulin release from pancreatic beta cells and reduced appetite, and might be good targets for treating NASH. A GLP-1 receptor agonist analog was shown to improve metabolic, biochemical, and histopathological indices

of NASH in mice by restoring hepatic lipid oxidation (Svegliati-Baroni et al. 2011). GLP-1 agonists (GLP-1 and GIP) are degraded by dipeptidyl peptidase-IV (DPP-IV). Therefore, inhibition
of DPP-IV extends the half-life of endogenous GLP-1 and GIP and results in diabetic control. The
long-term administration of a DPP-IV inhibitor has reduced liver fat content in animals with diet-
induced hepatic steatosis and insulin resistance (Kern et al. 2012). A meta-analysis of GLP-1 agonists and DPP-IV inhibitors revealed a mean reduction of alanine aminotransferase (ALT) levels to
14.1 IU/L after 16–48.7 weeks of treatment (Carbone et al. 2015). These drugs provide promising
opportunities for further large-scale trials.

25.4.3 Other Candidate Treatments

There are several basic or clinical trials for other candidates, such as lipid-lowering agents,
ursodeoxycholic acid (UDCA), and angiotensin receptor blockers (ARBs). Lipid-lowering drugs
such as statins can also improve ALT and radiological steatosis in hyperlipidemic patients with
NAFLD; however, histological improvements are not evident (Nelson et al. 2009). Ezetimibe
is a Niemann-Pick C1-like protein inhibitor that can reduce the intestinal accumulation of free
cholesterol. Ezetimibe has reduced histological NASH development in mice and in 10 patients
with NASH (Deushi et al. 2007, Yoneda et al. 2010). A large randomized control study (RCT)
is needed to further explore these results. Experiments in mice have revealed that ezetimibe
does not improve hepatic steatosis induced by a high-fructose diet, which results from hepatic
lipogenesis, yet was able to improve high-fat diet-induced steatosis, which partly results from
lipid trafficking via the small intestine (Ushio et al. 2013). This drug may have favorable effects
on NAFLD.

UDCA is also reportedly effective in some instances. Several RCTs have found improvements in
ALT but not in liver histology, even at high doses (Leuschner et al. 2010).

ARBs are widely used antihypertensive drugs and also have anti-fibrotic effects in many organs.
There are several pilot studies indicating their efficacy on NASH serum biochemistry and histology
(Yokohama et al. 2004, Georgescu et al. 2009). ARBs would be the first antihypertensive drugs to
be used for the treatment of NAFLD in patients with concomitant hypertension.

25.4.4 Newly Emerging Antioxidant Candidate

L-Carnitine is a precursor of carnitine palmitoyltransferase 1 (CPT-1), the rate-limiting enzyme for
mitochondrial β-oxidation that is central to mitochondrial function. Deficiencies in the mitochondrial carnitine-dependent transport system result in curtailed fatty acid oxidation. L-Carnitine is not
simply an inhibitor of oxidative stress that must be preserved in certain conditions but also has the
ability to reverse the mitochondrial function. L-Carnitine supplementation reduces NASs, hepatic
oxidative stress, intestinal unfavorable microbiota, and hepatocarcinogenesis in a mouse model of
NASH–hepatocarcinogenesis (Ishikawa et al. 2014).

Dietary intake of tomatoes has been reported to reduce the risk of human cancers (Ip et al.
2014). Lycopene is the most abundant carotenoid found in tomatoes, tomato products, and other
red fruits. The intake of lycopene inhibits NASH-promoted rat hepatic preneoplastic lesions (Ip
et al. 2013). Although clinical trials remain necessary, plasma carotenoid levels in NASH patients
have been shown to be low, thus suggesting the possibility of a role for tomato intake in NASH
prevention.

Molecular hydrogen has been shown to have powerful antioxidant effects with unique features
(Ohsawa et al. 2007). In cultured cells, hydrogen scavenges hydroxyl radicals but not superoxide, H_2O_2, or nitric oxide (NO) and prevents the decline in the mitochondrial membrane potential and the subsequent decrease in cellular ATP synthesis consistent with antioxidative effects.
Kawai et al. reported that drinking hydrogen-rich water has favorable effects in NASH models

(Kawai et al. 2012). Plasma transaminase levels, histological NASH, hepatic TNF-α, IL-6, and fatty acid synthesis–related gene expression as well as the oxidative stress biomarker 8-OHdG were decreased in the livers of established MCD diet–induced NASH models administered either hydrogen-rich water or the antioxidant pioglitazone. Although the decrease in hepatic cholesterol was smaller in the group given hydrogen-rich water, serum oxidative stress was reduced and antioxidant function was higher than in the pioglitazone group. Hydrogen administration also induced a reduction in hepatocarcinogenesis in the NASH–HCC mouse model.

25.5 CONCLUSION

Oxidative stress is definitely involved in the pathogenesis of NAFLD. How to target oxidative stress in NASH is still controversial. Long-term follow-up results of the patients treated with antioxidants are required. New antioxidants may be effective in controlling NAFLD progression. More basic and clinical experiments into this novel potential treatment option are required.

GWASs have confirmed the importance of the PNPLA3 gene polymorphism in NAFLD. Insulin resistance is an important factor that could be controlled by several agents. The gut microbiome has recently been recognized as being involved in NAFLD development. The pattern of microbiome diversity can induce intestinal mucosal permeability and result in lipopolysaccharidemia, which correlates with NASH progression. Macrophages play an important role in the induction of inflammation and insulin resistance. Ingestion of free fatty acids and free cholesterol induces ER stress and oxidative stress resulting in hepatic inflammation and fibrogenesis, which in turn induces progression to NASH. Antioxidant treatment is the gold standard for NASH; however, monitoring for unexpected side effects is necessary. ARBs might have a role in the control of fibrogenetic HSCs. Lipid-lowering agents could also be a candidate treatment. Adipokines such as IL-6 and TNF-α produced by adipocytes affect hepatocyte fat content and the inflammatory environment in the liver.

REFERENCES

Aharoni-Simon, M., M. Hann-Obercyger, S. Pen, Z. Madar, and O. Tirosh. 2011. Fatty liver is associated with impaired activity of PPARgamma-coactivator 1alpha (PGC1alpha) and mitochondrial biogenesis in mice. *Lab Invest* 91(7):1018–1028. doi: 10.1038/labinvest.2011.55 labinvest201155 [pii].

Bast, A. and G. R. Haenen. 2013. Ten misconceptions about antioxidants. *Trends Pharmacol Sci* 34(8): 430–436. doi: 10.1016/j.tips.2013.05.010.

Blaner, W. S., S. M. O'Byrne, N. Wongsiriroj et al. 2009. Hepatic stellate cell lipid droplets: A specialized lipid droplet for retinoid storage. *Biochim Biophys Acta* 1791(6):467–473. doi: 10.1016/j.bbalip.2008.11.001 S1388-1981(08)00207-2 [pii].

Brand, M. D. 2010. The sites and topology of mitochondrial superoxide production. *Exp Gerontol* 45(7–8): 466–472. doi: 10.1016/j.exger.2010.01.003.

Brunt, E. M., D. E. Kleiner, L. A. Wilson et al. 2009. Portal chronic inflammation in nonalcoholic fatty liver disease (NAFLD): A histologic marker of advanced NAFLD-clinicopathologic correlations from the nonalcoholic steatohepatitis clinical research network. *Hepatology* 49(3):809–820. doi: 10.1002/hep.22724.

Bugianesi, E. 2007. Non-alcoholic steatohepatitis and cancer. *Clin Liver Dis* 11(1):191–207, x–xi. doi: S1089-3261(07)00007-4 [pii] 10.1016/j.cld.2007.02.006.

Carbone, L. J., P. W. Angus, and N. D. Yeomans. 2015. Incretin-based therapies for the treatment of nonalcoholic fatty liver disease: A systematic review and meta-analysis. *J Gastroenterol Hepatol*. doi: 10.1111/jgh.13026.

Chalasani, N., Z. Younossi, J. E. Lavine et al. 2012. The diagnosis and management of non-alcoholic fatty liver disease: Practice guideline by the American Gastroenterological Association, American Association for the Study of Liver Diseases, and American College of Gastroenterology. *Gastroenterology* 142(7): 1592–1609. doi: 10.1053/j.gastro.2012.04.001.

Chiappini, F., A. Barrier, R. Saffroy et al. 2006. Exploration of global gene expression in human liver steatosis by high-density oligonucleotide microarray. *Lab Invest* 86(2):154–165. doi: 3700374 [pii] 10.1038/labinvest.3700374.

Cortez-Pinto, H., J. Chatham, V. P. Chacko, C. Arnold, A. Rashid, and A. M. Diehl. 1999. Alterations in liver ATP homeostasis in human nonalcoholic steatohepatitis: A pilot study. *JAMA* 282(17):1659–1664.

Csak, T., M. Ganz, J. Pespisa, K. Kodys, A. Dolganiuc, and G. Szabo. 2011. Fatty acids and endotoxin activate inflammasome in hepatocytes which release danger signals to activate immune cells in steatohepatitis. *Hepatology* 54(1):133–144. doi: 10.1002/hep.24341.

Day, C. P. and O. F. James. 1998. Steatohepatitis: A tale of two "hits"? *Gastroenterology* 114(4):842–845.

Deushi, M., M. Nomura, A. Kawakami et al. 2007. Ezetimibe improves liver steatosis and insulin resistance in obese rat model of metabolic syndrome. *FEBS Lett* 581(29):5664–5670. doi: 10.1016/j.febslet.2007.11.023.

Fassio, E., E. Alvarez, N. Dominguez, G. Landeira, and C. Longo. 2004. Natural history of nonalcoholic steatohepatitis: A longitudinal study of repeat liver biopsies. *Hepatology* 40(4):820–826. doi: 10.1002/hep.20410.

Fu, S., L. Yang, P. Li et al. 2011. Aberrant lipid metabolism disrupts calcium homeostasis causing liver endoplasmic reticulum stress in obesity. *Nature* 473(7348):528–531. doi: 10.1038/nature09968 nature09968 [pii].

Gadd, V. L., R. Skoien, E. E. Powell et al. 2014. The portal inflammatory infiltrate and ductular reaction in human non-alcoholic fatty liver disease. *Hepatology* 59(4):1393–1405. doi: 10.1002/hep.26937.

Georgescu, E. F., R. Ionescu, M. Niculescu, L. Mogoanta, and L. Vancica. 2009. Angiotensin-receptor blockers as therapy for mild-to-moderate hypertension-associated non-alcoholic steatohepatitis. *World J Gastroenterol* 15(8):942–954.

Haas, J. T., J. Miao, D. Chanda et al. 2012. Hepatic insulin signaling is required for obesity-dependent expression of SREBP-1c mRNA but not for feeding-dependent expression. *Cell Metab* 15(6):873–884. doi: 10.1016/j.cmet.2012.05.002.

Hernandez-Gea, V., Z. Ghiassi-Nejad, R. Rozenfeld et al. 2012. Autophagy releases lipid that promotes fibrogenesis by activated hepatic stellate cells in mice and in human tissues. *Gastroenterology* 142(4):938–946. doi: 10.1053/j.gastro.2011.12.044S0016-5085(12)00012-1 [pii].

Hernandez-Gea, V., M. Hilscher, R. Rozenfeld et al. 2013. Endoplasmic reticulum stress induces fibrogenic activity in hepatic stellate cells through autophagy. *J Hepatol* 59(1):98–104. doi: 10.1016/j.jhep.2013.02.016S0168-8278(13)00138-4 [pii].

Hoofnagle, J. H., M. L. Van Natta, D. E. Kleiner et al. 2013. Vitamin E and changes in serum alanine aminotransferase levels in patients with non-alcoholic steatohepatitis. *Aliment Pharmacol Ther* 38(2):134–143. doi: 10.1111/apt.12352.

Hui, J. M., A. Hodge, G. C. Farrell, J. G. Kench, A. Kriketos, and J. George. 2004. Beyond insulin resistance in NASH: TNF-alpha or adiponectin? *Hepatology* 40(1):46–54. doi: 10.1002/hep.20280.

Ip, B. C., K. Q. Hu, C. Liu et al. 2013. Lycopene metabolite, apo-10′-lycopenoic acid, inhibits diethylnitrosamine-initiated, high fat diet-promoted hepatic inflammation and tumorigenesis in mice. *Cancer Prev Res (Phila)* 6(12):1304–1316. doi: 10.1158/1940-6207.CAPR-13-0178.

Ip, B. C., C. Liu, L. M. Ausman, J. von Lintig, and X. D. Wang. 2014. Lycopene attenuated hepatic tumorigenesis via differential mechanisms depending on carotenoid cleavage enzyme in mice. *Cancer Prev Res (Phila)* 7(12):1219–1227. doi: 10.1158/1940-6207.CAPR-14-0154.

Ishikawa, H., A. Takaki, R. Tsuzaki et al. 2014. L-Carnitine prevents progression of non-alcoholic steatohepatitis in a mouse model with upregulation of mitochondrial pathway. *PLoS One* 9(7):e100627. doi: 10.1371/journal.pone.0100627.

Itoh, S., Y. Tsukada, Y. Motomura, and A. Ichinoe. 1979. Five patients with nonalcoholic diabetic cirrhosis. *Acta Hepatogastroenterol (Stuttg)* 26(2):90–97.

Kantartzis, K., F. Machicao, J. Machann et al. 2009. The DGAT2 gene is a candidate for the dissociation between fatty liver and insulin resistance in humans. *Clin Sci (Lond)* 116(6):531–537. doi: 10.1042/CS20080306 CS20080306 [pii].

Kawaguchi, T., Y. Sumida, A. Umemura et al. 2012. Genetic polymorphisms of the human PNPLA3 gene are strongly associated with severity of non-alcoholic fatty liver disease in Japanese. *PLoS One* 7(6):e38322. doi: 10.1371/journal.pone.0038322 PONE-D-12-07060 [pii].

Kawai, D., A. Takaki, A. Nakatsuka et al. 2012. Hydrogen-rich water prevents progression of nonalcoholic steatohepatitis and accompanying hepatocarcinogenesis in mice. *Hepatology* 56(3):912–921. doi: 10.1002/hep.25782.

Kern, M., N. Kloting, H. G. Niessen et al. 2012. Linagliptin improves insulin sensitivity and hepatic steatosis in diet-induced obesity. *PLoS One* 7(6):e38744. doi: 10.1371/journal.pone.0038744.

Lallukka, S., K. Sevastianova, J. Perttila et al. 2013. Adipose tissue is inflamed in NAFLD due to obesity but not in NAFLD due to genetic variation in PNPLA3. *Diabetologia* 56(4):886–892. doi: 10.1007/s00125-013-2829-9.

Lanthier, N., O. Molendi-Coste, Y. Horsmans, N. van Rooijen, P. D. Cani, and I. A. Leclercq. 2010. Kupffer cell activation is a causal factor for hepatic insulin resistance. *Am J Physiol Gastrointest Liver Physiol* 298(1):G107–G116. doi: 10.1152/ajpgi.00391.2009 ajpgi.00391.2009 [pii].

Larter, C. Z., S. Chitturi, D. Heydet, and G. C. Farrell. 2010. A fresh look at NASH pathogenesis. Part 1: The metabolic movers. *J Gastroenterol Hepatol* 25(4):672–690. doi: 10.1111/j.1440-1746.2010.06253.x.

Larter, C. Z. and G. C. Farrell. 2006. Insulin resistance, adiponectin, cytokines in NASH: Which is the best target to treat? *J Hepatol* 44(2):253–261. doi: 10.1016/j.jhep.2005.11.030.

Leuschner, U. F., B. Lindenthal, G. Herrmann et al. 2010. High-dose ursodeoxycholic acid therapy for non-alcoholic steatohepatitis: A double-blind, randomized, placebo-controlled trial. *Hepatology* 52(2):472–479. doi: 10.1002/hep.23727.

Li, Z., S. Yang, H. Lin et al. 2003. Probiotics and antibodies to TNF inhibit inflammatory activity and improve nonalcoholic fatty liver disease. *Hepatology* 37(2):343–350. doi: 10.1053/jhep.2003.50048 S0270913902141620 [pii].

Liu, H. Y., J. Han, S. Y. Cao et al. 2009. Hepatic autophagy is suppressed in the presence of insulin resistance and hyperinsulinemia: Inhibition of FoxO1-dependent expression of key autophagy genes by insulin. *J Biol Chem* 284(45):31484–31492. doi: 10.1074/jbc.M109.033936.

Ludwig, J., T. R. Viggiano, D. B. McGill, and B. J. Oh. 1980. Nonalcoholic steatohepatitis: Mayo Clinic experiences with a hitherto unnamed disease. *Mayo Clin Proc* 55(7):434–438.

Matteoni, C. A., Z. M. Younossi, T. Gramlich, N. Boparai, Y. C. Liu, and A. J. McCullough. 1999. Nonalcoholic fatty liver disease: A spectrum of clinical and pathological severity. *Gastroenterology* 116(6):1413–1419. doi: S0016508599005636 [pii].

Mittler, R., S. Vanderauwera, N. Suzuki et al. 2011. ROS signaling: The new wave? *Trends Plant Sci* 16(6):300–309. doi: 10.1016/j.tplants.2011.03.007 S1360-1385(11)00055-0 [pii].

Musso, G., M. Cassader, F. Rosina, and R. Gambino. 2012. Impact of current treatments on liver disease, glucose metabolism and cardiovascular risk in non-alcoholic fatty liver disease (NAFLD): A systematic review and meta-analysis of randomised trials. *Diabetologia* 55(4):885–904. doi: 10.1007/s00125-011-2446-4.

Nakamura, A., M. Yoneda, Y. Sumida et al. 2013. Modification of a simple clinical scoring system as a diagnostic screening tool for non-alcoholic steatohepatitis in Japanese patients with non-alcoholic fatty liver disease. *J Diabetes Investig* 4(6):651–658. doi: 10.1111/jdi.12101.

Nascimbeni, F., R. Pais, S. Bellentani et al. 2013. From NAFLD in clinical practice to answers from guidelines. *J Hepatol* 59(4):859–871. doi: 10.1016/j.jhep.2013.05.044.

Navarro, L. A., A. Wree, D. Povero et al. 2015. Arginase 2 deficiency results in spontaneous steatohepatitis: A novel link between innate immune activation and hepatic de novo lipogenesis. *J Hepatol* 62(2):412–420. doi: 10.1016/j.jhep.2014.09.015.

Nelson, A., D. M. Torres, A. E. Morgan, C. Fincke, and S. A. Harrison. 2009. A pilot study using simvastatin in the treatment of nonalcoholic steatohepatitis: A randomized placebo-controlled trial. *J Clin Gastroenterol* 43(10):990–994. doi: 10.1097/MCG.0b013e31819c392e.

Nelson, J. E., L. Wilson, E. M. Brunt et al. 2011. Relationship between the pattern of hepatic iron deposition and histological severity in nonalcoholic fatty liver disease. *Hepatology* 53(2):448–457. doi: 10.1002/hep.24038.

Novo, E., C. Busletta, L. V. Bonzo et al. 2011. Intracellular reactive oxygen species are required for directional migration of resident and bone marrow-derived hepatic pro-fibrogenic cells. *J Hepatol* 54(5):964–974. doi: S0168-8278(10)00923-2 [pii] 10.1016/j.jhep.2010.09.022.

Ohsawa, I., M. Ishikawa, K. Takahashi et al. 2007. Hydrogen acts as a therapeutic antioxidant by selectively reducing cytotoxic oxygen radicals. *Nat Med* 13(6):688–694. doi: 10.1038/nm1577.

Ono, M. and T. Saibara. 2006. Clinical features of nonalcoholic steatohepatitis in Japan: Evidence from the literature. *J Gastroenterol* 41(8):725–732. doi: 10.1007/s00535-006-1876-0.

Pacifico, L., C. Anania, F. Martino et al. 2011. Management of metabolic syndrome in children and adolescents. *Nutr Metab Cardiovasc Dis* 21(6):455–466. doi: S0939-4753(11)00039-1 [pii] 10.1016/j.numecd.2011.01.011.

Park, S. W., Y. Zhou, J. Lee et al. 2010. The regulatory subunits of PI3K, p85alpha and p85beta, interact with XBP-1 and increase its nuclear translocation. *Nat Med* 16(4):429–437. doi: 10.1038/nm.2099 nm.2099 [pii].

Pessayre, D. 2007. Role of mitochondria in non-alcoholic fatty liver disease. *J Gastroenterol Hepatol* 22(Suppl. 1):S20–S27. doi: JGH4640 [pii] 10.1111/j.1440-1746.2006.04640.x.

Qiao, A., J. Liang, Y. Ke et al. 2011. Mouse patatin-like phospholipase domain-containing 3 influences systemic lipid and glucose homeostasis. *Hepatology* 54(2):509–521. doi: 10.1002/hep.24402.

Qiao, S., M. Dennis, X. Song et al. 2015. A REDD1/TXNIP pro-oxidant complex regulates ATG4B activity to control stress-induced autophagy and sustain exercise capacity. *Nat Commun* 6:7014. doi: 10.1038/ncomms8014.

Rakoski, M. O., A. G. Singal, M. A. Rogers, and H. Conjeevaram. 2010. Meta-analysis: Insulin sensitizers for the treatment of non-alcoholic steatohepatitis. *Aliment Pharmacol Ther* 32(10):1211–1221. doi: 10.1111/j.1365-2036.2010.04467.x.

Rolo, A. P., J. S. Teodoro, and C. M. Palmeira. 2012. Role of oxidative stress in the pathogenesis of non-alcoholic steatohepatitis. *Free Radic Biol Med* 52(1):59–69. doi: 10.1016/j.freeradbiomed.2011.10.0 03S0891-5849(11)00632-0 [pii].

Romeo, S., J. Kozlitina, C. Xing et al. 2008. Genetic variation in PNPLA3 confers susceptibility to nonalcoholic fatty liver disease. *Nat Genet* 40(12):1461–1465. doi: 10.1038/ng.257ng.257 [pii].

Sanyal, A. J., N. Chalasani, K. V. Kowdley et al. 2010. Pioglitazone, vitamin E, or placebo for nonalcoholic steatohepatitis. *N Engl J Med* 362(18):1675–1685. doi: 10.1056/NEJMoa0907929.

Scarpulla, R. C. 2011. Metabolic control of mitochondrial biogenesis through the PGC-1 family regulatory network. *Biochim Biophys Acta* 1813(7):1269–1278. doi: 10.1016/j.bbamcr.2010.09.0 19S0167-4889(10)00260-0 [pii].

Servidddio, G., F. Bellanti, R. Tamborra et al. 2008. Alterations of hepatic ATP homeostasis and respiratory chain during development of non-alcoholic steatohepatitis in a rodent model. *Eur J Clin Invest* 38(4):245–252. doi: 10.1111/j.1365-2362.2008.01936.x.

Shyangdan, D., C. Clar, N. Ghouri et al. 2011. Insulin sensitisers in the treatment of non-alcoholic fatty liver disease: A systematic review. *Health Technol Assess* 15(38):1–110. doi: 10.3310/hta15380.

Siddique, A., J. E. Nelson, B. Aouizerat, M. M. Yeh, K. V. Kowdley, and Nash Clinical Research Network. 2013. Iron deficiency in patients with nonalcoholic fatty liver disease is associated with obesity, female gender, and low serum hepcidin. *Clin Gastroenterol Hepatol*. doi: 10.1016/j.cgh.2013.11.017.

Singal, A. G., H. Manjunath, A. C. Yopp et al. 2014. The effect of PNPLA3 on fibrosis progression and development of hepatocellular carcinoma: A meta-analysis. *Am J Gastroenterol* 109(3):325–334. doi: 10.1038/ajg.2013.476.

Sookoian, S. and C. J. Pirola. 2011. Meta-analysis of the influence of I148M variant of patatin-like phospholipase domain containing 3 gene (PNPLA3) on the susceptibility and histological severity of nonalcoholic fatty liver disease. *Hepatology* 53(6):1883–1894. doi: 10.1002/hep.24283.

Sumida, Y., M. Yoneda, H. Hyogo et al. 2011. A simple clinical scoring system using ferritin, fasting insulin, and type IV collagen 7S for predicting steatohepatitis in nonalcoholic fatty liver disease. *J Gastroenterol* 46(2):257–268. doi: 10.1007/s00535-010-0305-6.

Sutti, S., A. Jindal, I. Locatelli et al. 2014. Adaptive immune responses triggered by oxidative stress contribute to hepatic inflammation in NASH. *Hepatology* 59(3):886–897. doi: 10.1002/hep.26749.

Svegliati-Baroni, G., S. Saccomanno, C. Rychlicki et al. 2011. Glucagon-like peptide-1 receptor activation stimulates hepatic lipid oxidation and restores hepatic signalling alteration induced by a high-fat diet in nonalcoholic steatohepatitis. *Liver Int* 31(9):1285–1297. doi: 10.1111/j.1478-3231.2011.02462.x.

Takaki, A., D. Kawai, and K. Yamamoto. 2013. Multiple hits, including oxidative stress, as pathogenesis and treatment target in non-alcoholic steatohepatitis (NASH). *Int J Mol Sci* 14(10):20704–20728. doi: 10.3390/ijms141020704.

Takaki, A., D. Kawai, and K. Yamamoto. 2014. Molecular mechanisms and new treatment strategies for non-alcoholic steatohepatitis (NASH). *Int J Mol Sci* 15(5):7352–7379. doi: 10.3390/ijms15057352.

Tilg, H. and A. R. Moschen. 2010. Evolution of inflammation in nonalcoholic fatty liver disease: The multiple parallel hits hypothesis. *Hepatology* 52(5):1836–1846. doi: 10.1002/hep.24001.

Trachootham, D., J. Alexandre, and P. Huang. 2009. Targeting cancer cells by ROS-mediated mechanisms: A radical therapeutic approach? *Nat Rev Drug Discov* 8(7):579–591. doi: 10.1038/nrd2803.

Ushio, M., Y. Nishio, O. Sekine et al. 2013. Ezetimibe prevents hepatic steatosis induced by a high-fat but not a high-fructose diet. *Am J Physiol Endocrinol Metab* 305(2):E293–E2304. doi: 10.1152/ajpendo.00442.2012.

Valenti, L., A. Alisi, E. Galmozzi et al. 2010. I148M patatin-like phospholipase domain-containing 3 gene variant and severity of pediatric nonalcoholic fatty liver disease. *Hepatology* 52(4):1274–1280. doi: 10.1002/hep.23823.

Wang, B., S. H. Hsu, W. Frankel, K. Ghoshal, and S. T. Jacob. 2012. Stat3-mediated activation of microRNA-23a suppresses gluconeogenesis in hepatocellular carcinoma by down-regulating glucose-6-phosphatase and peroxisome proliferator-activated receptor gamma, coactivator 1 alpha. *Hepatology* 56(1):186–197. doi: 10.1002/hep.25632.

Watanabe, S., E. Hashimoto, K. Ikejima et al. 2015. Evidence-based clinical practice guidelines for nonalcoholic fatty liver disease/nonalcoholic steatohepatitis. *Hepatol Res* 45(4):363–377. doi: 10.1111/hepr.12511.

Watson, J. 2013. Oxidants, antioxidants and the current incurability of metastatic cancers. *Open Biol* 3(1):120144. doi: 10.1098/rsob.120144.

Yae, T., K. Tsuchihashi, T. Ishimoto et al. 2012. Alternative splicing of CD44 mRNA by ESRP1 enhances lung colonization of metastatic cancer cell. *Nat Commun* 3:883. doi: 10.1038/ncomms1892.

Yamaguchi, K., L. Yang, S. McCall et al. 2007. Inhibiting triglyceride synthesis improves hepatic steatosis but exacerbates liver damage and fibrosis in obese mice with nonalcoholic steatohepatitis. *Hepatology* 45(6):1366–1374. doi: 10.1002/hep.21655.

Yang, L., P. Li, S. Fu, E. S. Calay, and G. S. Hotamisligil. 2010. Defective hepatic autophagy in obesity promotes ER stress and causes insulin resistance. *Cell Metab* 11(6):467–478. doi: 10.1016/j.cmet.2010.04.005.

Yatsuji, S., E. Hashimoto, M. Tobari, M. Taniai, K. Tokushige, and K. Shiratori. 2009. Clinical features and outcomes of cirrhosis due to non-alcoholic steatohepatitis compared with cirrhosis caused by chronic hepatitis C. *J Gastroenterol Hepatol* 24(2):248–254. doi: JGH5640 [pii]10.1111/j.1440-1746.2008.05640.x.

Yokohama, S., M. Yoneda, M. Haneda et al. 2004. Therapeutic efficacy of an angiotensin II receptor antagonist in patients with nonalcoholic steatohepatitis. *Hepatology* 40(5):1222–1225. doi: 10.1002/hep.20420.

Yoneda, M., K. Fujita, Y. Nozaki et al. 2010. Efficacy of ezetimibe for the treatment of non-alcoholic steatohepatitis: An open-label, pilot study. *Hepatol Res* 40(6):566–573. doi: 10.1111/j.1872-034X.2010.00644.x.

Yu, X. X., S. F. Murray, S. K. Pandey et al. 2005. Antisense oligonucleotide reduction of DGAT2 expression improves hepatic steatosis and hyperlipidemia in obese mice. *Hepatology* 42(2):362–371. doi: 10.1002/hep.20783.

Zein, C. O., R. Lopez, X. Fu et al. 2012. Pentoxifylline decreases oxidized lipid products in nonalcoholic steatohepatitis: New evidence on the potential therapeutic mechanism. *Hepatology* 56(4):1291–1299. doi: 10.1002/hep.25778.

Zeng, T. S., F. M. Liu, J. Zhou, S. X. Pan, W. F. Xia, and L. L. Chen. 2015. Depletion of Kupffer cells attenuates systemic insulin resistance, inflammation and improves liver autophagy in high-fat diet fed mice. *Endocr J.* 62(7):615–626 doi: 10.1507/endocrj.EJ15-0046.

26 Reactive Oxygen Species and Diabetic Retinopathy

Mohammad Shamsul Ola and Haseeb Ahsan

CONTENTS

ABSTRACT

Oxidative stress is considered as a central factor in the pathophysiology of many diseases, including diabetic retinopathy (DR). Oxidative stress is caused by an imbalance between the production of reactive oxygen/nitrogen species (ROS/RNS) and its antioxidant level to detoxify these species. Increased levels of ROS/RNS reported in the retinas of diabetic animals have been found to be strongly associated with DR. The generation of ROS/RNS contributes to DR through a number of pathological processes, including increased retinal vascular permeability, reduced endothelial and neuronal cell survival, and retinal inflammation. A number of metabolic stresses have been implicated in the pathophysiology of DR. The retina responds to dysregulated metabolites, including hyperglycemia, through a number of biochemical changes, including the activation of protein kinase C (PKC) and polyol pathway, increased advanced glycation end products (AGEs) formation, and oxidative stress. In an environment of increased oxidative stress, free radicals cause the dysfunction of retinal cells and proteins. Thus, the inhibition of oxidative stress in retinal cells by antioxidants may augment oxidative damage in the retina. Therefore, antioxidant therapy may be an important step in combating sight-threatening complications, including blindness, in DR.

26.1 INTRODUCTION

Diabetes has become an epidemic in the society. It is a metabolic disease in which the levels of a number of metabolites, including glucose, hormones, lipids, nutrients and their metabolic pathways, are dysregulated, thus implicating in the pathophysiology of complications of diabetes, including diabetic retinopathy (DR) (Ola et al., 2012, 2013a; Ola and Alhomida, 2014; Ahsan, 2015). DR is one of the major complications of diabetes and the leading cause of vision loss or blindness among working adults. In DR, the imbalance in metabolites changes the production of both neurotrophic and vasoactive factors, which damages both neuronal and vascular cells, leading to DR.

Over the last 20 years, numerous studies investigating pathophysiological mechanisms have been conducted; however, the precise mechanism of how dysregulated metabolites damage the retina is elusive. A large number of studies have suggested hyperglycemia-induced imbalance in metabolic pathways as a major factor in damaging retina in diabetes. Landmark studies of Brownlee (2005) suggested a unifying mechanism of mitochondrial superoxide ($O_2^{\bullet-}$) production in diabetic retina under hyperglycemic conditions as the central player in damaging the retina (Nishikawa et al., 2000; Brownlee, 2005). An enhanced mitochondrial source of $O_2^{\bullet-}$ as a driving force for diabetes complications has been widely accepted, despite inconclusive support from several antioxidant clinical trials and some new studies questioning the mitochondrial production of $O_2^{\bullet-}$ and hyperglycemic conditions (Ola et al., 2006; Sharma, 2015a,b). Moreover, there are other endogenous cytosolic sources of $O_2^{\bullet-}$ besides mitochondria, such as NADPH oxidase (NOX), nitric oxide synthase (NOS), lipoxygenases, and xanthine oxidase (XO). Numerous studies reported that these activated enzymes increase $O_2^{\bullet-}$ production and damage the diabetic retina. In this chapter, we will critically review and discuss different sources of oxidative stress in diabetes and its implication in damaging retinal cells, thus leading to DR.

26.2   OXIDATIVE STRESS IN DIABETES

Oxidative stress is considered as a central factor involved in the pathophysiology of many diseases, including DR. Oxidative stress is caused by an imbalance between the production of reactive oxygen/nitrogen species (ROS/RNS) and its antioxidant level to detoxify these species. ROS consists of $O_2^{\bullet-}$, hydrogen peroxide (H_2O_2), and hydroxyl radicals ($^{\bullet}OH$). RNS consists of peroxynitrite, which results from the interaction between nitric oxide (NO) and the $O_2^{\bullet-}$. Peroxynitrite is a powerful oxidizing and nitrating agent. Increased ROS/RNS result in metabolic abnormalities, and these also can produce more ROS. Exposure to ROS/RNS damages the electron transport chain (ETC) system and mitochondrial DNA (mtDNA) within the mitochondria. Damaged mtDNA and ETC produce more ROS/RNS, and the vicious cycle propagates. Many genes important in the generation and neutralization of ROS are epigenetically modified further, increasing ROS/RNS levels. Increased levels of ROS/RNS reported in the retinas of diabetic animals have been suggested to be strongly associated with DR. The generation of ROS/RNS also contributes to DR through a number of pathological processes, including increased retinal vascular permeability, reduced endothelial and neuronal cell survival, and retinal inflammation (Coucha et al., 2015).

26.3   MECHANISMS OF OXIDATIVE STRESS IN DIABETIC RETINOPATHY

26.3.1   Hyperglycemia-Induced Oxidative Stress

Hyperglycemia-induced oxidative stress in the retina plays a pivotal role in the development of DR by damaging retinal cells (Sato et al., 2005). Numerous studies showed that high glucose increases the flux through glycolytic pathways, resulting in increased cytosolic NADH and tissue lactate to pyruvate ratios. This increases the tricarboxylic acid (TCA) cycle flux putting extra pressure on mitochondria to generate more electrons, thereby producing excess levels of $O_2^{\bullet-}$ (Ido et al., 1997; Madsen-Bouterse and Kowluru, 2008). This process of $O_2^{\bullet-}$ production due to excess glucose in diabetic patients is considered as the central mechanism causing tissue damage.

Nyengaard et al. (2004) proposed the concept of hyperglycemia-induced pseudohypoxia, which generates oxidative stress by changing the redox status in the retina; this was refuted by us and by Diederen et al. (2006b) and Ola et al. (2006). We have reported that hyperglycemia per se might not be increasing ROS production in mitochondria since there was no increased flux through glycolytic pathway or tricarboxylic acid cycle (Ola et al., 2006). Further, recent data suggest that excess glucose can cause a reduction in $O_2^{\bullet-}$ production in diabetic organs such as the heart, kidney, and liver (Herlein et al., 2009; Dugan et al., 2013). Another study revealed that excess intracellular glucose blocked pyruvate entry into mitochondria, which results in the reduced activity of the ETC and less $O_2^{\bullet-}$ production (Dugan et al., 2013). Ola et al.'s unpublished study, however, suggests that exposure of cultured retinal cells to high glucose could not increase the ROS level. Therefore, the role of mitochondria as the driving force for diabetic complications, including DR, needs to be reevaluated.

Other sources of ROS production by hyperglycemia include flux through the activation of AGEs, aldose reductase (AR), hexosamine, PKC, and poly(ADP-ribose) polymerase (PARP) pathways. One of the major pathways by which hyperglycemia contributes to cellular damage is through the generation of AGEs. Chronic hyperglycemia creates a favorable environment for nonenzymatic condensation reactions between reduced glucose and amine residues of proteins, nucleic acids, and lipids resulting in an irreversibly cross-linked complex of compounds collectively termed "advanced glycation end products" (Yamagishi et al., 2005; Ola et al., 2012). Elevated levels of AGEs have been found in retinal blood vessels, serum, and vitreous of diabetic patients and contribute to DR pathophysiology mainly by disturbing microvascular homeostasis through binding of AGEs to the receptor of AGEs (RAGEs) (Lu et al., 1998; Stitt, 2003; Goldin and Beckman, 2006). AGEs also cause endothelial cells to express different adhesion and chemoattraction factors through intracellular ROS generation (Ola et al., 2012). Furthermore, AGEs activate nuclear factor-κB (NF-κB) and NOX, which in turn increases ROS production and apoptosis of retinal cells (Zheng et al., 2004; Ola et al., 2012).

Hyperglycemia increases de novo synthesis of diacylglycerol, which activates PKC that in turn phosphorylates target proteins. Studies indicate that PKC activates the transcription factor NF-κB that alters NADPH-dependent oxidases and increases ROS formation (Frey et al., 2002). The effects of PKC activation on microvasculature include increased permeability, increased flow, apoptosis, and proliferation of endothelial cells (Kowluru and Chan, 2007). The cytosolic enzyme, aldose reductase, is a rate-limiting enzyme of polyol pathway that converts high intracellular glucose concentrations to sorbitol followed by oxidation to fructose using NADPH. The consumption of NADPH by this reaction inhibits glutathione level, which increases oxidative stress (Chung and Chung, 2005; Lorenzi, 2007).

Hyperglycemia induces the hexosamine pathway, which is implicated in the pathology of diabetes-induced oxidative stress and its complications. Flux through the hexosamine pathway results in an increased activation of transcription factors implicated in diabetic complications (Edwards et al., 2008). Hyperglycemia also activates PARP, which further cleaves nicotinamide adenine dinucleotide (NAD^+) to nicotinamide and ADP-ribose residues. Excessive amounts of PARP binding to NAD^+ ultimately result in decreased glycolysis and eventually cell death. Second, PARP inhibits glyceraldehyde-3-phosphate dehydrogenase (GAPDH), further increasing ROS and RNS production and causing DNA strand breaks and endothelial and neuronal dysfunction in DR (Ola et al., 2012).

26.3.2 Other Metabolic Sources of Oxidative Stress

An increased level of glutamate has been reported in the retina of diabetic patients, which may be damaging the retinal neurons by excitotoxicity (Diederen et al., 2006a). Increased extracellular level of glutamate in the neuronal tissue activates N-methyl D-Aspartate (NMDA) receptors, depolarizing the neuronal cells, which increases the influx of calcium and sodium ions into the cell and in turn generates free radicals and induces apoptosis (Zhang and Bhavnani, 2006). High levels of branched-chain amino acids found in the retina of diabetic rats may increase extracellular glutamate levels that cause excitotoxicity and oxidative stress (Gowda et al., 2011). In addition, an increased level of D-serine in diabetic eyes can also cause excitotoxicity by activating the NMDA receptor, thereby damaging neurons (Jiang et al., 2011).

Homocysteine is another potential metabolite whose elevated level has been associated with DR (Brazionis et al., 2008). Homocysteine level is reduced by the enzyme methionine synthase in the presence of vitamin B12 and folate as cofactors (Wright et al., 2008). We had reported a reduced expression of the folate transporter and a decreased folate level in the diabetic retina (Naggar et al., 2002). Thus, a lower level of folate in a diabetic retina may cause an increase in homocysteine levels. An elevated homocysteine level has been found to induce apoptosis in retinal ganglion cells (RGC)

(Ganapathy et al., 2011a). Homocysteine has also been shown to activate NMDA receptors and thereby has the potential to cause excitotoxicity of RGCs in diabetic retina (Ganapathy et al., 2011b).

Kynurenic acid is the product of tryptophan metabolism, which is suggested to play an important role in neurodegeneration. A correlation between decreased levels of kynurenic acid and glutamate excitotoxicity and free radical generation has been found. Kynurenic acid has been found to influence the excitotoxicity of neuronal cells by homocysteine (Chmiel-Perzyκska et al., 2007). Diabetes-induced neurodegenerative metabolites may attenuate neurotrophic factors, including brain-derived neurotrophic factor (BDNF) and nerve growth factor (NGF) by increasing oxidants in the diabetic retina (Al-Gayyar et al., 2011; Ozawa et al., 2012; Ola and Alhomida, 2014). These neurotrophic factors are known to regulate axonal growth and synaptic activity, as well as neuronal survival (Cunha et al., 2010). We and others reported a decreased level of neurotrophic factors in the diabetic retina; however, antioxidant treatments ameliorated their levels (Sasaki et al., 2010; Ola et al., 2015). Therefore, a strategy should be devised to preserve neurotrophic factors by antioxidants in the retina of diabetic rats to protect neurons.

High content of lipid in diabetes increases the risk of DR (Ansquer et al., 2009). Although, it is not fully understood how increased lipid levels affect the onset and progression of DR. However, it is suggested to occur through alterations in the levels of compounds such as ketone bodies, acylcarnitine, oxidized fatty acids, polyunsaturated fatty acids, sphingolipids, and ceramides (Adibhatla and Hatcher, 2007; Fox and Kester, 2010; Al-Shabrawey et al., 2011). In diabetes, a high-fat diet may increase oxidative stress (Kowluru and Chan, 2007) and contribute to the inflammatory response in the retina (Antonetti et al., 2006). Increased levels of lipid peroxidation have been observed both in diabetic rodents and in diabetic specimens of humans as measured by the products of lipid peroxidation, such as thiobarbituric acid reactive species (TBARS) and malondialdehyde (MDA). MDA metabolite serves as a marker for the DR severity (Gupta and Chari, 2005). Supplementation of vitamin E to diabetic animals and humans exhibited significantly lowered lipid peroxidation products in plasma and several tissues, including the retina and lens (Naziroglu et al., 1999; Di Leo et al., 2003; Devaraj et al., 2008). Thus, it suggests a potential role of vitamin E as an antioxidant in inhibiting lipid peroxidation and, thereby, retinopathy in diabetes (Pazdro and Burgess, 2010).

A large body of evidence suggests that activated metabolites of the renin–angiotensin system (RAS) in diabetic retina play a significant role in DR. Angiotensin II, a component of RAS that activates angiotensin type 1 receptor (AT1R), produces ROS, which damages retinal cells and particularly retinal ganglion cells in the diabetic retina (Wilkinson-Berka, 2006; Silva et al., 2009). Increased levels of AT1R in diabetic retina resulted in impaired neuronal function and the AT1R blocker telmisartan suppressed the impaired inner retinal function (Kurihara et al., 2008). Recently, we found a beneficial effect of AT1R blocker, telmisartan, toward neuroprotection in the retina of diabetic rats (Ola et al., 2013b). Thus, the role of RAS in increased reactivity and its therapeutic target may have an important role in neuroprotection in DR.

Other sources of oxidative stress are impaired activities of antioxidant defense enzymes such as glutathione reductase, glutathione peroxidase, superoxide dismutase, and catalase in diabetic retina, which provide strong evidence that oxidative stress potentially contributes to the pathogenesis of diabetes retinopathy (Hartnett et al., 2000; Kowluru and Chan, 2007; Al-Shabrawey et al., 2008; Madsen-Bouterse and Kowluru 2008). Therefore, the targeted use of antioxidant or agents to activate antioxidant enzymes will be an important therapeutic strategy to achieve optimum levels of ROS required for attenuating the pathology of DR.

26.4 ANTIOXIDANTS

Antioxidants that readily detoxify ROS/RNS and repair damage to biomolecules, including vitamins A, C, and E; glutathione; minerals; cofactors; and enzymes such as superoxide dismutase, catalase, glutathione peroxidase; and glutathione reductase are compromised. Insufficient antioxidant defenses damages protein, lipids, and DNA, resulting in altered function and, ultimately, apoptosis (Birben et al., 2012). In diabetic retina, these antioxidant systems are not effective in balancing the levels of oxidants, which makes retinal cells vulnerable to damage. Antioxidants have generated beneficial effects in ameliorating retinopathy in diabetic rodents, but clinical studies have been limited and are not very encouraging. With the ongoing use of antioxidants for various chronic diseases, there is a need for a controlled trial to recognize their potential in ameliorating the development of DR (Kowluru and Mishra, 2015). For DR, antioxidant supplements may have a small benefit, if any, but only as an adjunct to glycemic control. In very high-risk premature retinopathy and retinitis pigmentosa, different antioxidant supplements may be beneficial (Grover and Samson, 2014). Therapies that prevent ROS accumulation, maintain mitochondrial homeostasis, and protect DNA appear to be the most likely strategies to prevent the development of DR. Thus, therapies that could target multiple steps of oxidative stress and mitochondrial damage should provide a hope for the prevention of this multifactorial blinding complication of diabetes (Santos et al., 2011).

26.5 CONCLUSION

DR remains the leading cause of blindness in young adults and the mechanism(s) of this blinding disease remains elusive. However, oxidative stress is considered as the major factor involved in the pathophysiology of DR. Increased levels of ROS/RNS reported in the retinas of diabetic animals have been found to be strongly associated with a number of dysregulated metabolites including hyperglycemia. Mitochondrial sources of oxidative stress may damage mitochondria and the electron transport system (ETC), ultimately leading to mitochondrial DNA (mtDNA) damage. Damaged mtDNA may impair the transcription of genes important in the neutralization of ROS, which may further increase ROS, and the vicious cycle continues (Kowluru and Mishra, 2015). In addition, as discussed earlier, there are various nonmitochondrial sources of oxidative stress that may also implicate damage to retinal cells in diabetes. Targeted antioxidant therapy to reduce ROS/RNS production may lead to beneficial effects on retinal vascular permeability, endothelial and neuronal damage, and retinal inflammation in DR.

ACKNOWLEDGMENTS

MSO acknowledges King Abdul Aziz City for Science and Technology (KACST) (ARP 30-23) and the Department of Biochemistry, College of Science, King Saud University, for the research support.

REFERENCES

Adibhatla RM, Hatcher JF. Role of lipids in brain injury and diseases. *Future Lipidol.* 2007;2(4):403–422.

Ahsan H. Diabetic retinopathy-biomolecules and multiple pathophysiology. *Diabetes Metab Syndr.* 2015; 9(1):51–54.

Al-Gayyar MM, Matragoon S, Pillai BA, Ali TK, Abdelsaid MA, El-Remessy AB. Epicatechin blocks pro-nerve growth factor (proNGF)-mediated retinal neurodegeneration via inhibition of p75 neurotrophin receptor expression in a rat model of diabetes [corrected]. *Diabetologia.* 2011;54(3):669–680. Erratum in: *Diabetologia.* 2011;54(3):713.

Al-Shabrawey M, Mussell R, Kahook K, Tawfik A, Eladl M, Sarthy V, Nussbaum J et al. Increased expression and activity of 12-lipoxygenase in oxygen-induced ischemic retinopathy and proliferative diabetic retinopathy: Implications in retinal neovascularization. *Diabetes.* 2011;60(2):614–624. Erratum in: *Diabetes.* 2013;62(3):998.

Al-Shabrawey M, Rojas M, Sanders T, Behzadian A, El-Remessy A, Bartoli M, Parpia AK, Liou G, Caldwell RB. Role of NADPH oxidase in retinal vascular inflammation. *Invest Ophthalmol Vis Sci.* 2008;49(7):3239–3244.

Ansquer JC, Foucher C, Aubonnet P, Le Malicot K. Fibrates and microvascular complications in diabetes—Insight from the FIELD study. *Curr Pharm Des.* 2009;15(5):537–552.

Antonetti DA, Barber AJ, Bronson SK, Freeman WM, Gardner TW, Jefferson LS, Kester M et al. Diabetic retinopathy: Seeing beyond glucose-induced microvascular disease. *Diabetes.* 2006;55(9):2401–2411.

Birben E, Sahiner UM, Sackesen C, Erzurum S, Kalayci O. Oxidative stress and antioxidant defense. *World Allergy Organ J.* 2012;5(1):9–19.

Brazionis L, Rowley K Sr., Itsiopoulos C, Harper CA, O'Dea K. Homocysteine and diabetic retinopathy. *Diabetes Care.* 2008;31(1):50–56.

Brownlee M. The pathobiology of diabetic complications: A unifying mechanism. *Diabetes.* 2005;54(6):1615–1625.

Chmiel-Perzyκska I, Perzyκski A, Wielosz M, Urbaκska EM. Hyperglycemia enhances the inhibitory effect of mitochondrial toxins and D,L-homocysteine on the brain production of kynurenic acid. *Pharmacol Rep.* 2007;59:268–273.

Chung SS, Chung SK. Aldose reductase in diabetic microvascular complications. *Curr Drug Targets.* 2005;6(4):475–486.

Coucha M, Elshaer SL, Eldahshan WS, Mysona BA, El-Remessy AB. Molecular mechanisms of diabetic retinopathy: Potential therapeutic targets. *Middle East Afr J Ophthalmol.* 2015;22(2):135–144.

Cunha C, Brambilla R, Thomas KL. A simple role for BDNF in learning and memory? *Front Mol Neurosci.* 2010;3:1.

Devaraj S, Leonard S, Traber MG, Jialal I. Gamma-tocopherol supplementation alone and in combination with alpha-tocopherol alters biomarkers of oxidative stress and inflammation in subjects with metabolic syndrome. *Free Radic Biol Med.* 2008;44:1203–1208.

Di Leo MA, Ghirlanda G, Gentiloni Silveri N, Giardina B, Franconi F, Santini SA. Potential therapeutic effect of antioxidants in experimental diabetic retina: A comparison between chronic taurine and vitamin E plus selenium supplementations. *Free Radic Res.* 2003;37:323–330.

Diederen RM, La Heij EC, Deutz NE, Kijlstra A, Kessels AG, van Eijk HM, Liem AT, Dieudonné S, Hendrikse F. Increased glutamate levels in the vitreous of patients with retinal detachment. *Exp Eye Res.* 2006a;83(1):45–50.

Diederen RM, Starnes CA, Berkowitz BA, Winkler BS. Reexamining the hyperglycemic pseudohypoxia hypothesis of diabetic oculopathy. *Invest Ophthalmol Vis Sci.* 2006b;47(6):2726–2731.

Dugan LL, You YH, Ali SS, Diamond-Stanic M, Miyamoto S, DeCleves AE, Andreyev A et al. AMPK dysregulation promotes diabetes-related reduction of superoxide and mitochondrial function. *J Clin Invest.* 2013;123(11):4888–4899.

Edwards JL, Vincent AM, Cheng HT, Feldman EL. Diabetic neuropathy: Mechanisms to management. *Pharmacol Ther.* 2008;120(1):1–34.

Fox TE, Kester M. Therapeutic strategies for diabetes and complications: A role for sphingolipids? *Adv Exp Med Biol.* 2010;688:206–216.

Frey RS, Rahman A, Kefer JC, Minshall RD, Malik AB. PKCzeta regulates TNF-alpha-induced activation of NADPH oxidase in endothelial cells. *Circ Res.* 2002;90(9):1012–1019.

Ganapathy PS, Perry RL, Tawfik A, Smith RM, Perry E, Roon P, Bozard BR, Ha Y, Smith SB. Homocysteine-mediated modulation of mitochondrial dynamics in retinal ganglion cells. *Invest Ophthalmol Vis Sci.* 2011a;52:5551–5558.

Ganapathy PS, White RE, Ha Y, Bozard BR, McNeil PL, Caldwell RW, Kumar S, Black SM, Smith SB. The role of N-methyl-D-aspartate receptor activation in homocysteine-induced death of retinal ganglion cells. *Invest Ophthalmol Vis Sci.* 2011b;52:5515–5524.

Goldin A, Beckman JA, Schmidt AM, Creager MA. Advanced glycation end products: Sparking the development of diabetic vascular injury. *Circulation.* 2006;114(6):597–605.

Gowda K, Zinnanti WJ, LaNoue KF. The influence of diabetes on glutamate metabolism in retinas. *J Neurochem.* 2011;117(2):309–320.

Grover AK, Samson SE. Antioxidants and vision health: Facts and fiction. *Mol Cell Biochem.* 2014;388(1–2):173–183.

Gupta MM, Chari S. Lipid peroxidation and antioxidant status in patients with diabetic retinopathy. *Indian J Physiol Pharmacol.* 2005;49(2):187–192.

Hartnett ME, Stratton RD, Browne RW, Rosner BA, Lanham RJ, Armstrong D. Serum markers of oxidative stress and severity of diabetic retinopathy. *Diabetes Care.* 2000;23(2):234–240.

Herlein JA, Fink BD, O'Malley Y, Sivitz WI. Superoxide and respiratory coupling in mitochondria of insulin-deficient diabetic rats. *Endocrinology.* 2009;150(1):46–55.

Ido Y, Kilo C, Williamson JR. Cytosolic NADH/NAD+, free radicals, and vascular dysfunction in early diabetes mellitus. *Diabetologia*. 1997;40(Suppl. 2):S115–S117.

Jiang H, Fang J, Wu B, Yin G, Sun L, Qu J, Barger SW, Wu S. Overexpression of serine racemase in retina and overproduction of D-serine in eyes of streptozotocin-induced diabetic retinopathy. *J Neuroinflammation*. 2011;8:119.

Kowluru RA, Chan PS. Oxidative stress and diabetic retinopathy. *Exp Diabetes Res*. 2007;2007:43603.

Kowluru RA, Mishra M. Oxidative stress, mitochondrial damage and diabetic retinopathy. *Biochim Biophys Acta*. 2015;1852(11):2474–2483.

Kurihara T, Ozawa Y, Nagai N, Shinoda K, Noda K, Imamura Y, Tsubota K, Okano H, Oike Y, Ishida S. Angiotensin II type 1 receptor signaling contributes to synaptophysin degradation and neuronal dysfunction in the diabetic retina. *Diabetes*. 2008;57(8):2191–2198.

Lorenzi M. The polyol pathway as a mechanism for diabetic retinopathy: Attractive, elusive, and resilient. *Exp Diabetes Res*. 2007;2007:61038.

Lu M, Kuroki M, Amano S, Tolentino M, Keough K, Kim I, Bucala R, Adamis AP. Advanced glycation end products increase retinal vascular endothelial growth factor expression. *J Clin Invest*. 1998;101(6):1219–1224.

Madsen-Bouterse SA, Kowluru RA. Oxidative stress and diabetic retinopathy: Pathophysiological mechanisms and treatment perspectives. *Rev Endocr Metab Disord*. 2008;9(4):315–327.

Naggar H, Ola MS, Moore P, Huang W, Bridges CC, Ganapathy V, Smith SB. Downregulation of reduced-folate transporter by glucose in cultured RPE cells and in RPE of diabetic mice. *Invest Ophthalmol Vis Sci*. 2002;43(2):556–563.

Naziroglu M, Dilsiz N, Cay M. Protective role of intraperitoneally administered vitamins C and E and selenium on the levels of lipid peroxidation in the lens of rats made diabetic with streptozotocin. *Biol Trace Elem Res*. 1999;70:223–232.

Nishikawa T, Edelstein D, Du XL, Yamagishi S, Matsumura T, Kaneda Y, Yorek MA et al. Normalizing mitochondrial superoxide production blocks three pathways of hyperglycaemic damage. *Nature*. 2000;404(6779):787–790.

Nyengaard JR, Ido Y, Kilo C, Williamson JR. Interactions between hyperglycemia and hypoxia: Implications for diabetic retinopathy. *Diabetes*. 2004;53(11):2931–2938.

Ola MS, Ahmed MM, Abuohashish HM, Al-Rejaie SS, Alhomida AS. Telmisartan ameliorates neurotrophic support and oxidative stress in the retina of streptozotocin-induced diabetic rats. *Neurochem Res*. 2013b;38(8):1572–1579.

Ola MS, Ahmed MM, Ahmad R, Abuohashish HM, Al-Rejaie SS, Alhomida AS. Neuroprotective effects of rutin in streptozotocin-induced diabetic rat retina. *J Mol Neurosci*. 2015;56(2):440–448.

Ola MS, Alhomida AS. Neurodegeneration in diabetic retina and its potential drug targets. *Curr Neuropharmacol*. 2014;12(4):380–386.

Ola MS, Berkich DA, Xu Y, King MT, Gardner TW, Simpson I, LaNoue KF. Analysis of glucose metabolism in diabetic rat retinas. *Am J Physiol Endocrinol Metab*. 2006;290(6) E1057–E1067.

Ola MS, Nawaz MI, Khan HA, Alhomida AS. Neurodegeneration and neuroprotection in diabetic retinopathy. *Int J Mol Sci*. 2013a;14(2):2559–2572.

Ola MS, Nawaz MI, Siddiquei MM, Al-Amro S, Abu El-Asrar AM. Recent advances in understanding the biochemical and molecular mechanism of diabetic retinopathy. *J Diabetes Complications* 2012;26(1):56–64.

Ozawa Y, Sasaki M, Takahashi N, Kamoshita M, Miyake S, Tsubota K. Neuroprotective effects of lutein in the retina. *Curr Pharm Des*. 2012;18(1):51–56.

Pazdro R, Burgess JR. The role of vitamin E and oxidative stress in diabetes complications. *Mech Ageing Dev*. 2010;131(4):276–286.

Santos JM, Mohammad G, Zhong Q, Kowluru RA. Diabetic retinopathy, superoxide damage and antioxidants. *Curr Pharm Biotechnol*. 2011;12(3):352–361.

Sasaki M, Ozawa Y, Kurihara T, Kubota S, Yuki K, Noda K, Kobayashi S, Ishida S, Tsubota K. Neurodegenerative influence of oxidative stress in the retina of a murine model of diabetes. *Diabetologia*. 2010;53(5):971–979.

Sato H, Sato AS, Kawasaki AR, Yamamoto AT, Yamashita BT, Yamashita H. A retinal cell damage due to oxidative stress in diabetic retinopathy. *Invest Ophthalmol Vis Sci*. 2005;46:443.

Sharma K. Response to comment on Sharma. Mitochondrial hormesis and diabetic complications. Diabetes. 2015;64:663–672. *Diabetes*. 2015a;64(9):e34.

Sharma K. Mitochondrial hormesis and diabetic complications. *Diabetes*. 2015b;64(3):663–672.

Silva KC, Rosales MA, Biswas SK, Lopes de Faria JB, Lopes de Faria JM. Diabetic retinal neurodegeneration is associated with mitochondrial oxidative stress and is improved by an angiotensin receptor blocker in a model combining hypertension and diabetes. *Diabetes.* 2009;58(6):1382–1390.

Stitt AW. The role of advanced glycation in the pathogenesis of diabetic retinopathy. *Exp Mol Pathol.* 2003;75(1):95–108.

Wilkinson-Berka JL. Angiotensin and diabetic retinopathy. *Int J Biochem Cell Biol.* 2006;38(5–6):752–765.

Wright AD, Martin N, Dodson PM. Homocysteine, folates, and the eye. *Eye (Lond).* 2008;22(8):989–993.

Yamagishi S, Nakamura K, Imaizumi T. Advanced glycation end products (AGEs) and diabetic vascular complications. *Curr Diabetes Rev.* 2005;1(1):93–106.

Zhang Y, Bhavnani BR. Glutamate-induced apoptosis in neuronal cells is mediated via caspase-dependent and independent mechanisms involving calpain and caspase-3 proteases as well as apoptosis inducing factor (AIF) and this process is inhibited by equine estrogens. *BMC Neurosci.* 2006;7:49.

Zheng L, Szabó C, Kern TS. Poly(ADP-ribose) polymerase is involved in the development of diabetic retinopathy via regulation of nuclear factor-kappaB. *Diabetes.* 2004;53(11):2960–2967.

27 Reactive Oxygen Species
Physiology and Pathogenesis in Systemic Lupus Erythematosus

Ashish Aggarwal, Nidhi Mahajan, Mansoor Ali Syed, Bishnuhari Paudyal, Swapan K. Nath, and Dilip Shah

CONTENTS

ABSTRACT

Systemic lupus erythematosus (SLE) is a complex autoimmune disease that can affect virtually any organ, though kidneys, skin, lungs, brain, and heart-related comorbidities are frequently reported. Uncontrolled generation of reactive oxygen species (ROS) causes an oxidative stress state and results in oxidative modifications of functional proteins, lipid, and DNA. These consequences are detrimental to the immune system, causing a break in immune tolerance, apoptosis, necrosis, autophagy, and increased tissue damage. This chapter addresses current information on ROS generation, framing it within the context of their physiological and pathological roles in SLE.

27.1 INTRODUCTION

Free radicals and oxidants can be both harmful and helpful to the body. They are produced in mitochondria, cellular membranes, and in the endoplasmic reticulum (ER) membrane as physiological responses to a variety of internal and external stressors. These free radicals are now well recognized for playing both beneficial and deleterious roles in the body's immune defenses. Reactive oxygen species (ROS) are beneficial at low/moderate concentrations via activation of redox-sensitive signaling pathways, phagocytosis of the infected cells, induction of mitogenic responses for wound healing, and clearance of abnormal or aging cells as a part of important immune surveillance mechanisms [1,2]. When an overload of free radicals cannot gradually be destroyed by antioxidant systems in the body, their accumulation produces a condition referred to as oxidative stress. This free radical overload results in the oxidization of lipids, proteins, and DNA, all of which are detrimental to normal immune system function. Thus, a tightly controlled balance between beneficial and harmful levels of free radical production is a very important aspect in the maintenance of normal health.

Recently, numerous clinical and animal studies have provided evidence implicating increased ROS production as a key element in the pathogenesis of systemic lupus erythematosus (SLE). Oxidative damage to proteins, lipids, and DNA by uncontrolled ROS production exacerbates inflammation and induces apoptotic cell death, which in turn alters the expression of a variety of immune and inflammatory molecules and causes the production of a flare of antibodies and associated tissue damage [3,4]. Although the mechanisms of neoantibody flare generation in SLE patients are largely unknown, several studies have pointed to ROS involvement both in their effect in increasing apoptosis and in their effect in delaying the clearance of apoptotic bodies. The delay in clearance of apoptotic cells may prolong interaction between ROS and nuclear debris and generate neoepitopes that subsequently stimulate a broad spectrum of autoantibody formation, which can then lead to inflammation and organ damage in SLE. This chapter describes the generation of ROS and their recognized role in human physiology when maintained at controlled levels and the pathological role of ROS when they are not maintained at appropriate levels. These concepts are interwoven with recent progress in antioxidant therapy as it relates to SLE and potentially to other disease states.

27.2 GENERATION AND SCAVENGER OF ROS

27.2.1 Reactive Oxygen Species

ROS is a broad term used for the chemical species formed by the incomplete reduction of molecular oxygen. They are short-lived molecules produced by normal cellular metabolism, and they aid in a multitude of physiopathological processes [1]. Although they are short-lived molecules, if they are not scavenged adequately they cause a variety of pathological consequences. ROS include superoxide radical ($O_2^{-\bullet}$), peroxy radical (ROO^-), hydrogen peroxide (H_2O_2), singlet oxygen ($^1O_2$), perhydroxyl radical ($HO_2^\bullet$), and extremely reactive hydroxyl radical ($^\bullet OH$). These reactive chemical species lead to the generation of secondary radicals by further reacting with biomolecules (lipids, proteins, and nucleic acids). Mitochondria are the primary site of ROS production, while peroxisomes and cytosol also have the machinery necessary to generate ROS (Table 27.1).

TABLE 27.1
Reactive Oxygen Intermediates, Their Sources, and Modes of Action

	Source	Modes of Action
Superoxide (SO)	NADPH oxidase	Oxidative stress
	Xanthine oxidase	Redox signaling
	Complex I/complex III (mitochondria)	
	5-Lipoxygenase	
	Cyclooxygenase	
	Uncoupled nitric oxide synthase	
Hydrogen peroxide (H_2O_2)	Peroxisomes	Redox signaling
	Superoxide dismutase	Oxidative stress
Hydroxyl radical OH	Fenton reaction	Oxidative stress
	Haber–Weiss reaction	
Hypochlorous acid (HOCl)	Myeloperoxidase (MPO)	Chlorination
		Oxidative stress

During evolution, organisms evolved different antioxidant-processing mechanisms. It is not inconceivable that these mechanisms moved from harnessing ROS pathways for the energy benefit of the organism to defense mechanisms that either curtail the chemical reactivity of ROS or convert them to harmless entities. The balance between the rate of production of ROS and antioxidant defense mechanisms inevitably leads to a point at which ROS move from beneficial to harmful. This condition is commonly known as oxidative stress.

The controlled production of ROS is often beneficial. Oxidative stress is a tool for phagocytic cells to eliminate pathogenic organisms. Cytotoxic drugs on the other hand (used with intent to eliminate auto-reactive cells and thus restraining immunity) induce oxidative stress and cell death. Categorically, the pathological role of ROS includes (1) shift of intracellular redox condition, (2) oxidative modification of biomolecules such as protein, lipid, and nucleic acids (DNA and RNA), (3) gene regulations (activation of oxidative stress and gene mutation related to antioxidant enzymes), (4) activation of apoptosis caused by altered redox state, and (5) increased ER stress and autophagy [5].

27.2.1.1 Source of ROS and Their Scavengers

Sources of oxygen radicals can be both endogenous and exogenous. Endogenous oxygen radicals are attributed primarily to nonenzymatic reactions occurring at the mitochondrial electron transport chain during oxidative phosphorylation in ATP generation [6]. It has been estimated that 3%–5% of total electron flux accounts for ROS formation, which in a typical person (even at rest) corresponds to the production of approximately 2 kg of $O_2^{-\bullet}$ per annum. Additional sources of $O_2^{-\bullet}$ generation come from Nicotinamide adenine dinucleotide phosphate (NADPH) [7], nitric oxide synthase in peroxisomes [8], neutrophils (oxidative burst) [9], lysosomes, and microsomes [10].

In SLE, ROS originate primarily from mitochondria and blood compartments including lymphocytes. ROS have been shown to cause hyperpolarization of mitochondria and activation of T lymphocytes [11]. It has been well established that increased production of ROS or diminished levels of intracellular reduced glutathione in various blood compartments (RBC and lymphocytes) are associated with disease activity in SLE patients [5,12]. In addition, ROS can be produced from other sources, including NADPH oxidase (NOX) enzyme in activated phagocytes [7] and to a lesser extent in endothelial cells, macrophages, polymononuclear cells [12], lysosomes, and microsomes [10]. $\bullet$OH can also be generated from peroxynitrite, which in turn rapidly forms through the reaction between $NO^\bullet$ and $O_2^{-\bullet}$ under stoichiometric conditions in an appropriate environment. H_2O_2 is formed through the dismutation of $O_2^{-\bullet}$ catalyzed by the enzyme superoxide dismutase (SOD) and is also produced through the action of several other oxidase enzymes (e.g., amino acid

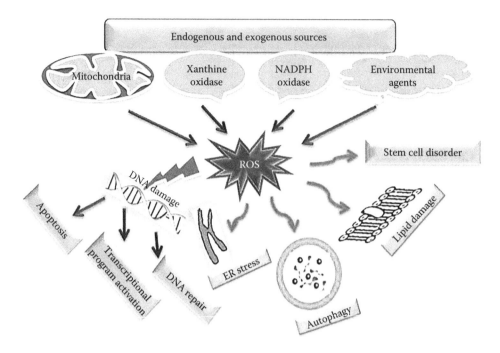

FIGURE 27.1   Overproduction of ROS and its contribution to various complications in SLE.

oxidases) (Figure 27.1). Tissue inflammation and chronic infection lead to the overproduction of $^{\bullet}NO$ and $O_2^{-\bullet}$, which rapidly combine to yield peroxynitrite in the reaction $O_2^{-\bullet} + {}^{\bullet}NO \rightarrow ONO_2^{-\bullet}$. In addition, ROS may amplify the inflammation process by stimulating the expression of genes involved in the inflammatory response, particularly via activation of nuclear transcription factor NF-kβ. This in turn upregulates the proinflammatory cytokines and leukocyte adhesion molecules. Exogenous sources include radiation, ultraviolet (UV) light, x-rays, gamma rays, chemicals that react to form peroxides, ozone, and $^1O_2$ (which promote $O_2^{-\bullet}$ formation), quinones, nitroaromatics, and chemicals that are metabolized to radicals, such as polyhalogenated alkanes, phenols, and aminophenols [13].

The most damaging ROS are $^{\bullet}OH$ and $O_2^{-\bullet}$. The latter can be converted by SOD into the relatively stable, nonradical H_2O_2, which is then reduced by three general mechanisms. First, it is the substrate for two enzymes, catalase (CAT) and glutathione peroxidase (GPx), which catalyze the conversion of H_2O_2 to $H_2O + O_2^{-\bullet}$. This is presumed to be a detoxification mechanism. Second, H_2O_2 can be converted by myeloperoxidase (MPO) in neutrophils to hypochlorous acid (HOCl). This appears to be a mechanism that forms a physiologically toxic agent, since HOCl is a strong oxidant that acts as a bactericidal agent in phagocytic cells. The reaction of HOCl with H_2O_2 yields $^1O_2$ and water. The biological significance of $^1O_2$ is unclear. Third, H_2O_2 can be converted in a spontaneous reaction catalyzed by Fe^{2+} (Fenton reaction) to the highly reactive $^{\bullet}OH$. As $^{\bullet}OH$ cannot be eliminated without causing oxidative damage, it reacts immediately with biomolecules such as lipid, protein, or DNA. This reaction causes severe consequences in SLE pathogenesis [14,15].

27.3   REACTIVE OXYGEN SPECIES IN NORMAL CELLULAR PROCESSES

Until the last decade, ROS were considered to be detrimental to living organisms, a perspective solely derived from toxicology studies as these species react with and modify the structure and function of cell components. Now, ROS are well recognized for playing a dual role as both deleterious and beneficial to living systems [16]. At low or moderate concentrations, ROS are required for the regulation of many cellular processes, including cell signaling, differentiation, proliferation, growth, apoptosis, and cytoskeletal regulation, and contribute to fundamental functions of the host defense system.

27.3.1 ROS in Immune Function

ROS are deeply involved in both arms of immune system function: innate and acquired responses. They are necessary for microbial killing, for limiting the specific immune response, and for termination of inflammation. Evidence for ROS involvement in immune function comes in part from studies on chronic granulomatous disease (CGD), caused by a lack of ROS-generating phagocyte NADPH oxidase 2 (NOX2). Lack of this phagocyte renders an individual immunodeficient due to recurrent infections associated with pneumonia, abscesses, or osteomyelitis. In response to some stimuli, phagocytes of CGD patients do not generate ROS, which is problematic for host defense because macrophages and neutrophils depend on ROS for efficient killing of bacteria by phagocytosis [17]. These phagocytic cells kill bacteria by a mechanism known as respiratory burst, which results in the generation of bactericidal ROS. Due to the activation of the membrane-bound NOX, oxygen consumption in polymorphonuclear leukocytes increases by at least 100-fold during phagocytosis, giving rise to the ROS burst, which damages the cell membrane of the invading organism. MPO from neutrophils also generates a bactericidal species HOCl from H_2O_2. Moreover, mononuclear phagocytes (macrophages) generate NO from L-arginine via nitric oxide (NO) synthase when activated. Cytotoxic effects of NO have been shown to be an important defense against parasitic fungi, protozoa, helminthes, and *Mycobacteria* but not extracellular pathogens [18]. Being a potential vasodilator, NO also slows down circulation, thus helping phagocytic cells, antibodies, and other factors necessary to mount an effective immune response at the site of the pathology, reflected by cardinal signs of inflammation: heat, redness, and swelling.

27.3.2 ROS in Thyroid Function

ROS have been shown to play an important role in patients with a rare form of hypothyroidism. H_2O_2 generated by dual oxidase 2 (DUOX2) acts as a necessary cofactor for thyroperoxidase, the enzyme participating in a final step of hormone production. This theory is well supported by congenital hypothyroid patients having mutations in the DUOX2 gene [19].

27.3.3 ROS in Skeletal Muscle

Skeletal muscle is a common target tissue for oxidative regulation and/or oxidative stress, since it requires a large supply of energy to ensure efficient contraction and is consequently liable to be exposed to excessive mitochondrial ROS. In the skeletal muscle, the production of ROS is promoted by multiple stimuli including muscle contraction, insulin, and hypoxia. Although under normal physiological conditions, antioxidant systems control the levels of ROS in skeletal muscle, oxidative stress can take place if ROS levels exceed the muscle antioxidant capabilities, and this can have deleterious effects. Recent evidence indicates that ROS can act as signaling intermediates in the regulation of skeletal muscle glucose uptake during contraction [20]. However, these results must be interpreted with caution as they have been inconsistent and depend on the specific model employed and the experimental design [21,22].

27.3.4 ROS in Signal Transduction

A purposeful role for ROS as regulators or secondary messenger molecules for various cellular functions has gained significant recognition over the past decades. An early study hinted at a role for ROS as secondary messengers, revealing that exogenous H_2O_2 could mimic the action of the insulin-like growth factor (IGF) and that IGF and nerve growth factor (NGF) stimulate endogenous H_2O_2

production [23]. Indeed, growth factors and cytokines are capable of generating ROS in a number of different cell types, and antioxidants and inhibitors of ROS-generating enzymatic pathways block specific growth factor–activated and/or cytokine-activated signaling events or physiological effects [24]. Moreover, the addition of oxidants activates the same cytokine-mediated and/or growth factor–mediated signaling pathways or recapitulates the physiological effects of the specific cytokines or growth factors [25]. Discoveries of the activation of several transcription factors by ROS, the role of NO as a signaling molecule that activates cyclic guanosine monophosphate (cGMP) production, and the presence of specific enzymes as a source of ROS formation in many tissues stimulated a revival in redox signaling studies [26]. The evidence of ROS/NOS as secondary messengers and signaling molecules at physiological concentrations has been reviewed comprehensively by D'Autreaux and Toledano [1].

27.3.4.1 Hydrogen Peroxide and NO˙ in Guanylate Cyclase Activation

Guanylate cyclase (GC) belongs to the family of heterodimeric heme proteins and catalyzes the formation of cGMP, which is utilized as an intracellular amplifier and a second messenger in a broad range of physiological responses [27]. The soluble guanylate cyclase (sGC) is known to be activated by both H_2O_2 and ˙NO. ˙NO binds to Fe^{2+}-heme groups in sGC, producing a conformational change at Fe^{2+}, which activates the enzyme. Its product, cGMP, modulates the function of protein kinases, ion channels, and other physiologically important targets, with the most important having roles in the regulation of smooth muscle tone [28] and inhibition of platelet adhesion [29]. Vascular smooth muscle relaxation is also mediated by a cGMP-dependent protein kinase that phosphorylates and activates a calcium-sensitive potassium channel [30].

27.3.4.2 NO-Mediated Activation of the GTP-Binding Protein p21 Ras and Protein Kinase Cascades

In human peripheral blood mononuclear cells and endothelial cells, NO was found to activate all three mitogen-activated protein kinase (MAPK) pathways [31,32]. The activation has been attributed to the NO-mediated stimulation of a membrane-associated protein tyrosine phosphatase activity, which may lead to the dephosphorylation and activation of the Src family protein tyrosine kinase p56lck [33]. Another Src family protein kinase, p60c-src, was also found to be activated by NO in fibroblasts [34]. This activation was associated with autophosphorylation at Tyr-416 and S–S bond–mediated aggregation of the kinase molecules. NO may also activate Ras by S-nitrosylation of cysteine-118 [32].

27.3.4.3 Redox Regulation of Cell Adhesion

Cell adhesion plays a key role in embryogenesis, cell growth, differentiation, wound healing, and other processes. Changes in the adhesive properties of cells are tightly redox regulated [35]. The expression of cell adhesion molecules is stimulated by bacterial lipopolysaccharides and by various cytokines such as tumor necrosis factor α (TNF-α), IL-1α, and IL-1β [36]. The adherence of leukocytes to endothelial cells is also significantly augmented by ROS. ROS generated from ligand receptor interactions can oxidize cysteine residues in phosphatases to inhibit their function and sustain signal transduction to the nucleus [37]. In endothelial cells, H_2O_2 or angiotensin II (Ang II) stimulation of ROS production activates eNOS to produce NO, which facilitates cell migration and proliferation [37,38]. The oxidant-induced adherence of neutrophils can be effectively blunted by ˙OH scavengers or iron chelators, suggesting that cell adherence is due in part to ˙OH generated from H_2O_2 within the cell. Adhesion of neutrophils to endothelial cells involves intercellular adhesion molecule-1 (ICAM-1), CD11b/CD18, and L-selectin [37]. In addition, ROS treatment of endothelial cells induces the phosphorylation of the focal adhesion kinase pp125FAK, a cytosolic tyrosine kinase that has been implicated in the oxidant-facilitated adhesion process [39].

27.3.4.4 ROS Production by Cyclooxygenase

Cyclooxygenase-1 has been implicated in ROS production in cells stimulated with TNF-α, IL-1, bacterial lipopolysaccharide, or the tumor promoter 4-O-tetradecanoylphorbol-13-acetate (TPA) [40]. A role for cyclooxygenase in the inflammatory response of 5-lipoxygenase defective mice was also suggested [41]; however, evidence for the participation of cyclooxygenase in redox signaling is still scarce.

27.4 UNCONTROLLED GENERATION OF ROS AND ITS ADVERSE EFFECTS

The production of ROS is an integral part of metabolism and originates primarily from mitochondria, NOX enzymes in phagocytes, and to a lesser extent from endothelial cells, T cells, and B cells [42]. The failure to check the excessive production of ROS produces several pathological consequences in cells. These include mitochondrial dysfunction, ER membrane dysfunction, autophagy, and cell death. Some of these deleterious effects of ROS are shown in Figure 27.1 and discussed in the following.

27.4.1 ROS AND MITOCHONDRIAL DYSFUNCTION

Mitochondria constantly metabolize oxygen and thereby produce ROS as a by-product. The most damaging ROS are $^{\bullet}OH$ and $O_2^{-\bullet}$, and if not eliminated, these ROS can cause oxidative damage. It has been demonstrated that ROS-mediated oxidative stress causes more damage to mtDNA than nuclear DNA due to the following reasons: (1) mtDNA lacks histone proteins, which are required to protect against ROS, (2) mtDNA lacks an adequate repair system, rendering it unable to cope with extensive damage, especially strand break, (3) mtDNA has very few noncoding sequences, therefore increasing the likelihood of a DNA mutation to affect gene function, and (4) mtDNA is located near the inner mitochondrial membrane, a major site of oxygen radical production. These changes in mitochondria homeostasis have been linked to several diseases including SLE.

Mounting genetic data implicate the disruption of mitochondrial function in SLE. Sle1c2 (a sublocus of the major lupus susceptibility locus Sle1 in mouse) is responsible for encoding an orphan nuclear receptor that regulates oxidative metabolism and mitochondrial function. Reduced expression of the gene in T cells caused greater mitochondrial mass, as estimated by increased voltage-dependent anion channel protein content. In humans, nonsynonymous polymorphisms in mitochondrial DNA encoding components of ETC complexes I and V have been associated with SLE. Furthermore, inactive alleles of uncoupling protein 2 (UCP2), which encodes mitochondrial UCP2, a protein that reduces oxidative stress, confer tendency toward several autoimmune diseases, including SLE. Increased biogenesis of mitochondria, which might be due to diminished turnover of lupus T cells, can also contribute to the increased production of ROS. Another consequence of overproduction of ROS is an increase in mitochondrial transmembrane potential, a process that plays a decisive role in cell survival by controlling the activity of redox-sensitive caspases. Peripheral blood T lymphocytes of patients with SLE exhibit mitochondrial hyperpolarization, mediate enhanced spontaneous and diminished activation-induced apoptosis, and sensitize lupus T cells to necrosis [12,43].

27.4.2 ROS AND APOPTOSIS

ROS are mostly generated in the mitochondria and play an important role in apoptosis induction under both physiological and pathological conditions. Several lines of evidence link intracellular ROS levels and apoptosis in numerous cell types. H_2O_2 has been shown to induce apoptosis in the blastocyst and neutrophils, and this apoptotic effect is diminished by CAT [44]. ROS generated by sodium arsenite activates the apoptosis program in eosinophils, and peroxynitrite causes DNA fragmentation and apoptosis in thymocytes, HL-60 cells, cultured cortical neurons, and PC12 cells [10].

It must be noted that contradictory results have been reported [45], which may be largely related to different levels of ROS production.

GSH depletion in response to oxidants has been widely reported to be associated with cell death [46]. GSH is essential for cell survival as suggested by data obtained from γ-glutamyl transpeptidase–deficient knockout mice. These mice exhibited apoptotic cell death [47]. Selective knockdown of GCL in distinct cell types also caused cell death in this model [48]. GSH levels have been shown to influence caspase activity, transcription factor activation, the expression and function of Bcl-2, ceramide production, thiol-redox signaling, and phosphatidylserine externalization. A remarkable feature of cells undergoing apoptosis is that they rapidly and selectively release a large fraction of their intracellular GSH into the extracellular space [49,50]. GPx has been shown to protect against apoptosis induced by Fas activation [51]. Replenishing GSH pools by glutathione and N-acetylcysteine (NAC) is known to protect against apoptosis [45]. On the contrary, glutathione depletion contributes to both extrinsic and intrinsic pathways of apoptosis [52,53].

In autoimmune diseases, depletion of glutathione is associated with various immune abnormalities including dysregulation of apoptosis, abnormal cytokine and chemokine production, and a variety of clinical features [54]. Several lines of evidence indicate that a depletion of intracellular glutathione related to apoptosis progression occurs in SLE patients [11,55,56]. The level of depletion correlates with the severity of the disease and also correlates with measurements of oxidative stress and apoptosis [55]. Specifically, glutathione levels were diminished in RBCs and total lymphocytes as well as lymphocyte subsets in these patients [55–58]. Diminished levels of glutathione in the RBCs and lymphocytes are positively associated with increased levels of oxidative stress markers such as ROS and lipid peroxidation in SLE patient [55,56,59]. Consistent with this, a negative association of GSH levels with apoptosis of T lymphocytes, $CD4^+$, $CD8^+$ T lymphocyte subsets, and intracellular activated caspase-3 was also found in these patients [49,100]. These results strongly suggest that glutathione is critical in the removal of activated $CD4^+$ T lymphocytes in these diseases. The role of glutathione as a therapeutic molecule to replenish depleted glutathione has been ascribed in part to reduction in the production of autoantibodies. Suwannaroj et al., have showed that the treatment of (NZB × NZW) F1 mice with NAC raised the ratio of GSH to GSSG and prolonged the survival of (NZB × NZW) F1 mice, concomitantly reducing autoantibody production and glomerulonephritis [60]. Intracellular glutathione has been shown to be involved in regulating several immune mechanisms in humans. While GSH scavenges ˙OH, $^1O_2$, and NO directly, it also catalytically detoxifies H_2O_2, $OONO^-$, and lipid peroxides by the activation of GPxs. It has been determined that modulation of intracellular glutathione inhibits complement-mediated damage in autoimmune diseases [61]. Glutathione is the major intracellular antioxidant defense within a cell it has been proposed that its depletion is a prerequisite for modulating the apoptotic machinery in autoimmune diseases like SLE. Replenishment of GSH levels by either high extracellular GSH or NAC may prevent increased ROS formation and place apoptosis under check in SLE patients.

27.4.3 ROS AND LIPID PEROXIDATION

ROS, mainly ˙OH, react with lipid membranes to generate reactive aldehydes. These can then *spread* oxidative damage via circulation in the case of SLE. Biochemically, in the initiation phase, a primary reactive radical abstracts a hydrogen atom from a methylene group to start peroxidation and form a conjugated diene, leaving an unpaired electron on the carbon. The carbon-centered fatty acid radicals combine with molecular oxygen in the propagation phase to yield highly reactive peroxyl radicals that react with additional lipid molecules to form hydroperoxides. Peroxyl radicals are capable of producing new fatty acid radicals, resulting in a radical chain reaction. The cascades of lipid peroxidation result in a variety of harmful end products including conjugated dienes, isoprostanes, 4-hydroxy-2-nonenal (HNE), HNE-modified proteins, malondialdehyde (MDA), MDA-modified proteins, protein-bound acrolein, and oxidized HDL (oxHDL), which are associated with disease activity in SLE. Oxidation by-products, such as HNE, could lead to neoantigens like HNE-modified 60 kDa Ro, which could in turn initiate autoimmunity or drive epitope spreading [62].

27.4.4 ROS AND PROTEIN OXIDATION

In addition to their role in lipid peroxidation, ROS can modify the structure and function of proteins. Metal-catalyzed protein oxidation results in the addition of carbonyl groups, cross-linking, or fragmentation of proteins. Lipid aldehydes resulting from peroxidation can react with sulfhydryl amino acids such as cysteine or basic amino acids such as histidine or lysine. It has been reported that many serum proteins are oxidatively modified, leading to the formation of neoantigens, which could in turn initiate autoimmunity in SLE [63]. However, the potential role of oxidative stress, especially the consequences of oxidative modification of proteins in the pathogenesis and progression of SLE, remains unresolved.

27.4.5 ROS AND DNA DAMAGE

Oxidative DNA damage is an inevitable consequence of cell metabolism. This damage is increased in SLE patients. Numerous modified forms of DNA have been found to be immunogenic and are involved in the generation of anti-DNA antibodies. The modification of DNA by ROS is also supported by the enhanced reactivity of SLE anti-DNA antibodies to ROS-denatured DNA [64]. In addition to its oxidative role in the generation of neo-DNA antibody formation in SLE, oxidative DNA damage causes several epigenetic changes such as DNA methylation, and methyl and acetyl modification of histones. In DNA damage mediated by oxidative stress, histone deacetylation levels increase in the nucleus and so DNA damage response decreases in SLE. Various histone deacetylase (HDAC) inhibitors (TSA, SAHAH, and ITF2357) have been shown to have therapeutic potential in animal models of SLE [65]. However, the mechanism through which HDAC inhibitors ameliorate disease remains to be elucidated.

27.4.6 ROS AND ENVIRONMENTAL AGENTS

A number of environmental agents, such as cigarette smoke, viral infection, and various chemicals, have been demonstrated to induce oxidative stress and exacerbate autoimmunity [66]. UV light, in particular UV-A1 and UV-B, can induce disease flares in patients with SLE and trigger disease onset [67]. It is evident that UV-B induces the apoptosis of keratinocytes and other dermal cells and releases a large amount of autoantigen and proinflammatory cytokines into the circulation, triggering autoimmune-related systemic inflammation [68]. Many toxic substances that activate alveolar macrophages, induce MPO activity, and produce free radicals are found in tobacco smoke [69]. In addition to the risk of development of SLE, smoking has been associated with skin flares in patients with SLE [70]. It is likely that tobacco smoke reduces the efficacy of antimalarials, which eventually induces the exacerbation of cutaneous lupus [71]. Both organic and elementary Hg can induce antinuclear antibody (ANA) in murine models and humans [72]. Both methyl-Hg and inorganic Hg induce the oxidative stress of T cells by depleting thiol-containing antioxidants and glutathione, leading to the production of ROS and enhancement of apoptosis and inactivation of PKC-δ of CD4 T cells [73]. Occupational exposure of Hg has been reported to increase the odds of developing SLE [74]. Lipstick, which contains a variety of chemical compounds, such as eosin and phthalates, has been shown to induce photosensitivity and lupus flares, as well as the production of anti-dsDNA antibodies and progression of renal disease in NZB/W F1 mice, respectively, partly and potentially due to the breach of immunological tolerance by molecular mimicry [75].

27.4.7 ROS AND AUTOPHAGY

Excessive oxidative stress and altered redox signaling are most commonly known to be involved in cell death signaling cascades. However, their role in the regulation of autophagy is largely unknown in autoimmune diseases. Autophagy is a persistent homeostatic process in which certain cell components are engulfed by autophagosomes and subsequently degraded to produce energy or preserve cellular viability and homeostasis [76]. Autophagy breaks down compromised cellular components, such as damaged organelles and aggregated proteins. Deposition of these components within cells can lead to toxic effects, resulting in the destruction of tissues, organisms, and biological systems [77]. Alterations in autophagic cycle rate (flux), initiating with the formation of phagophores and terminating with the degradation of autophagosome cargo after fusion with lysosomes, are generally observed in response to stress [78]. Elevated ROS resulting in autophagy promotes either cell survival or cell death, the fate depending upon the severity of stress occurring with a particular disease.

Several studies have shown that ROS accumulation activates the autophagy process. For example, a mutation in an antioxidative superoxide dismutase-1 (SOD1) gene modulates autophagy in cells. Reports from several laboratories have described the activation of autophagy in transgenic mice expressing mutant SOD1 [79]. In one report, SOD1^{G93A} transgenic mice displayed inhibition of the mTOR gene and accumulation of lipid-conjugated LC3, the mammalian homologue of Atg8 [79]. A recent report showed that mutant SOD1 interacts directly with p62 (also called SQSTM1), an LC3-binding partner known to target protein aggregates for autophagic degradation. Indeed, this interaction is proposed to mediate autophagic degradation of mutant SOD1 [80]. It has been suggested that two major ROS, H_2O_2 and $O_2^{-\cdot}$, function as the main regulators of autophagy [81]. H_2O_2 is an attractive candidate for signaling because it is relatively stable and long-lived compared with other ROS, and its neutral ionic state enables it to easily exit mitochondria. It has been implicated as a signaling molecule in various signal transduction pathways, including autophagy [82]. H_2O_2 can act as a direct modifier of thiol-containing proteins. Indeed, the product of ATG4 (autophagy-related gene 4), an essential protease in the autophagic pathway, has been identified as a direct target for oxidation by H_2O_2 during starvation [83]. Other studies report autophagy activation in response to exogenous H_2O_2 treatment. In most cases, this treatment leads to oxidative stress and mitochondrial damage, which induce autophagy. Thus, this evidence supports the vital role of oxidative stress in the induction of autophagy.

27.4.8 STEM CELL DISORDER AND ROS

It has been recently postulated in several studies that SLE is a disease that relies fundamentally on the development of stem cell disorders. A better understanding of the characteristics of bone marrow mesenchymal stem cell (BMSC) dysfunction in SLE could have major clinical implications for SLE treatment. It has been demonstrated that SLE BMSCs appear to show senescence in early passages in tissue culture [84]. In addition, more BMSCs undergo apoptosis in SLE patients than in normal controls. There are many factors related to senescence, including irreversible DNA damage, ROS levels, and shortening of telomeres [85]. PI3K/AKT has been recognized as an important signaling pathway that upregulates intracellular ROS levels by stimulating oxidative metabolism in the mitochondria [86]. AKT hyperactivation promotes metabolic activity in the mitochondria and inhibits FoxO transcriptional activity, resulting in increased ROS levels. FoxO3 is essential for the regulation of HSC fate by maintaining ROS levels below a harmful threshold [87].

27.5 ROS-MEDIATED BIOMARKERS

The production of free radicals is an integral part of human metabolism. However, if it is not tightly controlled, free radical production can lead to oxidative stress. The increased oxidative stress found in pathologic conditions of autoimmune diseases offers the possibility of identifying biomarkers of oxidative stress for use in new diagnostic, therapeutic, and preventive strategies for the management of systemic autoimmune diseases, as sustained interactions between ROS and apoptotic cell macromolecules may generate neoepitopes causing autoimmunity. This holds true for ROS-mediated alterations in other molecules such as membrane lipids, various proteins, and DNA (Figure 27.1). A series of biomarkers, each validated in sequence, are required for early detection and preventive treatments of the diseases. The most intuitive goals for a biomarker are to help diagnose symptomatic and presymptomatic disease and to provide surrogate end points to demonstrate clinical efficacy of new treatments.

The identification of biomarkers that allow an accurate assessment of the degree of oxidative stress will become important in clinical trials aimed at investigating the efficacy of antioxidant therapies for preventing/alleviating complications. A variety of functional assays both *in vivo* and *ex vivo* include measurements of total ROS in blood cells, lipid oxidation, protein oxidation, and DNA oxidation (Table 27.1). ROS/RNS have been detected *in vitro* by electron spin resonance with or without spin-trapping reagents or by chemiluminescence. These direct detection methods are not yet applicable for clinical examinations because of the instability of many reactive species (some half-lives being much shorter than seconds), as well as the requirement for sophisticated analytical techniques. Recently, however, 2′,7′-dichlorofluorescein (H2DCF) and dihydroethidium (DHE) have been used extensively to evaluate ROS production in various tissues by flow cytometry [88,89]. The products formed from ROS/RNS are considered more stable than ROS/RNS and are well suited for evaluating oxidative stress in biological samples. These metabolites include nitrate/nitrite markers for analyzing the roles of RNS [90,91] and/or concentrations of their oxidation target products, including lipid peroxidation end products and oxidized proteins for analyzing the roles of ROS in disease [92].

Lipids are susceptible to oxidation, and lipid peroxidation products are potential biomarkers for oxidative stress status *in vivo* and for related disorders [93]. Lipid peroxidation generates a variety of relatively stable decomposition end products. These include unsaturated reactive aldehydes, such as MDA [94], hexanoyl-Lys adduct (HEL), HNE, and 2-propenal (acrolein) [95], and isoprostanes (8-iso-PGF$_2$) [96], which can be measured in a variety of biological samples (serum/plasma and urine) as an indirect index of oxidative stress. Levels of MDA are often measured spectrophotometrically by the thiobarbituric acid–reacting substance (TBARS) assay. Recently, a highly sensitive ELISA method has been developed for measuring MDA levels in serum/plasma or other biological fluids [97]. F2-isoprostanes, especially 8-iso-PGF2, have been proposed as specific, reliable, and noninvasive markers of lipid peroxidation *in vivo* [98]. The mass chromatography technique (GC-MS) and ELISA ([99] can accurately and sensitively measure F2-isoprostanes in biological samples. 8-iso-PGF2-alpha has been used as a sensitive, noninvasive, and reliable marker of oxidative stress *in vivo* in vascular involved autoimmune diseases like SLE and MS [98]. HNE is a major and toxic aldehyde generated by free radical attack on polyunsaturated fatty acids (arachidonic, linoleic, and linoleic acids) and is considered as a second toxic messenger of oxygen free radicals [100]. HNE has been shown to exhibit facile reactivity with various biomolecules, including proteins and DNA, and is regarded as a sensitive marker for evaluating oxidative stress in the diseased state [101].

Proteins are initial targets of ROS. This leads to early formation of protein carbonyls in biological systems, and detection of increased levels of protein carbonyls has been proposed as a marker of disease-associated dysfunction [102]. Protein carbonyls are a widely used and chemically stable class of biomarkers for oxidative stress. They circulate for longer periods in the blood compared to other oxidized products and can be measured by spectrophotometer, HPLC, Western blotting, and ELISA [103] (Figure 27.2).

DNA damage can be caused by ROS generated under a variety of conditions, and several techniques have been developed to measure oxidatively modified DNA to evaluate disease state [103]. The most representative product that may reflect oxidative damage to DNA in the cell is 8-hydroxy-2-deoxyguanosine (8-OHdG), a product of oxidatively modified DNA base guanine [103,104]. The levels of 8-OHdG in various biological samples are correlated with disease activity in SLE [105].

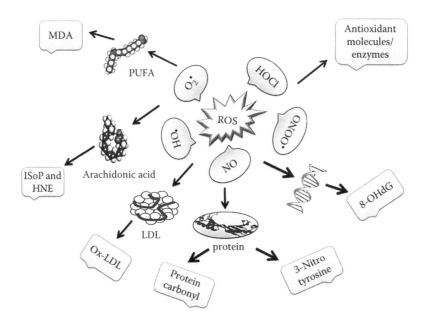

FIGURE 27.2 Formation of oxidative modified biomarkers by reactive oxygen species. Lipid peroxidation biomarkers: malondialdehyde, F2-isoprostane, acrolein, and Ox-LDL. Protein oxidation markers: protein carbonyl and protein nitration. Oxidative DNA damage biomarkers: 8-hydroxy-2′-deoxyguanosine (8-OHdG). Antioxidant enzymes and molecules: superoxide dismutase, catalase, glutathione peroxidase, oxidized glutathione, and total antioxidant capacity.

Many methods such as HPLC, GC-MS, LC-MS, and ELISA have been established to measure biological samples and are reviewed in detail by several articles [106]. HLPC is a frequently used method to measure 8-OHdG with high accuracy and sensitivity, but the procedure is complex and time-consuming, making it less promising for clinical use than the method such as ELISA [106,107]. The measurement of urinary 8-OHdG has been found to reflect whole-body oxidative damage [108] and was shown in humans to be independent of dietary influence. It is known that RNS such as oxides of nitrogen and peroxynitrite generated in various pathophysiological conditions can nitrate guanine and its related nucleosides and nucleotides in the free form or in DNA and RNA [109]. The 8-nitroguanine is a representative DNA nucleobase product of nitrative lesion by RNS. It is a suitable RNS marker for inflammatory diseases like RA [110]. Besides oxidative stress markers of biomolecules (lipid, protein, and DNA), antioxidant enzymes/molecules have been regularly used to evaluate antioxidant defense systems in humans [111].

Among these antioxidant enzymes, which include SOD, CAT, GPx, and xanthine oxidase (XO), and molecules such as ascorbic acid, β-carotene, Zn, selenium, Cu, and Fe, glutathione redox (GSH/GSSG) has been identified as a reliable marker for whole-body antioxidant index in humans [112–114]. It has been well established that a decrease in GSH concentration can be associated with the pathogenesis of SLE [5,53,115]. Several methods have been optimized to measure glutathione forms in human samples, including spectrophotometry, HPLC, and GC-MS. Among these markers of oxidative stress, the intracellular levels of glutathione are considered to be a reliable marker for measuring oxidative stress due to its close association with immune regulation of disease (Table 27.2)

27.6 ANTIOXIDANT THERAPY FOR SLE

Extensive research from several laboratories suggests that restoration of the redox balance using the antioxidant agent, N-acetylcysteamine (NAC), or diminishing the effect of oxidative stress by the

TABLE 27.2
A Summary of Reported Oxidant and Antioxidant Biomarkers in SLE

Study	ROS	Lipid Peroxidation	SOD	Catalase	Glutathione Peroxidase	GSH	Nitric Oxide	Protein Oxidation	DNA Oxidation	Tissue/Cell Studied
Shah et al. [14,23,49,54,102]	↑	↑	↓	↓	↓	↓				RBC, serum, lymphocyte
Perl et al. [22,24]	↑	↑								Lymphocyte
Turi et al. [69]		↑	↓		↓	↓				RBC
Hassan et al. [70]		↑	↓	↓	↓	↓				Serum
Kurient and Scofied [4,35]		↑								Serum
Taysi et al. [101]		↑	↓	↓	↓	↓				Serum
Serban et al. [45]		↑			↓					RBC, plasma
Turgay et al. [53]		↑		↑	↓					Plasma
Segal et al. [80]		↑			↓					Plasma
Bae et al. [100]		↑	↓							Plasma
Jovanovic et al. [13]		↑								Plasma
Abou-raya et al. [78]		↑	↓							Serum
Vipartene et al. [101]		↑	↓		↓					RBC
Mohan and Das [65]		↑	↓		↓					Plasma
Tewthanan et al. [16,51]		↑	↓	↓	↓	↓				Plasma
Morgan et. al. [32,84]						↓		↑		Serum
Zhang et al. [83,86]		↑						↑		Serum, blood
Ahsan et al. [87]								↑		Serum
Lunec et al. [91]									↑	Urine
Evan et al [90]									↑	Serum
Maeshima et al. [88]									↑	Urine
Ho et al. [62]							↑			Plasma
Gilkeson et al. [17]							↑			Serum
Wanchu et al. [63]							↑			Serum

Note: ↑, Significantly elevated levels; ↓, Significantly diminished levels.

intake of antioxidant nutrients, vitamins A, C, and E, carotene, lycopene, etc., may attenuate various oxidative stress–induced complications in SLE. NAC has been shown to be a promising candidate for the restoration of redox balance in both SLE patients and animal models. Murine models of lupus showed that NAC treatment suppressed not only the autoantibody formation but also reduced nephritis and prolonged survival. Administration of NAC has been shown to produce beneficial effects in mild SLE patients in terms of decreasing lipid peroxidation and improving CNS complication and endothelial function in patients with cerebrovascular involvement. Other antioxidants like cysteamine (CYST) have been shown to be beneficial in the treatment of (NZB × NZW) F1 lupus-prone mice. However, an effect of CYST in the treatment of SLE patients has not been shown. Collectively, therapeutic interventions that replenish the redox balance or decrease exposure to ROS and/or augment antioxidant defenses might be beneficial adjunctive therapies in the treatment of oxidative damage in SLE.

Despite the power of modern molecular approaches and persistent investigative efforts, lupus remains an enigmatic disorder and the agent triggering this autoimmune response remains to be identified. Patients with SLE have a diverse array of antinuclear autoantibodies, but the cellular and molecular mechanisms responsible for the production of antinuclear antibodies in SLE and the way by which these antibodies participate in tissue destruction remain highly controversial.

27.7 CONCLUSION

A body of evidence shows that ROS behave as a *double-edged sword*. On one hand, the species contribute to normal physiological functions and provide pivotal defenses against invading organisms. On the other hand, excessive production of ROS causes oxidative stress that can lead to cell damage, senescence, or death. These attributes make ROS an essential investigative target in the biochemistry and physiology of health and pathological mechanisms of diseases. Recently, NAC has shown great potential for treating systemic autoimmune diseases and has been in clinical trials for the therapeutic management of several other diseases. Further studies should be directed at a more specific role of antioxidant and other therapeutic molecules (inhibitors of histone deacetylation and the mTOR pathway) in the alleviation of oxidative stress, apoptosis, autophagy, and/or in increasing the severity of a given disease.

REFERENCES

1. D'Autreaux, B. and M.B. Toledano, ROS as signalling molecules: Mechanisms that generate specificity in ROS homeostasis. *Nat Rev Mol Cell Biol*, 2007. **8**(10): 813–824.
2. Gutowski, M. and S. Kowalczyk, A study of free radical chemistry: Their role and pathophysiological significance. *Acta Biochim Pol*, 2013. **60**(1): 1–16.
3. Kurien, B.T. and R.H. Scofield, Autoimmunity and oxidatively modified autoantigens. *Autoimmun Rev*, 2008. **7**(7): 567–573.
4. Sundaresan, M. et al., Requirement for generation of H_2O_2 for platelet-derived growth factor signal transduction. *Science*, 1995. **270**(5234): 296–299.
5. Shah, D. et al., Altered redox state and apoptosis in the pathogenesis of systemic lupus erythematosus. *Immunobiology*, 2013. **218**(4): 620–627.
6. Suski, J.M. et al., Relation between mitochondrial membrane potential and ROS formation. *Methods Mol Biol*, 2012. **810**: 183–205.
7. Vignais, P.V., The superoxide-generating NADPH oxidase: Structural aspects and activation mechanism. *Cell Mol Life Sci*, 2002. **59**(9): 1428–1459.
8. Zahrt, T.C. and V. Deretic, Reactive nitrogen and oxygen intermediates and bacterial defenses: Unusual adaptations in mycobacterium tuberculosis. *Antioxid Redox Signal*, 2002. **4**(1): 141–159.
9. Nagy, G. et al., Central role of nitric oxide in the pathogenesis of rheumatoid arthritis and systemic lupus erythematosus. *Arthritis Res Ther*, 2010. **12**(3): 210.
10. Ahmad, R., Z. Rasheed, and H. Ahsan, Biochemical and cellular toxicology of peroxynitrite: Implications in cell death and autoimmune phenomenon. *Immunopharmacol Immunotoxicol*, 2009. **31**(3): 388–396.

11. Perl, A., P. Gergely, Jr., and K. Banki, Mitochondrial dysfunction in T cells of patients with systemic lupus erythematosus. *Int Rev Immunol*, 2004. **23**(3–4): 293–313.

12. Perl, A., Oxidative stress in the pathology and treatment of systemic lupus erythematosus. *Nat Rev Rheumatol*, 2013. **9**(11): 674–686.

13. Ortona, E. et al., Redox state, cell death and autoimmune diseases: A gender perspective. *Autoimmun Rev*, 2008. **7**(7): 579–584.

14. Al-Shobaili, H.A. et al., Hydroxyl radical modification of immunoglobulin g generated cross-reactive antibodies: Its potential role in systemic lupus erythematosus. *Clin Med Insights Arthritis Musculoskelet Disord*, 2011. **4**: 11–19.

15. Kurien, B.T. and R.H. Scofield, Lipid peroxidation in systemic lupus erythematosus. *Indian J Exp Biol*, 2006. **44**(5): 349–356.

16. Valko, M. et al., Free radicals and antioxidants in normal physiological functions and human disease. *Int J Biochem Cell Biol*, 2007. **39**(1): 44–84.

17. Holland, S.M., Chronic granulomatous disease. *Clin Rev Allergy Immunol*, 2010. **38**(1): 3–10.

18. Moncada, S., R.M. Palmer, and E.A. Higgs, Nitric oxide: Physiology, pathophysiology, and pharmacology. *Pharmacol Rev*, 1991. **43**(2): 109–142.

19. Moreno, J.C. et al., Inactivating mutations in the gene for thyroid oxidase 2 (THOX2) and congenital hypothyroidism. *N Engl J Med*, 2002. **347**(2): 95–102.

20. Barbieri, E. and P. Sestili, Reactive oxygen species in skeletal muscle signaling. *J Signal Transduct*, 2012. **2012**: 982794.

21. Merry, T.L. et al., Skeletal muscle glucose uptake during contraction is regulated by nitric oxide and ROS independently of AMPK. *Am J Physiol Endocrinol Metab*, 2010. **298**(3): E577–E585.

22. Sandstrom, M.E. et al., Role of reactive oxygen species in contraction-mediated glucose transport in mouse skeletal muscle. *J Physiol*, 2006. **575**(Pt 1): 251–262.

23. Mukherjee, S.P. and C. Mukherjee, Similar activities of nerve growth factor and its homologue pro-insulin in intracellular hydrogen peroxide production and metabolism in adipocytes. Transmembrane signalling relative to insulin-mimicking cellular effects. *Biochem Pharmacol*, 1982. **31**(20): 3163–3172.

24. Lo, Y.Y. and T.F. Cruz, Involvement of reactive oxygen species in cytokine and growth factor induction of c-fos expression in chondrocytes. *J Biol Chem*, 1995. **270**(20): 11727–11730.

25. Thannickal, V.J. and B.L. Fanburg, Reactive oxygen species in cell signaling. *Am J Physiol Lung Cell Mol Physiol*, 2000. **279**(6): L1005–L1028.

26. Schreck, R., P. Rieber, and P.A. Baeuerle, Reactive oxygen intermediates as apparently widely used messengers in the activation of the NF-kappa B transcription factor and HIV-1. *EMBO J*, 1991. **10**(8): 2247–2258.

27. Gerzer, R. et al., Soluble guanylate cyclase purified from bovine lung contains heme and copper. *FEBS Lett*, 1981. **132**(1): 71–74.

28. Tiritilli, A., (Nitric oxide [NO]), vascular protection factor. Biology, physiological role and biochemistry of NO). *Presse Med*, 1998. **27**(21): 1061–1064.

29. Radomski, M.W., R.M. Palmer, and S. Moncada, The anti-aggregating properties of vascular endothelium: Interactions between prostacyclin and nitric oxide. *Br J Pharmacol*, 1987. **92**(3): 639–646.

30. Archer, S.L. et al., Nitric oxide and cGMP cause vasorelaxation by activation of a charybdotoxin-sensitive K channel by cGMP-dependent protein kinase. *Proc Natl Acad Sci USA*, 1994. **91**(16): 7583–7587.

31. Lander, H.M. et al., Differential activation of mitogen-activated protein kinases by nitric oxide-related species. *J Biol Chem*, 1996. **271**(33): 19705–19709.

32. Droge, W., Free radicals in the physiological control of cell function. *Physiol Rev*, 2002. **82**(1): 47–95.

33. Lander, H.M. et al., Activation of human peripheral blood mononuclear cells by nitric oxide-generating compounds. *J Immunol*, 1993. **150**(4): 1509–1516.

34. Akhand, A.A. et al., Nitric oxide controls Src kinase activity through a sulfhydryl group modification-mediated Tyr-527-independent and Tyr-416-linked mechanism. *J Biol Chem*, 1999. **274**(36): 25821–25826.

35. Frenette, P.S. and D.D. Wagner, Adhesion molecules—Part 1. *N Engl J Med*, 1996. **334**(23): 1526–1529.

36. Albelda, S.M., C.W. Smith, and P.A. Ward, Adhesion molecules and inflammatory injury. *FASEB J*, 1994. **8**(8): 504–512.

37. Sellak, H. et al., Reactive oxygen species rapidly increase endothelial ICAM-1 ability to bind neutrophils without detectable upregulation. *Blood*, 1994. **83**(9): 2669–2677.

38. Diep, Q.N. et al., PPARalpha activator effects on Ang II-induced vascular oxidative stress and inflammation. *Hypertension*, 2002. **40**(6): 866–871.

39. Schaller, M.D. et al., pp125FAK a structurally distinctive protein-tyrosine kinase associated with focal adhesions. *Proc Natl Acad Sci USA*, 1992. **89**(11): 5192–5196.

40. Feng, L. et al., Involvement of reactive oxygen intermediates in cyclooxygenase-2 expression induced by interleukin-1, tumor necrosis factor-alpha, and lipopolysaccharide. *J Clin Invest*, 1995. **95**(4): 1669–1675.

41. Goulet, J.L. et al., Altered inflammatory responses in leukotriene-deficient mice. *Proc Natl Acad Sci USA*, 1994. **91**(26): 12852–12856.

42. Ahsan, H., A. Ali, and R. Ali, Oxygen free radicals and systemic autoimmunity. *Clin Exp Immunol*, 2003. **131**(3): 398–404.

43. Tewthanom, K. et al., The effect of high dose of N-acetylcysteine in lupus nephritis: A case report and literature review. *J Clin Pharm Ther*, 2010. **35**(4): 483–485.

44. Pierce, G.B., R.E. Parchment, and A.L. Lewellyn, Hydrogen peroxide as a mediator of programmed cell death in the blastocyst. *Differentiation*, 1991. **46**(3): 181–186.

45. Devadas, S. et al., Fas-stimulated generation of reactive oxygen species or exogenous oxidative stress sensitize cells to Fas-mediated apoptosis. *Free Radic Biol Med*, 2003. **35**(6): 648–661.

46. Perricone, C., C. De Carolis, and R. Perricone, Glutathione: A key player in autoimmunity. *Autoimmun Rev*, 2009. **8**(8): 697–701.

47. Dalton, T.P. et al., Genetically altered mice to evaluate glutathione homeostasis in health and disease. *Free Radic Biol Med*, 2004. **37**(10): 1511–1526.

48. Valverde, M. et al., Survival and cell death in cells constitutively unable to synthesize glutathione. *Mutat Res*, 2006. **594**(1–2): 172–180.

49. Hammond, C.L., T.K. Lee, and N. Ballatori, Novel roles for glutathione in gene expression, cell death, and membrane transport of organic solutes. *J Hepatol*, 2001. **34**(6): 946–954.

50. Ghibelli, L. et al., Rescue of cells from apoptosis by inhibition of active GSH extrusion. *FASEB J*, 1998. **12**(6): 479–486.

51. Gouaze, V. et al., Glutathione peroxidase-1 protects from CD95-induced apoptosis. *J Biol Chem*, 2002. **277**(45): 42867–42874.

52. Franco, R. and J.A. Cidlowski, Apoptosis and glutathione: Beyond an antioxidant. *Cell Death Differ*, 2009. **16**(10): 1303–1314.

53. Shah, D., S. Sah, and S.K. Nath, Interaction between glutathione and apoptosis in systemic lupus erythematosus. *Autoimmun Rev*, 2013. **12**(7): 741–751.

54. Ballatori, N. et al., Glutathione dysregulation and the etiology and progression of human diseases. *Biol Chem*, 2009. **390**(3): 191–214.

55. Shah, D. et al., Association between T lymphocyte sub-sets apoptosis and peripheral blood mononuclear cells oxidative stress in systemic lupus erythematosus. *Free Radic Res*, 2011. **45**(5): 559–567.

56. Shah, D., A. Wanchu, and A. Bhatnagar, Interaction between oxidative stress and chemokines: Possible pathogenic role in systemic lupus erythematosus and rheumatoid arthritis. *Immunobiology*, 2011. **216**(9): 1010–1017.

57. Shah, D. et al., Oxidative stress in systemic lupus erythematosus: Relationship to Th1 cytokine and disease activity. *Immunol Lett*, 2010. **129**(1): 7–12.

58. Hassan, M.Q. et al., The glutathione defense system in the pathogenesis of rheumatoid arthritis. *J Appl Toxicol*, 2001. **21**(1): 69–73.

59. Jovanovic, V. et al., Lipid anti-lipid antibody responses correlate with disease activity in systemic lupus erythematosus. *PLoS One*, 2013. **8**(2): e55639.

60. Suwannaroj, S. et al., Antioxidants suppress mortality in the female NZB × NZW F1 mouse model of systemic lupus erythematosus (SLE). *Lupus*, 2001. **10**(4): 258–265.

61. Perricone, C. et al., Inhibition of the complement system by glutathione: Molecular mechanisms and potential therapeutic implications. *Int J Immunopathol Pharmacol*, 2011. **24**(1): 63–68.

62. Scofield, R.H. et al., Modification of lupus-associated 60-kDa Ro protein with the lipid oxidation product 4-hydroxy-2-nonenal increases antigenicity and facilitates epitope spreading. *Free Radic Biol Med*, 2005. **38**(6): 719–728.

63. Morgan, P.E., A.D. Sturgess, and M.J. Davies, Evidence for chronically elevated serum protein oxidation in systemic lupus erythematosus patients. *Free Radic Res*, 2009. **43**(2): 117–127.

64. Evans, M.D. et al., Aberrant processing of oxidative DNA damage in systemic lupus erythematosus. *Biochem Biophys Res Commun*, 2000. **273**(3): 894–898.

65. Regna, N.L. et al., Class I and II histone deacetylase inhibition by ITF2357 reduces SLE pathogenesis in vivo. *Clin Immunol*, 2014. **151**(1): 29–42.

66. Somers, E.C. and B.C. Richardson, Environmental exposures, epigenetic changes and the risk of lupus. *Lupus*, 2014. **23**(6): 568–576.

67. Zahn, S. et al., Ultraviolet light protection by a sunscreen prevents interferon-driven skin inflammation in cutaneous lupus erythematosus. *Exp Dermatol*, 2014. **23**(7): 516–518.

68. Caricchio, R., L. McPhie, and P.L. Cohen, Ultraviolet B radiation-induced cell death: Critical role of ultraviolet dose in inflammation and lupus autoantigen redistribution. *J Immunol*, 2003. **171**(11): 5778–5786.

69. Cohen, A.B., D.E. Chenoweth, and T.E. Hugli, The release of elastase, myeloperoxidase, and lysozyme from human alveolar macrophages. *Am Rev Respir Dis*, 1982. **126**(2): 241–247.

70. Takvorian, S.U., J.F. Merola, and K.H. Costenbader, Cigarette smoking, alcohol consumption and risk of systemic lupus erythematosus. *Lupus*, 2014. **23**(6): 537–544.

71. Ezra, N. and J. Jorizzo, Hydroxychloroquine and smoking in patients with cutaneous lupus erythematosus. *Clin Exp Dermatol*, 2012. **37**(4): 327–334.

72. Gardner, R.M. et al., Mercury exposure, serum antinuclear/antinucleolar antibodies, and serum cytokine levels in mining populations in Amazonian Brazil: A cross-sectional study. *Environ Res*, 2010. **110**(4): 345–354.

73. Shenker, B.J. et al., Induction of apoptosis in human T-cells by methyl mercury: Temporal relationship between mitochondrial dysfunction and loss of reductive reserve. *Toxicol Appl Pharmacol*, 1999. **157**(1): 23–35.

74. Cooper, G.S. et al., Occupational risk factors for the development of systemic lupus erythematosus. *J Rheumatol*, 2004. **31**(10): 1928–1933.

75. Lim, S.Y. and S.K. Ghosh, Autoreactive responses to environmental factors: 3. Mouse strain-specific differences in induction and regulation of anti-DNA antibody responses due to phthalate-isomers. *J Autoimmun*, 2005. **25**(1): 33–45.

76. Zhou, X.J., F.J. Cheng, and H. Zhang, Emerging view of autophagy in systemic lupus erythematosus. *Int Rev Immunol*, 2015. **34**(3): 280–292.

77. Doria, A., M. Gatto, and L. Punzi, Autophagy in human health and disease. *N Engl J Med*, 2013. **368**(19): 1845.

78. Bhattacharya, A. and N.T. Eissa, Autophagy and autoimmunity crosstalks. *Front Immunol*, 2013. **4**: 88.

79. Morimoto, N. et al., Increased autophagy in transgenic mice with a G93A mutant SOD1 gene. *Brain Res*, 2007. **1167**: 112–117.

80. Gal, J. et al., Sequestosome 1/p62 links familial ALS mutant SOD1 to LC3 via an ubiquitin-independent mechanism. *J Neurochem*, 2009. **111**(4): 1062–1073.

81. Scherz-Shouval, R. and Z. Elazar, Regulation of autophagy by ROS: Physiology and pathology. *Trends Biochem Sci*, 2010. **36**(1): 30–38.

82. Zhang, H. et al., Oxidative stress induces parallel autophagy and mitochondria dysfunction in human glioma U251 cells. *Toxicol Sci*, 2009. **110**(2): 376–388.

83. Scherz-Shouval, R. et al., Reactive oxygen species are essential for autophagy and specifically regulate the activity of Atg4. *EMBO J*, 2007. **26**(7): 1749–1760.

84. Li, X. et al., Enhanced apoptosis and senescence of bone-marrow-derived mesenchymal stem cells in patients with systemic lupus erythematosus. *Stem Cells Dev*, 2012. **21**(13): 2387–2394.

85. Giorgio, M. et al., Hydrogen peroxide: A metabolic by-product or a common mediator of ageing signals? *Nat Rev Mol Cell Biol*, 2007. **8**(9): 722–728.

86. Dolado, I. and A.R. Nebreda, AKT and oxidative stress team up to kill cancer cells. *Cancer Cell*, 2008. **14**(6): 427–429.

87. Yalcin, S. et al., FoxO3 is essential for the regulation of ataxia telangiectasia mutated and oxidative stress-mediated homeostasis of hematopoietic stem cells. *J Biol Chem*, 2008. **283**(37): 25692–25705.

88. Amer, J., A. Goldfarb, and E. Fibach, Flow cytometric measurement of reactive oxygen species production by normal and thalassaemic red blood cells. *Eur J Haematol*, 2003. **70**(2): 84–90.

89. Eruslanov, E. and S. Kusmartsev, Identification of ROS using oxidized DCFDA and flow-cytometry. *Methods Mol Biol*, 2010. **594**: 57–72.

90. Ho, C.Y. et al., Elevated plasma concentrations of nitric oxide, soluble thrombomodulin and soluble vascular cell adhesion molecule-1 in patients with systemic lupus erythematosus. *Rheumatology (Oxford)*, 2003. **42**(1): 117–122.

91. Wanchu, A. et al., Nitric oxide synthesis is increased in patients with systemic lupus erythematosus. *Rheumatol Int*, 1998. **18**(2): 41–43.

92. Offord, E., G. van Poppel, and R. Tyrrell, Markers of oxidative damage and antioxidant protection: Current status and relevance to disease. *Free Radic Res*, 2000. **33**(Suppl): S5–S19.

93. Taysi, S. et al., Serum oxidant/antioxidant status of patients with systemic lupus erythematosus. *Clin Chem Lab Med*, 2002. **40**(7): 684–648.

94. Segal, B.M. et al., Oxidative stress and fatigue in systemic lupus erythematosus. *Lupus*, 2012. **21**(9): 984–992.

95. Uchida, K., 4-Hydroxy-2-nonenal: A product and mediator of oxidative stress. *Prog Lipid Res*, 2003. **42**(4): 318–343.

96. Cracowski, J.L., T. Durand, and G. Bessard, Isoprostanes as a biomarker of lipid peroxidation in humans: Physiology, pharmacology and clinical implications. *Trends Pharmacol Sci*, 2002. **23**(8): 360–366.

97. Turi, S. et al., Oxidative stress and antioxidant defense mechanism in glomerular diseases. *Free Radic Biol Med*, 1997. **22**(1–2): 161–168.

98. Abou-Raya, A., D. el-Hallous, and H. Fayed, 8-Isoprostaglandin F2 alpha: A potential index of lipid peroxidation in systemic lupus erythematosus. *Clin Invest Med*, 2004. **27**(6): 306–311.

99. Basu, S., Isoprostanes: Novel bioactive products of lipid peroxidation. *Free Radic Res*, 2004. **38**(2): 105–122.

100. Parola, M. et al., 4-Hydroxynonenal as a biological signal: Molecular basis and pathophysiological implications. *Antioxid Redox Signal*, 1999. **1**(3): 255–284.

101. Wang, G. et al., Markers of oxidative and nitrosative stress in systemic lupus erythematosus: Correlation with disease activity. *Arthritis Rheum*, 2010. **62**(7): 2064–2072.

102. Shacter, E., Quantification and significance of protein oxidation in biological samples. *Drug Metab Rev*, 2000. **32**(3–4): 307–326.

103. Dalle-Donne, I. et al., Protein carbonyl groups as biomarkers of oxidative stress. *Clin Chim Acta*, 2003. **329**(1–2): 23–38.

104. Zaremba, T. and R. Olinski, Oxidative DNA damage—analysis and clinical significance. *Postepy Biochem*, 2010. **56**(2): 124–138.

105. Lunec, J. et al., 8-Hydroxydeoxyguanosine. A marker of oxidative DNA damage in systemic lupus erythematosus. *FEBS Lett*, 1994. **348**(2): 131–138.

106. Saito, S. et al., Quantitative determination of urinary 8-hydroxydeoxyguanosine (8-OH-dg) by using ELISA. *Res Commun Mol Pathol Pharmacol*, 2000. **107**(1–2): 39–44.

107. Shimoi, K. et al., Comparison between high-performance liquid chromatography and enzyme-linked immunosorbent assay for the determination of 8-hydroxy-2′-deoxyguanosine in human urine. *Cancer Epidemiol Biomarkers Prev*, 2002. **11**(8): 767–770.

108. Halliwell, B. and M. Whiteman, Measuring reactive species and oxidative damage in vivo and in cell culture: How should you do it and what do the results mean? *Br J Pharmacol*, 2004. **142**(2): 231–255.

109. Ohshima, H., T. Sawa, and T. Akaike, 8-nitroguanine, a product of nitrative DNA damage caused by reactive nitrogen species: Formation, occurrence, and implications in inflammation and carcinogenesis. *Antioxid Redox Signal*, 2006. **8**(5–6): 1033–1045.

110. Chang, H.R. et al., Formation of 8-nitroguanine in blood of patients with inflammatory gouty arthritis. *Clin Chim Acta*, 2005. **362**(1–2): 170–175.

111. Wei, Y.H. et al., Oxidative stress in human aging and mitochondrial disease-consequences of defective mitochondrial respiration and impaired antioxidant enzyme system. *Chin J Physiol*, 2001. **44**(1): 1–11.

112. Townsend, D.M., K.D. Tew, and H. Tapiero, The importance of glutathione in human disease. *Biomed Pharmacother*, 2003. **57**(3–4): 145–155.

113. Turgay, M. et al., Oxidative stress and antioxidant parameters in a Turkish group of patients with active and inactive systemic lupus erythematosus. *APLAR J Rheumatol*, 2007. **10**(2): 101–106.

114. Bae, S.C., S.J. Kim, and M.K. Sung, Impaired antioxidant status and decreased dietary intake of antioxidants in patients with systemic lupus erythematosus. *Rheumatol Int*, 2002. **22**(6): 238–243.

115. Serban, M.G., E. Balanescu, and V. Nita, Lipid peroxidase and erythrocyte redox system in systemic vasculitides treated with corticoids. Effect of vitamin E administration. *Rom J Intern Med*, 1994. **32**(4): 283–289.

28 Oxidative Stress in Fibromyalgia Syndrome

Ghizal Fatima

CONTENTS

ABSTRACT

Fibromyalgia syndrome (FMS) is one of the least understood pain syndromes in medicine today, which is a chronic disorder characterized by persistent and widespread pain. Other symptoms include tingling of the skin, prolonged muscle spasms, weakness in the limbs, nerve pain, muscle twitching, and chronic sleep disturbances. FMS is a common chronic pain syndrome of unknown etiology and limited treatment options. In addition to this, FMS is perceived as a disorder of central sensitization. Proposed initiating factors include genetic predisposition, physical trauma, and infection, but free radical–mediated oxidative stress and inflammatory cytokines may also play important roles in its pathogenesis. Previous studies have reported oxidative stress in the patients with FMS, but the results were inconsistent. Oxidative stress and nitric oxide (NO) are involved in FMS pathophysiology; however, it is still not clear whether oxidative stress abnormalities are the cause of FMS. There are several studies indicating oxidative stress in patients with FMS. Oxidant (malondialdehyde [MDA]) and antioxidant (superoxide dismutase [SOD]) balances were found to be changed in fibromyalgia patients. Furthermore, increased free radical levels may be responsible for the development of FMS, and free radical–mediated oxidative stress may also play important roles in its pathogenesis. There are studies supporting the hypothesis of FMS as an oxidative disorder. Moreover, oxidative stress is supposed to be increased in patients with FMS, which is related to the severity of FMS symptoms. Therefore, it is important to understand whether the oxidative stress parameters are involved in FMS and the relationship between these and antioxidants in FMS patients. In this chapter, we will elucidate the importance of oxidative stress and its possible relationship with FMS.

28.1 INTRODUCTION

Fibromyalgia is derived from Latin word *fibro*, meaning "fibrous tissues," Greek word *myo*, meaning "muscle," and Greek word *algos*, meaning "pain"; thus, the term "fibromyalgia" means "muscle and connective tissue pain." Fibromyalgia symptoms are not restricted to pain only; therefore, the alternative term "fibromyalgia syndrome" (FMS) is used for the condition. Epidemiological studies indicate that poor sleep quality is a risk factor for the development of chronic widespread pain, and sleep deprivation impairs descending pain-inhibition pathways that are important in controlling and coping with pain (Choy, 2015). FMS is considered to be a disorder of central pain processing that produces heightened responses to painful stimuli (hyperalgesia) and painful responses to nonpainful stimuli (allodynia). Therefore, the heightened state of pain transmission may be due to increase in nociceptive neurotransmitters, such as substance P and glutamate. Other symptoms of FMS include debilitating fatigue, sleep disturbance, headaches, anxiety, depression, cold hands and feet, lowered immune function, chemical sensitivities, and joint stiffness (Lukkahatai et al., 2016). Few patients with FMS also report difficulty in swallowing, bowel and bladder abnormalities, numbness, tingling, and cognitive dysfunction (Clauw, 2015). FMS is a persistent and potentially debilitating disorder that can have a devastating effect on the quality of life, impairing the patient's ability to work and participate in everyday activities, as well as affecting relationships with family, friends, and employers. It imposes heavy economic burdens on society and on the patient (Clauw, 2014).

There is as yet no cure for FMS. Some treatments have been shown by controlled clinical trials to effectively reduce symptoms, including medications, behavioral interventions, patient education, and exercise. FMS is one of the least understood pain syndromes in medicine today, which is a chronic disorder characterized by persistent and widespread pain, with an estimated prevalence of 2%–4% in the adult general population (3.4% for women and 0.5% for men) (Jay and Barkin, 2015). The initial FMS criteria included tenderness on pressure (tender points) in at least 11 of 18 defined anatomic sites with the presence of widespread pain (Figure 28.1) (Wolfe et al., 1990). However, according to recent criteria, it is clear that apart from the pain, other seminal features of the disorder such as cognitive dysfunction, unrefreshing sleep, fatigue, and mood disorders also play an important role in the diagnosis (Wolfe et al., 2010).

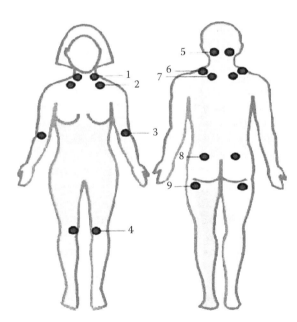

Tender points in fibromyalgia syndrome

ACR Criteria (1990)

Widespread pain history: (>3 months = 1, >6 months = 2, >1 year = 3)

- All four quadrants of the body.
- Both sides of the body.
- Above and below the waist.

Tender Points Examination: (11/18)

1) Low cervical: Pain at the anterior aspect of intertransverse spaces at C5-C7.
2) Second rib: Pain at costochondral junction just lateral to the junction on the upper surface.
3) Lateral epicondyle: Pain at point 2 cm distal to epicondyles.
4) Knee: Pain at medial fat pad proximal to joint line.
5) Occiput: Pain at suboccipital muscle insertion.
6) Trapezius: Pain at the midpoint of upper border.
7) Supraspinatus: Pain at above the scapula spine near the medial border.
8) Gluteal: Pain in the upper outer quadrants of buttocks in anterior fold of muscle.
9) Greater trochanter: Pain at the posterior region of trochanteric prominance.

For FMS diagnosis:

- History of chronic widespread pain ≥3 months.
- Patients must exhibit ≥11 of 18 tender points.

FIGURE 28.1   Diagnostic criteria of FMS.

Recently, oxidative stress has been proposed as a relevant event in the pathogenesis of this disorder (Fatima et al., 2013). Most of the body's energy is produced by the enzymatically controlled reaction of oxygen with hydrogen in oxidative phosphorylation occurring within the mitochondria during oxidative metabolism. During this enzymatic reduction of oxygen to produce energy, free radicals are formed (Valko et al., 2007). In the normal condition, there is a balance between reactive oxygen species (ROS) and antioxidants within the cell, in the membranes, and in the extracellular space. Endogenous free radical scavengers are overwhelmed by excessive production of ROS. The ROS attack the polyunsaturated fatty acids (PUFAs) in the membrane lipids, thereby leading to lipid peroxidation (LP) resulting in the loss of fluidity of the membranes, changes in membrane potentials, and eventually rupture leading to the release of the cell and organelle contents.

Therefore, increased oxidative stress results from an imbalance between products of oxidation and antioxidant defenses. These toxic molecules become highly reactive in their formation because of their altered number of unpaired valence electrons (Mahdi and Fatima, 2014). There are several inflammatory clinical conditions associated with increased oxidative stress, but novel data suggest a relationship between oxidative stress and pain perception (Altindag and Celik, 2006). Furthermore, oxidative stress is increased in patients with chronic fatigue syndrome (CFS). There is little information about oxidative stress in FMS. Several disorders are associated with oxidative stress that is manifested by LP, protein oxidation, and other markers. Few studies have shown some evidence that oxidative stress may have a role in the pathophysiology of FMS (Neyal et al., 2012). Therefore, in this chapter, we will elucidate the complex network of oxidative and antioxidative imbalance in patients with FMS.

28.2 REACTIVE OXYGEN SPECIES

ROS are highly reactive chemical species with an unpaired electron and are formed by catalyzing transition metals like iron, copper, or manganese. These ROS have been suggested to play important roles in various neurodegenerative and neuropsychiatric diseases, including Alzheimer's disease and Parkinson's disease, as well as depression and rheumatologic conditions such as rheumatoid arthritis (RA), ankylosing spondylitis (AS), and FMS.

A major source of radicals in biological systems is dioxygen (O_2). The radicals originating from molecular oxygen are generally named ROS. One of the important enzymatic sources of superoxide anion radical ($O_2^{\cdot-}$) is xanthine oxidase (XO). Purine nucleotides are degraded by a pathway in which the phosphate group is lost by the action of $5'$ nucleotidase. Adenosine is then deaminated to inosine by adenosine deaminase (ADA). Inosine is hydrolyzed to yield its purine base hypoxanthine. Hypoxanthine is oxidized successively to xanthine and then to uric acid by XO. XO is an important iron and molybdenum-containing enzyme. The enzyme exists primarily in the xanthine dehydrogenase (desulfo) form and can be converted to XO (D-to-O conversion) by a variety of conditions including proteolysis, homogenization, and sulfhydryl oxidation (Neill et al., 2015). Current interest in XO systems results from its proposed role in post-ischemic reperfusion injury; in such cases, the activity of this $O_2^{\cdot-}$ producer enzyme may increase through the proteolytic conversion of xanthine dehydrogenase to XO and produce an enormous amount of $O_2^{\cdot-}$ (Phaniendra et al., 2015).

The major intracellular antioxidant enzymes, Cu,Zn-SOD in the cytoplasm and manganese-superoxide dismutase (Mn-SOD) in the mitochondria, rapidly and specifically reduce superoxide radicals to hydrogen peroxide (H_2O_2). The other endogenous antioxidant enzyme, glutathione peroxidase (GPx), acts to decompose H_2O_2 to water. The assessment of thiobarbituric acid reactive substances (TBARSs), MDA, or 4-hydroxynonenal is probably the most commonly applied method for the measurement of LP. The central nervous system (CNS) is more sensitive to ROS attack than the other body compartments and organs, because of its high lipid content.

28.2.1 Effects of Oxidative Stress

Oxidative stress occurs as a result of an imbalance between free radical production and antioxidant defenses and is associated with damage to a wide range of molecular species including lipids, proteins, and nucleic acids. Lipoprotein particles or membranes characteristically undergo the process of LP, giving rise to a variety of products including short-chain aldehydes such as MDA or 4-hydroxynonenal, alkanes, and alkenes, conjugated dienes, and a variety of hydroxides and hydroperoxides. Many of these products can be measured as markers of LP. Oxidative damage to proteins and nucleic acids similarly gives rise to a variety of specific damage products as a result of modifications of amino acids or nucleotides. Such oxidative damage might also lead to cellular dysfunction, and it might contribute to the pathophysiology of a wide variety of diseases including FMS.

28.2.2 Oxidative Stress in FMS

LP and carbonylated proteins, the end products of membrane damage that are induced by ROS, are increased in the plasma of patients with FMS (Ozgocmen et al., 2006). The role of free radical–mediated oxidative damage was investigated in the etiopathogenesis of FMS by Fassbender and Wegner (1973) who suggested that muscle tender points in FMS result from local hypoxia. Another study showed abnormal oxygen pressure at the muscle surface above trigger points. Bengtsson et al. (1986) investigated oxidative metabolism and found that adenosine diphosphate and phosphoryl creatine levels decreased and adenosine monophosphate and creatine levels increased in FMS patients. Furthermore, total antioxidant capacity or antioxidant enzymes, such as SOD and catalase, are decreased in the plasma of patients with FMS (Sendur et al., 2009). Elevated total SOD activity and unchanging total GPX1 activity in FMS patients could be the reason for increased oxidative stress and LP observed in FMS patients (Akbas et al., 2014). Recent studies have reported some evidence demonstrating that oxidative stress, mitochondrial dysfunction, and inflammation may have a role in the pathophysiology of FMS (Ali, 2003). Skin biopsies from patients showed a significant mitochondrial dysfunction with reduced mitochondrial chain activities and bioenergetics levels and increased levels of oxidative stress. These data were related to increased levels of inflammation and correlated with pain, the principal symptom of FMS. A study by Sánchez et al. (2015) supports the role of oxidative stress, mitochondrial dysfunction, and inflammation as interdependent events in the pathophysiology of FMS with a special role in the peripheral alterations.

Research has been directed to the plasma or serum of patients as a study model, with a need for cellular models, as this is the place where activation and control of the ROS-producing machinery occur. In this regard, H_2O_2, as one of the free oxygen radicals that results from the oxygen of the ROS, has been found to increase in the neutrophils of patients with FMS. Similarly, high levels of $O_2^{\bullet-}$ of mitochondrial origin have been observed in the peripheral blood mononuclear cells of patients with FMS. In this model, patients had low levels of CoQ_{10}, a vital element in the mitochondrial respiratory chain whose primary mission is the electron transport from complexes I and II to III, in addition to regulating the coupling of proteins, the pore transition, and mitochondrial oxidation of fatty acids, so that a deficiency of the cell induces a drop in the activity of complex II + III, complex III, and complex IV, plus reduces the expression of mitochondrial proteins involved in oxidative phosphorylation, decreases mitochondrial membrane potential, and increases the production of ROS (Cordero et al., 2010). But, from a physiological point of view, what relationship exists between oxidative stress and the symptoms of FMS? It is known that ROS are involved in the etiology of one of the major symptoms of FMS, which is pain.

The $O_2^{\bullet-}$ plays an important role in the development of pain on one side by peripheral nervous system and CNS sensitization and thus induces an alteration of nociception and, on the other hand, contributes to it through the activation of several cytokines such as TNF-α, IL-1, and IL-6 (Menzies and Lyon, 2010). The role of cytokines in FMS has been widely discussed, although not as an etiologic mechanism but as a factor in the worsening of symptoms (Cordero et al., 2013).

Although the mechanisms by which oxidative stress can alter muscle sensitivity are still unknown, it is possible that oxidative damage interferes with the muscles by reducing local nociceptors, which causes a decrease in the pain threshold. On the other hand, LP has been associated with a typical symptom of FMS: fatigue. High levels of LP, as well as an interesting correlation with this symptom, have been demonstrated in CFS, a disease with a high rate of comorbidity with FMS (Cordero et al., 2010).

NO is a highly diffusible and labile, gaseous messenger molecule involved in various biological functions such as vasodilatation or vascular regulation, modulation of nociception, immune function, neurotransmission, and excitation–contraction coupling. NO also acts as a metabolic regulator during exercise. The production of NO from L-arginine is catalyzed by the dioxygenase, nitric oxide synthase (NOS), which has three isoforms: neuronal (nNOS), inducible (iNOS), and endothelial (eNOS). Recently, NO has been shown to modulate the levels of ROS in a variety of cells. Larson et al. (2000) suggested the possible role of NO in pain modulation in FMS. Interestingly, significant correlations are observed between the levels of antioxidants in both plasma and serum with respect to the score on a visual analogue pain scale and the degree of morning stiffness by patients. NO is known to be both a ROS and a neurotransmitter in the CNS and the peripheral nervous system. However, NO is described as an atypical neurotransmitter in the nervous system, and it may be much more suitable to consider it as a second messenger or hormone. Most of the effects of NO are mediated through the activation of the enzyme guanylate cyclase, which produces guanosine 3,5-cyclic monophosphate (cGMP). Additionally, several reports have shown that the permeability of the blood–brain barrier is affected by free radicals and peroxynitrite. NO is widely assessed in various inflammatory conditions, and studies in RA have demonstrated increased production of NO and decreased levels after treatment with prednisolone. Therefore, it is very valuable to determine the interference of NO generation in neuropsychiatric and rheumatologic disorders. On the other hand, LP in serum has demonstrated a high degree of correlation with the level of depression presented by patients with FMS, which shows the relationship between the balance of oxidants/antioxidants and symptoms of FMS. Signs of oxidative stress in FMS include a high level of oxidative damage to DNA, as seen in the biopsy samples of patients with FMS.

Reduced oxidative metabolism and mitochondrial abnormalities in FMS also support a mitochondrial defect as a contributor. Moreover, since the mitochondria supply energy to the cell through oxidative phosphorylation, the lower level of ATP that results from a low mitochondrial activity may explain the low exercise capacity and fatigue reported in patients with FMS. These results suggest that ROS production in the mitochondria may be involved in oxidative stress, and CoQ_{10} deficiency and mitochondrial dysfunction could also be involved in the pathophysiology of FMS. These results confirm the oxidative stress background of FMS, probably due to a defect in the antioxidant system (SOD, CoQ_{10}) and a high production of ROS. Finding the origin of oxidative stress could help us to understand the pathophysiology of FMS and to offer new therapeutic strategies for this disease.

28.2.3 ANTIOXIDANT DEFENSE SYSTEMS

An antioxidant can be defined as "any substance present in low concentrations compared to that of an oxidizable substrate, which significantly delays or inhibits the oxidation of that substrate." The physiological role of antioxidants, as this definition suggests, is to prevent damage to cellular components arising as a consequence of chemical reactions involving free radicals. In recent years, a substantial body of evidence has developed supporting a key role for free radicals in many fundamental cellular reactions and suggesting that oxidative stress might be important in the pathophysiology of common diseases including atherosclerosis, chronic renal failure, and diabetes mellitus. As radicals have the capacity to react in an indiscriminate manner leading to damage to almost any cellular component, an extensive range of antioxidant defenses, both endogenous and exogenous, are present to protect cellular components from free radical–induced damage.

28.2.4 Antioxidant in FMS

ROS, such as $O_2^{\cdot-}$, $^{\cdot}OH$, and H_2O_2, are produced in metabolic and physiological processes, and harmful oxidative reactions may occur in organisms. The oxidative effects of ROS are controlled by exogenous antioxidants such as vitamins E and C and also by endogenous antioxidants such as scavenger enzymes SOD, GPx, bilirubin, and uric acid. When the production of damaging free radicals exceeds the capacity of the body's antioxidant defense to detoxify them, a condition known as oxidative stress occurs, such that its increase in oxidant and decrease in antioxidant cannot be prevented, and the oxidative/antioxidative balance shifts toward the oxidative status. Cellular injury can be caused by oxidative stress, which has been implicated in over a hundred disorders, including FMS. Inherent to its function, the skeletal muscle is continuously exposed to fluctuations in its redox environment as, during exercise, ROS production by the mitochondrial respiratory chain increases. Therefore, the capability of the skeletal muscle to respond to oxidative stress may not be surprising. In vitro experiments have demonstrated that skeletal myocytes adapt to oxidative stress by the upregulation of antioxidant enzymes such as Cu, Zn-SOD, catalase, and GPx (Wang et al., 2004). This suggests that under some conditions (such as aging, poor sleep, or microtrauma), antioxidant capacity may be impaired, as it may result in more oxidative stress in the skeletal muscle. Potential triggers of oxidative stress in the muscle compartment include hypoxia and local sources of ROS and reactive nitrogen species; skeletal muscle tropical state, contractility, and fatigability may be affected by oxidative stress, resulting in skeletal muscle dysfunction.

Although the etiology of FMS remains unknown, recent data suggest that the oxidant/antioxidant balance may play a role in its development. It is suggested that oxidant/antioxidant imbalance is related to the disease process, and the increase in free radical levels may be responsible for the development of FMS. Bramwell et al. (2000) suggested that ascorbic acid treatment may be provided to improve on clinical findings and the quality of life in FMS patients. It has been suggested that the modified "Myers cocktail," which consists of magnesium, calcium, vitamin B, and vitamin C, has been found to be effective against acute asthma attacks, migraines, CFS, FMS, acute muscle spasm, upper respiratory tract infections, chronic sinusitis, seasonal allergic rhinitis, cardiovascular disease, and other disorders.

28.2.5 Antioxidants in FMS Treatment

To focus on the clinical aspect of FMS, which still largely is uncertain and is one of the main problems of this disease that is the lack of effective treatment. This leads the specialists to treat the symptoms of the disease rather than its causes and cure sometimes leading to worsening of the side effects of the disorders; and in many cases, these drugs induce an increase in oxidative stress. CoQ_{10} has been shown in in vitro experiments with peripheral blood mononuclear cells of patients with FMS, either through its antioxidant role or by offsetting the deficit significantly, to reduce ROS levels and to induce a mitochondrial degradation pathway known as mitophagy. This result could provide insights into the beneficial effect obtained in patients after administration of CoQ_{10} along with *Ginkgo biloba* shown by a pilot study in which there was a significant improvement in the quality of life of patients (Lister, 2002). Fatigue, one of the most typical symptoms of FMS, has been reduced by treatment with CoQ_{10} in both animal and human physical fatigue models after exercise. It should also be noted that CoQ_{10} has been shown to reduce muscle pain induced by statins in patients, and animal models have proven an anti-inflammatory and antinociceptive effect, and CoQ_{10} has recently been observed to regulate the expression of certain proinflammatory cytokine genes such as TNF-α (Schmelzer et al., 2008), whose role has already been described in FMS.

Antioxidant therapies have proven effective in many pathological processes in which oxidative stress plays an important role. CoQ_{10}, vitamin E or α-tocopherol, vitamin C or ascorbic acid, melatonin, SOD, vitamin A or retinol, glutathione (GSH), *N*-acetylcysteine, etc., are some of the antioxidants used in randomized trials of patients with a variety of diseases. However, in the case of FMS, there are still no double-blind and placebo-controlled trials in which the possible mechanisms demonstrate the benefits of these therapies in general and of CoQ_{10} in particular (Iqbal et al., 2011). The complexity of this disease makes it difficult to assess the effectiveness of a single treatment, thus requiring a multidisciplinary therapeutic approach in which the use of antioxidants would acquire a role as cotreatment. Although oxidative stress in FMS is an accepted fact, its role in the disease from a physiological point of view is not yet clear, and the mechanism by which high levels of free radicals, low levels of antioxidants, or both processes simultaneously can have effects on the worsening of symptoms is still unknown. Therefore, further studies are necessary in this regard and in the design of controlled trials on the therapeutic effect of antioxidants.

Oxidative stress parameter like LP levels has been implicated in the severity of the clinical symptoms in FMS, and it has been suggested that antioxidant therapy could be beneficial in FMS. In addition, prevention of LP has been demonstrated to be neuroprotective in a variety of neuropathological paradigms. It is known that LP, as a consequence of oxidative stress, indirectly reflects intracellular ROS generation, and ROS are known to be implicated in the etiology of pain, one of the most prominent symptoms in FMS, by inducing peripheral and central hyperalgesia. The possible link between FMS and oxidative stress presents a logical proposal that antioxidant supplementation (vitamins, omega-3 and omega-6 fatty acids, and other dietary supplements) may have value in the management. Furthermore, the use of plant-based diets or omega-3 fatty acids, which are also potent antioxidants, has been suggested to cause improvement in serum peroxidation products, pain scores, stiffness, depression, and sleep quality in cases with FMS. But these reports are usually pilot results or uncontrolled open-label studies that need further additional researches. All these data support the idea that antioxidant therapy may be beneficial in FMS patients. Nevertheless, although oxidative stress is accepted to be involved in the pathophysiology of FMS and the mitochondrial dysfunction could be involved in this disease (Sánchez et al., 2015), more studies are necessary to elucidate the origin of this oxidative disorder and its role in the etiology of FMS. In conclusion, the hypothesis that mitochondrial dysfunction is the origin of oxidative stress in FMS patients could help to understand the complex pathophysiology of this disorder and may lead to the development of new therapeutic strategies for the prevention and treatment of this disease.

28.2.6 Toxic Metals and Oxidative Stress in FMS

Toxic metals like lead, aluminum, magnesium, and zinc are neurotoxin, causing cognitive dysfunction, and interfere with energy metabolism resulting in fatigue, paresthesia, and weakness, all hallmarks of FMS. Heavy metal toxicity caused by the increased levels of pollution and the use of chemicals in industry is a growing threat to health and development. Distribution of the metal in different target organs varies with route, dose, and duration of exposure. A number of FMS patients are unknowingly suffering from heavy metal poisoning (i.e., aluminum, lead, magnesium, zinc, and arsenic). However, muscle pain in FMS has been associated with magnesium (Mg) and selenium (Se) deficiency (Eisinger et al., 1994). The metal aluminum is present in our food, water supply, and soil; most people suffer from some degree of aluminum toxicity. After years of accumulated exposure and storage of it in body tissues, it is clear that aluminum causes poisoning of the nervous system with a range of symptoms that are similar to FMS like disturbed sleep, nervousness, emotional instability, memory loss, headaches, and impaired intellect. Toxic metals trigger the production of free radicals, leading to oxidative stress and depletion of the body's master antioxidant as well as influencing the metabolism of metallothioneins (small metal-binding proteins rich in sulfur). Toxic metals are also known to stimulate the production of inflammatory messengers known as cytokines in the immune system causing immense pain (Fournié et al., 2001). These effects and more are known to be important in FMS. Thus, the symptoms of FMS alone are suggestive of metal toxicity, and symptoms such as fatigue, poor memory and concentration, and "brain fog" could all be caused or exacerbated by toxic metals like aluminum and lead. Heavy metal toxicity is a growing threat to health and development. Research has suggested that serum magnesium and zinc levels may play an important role in the pathophysiology of FMS (Sendur et al., 2008).

An increased susceptibility to allergies and HPA-axis dysfunction has long been thought to be central to CFS and FMS. Metal toxicity (aluminum, lead, etc.) is characterized by diffuse muscle pain, poor sleep, chronic malaise, brain fog, dizziness, headaches, anxiety, and numbness, which are the same as that of symptoms of FMS. Our bodies need some trace amounts of metals (aluminum, mercury, lead, copper, arsenic, and others), but at high levels, these metal toxins can be quite serious, even fatal. They are found in the air, soil, drinking water, fuel, medicines, cosmetics, etc., and they produce many FMS-like symptoms. The immune and nervous systems are closely linked, and it has been suggested that exposure to toxic metals could upset the delicate balance between them and lead to diseases such as CFS and FMS, where neuroimmune dysfunction is present. Through increased oxidative stress, toxic metals can damage mitochondrial DNA and cause the mitochondrial dysfunction, which inevitably leads to fatigue and other symptoms associated with FMS. Toxic metals not only increase oxidative stress but also deplete GSH and the amino acid cysteine vital for its production, thus lowering the body's defenses against free radicals.

28.3   CONCLUSIONS

Free radicals have been implicated in the etiology of a large number of major diseases. They can adversely alter many crucial biological molecules, leading to the loss of form and function. Such undesirable changes in the body can lead to diseased conditions. Antioxidants can protect against the damage induced by free radicals acting at various levels. Dietary and other components of plants form major sources of antioxidants. The relationship among free radicals, antioxidants, and functioning of various organs and organ systems is highly complex, and the discovery of "redox signaling" is a milestone in this crucial relationship. Recent research centers on various strategies to protect crucial tissues and organs against oxidative damage induced by free radicals. Many novel approaches are made, and significant findings have come to light in the last few years. There is overwhelming evidence that oxidative stress occurs in cells as a consequence of normal physiological processes and environmental interactions and that the complex web of antioxidant defense systems plays a key role in protecting against oxidative damage. These processes appear to be disordered in many conditions, and a plausible hypothesis may be constructed implicating oxidative stress as a cause of tissue damage. The toxicity of metals like aluminum, lead, and arsenic is mediated through free radical generation in the cells that alters the oxidative stress parameters and may lead to FMS. Thus, a clear understanding on the oxidative stress, antioxidant parameters, and disturbed metal ion level in patients with FMS may provide useful information to augment the understanding of pathophysiology and may also help in routine diagnosis of the patients. However, complete understanding of the biochemical events occurring at a cellular level to influence oxidative damage is required to guide future therapeutic advances.

REFERENCES

Akbas A, Inanir A, Benli I, Onder Y, Aydogan L. Evaluation of some antioxidant enzyme activities (SOD and GPX) and their polymorphisms (MnSOD2 Ala9Val, GPX1 Pro198Leu) in fibromyalgia. *Eur Rev Med Pharmacol Sci.* 2014; 18: 1199–1203.
Ali M. The cause of fibromyalgia: The respiratory-to-fermentative shift (the DysOx State) in ATP production. *J Integr Med.* 2003; 8: 135–140.
Altindag O, Celik H. Total antioxidant capacity and the severity of the pain in patients with fibromyalgia. *Redox Rep.* 2006; 11: 131–135.

Bengtsson A, Henriksson KG, Larsson J. Reduced high energy phosphate levels in the painful muscles of patients with primary fibromyalgia. *Arthritis Rheum.* 1986; 15: 1–6.

Bramwell B, Ferguson S, Scarlett N, Macintosh A. The use of ascorbigen in the treatment of fibromyalgia patients: A preliminary trial. *Altern Med Rev.* 2000; 5: 455–462.

Choy EH. The role of sleep in pain and fibromyalgia. *Nat Rev Rheumatol.* 2015; 11: 513–520.

Clauw DJ. Fibromyalgia: A clinical review. *JAMA.* 2014; 311: 1547–1555.

Clauw DJ. Fibromyalgia and related conditions. *Mayo Clin Proc.* 2015; 90: 680–692.

Cordero MD, DeMiguel M, Moreno Fernández AM et al. Mitochondrial dysfunction and mitophagy activation in blood mononuclear cells of fibromyalgia patients: Implications in the pathogenesis of the disease. *Arthritis Res Ther.* 2010; 12: R17.

Cordero MD, Díaz-Parrado E, Carrión AM et al. Is inflammation a mitochondrial dysfunction-dependent event in fibromyalgia? *Antioxid Redox Signal.* 2013; 18: 800–807.

Eisinger J, Plantamura A, Marie PA, Ayavou T. Selenium and magnesium status in fibromyalgia. *Magnes Res.* 1994; 7: 285–258.

Fassbender HG, Wegner K. Morphology and pathogenesis of soft tissue rheumatism. *Z Rheumaforsch.* 1973; 32: 355–374.

Fatima G, Das SK, Mahdi AA. Oxidative stress and antioxidative parameters and metal ion content in patients with fibromyalgia syndrome: Implications in the pathogenesis of the disease. *Clin Exp Rheumatol.* 2013; 31: 128–133.

Fournié GJ, Mas M, Cautain B, Savignac M, Subra JF, Pelletier L, Saoudi A, Lagrange D, Calise M, Druet P. Induction of autoimmunity through bystander effects: Lessons from immunological disorders induced by heavy metals. *J Autoimmun.* 2001; 16: 319–326.

Iqbal R, Mughal MS, Arshad N, Arshad M. Pathophysiology and antioxidant status of patients with fibromyalgia. *Rheumatol Int.* 2011; 31: 149–152.

Jay GW, Barkin RL. Fibromyalgia. *Dis Mon.* 2015; 61: 66–111.

Larson AA, Giovengo SL, Russell IJ et al. Changes in the concentrations of amino acids in the cerebrospinal fluid that correlate with pain in patients with fibromyalgia: Implications for nitric oxide pathways. *Pain.* 2000; 87: 201–211.

Lister RE. An open, pilot study to evaluate the potential benefits of coenzyme Q_{10} combined with Ginkgo biloba extract in fibromyalgia syndrome. *J Int Med Res.* 2002; 30: 195–199.

Lukkahatai N, Walitt B, Espina A, Gelio A, Saligan LN. Understanding the association of fatigue with other symptoms of fibromyalgia: Development of a cluster model. *Arthritis Care Res (Hoboken).* 2016; 68(1): 99–107.

Mahdi AA, Fatima G. A quest for better understanding of biochemical changes in fibromyalgia syndrome. *Indian J Clin Biochem.* 2014; 29: 1–2.

Menzies V, Lyon DE. Integrated review of the association of cytokines with fibromyalgia and fibromyalgia core symptoms. *Biol Res Nurs.* 2010; 11: 387–394.

Neyal M, Yimenicioglu F, Aydeniz A et al. Plasma nitrite levels, total antioxidant status, total oxidant status, and oxidative stress index in patients with tension-type headache and fibromyalgia. *Clin Neurol Neurosurg.* 2012; 12: 453–452.

O'Neill S, Brault J, Stasia MJ, Knaus UG. Genetic disorders coupled to ROS deficiency. *Redox Biol.* 2015; 6: 135–156.

Ozgocmen S, Ozyurt H, Sogut S, Akyol O, Ardicoglu O, Yildizhan H. Antioxidant status, lipid peroxidation and nitric oxide in fibromyalgia: Etiologic and therapeutic concerns. *Rheumatol Int.* 2006; 26: 598–603.

Phaniendra A, Jestadi DB, Periyasamy L. Free radicals: Properties, sources, targets, and their implication in various diseases. *Indian J Clin Biochem.* 2015; 30: 11–26.

Sánchez-Domínguez B, Bullón P, Román-Malo L, Marín-Aguilar F, Alcocer-Gómez E, Carrión AM, Sánchez-Alcazar JA, Cordero MD. Oxidative stress, mitochondrial dysfunction and, inflammation common events in skin of patients with Fibromyalgia. *Mitochondrion.* March 2015; 21: 69–75.

Schmelzer C, Lindner I, Rimbach G et al. Functions of coenzyme Q_{10} in inflammation and gene expression. *Biofactors.* 2008; 32: 179–183.

Sendur OF, Tastaban E, Turan Y et al. The relationship between serum trace element levels and clinical parameters in patients with fibromyalgia. *Rheumatol Int.* 2008; 28: 1117–1121.

Sendur OF, Turan Y, Tastaban E, Yenisey C, Serter M. Serum antioxidants and nitric oxide levels in fibromyalgia: A controlled study. *Rheumatol Int.* 2009; 29: 629–633.

Valko M, Leibfritz D, Moncol J et al. Free radicals and antioxidants in normal physiological functions and human disease. *Int J Biochem Cell Biol.* 2007; 39: 44–84.

Wang ZQ, Porreca F, Cuzzocrea S et al. A newly identified role for superoxide in inflammatory pain. *J Pharmacol Exp Ther.* 2004; 309: 869–878.

Wolfe F, Clauw DJ, Fitzcharles MA et al. The American College of Rheumatology preliminary diagnostic criteria for fibromyalgia and measurement of symptom severity. *Arthritis Care Res (Hoboken)* 2010; 62: 600–610.

Wolfe F, Smythe HA, Yunus MB et al. The American College of Rheumatology 1990 criteria for the classification of fibromyalgia. *Arthritis Rheum.* 1990; 33: 160–172.

29 Redox Mechanisms in Pulmonary Diseases

Li Zuo, Chia-Chen Chuang, and Tingyang Zhou

CONTENTS

ABSTRACT

Excessive production of reactive oxygen species (ROS), either from endogenous or exogenous sources, can compromise the body's antioxidant defenses leading to disrupted redox homeostasis. In particular, this redox imbalance (termed "oxidative stress" [OS]) has been manifested in the pathogenesis of many pulmonary diseases, such as chronic obstructive pulmonary disease (COPD) and asthma, due to the fact that lungs are highly susceptible to ROS-induced injuries. OS damages important cellular constituents, deters normal cell function, and contributes to the detrimental pathological characteristics often seen in pulmonary diseases. Although antioxidant or redox therapies targeting redox restoration are feasible approaches in alleviating OS-related lung impairment, various results are yielded. Thus, a thorough understanding of the role of ROS in pulmonary diseases is essential for the development of optimal treatments. In this chapter, we review the major pulmonary diseases—including pulmonary hypertension, COPD, and asthma—as well as their association with OS.

29.1 INTRODUCTION

Due to constant exposure to atmospheric oxygen, pollutants, and endogenous oxygen free radicals produced via metabolism, lungs are highly susceptible to reactive oxygen species (ROS)-mediated injuries (Rahman and MacNee 2012, Valavanidis et al. 2013). Excessive ROS overwhelm the body's naturally equipped antioxidant mechanisms, disrupting redox homeostasis leading to oxidative stress (OS) (Park et al. 2009). It has been reported that OS is associated with deleterious pathological characteristics, including inflammation, which often occurs in pulmonary diseases

(Lee and Yang 2012). The presence of OS is involved in the development of multiple pulmonary diseases, such as chronic obstructive pulmonary disease (COPD) and lung cancer. OS also causes damage to important cellular constituents (e.g., lipids, proteins, and DNA), affecting normal cell function and influencing signal transduction (Kirkham and Rahman 2006). Various approaches encompassing antioxidant or redox therapy, in attempts to restore the redox balance, have yielded positive results (Park et al. 2009). However, a thorough understanding of exact mechanisms of specific pulmonary diseases is paramount for successful treatments. In this chapter, we aim to present the pathogenesis of major pulmonary diseases that have been highly associated with OS, notably COPD, asthma, and pulmonary hypertension (PH), and the roles that ROS play in each disease.

29.2   EFFECTS OF ROS IN PULMONARY DISEASES

In addition to the production of ROS from endogenous sources such as mitochondria and inflammatory cells, environment-derived ROS are prominent in the lungs. Ranging from oxidant gases to airborne particulate matter, inhalable particles can promote ROS generation and initiate inflammatory responses (Kirkham and Rahman 2006, Rahman and MacNee 2012). Although inflammation occurs as a necessary defense against insults such as infection, excessive inflammation, as an intense response to eliminate invaders, may be regarded harmful (Moldoveanu et al. 2009). Considering the relevance of chronic inflammation in multiple pulmonary diseases, the lungs contain several antioxidants in order to effectively combat OS and lessen OS-induced tissue damage (Zuo et al. 2013).

NADPH oxidases (Nox) reside in various cells, particularly in endothelial cells and phagocytes. When stimulated, Nox can produce ROS, which participate in the lung's host defense. Indeed, the lack of Nox, due to genetic defect in ROS-forming capability, results in poor eradication of pathogens and toxicants (Forman and Torres 2002, Bedard and Krause 2007). ROS released by elevated levels of phagocytes may generate end products of lipid peroxidation, such as thiobarbituric acid reactive products and isoprostanes (Fischer et al. 2015). In addition, Nox activities can regulate other ROS-generating oxidases including xanthine oxidase (XO) (Demarco et al. 2010). Small amounts of ROS can be formed as by-products of cellular metabolism in mitochondrial electron transport chain (Mittal et al. 2014). Normally, ROS generated in Nox and mitochondria are kept at basal levels, functioning as signaling mediators for the control of ion channels and other essential cellular components. When pulmonary antioxidants are unable to counterbalance both endogenous and exogenous ROS, increased oxidant burden may lead to deleterious cellular alterations in the lungs (Demarco et al. 2010).

As a cell-signaling messenger, ROS also facilitate inflammatory cell influx and cytokine production via nuclear factor kappa-light-chain-enhancer of activated B cells (NF-κB)-dependent pathways (Forman and Torres 2002). NF-κB is a redox-sensitive transcription factor that is essential in the inflammatory and immune responses. Specifically, its activation has been implicated in the pathogenesis of asthma and COPD, two commonly known chronic inflammatory diseases (Henderson et al. 2002, Park et al. 2009). In addition, ROS may induce activator protein-1 (AP-1)-related gene expression, which is likely involved in hyperoxic inflammatory lung injury (Pepperl et al. 2001).

29.3   REDOX MECHANISMS IN PULMONARY PATHOGENESIS

29.3.1   COPD

Increased rate of cigarette smoking and industrial waste including gases has been shown to elevate environmental pollution and result in high prevalence of COPD in many urbanized countries. Commonly characterized as a preventable and treatable inflammatory disease, COPD exhibits persistent airflow limitation and several respiratory structural alterations such as parenchymal destruction (Decramer and Vestbo 2015). COPD continues to impose significant economic and social burdens worldwide. In particular, comorbidities often associated with increased age may contribute to the exacerbation of COPD and frequent hospitalization (Decramer and Vestbo 2015, Mannino et al. 2015). The abnormal inflammatory response to inhaled noxious particles/gases cumulatively induces subsequent pathological changes, highlighting the pathogenesis of COPD (Decramer and Vestbo 2015). In particular, cigarette smoke is considered as a direct contributing factor for COPD. It has been found that cigarette smoke contains high concentrations of short-lived oxidants, such as nitric oxide (NO) and superoxide anion ($O_2^{\cdot-}$), and long-lived organic radicals (Kirkham and Rahman 2006). Schamberger et al. reported that cigarette smoke extract can alter cell differentiation in human bronchial epithelium, significantly reduce ciliated cell numbers, and hamper the function of pathogen clearance (Schamberger et al. 2015). However, smoking cessation does not effectively relieve COPD symptoms, suggesting that more complex mechanisms are involved in the development of COPD (Zuo et al. 2014a).

Besides persistent inflammation, recurrent OS in the lungs due to environmental exposures has been implicated in COPD development. An endogenous imbalance between oxidants and antioxidants may be an essential factor underlying the pathogenesis of COPD (Perera et al. 2007). Cigarette smoking generates a complex aerosol via incomplete combustion, which includes a mixture of various oxidant compounds that can directly damage tissues and cells, elevate OS, and initiate inflammation. Commonly recognized ROS, such as $O_2^{\cdot-}$, hydrogen peroxide (H_2O_2), NO, and peroxynitrite (ONOO$^-$), have been found in the particulate of cigarette smoke (Fischer et al. 2015). On the other hand, the concentration of antioxidants (e.g., vitamins) is lowered in the serum of COPD patients, which likely corresponds to the increased OS (Tsiligianni and van der Molen 2010, Lehouck et al. 2012). Notably, glutathione (GSH) metabolism is particularly induced by smoking. Therefore, the supplementation of exogenous antioxidants is a potential approach in alleviating OS-associated injuries (Fischer et al. 2015).

Cigarette smoke–induced OS triggers a cascade involving the activation of key redox-sensitive transcription factors such as NF-κB and nuclear factor erythroid 2-related factor 2 (Nrf2), followed by inflammation and apoptosis (Fischer et al. 2015). As mentioned previously, smoking cessation does not alleviate COPD symptoms. In fact, OS continues to exist after smoking cessation. The retained particulates of cigarette smoke may continue to participate in the generation of oxidants (Louhelainen et al. 2009). Another possible explanation describes the disturbed iron homeostasis during smoking, which may produce additional oxidants upon the interaction with smoking particles (Fischer et al. 2015). In addition, exposure to other noxious particles, such as ambient air pollution particles, also leads to OS (Tao et al. 2003). OS can regulate mucin gene expression, which is upregulated when exposed to tobacco smoke constituents (e.g., H_2O_2) (Fischer et al. 2015). Mucins comprise epithelial mucus in the respiratory tract and function to provide mucociliary clearance and lung defense against environmental pathogens and toxins. The overproduction of mucus and mucins has been observed in patients with COPD, which is highly associated with airway obstruction (Rose and Voynow 2006). In particular, *MUC5AC* expression is upregulated, indicating the involvement of mucous cell hyperplasia and hypertrophy in COPD pathogenesis (Caramori et al. 2009).

29.3.2 ASTHMA

Asthma is a long-term airway inflammatory disorder, defined by symptoms including wheezing, coughing, and shortness of breath. Typical pathophysiological changes in asthma are associated with reversible airflow limitation and airway hyperreactivity. However, long-standing asthma results in permanent airway remodeling (Comhair and Erzurum 2010, FitzGerald and Reddel 2015). Reversible symptoms vary with time and may become undetectable without medication. The underlying pathological presentations such as inflammation and hyperresponsiveness usually persist, even in the absence of symptoms or with normal lung function (FitzGerald and Reddel 2015).

Based on the report by the Asthma and Allergy Foundation of America (AAFA), over 300 million people are suffering from this disorder worldwide. In the United States, asthma accounts for nine deaths per day and causes a yearly economic loss of $56 billion (Jiang et al. 2014).

Several types of asthma have been defined by their phenotypes including allergic asthma, nonallergic asthma, and obesity-related asthma (FitzGerald and Reddel 2015). Allergic asthma, most commonly found in childhood, is characterized by eosinophilic airway inflammation, usually correlated with a familial history of allergic diseases. On the other hand, nonallergic asthma is often seen in adults. The inflammatory cells present in these asthmatics include neutrophils and eosinophils. Certain asthma patients with obesity demonstrate significant respiratory symptoms while maintaining low levels of eosinophilic airway inflammation (FitzGerald and Reddel 2015).

During acute severe asthma, interleukin (IL)-8 mediates the recruitment of neutrophils to the airway, leading to substantial inflammation (Ordonez et al. 2000). Other inflammatory cells including mast cells, eosinophils, and macrophages are also stimulated upon exposure to allergens (Jiang et al. 2014). The accumulation of these inflammatory cells initiates excessive production of $O_2^{\cdot-}$ and H_2O_2, which may be implicated in the signaling cascade of asthma (Pelaia et al. 2004, Zuo et al. 2013, Jiang et al. 2014). For instance, ROS can activate NF-κB in epithelial cells, promoting the expression of proinflammatory cytokines and adhesion molecules (Zuo et al. 2013, Jiang et al. 2014). Increased ROS formation plays a critical role in mediating the activity of tumor necrosis factor-α (TNF-α) and IL-1 (Jiang et al. 2014). These mediators can further modulate other proinflammatory cytokines such as vascular cell adhesion molecules and intercellular adhesion molecule-1, which are implicated in the pathogenesis of asthma (Kim et al. 2008, Jiang et al. 2014). Moreover, Nox is a critical source of ROS in asthma. Nox-derived ROS can trigger the production of immunoglobulin E (IgE), which acts as an important mediator to initiate the hypersensitive responses in the asthmatic airway (Jiang et al. 2014).

Healthy lungs are normally characterized toward a more reducing environment, as a result of a robust antioxidant system (Comhair and Erzurum 2010). However, the redox balance is destabilized in asthma patients because of the excessive ROS accumulation and compromised antioxidant activities (Comhair and Erzurum 2010, Jiang et al. 2014). The ratio of GSH/glutathione disulfide (GSSG), a critical redox indicator, is markedly lowered in epithelial lining fluid (Comhair and Erzurum 2010). Substantial oxidative damages such as DNA mutation, protein modification, and lipid peroxidation are observed in asthma lungs (Jiang et al. 2014). Specifically, a highly oxidative environment can lead to the inefficacy of antioxidant enzymes thus exacerbating OS conditions (Comhair and Erzurum 2010). Studies have revealed markedly decreased superoxide dismutase (SOD) and catalase activities in asthma subjects (Comhair and Erzurum 2010). Altered intracellular redox conditions can stimulate mitogen-activated protein kinase (MAPK) signaling including c-jun N-terminal kinase (JNK) and extracellular signal–regulated kinase (ERK), further facilitating the inflammatory process (Comhair and Erzurum 2010).

Reactive nitrogen species (RNS) also play a role in the pathogenesis of asthma (Zuo et al. 2014b). Increased NO production is associated with enhanced nitric oxide synthase (NOS) activity in the lung (Comhair and Erzurum 2010, Zuo et al. 2014b). NO can react with $O_2^{\cdot-}$ to form $ONOO^-$, a highly toxic molecule that results in tyrosine nitration and subsequent respiratory tract damage (Comhair and Erzurum 2010, Zuo et al. 2014b). Accordingly, therapies regulating the redox state in the airway may be beneficial in asthma control.

29.3.3 PULMONARY HYPERTENSION

A total of six groups of PH have been classified clinically, which altogether exhibit a higher mean pulmonary arterial pressure at rest (≥25 mmHg) and histological abnormalities of pulmonary vasculature (Galie et al. 2009, Wong et al. 2013). In particular, pulmonary arterial hypertension (PAH; group 1) is a clinical condition of PH highlighted by aberrant smooth muscle and endothelial dysfunction, leading to profound vascular remodeling and vasoconstriction. Specifically, overexpressed vasoconstrictors (e.g., endothelin-1 and thromboxane A_2) and reduced vasodilator bioavailability (e.g., NO and prostacyclin) partly contribute to the vasoconstriction induced by endothelium dysfunction (Demarco et al. 2010, Hoeper et al. 2013).

Studies have observed enhanced levels of lipid peroxidation–associated by-products such as urinary F_2-isoprostane and plasma malonic dialdehyde in patients with PH, suggesting the involvement of OS in PH pathogenesis (Cracowski et al. 2001, Irodova et al. 2002). In a disease state, NOS along with Nox and XO become major contributors for ROS formation in the vasculature. OS impairs vasodilator synthesis by damaging endothelial and prostacyclin synthase. ROS promote smooth muscle proliferation, leading to vascular remodeling (Demarco et al. 2010). Bowers et al. supported the occurrence of OS by examining the samples of lung tissue from severe PH patients. An elevated expression of 8-hydroxydeoxyguanosine (8-OHdG) and arachidonic acid metabolites was detected in the immunohistological staining of lung sections, indicating the production of ROS in the lungs of PH subjects (Bowers et al. 2004). The 8-OHdG serves as a reliable biomarker for DNA oxidation and has been used in various studies (Kasai 2002). Interestingly, high nitrotyrosine levels are also observed, which may be attributed to $ONOO^-$ formation in the endothelial and smooth muscle cells. This leads to one possible explanation of NO depletion frequently observed in PH patients, that is, the consumption of NO via an $ONOO^-$-generating cascade rather than the cessation of NO production (Bowers et al. 2004). Endogenous antioxidant amounts, such as manganese SOD, are also significantly impacted by the OS in PH (Bowers et al. 2004). Reis et al. showed lower levels of vitamin E and GSH in patients with PAH. Additionally, the increment of circulating cytokines including TNF-α and IL-6 further confirms the involvement of inflammation in PH (Reis et al. 2013).

Animal models of PH help to elucidate the complex interactions between OS and the pathological features presented in PH. In a study using rat models, the activated local renin–angiotensin system (RAS) within the lungs and pulmonary vasculature induces Nox-associated OS, which likely contributes to the PAH pathogenesis. Thus, targeting the RAS may be a potential gene therapy for PAH (Ferreira et al. 2009, Demarco et al. 2010). ROS are also generated under stress, such as hypoxia and monocrotaline (pyrrolizidine alkaloid), in part due to an influx of inflammatory cells and acute inflammation. Particularly, bone marrow–derived monocytic progenitor cells aggregate in the pulmonary vasculature. Monocytes then activate vascular wall fibrosis via the secretion of mitogenic and fibrogenic cytokines (Wilson et al. 1989, Demarco et al. 2010). Moreover, through the activation of ERK in the presence of H_2O_2, the downstream GATA-binding factor 4 (GATA4) transcription factor promotes the hyperplasia of smooth muscle, thereby leading to vasoconstriction (Wong et al. 2013, Zuo et al. 2014c).

29.4 REDOX THERAPIES

29.4.1 COPD

Cigarette smoke is a major etiological factor for the oxidative burden in COPD. Other than smoking cessation, various pharmacological therapies including antioxidant agents have been tested in the clinical setting. Small molecule thiol antioxidants, such as N-acetyl-L-cysteine (NAC), can neutralize oxidative states by elevating intracellular GSH levels (Biswas et al. 2013). Oral administration of NAC has been shown to alleviate pulmonary emphysema in a rat model (Rubio et al. 2004). The beneficial clinical effects of NAC and other thiol agents may be promising in the treatment of COPD (Biswas et al. 2013). The supplementation of antioxidant enzyme mimetics, such as synthetic SOD/catalase, and dietary vitamins demonstrates improved redox status and reduced OS-mediated injuries (Zhang et al. 2004, Nyunoya et al. 2011). Moreover, the suppression of redox effector factor-1 (Ref-1) and thioredoxin, both redox sensors, can attenuate cigarette smoke–induced OS and inflammation via the inhibition of NF-κB and AP-1 signaling, respectively (Nguyen et al. 2003, Souza et al. 2005, Rahman 2012).

29.4.2 Asthma

Studies have attempted to explore the possibilities of alleviating asthmatic symptoms by interrupting related inflammatory pathways (Comhair and Erzurum 2010). For instance, therapeutic strategies have been developed via the inhibition of NF-κB and AP-1, which have been implicated in the signaling of airway inflammation (Comhair and Erzurum 2010). The application of PNRI-299, a potent and exclusive inhibitor of AP-1, dramatically attenuated both airway eosinophil infiltration and IL-4 release (Comhair and Erzurum 2010). Furthermore, MOL 294 also effectively reduced airway hyperreactivity and inflammation in asthmatic mice by inhibiting the activity of NF-κB and AP-1 (Comhair and Erzurum 2010). Other studies have also provided evidence that treatment with SOD mimics may reduce airway inflammation in asthma patients (Comhair and Erzurum 2010).

29.4.3 PH

According to the European Respiratory Society (ERS) guideline, treatments of PH encompass endothelium receptor antagonists, prostacyclin, and phosphodiesterase type 5 inhibitors, which mainly function to increase NO, a vasodilator, in order to relieve pulmonary vascular resistance (Galie et al. 2009). However, the effectiveness of antioxidants has been gradually recognized in various PH animal models. For instance, SOD was shown to reduce oxidation and improve the oxygenation of neonatal lambs with sustained PH (Lakshminrusimha et al. 2006). Accordingly, Redout et al. showed that a SOD/catalase mimetic can attenuate OS and the development of interstitial fibrosis in a monocrotaline-induced PH rat model (Redout et al. 2010).

29.5 CONCLUSION

Disrupted redox homeostasis in the lung either through excess ROS generation or compromised antioxidant defenses is implicated in the pathogenesis of pulmonary diseases, notably COPD, asthma, and PH. Along with OS-associated injuries, prolonged lung inflammation further damages the tissue and has been markedly associated with the development of pulmonary diseases. Dietary antioxidants and antioxidant enzyme mimetics have beneficial effects in restoring oxidative imbalance, thus serving as feasible approaches in managing OS-mediated lung impairment. Taken together, an understanding of the exact mechanisms involved in antioxidant or redox therapy may certainly help in elucidating novel methods to treat pulmonary lung diseases.

ACKNOWLEDGMENTS

The authors thank Benjamin Pannell for his assistance during the manuscript preparation.

REFERENCES

Bedard, K. and K. H. Krause. 2007. The NOX family of ROS-generating NADPH oxidases: Physiology and pathophysiology. *Physiol Rev* 87 (1):245–313.
Biswas, S., J. W. Hwang, P. A. Kirkham, and I. Rahman. 2013. Pharmacological and dietary antioxidant therapies for chronic obstructive pulmonary disease. *Curr Med Chem* 20 (12):1496–1530.
Bowers, R., C. Cool, R. C. Murphy, R. M. Tuder, M. W. Hopken, S. C. Flores, and N. F. Voelkel. 2004. Oxidative stress in severe pulmonary hypertension. *Am J Respir Crit Care Med* 169 (6):764–769.
Caramori, G., P. Casolari, C. Di Gregorio, M. Saetta, S. Baraldo, P. Boschetto, K. Ito et al. 2009. MUC5AC expression is increased in bronchial submucosal glands of stable COPD patients. *Histopathology* 55 (3):321–331.

Comhair, S. A. and S. C. Erzurum. 2010. Redox control of asthma: Molecular mechanisms and therapeutic opportunities. *Antioxid Redox Signal* 12 (1):93–124.

Cracowski, J. L., C. Cracowski, G. Bessard, J. L. Pepin, J. Bessard, C. Schwebel, F. Stanke-Labesque, and C. Pison. 2001. Increased lipid peroxidation in patients with pulmonary hypertension. *Am J Respir Crit Care Med* 164 (6):1038–1042.

Decramer, M. and J Vestbo. 2015. Global strategy for the diagnosis, management and prevention of COPD. Global Initiative for Chronic Obstructive Lung Disease.

Demarco, V. G., A. T. Whaley-Connell, J. R. Sowers, J. Habibi, and K. C. Dellsperger. 2010. Contribution of oxidative stress to pulmonary arterial hypertension. *World J Cardiol* 2 (10):316–324.

Ferreira, A. J., V. Shenoy, Y. Yamazato, S. Sriramula, J. Francis, L. Yuan, R. K. Castellano et al. 2009. Evidence for angiotensin-converting enzyme 2 as a therapeutic target for the prevention of pulmonary hypertension. *Am J Respir Crit Care Med* 179 (11):1048–1054.

Fischer, B. M., J. A. Voynow, and A. J. Ghio. 2015. COPD: Balancing oxidants and antioxidants. *Int J Chron Obstruct Pulmon Dis* 10:261–276.

FitzGerald, J. M. and H. K. Reddel. 2015. Global initiative for asthma (GINA): Global strategy for asthma management and prevention. NIH Publication (02-3659).

Forman, H. J. and M. Torres. 2002. Reactive oxygen species and cell signaling: Respiratory burst in macrophage signaling. *Am J Respir Crit Care Med* 166 (12 Pt 2):S4–S8.

Galie, N., M. M. Hoeper, M. Humbert, A. Torbicki, J. L. Vachiery, J. A. Barbera, M. Beghetti et al. 2009. Guidelines for the diagnosis and treatment of pulmonary hypertension: The Task Force for the Diagnosis and Treatment of Pulmonary Hypertension of the European Society of Cardiology (ESC) and the European Respiratory Society (ERS), endorsed by the International Society of Heart and Lung Transplantation (ISHLT). *Eur Heart J* 30 (20):2493–2537.

Henderson, W. R., Jr., E. Y. Chi, J. L. Teo, C. Nguyen, and M. Kahn. 2002. A small molecule inhibitor of redox-regulated NF-kappa B and activator protein-1 transcription blocks allergic airway inflammation in a mouse asthma model. *J Immunol* 169 (9):5294–5299.

Hoeper, M. M., H. J. Bogaard, R. Condliffe, R. Frantz, D. Khanna, M. Kurzyna, D. Langleben et al. 2013. Definitions and diagnosis of pulmonary hypertension. *J Am Coll Cardiol* 62 (25 Suppl):D42–D50.

Irodova, N. L., V. Z. Lankin, G. K. Konovalova, A. G. Kochetov, and I. E. Chazova. 2002. Oxidative stress in patients with primary pulmonary hypertension. *Bull Exp Biol Med* 133 (6):580–582.

Jiang, L., P. T. Diaz, T. M. Best, J. N. Stimpfl, F. He, and L. Zuo. 2014. Molecular characterization of redox mechanisms in allergic asthma. *Ann Allergy Asthma Immunol* 113 (2):137–142.

Kasai, H. 2002. Chemistry-based studies on oxidative DNA damage: Formation, repair, and mutagenesis. *Free Radic Biol Med* 33 (4):450–456.

Kim, H., J. S. Hwang, C. H. Woo, E. Y. Kim, T. H. Kim, K. J. Cho, J. H. Kim, J. M. Seo, and S. S. Lee. 2008. TNF-alpha-induced up-regulation of intercellular adhesion molecule-1 is regulated by a Rac-ROS-dependent cascade in human airway epithelial cells. *Exp Mol Med* 40 (2):167–175.

Kirkham, P. and I. Rahman. 2006. Oxidative stress in asthma and COPD: Antioxidants as a therapeutic strategy. *Pharmacol Ther* 111 (2):476–494.

Lakshminrusimha, S., J. A. Russell, S. Wedgwood, S. F. Gugino, J. A. Kazzaz, J. M. Davis, and R. H. Steinhorn. 2006. Superoxide dismutase improves oxygenation and reduces oxidation in neonatal pulmonary hypertension. *Am J Respir Crit Care Med* 174 (12):1370–1377.

Lee, I. T. and C. M. Yang. 2012. Role of NADPH oxidase/ROS in pro-inflammatory mediators-induced airway and pulmonary diseases. *Biochem Pharmacol* 84 (5):581–590.

Lehouck, A., C. Mathieu, C. Carremans, F. Baeke, J. Verhaegen, J. Van Eldere, B. Decallonne, R. Bouillon, M. Decramer, and W. Janssens. 2012. High doses of vitamin D to reduce exacerbations in chronic obstructive pulmonary disease: A randomized trial. *Ann Intern Med* 156 (2):105–114.

Louhelainen, N., P. Rytila, T. Haahtela, V. L. Kinnula, and R. Djukanovic. 2009. Persistence of oxidant and protease burden in the airways after smoking cessation. *BMC Pulm Med* 9:25.

Mannino, D. M., K. Higuchi, T. C. Yu, H. Zhou, Y. Li, H. Tian, and K. Suh. 2015. Economic burden of COPD in the presence of comorbidities. *Chest* 148 (1):138–150.

Mittal, M., M. R. Siddiqui, K. Tran, S. P. Reddy, and A. B. Malik. 2014. Reactive oxygen species in inflammation and tissue injury. *Antioxid Redox Signal* 20 (7):1126–1167.

Moldoveanu, B., P. Otmishi, P. Jani, J. Walker, X. Sarmiento, J. Guardiola, M. Saad, and J. Yu. 2009. Inflammatory mechanisms in the lung. *J Inflamm Res* 2:1–11.

Nguyen, C., J. L. Teo, A. Matsuda, M. Eguchi, E. Y. Chi, W. R. Henderson, Jr., and M. Kahn. 2003. Chemogenomic identification of Ref-1/AP-1 as a therapeutic target for asthma. *Proc Natl Acad Sci USA* 100 (3):1169–1173.

Nyunoya, T., T. H. March, Y. Tesfaigzi, and J. Seagrave. 2011. Antioxidant diet protects against emphysema, but increases mortality in cigarette smoke-exposed mice. *COPD* 8 (5):362–328.

Ordonez, C. L., T. E. Shaughnessy, M. A. Matthay, and J. V. Fahy. 2000. Increased neutrophil numbers and IL-8 levels in airway secretions in acute severe asthma: Clinical and biologic significance. *Am J Respir Crit Care Med* 161 (4 Pt 1):1185–1190.

Park, H. S., S. R. Kim, and Y. C. Lee. 2009. Impact of oxidative stress on lung diseases. *Respirology* 14 (1): 27–38.

Pelaia, G., G. Cuda, A. Vatrella, L. Gallelli, D. Fratto, V. Gioffre, B. D'Agostino et al. 2004. Effects of hydrogen peroxide on MAPK activation, IL-8 production and cell viability in primary cultures of human bronchial epithelial cells. *J Cell Biochem* 93 (1):142–152.

Pepperl, S., M. Dorger, F. Ringel, C. Kupatt, and F. Krombach. 2001. Hyperoxia upregulates the NO pathway in alveolar macrophages in vitro: Role of AP-1 and NF-kappaB. *Am J Physiol Lung Cell Mol Physiol* 280 (5):L905–L913.

Perera, W. R., J. R. Hurst, T. M. Wilkinson, R. J. Sapsford, H. Mullerova, G. C. Donaldson, and J. A. Wedzicha. 2007. Inflammatory changes, recovery and recurrence at COPD exacerbation. *Eur Respir J* 29 (3):527–534.

Rahman, I. 2012. Pharmacological antioxidant strategies as therapeutic interventions for COPD. *Biochim Biophys Acta Mol Basis Dis* 1822 (5):714–728.

Rahman, I. and W. MacNee. 2012. Antioxidant pharmacological therapies for COPD. *Curr Opin Pharmacol* 12 (3):256–265.

Redout, E. M., A. van der Toorn, M. J. Zuidwijk, C. W. van de Kolk, C. J. van Echteld, R. J. Musters, C. van Hardeveld, W. J. Paulus, and W. S. Simonides. 2010. Antioxidant treatment attenuates pulmonary arterial hypertension-induced heart failure. *Am J Physiol Heart Circ Physiol* 298 (3):H1038–H1047.

Reis, G. S., V. S. Augusto, A. P. Silveira, A. A. Jordao, Jr., J. Baddini-Martinez, O. Poli Neto, A. J. Rodrigues, and P. R. Evora. 2013. Oxidative-stress biomarkers in patients with pulmonary hypertension. *Pulm Circ* 3 (4):856–861.

Rose, M. C. and J. A. Voynow. 2006. Respiratory tract mucin genes and mucin glycoproteins in health and disease. *Physiol Rev* 86 (1):245–278.

Rubio, M. L., M. C. Martin-Mosquero, M. Ortega, G. Peces-Barba, and N. Gonzalez-Mangado. 2004. Oral N-acetylcysteine attenuates elastase-induced pulmonary emphysema in rats. *Chest* 125 (4):1500–1506.

Schamberger, A. C., C. A. Staab-Weijnitz, N. Mise-Racek, and O. Eickelberg. 2015. Cigarette smoke alters primary human bronchial epithelial cell differentiation at the air-liquid interface. *Sci Rep* 5:8163.

Souza, D. G., A. T. Vieira, V. Pinho, L. P. Sousa, A. A. Andrade, C. A. Bonjardim, M. McMillan, M. Kahn, and M. M. Teixeira. 2005. NF-kappaB plays a major role during the systemic and local acute inflammatory response following intestinal reperfusion injury. *Br J Pharmacol* 145 (2):246–254.

Tao, F., B. Gonzalez-Flecha, and L. Kobzik. 2003. Reactive oxygen species in pulmonary inflammation by ambient particulates. *Free Radic Biol Med* 35 (4):327–340.

Tsiligianni, I. G. and T. van der Molen. 2010. A systematic review of the role of vitamin insufficiencies and supplementation in COPD. *Respir Res* 11:171.

Valavanidis, A., T. Vlachogianni, K. Fiotakis, and S. Loridas. 2013. Pulmonary oxidative stress, inflammation and cancer: Respirable particulate matter, fibrous dusts and ozone as major causes of lung carcinogenesis through reactive oxygen species mechanisms. *Int J Environ Res Public Health* 10 (9):3886–3907.

Wilson, D. W., H. J. Segall, L. C. Pan, and S. K. Dunston. 1989. Progressive inflammatory and structural changes in the pulmonary vasculature of monocrotaline-treated rats. *Microvasc Res* 38 (1):57–80.

Wong, C. M., G. Bansal, L. Pavlickova, L. Marcocci, and Y. J. Suzuki. 2013. Reactive oxygen species and antioxidants in pulmonary hypertension. *Antioxid Redox Signal* 18 (14):1789–1796.

Zhang, H. J., S. R. Doctrow, L. Xu, L. W. Oberley, B. Beecher, J. Morrison, T. D. Oberley, and K. C. Kregel. 2004. Redox modulation of the liver with chronic antioxidant enzyme mimetic treatment prevents age-related oxidative damage associated with environmental stress. *FASEB J* 18 (13):1547–1549.

Zuo, L., F. He, G. G. Sergakis, M. S. Koozehchian, J. N. Stimpfl, Y. Rong, P. T. Diaz, and T. M. Best. 2014a. Interrelated role of cigarette smoking, oxidative stress, and immune response in COPD and corresponding treatments. *Am J Physiol Lung Cell Mol Physiol* 307 (3):L205–L218.

Zuo, L., M. S. Koozechian, and L. L. Chen. 2014b. Characterization of reactive nitrogen species in allergic asthma. *Ann Allergy Asthma Immunol* 112 (1):18–22.

Zuo, L., N. P. Otenbaker, B. A. Rose, and K. S. Salisbury. 2013. Molecular mechanisms of reactive oxygen species-related pulmonary inflammation and asthma. *Mol Immunol* 56 (1–2):57–63.

Zuo, L., B. A. Rose, W. J. Roberts, F. He, and A. K. Banes-Berceli. 2014c. Molecular characterization of reactive oxygen species in systemic and pulmonary hypertension. *Am J Hypertens* 27 (5):643–650.

30 Pathological Effects of Elevated Reactive Oxygen Species on Sperm Function

Gulfam Ahmad and Ashok Agarwal

CONTENTS

ABSTRACT

Reactive oxygen species (ROS) are produced during the normal cellular metabolism and are essential in carrying out the cell functions. In semen, ROS are produced by spermatozoa and leukocytes. This chapter discusses the cellular and subcellular sources of ROS in semen and their regulation at the molecular level. Physiological levels of ROS are essential for normal sperm function; however, an imbalance between ROS and antioxidant levels causes oxidative stress (OS), which affects sperm function. The pathological effects of elevated levels of ROS on sperm parameters such as membrane function, motility, capacitation, acrosome reaction, and sperm–zona binding capacity are discussed in great details. Further, the OS-induced sperm apoptotic pathway explaining the key factors involved in the cell death has been deliberated. Finally, future directions and recommendations regarding the treatment of OS are discussed.

30.1 INTRODUCTION

Oxygen is essential to sustain life, but its harmful metabolites such as reactive oxygen species (ROS) pose a potential threat to cellular functions and cell survival. The abbreviation is used to describe several free radicals and highly reactive entities that are derived from the molecular oxygen. ROS are exceedingly reactive with one or more unpaired electrons. The production of oxygen-derived free radicals is the bane to every aerobic species. These are produced during the normal metabolism of the cells. Principal ROS agents involved in the biology of reproduction include hydrogen peroxide (H_2O_2), hydroxyl ion (OH^-), superoxide anion ($O_2^{\cdot-}$), hydroxyl radical ($^\cdot OH$), and peroxide. Nitroxyl radicals, on the other hand, which are derived from nitrogen, termed "reactive nitrogen species," such as nitric oxide (NO^-) and peroxynitrite ($ONOO^-$), also play an important role in reproduction (Agarwal et al., 2003; Maneesh and Jayalekshmi, 2006).

In semen, ROS are produced mainly by spermatozoa and leukocytes. Seminal plasma acts as an antioxidant reservoir and possesses strong antioxidant capacity. Indeed, a physiological balance exists between the production of ROS and antioxidants. Evidence supports the beneficial effects of physiological levels of ROS on the normal sperm function. However, an overproduction of ROS, which exceeds the antioxidant capacity of seminal plasma, results in oxidative stress (OS). This imbalance escalates the deleterious effects of ROS on spermatozoa function that leads to male infertility (Agarwal et al., 2008; Mahfouz et al., 2009). The primary objective of this chapter is to describe the production and molecular regulation of ROS in semen, to describe the impact of ROS on sperm function, and to shed light on OS-induced sperm apoptosis.

30.1.1 CELLULAR AND SUBCELLULAR ROS SOURCES IN SEMEN: FRIEND OR FOE?

Semen is an organic fluid that contains spermatozoa as its major constituent, along with secretions from the testes, epididymis, seminal vesicles, prostate, and to some extent the Cowper's gland. In cases of male genital tract infection, leukocytes, macrophages, and sometimes Sertoli cells are also seen in seminal ejaculates (Fisher and Aitken, 1997). In semen, among all the key players, it is the spermatozoa, mainly the immature ones, and the leukocytes that contribute significantly to ROS production.

30.1.1.1 Spermatozoa-Mediated ROS Production

Immature spermatozoa with abnormal morphology particularly in the head and retained cytoplasm are substantial sources of ROS. Any abnormalities in cytoplasm extrusion during the differentiation phase of spermatogenesis deploy enzymes such as glucose-6-phosphate dehydrogenase (G6PD) and nicotinamide adenine dinucleotide phosphate (NADPH) oxidase (NOX2), which result in the production of ROS through NADPH formation (Aitken et al., 1997; Agarwal et al., 2014). There are two main ways of spermatozoa-mediated ROS production: (1) the NOX2 system in the spermatozoa plasma membrane and (2) the NADPH-dependent oxidoreductase (diaphorase) present in the sperm mitochondria (Aitken et al., 1992; Gavella and Lipovac, 1992). The production of ROS by spermatozoa is further related to the quality of semen. If the semen contains high number of immature spermatozoa, the rate of production of ROS is correspondingly higher. When different subpopulations of spermatozoa were separated by density gradients, ROS production was found to be highest in the immature population and lowest in mature and morphologically normal spermatozoa (Gil-Guzman et al., 2001). Studies have also shown that mature spermatozoa produce ROS during their normal metabolism. The generation of ROS by mature spermatozoa is considered as a normal physiological process, which is required for downstream spermatozoa function (Du Plessis et al., 2015). Increased ROS production has also been reported during the preparation of spermatozoa for assisted reproductive techniques (ARTs) such as the density gradient method when spermatozoa were exposed to repeated cycles of centrifugation. The length of time to which spermatozoa were exposed to centrifugation causes more ROS production compared to the force of centrifugation (Agarwal et al., 2003).

30.1.1.2 Leukocytes

The presence of leukocytes at concentrations of $>10^6$ mL^{-1} of semen is defined as leukocytospermia by World Health Organization (WHO). Peroxidase-positive leukocytes, contributed mainly by the prostate and seminal vesicles, are the substantial generators of ROS in the semen. Myeloperoxidase staining is used to quantify their presence in the neat semen. Several controversial studies have reported the association of leukocytospermia, ROS production, and male infertility. Nonetheless, with the exception of a few reports, most studies indicate a strong relation between leukocytospermia and ROS production that lead to impaired sperm function (Wolff et al., 1990; Sharma et al., 2001; Mahfouz et al., 2009). Polymorphonuclear (PMN) leukocytes and macrophages are the main subpopulations of leukocytes in semen, which contribute 50%–60% and 20%–30% of ROS production, respectively (Wolff, 1995). ROS production depends on the activity of leukocytes because activated forms of cells generate 100-fold more oxygen species compared to nonactivated cells. Activation of leukocytes depends on several elements such as infection and inflammatory responses (Potts and Pasqualotto, 2003). These factors cause the activation of a cascade of reactions involved in NOX2 catalysis and monophosphate shunt in fighting against the invading antigens and infection. Once these responses are triggered, it leads to the production of free radicals (Babior, 1999). The PMN leukocytes, being the main subpopulation of white blood cells in the semen, contribute a larger share of ROS production. Nonetheless, once the macrophages are activated, it too causes a surge in ROS production (Ochsendorf, 1999). Direct cell-to-cell contact of leukocytes with spermatozoa or the liberation of cellular by-products also causes the production of ROS by spermatozoa themselves (Saleh et al., 2002). The subcellular components involved in ROS production include some cell organelles, different enzymes, and long-chain fatty acids (LCFA) (Figure 30.1).

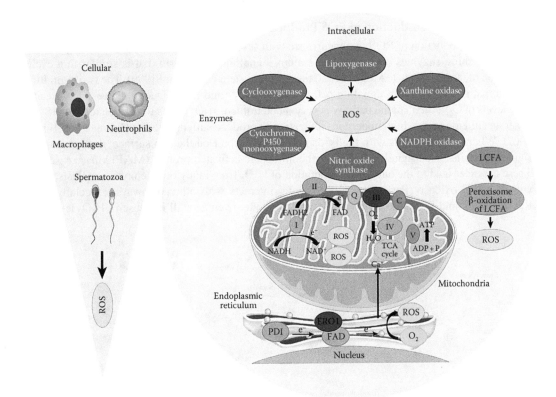

FIGURE 30.1   (See color insert.) Major cellular and subcellular components responsible for ROS production in semen. (Reprinted with permission, Cleveland Clinic Center for Medical Art & Photography © 2015. All Rights Reserved.)

30.1.2 MOLECULAR REGULATION OF ROS PRODUCTION

30.1.2.1 Activation of NADPH Oxidase

The respiratory burst occurs as a result of enhanced production of ROS by activated leukocytes due to the much higher consumption of oxygen by these cells. This process is catalyzed by the membrane-bound enzyme complex called NOX2 (Babior et al., 1973). Among several enzymes known for the production of different moieties of ROS, NOX2 is the most significant enzyme. The NOX2 complex of neutrophils and other immune responsive cells comprise membrane-linked cytochrome b_{558} and several components of cytosol (p40^{phox}, P47^{phox}, and P67^{phox}). The functional activity of the NOX2 is regulated by G-protein Rac, which is part of a complex regulatory system (Figure 30.2) (Hancock et al., 2001; Bokoch and Diebold, 2002).

When immune cells (phagocytes and PMN) are in the resting state, a heterodimer consisting of two polypeptides, p22-phox and gp91-phox, two heme molecules, and one flavin adenine dinucleotide (FAD) group empowers the electron transmission from cytosolic NADPH to the molecular oxygen without the involvement of NOX2 activity (Hancock et al., 2001). The behavior of gp91-phox polypeptide as the H^+ ion channel results in charge compensation. Once a transmembrane stimulatory response is initiated, several cytosolic polypeptides (p40^{phox}, p47^{phox}, and p67^{phox}) assemble at the interior face of the cell membrane to form the fully activated NOX2 enzyme complex. The active NOX2 also requires the simultaneous translocation of Rac GTPase. The whole complex of membrane-associated heterodimer cytochrome b_{558}, cytosolic polypeptides, two heme groups, and one FAD allows the flow of electrons from NADPH to FAD and then to heme and finally to molecular oxygen. As a result of activation of the NOX2 and transfer of electrons, $O_2^{\bullet-}$ are produced by the reduction of molecular oxygen (Figure 30.2) (Hancock et al., 2001).

30.1.2.2 Signal Transduction in ROS Production

Intracellular production of ROS in concurrence with several antioxidant enzymes plays a pertinent role in switching enzymes *on and off* by a redox signaling mechanism that is similar to a cyclic adenosine monophosphate (cAMP) second messenger system (Hou et al., 1999). The two main ROS species that act through this signaling pathway are the $O_2^{\bullet-}$ and H_2O_2. Due to the very low steady-state levels of $O_2^{\bullet-}$, their spatial activity is very much limited. H_2O_2 reacts with thiolate anion (S^-) to generate sulfenic acid, which undergoes ionization to produce sulfenate (SO^-) (Forman and Torres, 2002). When a ligand binds with the tyrosine kinase receptor on the cell surface, the downstream signaling begins. This signaling activates the mitogenic-activated protein (MAP) kinase cascades. These cascades lead to the enormous generation of H_2O_2 from many other enzyme catalysts, along with NADPH oxidase (Park et al., 2006). H_2O_2 interacts with other pathways, especially SOS-Ras-Raf-ERK and PI3K/Akt in a dose-dependent mode. Even a small increase in H_2O_2 levels can

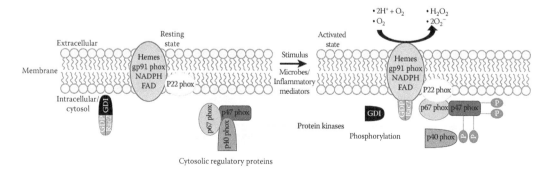

FIGURE 30.2 **(See color insert.)** Mechanistic activation of NADPH oxidase complex and production of superoxide anions. (Reprinted with permission, Cleveland Clinic Center for Medical Art & Photography © 2015. All Rights Reserved.)

cause tremendous reentry into the cell as occurs in the case of NOX1 expression. However, permanent elevation in cellular H_2O_2 levels causes cell cycle arrest and, eventually, apoptosis (Burch and Heintz, 2005).

Peroxiredoxins, a novel family of peroxidases, in conjunction with reducing equivalents delivered by thioredoxin, serve as the key regulators of H_2O_2 and mitogenic signaling. As the mitogenic signaling begins and the ROS production commences, these peroxiredoxins are activated and interacted with mitogenic signaling to control further signaling and to limit the effect of ROS-mediated activation on downstream mitogenic cascade targets (Choi et al., 2005).

30.1.3 REDOX SIGNALING

Spermatozoa are nondividing cells and are transcriptionally inactive. However, it is important to understand the basic mechanism of cell division, as the division in terminal differentiated cells is controlled by oxidants (Latella et al., 2001). Cell cycle is an extremely ordered and sequential process that ensures the truthful replication of cellular genome and division into two daughter cells. The cell cycle consists of four active stages: that is, Gap 1(G_1), Synthesis (S), Gap 2 (G_2), and Mitosis (M), and one quiescent state (G_0). In order to guarantee the faithful progression through the cell cycle, there are three checkpoints that play a strategic role. The first, the G_1 checkpoint safeguards cell nutrients, cell size, and growth factors required for DNA synthesis. The second, the G_2 checkpoint ensures that an appropriate cell size has been achieved and that DNA damage has been repaired before it confirms the correct entry of the cell into M phase. The third, the metaphase checkpoint or spindle assembly checkpoint controls the accurate attachment of spindles with chromosomes and ensures that the cell is ready for division (Lilly and Duronio, 2005).

In the cells that withdraw from the cell cycle and enter in G_0 state, their reentry into G_1 is regulated by oxidants. The expression of cyclin D1, the main protein required for reentry into the cell cycle, is promoted by the redox signaling pathway. The expression of this protein is further regulated and is considered as a marker for mitogenic stimulation (Burch and Heintz, 2005). The cyclin D/CDK complex phosphorylates the retinoblastoma (pRBb) protein with a redox potential of around -207 mV at G_1 phase. At higher redox potentials, the pRb protein undergoes dephosphorylation and the cell cycle ceases. However, maximum ROS production occurs during the G_2/M phase of the cell cycle (Havens et al., 2006).

30.1.4 CELLULAR GLUTATHIONE REDOX SYSTEM AND SPERMATOGENESIS

In testes and epididymal tissues, cellular levels of glutathione (GSH) are very low. During sperm elongation, more than 90% of GSH is removed (Conrad et al., 2015). Such low GSH levels are required to allow the oxidative phases to shape the spermatozoa acrosome, nucleus, midpiece, and flagellum. Despite the low abundance, some of the glutathione peroxidase (Gpx) family members exercise an essential role in the maturation process of mammalian spermatozoa (Noblanc et al., 2011). Mammals express eight Gpx; nonetheless, the presence of all these in spermatozoa has not been confirmed. Glutathione peroxidase 1 (Gpx1) is expressed in testes, Gpx2 in intestinal tissue, and Gpx3 and Gpx5 in epididymal tissue. Gpx4 has the most abundant expression including all essential compartments of the testes and in mature spermatozoa. Gpx6 is expressed in olfactory epithelium, while scanty information is available about the expression of Gpx7 and Gpx8, which are known as endoplasmic reticulum resident (ER resident) (Conrad et al., 2015).

Among the eight GPxs, Gpx3, Gpx4, and Gpx5 have the more important role in sperm function. Gpx3 and Gpx5, being present throughout the epididymis and caput, respectively, protect the male gamete genome from deleterious effects of increased ROS. Although there is a lack of evidence of any phenotype in the absence of Gpx3, elevated levels of ROS were observed in spermatozoa and cauda epididymis epithelium in mice lacking Gpx5. When the offsprings of these mice that were older than a year were allowed to breed, significant miscarriages and fetal developmental defects were seen. These findings were in concurrence with the fact that Gpx5$^{-/-}$ sperm manifest higher DNA oxidation (Chabory et al., 2009).

Gpx4 is considered the ideal target for the testicular selenium and is rendered responsible for male infertility (Ursini et al., 1999). Gpx4 was identified as the most copious selenoprotein in the capsule of sperm mitochondria, which provides stability to the sperm midpiece. Gpx4 is the key enzyme involved in sperm capsule protein oxidation. The involvement of Gpx4 has also been confirmed with other capsular proteins, such as sperm mitochondria–associated cysteine-rich protein (SMCP). When the Gpx4 gene was disrupted, increased embryonic mortality was observed. Additionally, the role of Gpx4 has also been associated with the sperm fibrous sheath and the acrosome, and its disruption causes male infertility (Maiorino et al., 2005; Schneider et al., 2009).

In the previous sections of this chapter, the major players involved in the production of ROS, the regulation of ROS production at molecular levels, and the important redox system implicated in normal spermatogenesis have been discussed. In the subsequent sections, the impact of OS on functional sperm parameters is emphasized.

30.1.5 OS AND SPERM DYSFUNCTION

The conditions when the levels of oxidants exceed antioxidants are called OS. During the state of OS, the levels of peroxidation increase and pathological effects cause the impairment of cellular function. All integral cellular components, that is, proteins, nucleic acids, lipids, and carbohydrates, become critical targets of OS (Agarwal et al., 2003). OS-induced cellular damage depends on the type and amount of ROS involved, the duration of exposure to ROS and several other factors such as antioxidants capacity, neighboring environment, available oxygen and the temperature.

30.1.5.1 Spermatozoa Membrane Lipid Peroxidation

The plasma membrane of spermatozoa is rich in polyunsaturated fatty acids (PUFAs). The fatty acids are highly susceptible to ROS attack, which leads to the activation of downstream chemical reactions. The oxidative deterioration of PUFA is called the lipid peroxidation (LPO). There are two kinds of LPO: (1) enzymatic LPO, which depends on the NADPH and adenosine diphosphate (ADP), and (2) nonenzymatic LPO, which is independent of such enzymes. The most common effect of LPO on cellular function is the perturbation in the structure and function of cellular and organelle membranes. The intracellular ion transport, metabolite gradients, and receptor-mediated signal transduction operate aberrantly. The transcriptionally inactive state of the sperm cell makes it much more susceptible to LPO. Unlike other cells, spermatozoa cannot repair the damage caused by ROS due to the lack of essential cytoplasmic enzymes involved in the repair mechanism (Alvarez et al., 1987; Krausz et al., 1994; Agarwal et al., 2003; Aitken and Baker, 2013).

30.1.5.2 Spermatozoa Motility

The impact of increased ROS on sperm motility has been clearly documented (de Lamirande and Gagnon, 1992; Agarwal et al., 1994; Armstrong et al., 1999; Desai et al., 2009). Increased ROS levels may hinder the normal function of the redox system that is required for sperm motility. As explained in previous sections, the systematic knockout of mitochondrial Gpx4 resulted in massive reduction in sperm motility and dramatic midpiece phenotypic abnormalities (Schneider et al., 2009). Another reason could be the intercellular diffusion of H_2O_2 across the plasma membrane that hinders the normal functions of certain enzymes, such as G6PD. The function of this enzyme is to control the efflux of glucose through the hexose monophosphate shunt, which eventually regulates intracellular NADPH. NOX2 system is the key player in ROS production (Aitken et al., 1997). Once G6PD activity is altered, NADPH becomes unavailable and the oxidized glutathione accumulates in the cells, which results in reduced levels of glutathione. Consequently, the antioxidant system of spermatozoa weakens, rendering the sperm membrane susceptible to LPO (Griveau et al., 1995).

Other speculation in reduced sperm motility is the decreased axonemal protein phosphorylation involved in sperm membrane fluidity (de Lamirande and Gagnon, 1995).

30.1.5.3 Spermatozoa Capacitation

Capacitation is the ability of spermatozoa to fertilize the oocyte, and to achieve this, it requires a series of morphological and metabolic changes of the sperms in the female reproductive tract. This process enables the spermatozoa to bind and penetrate the zona pellucida. Both biochemical and molecular events are involved in sperm capacitation; production of cAMP through the activation of adenylyl cyclase (AC), activation of calcium channels, production of ROS, efflux of cholesterol from the plasma membrane, increased intracellular pH, and activation of protein kinases (O'Flaherty, 2015). Sperm capacitation in mammals is an oxidative phenomenon. Although normal levels of ROS are a prerequisite for sperm capacitation, higher and sustained ROS levels, such as during OS, can lead to premature capacitation (O'Flaherty, 2015) or prevent spermatozoa from undergoing capacitation (Morielli and O'Flaherty, 2015). Since phosphorylation events are mandatory during sperm capacitation, any hindrance in these events may lead to failure in sperm capacitation. In infertile men, it was observed that spermatozoa were unable to undergo tyrosine phosphorylation (Buffone et al., 2005). When spermatozoa were treated with H_2O_2 at a concentration of 1 mM before undergoing capacitation, decreased tyrosine phosphorylation was noted. These findings propose a delayed sperm capacitation under OS conditions. On the other side, multifaceted redox changes of protein sulfhydryl (SH) occur during the events of sperm capacitation and the levels of SH are changed. During the initial 30–60 min of capacitation, the SH content of detergent-soluble proteins (Triton X-10) increases (de Lamirande and Gagnon, 2003), which on one hand is beneficial, but at the same time, it also creates a situation of OS and, ultimately, early capacitation. This is similar to sperm cryopreservation, where increased SH content of Triton-soluble proteins is known to cause OS (Cormier et al., 1997), leading to premature sperm capacitation. Unlike sperm motility, the effects of OS on sperm capacitation are scarcely reported and those reported remain controversial (Morielli and O'Flaherty, 2015). Furthermore, the detailed underlying molecular mechanism of sperm capacitation is yet to be elucidated. Likewise, exogenous ROS causes hyperactivation of spermatozoa but is prevented by ROS scavengers, that is, catalase, SOD, and NOX2. However, the exact mechanism with regard to the impact of ROS on sperm hyperactivation is not known.

30.1.5.4 Acrosome Reaction

Sperm acrosome is a cap-shaped structure derived from the Golgi apparatus. It develops over the anterior part of the sperm head. It contains several enzymes playing a crucial role in fertilization. The intact acrosome is important for fertilization, and if the acrosome reaction (release of enzymes) has occurred, spermatozoa are unable to fertilize the oocyte. ROS have a negative correlation with acrosome function. The higher the OS, the lower the percentage of spermatozoa with intact acrosomes. Currently, there is a lack of consensus as to which free radical or ROS are the major contributors in the regulation of acrosome function. Nonetheless, the significant stimulatory effect of H_2O_2 on spontaneous acrosome reaction was observed compared to controls. The effects were evident even at 0.01 mM concentration of H_2O_2. On the other hand, H_2O_2 showed inhibitory effects on acrosome reaction in spermatozoa treated with progesterone (Oehninger et al., 1995). Dose-dependent decrease in lysophosphatidylcholine-induced acrosome reaction was also observed in spermatozoa treated with H_2O_2 (Morielli and O'Flaherty, 2015). Likewise, in previous experiments, a significant increase was observed in the percentage of acrosome-reacted (loss of acrosome function) spermatozoa when treated with H_2O_2 compared to controls (Aitken et al., 1993).

30.1.5.5 Spermatozoa–Zona Pellucida Binding

Sperm–zona pellucida binding is a crucial step in achieving successful fertilization and eventually pregnancy. Physiological levels of ROS are required for the effective binding of spermatozoa to the zona pellucida. However, raised ROS levels have a deleterious effect on sperm–zona binding and result in fertilization failure. A significant inhibitory effect of H_2O_2 on sperm–zona binding has been reported at doses 0.2 mM and above. Lower doses of H_2O_2, that is, 0.05 and 0.1 mM, did not reveal any negative effect, and this demonstrated that spermatozoa were bound to the zona pellucida in a similar fashion as that of the controls (Oehninger et al., 1995). This again reveals the importance of physiological levels of ROS in the normal sperm function. Whenever OS occurs as result of higher ROS, it gives a negative impact on sperm–zona binding. Studies have also shown a negative association between the levels of ROS and the decreased capacity of sperm–egg fusion (zona-free hamster egg) (Aitken et al., 1991). Similarly, a significant reduction in sperm–zona binding and penetration was observed in spermatozoa treated with xanthine oxidase and directly with H_2O_2 (0.2 mM) (Aitken et al., 1993).

30.1.6 OS AND SPERM APOPTOSIS

Apoptosis is a genetically determined process of programmed cell death. Indeed, each spermatozoon is destined to undergo apoptosis except for a few, which achieves immortality through fertilization (Aitken and Baker, 2013). The majority of the ejaculated spermatozoa will undergo apoptosis during their strenuous effort to embrace the oocyte for fertilization. Therefore, apoptosis is not induced in spermatozoa, rather it results because of the spontaneous attempts the spermatozoa have to make to engage with the oocyte. In order to recognize the oocyte, inseminated spermatozoa will undergo a cascade of cellular events and then achieve capacitation in the female reproductive tract. During the course of these events, the numerous surface receptors present on the spermatozoa help in the recognition of the zona pellucida (Reid et al., 2011; Baker et al., 2012). The hyperactivity achieved through the capacitation of spermatozoa plays a key role in defining the spermatozoa with the greatest motility to successfully fertilize the oocyte.

The ROS generated during the sperm capacitation significantly deplete the cholesterol from the plasma membrane, thus increasing its fluidity, in order to facilitate the fertilization process (Aitken and Baker, 2013). However, the continuous production of ROS by spermatozoa eventually overcomes their limited defensive mechanisms and the cell undergoes OS, which in turn activates the intrinsic cascade of apoptosis (Aitken, 2011). Unlike somatic cells, apoptosis in spermatozoa is not triggered during the cell cycle checkpoints but is instead a programmed manifestation.

Human spermatozoa face different challenges compared to other mammalian species. Unlike mammals, humans do not exhibit a behavioral estrous pattern, which guides the synchronized insemination just before ovulation. Rather, human spermatozoa have to wait for the ovulation to occur, and the waiting may prolong for several days. A pertinent question then arises as to what prevents the spermatozoa from undergoing apoptosis while waiting for such a long time in the female reproductive tract? Indeed, several factors such as glucose regulation and hormones such as insulin or prolactin play a prosurvival role in protecting the spermatozoa from undergoing apoptosis. These factors play a significant antiapoptotic role by activating the phosphatidylinositide 3-kinase (PI3 kinase), which produces phosphatidylinositol triphosphate (PIP3) (Pujianto et al., 2010; Aitken and Baker, 2013). However, the physiological functioning of these factors is hampered when there is OS due to persistent ROS production.

Sustained generation of ROS by sperm mitochondria is crucial in triggering sperm apoptosis. One of the best known reasons of continuous mitochondrial ROS production is LPO, an unavoidable consequence of OS, which causes interruption in the mitochondrial electron transport chain (mETC). Thus, the first sign of activation of intrinsic cascades in sperm apoptosis is the induction of OS, further synchronized by several other factors, that is, age, lifestyle exposure, smoking, varicocele, leukocytes, and environmental factors (Aitken and Baker, 2013).

Spermatozoa contain a high number of unsaturated fatty acids, which make these cells vulnerable to free radicals and consequently trigger increased LPO, thus placing the spermatozoa under immense OS. As a result of such stress, electrophilic lipid aldehydes such as 4-hydroxynonenal (4HNE) are produced by making adducts with susceptible proteins more commonly at lysine,

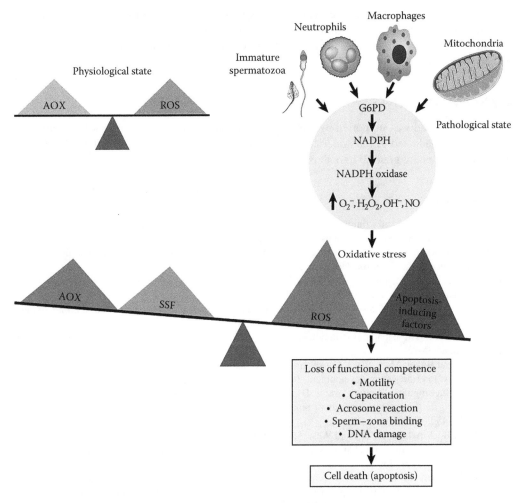

FIGURE 30.3 **(See color insert.)** Key players involved in oxidative stress and sperm apoptosis. (Reprinted with permission, Cleveland Clinic Center for Medical Art & Photography © 2015. All Rights Reserved.)

histidine, and cysteine residues. Such adducts impair the mETC function leading to leakage of electrons and formation of oxides. ROS generation is considered a self-disseminating phenomenon, which once initiated cannot be broken or reversed and leads to cell death (Figure 30.3) (du Plessis et al., 2010).

30.2 CONCLUSION

A conclusive body of evidence supports the drastic effects of elevated ROS/OS on sperm function, which ultimately leads to male subfertility or infertility. Despite the fact that common conditions such as age, smoking, heat, pollution, toxins, diet, obesity lead to OS, little attention is given to the prevention of this cause by physicians specializing in infertility. In infertile couples where male has high ROS/OS, many clinicians treat with ART to achieve pregnancy. However, the impact of OS on the quality of sperm (sperm DNA) should not be neglected. Sperm with damaged DNA can fertilize the oocyte but cannot maintain the pregnancy and results in poor quality of embryos/blastocysts, which consequently end up in a miscarriage. Therefore, as a priority, OS should be addressed before attempting pregnancy to avoid financial and psychological distress. Though controversial, antioxidant therapy in men with OS has proven to be beneficial, and almost one-third of men experiencing infertility are prescribing to this therapy. Besides nutritional changes, it is also imperative to modify the lifestyle of the couples to minimize the risk of OS. However, further large randomized controlled trials are needed in order to validate the best dose and combination of antioxidant supplements.

ACKNOWLEDGMENTS

Dr. Gulfam Ahmad was supported by a Fulbright Scholar award by the Fulbright Foundation. His research was conducted at the American Center for Reproductive Medicine at Cleveland Clinic, Ohio. The authors are grateful to both organizations for their support.

REFERENCES

Agarwal, A., Ikemoto, I., and Loughlin, K. R. 1994. Relationship of sperm parameters with levels of reactive oxygen species in semen specimens. *J Urol*, 152, 107–110.

Agarwal, A., Makker, K., and Sharma, R. 2008. Clinical relevance of oxidative stress in male factor infertility: An update. *Am J Reprod Immunol*, 59, 2–11.

Agarwal, A., Saleh, R. A., and Bedaiwy, M. A. 2003. Role of reactive oxygen species in the pathophysiology of human reproduction. *Fertil Steril*, 79, 829–843.

Agarwal, A., Tvrda, E., and Sharma, R. 2014. Relationship amongst teratozoospermia, seminal oxidative stress and male infertility. *Reprod Biol Endocrinol*, 12, 45.

Aitken, R. J. 2011. The capacitation-apoptosis highway: Oxysterols and mammalian sperm function. *Biol Reprod*, 85, 9–12.

Aitken, R. J. and Baker, M. A. 2013. Causes and consequences of apoptosis in spermatozoa; contributions to infertility and impacts on development. *Int J Dev Biol*, 57, 265–272.

Aitken, R. J., Buckingham, D., and Harkiss, D. 1993. Use of a xanthine oxidase free radical generating system to investigate the cytotoxic effects of reactive oxygen species on human spermatozoa. *J Reprod Fertil*, 97, 441–450.

Aitken, R. J., Buckingham, D. W., and West, K. M. 1992. Reactive oxygen species and human spermatozoa: Analysis of the cellular mechanisms involved in luminol- and lucigenin-dependent chemiluminescence. *J Cell Physiol*, 151, 466–477.

Aitken, R. J., Fisher, H. M., Fulton, N., Gomez, E., Knox, W., Lewis, B., and Irvine, S. 1997. Reactive oxygen species generation by human spermatozoa is induced by exogenous NADPH and inhibited by the flavoprotein inhibitors diphenylene iodonium and quinacrine. *Mol Reprod Dev*, 47, 468–482.

Aitken, R. J., Irvine, D. S., and Wu, F. C. 1991. Prospective analysis of sperm-oocyte fusion and reactive oxygen species generation as criteria for the diagnosis of infertility. *Am J Obstet Gynecol*, 164, 542–551.

Alvarez, J. G., Touchstone, J. C., Blasco, L., and Storey, B. T. 1987. Spontaneous lipid peroxidation and production of hydrogen peroxide and superoxide in human spermatozoa. Superoxide dismutase as major enzyme protectant against oxygen toxicity. *J Androl*, 8, 338–348.

Armstrong, J. S., Rajasekaran, M., Chamulitrat, W., Gatti, P., Hellstrom, W. J., and Sikka, S. C. 1999. Characterization of reactive oxygen species induced effects on human spermatozoa movement and energy metabolism. *Free Radic Biol Med*, 26, 869–880.

Babior, B. M. 1999. NADPH oxidase: An update. *Blood*, 93, 1464–1476.

Babior, B. M., Kipnes, R. S., and Curnutte, J. T. 1973. Biological defense mechanisms. The production by leukocytes of superoxide, a potential bactericidal agent. *J Clin Invest*, 52, 741–744.

Baker, M. A., Nixon, B., Naumovski, N., and Aitken, R. J. 2012. Proteomic insights into the maturation and capacitation of mammalian spermatozoa. *Syst Biol Reprod Med*, 58, 211–217.

Bokoch, G. M. and Diebold, B. A. 2002. Current molecular models for NADPH oxidase regulation by Rac GTPase. *Blood*, 100, 2692–2696.

Buffone, M. G., Calamera, J. C., Verstraeten, S. V., and Doncel, G. F. 2005. Capacitation-associated protein tyrosine phosphorylation and membrane fluidity changes are impaired in the spermatozoa of asthenozoospermic patients. *Reproduction*, 129, 697–705.

Burch, P. M. and Heintz, N. H. 2005. Redox regulation of cell-cycle re-entry: Cyclin D1 as a primary target for the mitogenic effects of reactive oxygen and nitrogen species. *Antioxid Redox Signal*, 7, 741–751.

Chabory, E., Damon, C., Lenoir, A., Kauselmann, G., Kern, H., Zevnik, B., Garrel, C. et al. 2009. Epididymis seleno-independent glutathione peroxidase 5 maintains sperm DNA integrity in mice. *J Clin Invest*, 119, 2074–2085.

Choi, M. H., Lee, I. K., Kim, G. W., Kim, B. U., Han, Y. H., Yu, D. Y., Park, H. S. et al. 2005. Regulation of PDGF signalling and vascular remodelling by peroxiredoxin II. *Nature*, 435, 347–353.

Conrad, M., Ingold, I., Buday, K., Kobayashi, S., and Angeli, J. P. 2015. ROS, thiols and thiol-regulating systems in male gametogenesis. *Biochim Biophys Acta*, 1850(8), 1566–1574.

Cormier, N., Sirard, M. A., and Bailey, J. L. 1997. Premature capacitation of bovine spermatozoa is initiated by cryopreservation. *J Androl*, 18, 461–468.

De Lamirande, E. and Gagnon, C. 1992. Reactive oxygen species and human spermatozoa. II. Depletion of adenosine triphosphate plays an important role in the inhibition of sperm motility. *J Androl*, 13, 379–386.

De Lamirande, E. and Gagnon, C. 1995. Impact of reactive oxygen species on spermatozoa: A balancing act between beneficial and detrimental effects. *Hum Reprod*, 10 (Suppl 1), 15–21.

De Lamirande, E. and Gagnon, C. 2003. Redox control of changes in protein sulfhydryl levels during human sperm capacitation. *Free Radic Biol Med*, 35, 1271–1285.

Desai, N., Sharma, R., Makker, K., Sabanegh, E., and Agarwal, A. 2009. Physiologic and pathologic levels of reactive oxygen species in neat semen of infertile men. *Fertil Steril*, 92, 1626–1631.

Du Plessis, S. S., Agarwal, A., Halabi, J., and Tvrda, E. 2015. Contemporary evidence on the physiological role of reactive oxygen species in human sperm function. *J Assist Reprod Genet*, 32, 509–520.

Du Plessis, S. S., Mcallister, D. A., Luu, A., Savia, J., Agarwal, A., and Lampiao, F. 2010. Effects of H_2O_2 exposure on human sperm motility parameters, reactive oxygen species levels and nitric oxide levels. *Andrologia*, 42, 206–210.

Fisher, H. M. and Aitken, R. J. 1997. Comparative analysis of the ability of precursor germ cells and epididymal spermatozoa to generate reactive oxygen metabolites. *J Exp Zool*, 277, 390–400.

Forman, H. J. and Torres, M. 2002. Reactive oxygen species and cell signaling: Respiratory burst in macrophage signaling. *Am J Respir Crit Care Med*, 166, S4–S8.

Gavella, M. and Lipovac, V. 1992. NADH-dependent oxidoreductase (diaphorase) activity and isozyme pattern of sperm in infertile men. *Arch Androl*, 28, 135–141.

Gil-Guzman, E., Ollero, M., Lopez, M. C., Sharma, R. K., Alvarez, J. G., Thomas, A. J., Jr. and Agarwal, A. 2001. Differential production of reactive oxygen species by subsets of human spermatozoa at different stages of maturation. *Hum Reprod*, 16, 1922–1930.

Griveau, J. F., Dumont, E., Renard, P., Callegari, J. P., and Le Lannou, D. 1995. Reactive oxygen species, lipid peroxidation and enzymatic defence systems in human spermatozoa. *J Reprod Fertil*, 103, 17–26.

Hancock, J. T., Desikan, R., and Neill, S. J. 2001. Role of reactive oxygen species in cell signalling pathways. *Biochem Soc Trans*, 29, 345–350.

Havens, C. G., Ho, A., Yoshioka, N., and Dowdy, S. F. 2006. Regulation of late G1/S phase transition and APC Cdh1 by reactive oxygen species. *Mol Cell Biol*, 26, 4701–4711.

Hou, Y. C., Janczuk, A., and Wang, P. G. 1999. Current trends in the development of nitric oxide donors. *Curr Pharm Des*, 5, 417–441.

Krausz, C., Mills, C., Rogers, S., Tan, S. L., and Aitken, R. J. 1994. Stimulation of oxidant generation by human sperm suspensions using phorbol esters and formyl peptides: Relationships with motility and fertilization in vitro. *Fertil Steril*, 62, 599–605.

Latella, L., Sacco, A., Pajalunga, D., Tiainen, M., Macera, D., D'angelo, M., Felici, A., Sacchi, A., and Crescenzi, M. 2001. Reconstitution of cyclin D1-associated kinase activity drives terminally differentiated cells into the cell cycle. *Mol Cell Biol*, 21, 5631–5643.

Lilly, M. A. and Duronio, R. J. 2005. New insights into cell cycle control from the Drosophila endocycle. *Oncogene*, 24, 2765–2775.

Mahfouz, R., Sharma, R., Sharma, D., Sabanegh, E., and Agarwal, A. 2009. Diagnostic value of the total antioxidant capacity (TAC) in human seminal plasma. *Fertil Steril*, 91, 805–811.

Maiorino, M., Roveri, A., Benazzi, L., Bosello, V., Mauri, P., Toppo, S., Tosatto, S. C., and Ursini, F. 2005. Functional interaction of phospholipid hydroperoxide glutathione peroxidase with sperm mitochondrion-associated cysteine-rich protein discloses the adjacent cysteine motif as a new substrate of the selenoperoxidase. *J Biol Chem*, 280, 38395–38402.

Maneesh, M. and Jayalekshmi, H. 2006. Role of reactive oxygen species and antioxidants on pathophysiology of male reproduction. *Indian J Clin Biochem*, 21, 80–89.

Morielli, T. and O'Flaherty, C. 2015. Oxidative stress impairs function and increases redox protein modifications in human spermatozoa. *Reproduction*, 149, 113–123.

Noblanc, A., Kocer, A., Chabory, E., Vernet, P., Saez, F., Cadet, R., Conrad, M., and, Drevet, J. R. 2011. Glutathione peroxidases at work on epididymal spermatozoa: An example of the dual effect of reactive oxygen species on mammalian male fertilizing ability. *J Androl*, 32, 641–650.

O'Flaherty, C. 2015. Redox regulation of mammalian sperm capacitation. *Asian J Androl*, 17(4), 583–590.

Ochsendorf, F. R. 1999. Infections in the male genital tract and reactive oxygen species. *Hum Reprod Update*, 5, 399–420.

Oehninger, S., Blackmore, P., Mahony, M., and Hodgen, G. 1995. Effects of hydrogen peroxide on human spermatozoa. *J Assist Reprod Genet*, 12, 41–47.

Park, H. S., Park, D., and Bae, Y. S. 2006. Molecular interaction of NADPH oxidase 1 with betaPix and Nox Organizer 1. *Biochem Biophys Res Commun*, 339, 985–990.

Potts, J. M. and Pasqualotto, F. F. 2003. Seminal oxidative stress in patients with chronic prostatitis. *Andrologia*, 35, 304–308.

Pujianto, D. A., Curry, B. J., and Aitken, R. J. 2010. Prolactin exerts a prosurvival effect on human spermatozoa via mechanisms that involve the stimulation of Akt phosphorylation and suppression of caspase activation and capacitation. *Endocrinology*, 151, 1269–1279.

Reid, A. T., Redgrove, K., Aitken, R. J., and Nixon, B. 2011. Cellular mechanisms regulating sperm-zona pellucida interaction. *Asian J Androl*, 13, 88–96.

Saleh, R. A., Agarwal, A., Kandirali, E., Sharma, R. K., Thomas, A. J., Nada, E. A., Evenson, D. P., and Alvarez, J. G. 2002. Leukocytospermia is associated with increased reactive oxygen species production by human spermatozoa. *Fertil Steril*, 78, 1215–1224.

Schneider, M., Forster, H., Boersma, A., Seiler, A., Wehnes, H., Sinowatz, F., Neumuller, C. et al. 2009. Mitochondrial glutathione peroxidase 4 disruption causes male infertility. *Faseb J*, 23, 3233–3242.

Sharma, R. K., Pasqualotto, A. E., Nelson, D. R., Thomas, A. J., Jr., and Agarwal, A. 2001. Relationship between seminal white blood cell counts and oxidative stress in men treated at an infertility clinic. *J Androl*, 22, 575–583.

Ursini, F., Heim, S., Kiess, M., Maiorino, M., Roveri, A., Wissing, J., and Flohe, L. 1999. Dual function of the selenoprotein PHGPx during sperm maturation. *Science*, 285, 1393–1396.

Wolff, H. 1995. The biologic significance of white blood cells in semen. *Fertil Steril*, 63, 1143–1157.

Wolff, H., Politch, J. A., Martinez, A., Haimovici, F., Hill, J. A., and Anderson, D. J. 1990. Leukocytospermia is associated with poor semen quality. *Fertil Steril*, 53, 528–536.

31 Reactive Oxygen Species and Male Fertility
The Physiological Role

Ibukun P. Oyeyipo, Bongekile Skosana, and Stefan S. Du Plessis

CONTENTS

ABSTRACT

Reactive oxygen species (ROS) are unstable, reactive by-products of oxygen metabolism with the ability to oxidize and modify nearby molecules. Mammalian sperm were the first cells discovered to produce ROS. Excessive ROS production has been shown to result in reduced sperm motility, increased DNA fragmentation, membrane lipid peroxidation, sperm apoptosis, and a reduction in sperm–oocyte fusion. The perception that ROS is damaging to male gametes was widely accepted until they were discovered to also provide numerous functions vital to the success of fertilization. The key is to keep their levels controlled within certain ranges to allow proper physiological function. Physiological levels of ROS are necessary for the activation of intracellular pathways leading to sperm maturation, capacitation, hyperactivation, acrosome reaction, and chemotactic processes important in sperm–oocyte fusion. This chapter will discuss various physiological roles of ROS and the mechanisms through which it stimulates essential activities of spermatozoa.

31.1 INTRODUCTION

Reactive oxygen species (ROS) play a vital role in male fertility at low concentrations. Evidence suggests that it is involved in the activation of intracellular pathways responsible for spermatozoal maturation, capacitation, hyperactivation, acrosomal reaction, chemotactic processes, and fusion with the female gamete. This chapter will discuss various physiological roles and mechanisms of action of ROS in the acquisition of structural integrity and the physiological activities of spermatozoa.

31.1.1 SPERM MATURATION

The maturation of spermatozoa consists of several critical physiological and morphological modifications in the testis and epididymis. These physiological processes include a series of highly organized steps such as proliferation and renewal of stem cells, genetic remodeling including reduction of chromosomes, and genetic exchange between sister chromatids and morphological alterations that lead to sperm activation. During the process of spermatogenesis, spermatogonial stem cells either go through proliferation so as to sustain the population of stem cells or undergo differentiation into primary spermatocytes. Primary spermatocytes undergo meiotic division resulting in secondary spermatocytes, which then undergo a second meiotic division to form spermatids. A further developmental process referred to as spermiogenesis entails morphological changes such as the formation of a tail, axoneme, and acrosome from centriole, microtubules, and Golgi apparatus, respectively, removal of excessive cytoplasm and organelles, which lead to the formation of residual bodies, and tight packaging of the male genome through the replacement of most of histones by transition proteins and protamines. Mature spermatids are then released from the supporting Sertoli cells into the lumen of the seminiferous tubule. At this stage, the spermatozoa are immobile but acquire motility, capability to fertilize, and further sperm oxidation as they are transported to the epididymis. During transit and storage of spermatozoa in the epididymis, they undergo membrane, nuclear, and enzyme-related remodeling. This includes the release, attachment, and rearrangement of surface proteins (Gil-Guzman et al. 2001; Vernet et al. 2004). It also involves assembling of the signal transduction machinery necessary for spermatozoa to undergo hyperactivation and capacitation (Aitken et al. 2004).

Spermatozoa were recognized as the first mammalian cell to generate H_2O_2 as early as 1946 (Tosic and Walton 1946). Before this time, it had been concluded that ROS adversely impact the motility of human spermatozoa (Macleod 1943). Nowadays, it is believed that ROS affect the function and quality of human spermatozoa, but this is subject to its concentration. Studies have demonstrated that a certain amount of ROS is necessary for specific developmental steps of spermatozoa including all oxidative steps essential for the formation of the rigid structure of the nucleus, midpiece, and tail; hence, it is evident that spermatozoa develop in a highly complex antioxidant network maintaining a tightly controlled frame of ROS generation, as well as preventing the adverse effect of excessive ROS on spermatozoa. Sperm development within the testis and epididymis causes expression and loss of chains of redox enzymes and exploitation of the antioxidant endowment found in the epididymis and other environments they are in contact with on their journey from the epididymis to the female duct (Vernet et al. 2004). As a matter of fact, the seminal fluid is highly enriched with antioxidants with low molecular weight (Jones et al. 1979; Rhemrev et al. 2000), and these enzymes contribute directly or indirectly to redox homeostasis, thereby playing an important role in male fertility.

The process of chromatin packaging, characterized by chromatin stability during the maturation of mammalian spermatozoa, is regulated via redox processes (Sabeur and Ball 2007). Responsible for the stability of chromatin are the inter- and intramolecular disulfide bonds between cysteine residues of protamines. The small, nuclear proteins that replace histones during spermiogenesis are known as protamines. As the spermatozoa travels from the caput to caudal epididymis, oxidation of thiol groups of protamines occurs (Saowaros and Panyim 1978). More recently, it has been concluded that an oxidation process is likely to occur in the caudal epididymis, where the sperm are stored prior to ejaculation (Aitken et al. 2004). ROS have been proposed to act as oxidizing agents in this process, thereby facilitating the formation of disulfide bonds, increasing chromatin stability, and protecting DNA from physical or chemical damage (Rousseaux and Rousseaux-Prevost 1995), while further chromatin condensation is a necessary protective process since spermatozoa possess minimal repair mechanisms (De Lamirande and Gagnon 1992a), and thus, in this way, ROS protect spermatozoa from future damage induced by oxidative stress (OS).

Similarly, peroxides have been shown to be involved in the formation of a protective coat known as the mitochondrial capsule. This coat confers the prevention of proteolytic degradation (Roveri et al. 2001). Mitochondria need such protection because they are crucial to cell metabolism and mediate apoptosis and ROS production, and any damage to the mitochondria may result in impaired function (Venkatesh et al. 2009).

The effects of oral antioxidant therapy over a 3-month period for treating male infertility have shown an improved DNA integrity and reduced ROS production. However, unusual decondensation of sperm DNA was reported (Menezo et al. 2007; Tunc et al. 2009). This indicates that high antioxidant levels may interfere with the oxidative conditions necessary for the proper formation of inter- and intramolecular disulfide bonds, resulting in improper DNA compaction and paternal DNA decondensation. It is well recognized that controlled amounts of ROS are essential for these processes and fertilization. Information in literature regarding the role of ROS and mechanisms that keep it at physiological levels is still limited due to the inability to study the development of spermatozoa in vitro, differences between highly divergent redox condition in vivo and in vitro, and also inadequate genetic animal models to clearly assign distinct redox-dependent processes to the individual steps of spermatogenesis in vivo.

31.1.2 Motility and Hyperactivation

Hyperactivation is poorly understood, but it is considered to be a subset of capacitation. Normally, spermatozoal motility is characterized by low-amplitude flagellar movement and linear velocity, while when hyperactivated, spermatozoa exhibit high-amplitude, asymmetric flagellar movement, pronounced lateral head displacement, and nonlinear trajectory (Burkman 1991; Baldi et al. 2000). This enables the spermatozoa to penetrate the cumulus cells and zona pellucida (ZP) surrounding the oocyte. (De Lamirande et al. 1997b). It may also facilitate progressive movement and prevent stagnation through the oviduct, thus adding another potential benefit to sperm function (Suarez 2008). Hyperactivation generally primes the spermatozoa for fertilization. It also helps in selecting only mature spermatozoa to reach the oocyte due to hyperactive motility and an increased responsiveness to chemotactic signaling (Eisenbach 1999).

A number of studies have shown that only hyperactivation in spermatozoa increases their motility and enables them to undergo acrosome reaction (AR), thus acquiring the features necessary to successfully fertilize an ovum. De Lamirande and Gagnon (1993a,b) investigated the involvement of ROS in the initiation of the hyperactivation process on human spermatozoa and demonstrated that O_2^-, but not H_2O_2, was the main species responsible for the enhanced hyperactivity exhibited by spermatozoa, while a later study concluded that a concentration of 25 μM H_2O_2 enhances hyperactivation (Griveau et al. 1997). The differences in hyperactivation criteria may explain the discrepancy obtained from these studies.

In contrast, other studies have differed, indicating that the exogenous addition of H_2O_2 to spermatozoa had no significant effect on motility (Aitken et al. 1995), while Oehninger et al. (1995) found that motility was significantly decreased with exogenous addition of H_2O_2. The discrepancy observed in the results may be reasonably associated with differences in concentrations employed by different studies. Aitken et al. (1995) used 50 and 100 μM H_2O_2, while concentrations of 10 and $\geq$50 μM used by Oehninger et al. (1995) bypassed the optimal concentration (25 μM) determined earlier by Griveau et al. (1997). These studies provide evidence for the impact of H_2O_2 on hyperactivation and suggest that the role of H_2O_2 is concentration specific. Taken together, these studies indicate that ROS can positively influence the process of hyperactivation in spermatozoa.

31.1.3 Capacitation

Capacitation occurs after ejaculation of spermatozoa into the female genital tract and is the penultimate process in the maturation of spermatozoa that is required to make them competent to fertilize the ovum successfully (Choudhary et al. 2010). The controlled ROS generation occurs in spermatozoa during this process, leading to various molecular modifications. The first step in the process is an increase in cyclic adenosine 3',5'-monophosphate (cAMP). The cAMP pathway has been shown to be essential for numerous biological processes including activation of enzyme and regulation of gene expression. In this pathway, there is activation of protein kinase A (PKA), phosphorylation of PKA substrates (arginine, threonine, and serine), which leads to phosphorylation of mitogen-activated protein kinase (MEK) (extracellular signal–regulated kinase)-like proteins and threonine–glutamate–tyrosine, and finally tyrosine phosphorylation of fibrous sheath proteins. The entire adenylate cyclase (AC)/cAMP/PKA cascade is sensitive to high pH (Aitken et al. 1998a,b). The free radical O_2^- was reported to enhance intracellular pH by acting on the Na^+/H^+ exchanger and stimulating H^+-conducting pathways (Demaurex et al. 1996).

The molecular process includes influx of Ca^{2+} and HCO_3^-, efflux of cholesterol, and an increase in cAMP activity, pH, ROS generation, protein phosphorylation (Ser/Thr and tyrosine), and membrane hyperpolarization (De Lamirande et al. 1997; Visconti et al. 2002). The first evidence showing the involvement of ROS in the physiology of human sperm concluded that exposure of spermatozoa to O_2^- resulted in increased capacitation and higher activation than those treated with Ham's F-10 medium alone and a capacitation inducer. These effects were reported to be prevented with superoxide dismutase (SOD) (de Lamirande and Gagnon 1993). In a later study by the same group of scientists, induction of capacitation with biological fluids such as follicular fluid or fetal cord serum was prevented by SOD, further indicating the importance of ROS in the process of sperm capacitation. Other experiments have also shown that the potentials of these fluids to scavenge O_2^- inversely correlated with the diverse rates of sperm capacitation induced by biological fluids (de Lamirande et al. 1993), while an increase in the induction of extracellular O_2^- production at the commencement of incubation over a period of 4 h has been shown to be essential for the expression of sperm capacitation (de Lamirande and Gagnon 1995). To show the involvement of O_2^-, de Lamirande et al. (1998) confirmed that a higher concentration of O_2^- is needed to initiate the hyperactivation of spermatozoa, and subsequent basal level of this free radical is required to ensure that the cell remains in this state. In fact, adding SOD to hyperactivated spermatozoa was shown to stop the process (Leclerc et al. 1998). The effects of apocynin (APO), a reduced form of nicotinamide adenine dinucleotide phosphate (NADPH) oxidase inhibitor, on capacitation showed that APO prevented the generation of free radical generation and led to cell death. De Lamirande et al. (2009) showed that nitric oxide (NO), O_2^-, and H_2O_2 play a significant role in the process of capacitation, and the mode of action of O_2^- and NO is similar. NO is generated during the whole process of capacitation, and it can activate P-Tyr through cyclic guanosine monophosphate (cGMP) and cAMP simultaneously (Herrero et al. 1999). There is a general consensus that the presence of ROS is essential for the amplification in P-Tyr of proteins (Leclerc et al. 1997; Dona et al. 2011). Inhibited ROS production in vitro has been shown to result in a decrease of tyrosine-phosphorylated proteins, justifying the hypotheses of ROS-induced phosphorylation (Aitken et al. 1998a,b).

Furthermore, Revelli et al. (1999) suggested that nitric oxide synthase (NOS) are present in human spermatozoa for the production of NO, which is necessary for fertilization. Interestingly, a new hypothesis of ROS-induced ROS generation implies that NO stimulates O_2^- formation and vice versa in sperm development, and free radicals can react together to form peroxynitrite, an effector in capacitation (De Lamirande and Lamothe 2009).

O_2^- is not the only ROS involved in the acquisition of fertilizing ability by spermatozoa, but H_2O_2 at physiological concentration also accelerated the initiation of capacitation and sperm hyperactivation by 43% and 37%, respectively, after 3 h of incubation (Griveau et al. 1995). There is still no direct evidence for the production of H_2O_2 during the process of capacitation in the human sperm, but it is conceivable that there is generation of H_2O_2 in vivo, which might originate from the $O_2^{\bullet-}$ dismutation, or even that spermatozoa are generally subjected to ROS generated from the cells or fluids of the female reproductive tract.

NO, also a free radical, has deleterious effects on human sperm motility at high concentrations, while at low concentration using diethylamine-NONOate and spermine-NONOate, it has been shown to significantly induce capacitation in human spermatozoa induced without affecting sperm motility (Zini et al. 1995). However, experiments performed using high concentrations of an NOS inhibitor suggest that NO may be involved in hamster sperm hyperactivation

(Yeoman 1994). NO is believed to originate from the female reproductive tract since it has been proven to show NO synthase activity (Sladek et al. 1993). The aforementioned studies demonstrate that sperm capacitation is an oxidative process that can be induced exogenously when low concentrations are added.

Ionophore A23187, which acts as a membrane Ca^{2+} transporter, has been used to activate both capacitation and AR. To distinguish whether the effect was on capacitation or AR, 25 μM H_2O_2 was added to the incubating medium and spermatozoa were exposed to catalase at various times of incubation. Catalase was added initially to the incubation medium and significantly decreased the AR when compared to the control, which had no catalase addition and late addition of catalase (5 h 45 min) groups, showing that H_2O_2 affects the long-term process of capacitation rather than the quick AR (Griveau et al. 1995).

Considering the role of tyrosine phosphorylation in capacitation, recent study has shown that catalase decreased and H_2O_2 increased tyrosine phosphorylation in dose-dependent manners, thereby indicating the involvement of H_2O_2 in the process of capacitation (Rivlin et al. 2004).

ROS have been shown to increase tyrosine phosphorylation by both activation of tyrosine kinases and inhibition of phosphotyrosine phosphatases (PTPase) via oxidation of cysteine residues (De Lamirande et al. 1997; Leclerc et al. 1997). Recent studies have suggested that O_2^- stimulates AC (Zhang and Zheng 1996; Lewis and Aitken 2001), while H_2O_2 directly acts on kinases and phosphatases to induce capacitation (Lewis and Aitken 2001). AC produces cAMP and cAMP stimulates PKA, a necessary component for capacitation (O'Flaherty et al. 2006). PKA acts on a variety of substrates to induce tyrosine phosphorylation, which primarily occurs in the principal piece of the flagellum and the midpiece (Aitken et al. 2007). Tyrosine phosphorylation may not be the final step in capacitation, but it is generally used as a measurement of capacitation progress since the ability to proceed to the AR in response to a variety of stimuli is subject to the amount of tyrosine phosphorylation (Aitken et al. 1995).

The different studies on the effects of ROS on capacitation indicate that ROS can positively enhance capacitation in human spermatozoa, but this depends on the specific ROS involved. O_2^- and H_2O_2 have been shown to stimulate different molecules in the biochemical pathway, but this is subject to the method of AR induction. Overall, despite the mixed results, H_2O_2 is recognized to be more centrally involved in the regulation of tyrosine phosphorylation (Aitken et al. 1995; Aitken 1997). Most studies to date on both human and other mammalian spermatozoa show that H_2O_2 is responsible for the activation of capacitation.

31.1.4 ACROSOME REACTION

The biochemistry of capacitation and AR is closely related in many respects. ROS have been found to be involved in the phosphorylation of the same tyrosine proteins in both capacitation and AR (De Lamirande et al. 1998).

After the passage of the hyperactivated spermatozoa through the cumulus oophorous of the oocyte, it binds to the ZP and initiates the exocytotic release of enzymes that are proteolytic in nature, which creates a pore in the ZP's extracellular matrix. This then enables sperm to penetrate these physical outer barriers and fuse with the oocyte (De Lamirande and O'Flaherty 2008). The events that result in the AR include phosphorylation of similar tyrosine proteins, influx of Ca^{2+}, and increased cAMP and PKA levels, which substantially overlap with those of capacitation. ROS may stimulate a variety of targets, including AC, protein kinase C (PKC), and phospholipase A2 (PLA2), which have been shown to play a role in the AR (Gopalakrishna et al. 2013; Korbecki et al. 2013). It was shown that both O_2^- and H_2O_2 may activate PLA2 (Sawada et al. 1991). Activation of PLA2 has been shown to lead to the increased fluidity of the sperm plasma membrane, implicating that it plays a vital role in the regulation of AR and contributes to successful sperm–oocyte fusion (Goldman et al. 1992; Griveau et al. 1995).

The AR is also a short, irreversible process associated with a rapid production of extracellular O_2^- leading to the increased tyrosine phosphorylation of specific proteins (Griveau et al. 1995; De Lamirande et al. 1998). The source of ROS in this process is through the dismutation of O_2^- into H_2O_2 by NADPH oxidase. The two ROS members, at physiological concentrations, have been shown to have a positive effect on the AR (Aitken et al. 1997; Griveau and Lannou 1997; O'Flaherty et al. 2006). In line with this, NO has been reported to increase the percentage of sperm undergoing the AR (Herrero and Gagnon 2001). Results regarding the specific ROS involved report the beneficial effect of H_2O_2 and the negative effect of catalase, thus suggesting that H_2O_2 is the major species responsible for the positive effect on the AR.

It has been proven that mild oxidative conditions are essential in the binding of spermatozoa to the ZP, and ROS is involved in the in vivo AR through its effect on the phosphorylation of the plasma membrane proteins. When O_2^-, H_2O_2, and NO were added to the seminal plasma in vitro at physiological concentrations, it was observed that there was activation of the AR. Lipid peroxidation also enhances the effect of spermatozoa on ZP (Aitken et al. 1989). When spermatozoa were incubated in conditions that increased O_2^- production and induced capacitation, there was an increase in tyrosine phosphorylation of two major human sperm proteins of 81 and 105 kDa (Leclerc et al. 1996). However, tyrosine phosphorylation of other proteins occurs during the AR induced by A23187 (Aitken et al. 1995). Tyrosine kinases and phosphatases, which are susceptible to redox regulation, are the two types of enzymes involved in the regulation of tyrosine phosphorylation. However, it is unclear whether an increase in kinase activities or a decrease in phosphatases leads to the increase in protein tyrosine phosphorylation observed during sperm capacitation and AR. Tyrosine phosphorylation of proteins is the only component of the signal transduction mechanism for which a redox regulation linked to sperm AR has been demonstrated (Aitken et al. 1995). It is also plausible that oxidative conditions activate or inhibit other enzymes or mechanisms essential for sperm capacitation and AR as observed in other systems such as in endothelial cells and coronary artery (Grover et al. 1992; Taher et al. 1993). It is, therefore, plausible that more than one component of the signal transduction pathways is modified by oxidation in order for capacitation and AR to proceed.

Animal experiments also showed that induction of lipid peroxidation, using low concentrations of iron, ascorbic acid, and albumin, enhanced the binding of mouse spermatozoa to ZP and fertilizing potential by 50% (Kodama et al. 1996). Studies have also shown that catalase prevents AR in an A23187-induced AR protocol and also disallows binding and penetration of the zona-free hamster oocyte. These processes were, however, reactivated by the addition of a low concentration of H_2O_2. In studies conducted on human spermatozoa, membrane fluidity and rates of sperm–oocyte fusion were increased by ROS, which occurs during the biochemical events of capacitation and AR. This mechanism involves the inhibition of protein tyrosine phosphatase activity and prevents deactivation and dephosphorylation of PLA2. PLA2 cleaves the secondary fatty acid from the triglycerol backbone of the membrane phospholipid and increases the membrane's fluidity (Calamera et al. 2003; Khosrowbeygi et al. 2007). It is also evident in literature that α-tocopherol (a fat-soluble antioxidant) causes a dose-dependent decrease in the rate of AR, thus indicating that some level of oxidation of membrane lipids may play a role in AR (Griveau et al. 1995). It is proven that the most important function of O_2^- in AR induction is in the stimulation of an increase in membrane fluidity through the release of unesterified fatty acids (Griveau et al. 1995).

Ichikawa et al. (1999) compared the effect of small amounts of free radicals in stimulating the exocytosis and activation of acrosin. They found a strong correlation between the presence of a small quantity of radical species and an increase in the number of fully mature spermatozoa, while high levels of ROS produced the opposite effect. No relationship was found between the acrosin activity and the level of free radicals. Currently, there is no consensus as to which free radical is fully implicated in regulating the AR.

The role H_2O_2 plays in sperm penetration of zona-free hamster eggs in the presence of Biggers, Whitten, and Whittingham (BWW) medium containing A23187 ionophore and bovine serum albumin (BSA) has been investigated. The addition of catalase, an antioxidant enzyme of H_2O_2, stopped the aforementioned process, while it is resumed by adding H_2O_2 (Aitken et al. 1998a,b). Similarly, SOD (an O_2^- scavenger) did not have any effect on the procedure, thereby inferring that H_2O_2, and not O_2^-, is required for the AR. In contrary, Griveau et al. (1995) showed a positive correlation between the O_2^- amplification and induction of the AR, and in the presence of SOD, no change in the free radical level or AR was observed, further strengthening the claim that O_2^- is the free radical involved in the ionophore-induced AR.

De Lamirande et al. (1998) endeavored to clarify this controversy by investigating the effect of inducers of the AR including fetal cord serum ultrafiltrate (FCSu), follicular fluid ultrafiltrate (FFu), lipid-disturbing agent (LDA), and A23187 ionophore on AR. They subsequently concluded that both O_2^- and H_2O_2 play a role in the AR.

31.1.5 SPERM–OOCYTE FUSION

The fusion of the spermatozoon with the oocyte, after degradation of the ZP upon acrosomal exocytosis, ends the long journey through the female genital tract. A strong correlation has been shown to exist between increased ROS levels within physiological level and increased sperm–oocyte fusion (Aitken et al. 1997; Griveau et al. 1997). Increased expression of phosphorylated tyrosine proteins correlated with high rates of sperm–oocyte fusion (Aitken et al. 1997), indicating that sperm–oocyte fusion is correlated with the events of capacitation and AR. Both H_2O_2 and O_2^- contribute to the increase in fertilization rates, as shown by several studies from the same group (Aitken et al. 1995, 1996, 1997).

31.2 CONCLUSION

ROS is involved in the activation of intracellular pathways responsible for spermatozoal maturation, capacitation, hyperactivation, acrosomal reaction, chemotactic processes, and fusion with the female gamete. ROS provide numerous functions vital to the success of fertilization. Sperm maturation is aided by the activation of AC and cAMP, which trigger downstream molecules that improve the efficacy of the maturation processes required to take place. ROS also have a hand in many postejaculatory processes. In order for fertilization capacity to be acquired, a balance is essential for the formation and degradation of ROS acting at certain instances during the existence of spermatozoa. Their presence also brings about beneficial effects in the formation of the mitochondrial capsule and in the condensation of sperm DNA. The main ROS molecule involved in physiological processes differs. O_2^- is the principle ROS required in hyperactivation, while H_2O_2 has been shown to play a more significant role in capacitation and AR.

Free radicals are neither exclusively beneficial nor exclusively detrimental, thus/therefore a balance between the risks and benefits of their presence is necessary for the functioning and survival of spermatozoa. ROS need to be maintained at suitable levels to assure proper physiological function, while ensuring that pathological damage is averted.

REFERENCES

Aitken J, Clarkson JS, Fishel J. Generation of reactive oxygen species, lipid peroxidation, and human sperm function. *Biology of Reproduction.* 1989; 40: 183–197.

Aitken JR. Molecular mechanisms regulating human sperm function. *Molecular Human Reproduction.* 1997; 3(3): 169–173.

Aitken RJ, Buckingham DW, Harkiss D, Paterson M, Fisher H, Irvine DS. The extragenomic action of progesterone on human spermatozoa is influenced by redox regulated changes in tyrosine phosphorylation during capacitation. *Molecular and Cellular Endocrinology.* 1996; 117: 83–93.

Aitken RJ, Harkiss D, Knox W, Paterson M, Irvine DS. A novel signal transduction cascade in capacitating human spermatozoa characterised by a redox-regulated, cAMP-mediated induction of tyrosine phosphorylation. *Journal of Cell Science* 1998a; 111(Pt 5): 645–656.

Aitken RJ, Harkiss D, Knox W, Paterson M, Irvine DS. On the cellular mechanisms by which the bicarbonate ion mediates the extragenomic action of progesterone on human spermatozoa. *Biology of Reproduction.* 1998b; 58(1): 186–196.

Aitken RJ, Nixon B, Lin M, Koppers AJ, Lee YH, Baker MA. Proteomic changes in mammalian spermatozoa during epididymal maturation. *Asian Journal of Andrology.* 2007; 9: 554–564.

Aitken RJ, Paterson M, Fisher H, Buckingham DW, van Duin M. Redox regulation of tyrosine phosphorylation in human spermatozoa and its role in the control of human sperm function. *Journal of Cell Science.* 1995; 180: 2017–2025.

Aitken RJ, Ryan AL, Baker MA, McLaughlin EA. Redox acitivity associated with the maturation and capacitation of mammalian spermatozoa. *Free Radical Biology and Medicine.* 2004; 36(8): 994–1010.

Baldi E, Luconi M, Bonaccorsi L, Muratori M, Forti G. Intracellular events and signaling pathways involved in sperm acquisition of fertilizing capacity and acrosome reaction. *Frontiers in Bioscience.* November 2000; 5: 110–123.

Burkman LJ. Discrimination between nonhyperactivated and classical hyperactivated motility patterns in human spermatozoa. *Fertility and Sterility.* 1991; 55(2): 363–371.

Calamera J, Buffone M, Ollero M, Alvarez J, Doncel GF. Superoxide dismutase content and fatty acid composition in subsets of human spermatozoa from normozoospermic, asthenozoospermic, and polyzoospermic semen samples. *Molecular Reproduction and Development.* 2003; 66: 422–430.

Choudhary R, Chawala VK, Soni ND, Kumar J, Vyas RK. Oxidative stress and role of antioxidants in male infertility. *Pakistan Journal of Physiology.* 2010; 6: 54–59.

De Lamirande E, Gagnon C. Reactive oxygen species and human spermatozoa II. Depletion of adenosine triphosphate plays an important role in the inhibition of sperm motility. *Journal of Andrology.* 1992; 13(5): 379–386.

De Lamirande E, Gagnon C. A positive role for the superoxide anion in the triggering of human sperm hyperactivation and capacitation. *International Journal of Andrology.* 1993a; 16: 21–25.

De Lamirande E, Gagnon C. Human sperm hyperactivation and capacitation as parts of an oxidative process. *Free Radical Biology and Medicine.* 1993b; 14(2): 157–166.

De Lamirande E, Gagnon C. Capacitation-associated production of superoxide anion by human spermatozoa. *Free Radical Biology and Medicine.* 1995; 18: 487–495.

De Lamirande E, Harakat A, Gagnon C. Human sperm capacitation induced by biological fluids and progesterone, but not by NADH or NADPH, is associated with the production of superoxide anion. *Journal of Andrology.* 1998a; 19(2): 215–225.

De Lamirande E, Jiang H, Zini A, Kodama H, Gagnon C. Reactive oxygen species and sperm physiology. *Reviews of Reprodroduction.* 1997a; 2(1): 48–54.

De Lamirande E, Lamothe G, Villemure M. Control of superoxide and nitric oxide formation during human sperm capacitation. *Free Radical Biology and Medicine.* 2009; 46(10): 1420–1427.

De Lamirande E, Leclerc P, Gagnon C. Capacitation as a regulatory event that primes spermatozoa for the acrosome reaction and fertilization. *Molecular Human Reproduction.* 1997b; 3(3): 175–194.

De Lamirande E, O'Flaherty C. Sperm activation: Role of reactive oxygen species and kinases. *Biochimica et Biophysica Acta.* 2008; 1784: 106–115.

De Lamirande E, Tsai C, Harakat A, Gagnon C. Involvement of reactive oxygen species in human sperm arcosome reaction induced by A23187, lysophosphatidylcholine, and biological fluid ultrafiltrates. *Journal of Andrology.* 1998b; 19(5): 585–594.

Demaurex N, Downey GP, Waddell TK, Grinstein S. Intracellular pH regulation during spreading of human neutrophils. *Journal of Cell Biology.* 1996; 133(6): 1391–1402.

Donà G, Fiore C, Tibaldi E, Frezzato F, Andrisani A, Ambrosini G et al. Endogenous reactive oxygen species content and modulation of tyrosine phosphorylation during sperm capacitation. *International Journal of Andrology.* 2011; 34(5 Pt 1): 411–419.

Eisenbach M. Mammalian sperm chemotaxis and its association with capacitation. *Developmental Genetics.* 1999; 25: 87–94.

Gil-Guzman E, Ollero M, Lopez MC et al. Differential production of reactive oxygen species by subsets of human spermatozoa at different stages of maturation. *Human Reproduction Update.* 2001; 16(9): 1922–1939.

Goldman R, Ferber E, Zort U. Reactive oxygen species are involved in the activation of cellular phospholipase A2. *FEBS Letters.* 1992; 309(2): 190–192.

Gopalakrishna R, McNeill TH, Elhiani AA, Gundimeda U. Methods for studying oxidative regulation of protein kinase C. *Methods Enzymology.* 2013; 528: 79–98.

Griveau JF, Lannou DL. Reactive oxygen species and human spermatozoa: Physiology and pathology. *International Journal of Andrology*. 1997; 20: 61–69.

Griveau JF, Renard P, Le Lannou D. Superoxide anion production by human spermatozoa as a part of the ionophore-induced acrosome reaction process. *International Journal of Andrology*. 1995; 18(2): 67–74.

Grover AK, Samson SE, Fomin VP. Peroxide inactivates calcium pumps in pig coronary artery. *American Journal of Physiology*. 1992; 263: H537–H543.

Herrero MB, de Lamirande E, Gagnon C. Nitric oxide regulates human sperm capacitation and protein-tyrosine phosphorylation in vitro. *Biology of Reproduction* 1999; 61(3): 575–581.

Herrero MB, Gagnon C. Nitric oxide: A novel mediator of sperm function. *Journal of Andrology*. 2001; 22(3): 349–356.

Ichikawa T, Oeda T, Ohmori H, Schill WB. Reactive oxygen species influence the acrosome reaction but not acrosin activity in human spermatozoa. *International Journal of Andrology*. 1999; 22(1): 37–42.

Jones R, Mann T, Sherins R. Peroxidative breakdown of phospholipids in human spermatozoa, spermicidal properties of fatty acid peroxides, and protective action of seminal plasma. *Fertility and Sterility*. 1979; 31: 531–537.

Khosrowbeygi A, Zarghami N. Fatty acid composition of human spermatozoa and seminal plasma levels of oxidative stress biomarkers in subfertile males. *Prostaglandins, Leukotrienes and Essential Fatty Acids*. 2007; 77: 117–121.

Kodama H, Kuribayashi Y, Gagnon C. Effect of sperm lipid peroxidation on fertilization *Journal of Andrology*. 1996; 17: 151–157.

Korbecki J, Baranowska-Bosiacka I, Gutowska I, Chlubek D. The effect of reactive oxygen species on the synthesis of prostanoids from arachidonic acid. *Journal of Physiology and Pharmacology*. 2013; 64(4): 409–421.

Leclerc P, de Lamirande E, Gagnon C. Cyclic adenosine 3',5' monophosphate-dependent regulation of protein tyrosine phosphorylation in relation to human sperm capacitation and motility. *Biology of Reproduction*. 1996; 55: 684–695.

Leclerc P, de Lamirande E, Gagnon C. Interaction between Ca^{2+}, cyclic 3', 5' adenosine monophosphate, the superoxide anion, and tyrosine phosphorylation pathways in the regulation of human sperm capacitation. *Journal of Andrology*. 1998; 19(4): 434–443.

Leclerc P, de Lamirande E, Gagnon C. Regulation of protein-tyrosine phosphorylation and human sperm capacitation by reactive oxygen derivatives. *Free Radical Biology and Medicine*. 1997; 22(4): 643–656.

Lewis B, Aitken RJ. A redox-regulated tyrosine phosphorylation cascade in rat spermatozoa. *Journal of Andrology*. 2001; 22(4): 611–622.

MacLeod J. The role of oxygen in the metabolism and motility of human spermatozoa. *The American Journal of Physiology*. 1943; 138: 512–518.

Menezo Y. Antioxidants to reduce sperm DNA fragmentation: An unexpected adverse effect. *Reproductive Biomedicine Online*. 2007; 14(4): 418–421.

Oehninger S, Blackmore P, Mahony M, Hodgen G. Effects of hydrogen peroxide on human spermatozoa. *Journal of Assisted Reproduction and Genetics*. 1995; 12(1): 41–47.

O'Flaherty C, Lamirande Ed, Gagnon C. Positive role of reactive oxygen species in mammalian sperm capacitation: Triggering and modulation of phosphorylation events. *Free Radical Biology and Medicine*. 2006; 41: 528–540.

Revelli A, Soldati G, Costamagna C, Pellerey O, Aldieri E, Massobrio M et al. Follicular fluid proteins stimulate nitric oxide (NO) synthesis in human sperm: A possible role for NO in acrosomal reaction. *The Journal of Cellular Physiology*. 1999; 178(1): 85–92.

Rhemrev JP, van Overveld FW, Haenen GR, Teerlink T, Bast A, Vermeiden JP. Quantification of the non-enzymatic fast and slow TRAP in a postaddition assay in human seminal plasma and the antioxidant contributions of various seminal compounds. *Journal of Andrology*. 2000; 21: 913–920.

Rivlin J, Mendel J, Rubinstein S, Etkovitz N, Breitbart H. Role of hydrogen peroxide in sperm capacitation and acrosome reaction. *Biology of Reproduction*. 2004; 70: 518–522.

Rousseaux J, Rousseaux-Prevost R. Molecular localization of free thiols in human sperm chromatin. *Biology of Reproduction*. 1995; 52: 1066–1072.

Roveri A, Ursini F, Flohe L, Maiorino M. PHGPx and spermatogenesis. *BioFactors*. 2001; 14: 213–222.

Sabeur K, Ball B. Characterization of NADPH oxidase 5 in equine testis and spermatozoa. *Society for Reproduction and Fertility*. 2007; 134: 263–270.

Saowaros W, Panyim S. The formation of disulfide bonds in human protamines during sperm maturation. *Experientia*. 1978; 35(2): 191–192.

Sawada M, Carlson JC. Rapid plasma membrane changes in superoxide radical formation, fluidity, and phospholipase A2 activity in the corpus luteum of the rat during induction of luteolysis. *Endocrinology.* 1991; 128(6): 2992–2998.

Sladek MS, Regenstein AC, Lykins D, Roberts JM. Nitric oxide synthase activity in pregnant rabbit uterus decreases on the last day of pregnancy. *American Journal of Obstetrics and Gynecology.* 1993; 169: 1285–1291.

Suarez SS. Control of hyperactivation in sperm. *Human Reproduction Update.* 2008; 14(6): 647–657.

Taher MM, Garcia JGN, Natarajan V. Hydroperoxide-induced diacylglycerol formation and protein kinase C activation in vascular endothelial cells. *Archives of Biochemistry and Biophysics.* 1993; 303: 260–266.

Tosic J, Walton A. Formation of hydrogen peroxide by spermatozoa and its inhibitory effect of respiration. *Nature.* 1946; 158: 485.

Tunc O, Thompson J, Tremellen K. Improvement in sperm DNA quality using an oral antioxidant therapy. *Reproductive Biomedicine Online.* 2009; 18(6): 761–768.

Venkatesh S, Deecaraman M, Kumar R, Shamsi MB, Dada R. Role of reactive oxygen species in the pathogenesis of mitochondrial DNA (mtDNA) mutations in male infertility. *Indian Journal of Medical Research.* 2009; 129: 127–137.

Vernet P, Aitken RJ, Drevet JR. Antioxidant strategies in the epididymis. *Molecular and Cellular Endocrinology.* 2004; 216: 31–39.

Visconti PE, Westbrook VA, Chertihin O, Demarco I, Sleight S, Diekman AB. Novel signaling pathways involved in sperm acquisition of fertilizing capacity. *Journal of Reproductive Immunology.* 2002; 53: 133–150.

Yeoman RR. Blockade of nitric oxide synthase inhibits hamster sperm hyperactivation. *Biology of Reproduction.* 1994; 50(1): 104.

Zhang H, Zheng R-L. Promotion of human sperm capacitation by superoxide anion. *Free Radical Research.* 1996; 24(4): 261–268.

Zini A, de Lamirande E, Gagnon C. Low levels of nitric oxide promotes human sperm capacitation *in vitro.* *Journal of Andrology.* 1995; 16: 424–431.

32 Impact of Oxidative Stress on DNA Damage in Human Spermatozoa

J.V. Villegas, P. Uribe, R. Boguen, and F. Treulen

CONTENTS

ABSTRACT

Oxidative stress (OS) results from elevated levels of reactive oxygen species (ROS) overpassing the antioxidant defenses. OS is the most frequent cause of functional loss of DNA in spermatozoa, and there is a high correlation between OS and DNA fragmentation (DF). High levels of DNA damage in sperm are correlated with male infertility, embryonic development disorders, and high rates of miscarriage. DNA damage has been associated with pathological conditions such as varicocele and obesity and lifestyle issues like smoking, exposure to xenobiotics, toxic metals, heat, and mobile phone radiation. Aging and antioxidant depletion have also been linked to elevated levels of sperm ROS and DNA damage. Considering that OS plays an important role in sperm DNA damage and infertility, it is becoming increasingly important to address research into the role of antioxidants *in vitro* and *in vivo* in order to protect spermatozoa against such oxidative damage and thus improve their fertilization potential.

In this chapter, the impact of OS on sperm DF is revised by analyzing in first place the particular acquisition of sperm DNA–specific features for genome protection and the evidences of the high susceptibility of abnormal spermatozoa to DNA damage. Next, methods to evaluate the sperm DNA integrity are briefly enumerated. Then, the following aspects are developed: evidences *in vivo* and *in vitro* of the damaging effects of ROS on sperm DNA integrity; the impact of oxidative DNA damage on assisted reproductive technologies (ARTs), and the oxidative DNA damage induced by ART procedures. Finally, some aspects of ROS and DNA damage during sperm cryopreservation and the antioxidant supplementation to avoid sperm DNA damage are analyzed.

32.1   INTRODUCTION

Oxidative stress (OS) is the result of elevated levels of reactive oxygen species (ROS) overpowering the antioxidant defenses. ROS in low levels are essential for processes achieving oocyte fecundation by spermatozoa. They are regulators of sperm function and also act as a double-edged sword since increased ROS levels can lead to a state of OS that, on the contrary, compromises sperm function (Aitken and Koppers 2011). Given the high correlation between OS and DNA fragmentation (DF), it can be stated that OS is the most frequent cause of integrity loss of DNA in spermatozoa (Aitken and De Iuliis 2010). High levels of DNA damage in the spermatozoa are frequently associated with male infertility and are also correlated with a variety of adverse clinical effects including infertility, embryonic development disorders, and high rates of miscarriage (Aitken and De Iuliis 2010). Sperm DNA integrity is crucial for maintaining paternal reproductive potential (Avendano and Oehninger 2011).

32.1.1   ACQUISITION OF SPECIFIC FEATURES OF SPERM DNA FOR GENOME PROTECTION

Sperm DNA acquires their features through a complex process during spermatogenesis that culminates in the epididymis. Spermatogenesis consists of three phases: mitotic amplification, meiotic and post-meiotic, or spermiogenesis. Following meiosis and during spermiogenesis in mammals, round spermatids differentiate into mature sperm and histones are replaced by protamines. The nuclear volume dramatically reduces in a process associated with histone hyperacetylation, and the DNA breaks. DNA strand breaks may eliminate DNA supercoils formed during histone removal. Mammalian protamines are arginine- and cysteine-rich proteins, and it is believed that they stabilize sperm chromatin through their assembly in the minor DNA groove. The arginine moieties within protamines mediate strong DNA binding, whereas the cysteine moieties facilitate the formation of disulfide bonds essential for highly compacted chromatin. It has been suggested that zinc ions contribute to DNA packaging by linking protamines with zinc bridges (Rathke et al. 2014). Finally, protamines replace 85% of the histones resulting in a supercoiled structure known as toroid (Rousseaux et al. 2008).

In the epididymis, protamines undergo cross-linking of disulfide bonds, further reducing the volume of the sperm nucleus (Barratt et al. 2010). However, 10%–15% of human sperm DNA retain a histone-based nucleosomal structure, suggesting that these paternally contributed regions are important for normal embryogenesis (Rathke et al. 2014). As a result, a normal sperm has an extremely condensed nucleus that confers protection from the effect of genotoxic factors and facilitates the transport of the paternal genome to the female tract (Manicardi et al. 1998).

32.1.2   ABNORMAL SPERMATOZOA ARE HIGHLY SUSCEPTIBLE TO DNA DAMAGE

In humans, a fraction of spermatozoa has a relatively high degree of DF. This fragmentation can be derived from aberrant chromatin packaging during spermatogenesis, defective apoptosis, or excessive production of ROS in the ejaculate. Spermatozoa with a higher than normal level of histone show defects in chromatin condensation with areas susceptible to DNA damage as reported by De Iuliis et al. (2009); they found that defective spermatozoa show a positive correlation between increased levels of histones and ROS, damaged DNA, and 8-hydroxy-2′-deoxyguanosine (8-OHdG). DNA damage occurs during sperm production or transport and storage (Sakkas and Alvarez 2010). In order to explain the origin of nuclear DNA damage, Aitken et al. (2009) suggested that faulty spermatogenesis can lead to defective chromatin remodeling and to a poorly protaminated nuclear DNA that is more susceptible to attack by ROS. Nucleases could contribute to sperm DNA damage by fragmenting the DNA of the spermatozoon for its final elimination (Aitken et al. 2014).

32.1.3 METHODS TO EVALUATE THE SPERM DNA INTEGRITY

In order to assess DNA integrity, methods that evaluate the chromatin structure such as sperm chromatin structure assay (SCSA) (Darzynkiewicz et al. 1975) and the acridine orange test (Tejada et al. 1984) have been used. Other methods analyze sperm DF including the terminal deoxyribonucleotidyl transferase–mediated deoxyuridine triphosphate (dUTP) nick end-labeling (TUNEL) assay (Gorczyca et al. 1993), the *in situ* nick translation assay (Manicardi et al. 1995), the single-cell gel electrophoresis (Comet) assay (McKelvey-Martin et al. 1997), and the sperm chromatin dispersion test (Fernandez et al. 2003). A more recent addition is the method based on detection of the oxidative base 8-OHdG, which is a marker of DNA damage by OS (De Iuliis et al. 2009). The most frequently used methods to detect DNA breaks are TUNEL and Comet and SCSA to evaluate the chromatin structure.

32.1.4 DAMAGING EFFECTS OF ROS ON SPERM DNA INTEGRITY

High concentrations of seminal ROS have been detected in men with elevated DNA damage (Wright et al. 2014). ROS cause DNA damage in the form of modification of bases, production of abasic sites, frame shifts in DNA sequence, inter- or intrastrand cross-links, and DNA–protein cross-links (Bennetts et al. 2008); all these alterations can cause adduct bases like 8-OHdG and O6-methylguanine to destabilize the DNA structure, generating single- or double-strand breaks, or interference with the action of the methyl transferases leading to global DNA hypomethylation (Aitken et al. 2012). DNA damage can give rise to mutations and even lethal genetic defects in the offspring (Moustafa et al. 2004, Tominaga et al. 2004). Sperm DNA can be damaged at both sites in nuclei or in mitochondria (Sakkas and Alvarez 2010). The mitochondrial DNA, unlike the nuclear DNA, is not protected by histones but physically associated with the inner mitochondrial membrane, where oxygen radicals are produced as a result of oxidative phosphorylation (Beckman and Ames 1998). Consequently, the mitochondrial DNA accumulates polymorphisms and mutations in larger number than the nuclear DNA (Wallace and Fan 2009).

In infertile subjects, DNA damage has been associated with endogenous ROS production by spermatozoa or exogenous ROS production by leukocytes, as well as with many pathological conditions such as varicocele and obesity and lifestyle issues like smoking, exposure to xenobiotics, toxic metals, heat, and mobile phone radiation. Aging and antioxidant depletion have also been linked to elevated levels of sperm ROS and DNA damage (Wright et al. 2014).

32.1.5 ROS-INDUCED DNA DAMAGE IN HUMAN SPERMATOZOA IN VIVO

Kodama et al. (1997) observed a link between ROS and DNA damage when they found that the sperm of infertile men presented higher levels of ROS and 8-OHdG than the sperm in fertile men. This association was confirmed after 2 months of antioxidant treatment, when the patients showed a significant reduction in ROS and DNA damage levels (Kodama et al. 1997). Individuals with idiopathic and normozoospermic infertility also show elevated ROS levels and DF, indicating that the association between increased ROS and DNA damage is independent of the spermiogram (Mayorga-Torres et al. 2013). It was reported that 25% increase in seminal ROS in infertile patients is associated with a 10% increase in sperm with DF (Mahfouz et al. 2010). It was also reported that ROS can reduce sperm DNA methylation, which together with DNA integrity has been recovered with antioxidant treatment (Tunc and Tremellen 2009). The link between excessive ROS and DNA damage has also been confirmed in smokers. Smoking is associated with high ROS levels and consumption of antioxidant capacity. Smokers present a marked increase in DNA oxidation, reflecting serious damage (Fraga et al. 1996). Patients with varicocele also have high ROS levels and DF, although they do not necessarily present decreased sperm quality (Saleh et al. 2003, Smith et al. 2006). A month after the varicocelectomy, ROS levels are significantly reduced, and in the sixth month, reduction in DF levels is observed (Dada et al. 2010). Synergistic action between smoking and varicocele has been observed by comparing the sperm damage observed in patients with the damage caused by each factor individually (Taha et al. 2014).

Other environmental factors that cause high ROS levels are ambient contaminants such as dichlorodiphenyldichloroethylene (p, p′-DDE) and radio frequency electromagnetic radiation (RFER). Exposure to p, p′-DDE is associated with increased ROS and loss of sperm quality as well as DF (Pant et al. 2014). The RFER emitted by mobile phones also causes increased sperm ROS. The use of mobile phones for more than 4 h per day is associated with higher levels of sperm DF (Rago et al. 2013). *In vitro* studies have been used to confirm directly the harmful effects of ROS on human sperm.

32.1.6 *In Vitro* Impact of ROS on Human Sperm DNA

Link between ROS and DNA damage has also been studied *in vitro*. Hughes et al. (1996) exposed human spermatozoa to H_2O_2 and observed the higher susceptibility of infertile patient's sperm DNA to oxidative damage. In a similar study, it was reported that damaged sperms are still capable of achieving normal rates of fertilization with intracytoplasmic sperm injection (ICSI) (Twigg et al. 1998). Aitken et al. (1998) found that purified populations of sperm with high motility obtained from normozoospermic donors resulted in a dose-dependent induction of high rates of DF when exposed to H_2O_2, indicating that normal spermatozoa are also susceptible to DNA damage by OS.

Also, by adding H_2O_2 to DNA, it can be shown that strand breaks in DNA occur (Duru et al. 2000). The DF produced *in vitro* is not affected by treatment with nuclease inhibitors but decreases with a glutathione peroxidase inhibitor (Muratori et al. 2003), suggesting that DF *in vitro* is produced by the direct effect of ROS. Recently, we reported that increasing levels of intracellular calcium by exposure of human spermatozoa to ionomycin increased the level of ROS and DFs, which were associated with the opening of pores in the inner mitochondrial membrane (Treulen et al. 2015). We also found that the ROS produced are mostly H_2O_2 and can be reduced by the addition of N-acetylcysteine but neither by antioxidants, such as 4-hydroxy-2,2,6,6,-tetramethyl piperidinoxyl and N-tert-butyl-α-phenylnitrone, nor by a nuclease inhibitor. These are in agreement with the fact that H_2O_2 is a small molecule and hence able to cross membranes and reach the sperm's nucleus and DNA.

Other *in vitro* systems for inducing OS include the use of progesterone, which has a similar effect of increasing ROS formation and fragment DNA (Lozano et al. 2009), phytoestrogens and estrogenic compounds, which can also damage DNA (Cemeli and Anderson 2011), and electrophilic aldehydes such as 4-hydroxynonenal and acrolein, which stimulate the formation of mitochondrial superoxide and involve in disrupt the respiratory chain to disrupt with the subsequent loss of mitochondrial membrane potential and end in oxidative DNA adduct formation, DF, and cell death. These effects are reverted by adding nucleophilic compounds like penicillamine to the incubation medium (Aitken et al. 2012b). The causative role of ROS on sperm DNA damage has also been demonstrated through the use of antioxidants *in vitro*. Antioxidants such as glutathione and hypotaurine have also been shown to protect against DNA damage induced by H_2O_2 (Donnelly et al. 2000). These results suggest that antioxidants can be used to improve the reproductive technologies facing OS.

32.1.7 Impact of Oxidative DNA Damage on ARTs

As a result of the increment in infertility due to the continued decline in sperm quality, couples increasingly require assisted reproductive technologies (ARTs) to conceive and give birth to baby of their own (Farquhar et al. 2014). Thus, intrauterine insemination (IUI), *in vitro* fertilization (IVF), and ICSI play an important role in treating infertility (Agarwal and Said 2005).

Despite numerous advances, the current success rate of ART procedures is inadequate, and the impact of OS on DNA integrity appears to be the most important factor that affects ART outcomes (Agarwal et al. 2006). DNA integrity is an important issue in ART procedures since spermatozoa with normal characteristics, according to WHO criteria, may carry damaged DNA, including high levels of DF (Benchaib et al. 2007). As previously mentioned, elevated levels of sperm DF are a well-recognized factor for negative pregnancy outcomes (Erenpreiss et al. 2006), and sperm genome abnormalities are often involved in the failure to obtain an embryo and/or pregnancy by ART techniques (Benchaib et al. 2003, 2007). High levels of DNA damage in human spermatozoa are usually associated with reduced fertilization rates and/or embryo cleavage after IVF and ICSI and may be associated with early embryo death (Agarwal and Said 2005). Therefore, sperm DNA damage could be a useful biomarker for predicting ART outcomes (Lewis et al. 2013). Specifically, it is affirmed that with sperm DNA damage above 20% as measured by the SCSA, there is a reduction in the chances of pregnancy by IUI, and the probability of fertilization seems to be close to zero if the proportion of sperm cells with DNA damage exceeds 30% (Erenpreiss et al. 2006). Similarly, when samples carrying more than 12% spermatozoa with fragmented DNA, as detected by TUNEL assay, are used for insemination, no pregnancies are achieved (Duran et al. 2002). If the test for 8-OHdG is employed, the results are even more sensitive, with a lower threshold value of 11.5% (Lewis et al. 2013). Moreover, when the DF index exceeds 15%, the miscarriage risk may increase fourfold (Benchaib et al. 2007). The oocytes are able to repair the damaged sperm DNA when it is low, but this ability is overwhelmed in cases of high levels of sperm DF (Benchaib et al. 2003).

Additionally, DNA damage has an even more serious impact because it is likely to increase the transmission of genetic diseases, childhood cancer, or infertility in the offspring (Cocuzza et al. 2007). The safety of the ICSI procedure has been questioned because this procedure is able to overcome the natural barrier for high sperm DNA damage and initiate a successful pregnancy when this would hardly be possible through natural conception, IUI, or even IVF (Erenpreiss et al. 2006). Some systematic reviews and meta-analyses addressing the prevalence of birth defects in children conceived by ARTs have revealed the increased risk of birth defects in children conceived through these procedures (Hansen et al. 2005, Wen et al. 2012).

Considering the importance of DNA integrity to the success of ARTs and the impact of DNA damage on both the procedure outcome and the well-being of conceived children, it would be useful to measure the level of sperm DF in order to provide the clinician with an opportunity to decrease the percentage of sperm DF prior to beginning any ART procedure (Benchaib et al. 2007). Although the Practice Committee of the American Society for Reproductive Medicine indicates that there is not enough evidence to recommend the routine assessment of sperm DNA damage (Practice Committee of the American Society for Reproductive Medicine 2013), several research groups support the inclusion of DNA analysis in infertility assessment (Wright et al. 2014).

32.1.8 Oxidative DNA Damage Induced by ART Procedure

The DNA can also be damaged during the laboratory procedure, and gametes and embryos when manipulated for ART procedures are exposed to various ROS-inducing factors, with the risk of OS being greater *in vitro* than *in vivo* mainly because of the lack of physiological antioxidant protection (Agarwal et al. 2014). ROS during ART procedures could be generated either endogenously from gametes or exogenously from the environmental factors. Sources of ROS in the ART protocol include contact with immature spermatozoa during centrifugation steps, contamination by leukocytes, and exposure of gametes to pO_2 since, *in vitro*, gametes are present in hyperoxic conditions compared with the *in vivo* environment (Agarwal et al. 2006). Also, spermatozoa display substantial DF following routine incubation *in vitro* prior to ICSI, and it is recommended that there be no delay in sperm injection so as to protect the offsprings for the genetic abnormalities (Dalzell et al. 2004).

32.1.9 Sperm Cryopreservation, ROS, and DNA Damage

In addition to ART, the techniques aimed at preserving gametes (eggs and sperms) are an emerging field that allows these reproductive elements to be preserved such as in cancer patients who need gonadotoxic treatment or in patients who want to preserve their gametes to postpone child-bearing. Cryopreservation of sperm is also considered a standard procedure prior to vasectomy, and men exposed to toxins or working conditions that may adversely affect spermatogenesis are also candidates for gamete preservation procedures (Gonzalez et al. 2012). Moreover, in developed countries, donating sperm to semen cryobanks is common for IVF treatment in women who want to get pregnant but are facing male factor infertility or when there is no male partner (Brezina et al. 2015). Despite the research conducted for optimizing semen cryopreservation, the post-thaw survival rate remains limited and fails to meet ideal expectations. Cell damage includes alteration of the plasma membrane, mitochondrial dysfunction, decreased motility, and oxidative DNA damage (Bateni et al. 2014). Although the use of cryoprotectants and the optimization of protocols have increased cell viability, the freeze–thaw process also induces considerable OS as a result of ROS formation, and this procedure increases oxidative DNA damage and fragmentation levels in spermatozoa (Thomson et al. 2009).

32.1.10 ANTIOXIDANT SUPPLEMENTATION AND SPERM DNA DAMAGE

Employment of antioxidants for the purpose of avoiding OS may consist of either oral supplementation for several months prior to the ART cycle or *in vitro* supplementation of media during the ART protocol (Agarwal et al. 2014). Patient treatment with oral antioxidant vitamins C and E is a standard practice for male infertility in an attempt to decrease ROS formation and improve fertility, assuming that the patient's antioxidant defenses are inadequate (Wright et al. 2014). In other interesting study, Devaraj et al. (2008) showed that a daily dose of 30 mg of lycopene was required for a 9% decrease in DNA damage. Lycopene supplementation decreases lipid peroxidation and DNA damage and increases antioxidants such as catalase and glutathione peroxidase (Durairajanayagam et al. 2014).

However, in the study by Menezo et al. (2007), mixed antioxidant treatment led to an increase in sperm decondensation, indicating that antioxidant therapy could interfere with paternal gene activity during preimplantation development (Menezo et al. 2007). Similarly, a recent study reported that the addition of ascorbic acid to the medium for swim-up improves several seminal quality parameters and DF but has a negative effect on acrosome reaction (Fanaei et al. 2014).

In order to protect spermatozoa from the OS generated during the freeze–thaw process, the supplementation of cryopreservation medium with several antioxidants has been addressed with promising results mainly in terms of motility, viability, and DNA integrity (Agarwal et al. 2014). Genistein is a naturally occurring phytoestrogen found in relatively high concentrations in soybean and is effective at reducing the level of DF and oxidative damage during cryopreservation (Thomson et al. 2009). Ganglioside antioxidants have also been used in human spermatozoa that are exposed to cryopreservation to prevent the damaging effects of DF (Gavella and Lipovac 2013).

Antioxidants can also be employed in the ART settings to improve the damaging effects of OS on gametes and embryos. There have been numerous studies about the effect of antioxidant therapy *in vivo* and *in vitro*, and as commented in the review by Wright et al. (2014), "antioxidant treatment may play a relevant role," but there are some contradictory results, and "large well-controlled trials are still required for conclusive evidence for their efficacy."

32.2 CONCLUSIONS

Despite the extreme compaction of sperm DNA, an increase in the susceptibility to damage by OS can be found in spermatozoa, especially when there are defects in the chromatin remodeling. The most frequent cause of damage to sperm integrity is OS. ART procedures expose sperm to OS, which is why the time for keeping the gametes *in vitro* must be reduced as far as possible. ART, like ICSI, skips over the natural barriers to avoid fertilization with gametes carrying DNA damage;

therefore, the analysis of DNA integrity prior to ART is advisable. Considering that OS plays such an important role in sperm DNA damage and infertility, it is becoming increasingly important to address research into the role of antioxidants *in vitro* and *in vivo* in order to protect spermatozoa against such oxidative damage and thus improve their fertilization potential.

ACKNOWLEDGMENTS

We thank the German Academic Exchange Service (DAAD) for financing J.V. Villegas.

We apologize to those authors who were not mentioned owing to space constraints, which often caused us to cite review articles instead.

ABBREVIATIONS

8-OHdG	8-Hydroxy-2′-deoxyguanosine
ART	Assisted reproductive technology
DF	DNA fragmentation
ICSI	Intracytoplasmatic sperm injection
OS	Oxidative stress
ROS	Reactive oxygen species

REFERENCES

Agarwal, A., D. Durairajanayagam, and S. S. du Plessis. 2014. Utility of antioxidants during assisted reproductive techniques: An evidence based review. *Reprod Biol Endocrinol* 12:112.

Agarwal, A. and T. M. Said. 2005. Oxidative stress, DNA damage and apoptosis in male infertility: A clinical approach. *BJU Int* 95 (4):503–507.

Agarwal, A., T. M. Said, M. A. Bedaiwy, J. Banerjee, and J. G. Alvarez. 2006. Oxidative stress in an assisted reproductive techniques setting. *Fertil Steril* 86 (3):503–512.

Aitken, R. J. and G. N. De Iuliis. 2010. On the possible origins of DNA damage in human spermatozoa. *Mol Hum Reprod* 16 (1):3–13.

Aitken, R. J., G. N. De Iuliis, and R. I. McLachlan. 2009. Biological and clinical significance of DNA damage in the male germ line. *Int J Androl* 32 (1):46–56.

Aitken, R. J., D. Harkiss, W. Knox, M. Paterson, and D. S. Irvine. 1998. A novel signal transduction cascade in capacitating human spermatozoa characterised by a redox-regulated, cAMP-mediated induction of tyrosine phosphorylation. *J Cell Sci* 111 (Pt 5):645–656.

Aitken, R. J., K. T. Jones, and S. A. Robertson. 2012a. Reactive oxygen species and sperm function—In sickness and in health. *J Androl* 33 (6):1096–1106.

Aitken, R. J. and A. J. Koppers. 2011. Apoptosis and DNA damage in human spermatozoa. *Asian J Androl* 13 (1):36–42.

Aitken, R. J., T. B. Smith, M. S. Jobling, M. A. Baker, and G. N. De Iuliis. 2014. Oxidative stress and male reproductive health. *Asian J Androl* 16 (1):31–38.

Aitken, R. J., S. Whiting, G. N. De Iuliis, S. McClymont, L. A. Mitchell, and M. A. Baker. 2012b. Electrophilic aldehydes generated by sperm metabolism activate mitochondrial reactive oxygen species generation and apoptosis by targeting succinate dehydrogenase. *J Biol Chem* 287 (39):33048–33060.

Avendano, C. and S. Oehninger. 2011. DNA fragmentation in morphologically normal spermatozoa: How much should we be concerned in the ICSI era? *J Androl* 32 (4):356–363.

Barratt, C. L., R. J. Aitken, L. Bjorndahl, D. T. Carrell, P. de Boer, U. Kvist, S. E. Lewis et al. 2010. Sperm DNA: Organization, protection and vulnerability: From basic science to clinical applications—A position report. *Hum Reprod* 25 (4):824–838.

Bateni, Z., L. Azadi, M. Tavalaee, A. Kiani-Esfahani, M. Fazilati, and M. H. Nasr-Esfahani. 2014. Addition of Tempol in semen cryopreservation medium improves the post-thaw sperm function. *Syst Biol Reprod Med* 60 (4):245–250.

Beckman, K. B. and B. N. Ames. 1998. The free radical theory of aging matures. *Physiol Rev* 78 (2):547–581.

Benchaib, M., V. Braun, J. Lornage, S. Hadj, B. Salle, H. Lejeune, and J. F. Guerin. 2003. Sperm DNA fragmentation decreases the pregnancy rate in an assisted reproductive technique. *Hum Reprod* 18 (5):1023–1028.

Benchaib, M., J. Lornage, C. Mazoyer, H. Lejeune, B. Salle, and J. Francois Guerin. 2007. Sperm deoxyribonucleic acid fragmentation as a prognostic indicator of assisted reproductive technology outcome. *Fertil Steril* 87 (1):93–100.

Bennetts, L. E., G. N. De Iuliis, B. Nixon, M. Kime, K. Zelski, C. M. McVicar, S. E. Lewis, and R. J. Aitken. 2008. Impact of estrogenic compounds on DNA integrity in human spermatozoa: Evidence for cross-linking and redox cycling activities. *Mutat Res* 641 (1–2):1–11.

Brezina, P. R., W. H. Kutteh, A. P. Bailey, J. Ding, R. W. Ke, and J. L. Klosky. 2015. Fertility preservation in the age of assisted reproductive technologies. *Obstet Gynecol Clin North Am* 42 (1):39–54.

Cemeli, E. and D. Anderson. 2011. Mechanistic investigation of ROS-induced DNA damage by oestrogenic compounds in lymphocytes and sperm using the comet assay. *Int J Mol Sci* 12 (5):2783–2796.

Cocuzza, M., S. C. Sikka, K. S. Athayde, and A. Agarwal. 2007. Clinical relevance of oxidative stress and sperm chromatin damage in male infertility: An evidence based analysis. *Int Braz J Urol* 33 (5):603–621.

Dada, R., M. B. Shamsi, S. Venkatesh, N. P. Gupta, and R. Kumar. 2010. Attenuation of oxidative stress & DNA damage in varicocelectomy: Implications in infertility management. *Indian J Med Res* 132:728–730.

Dalzell, L. H., C. M. McVicar, N. McClure, D. Lutton, and S. E. Lewis. 2004. Effects of short and long incubations on DNA fragmentation of testicular sperm. *Fertil Steril* 82 (5):1443–1445.

Darzynkiewicz, Z., F. Traganos, T. Sharpless, and M. R. Melamed. 1975. Thermal denaturation of DNA in situ as studied by acridine orange staining and automated cytofluorometry. *Exp Cell Res* 90 (2):411–428.

De Iuliis, G. N., L. K. Thomson, L. A. Mitchell, J. M. Finnie, A. J. Koppers, A. Hedges, B. Nixon, and R. J. Aitken. 2009. DNA damage in human spermatozoa is highly correlated with the efficiency of chromatin remodeling and the formation of 8-hydroxy-2′-deoxyguanosine, a marker of oxidative stress. *Biol Reprod* 81 (3):517–524.

Devaraj, S., S. Mathur, A. Basu, H. H. Aung, V. T. Vasu, S. Meyers, and I. Jialal. 2008. A dose-response study on the effects of purified lycopene supplementation on biomarkers of oxidative stress. *J Am Coll Nutr* 27 (2):267–273.

Donnelly, E. T., M. O'Connell, N. McClure, and S. E. Lewis. 2000. Differences in nuclear DNA fragmentation and mitochondrial integrity of semen and prepared human spermatozoa. *Hum Reprod* 15 (7):1552–1561.

Durairajanayagam, D., A. Agarwal, C. Ong, and P. Prashast. 2014. Lycopene and male infertility. *Asian J Androl* 16 (3):420–425.

Duran, E. H., M. Morshedi, S. Taylor, and S. Oehninger. 2002. Sperm DNA quality predicts intrauterine insemination outcome: A prospective cohort study. *Hum Reprod* 17 (12):3122–3128.

Duru, N. K., M. Morshedi, and S. Oehninger. 2000. Effects of hydrogen peroxide on DNA and plasma membrane integrity of human spermatozoa. *Fertil Steril* 74 (6):1200–1207.

Erenpreiss, J., M. Spano, J. Erenpreisa, M. Bungum, and A. Giwercman. 2006. Sperm chromatin structure and male fertility: Biological and clinical aspects. *Asian J Androl* 8 (1):11–29.

Fanaei, H., S. Khayat, I. Halvaei, V. Ramezani, Y. Azizi, A. Kasaeian, J. Mardaneh, M. R. Parvizi, and M. Akrami. 2014. Effects of ascorbic acid on sperm motility, viability, acrosome reaction and DNA integrity in teratozoospermic samples. *Iran J Reprod Med* 12 (2):103–110.

Farquhar, C., J. R. Rishworth, J. Brown, W. L. Nelen, and J. Marjoribanks. 2014. Assisted reproductive technology: An overview of Cochrane reviews. *Cochrane Database Syst Rev* 12:CD010537.

Fernandez, J. L., L. Muriel, M. T. Rivero, V. Goyanes, R. Vazquez, and J. G. Alvarez. 2003. The sperm chromatin dispersion test: A simple method for the determination of sperm DNA fragmentation. *J Androl* 24 (1):59–66.

Fraga, C. G., P. A. Motchnik, A. J. Wyrobek, D. M. Rempel, and B. N. Ames. 1996. Smoking and low antioxidant levels increase oxidative damage to sperm DNA. *Mutat Res* 351 (2):199–203.

Gavella, M. and V. Lipovac. 2013. Protective effects of exogenous gangliosides on ROS-induced changes in human spermatozoa. *Asian J Androl* 15 (3):375–381.

Gonzalez, C., M. Boada, M. Devesa, and A. Veiga. 2012. Concise review: Fertility preservation: An update. *Stem Cells Transl Med* 1 (9):668–672.

Gorczyca, W., J. Gong, and Z. Darzynkiewicz. 1993. Detection of DNA strand breaks in individual apoptotic cells by the in situ terminal deoxynucleotidyl transferase and nick translation assays. *Cancer Res* 53 (8):1945–1951.

Hansen, M., C. Bower, E. Milne, N. de Klerk, and J. J. Kurinczuk. 2005. Assisted reproductive technologies and the risk of birth defects—A systematic review. *Hum Reprod* 20 (2):328–338.

Hughes, C. M., S. E. Lewis, V. J. McKelvey-Martin, and W. Thompson. 1996. A comparison of baseline and induced DNA damage in human spermatozoa from fertile and infertile men, using a modified comet assay. *Mol Hum Reprod* 2 (8):613–619.

Kodama, H., R. Yamaguchi, J. Fukuda, H. Kasai, and T. Tanaka. 1997. Increased oxidative deoxyribonucleic acid damage in the spermatozoa of infertile male patients. *Fertil Steril* 68 (3):519–524.

Lewis, S. E., R. John Aitken, S. J. Conner, G. D. Iuliis, D. P. Evenson, R. Henkel, A. Giwercman, and P. Gharagozloo. 2013. The impact of sperm DNA damage in assisted conception and beyond: Recent advances in diagnosis and treatment. *Reprod Biomed Online* 27 (4):325–337.

Lozano, G. M., I. Bejarano, J. Espino, D. Gonzalez, A. Ortiz, J. F. Garcia, A. B. Rodriguez, and J. A. Pariente. 2009. Relationship between caspase activity and apoptotic markers in human sperm in response to hydrogen peroxide and progesterone. *J Reprod Dev* 55 (6):615–621.

Mahfouz, R., R. Sharma, A. Thiyagarajan, V. Kale, S. Gupta, E. Sabanegh, and A. Agarwal. 2010. Semen characteristics and sperm DNA fragmentation in infertile men with low and high levels of seminal reactive oxygen species. *Fertil Steril* 94 (6):2141–2146.

Manicardi, G. C., P. G. Bianchi, S. Pantano, P. Azzoni, D. Bizzaro, U. Bianchi, and D. Sakkas. 1995. Presence of endogenous nicks in DNA of ejaculated human spermatozoa and its relationship to chromomycin A3 accessibility. *Biol Reprod* 52 (4):864–867.

Manicardi, G. C., D. Bizzaro, M. Mandrioli, and U. Bianchi. 1998. Silver staining as a new banding technique to identify aphid chromosomes. *Chromosome Res* 6 (1):55–57.

Mayorga-Torres, B. J., W. Cardona-Maya, A. Cadavid, and M. Camargo. 2013. [Evaluation of sperm functional parameters in normozoospermic infertile individuals]. *Actas Urol Esp* 37 (4):221–227.

McKelvey-Martin, V. J., N. Melia, I. K. Walsh, S. R. Johnston, C. M. Hughes, S. E. Lewis, and W. Thompson. 1997. Two potential clinical applications of the alkaline single-cell gel electrophoresis assay: (1). Human bladder washings and transitional cell carcinoma of the bladder; and (2). Human sperm and male infertility. *Mutat Res* 375 (2):93–104.

Menezo, Y. J., A. Hazout, G. Panteix, F. Robert, J. Rollet, P. Cohen-Bacrie, F. Chapuis, P. Clement, and M. Benkhalifa. 2007. Antioxidants to reduce sperm DNA fragmentation: An unexpected adverse effect. *Reprod Biomed Online* 14 (4):418–421.

Moustafa, M. H., R. K. Sharma, J. Thornton, E. Mascha, M. A. Abdel-Hafez, A. J. Thomas, Jr., and A. Agarwal. 2004. Relationship between ROS production, apoptosis and DNA denaturation in spermatozoa from patients examined for infertility. *Hum Reprod* 19 (1):129–138.

Muratori, M., M. Maggi, S. Spinelli, E. Filimberti, G. Forti, and E. Baldi. 2003. Spontaneous DNA fragmentation in swim-up selected human spermatozoa during long term incubation. *J Androl* 24 (2):253–262.

Pant, N., M. Shukla, A. D. Upadhyay, P. K. Chaturvedi, D. K. Saxena, and Y. K. Gupta. 2014. Association between environmental exposure to p, p'-DDE and lindane and semen quality. *Environ Sci Pollut Res Int* 21 (18):11009–11016.

Practice Committee of the American Society for Reproductive Medicine. 2013. The clinical utility of sperm DNA integrity testing: A guideline. *Fertil Steril* 99 (3):673–677.

Rago, R., P. Salacone, L. Caponecchia, A. Sebastianelli, I. Marcucci, A. E. Calogero, R. Condorelli et al. 2013. The semen quality of the mobile phone users. *J Endocrinol Invest* 36 (11):970–974.

Rathke, C., W. M. Baarends, S. Awe, and R. Renkawitz-Pohl. 2014. Chromatin dynamics during spermiogenesis. *Biochim Biophys Acta* 1839 (3):155–168.

Rousseaux, S., N. Reynoird, E. Escoffier, J. Thevenon, C. Caron, and S. Khochbin. 2008. Epigenetic reprogramming of the male genome during gametogenesis and in the zygote. *Reprod Biomed Online* 16 (4):492–503.

Sakkas, D. and J. G. Alvarez. 2010. Sperm DNA fragmentation: Mechanisms of origin, impact on reproductive outcome, and analysis. *Fertil Steril* 93 (4):1027–1036.

Saleh, R. A., A. Agarwal, R. K. Sharma, T. M. Said, S. C. Sikka, and A. J. Thomas, Jr. 2003. Evaluation of nuclear DNA damage in spermatozoa from infertile men with varicocele. *Fertil Steril* 80 (6):1431–1436.

Smith, R., H. Kaune, D. Parodi, M. Madariaga, R. Rios, I. Morales, and A. Castro. 2006. Increased sperm DNA damage in patients with varicocele: Relationship with seminal oxidative stress. *Hum Reprod* 21 (4):986–993.

Taha, E. A., A. M. Ezz-Aldin, S. K. Sayed, N. M. Ghandour, and T. Mostafa. 2014. Smoking influence on sperm vitality, DNA fragmentation, reactive oxygen species and zinc in oligoasthenoteratozoospermic men with varicocele. *Andrologia* 46 (6):687–691.

Tejada, R. I., J. C. Mitchell, A. Norman, J. J. Marik, and S. Friedman. 1984. A test for the practical evaluation of male fertility by acridine orange (AO) fluorescence. *Fertil Steril* 42 (1):87–91.

Thomson, L. K., S. D. Fleming, R. J. Aitken, G. N. De Iuliis, J. A. Zieschang, and A. M. Clark. 2009. Cryopreservation-induced human sperm DNA damage is predominantly mediated by oxidative stress rather than apoptosis. *Hum Reprod* 24 (9):2061–2070.

Tominaga, H., S. Kodama, N. Matsuda, K. Suzuki, and M. Watanabe. 2004. Involvement of reactive oxygen species (ROS) in the induction of genetic instability by radiation. *J Radiat Res* 45 (2):181–188.

Treulen, F., P. Uribe, R. Boguen, and J. V. Villegas. 2015. Mitochondrial permeability transition increases reactive oxygen species production and induces DNA fragmentation in human spermatozoa. *Hum Reprod* 30 (4):767–776.

Tunc, O. and K. Tremellen. 2009. Oxidative DNA damage impairs global sperm DNA methylation in infertile men. *J Assist Reprod Genet* 26 (9–10):537–544.

Twigg, J. P., D. S. Irvine, and R. J. Aitken. 1998. Oxidative damage to DNA in human spermatozoa does not preclude pronucleus formation at intracytoplasmic sperm injection. *Hum Reprod* 13 (7):1864–1871.

Wallace, D. C. and W. Fan. 2009. The pathophysiology of mitochondrial disease as modeled in the mouse. *Genes Dev* 23 (15):1714–1736.

Wen, J., J. Jiang, C. Ding, J. Dai, Y. Liu, Y. Xia, J. Liu, and Z. Hu. 2012. Birth defects in children conceived by in vitro fertilization and intracytoplasmic sperm injection: A meta-analysis. *Fertil Steril* 97 (6):1331–1337 e1–e4.

Wright, C., S. Milne, and H. Leeson. 2014. Sperm DNA damage caused by oxidative stress: Modifiable clinical, lifestyle and nutritional factors in male infertility. *Reprod Biomed Online* 28 (6):684–703.

33 Role of Reactive Oxygen Species in Skin Aging and Its Prevention

Vibha Rani

CONTENTS

ABSTRACT

Skin aging is triggered by many intrinsic and extrinsic factors. These factors cause the increased production of reactive oxygen species (ROS) leading to oxidative stress due to imbalance between ROS production and the action of several enzymatic and nonenzymatic antioxidants present in the tissue. Oxidative stress plays a central role in aging by activating other factors responsible for aging—matrix metalloproteinases (MMPs) and inflammatory responses that further damage skin cells. To counteract this, there is a need for identifying and supplementing external, alternative sources of antioxidants. Synthetic antioxidants sometimes have harmful side effects, hence natural antioxidants are preferred. This chapter describes and discusses various factors causing skin aging along with the natural antiaging strategies widely used by the skin care industry.

33.1 INTRODUCTION

The accumulation of damages to cells is responsible for sequential alterations that lead to aging and associated progressive increase in the chance of disease development. Skin, the largest organ in the body, performs a wide range of functions, including protection, sensation, heat regulation, control of evaporation, storage, synthesis of various molecules, absorption, and water resistance. It is one of the most important parts of the human body as it regularly interfaces with the environment and is the first line of defense from external, harmful factors. Skin is exposed to the external environment and is prone to a wide range of harmful compounds leading to damage that causes wrinkles, dark patches, dry and sagging skin, and skin cancers [1].

33.2   SKIN: ANATOMY AND FUNCTION

Human skin has two major components separated by the basement membrane. The outer membrane or outermost layer is called epidermis while the inner layer is called dermis. Between these two layers is the hypodermis. The topmost layer of the epidermis is called the stratum corneum, which is composed of overlapping cells that create a network of protection from the environment. The main cell type of the epidermis is the keratinocyte, which proliferates by division in the basal layer lying above the basement membrane [2]. In addition to keratinocytes, other cell types in the epidermis include melanocytes and Langerhans cells (LCs). Most of the melanocytes are distributed in the basal layer, synthesizing and transferring melanin to adjacent keratinocytes, giving skin color and photoprotection. LCs are antigen-presenting cells (APCs) in the epidermis, and their long dendritic structures comprise the first line of immunological barriers. Once activated by endogenous or exogenous antigens, they can migrate to the skin and then pass into the lymph nodes and thereafter activate T cells [3]. T cells then traffic back to the skin to trigger an immune response against the antigens. The epidermis is replaced approximately every 28 days as older cells shed from the surface and newer cells move up from the lower-level basal cell layer. As basal cells divide and move upward from the basal cell layer toward the surface of the skin, they become gradually flatter and harder. After that, they move towards the upper layer of the epidermis called stratum corneum, their nucleus begins to degenerate. Under the programmed activities of specific enzymes, these cells lose their cohesion, separate from the surface and then shed [4].

Apart from these cells, the epidermis also contains several chromophores. Chromophores are molecules that absorb light rays that fall on the skin cells. Examples include urocanic acid, aromatic amino acids in proteins, and melanins and their precursors. Chromophores in solution have very characteristic absorption spectra, and the energy from such absorption may enable a chromophore to undergo chemical interactions within itself or with other molecules in its immediate vicinity. Thus, a chromophore may be a target biomolecule, such as nucleic acids, the structural alteration of which, as a result of ultraviolet ray (UVR) absorption, triggers a biological response. The second layer of dermis includes blood vessels, lymph vessels, and nerves. The dermis has two sections: an upper layer called the papillary dermis and a lower reticular dermis. The dermis is composed primarily of collagen and elastin. It is also rich in collagen fibers (mainly types I and III). Also, the extracellular matrix (ECM) found in the dermis plays a major role in skin aging.

Aging can be caused by exposure to UV rays from the sun, environmental pollutants like cigarette smoke, ionizing radiation, alcohol intake, poor nutrition, overeating, stress, lack of sleep, infections, and reduced or altered levels of some of the hormones that change with the age [5,6].

Two types of skin aging exist: (1) "intrinsic aging," which is mainly genetically determined and includes factors such as telomerase activity and hormonal expression, etc., and is similar to aging of other organs, and (2) "extrinsic aging," which is caused by environmental stress such as UV rays, pollutants, etc. [7]. Although the factors inducing aging are classified as *intrinsic* and *extrinsic*, their pathways and responses are mostly common. The visible signs of aging are mostly fine wrinkles, loss of elasticity, spots, and reduced epidermal and dermal thickness. Physiological changes include epidermal atrophy, enhanced number of dead cells in the keratinocytes due to its decreased mitotic rate, decreased proliferative capacity, and cellular senescence (Figure 33.1). In this chapter, we compile mechanisms of skin aging and the strategies to reduce the aging process.

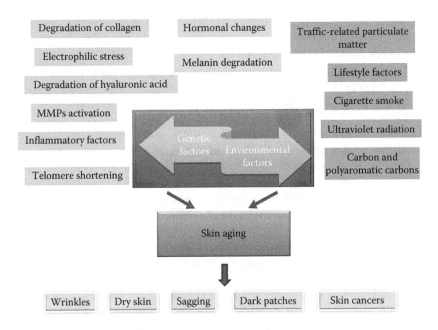

FIGURE 33.1 **(See color insert.)** Factors affecting skin aging.

33.3 FACTORS AFFECTING SKIN AGING

33.3.1 EXTRINSIC FACTORS

Solar UVR plays the most important role in skin aging via the production of reactive oxygen species (ROS) in the skin cells. Solar UVR has been classified as follows: UVA (320–400 nm), UVB (280–320 nm), and UVC (200–280 nm). As UVC is primarily absorbed by the ozone layer it cannot reach the earth, while some UVB and all UVA can penetrate the ozone layer and cause erythema (sunburn), skin cancer, and photoaging. Skin wrinkling and sagging are mostly caused by UVA as they can penetrate deeper into the skin layer.

DNA in the keratinocytes is mainly affected by UVB rays where two main photochemical reactions occur, the formation of cyclobutane pyrimidine dimers (CPDs) and (6–4) pyrimidine–pyrimidone photoproducts [(6–4)PPs]. DNA damage leads to the activation of an intrinsic pathway that includes cytochrome c release from the mitochondria. The p53 is thus released, causing permeation of the mitochondrial outer membrane, which then brings about the leakage of cytochrome c into the cytosol that further triggers a caspase cascade, leading to the apoptosis of the skin cells—a reason for skin pigmentation. Exposure to UVR also leads to the activation of chromophores in the epidermis. *Trans-urocanic* acid (UCA), a chromophore present at high concentrations in the stratum corneum, also absorbs UVB, and this isomerizes *trans*-UCA to *cis*-UCA, which produces the ROS in keratinocytes, leading to oxidative damage, another major reason for skin aging.

The skin morphology differs depending on whether it is photoaged or chronologically aged. Photoaged skin has more wrinkles, spots, pigmentation, and a leathery appearance, while chronologically aged skin is thin, sustains reduced elasticity, has less spots, and remains somewhat smooth [8,9].

In addition to its role in ROS production, leading to oxidative stress, UVR causes collagen fragmentation and depletion of hyaluronic acid (HA). It also plays an important part in the activation of MMPs that degrade the ECM [9]. MMPs production by UV is linked with ROS production that triggers the signaling pathway, activating the transcription factor AP-1, which leads to a decreased synthesis of collagen and increased expression of MMPs [10]. MMPs are a family of Zn^{++}-dependent endopeptidases that degrade the ECM by cleaving most of its constituents. Type I collagen is the major constituent of dermal ECM with lesser amounts of type III collagen, elastin, proteoglycans, and fibronectin. Collagen fibers are responsible for strengthening and providing elasticity to the skin, as well as for protecting its morphology and structure from damage. It has been found that MMPs like collagenase, 92-kd gelatinase, and stromelysin-1 can jointly degrade skin collagen in the dermis [11].

Other factors that cause premature skin aging are the pollutants present in our environment, including cigarette smoke, traffic-related particulate matters, carbon and polyaromatic carbons (PAC), etc. These small particles carry with them some metals, organic chemicals, and other compounds that can enter the cell and initiate the signaling cascade triggering aging. These nanoparticles can penetrate the skin through the pores and can activate the signaling pathways that cause the degradation of the ECM and hence affect the elasticity of the skin; they are also responsible for age spots [12]. Other compounds causing premature aging are polychlorinated benzene (PCB) used in transformers, plasticizer in paints, cements, etc. These compounds cause skin aging via DNA damage, oxidative stress, and telomere shortening by the inactivation of the enzyme telomerase [13].

Cigarette smoke affects not just smokers but also the people around, passive smokers. Nicotine, alkaloid, and a volatile gas in cigarette smoke cause unfavorable skin changes that enhance the aging process [14]. When compared, the skin of a 40-year-old smoker looks like that of a 70-year-old nonsmoker. Nicotine in the smoke gets absorbed by the skin and the mucous membrane in dose-dependent manner, which damage the skin irreversibly by affecting both directly on the epidermis and indirectly through the bloodstream. It has been found that due to exposure to smoke, squalene, a biochemical precursor of steroids, gets converted to squalene monohydroperoxide and also that there is a dose-dependent increase in other hydroperoxides derived from the smoke. This further leads to lipid peroxidation [15]. The acceleration of the production of ROS by smoke leads to damages to the repair and control functions of cells. These compounds in the smoke further upregulate the synthesis and expression of MMPs that further damage the ECM [16].

33.3.2 Intrinsic Factors

The intrinsic factors and conditions in the body are responsible for a person's appearance. These factors include hormonal changes, electrophilic stress due to the different metabolic pathways occurring inside the body, and the genetic makeup. All these factors contribute toward the aging process. They result in many biochemical changes in the body like the degradation of HA, which is responsible for the maintenance of the moisture content of the skin and collagen degradation by the MMPs activated by ROS. The ongoing normal process of telomere shortening is also accelerated by these factors.

Oxidative stress is one of the major intrinsic factors for skin aging. ROS often include not only free radicals but also nonradicals such as singlet oxygen ($^{1}O_2$), peroxynitrite ($ONOO^-$), hydrogen peroxide (H_2O_2), and ozone (O_3). A free radical is a chemical species possessing an unpaired electron. Free radicals include hydroxyl ($OH^•$), superoxide ($O_2^{•-}$), nitric oxide ($NO^•$), thyl ($RS^•$), and peroxyl ($RO_2^•$). When in excess, these radicals can cause many health problems, including aging. To combat them, antioxidants are present in the body, such as vitamin E isoforms, vitamin C, glutathione (GSH), uric acid, ubiquinol, and antioxidant enzymes such as glutathione peroxidase, superoxide dismutase, and catalase. These antioxidants have the ability to scavenge these radicals. Other potent antioxidants include ascorbate, carotenoids, and sulfhydryls.

Oxidative stress refers to the imbalance between increased ROS production and decreased antioxidant activity. Also produced in our bodies are reactive nitrogen species (RNS) [17]. Nitric oxide (an RNS) contains one unpaired electron on the antibonding orbital and is, therefore, a radical. $NO^•$ is generated by the enzyme nitric oxide synthases (NOSs), which metabolize arginine to citrulline forming $NO^•$ [18]. $NO^•$ has a half-life of only a few seconds in an aqueous environment but has greater stability in an environment with a lower oxygen concentration. Nitric oxide (NO) has also been reported to be produced by melanocytes and keratinocytes in response to pro-inflammatory cytokines [19]. NO, in response to UVB, increases blood flow in human skin and contributes, to erythema involving melanocytes [19]. Overproduction of RNS is called nitrosative stress [20]. This stress further leads to nitrosylation reactions that damage the proteins.

Superoxide dismutase is located in the mitochondrial matrix where it forms the first line of antioxidant defense against $O_2^{-\bullet}$ produced as by-products of oxidative phosphorylation. These ROS are short-lived and cause damage to DNA, proteins, and lipids and also react with other reactive oxygen/nitrogen species like nitric oxide to form highly reactive peroxynitrite. An overexpression of SOD results in enhanced H_2O_2 concentration, which leads to the activation of distinct signaling pathways and transcription factors, among them is the heterodimeric AP-1, which induces interstitial collagenase (MMP-1) and degradation of interstitial collagen, leading to skin aging.

Another harmful ROS, H_2O_2, is scavenged by catalase, which is localized in peroxisomes. The reaction generates water and oxygen. Glutathione peroxidase (GPx) is another oxidant for H_2O_2, which catalyzes the reduction of H_2O_2, and lipid hydroperoxide generated during lipid peroxidation to water. Glutathione is found abundantly in the cytosol (1–11 mM), nuclei (3–15 mM), and mitochondria (5–11 mM) and is the major soluble antioxidant in these organelles [21]. Glutathione's (GSH) main function in the nucleus is to maintain the redox state of critical protein sulfhydryl, which is necessary for DNA repair and expression. If glutathione disulfide (GSSG) is imbalanced and the ratio of GSH/GSSG is in favor of GSSG, then it can lead to oxidative stress [22]. During the aging process, the free radical–induced damage in DNA, especially 8OH-dG, accumulates in the nuclear and mitochondrial DNA. Also, repair activities appear to decline with age.

Antioxidant status does not change significantly with age; human studies have shown that the levels of SOD, GPx, catalase, and ceruloplasmin among the age groups of 35–39, 50–54, and 65–69 years did not change [23]. ROS production occurring, both, via extrinsic or intrinsic pathways of skin aging results in the activation of the transcription factor c-Jun via mitogen-activated protein kinases (MAPK), leading to the overexpression of the matrix metalloproteinase MMP-1, MMP-3, and MMP-9, thereby preventing the expression of procollagen-1. This decreases the collagen level, which is a characteristic of the aged skin [24].

Electrophilic stress also contributes to skin aging. Most biological macromolecules react with each other inside the body and form compounds that are either harmful or useful to the body. These molecules are either nucleophile or electrophile [25]. Electrophilic compounds (or electrophilic centers) are electron-deficient compounds and they accept electrons to attain stability. The reactivity of an electrophile depends not only on its own structure but also on that of its electron donor partner, that is, the nucleophile [25].

Two major types of pathways lead to aging by electrophiles: (1) pro-aging factors that can cause toxicity and lead to destabilization of biological systems, which further leads to a gradual decrease in homeostasis and eventually to cellular death, and (2) change in signaling pathways that could weaken cell survival mechanisms or, perhaps, trigger pro-aging damage via different processes [18]. For example, TOR signaling is essential during growth but, if overexpressed, it may cause inappropriate growth/proliferation, which would further lead to loss of homeostasis and ultimately to aging [26,27]. Majority of cellular electrophiles that are generated in the human body are from polyunsaturated fatty acids (PUFAs) that are formed by a peroxidation chain reaction triggered by free radicals and spread without involving ROS. One such example is the formation of lipid-derived electrophiles such as 4-hydroxynon-2-enal (4-HNE). It is produced in the membrane by reactions of PUFAs present there, and concentration could exceed 300 µM [28]. Thus, 4-HNE not only forms within or near membranes but also appears to concentrate in such places. Prolonged exposure of integral membrane proteins and amino-phospholipids to high local 4-HNE concentrations could cause untargeted damage and accelerate aging [27]. This electrophile is also responsible for (1) destabilizing biological systems and (2) modulating signaling pathways that control longevity assurance mechanisms.

Certain hormones, such as sex hormones, are other crucial factors that have the ability to activate the inflammatory pathways that lead to MMP activation and also contribute to the structural changes observed in aged skin [29]. Changes also include collagen degradation, dryness, loss of elasticity, epidermal atrophy, and wrinkling [30].

Extracellular matrix components are important players in skin aging. Hyaluronic acid (HA), a high-molecular-weight polysaccharide, having a net negative charge, is present in the dermis responsible for the moisture retention in the skin and acts as a lubricant and shock absorber to maintain skin morphology; with age, its level and its physiological functioning decreases, which decreases its ability to bind water, making the skin rough with decreased elasticity, leading to wrinkle formation [31]. The HA content of the dermis is significantly higher than that of the epidermis. During the aging process, the level of this epidermal HA decreases. This decrease in the HA content is also attributed to extrinsic factors such as UVR that causes the accumulation and abnormal localization of the glycosaminoglycan (GAG), a chief component of the dermis [32], leading to the reduction in HA. The decrease in HA results in the cross-linking of collagen and hence the loss of extractability. Thus, it causes dehydration, atrophy, and loss of elasticity of the skin [33]. Collagen degradation due to UVR exposure also decreases HA synthesis by blocking the activity of hyaluronic acid synthases (HAS) [34].

The primary component of the dermis and the most abundant protein in the human body, collagen, is also responsible for providing strength, rigidity, and support to human skin, but as with age, it starts to deteriorate and results in natural aging signs like wrinkles and loosening of skin. Of the dry skin mass, 70% is composed of collagen [35]. In aged skin, collagen is categorized by thickened fibrils, organized in rope-like bundles that appear to be in disorder in comparison to the pattern observed in younger skin; also, the ratio varies between collagen types in skin changes with aging [36]. In young skin, collagen I comprises 80% and collagen III comprises about 15% of total skin collagen but in older skin, the ratio of Type III to Type I collagen has been shown to increase due to loss of collagen I [37]. In skin affected by UVR or other radiations, collagen I levels have been shown to be reduced by 59% [38]. Yet another type of collagen present between the dermal/epidermal layers is collagen IV, which imparts a structural framework for other molecules and plays a key role in maintaining mechanical stability. This collagen type is not much affected by radiations but plays an important role in wrinkle formation. The mechanical stability of the junction between dermal and epidermal layer may be adversely affected by the loss of collagen IV, resulting in wrinkle formation [39]. Collagen VII is the primary constituent in anchoring fibrils that attach the basement membrane zone to the underlying papillary dermis. This collagen is also affected by radiations. Photodamaged skin has a significantly lower number of fibrils as compared to normal skin [40].

Collagens are degraded by enzymes called MMPs. Exposure of skin to radiations upregulates the synthesis of these enzymes further leading to progressive degradation of collagen. The upregulated activity of certain transcription factors such as c-jun and c-fos is reported to be responsible for the increased MMP activity. UVR exposure leads to an increase in the amount of transcription factors. Activator protein-1 (AP-1) is then formed by the combination of c-jun and c-fos. In turn, AP-1 activates the MMP genes, which stimulate the production of collagenase, gelatinase, and stromelysin (types of MMPs). Collagen degradation is mediated by AP-1 activation and by the inhibition of transforming growth factor-β (TGF-β) signaling pathway [35]. TGF-β is involved in the formation of collagen by breaking down the pro-collagen form into collagen when required. As collagenase attacks and degrades collagen, long-term elevations in the levels of collagenase and other MMPs result in the disorganized and clumped collagen identified in photoaged skin.

Elastin is also a part of ECM in the skin that is present in the dermal layer. Its main function is to provide flexibility to the skin. Alterations in elastic fibers are also associated with UVR radiations, and "elastosis," an accumulation of amorphous elastin material, is considered to be caused by exposure to radiations of photoaged skin. UVR exposure induces a thickening and coiling of elastic fibers in the papillary dermis [41]. In addition, small amounts of sugar (GAG, hyaluronic acid), lipids, and abnormally high levels of polar amino acids (aspartic acid, threonine, serine, glutamic acid, lysine, histidine) have been found in elastin obtained from elderly or aged skin [42]. The mechanism behind age-related changes in elastin is not yet well worked out, but MMPs are thought to play a role in this process. It is also responsible for elastin degradation [43]. Older skin experiences a degenerative reaction characterized by changes in the normal pattern of immature elastic fibers, called oxytalan, that are located in the papillary dermis. These fibers form a network in young skin that starts perpendicularly from the uppermost section of the papillary dermis to just beneath the basement membrane. This network gradually disappears with age [44]. Consequently, skin elasticity is also gradually lost with age [45]. The sagging skin observed in older people is due to loss of elasticity; UV radiations accelerate these mechanisms leading to premature aging with saggy skin, loss in elasticity, and compact elastin networks.

Melanin degradation induced by various factors also contributes to premature aging. Skin color is mainly determined by the combined effects of carotenoids, oxy/deoxy-hemoglobin and, most importantly, by different types of melanin and also the way that melanin is packaged and distributed in melanosomes [46].

The synthesis of melanin takes place in specific ovoid organelles known as melanosomes, which are produced in dendritic melanocytes present in the epidermal layer. Melanin synthesized within melanosomes is transported via dendrites to adjacent keratinocytes, and it accumulates within keratinocytes and melanocytes that shield DNA from UV light [47]. Melanin is a complex of lighter red/yellow, alkali soluble sulfur containing pheomelanin and darker brown/black insoluble eumelanin [48,49]. Differences in skin pigmentation do not result from differences in the number of melanocytes in the skin, but from differences in the melanogenic activity, the type of melanin produced, and the size, number, and packaging of melanosomes [50,51]. Melanin provides protection against UV-induced photodamage. However, melanin can also have toxic properties, especially after exposure to UVR [52]. *In vitro* studies have shown that melanin reacts with DNA and acts as a photosensitizer that produces ROS after UVA exposure, resulting in single-strand DNA breaks in skin cells. Pheomelanin is especially prone to photodegradation and is thought to contribute to the damaging effects of UVR [53–56]. It can generate H_2O_2 and $O_2^{\cdot-}$ and might cause mutations in melanocytes or other cells. In addition, pheomelanin has been associated with higher rates of apoptotic cell death after UV radiation [10,57].

The most common reason for the aging of every multicellular organism is the telomere shortening. Telomeres are the repetitive nucleotide sequences present at the end of the chromatids. Intrinsic aging occurs mostly as a result of telomere shortening, since it cannot be replicated fully leading to a loss of 100–200 base pairs with every replication and division, thus limiting the number of divisions that a cell can undergo [58,59]. Thus, after a certain number of divisions, the cell stops dividing and becomes aged and undergoes senescence. Telomerase plays an important role in overcoming this problem. Telomerase is an enzyme that adds telomeric repeats at the end, thus saving telomeres from getting shorter. UV radiations trigger telomerase activity in the epidermis, which further leads to skin aging and cancer. Also, the ROS, produced in the cell from UV radiation exposure, cause genetic alterations leading to mutations and/or senescence in the telomeres by converting guanine into 8-oxoguanine [60].

Genetic regulation is another important factor responsible for skin aging. As the skin undergoes aging, there is a decrease in lipid synthesis. Some genes, such as those involved in fatty acid synthesis, collagen synthesis, cholesterol synthesis, and those for epidermal differentiation, get downregulated with age while there is an upregulation of the elastin gene expression [61]. Aging is further induced by the disturbed lipid metabolism, changes in insulin and STAT3 signaling, upregulation of apoptotic pathways, and downregulation of the AP-1 family. There is also deregulation of cytoskeletal proteins such as keratin 2A, 6A, and 16A that are responsible for maintaining the structure of skin cells, extracellular matrix components (e.g., PI3, S100A2, A7, A9, SPRR2B), and deregulation of the cell cycle by downregulation of the genes responsible for the control of the cell cycle (e.g., CDKs, GOS2) [60,61]. The expression and downregulation of these genes are the major intrinsic factors for skin aging.

Inflammatory factors act as catalysts that aggravate skin aging. Some factors, directly or indirectly, induce the micro-inflammatory cycle leading to the expression of intercellular adhesion molecule-1 (ICAM-1) in endothelial cells [5]. ICAM1 is an endothelial and leukocyte-associated protein present at a very low concentration in normal conditions, but it increases upon cytokine stimulation. It plays an important role in the pro-inflammatory effect and response. It recruits immune cells that digest the ECM by secreting enzymes like collagenases, myeloperoxidases, and ROS. In this process, neighboring cells also get damaged, thus secreting prostaglandins and leukotrienes. This signals the induction of the degranulation of resident mast cells, resulting in the release of histamine and the cytokine TNF-α and, thus, activating endothelial cells lining adjacent capillaries that release P-selectin and synthesize ICAM-1. This further results in the accumulation of ECM damage and hence skin aging. UV-induced DNA breaks in LCs induce suppressive cytokines such as IL-10, resulting in the inhibition of contact hypersensitivity responses [62].

33.4 ANTIAGING: ANTIOXIDATIVE STRATEGIES AGAINST SKIN AGING

Evidences exist that free radicals and ROS are involved in a number of diseases including skin aging [3]. Though the body's defense mechanisms are present to overcome stress, due to exposure to some extrinsic factors like UV rays, the defense mechanisms often fall short of their full activities. Hence, there is a need to explore possible antiaging solutions.

Antioxidants are chemical compounds that can provide protection from endogenous and exogenous oxidative stresses by blocking or delaying the activity of free radicals by inhibiting either the initiation or propagation of oxidizing chain reactions. They scavenge and neutralize free radicals ROS/RNS rather than eliminate them [63]. Antioxidants such as vitamins (vitamin C, vitamin E), flavonoids, and phenolic acids play the main role in fighting against free radicals that are the main cause of numerous skin changes.

Antioxidants may act in four ways:

1. Break chain reaction resulting in reduction in the levels of reactive radicals
2. Chelate/deactivate metals—rendering radicals inactive
3. Scavenge singlet oxygen (highly toxic)
4. Scavenge ROS

33.4.1 CLASSIFICATION OF ANTIOXIDANTS

Mammalian cells possess a number of defense mechanisms [64].

1. *Endogenous antioxidants*: Generated naturally
   a. *Enzymatic*: SOD, CAT, and GPx, which can destroy toxic peroxides. There are five major families of internal antioxidant enzymes found in most organisms [65,66]:
      i. *Superoxide dismutases* (SODs): A family of closely related enzymes that catalyze the conversion of $O_2^{\cdot-}$ into oxygen and H_2O_2.
      ii. *Catalase*: Catalyzes the conversion of H_2O_2 to water and oxygen; also oxidizes toxins such as formaldehyde, formic acid, and alcohols. Catalase becomes activated when the glutathione peroxide pathway approaches saturation.
      iii. *Glutathione system*: Consists of glutathione, glutathione reductase, and glutathione peroxidase. Glutathione peroxidase catalyzes the breakdown of H_2O_2 to water and reduces lipid peroxides. The role of glutathione reductase is to reduce oxidized glutathione (glutathione disulfide).
      iv. *Thioredoxin system*: Consists of thioredoxin and thioredoxin reductase. Thioredoxin acts as a proficient reducing agent by scavenging ROS and maintaining other food proteins, allergenic proteins, albumin proteins, etc., in a reduced state.
      v. *Peroxiredoxins*: Catalyze the reduction of peroxides.

b. *Nonenzymatic molecules*: These *include* thioredoxin, thiols, and disulfide bonding also plays important roles in antioxidant defense systems.
   i. *Metabolic antioxidants*: These are produced by metabolism in the body, for example, lipoid acid, glutathione, L-arginine, melatonin, uric acid, bilirubin, and metal-chelating proteins.
   ii. *Nutrient antioxidants*: Compounds that cannot be produced in the body and must be provided through foods or supplements, for example, vitamins E and C, carotenoids, trace metals (selenium, zinc), flavonoids, and omega-3 and omega-6 fatty acids.
2. *Exogenous antioxidants*: Externally supplied through foods such as α-tocopherol, β-carotene, ascorbic acid, and micronutrient elements such as zinc and selenium.

There is a second system of classification into two basic categories: synthetic and natural.

1. *Synthetic antioxidants*: Compounds with phenolic structures consisting of different degrees of alkyl substitution
2. *Natural antioxidants*: Naturally occurring compounds found in many animals and plants. They can be
   Phenolic compounds: Tocopherols—Vitamin E, flavonoids, nonflavonoid phenolic, polyphenols, and phenolic acids)
   Nitrogen compounds: Alkaloids, chlorophyll derivatives, amino acids, and amines
   Carotenoids and terpenoids: Found in deep orange and dark green vegetables, β-carotene, and other carotenoids, such as γ-carotene, α-carotene, and β-criptoxanthin, which are potent antioxidants of plant origins
   Ascorbic acid: Found in citrus and other fruits and vegetables
3. *Synthetic antioxidants*: Such as butylated hydroxyanisole (BHA), propyl gallate (PG), tertiary butyl hydroquinone (TBHQ), and butylated hydroxytoluene (BHT) have been used as antioxidants since the beginning of this century; yet, restrictions have been imposed on the use of these compounds because of their carcinogenicity. Thus, the interest in natural antioxidants has increased considerably [67].

33.4.2 PLANTS AS A SOURCE OF ANTIOXIDANTS: PHYTOCHEMICALS

Antioxidants are oxidized during the process of destroying free radicals. Therefore, antioxidant resources must be continuously replenished in the body. Hence depending on the frequency of use, in one system, an antioxidant might be very effective against free radicals but it could become ineffective in other systems [68]. Hence, there is a need to turn toward exogenous, natural sources of antioxidants.

Globally, conventional medicines are now being revaluated by carrying out extensive research on different plant species and their therapeutic principles. Plants, herbs, and herbal preparations have high potentials due to their antioxidant activities. As plants are constantly exposed to sunlight, which causes oxidative stress in them, they produce a lot of antioxidants to counter this stress. Hence, they can represent a source of new compounds with antioxidant activities. Many plants have now been discovered that contain a wide variety of compounds that have antioxidant activities, including polyphenols such as flavonoids and tocopherols. Phenolic compounds in plants are widely accepted to confer resistance against oxidation. Secondary metabolites and whole extracts from plants have also been widely investigated and some have been found to have anticollagenase and anti-elastase activities, which can contribute to antiaging (Table 33.1) [69].

Although isolated plant compounds have high potential to protect skin from being damaged, whole herb extracts and herbal preparations also showed higher potentials due to their complex

TABLE 33.1

Some Common Plants with Well-Defined Antioxidant and Antiaging Potential

S. No.	Name of Plant	Part Used	Major Chemical Constituent
1.	*Aloe barbadensis* (Aloe vera)	Gel	Acemannan
2.	*Camellia sinensis* (Green tea)	Leaves	Epigallocatechin-3-gallate
3.	*Curcuma longa* (Turmeric)	Roots	Curcumin
4.	*Rubus fruticosus* (Berries)	Skin	Resveratrol
5.	*Camellia sinensis* (Green/Black Tea)	Leaves	ECGC
6.	*Juglans regia* (Walnut)	Shell	Juglone
7.	*Panax ginseng* L. Ginseng (Red ginseng extract)	Roots	Ginsenosides
8.	*Solanum persicum* (Tomato)	Fruit	Lycopene
9.	*Persea americana* (Avocado oil)	Fruit	Vitamins D and E, beta carotene, protein, lecithin fatty acid
10.	*Borago officinalis* (Borage oil)	Whole	Gamma linoleic acid
11.	*Citrus aurantium* (Orange)	Peel	Essential oils, triterpenes
12.	*Brassica oleracea* (Broccoli)	Whole	Glucorapharin mixes with myrosinase (a native enzyme of broccoli) to form a new substance called sulforaphane

Sources: Pulok, K.M., *Phytomedicine*, 19(1), 64, 2011; Ashok, D.B. et al., *J. Clin. Biochem. Nutr.*, 41(1), 1, 2007.

composition and antioxidant activities. Hence, for centuries, Ayurveda has been used in medicines and cosmetics as it has shown potential to treat different kinds of skin diseases, improve skin appearance, and reduce wrinkles and other signs of aging. Many studies showed that green and black tea (polyphenols) can restructure adverse skin reactions following UV exposure. Also, the gel isolated from aloe vera is believed to stimulate skin rejuvenation by assisting in new cell growth. The traditional use of plant in medication or beautification has become an ideal area for research [70].

33.4.2.1 Green Tea

Green tea, extracted from the plant *Camellia sinensis*, has been identified to retain potent antioxidant and anti-inflammatory activities in both human and animal skins. Green tea extract contains catechins, which help prevent and repair skin damage and may even help in preventing chemically and radiation-induced skin cancers. There are four major polyphenolic catechins present in green tea (Figure 33.2) [73]:

1. (−)-Epicatechin (EC)
2. (−)-Epigallocatechin (EGC)
3. (−)-Epicatechin-3-gallate (ECG)
4. (−)-Epigallocatechin-3-gallate (EGCG)

EGCG is the most abundant of the four. The main biologically active ingredient in green tea has anti-inflammatory, antioxidant, and sunscreen properties. Green tea polyphenols (GTPs) possess antioxidant activity and anti-inflammatory and anti-carcinogenic agents. GTP can be delivered either orally or topically. In human skin, GTPs have been shown to protect epidermal Langerhans cells from UV damage, reduce the number of sunburn cells, protect from UV-induced erythema, immunosuppression, and DNA damage. The polyphenolic compounds in tea have been shown to influence biochemical pathways that play a role in cell proliferation and inflammatory responses.

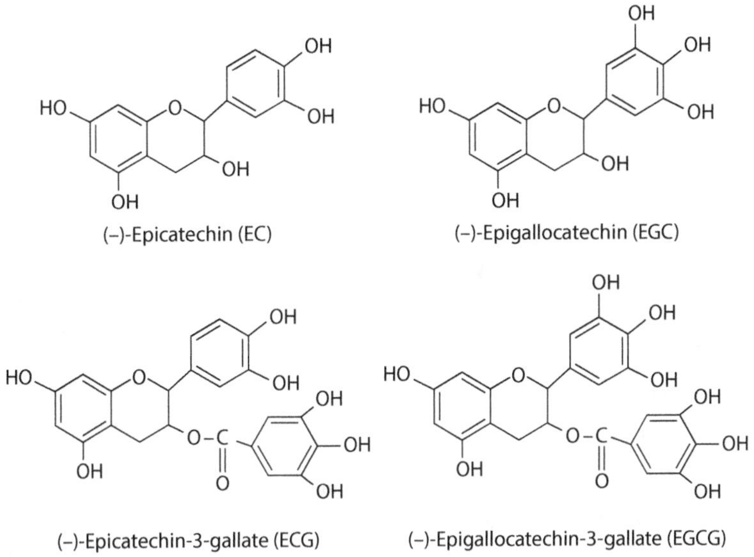

FIGURE 33.2 Principal polyphenolic catechins from green tea.

GTPs get rapidly metabolized when they react with ROS and provide the first line of defense against ROS *in vivo*, and they also inhibit pro-oxidant enzymes, which provide the second line of defense against free radicals. Oxidative stress activates many pathways and molecules like MAPK, AP-1, p-53, JNK, and MMP, resulting in the activation of inflammatory pathways that are involved in the pathogenesis of various skin diseases [74].

Research indicates that GTPs do not only scavenge ROS and act as antioxidants in the epidermis but also act as modulators of different gene groups and signaling pathways. The molecular targets for GTPs include Ras and AP-1, elements of the mitogen-activated protein kinase (MAPK) signaling pathway (Figure 33.3) [73].

33.4.2.1.1 Possible Mechanisms for Green Tea Counteracting Photoaging

Treatment with GTPs exerts the protective effects on oxidatively stressed cells [75–78]. It protects UVB-induced depletion of GPx, catalase and GSH in the skin.

EGCG, when applied to SKH-1 hairless mice before UVB exposure, significantly prevented UVB-induced depletion of the antioxidant enzymes. It also inhibits UVB-induced lipid peroxidation and protein oxidation and induced the expression of MAPK, ERK 1 and 2, c-jun N-terminal kinase (JNK), and p38 [79]. Green tea also enhances the production of antioxidant enzymes and depletes the production of oxidizing enzymes such as nitric oxide synthase, lipoxygenases, COX, and xanthine oxidase. In mouse models, EGCG has been effective in inhibiting AP-1 and NF-kB activities in the epidermis, which is responsible for collagen breakdown. Oral administration of GTP also resulted in the inhibition of UVB-induced expression of matrix degrading MMPs, such as MMP-2, MMP-3, MMP-7, and MMP-9, in hairless mouse skin. It also inhibits stimulated granulocytes.

33.4.2.2 Aloe Vera

For centuries, the *aloe vera* plant has been used for its curative and therapeutic properties for cosmetic and medicinal purposes. Seventy-five active ingredients have been identified from the inner gel, but most of their therapeutic effects are yet to be discovered. Some of the major constituents of *aloe vera* are as follows: anthraquinones/anthrones, essential amino acids (isoleucine, leucine, lysine, methionine, phenylalanine, threonine, valine, and tryptophan), nonessential amino acids (arginine, asparagine, cysteine, glutamic acid, glycine, histidine, proline, serine, tyrosine, glutamine,

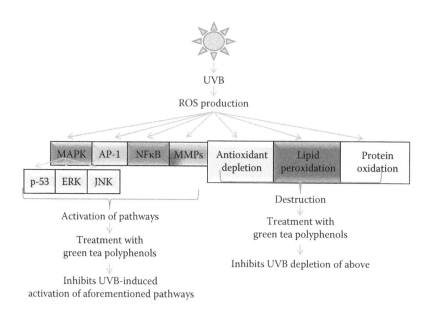

FIGURE 33.3 **(See color insert.)** Endogenous ROS production due to UVB has a wide range of harmful effects on human health. It results in the activation of various pathways like mitogen-activated protein kinases (MAPK) and its family members extracellular signal–regulated kinase (ERK) 1 and 2, c-jun N-terminal kinase (JNK), and p38; activating protein-1 (AP-1); nuclear factor kappa B (NFκB); and matrix metalloproteins (MMPs); all of these ultimately lead to oxidative stress, aging, and disease. They also reduce antioxidants, enhance the release of pro-oxidants, and cause cell membrane and ECM breakdown by lipid peroxidation and protein oxidation. GTPs have been shown to inhibit all these pathways when applied to epidermal cells before or after UVB exposure. They enhance the production of antioxidant enzymes and deplete the production of pro-oxidant enzymes. They also prevent lipid peroxidation and protein oxidation.

and aspartic acid) chromone, enzymes (amylase, bradykinase, catalase, cellulase, oxidase, alkaline phosphatase, proteolyase, creatine phosphokinase, and carboxypeptidase), lignin, mono- and poly-saccharides (glucose and acemannan, respectively), vitamins (A, C, E, B1, B2, B3, B5, B6, and B12), salicylic acid, saponins, and sterols that can help reverse damage caused by sun exposure [80].

There are more than 250 species of *Aloe* grown around the world. Five of them, *Aloe barbadensis*, *Aloe perryi*, *Aloe ferox*, *Aloe arborescens*, and *Aloe saponaria*, have been identified to possess medicinal benefits [81]. It is observed that the *aloe vera* plant, at various development stages, contains different active components and hence confers different levels of antioxidant activities. Adult mature plant activity provided equivalent or higher antioxidant activity as compared to BHT and R-tocopherol [82] (Figure 33.4). Some possible mechanisms of skin protection after UV exposure and added skin benefits are listed as follows [81–84]:

1. *Aloe vera* gel components can prevent UV or γ-radiation-induced skin reactions in mice and humans. Mannose-6-phosphate plays a role in this.
2. Polysaccharides enhance the release of cytokines, which increase the proliferation of fibroblasts, which ultimately cause skin repair.
3. Glucomannan, a mannose-rich polysaccharide, and gibberellin, a growth hormone, interact with growth factor receptors on the fibroblast, thereby stimulating its activity and proliferation, which in turn significantly increase collagen synthesis after topical and oral application of aloe vera. Aloe gel not only increases collagen content of the wound but also changes collagen composition and increases the degree of collagen cross-linking. Due to this, it accelerates wound contraction and increases the breaking strength of the resulting scar tissue.

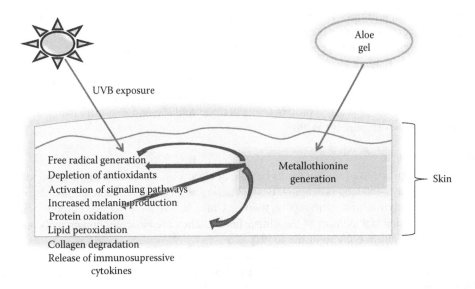

FIGURE 33.4 **(See color insert.)** The exact mechanism by which aloe vera protects skin from sun damage is unknown. But following the application of aloe vera gel, an antioxidant protein, metallothionein, is generated in the skin, which scavenges hydroxyl radicals and regulates SOD and glutathione peroxidase in the skin. It reduces the production and liberation of immunosuppressive cytokines (IL-10) by skin keratinocyte and, hence, prevents UV-induced suppression of delayed type hypersensitivity (DTH). (From Tanuja, Y. et al., *Environ. Toxicol. Pharmacol.*, 39(1), 384, 2015. With permission.)

4. Gel components in the range of 500 to 1000 Da are reported to suppress UVB induced function of Langerhans cells of the epidermis.
5. Topical application of aloe gel has shown to increase collagen production and deposition.
6. Glycoprotein fraction from aloe vera was found to accelerate wound healing. This 5.5 kDa glycoprotein isolated from aloe vera showed an increase in cell migration and sped up wound healing in human keratinocytes monolayer. This glycol protein also stimulated epidermal tissue formation.
7. Aloe vera has antioxidant effects. Some oligo elements like manganese and selenium are constituents of antioxidant enzymes that enhance its radical scavenging activity.
8. Aloe vera contains the enzyme bradykinase that stops inflammatory response caused by sunburn.
9. Acemannan, a D-isomer mucopolysaccharide found in the inner leaf gel of the aloe vera plant, speeds up the regeneration of skin tissue by intervening in the stimulation of macrophages and increased production of collagen and fibroblasts.
10. Aloe extracts and aloin, an anthraquinone from aloe vera, can act as UV absorbers as they have been reported to block 20%–30% of the sun's rays.
11. Aloe vera also protects against pro-oxidant-induced membrane and cellular damage by significantly reducing the levels of cytochrome P450 and cytochrome b5.

33.4.2.3 Berries

Berry fruits are rich sources of phenolics such as phenolic acids, flavonoids, stilbenes, and tannins. Their phenolics possess one or more aromatic rings with hydroxyl groups. The skin of fresh grapes contains resveratrol, which shows a broad variety of beneficial effects, including cardioprotection and neuroprotection; it also possesses antimicrobial and chemopreventive properties [86].

Resveratrol has high-affinity receptors present in the human skin [87] and is known to express significant antiaging protection properties in the following ways.

It affects the sirtuin (SIRT) gene activation, which is associated with aging and longevity, which in turn reduces photoaging by inhibiting the activation of matrix metalloproteins [88]. It also inhibits the NF-kB signaling pathway and its nuclear translocation [89]. Resveratrol is known to increase the level of NF-E2-related factor-2 (Nrf2) in the nucleus by decreasing its repressor Kelch-like ECH-associated protein 1 (Keap1) protein levels in the cytoplasm and also increasing the activity of antioxidant enzymes (Figure 33.5) [90].

33.4.2.4 Turmeric

Turmeric has been accepted widely for its beneficial effects on the skin. It plays an important role in reducing UV rays–induced skin aging. Besides acting as an antiaging factor, it also displays anti-inflammatory, antiviral, and antifungal activities. Curcumin is the principal curcuminoid of this popular spice, which is a member of the ginger family. They are basically polyphenols responsible for imparting the yellow color to turmeric [91].

Curcumin acts as an antiaging compound by inhibiting the NF-kB pathway that is responsible for MMPs activation and subsequent collagen degradation. It acts by inhibiting TNF, H_2O_2, and PMA responsible for NF-kB activation (Figure 33.6). Curcumin also inhibits the phosphorylation of IkBα, an inhibitor bound to NF-kB in cytoplasm making it inactive. However, phosphorylation and subsequent degradation of IkBα release NF-kB into the nucleus and activate MMP genes such as MMP-1, 3, 9, etc. [92].

33.4.2.5 Tomatoes

Tomatoes are rich in lycopenes, an acyclic hydrocarbon and a potent antioxidant. Lycopene belongs to a group of fat-soluble plant compounds known as carotenoids. It gives many fruits and vegetables their bright yellow, red, and orange colors. Besides performing important functions during

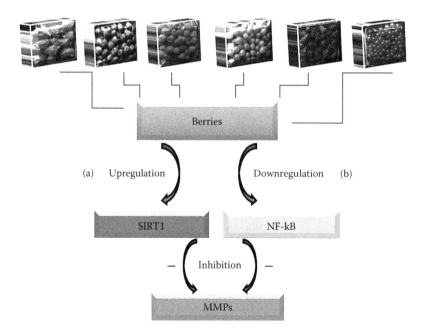

FIGURE 33.5 **(See color insert.)** Berries show antiaging property by inhibiting the activation of MMPs, which are known to degrade collagen by two main pathways: (a) by upregulating SIRT1 gene, which inhibits MMPs, and (b) by downregulating the NF-kB transcription factor, which inhibits MMPs production.

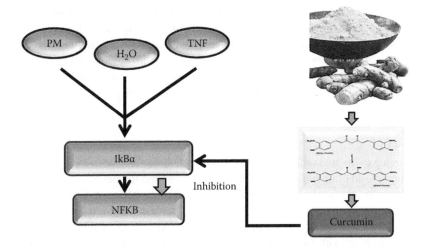

FIGURE 33.6 **(See color insert.)** Mechanism of action of curcumin: Curcumin inhibits the phosphorylation and degradation of IkBα and subsequent translocation in the nucleus.

photosynthesis, they also act as free radical scavengers that protect the organism from harmful effects of UV ray exposure [93]. It confers protection from photoaging in the following ways.

Lycopene suppresses UVB-induced activity of an enzyme called ornithine decarboxylase, an important initiating and rate-controlling factor involved in stabilizing DNA structure in the nucleus of the skin cells and maintaining DNA double-strand break repair pathways [94]. UVB radiation is also found to deplete PCNA (proliferating cell nuclear antigen), which is vital for DNA synthesis and repair. Lycopene significantly reverses the reduction of PCNA caused by UVB exposure. Besides protecting skin cells from free radical damage, lycopene also produces beneficial effects by improving the formation of cell-to-cell junctions [95]. Lycopene may also strengthen skin by inhibiting the activity of enzymes involved in the destruction of collagen, downregulate collagenase, and help to maintain natural firmness (Figure 33.7) [96].

33.4.2.6 Broccoli

Concentrated extracts from cruciferous vegetables, such as broccoli, containing bioactive phytonutrients can help to reduce skin damage leading to photoaging. Broccoli seeds are rich in

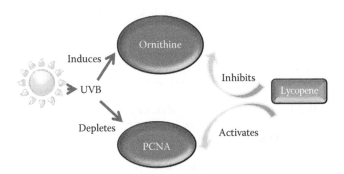

FIGURE 33.7 **(See color insert.)** UVB induces ornithine decarboxylase and MMPs, and, on the other hand, it depletes PCNA. Lycopene reverses the induction and depletion of ornithine decarboxylase and PCNA, respectively.

glucoraphanin, a glucosinolate and the bio-precursor of the widely extolled chemopreventive agent sulforaphane. Apart from preventing cancer and inflammation, it also has a role in maintaining youthful and radiating skin. Glucoraphanin, abundantly present especially in newly sprouted broccoli (10–100 times higher levels), gets released and mixes with myrosinase (a native enzyme of broccoli) to form a substance called sulforaphane, a sulfur containing anticancerous compound [97]. Sulforaphane exhibits UV protection properties by stimulating antioxidant enzymes like SOD, GST, etc. [98,99].

33.4.2.7 Walnut

Walnuts, belonging to the Juglandaceae family, are rounded, single-seeded stone fruits. The shuck of the walnut is composed of involucral bracts, perianth tissue, and the outer layer of the pericarp. The hard shell surrounding the seed of walnut is the inner layer of the pericarp. The fruit has been known to have antiaging potential because of its antioxidant property due to high content of phenolic acid, flavonoids, and melatonin [100]. Its main component is juglone (5-hydroxy-1,4-naphthoquinone), which is known to react with the keratin proteins present in the skin to form sclerojuglonic compounds that possess UV protection properties [101]. Another compound present in walnuts is melatonin, a potent radical scavenger that not only can scavenge the hydroxyl radicals but also perform indirect antioxidant actions by stimulating antioxidative enzymes such as superoxide dismutase and catalase. It has been postulated that increasing the levels of melatonin in diet can reduce oxidative stress, which in turn may reduce general aging [102].

33.5 CONCLUSION

With time, people become more concerned about their skin, the window of human health that shows the most obvious signs of aging. Research in the area of skin aging is gaining considerable attention. Research has produced good information in this field. Based on these results, many ways to prevent, even revert, skin aging have been developed. Supplementing bioactive compounds in food and methods to reduce UVR exposure may be promising in overcoming this problem more effectively in the future. More research, especially on promising plant extracts, may add more powerful ways to overcome this problem and may provide the opportunity to people to keep their skin healthy for longer. Hence, research should focus more on identifying and evaluating plants and herbs from natural sources that can be used to reduce the process of skin aging.

REFERENCES

1. Swann G. 2010. The skin is the body's largest organ. *J Vis Commun Med* 33:148–149.
2. Pincelli C, Marconi A. 2010. Keratinocyte stem cells: Friends and foes. *J Cell Physiol* 225(2):310–315.
3. Chih HL et al. 2013. Molecular mechanisms of UV-induced apoptosis and its effects on skin residential cells: The implication in UV-based phototherapy. *Int J Mol Sci* 14(3): 6414–6435.
4. Stern R. 2010. Hyaluronan and the process of aging in skin. In: Farage M, Miller K, Maibach H, eds. *Textbook of Aging Skin*. Berlin, Heidelberg: Springer, pp. 225–238.
5. Giacomoni PU, Rein G. 2001. Factors of skin aging share common mechanisms. *Biogerontology* 2(4):219–229.
6. Borut P, Raja D. 2012. Free radicals and extrinsic skin aging. *Dermatol Res Pract* 2012:135206.
7. Florence DC et al. 2012. UV, stress and aging. *Dermato-Endocrinol* 4(3):236–240.
8. Ivana B et al. 2013. Skin aging: Natural weapons and strategies. *Evid Based Complement Alternat Med* 2013:827248.
9. Fisher GJ. 1997. Pathophysiology of premature skin aging induced by ultraviolet light. *N Engl J Med* 337(20):1419–1428.
10. Andrea V, Jean K. 2012. Environmental influences on skin aging and ethnic-specific manifestations. *Dermato-Endocrinol* 4(3):227–231.
11. Jaiswal A et al. 2011. Comparative analysis of human matrix metalloproteinases: Emerging therapeutic targets in diseases. *Bioinformation* 6(1):23–30.

12. Andrea V et al. 2010. Air borne particle exposure and extrinsic skin aging. *J Invest Dermatol* 130:2719–2726.

13. Senthilkumar PK et al. 2011. Air borne polychlorinated biphenyls (PCBs) reduce telomerase activity and shorten telomere length in immortal human skin keratinocytes (HaCat). *Toxicol Lett* 204(1):64–70.

14. Urbanska M et al. 2012. Cigarette smoking and its influence on skin aging. *Przeql Lek* 69(10):1111–1114.

15. Egawa M et al. 1999. Oxidative effects of cigarette smoke on the human skin. *Int J Cosmet Sci* 21(2):83–98.

16. Arisa O et al. 2012. Smoking and the skin. *Int J Dermatol* 51(3):250–262.

17. Marian VA et al. 2007. Free radicals and antioxidants in normal physiological functions and human disease. *Int J Biochem Cell Biol* 39:44–84.

18. Piotr Z. 2011. Relationship of electrophilic stress to aging. *Free Radic Biol Med* 51(6):1087–1105.

19. Christine RG et al. 1997. Nitric oxide produced by ultraviolet-irradiated keratinocytes stimulates melanogenesis. *J Clin Invest* 99(4):635–642.

20. Klatt P, Lamas S. 2000. Regulation of protein function by S glutathiolation in response to oxidative and nitrosative stress. *Eur J Biochem* 267(16):4928–4944.

21. Shen D et al. 2005. Glutathione redox state regulates mitochondrial reactive oxygen production. *J Biol Chem* 280(27):25305–25312.

22. Nogueira CW et al. 2004. Organo-selenium and organo-tellurium compounds. *Toxicol Pharmacol Chem Rev* 104(12):6255–6285.

23. Barnett YA, King CM. 1995. Investigation of antioxidant status, DNA-repair capacity and mutation as a function of age in humans. *Mut Res* 338:115–128.

24. Zouboulis CC, Makrantonaki E. 2012. Hormonal therapy of intrinsic aging. *Rejuvenation Res* 15(3):302–312.

25. Mayr H, Ofial AR. 2005. Kinetics of electrophile-nucleophile combinations: A general approach to polar organic reactivity. *Pure Appl Chem* 77:1807–1821.

26. Shaw RJ, Cantley LC. 2006. Ras, PI(3)K and mTOR signalling controls tumour cell growth. *Nature* 441(7092):424–430.

27. Halliwell B. 2006. Oxidative stress and neuro-degeneration: Where are we now? *J Neurochem* 97(6):1634–1658.

28. Koster JF et al. 1986. Comparison of the inactivation of microsomal glucose-6-phosphatase by in situ lipid peroxidation-derived 4-hydroxynonenal and exogenous 4-hydroxynonenal. *Free Radic Res Commun* 1(4):273–287.

29. Stefanie W et al. 2009. Changes in skin physiology and clinical appearance after micro-droplet placement of hyaluronic acid in aging hands. *J Cosmet Dermatol* 8:216–225.

30. Bernstein EF et al. 1996. Chronic sun exposure alters both the content and distribution of dermal glycosaminoglycans. *Br J Dermatol* 135(2):255–262.

31. Eleni P et al. 2012. Hyaluronic acid: A key molecule in skin aging. *Dermato-Endocrinol* 4(3):253–258.

32. Röck K et al. 2011. Collagen fragments inhibit hyaluronan synthesis in skin fibroblasts in response to ultraviolet B (UVB): New insights into mechanisms of matrix remodeling. *J Biol Chem* 286(20):18268–18276.

33. Gniadecka M et al. 1998. Water and protein structure in photoaged and chronically aged skin. *J Invest Dermatol* 111(6):1129–1133.

34. Fenske NA, Lober CW. 1986. Structural and functional changes of normal aging skin. *J Am Acad Dermatol* 15(4):571–585.

35. Oikarinen A. 1990. The aging of skin: Chronoaging versus photoaging. *Photodermatol Photoimmunol Photomed* 7(1):3–4.

36. Fisher GJ et al. 1997. Pathophysiology of premature skin aging induced by ultraviolet light. *N Engl J Med* 337(20):1419–1428.

37. Contet-Audonneau JL et al. 1999. A histological study of human wrinkle structures: Comparison between sun exposed areas of the face, with or without wrinkles, and sun protected areas. *Br J Dermatol* 140(6):1038–1047.

38. Craven NM et al. 1997. Clinical features of photodamaged human skin are associated with a reduction in collagen VII. *Br J Dermatol* 137(3):344–350.

39. Rittie L, Fisher GJ. 2002. UV-light-induced signal cascades and skin aging. *Ageing Res Rev* 1(4):705–720.

40. Mitchel RE. 1967. Chronic solar dermatosis: A light and electron microscopic study of the dermis. *J Invest Dermatol* 48(3):203–220.

41. Scharffetter-KK et al. 2000. Photoaging of the skin from phenotype to mechanisms. *Exp Gerontol* 35(3):307–316.

42. Montagna W, Carlisle K. 1979. Structural changes in aging human skin. *J Invest Dermatol* 73(1):47–53.

43. Escoffier C et al. 1989. Age-related mechanical properties of human skin: An *in vivo* study. *J Invest Dermatol* 93(3):353–357.

44. Stamatas GN et al. 2004. Non-invasive measurements of skin pigmentation *in situ*. *Pigment Cell Res* 17(6):618–626.

45. Montagna W, Carlisle K. 1991. The architecture of black and white facial skin. *J Am Acad Dermatol* 24(6):929–937.

46. Thody AJ et al. 1991. Pheomelanin as well as eumelanin is present in human epidermis. *J Invest Dermatol* 97:340–344.

47. Zanetti R et al. 2001. Development of an integrated method of skin phenotype measurement using the melanin. *Melanoma Res* 11(6):551–557.

48. Duchon J. 1970. Chemical composition of melanosomes. *Dermatol Monatsschr* 156(5):371.

49. Pathak MA, Jimbow K, Fitzpatrick T. 1980. Photobiology of pigment cells. In: Seiji M, ed. *Phenotypic Expression in Pigment Cells*. Tokyo, Japan: University of Tokyo Press, pp. 655–670.

50. Kipp C, Young AR. 1999. The soluble eumelanin precursor 5,6-dihydroxyindole-2-carboxylic acid enhances oxidative damage in human keratinocyte DNA after UVA irradiation. *Photochem Photobiol* 70:191–198.

51. Kvam E, Dahle J. 2004. Melanin synthesis may sensitize melanocytes to oxidative DNA damage by ultraviolet A radiation and protect melanocytes from direct DNA damage by ultraviolet B radiation. *Pigment Cell Res* 17(5):549–550.

52. Marrot L et al. 1999. The human melanocyte as a particular target for UVA radiation and an endpoint for photoprotection assessment. *Photochem Photobiol* 69(6):686–693.

53. Chedekel MR. 1982. Photochemistry and photobiology of epidermal melanins. *Photochem Photobiol* 35(6):881–885.

54. Chedekel MR et al. 1977. Photodestruction of phaeomelanin. *Photochem Photobiol* 26:651–653.

55. Chedekel MR et al. 1978. Photodestruction of pheomelanin: Role of oxygen. *Proc Natl Acad Sci USA* 75(11):5395–5399.

56. Hill HZ, Hill GJ. 2000. UVA, pheomelanin and the carcinogenesis of melanoma. *Pigment Cell Res* 13(s8):140–144.

57. Kosmadaki MG, Gilchrest BA. 2004. The role of telomeres in skin aging/photoaging. *Micron* 35(3):155–159.

58. Erin MB, Aloysius JK. 2011. The role of telomeres in the aging of human skin. *Exp Dermatol* 20(4):297–302.

59. Paraskevi G, Markus B. 2012. Advanced glycation end products. *Dermato-Endocrinol* 4(3):259–270.

60. Makrantonaki E, Zouboulis CC. 2009. Androgens and aging of the skin. *Curr Opin Endocrinol Diabetes Obes* 16:240–245.

61. Evgenia M et al. 2012. Genetics and skin aging. *Dermato-Endocrinol* 4(3):280–284.

62. Lin DS. 2011. A novel mutation in PYCR1 causes an autosomal recessive cutis laxa with premature aging features in a family. *Am J Med Genet A* 155A(6):1285–1289.

63. Lien Ai PH. 2008. Free radicals, antioxidants in disease and health. *Int J Biomed Sci* 4(2):89–96.

64. Halliwell B. 2007. Biochemistry of oxidative stress. *Biochem Soc Trans* 35(Pt 5):1147–1150.

65. Farrukh A. 2006. Antioxidant and free radical scavenging properties of twelve traditionally used Indian medicinal plants. *Turk J Biol* 30:177–183.

66. Noori S. 2012. An overview of oxidative stress and antioxidant defensive system. *Open Access Sci Rep* 1:413.

67. Velioglu YS et al. 1998. Antioxidant activity and total phenolics in selected fruits, vegetables, and grain products. *J Agric Food Chem* 46:4113–4117.

68. Young I, Woodside J. 2001. Antioxidants in health and disease. *J Clin Pathol* 54(3):176–186.

69. Tamsyn SA et al. 2009. Anti-collagenase, anti-elastase and anti-oxidant activities of extracts from 21 plants. *BMC Complement Alternat Med* 9(1):27.

70. Radava RK, Kapil MK. 2011. Potential of herbs in skin protection from ultraviolet radiation. *Pharmacogn Rev* 5(10):164–173.

71. Pulok KM. 2011. Bioactive compounds from natural resources against skin aging. *Phytomedicine* 19(1):64–73.

72. Ashok DB et al. 2007. Current status of herbal drugs in India: An overview. *J Clin Biochem Nutr* 41(1):1–11.

73. Stephen H. 2005. Green tea and the skin. *J Am Acad Dermatol* 52(6):1049–1059.

74. Toren F, Nikki JH. 2000. Oxidants, oxidative stress and the biology of aging. *Nature* 408(6809): 239–247.

75. Margaret B et al. 1998. Epigallocatechin 3-gallate inhibition of ultraviolet B-induced AP-1 activity. *Carcinogenesis* 19(12):2201–2201.

76. Farrukh A et al. 2003. Suppression of UVB-induced phosphorylation of mitogen-activated protein kinases and nuclear factor kappa B by green tea polyphenol in SKH-1 hairless mice. *Oncogene* 22(58):9254–9264.

77. Vayalil PK. 2004. Green tea polyphenols prevent ultraviolet light-induced oxidative damage and matrix metalloproteinases expression in mouse skin. *J Invest Dermatol* 122(6):1480–1487.

78. Karin M. 1995. The regulation of AP-1 activity by mitogen-activated protein kinases. *J Biol Chem* 270(28):16483–16486.

79. Chen C, Yu R, Owuor ED, Kong AN. 2000. Activation of antioxidant-response element (ARE), mitogen-activated protein kinases (MAPKs) and caspases by major green tea polyphenol components during cell survival and death. *Arch Pharm Res* 23(6):605–612.

80. Verma SK et al. 2011. *Aloe vera*: Their chemicals composition and applications: A review. *Int J Biol Med Res* 2(1):466–471.

81. Basmatker G et al. 2011. *Aloe vera*: A valuable multifunctional cosmetic ingredient. *Int J Med Arom Plant* 1(3):338–341.

82. Yun Hu et al. 2003. Evaluation of antioxidant potential of aloe vera (*Aloe barbadensis* miller) extracts. *J Agric Food Chem* 51(26):7788–7791.

83. Josias HH. 2008. Composition and applications of *Aloe vera* leaf gel. *Molecules* 13(8):1599–1616.

84. Reynolds T, Dweck AC. 1999. *Aloe vera* leaf gel: A review update. *J Ethnopharmacol* 68(1–3):3–37.

85. Tanuja Y et al. 2015. Anticedants and natural prevention of environmental toxicants induced accelerated aging of skin. *Environ Toxicol Pharmacol* 39(1):384–391.

86. Octavio PL et al. 2010. Berries: Improving human health and healthy aging, and promoting quality life—A review. *Plant Foods Hum Nutr* 65(3):299–308.

87. Bastianetto S et al. 2010. Protective action of resveratrol in human skin possible involvement of specific receptor binding sites. *PLoS One* 5(9):1–12.

88. Lee JS et al. 2010. Negative regulation of stress-induced matrix metalloproteinase-9 by Sirt1 in skin tissue. *Exp Dermatol* 19(12):1060–1066.

89. Vaqar M et al. 2003. Suppression of ultraviolet b exposure-mediated activation of NF-κB in normal human keratinocytes by resveratrol. *Neoplasmia* 5(1):74–82.

90. Liu Y et al. 2011. Resveratrol protects human keratinocytes HaCaT cells from UVA-induced oxidative stress damage by downregulating Keap1 expression. *Eur J Pharmacol* 650(1):130–137.

91. Akram M et al. 2010. *Curcuma longa* and Curcumin: A review article. *Plant Biol* 55(2):65–70.

92. Singh S, Agarwal BB. 1995. Activation of transcription factor NF-kB is suppressed by curcumin. *J Biol Chem* 270:24995–25000.

93. Rao AV, Rao LG. 2007. Carotenoids and human health. *Pharmacol Res* 55(3):207–216.

94. Fazekas Z. 2003. Protective effects of lycopene against ultraviolet B induced photodamage. *Nutr Cancer* 47(2):181–187.

95. Aust O et al. 2003. Lycopene oxidation product enhances gap junctional communication. *Food Chem Toxicol* 41(10):1399–1407.

96. Huang CS et al. 2007. Lycopene inhibits matrix metalloproteinase-9 expression and down-regulates the binding activity of nuclear factor-kappa B and stimulatory protein-1. *J Nutr Biochem* 18(7):449–456.

97. Fahey JW et al. 1997. Broccoli sprouts: An exceptionally rich source of inducers of enzymes that protect against chemical carcinogens. *Proc Natl Acad Sci USA* 94(19):10367–10372.

98. Jenkins SN et al. 2006. Protection against UV-light-induced skin carcinogenesis in SKH-1 high-risk mice by sulforaphane-containing broccoli sprout extracts. *Cancer Lett* 240(2):243–252.

99. Talalay P et al. 2007. Sulforaphane mobilizes cellular defenses that protect skin against damage by UV radiation. *Proc Natl Acad Sci USA* 104(44):17500–17505.

100. Muthaiyah B et al. 2011. Protective effects of walnut extract against amyloid beta peptide induced cell death and oxidative stress in PC12 cells. *Neurochem Res* 36(11):2096–2103.

101. Korac RR, Khambhlja KM. 2011. Potential of herbs in skin protection from ultraviolet radiation. *Pharmacogn Rev* 5(10):164–173.

102. Reiter RJ et al. 2005. Melatonin in walnuts: Influence on levels of melatonin and total antioxidant capacity of blood. *Nutrition* 21(9):920–924.

34 ROS and Ataxia Telangiectasia

Tetsuo Nakajima

CONTENTS

ABSTRACT

Reactive oxygen species (ROS) are known to be involved in the development of many hereditary and nonhereditary diseases. Mutations in the Ataxia-telangiectasia mutated (*ATM*) gene cause Ataxia-telangiectasia (AT). Although ATM protein has been investigated from the viewpoint of the functions in DNA repair system so far, it has also been demonstrated that ATM is directly regulated by oxidative stress. In addition, some AT phenotypes seem to be due to the induction of oxidative stress, and studies of ROS and ATM contribute to understanding lifestyle-related diseases. Here, the recent advances on this subject are highlighted and discussed.

34.1   INTRODUCTION

Ataxia telangiectasia (AT) is a hereditary human disorder that is caused by mutations in the *ATM* (ataxia telangiectasia mutated) gene. As ATM plays a key role in the DNA repair system, the functions of ATM in DNA repair have been extensively investigated. AT phenotypes also include several metabolic syndromes that do not appear to be directly related to impaired DNA damage and repair. Although ATM is thought to have cytoplasm-specific functions that influence the AT phenotypes, subcellular localization and functions of ATM have not been sufficiently clarified. These functions are likely to be related to oxidative stress and induced production of reactive oxygen species (ROS); however, the underlying mechanisms of ATM activation remain unclear.

Recently, it was demonstrated that oxidative stress induces ATM activation independent of DNA damage. In addition, antioxidants were shown to attenuate cancer progression and to increase the shortened life-span of ATM-deficient mice. These findings suggest that ATM has specific functions related to oxidative stress. In this chapter, the relationship between ATM function and ROS-inducing oxidative stress is examined with references to the phenotypes of AT patients and ATM-deficient mice. The findings presented here for AT may also shed light on the mechanisms underlying metabolic syndrome.

34.2 ATM AS THE GENE RESPONSIBLE FOR AT

AT is an autosomal recessive disorder caused by mutations in the *ATM* gene, which was first cloned in 1995 (Savitsky et al. 1995). AT patients either lack the ATM protein or have impaired ATM kinase activity; typically suffer from progressive neurological dysfunction, oculocutaneous telangiectasia, and immunodeficiency; and are at high risk for cancer and pulmonary infection (Boder and Sedgwick 1958, McKinnon 2004, Taylor and Byrd 2005). As ATM plays a key role in DNA damage repair signaling systems, AT patients are also hypersensitive to DNA-damaging factors, such as radiation. In addition, AT patients have an increased risk of developing insulin resistance and type 2 diabetes (Bar et al. 1978).

In the general population, the estimated frequency of heterozygous AT carriers is about 1.4% (Swift et al. 1987). Polymorphisms of ATM are associated with an increased risk of cancer (Gao et al. 2011, Shen et al. 2012, Liu et al. 2014). The phenotypes of AT cells have been well characterized and include ataxia, immunodeficiency, and radiosensitivity (Chun and Gatti 2004). Several comprehensive reviews have been published on the phenotypic characterization of AT (Boder and Sedgwick 1958, Shiloh and Ziv 2013, Stagni et al. 2014, Taylor et al. 2015).

The *ATM* gene (9168 bp) is located on chromosome 11q22-23, which encodes an approximately 350 kDa serine threonine kinase. ATM belongs to the PI3-kinase family, which includes DNA-protein kinase (DNA-PK) and the ATM- and Rad3-related kinase (ATR) (McKinnon 2012). Members of this family share common protein motifs, including FAT, FATC (FAT-C terminal), and a kinase domain (McKinnon 2012). ATM is constitutively expressed and is activated upon DNA damage. Similar to DNA-PK and ATR, ATM is a main functional protein in DNA damage–repair systems. Although the role of ATM in DNA damage–repair responses, particularly with respect to its involvement in the underlying mechanisms of DNA damage–related diseases, has been well studied, recent findings suggest that ATM is activated by oxidative stress (Guo et al. 2010a) and impaired ATM function is associated with the regulation of ROS production. For example, ATM deficiency shortens the life-span of mice and causes an increase in lymphoma incidence and hematopoietic stem cells (HSC) loss (Ito et al. 2004, Reliene and Schiestl 2006, 2007, Reliene et al. 2008). The administration of the antioxidant N-acetyl-L-cysteine (NAC) to ATM-deficient mice leads to an increased life-span and a reduction in the incidence and multiplicity of lymphoma and loss of HSC. These findings suggest that the predisposition to cancer and premature aging observed in ATM-deficient mice are related to oxidative stress and indicate that ATM may play a key role in homeostatic functions, such as stem cell maintenance, via oxidative stress regulation.

34.3 ATM ACTIVATION AND OXIDATIVE STRESS

AT patients display multiple phenotypes that are generally considered to be related to impairment of DNA repair systems. ATM normally exists in cells in an inactive state as a dimer or tetramer. In response to DNA double-strand breaks (DSBs), ATM activation is mediated by the Mre11-Rad50-NBS1 (MRN) complex, which recruits ATM to the site of DSBs, where ATM becomes activated in monomeric or dimeric form. In addition to DNA repair, the involvement of ATM in oxidative stress responses has also been widely investigated (Barzilai et al. 2002, Nakajima 2009, Pagano et al. 2009). Recent evidence suggests that the oxidative stress–induced regulation of ATM differs from that of DSBs-induced systems. Specifically, ATM activation by H_2O_2-induced oxidative stress occurs via the formation of disulfide bridges involving multiple cysteine residues, including cysteine 2991, in the FATC domains of ATM dimers (Guo et al. 2010a,b). Is this activation mechanism related to the DSB-induced system? p53 and Chk2 are representative targets, both of which

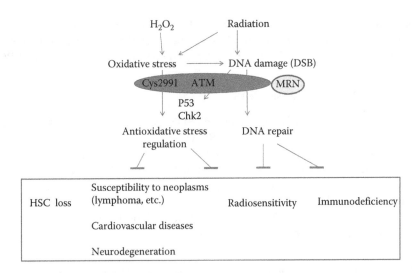

FIGURE 34.1 Involvement of oxidative stress and DNA damage in ATM-related diseases and metabolism.

are phosphorylated by ATM. The phosphorylation of these proteins is also induced by hydrogen peroxide (H_2O_2) and DSBs (Guo et al. 2010a). However, the phosphorylation of γH2AX and Kap1, which are substrates phosphorylated by ATM in response to DSBs for DNA damage repair is induced only by DSB. In addition, p53 and Chk2 phosphorylation by ATM activation in response to H_2O_2 does not require MRN, which is an important complex for DSB-induced phosphorylation of these proteins. The antioxidant NAC inhibits H_2O_2-induced p53 phosphorylation but has no effects of p53 phosphorylation induced by MRN and DSBs. Taken together, these findings suggest that the H_2O_2-induced signaling pathway has distinct roles from the DSB-induced pathway, even though both DSB and H_2O_2 lead to ATM activation (Guo et al. 2010a).

ATM oxidation inhibits the binding of ATM to MRN (Guo et al. 2010b). ATM proteins lacking an FATC domain respond to MRN, but not to H_2O_2, for the phosphorylation of p53, suggesting that ATM oxidation in the FATC domain has specific roles in p53 phosphorylation. A global proteome analysis to identify components of the ATM signaling pathway detected numerous substrates, including p53, Chk2, and BRCA1 (Matsuoka et al. 2007). The analysis also revealed that many substrates of ATM do not appear to be directly linked to DSB repair. Cells with the ATM R3047X mutation in the FATC domain, which is deleted in the last 10 amino acids, are less radiosensitive than more common AT phenotypes, and a patient with the mutation was reported to have no immunodeficiency (Chessa et al. 1992, Gilad et al. 1998, Toyoshima et al. 1998). Although radiation is able to cause DNA damage, low-linear energy transfer (LET) radiation can also induce oxidative stress. In addition to the role of ATM in DNA damage repair systems (McKinnon 2012), the reduced radiosensitivity associated with the R3047X mutation suggests that ATM has distinct protective roles against oxidative stress and DNA-damaging factors. In addition, the FATC domain has only been observed in terrestrial vertebrates, suggesting the domain was acquired in the process of evolution (Guo et al. 2010a,b).

A schematic illustration of the proposed functions of ATM is presented in Figure 34.1.

34.4 ROS INVOLVEMENT IN ATM FUNCTION AND AT CHARACTERISTICS

34.4.1 Cancer Incidence and ROS Regulation

In AT patients and ATM-deficient mice, increases in ROS appear to induce the progression of AT-related diseases. For example, the inhibition of ROS in ATM-deficient mice reduces the incidence of thymic lymphomas. The antioxidant NAC inhibits lymphoma progenitor cells and also restores normal T-cell development and inhibits aberrant V(D)J recombination (Ito et al. 2007). Evidence suggests that ATM controls proper DNA recombination via the regulation of ROS.

34.4.2 STEM CELL REGULATION AND ROS

In ATM-deficient mice, increased DNA damage and ROS levels are observed in HSC, whereas only increased DNA damage has been detected in BRCA1-deficient mice (Santos et al. 2014). In HSC regulation, ATM appears to delay progression toward senescence. Deficiency of ATM function in the regulation of the progression increases ROS formation and DNA damage and causes defects in HSC repopulating capacity (Ito et al. 2006). As NAC treatment improves HSC maintenance in ATM-deficient mice, ATM appears to play an important role in HSC maintenance through the regulation of ROS (Ito et al. 2006). More recently, it was demonstrated that the ATM-BID signal pathway participates in this process (Maryanovich et al. 2012).

34.4.3 NEURAL STEM CELL, NEURAL DIFFERENTIATION, AND ATM FUNCTION

The function of ATM in the regulation of neural stem cells has also attracted considerable attention. In ATM-deficient mice, mild neurodegeneration appears to result from the abnormal proliferation of neural stem cells due to increased cellular levels of ROS (Kim and Wong 2009a). Treatment with NAC or p38 MAPK inhibitor restores normal proliferation of neural stem cells, and it was also demonstrated that the Akt and Erk1/2 signaling pathways are involved in this process (Kim and Wong 2009a). Thus, ATM might participate in neural stem cell control via the regulation of ROS and p38 (Kim and Wong 2009b). Moreover, ATM may also function in redox regulation in astrocytes to control microenvironments for neural differentiation. Indeed, with respect to the involvement of ATM in neural cell differentiation, evidence suggests that ATM play a role in the dendritogenesis of Purkinje cells via ROS regulation and astrocyte function (Chen et al. 2003, Liu et al. 2005). In addition, the increased sensitivity to oxidative stress in differentiated oligodendrocytes lacking functional ATM appears to occur in both humans and mice, although ATM may have different functions in the differentiation of neurons in these two species (Allen et al. 2001, Carlessi et al. 2009).

34.4.4 NEURODEGENERATION AND ROS

Neurodegeneration in AT patients may be induced by oxidative stress due to the loss of ATM function. The ATM protein is distributed in both the nucleus and cytoplasm in neural cells. In ATM-deficient mice, although severe neurodegeneration is not observed, neurological defects and neuropathological changes are observed and provide important insights into the understanding of AT mechanisms (Lavin 2013).

The antioxidant 5-carboxy-1,1,3,3-tetramethylisoindolin-2-yloxyl (CTMIO) inhibits cell death and reduces oxidative stress in Purkinje cells in ATM-deficient mice (Chen et al. 2003, Gueven et al. 2006). In addition, ATM participates in neural stem cell maintenance via the regulation of ROS, as mentioned earlier. These findings strongly indicate that ROS are involved in AT-related diseases and may be related to the cytoplasmic function(s) of ATM, as described in the following section (Chen et al. 2003, Gueven et al. 2006).

34.5 ATM DYSFUNCTION AND ROS PRODUCTION

How does ATM deficiency induce ROS in cells? Cells from AT patients have low antioxidative capacity (Reichenbach et al. 1999). AT cells and ATM-deficient mouse cells are highly sensitive to oxidative stress and exhibit characteristics similar to cells chronically exposed to oxidative stress

(Takao et al. 2000, Gatei et al. 2001). ATM might function to control ROS in cells and ATM dysfunction appears to lead to ROS production.

Recently, ATM was reported to play a role in the regulation of the pentose phosphate pathway (PPP). PPP induction links to NADH production and the induction of glutathione (GSH) synthesis (Cosentino et al. 2011). The loss of ATM reduces the transcriptional activity of genes related to antioxidative activity, leading to decreases in GSH and catalase levels (Meredith and Dodson 1987, Kamsler et al. 2001, Li et al. 2004). Recently, it was demonstrated that NADPH oxidase 4 is activated in the brains of AT patients and may be involved in neurodegeneration (Weyemi et al. 2015). Treatment with an NADPH inhibitor prevents the formation of thymic lymphomas in ATM-deficient mice. NADPH likely produces ROS in the vicinity of the nucleus. NADPH oxidase 4 may be involved in the phenotypes of AT via the generation of ROS.

34.6 CYTOPLASMIC ATM FUNCTION AND ROS

ATM may be linked to mitochondrial function because the induction of mitochondria dysfunction leads to ATM activation (Valentin-Vega et al. 2012). In addition, mitochondria dysfunction resulting from the loss of ATM leads to increased ROS levels (Ambrose et al. 2007, Valentin-Vega et al. 2012, Maryanovich and Gross 2013). Mitochondria contribute to endogenous ROS production through the electron transport chain/oxidative phosphorylation system (Kowaltowski et al. 2009). In AT cells from the patients, the antioxidant alpha-lipoic acid increases mitochondrial respiration rates (Ambrose et al. 2007). In contrast, when wild-type cells are transfected with siRNA against ATM, the respiration rates of mitochondria are reduced. Moreover, the stable expression of full-length ATM improves the respiration rates of mitochondria in AT cells. Together, these findings indicate that ATM is required for normal mitochondrial regulation. On the other hand, it was also demonstrated that the loss of ATM decreases mitophagy (a type of autophagy), which likely results in an increase in the number of abnormal mitochondria (Valentin-Vega et al. 2012). AT might be considered in part as a mitochondrial dysfunction and associated disease. In addition, ATM function in mitochondria is implicated in liver disease induced by a high fat diet (HFD) in ATM (+/−) ApoE (−/−) mice, as mitochondrial dysfunction is easily caused by HFD-induced ROS formation in mice (Mercer et al. 2012). It was also reported that protein phosphatase 2A (PP2A) dysregulation resulting from the loss of ATM induces nuclear accumulation of HDAC4 in neurons, leading to neurodegeneration (Li et al. 2012). Although the relationship between ATM function and the regulation of ROS production in PP2A regulation remains unclear, as endogenous PP2A is predominantly cytoplasmic, ATM has been suggested to have cytoplasm-specific functions in the regulation.

Outside of the nucleus, ATM may also function specifically within peroxisomes. Oxidative stress regulation in peroxisomes is related to many diseases, including cancer and neurodegeneration. As ATM is found in peroxisomes, further studies examining the function of ATM in this organelle are warranted (Watters et al. 1999, Fransen et al. 2012).

34.7 ATM AND ROS IN HIF-1 AND AUTOPHAGY REGULATION

How does ATM signaling, particularly in the cytoplasm, regulate cell function? It was demonstrated that ATM activation leads to autophagy by the repression of mammalian target of rapamycin (mTOR) C1 via TSC2 activation (Alexander et al. 2010a). Activation of these signaling pathways might result in the removal of severely ROS-damaged cells. mTOR belongs to the PI3K family and forms a complex (mTORC1) in association with raptor and mLST8 (Cam et al. 2010). mTOR is a nutrient sensor and is involved in cell growth, but is not known to be involved in DNA repair systems. Interestingly, mTOR is suggested to be related to ROS and ATM signaling (Alexander et al. 2010a,b). Rapamycin inhibits ROS production and mTORC1 in ATM-deficient cells and suppresses lymphogenesis probably via autophagy regulation in ATM-deficient mice (Alexander et al. 2010a).

It suggests that ATM participation in autophagy regulation is needed for preventing cancer progression (Alexander et al. 2010a). With respect to cancer progression, ATM dysfunction also affects hypoxia-inducible factor 1 (HIF-1) pathways. It was reported that increases in ROS levels in ATM-deficient cells may induce angiogenesis via increased expression of HIF-1 (Ousset et al. 2014). In addition, ATM activation in response to hypoxia (low oxygen levels) results in phosphorylation of HIF-1α, mediating downregulation of mTORC1. Though the involvement of ROS in this regulation remains unknown, the response of ATM to hypoxia is independent of DSB (Cam et al. 2010). The dysregulation by loss of ATM would lead to cancer progression via mTORC1 activation.

34.8 ATM AND LIFESTYLE-RELATED DISEASES

ATM is involved in ROS regulation, and many lifestyle-related diseases, including diabetes and atherosclerosis, are initiated or mediated by ROS (Roberts and Sindhu 2009). ROS may contribute to the development of insulin resistance, which frequently occurs in AT patients (Bar et al. 1978), who often develop type 2 diabetes mellitus (T2DM) (Schalch 1970). Interestingly, metformin, an antihyperglycemic agent used to treat T2DM, may be effective via the control of ATM function (Zhou et al. 2011). As the relationship between ATM and T2DM may involve ROS (Armata et al. 2010), AMP-activated protein kinase (AMPK), which is a major regulator in energy homeostasis and also activated in response to oxidative stress, is likely to be involved in the regulation of ATM in diabetes (Xie et al. 2011). Evidence suggests that ATM activates AMPK either directly or indirectly via LKB1, leading to TSC2 activation and mTORC1 suppression (Alexander et al. 2010a,b). AMPK-mTOR pathway is an important mechanism in autophagy regulation in response to energy stress and glucose starvation. ATM may have a key role in energy control. In addition, the impairment of p53 Ser18 phosphorylation by ATM leads to severe defects in glucose homeostasis, and NAC treatment renders p53 Ser18-deficient mice glucose tolerant. Together, these findings suggest that the ATM-p53 pathway participates in the regulation of glucose homeostasis and is related to ROS regulation (Armata et al. 2010). In addition to T2DM, ATM may be involved in the development of nonalcoholic steatohepatitis (NASH), as ROS are known to play a major role in the genesis of NASH (Caballero et al. 2014). In ATM (+/−) ApoE (−/−) mice fed a HFD, which induce obesity and oxidative stress (Mercer et al. 2012, Vares et al. 2014, Marseglia et al. 2015), atherosclerosis is promoted (Mercer et al. 2012) and may be due to the reduced DNA repair capability associated with the ATM haplotype. DNA damage caused by ROS induced by a HFD promotes plaque formation. On the other hand, mitochondria dysfunction caused by ROS production and DNA damage is also considered to promote atherosclerosis. Roles of ATM involve ROS regulation and its impairment may promote lifestyle-related diseases.

34.9 INVOLVEMENT OF ROS IN THE DEVELOPMENT OF LIFESTYLE-RELATED DISEASES ASSOCIATED WITH ATM

As described earlier, liver disease and atherosclerosis are lifestyle-related diseases that may be mediated by ATM. The relationships between ATM, ROS, and selected lifestyle-related diseases are presented in Table 34.1. Cigarette smoking induces oxidative stress (van der Vaart et al. 2004). Although cancer predisposition induced by smoking habit is likely related to the DNA repair–associated function of ATM (Jiang et al. 2007, Gao et al. 2011), cardiovascular disease (Paschalaki et al. 2013) and pulmonary emphysema (Volonte and Galbiati 2009, Volonte et al. 2009) may be related to ATM-mediated ROS regulation (Zhan et al. 2010, Paschalaki et al. 2013). Evidence suggests that expression of caveolin-1, the major protein component of specialized plasma membrane invaginations called caveolae involved in signaling and endocytic trafficking, participates in ROS regulation related to ATM and is required for the development of pulmonary emphysema (Volonte and Galbiati 2009, Volonte et al. 2009).

TABLE 34.1

Involvement of ROS and ATM in Lifestyle-Related Diseases

Lifestyle-Related Factors	Smoking Habit	Alcohol Consumption	Obesity or High-Fat Diet
Involvement of ROS production	Yes (van der Vaart et al. 2004)	Yes (Das and Vasudevan 2007)	Yes (Vares et al. 2014, Marseglia et al. 2015)
Relationship with ATM	Positive (Gao et al. 2011, Jiang et al. 2007, Paschalaki et al. 2013, Volonte et al. 2009)	Positive (Clemens et al. 2011)	Positive (Mercer et al. 2012, Daugherity et al. 2012)
Diseases possibly related to ATM and ROS	Cardiovascular disease	Liver disease (Clemens et al. 2011)	Metabolic syndrome
	Pulmonary emphysema (Paschalaki et al. 2013, Volonte and Galbiati 2009a)		Cardiovascular diseases (Mercer et al. 2012, Daugherity et al. 2012)

Consumption of alcohol is also an oxidative stress inducer (Das and Vasudevan 2007). The finding that ethanol, which is not mutagenic (Poschl and Seitz 2004), can induce ATM activation suggests that ROS is directly involved in ATM regulation. Ethanol may influence G2/M arrest via ATM and lead to the initiation and progression of alcohol-related liver diseases (Clemens et al. 2011).

In HFD-administered mice, the progression of liver fibrosis and apoptosis is likely mediated by ATM (Daugherity et al. 2012). As mentioned earlier, the frequency of metabolic syndromes is increased in ATM (+/−) and ApoE (−/−) mice fed with a HFD. Thus ATM may have protective functions in conditions promoting lifestyle-related diseases and induction of the diseases may be associated with the function of ATM in the regulation of ROS production (Mercer et al. 2012).

34.10 CONCLUSION

From the viewpoint of DNA damage responses and ROS regulation, ATM has been extensively studied to determine its protective function against various stresses. ROS regulation appears to be directly involved in multiple functions of ATM. ATM itself is not an antioxidant and would be more aptly described as a type of sensor. The fact that ATM functions to maintain HSC indicates that the regulation of oxidative stress by ATM requires systematic orchestration for normal homeostasis function. Studies on the molecular evolution of ATM (Guo et al. 2010a,b) suggest that oxidative stress not only overwhelms living organisms, but also may have been an indispensable stimulator for the evolution of protective functions against environmental stresses. It is likely that ATM has a central role in the regulation and utilization of ROS for homeostasis regulation and stress responses. Researching ROS regulation would clarify ATM functions, leading to the development of medical applications for lifestyle-related diseases such as diabetes and atherosclerosis, as well as for the understanding of AT.

REFERENCES

Alexander, A., S. L. Cai, J. Kim et al. 2010a. ATM signals to TSC2 in the cytoplasm to regulate mTORC1 in response to ROS. *Proc Natl Acad Sci USA* 107(9):4153–4158. doi: 10.1073/pnas.0913860107.

Alexander, A., J. Kim, and C. L. Walker. 2010b. ATM engages the TSC2/mTORC1 signaling node to regulate autophagy. *Autophagy* 6(5):672–673. doi: 10.4161/auto.6.5.12509.

Allen, D. M., H. van Praag, J. Ray et al. 2001. Ataxia telangiectasia mutated is essential during adult neurogenesis. *Genes Dev* 15:554–566.

Ambrose, M., J. V. Goldstine, and R. A. Gatti. 2007. Intrinsic mitochondrial dysfunction in ATM-deficient lymphoblastoid cells. *Hum Mol Genet* 16(18):2154–2164. doi: 10.1093/hmg/ddm166.

Armata, H. L., D. Golebiowski, D. Y. Jung, H. J. Ko, J. K. Kim, and H. K. Sluss. 2010. Requirement of the ATM/p53 tumor suppressor pathway for glucose homeostasis. *Mol Cell Biol* 30(24):5787–5794. doi: 10.1128/MCB.00347-10.

Bar, R. S., W. R. Levis, M. M. Rechler et al. 1978. Extreme insulin resistance in ataxia telangiectasia: Defect in affinity of insulin receptors. *N Engl J Med* 298(21):1164–1171.

Barzilai, A., G. Rotmanb, and Y. Shiloh. 2002. ATM deficiency and oxidative stress: A new dimension of defective response to DNA damage. *DNA Repair (Amst)* 1:3–25.

Boder, E. and R. P. Sedgwick. 1958. Ataxia-telangiectasia; a familial syndrome of progressive cerebellar ataxia, oculocutaneous telangiectasia and frequent pulmonary infection. *Pediatrics* 21(4):526–554.

Caballero, T., M. Caba-Molina, J. Salmeron, and M. Gomez-Morales. 2014. Nonalcoholic steatohepatitis in a patient with ataxia-telangiectasia. *Case Reports Hepatol* 2014:761250. doi: 10.1155/2014/761250.

Cam, H., J. B. Easton, A. High, and P. J. Houghton. 2010. mTORC1 signaling under hypoxic conditions is controlled by ATM-dependent phosphorylation of HIF-1alpha. *Mol Cell* 40(4):509–520. doi: 10.1016/j.molcel.2010.10.030.

Carlessi, L., L. De Filippis, D. Lecis, A. Vescovi, and D. Delia. 2009. DNA-damage response, survival and differentiation in vitro of a human neural stem cell line in relation to ATM expression. *Cell Death Differ* 16(6):795–806. doi: 10.1038/cdd.2009.10.

Chen, P., C. Peng, J. Luff et al. 2003. Oxidative stress is responsible for deficient survival and dendritogenesis in purkinje neurons from ataxia-telangiectasia mutated mutant Mice. *J Neurosci* 23(36):11453–11460.

Chessa, L., P. Petrinelli, A. Antonelli et al. 1992. Heterogeneity in ataxia-telangiectasia: Classical phenotype associated with intermediate cellular radiosensitivity. *Am J Med Genet* 42(5):741–746.

Chun, H. H. and R. A. Gatti. 2004. Ataxia-telangiectasia, an evolving phenotype. *DNA Repair (Amst)* 3(8–9):1187–1196. doi: 10.1016/j.dnarep.2004.04.010.

Clemens, D. L., K. J. Schneider, and R. F. Nuss. 2011. Ethanol metabolism activates cell cycle checkpoint kinase, Chk2. *Alcohol* 45(8):785–793. doi: 10.1016/j.alcohol.2011.07.005.

Cosentino, C., D. Grieco, and V. Costanzo. 2011. ATM activates the pentose phosphate pathway promoting anti-oxidant defence and DNA repair. *EMBO J* 30(3):546–555. doi: 10.1038/emboj.2010.330.

Das, S. K. and D. M. Vasudevan. 2007. Alcohol-induced oxidative stress. *Life Sci* 81(3):177–187. doi: 10.1016/j.lfs.2007.05.005.

Daugherity, E. K., G. Balmus, A. Al Saei et al. 2012. The DNA damage checkpoint protein ATM promotes hepatocellular apoptosis and fibrosis in a mouse model of non-alcoholic fatty liver disease. *Cell Cycle* 11(10):1918–1928. doi: 10.4161/cc.20259.

Fransen, M., M. Nordgren, B. Wang, and O. Apanasets. 2012. Role of peroxisomes in ROS/RNS-metabolism: Implications for human disease. *Biochim Biophys Acta* 1822(9):1363–1373. doi: 10.1016/j.bbadis.2011.12.001.

Gao, Y., R. B. Hayes, W. Y. Huang et al. 2011. DNA repair gene polymorphisms and tobacco smoking in the risk for colorectal adenomas. *Carcinogenesis* 32(6):882–887. doi: 10.1093/carcin/bgr071.

Gatei, M., D. Shkedy, K. K. Khanna et al. 2001. Ataxia-telangiectasia: Chronic activation of damage-responsive functions is reduced by a-lipoic acid. *Oncogene* 20:289–294.

Gilad, S., L. Chessa, R. Khosravi et al. 1998. Genotype-phenotype relationships in ataxia-telangiectasia and variants. *Am J Hum Genet* 62(3):551–561.

Gueven, N., J. Luff, C. Peng, K. Hosokawa, S. E. Bottle, and M. F. Lavin. 2006. Dramatic extension of tumor latency and correction of neurobehavioral phenotype in Atm-mutant mice with a nitroxide antioxidant. *Free Radic Biol Med* 41(6):992–1000. doi: 10.1016/j.freeradbiomed.2006.06.018.

Guo, Z., R. Deshpande, and T. T. Paull. 2010b. ATM activation in the presence of oxidative stress. *Cell Cycle* 9(24):4805–4811. doi: 10.4161/cc.9.24.14323.

Guo, Z., S. Kozlov, M. F. Lavin, M. D. Person, and T. T. Paull. 2010a. ATM activation by oxidative stress. *Science* 330:517–521.

Ito, K., A. Hirao, F. Arai et al. 2004. Regulation of oxidative stress by ATM is required for self-renewal of haematopoietic stem cells. *Nature* 431:997–1002.

Ito, K., A. Hirao, F. Arai et al. 2006. Reactive oxygen species act through p38 MAPK to limit the lifespan of hematopoietic stem cells. *Nat Med* 12(4):446–451. doi: 10.1038/nm1388.

Ito, K., K. Takubo, F. Arai et al. 2007. Regulation of reactive oxygen species by atm is essential for proper response to DNA double-strand breaks in lymphocytes. *J Immunol* 178(1):103–110. doi: 10.4049/jimmunol.178.1.103.

Jiang, Y., Z. D. Liang, T. T. Wu, L. Cao, H. Zhang, and X. C. Xu. 2007. Ataxia-telangiectasia mutated expression is associated with tobacco smoke exposure in esophageal cancer tissues and benzo[a]pyrene diol epoxide in cell lines. *Int J Cancer* 120(1):91–95. doi: 10.1002/ijc.22121.

Kamsler, A., D. Daily, A. Hochman et al. 2001. Increased oxidative stress ataxia telangiectasia evidenced by alterations in redox state of brains from atm-deficient mice1. *Cancer Res* 61:1849–1854.

Kim, J. and P. K. Wong. 2009a. Loss of ATM impairs proliferation of neural stem cells through oxidative stress-mediated p38 MAPK signaling. *Stem Cells* 27(8):1987–1998. doi: 10.1002/stem.125.

Kim, J. and P. K. Wong. 2009b. Oxidative stress is linked to ERK1/2-p16 signaling-mediated growth defect in ATM-deficient astrocytes. *J Biol Chem* 284(21):14396–14404. doi: 10.1074/jbc.M808116200.

Kowaltowski, A. J., N. C. de Souza-Pinto, R. F. Castilho, and A. E. Vercesi. 2009. Mitochondria and reactive oxygen species. *Free Radic Biol Med* 47(4):333–343. doi: 10.1016/j.freeradbiomed.2009.05.004.

Lavin, M. F. 2013. The appropriateness of the mouse model for ataxia-telangiectasia: Neurological defects but no neurodegeneration. *DNA Repair (Amst)* 12(8):612–619. doi: 10.1016/j.dnarep.2013.04.014.

Li, B., X. Wang, N. Rasheed et al. 2004. Distinct roles of c-Abl and Atm in oxidative stress response are mediated by protein kinase C. *Genes Dev* 18:1824–1837.

Li, J., J. Chen, C. L. Ricupero et al. 2012. Nuclear accumulation of HDAC4 in ATM deficiency promotes neurodegeneration in ataxia telangiectasia. *Nat Med* 18(5):783–790. doi: 10.1038/nm.2709.

Liu, J., X. Wang, Y. Ren, X. Li, X. Zhang, and B. Zhou. 2014. Effect of single nucleotide polymorphism Rs189037 in ATM gene on risk of lung cancer in Chinese: A case-control study. *PLoS One* 9(12):e115845. doi: 10.1371/journal.pone.0115845.

Liu, N., G. Stoica, M. Yan et al. 2005. ATM deficiency induces oxidative stress and endoplasmic reticulum stress in astrocytes. *Lab Invest* 85(12):1471–1480.

Marseglia, L., S. Manti, G. D'Angelo et al. 2015. Oxidative stress in obesity: A critical component in human diseases. *Int J Mol Sci* 16(1):378–400. doi: 10.3390/ijms16010378.

Maryanovich, M. and A. Gross. 2013. A ROS rheostat for cell fate regulation. *Trends Cell Biol* 23(3):129–134. doi: 10.1016/j.tcb.2012.09.007.

Maryanovich, M., G. Oberkovitz, H. Niv, L. Vorobiyov, Y. Zaltsman, O. Brenner, T. Lapidot, S. Jung, and A. Gross. 2012. The ATM-BID pathway regulates quiescence and survival of haematopoietic stem cells. *Nat Cell Biol* 14(5):535–541. doi: 10.1038/ncb2468.

Matsuoka, S., B. A. Ballif, A. Smogorzewska et al. 2007. ATM and ATR substrate analysis reveals extensive protein networks responsive to DNA damage. *Science* 316:1160–1166.

McKinnon, P. J. 2004. ATM and ataxia telangiectasia. *EMBO Rep* 5(8):772–776. doi: 10.1038/sj.embor.7400210.

McKinnon, P. J. 2012. ATM and the molecular pathogenesis of ataxia telangiectasia. *Annu Rev Pathol* 7:303–321.

Mercer, J. R., E. M. Yu, N. Figg et al. 2012. The mitochondria-targeted antioxidant MitoQ decreases features of the metabolic syndrome in ATM(+/–)/ApoE(–/–) mice. *Free Radic Biol Med* 52(5):841–849. doi: 10.1016/j.freeradbiomed.2011.11.026.

Meredith, M. J. and M. L. Dodson. 1987. Impaired glutathione biosynthesis in cultured human ataxia-telangiectasia cells. *Cancer Res* 47:4576–4581.

Nakajima, T. 2009. Cell signaling in ataxia telangiectasia. In *Molecular Mechanisms of Ataxia Telangiectasia*, Ed. S. I. Ahmad, pp. 14–22. Austin, TX: Landes Bioscience.

Ousset, M., F. Bouquet, F. Fallone et al. 2014. Loss of ATM positively regulates the expression of hypoxia inducible factor 1 (HIF-1) through oxidative stress: Role in the physiopathology of the disease. *Cell Cycle* 9(14):2886–2894. doi: 10.4161/cc.9.14.12253.

Pagano, G., D. Paolo, and G. Castello. 2009. Ataxia telangiectasia: An oxidative stress-related disease. In *Molecular Mechanisms of Ataxia Telangiectasia*, Ed. S. I. Ahmad, pp. 72–77. Austin, TX: Landes Bioscience.

Paschalaki, K. E., R. D. Starke, Y. Hu et al. 2013. Dysfunction of endothelial progenitor cells from smokers and chronic obstructive pulmonary disease patients due to increased DNA damage and senescence. *Stem Cells* 31(12):2813–2826. doi: 10.1002/stem.1488.

Poschl, G. and H. K. Seitz. 2004. Alcohol and cancer. *Alcohol Alcohol* 39(3):155–165. doi: 10.1093/alcalc/agh057.

Reichenbach, J., R. Schubert, C. Schwan, K. Müller, H. J. Böhles, and S. Zielen. 1999. Anti-oxidative capacity in patients with ataxia telangiectasia. *Clin Exp Immunol* 117(3):535–539.

Reliene, R., S. M. Fleming, M. F. Chesselet, and R. H. Schiestl. 2008. Effects of antioxidants on cancer prevention and neuromotor performance in Atm deficient mice. *Food Chem Toxicol* 46(4):1371–1377. doi: 10.1016/j.fct.2007.08.028.

Reliene, R. and R. H. Schiestl. 2006. Antioxidant *N*-acetyl cysteine reduces incidence and multiplicity of lymphoma in Atm deficient mice. *DNA Repair (Amst)* 5(7):852–859. doi: 10.1016/j.dnarep.2006.05.003.

Reliene, R. and R. H. Schiestl. 2007. Antioxidants suppress lymphoma and increase longevity in atm-deficient mice. *J Nutr* 137:229S–232S.

Roberts, C. K. and K. K. Sindhu. 2009. Oxidative stress and metabolic syndrome. *Life Sci* 84(21–22):705–712. doi: 10.1016/j.lfs.2009.02.026.

Santos, M. A., R. B. Faryabi, A. V. Ergen et al. 2014. DNA-damage-induced differentiation of leukaemic cells as an anti-cancer barrier. *Nature* 514(7520):107–111. doi: 10.1038/nature13483.

Savitsky, K., A. Bar-Shira, S. Gilad et al. 1995. A single ataxia telangiectasia gene with a product similar to P1–3 kinase. *Science* 268:1749–1753.

Schalch, D. S., D. E. McFarlin, and M. H. Barlow. 1970. An unusual form of diabetes mellitus in ataxia telangiectasia. *N Engl J Med* 282(25):1396–1402.

Shen, L., Z. H. Yin, Y. Wan, Y. Zhang, K. Li, and B. S. Zhou. 2012. Association between ATM polymorphisms and cancer risk: A meta-analysis. *Mol Biol Rep* 39(5):5719–5725.

Shiloh, Y. and Y. Ziv. 2013. The ATM protein kinase: Regulating the cellular response to genotoxic stress, and more. *Nat Rev Mol Cell Biol* 14(4):197–210. doi: 10.1038/nrm3546.

Stagni, V., V. Oropallo, G. Fianco, M. Antonelli, I. Cina, and D. Barila. 2014. Tug of war between survival and death: Exploring ATM function in cancer. *Int J Mol Sci* 15(4):5388–5409. doi: 10.3390/ijms15045388.

Swift, M., P. J. Reitnauer, D. Morrell, and C. L. Chase. 1987. Breast and other cancers in families with ataxia-telangiectasia. *N Engl J Med* 316(21):1289–1294.

Takao, N., Y. Li, and K. Yamamoto. 2000. Protective roles for ATM in cellular response to oxidative stress. *FEBS Lett* 472:133–136.

Taylor, A. M. and P. J. Byrd. 2005. Molecular pathology of ataxia telangiectasia. *J Clin Pathol* 58(10):1009–1015. doi: 10.1136/jcp.2005.026062.

Taylor, A. M., Z. Lam, J. I. Last, and P. J. Byrd. 2015. Ataxia telangiectasia: More variation at clinical and cellular levels. *Clin Genet* 87(3):199–208.

Toyoshima, M., T. Hara, H. Zhang et al. 1998. Ataxia-telangiectasia without immunodeficiency: Novel point mutations within and adjacent to the phosphatidylinositol 3-kinase-like domain. *Am J Med Genet* 75(2):141–144.

Valentin-Vega, Y. A., K. H. MacLean, J. Tait-Mulder et al. 2012. Mitochondrial dysfunction in ataxia-telangiectasia. *Blood* 119(6):1490–1500. doi: 10.1182/blood-2011-08373639.

van der Vaart, H., D. S. Postma, W. Timens, and N. H. ten Hacken. 2004. Acute effects of cigarette smoke on inflammation and oxidative stress: A review. *Thorax* 59(8):713–721. doi: 10.1136/thx.2003.012468.

Vares, G., B. Wang, H. Ishii-Ohba, M. Nenoi, and T. Nakajima. 2014. Diet-induced obesity modulates epigenetic responses to ionizing radiation in mice. *PLoS One* 9(8):e106277. doi: 10.1371/journal.pone.0106277.

Volonte, D. and F. Galbiati. 2009. Caveolin-1, cellular senescence and pulmonary emphysema. *Aging* 1(9):831–835.

Volonte, D., B. Kahkonen, S. Shapiro, Y. Di, and F. Galbiati. 2009. Caveolin-1 expression is required for the development of pulmonary emphysema through activation of the ATM-p53-p21 pathway. *J Biol Chem* 284(9):5462–5466. doi: 10.1074/jbc.C800225200.

Watters, D., P. Kedar, K. Spring et al. 1999. Localization of a portion of extranuclear ATM to peroxisomes. *J Biol Chem* 274(48):34277–34282.

Weyemi, U., C. E. Redon, T. Aziz et al. 2015. NADPH oxidase 4 is a critical mediator in Ataxia telangiectasia disease. *Proc Natl Acad Sci USA* 112(7):2121–2126. doi: 10.1073/pnas.1418139112.

Xie, Z., C. He, and M.-H. Zou. 2011. AMP-activated protein kinase modulates cardiac autophagy in diabetic cardiomyopathy. *Autophagy* 7(10):1254–1255. doi: 10.4161/auto.7.10.16740.

Zhan, H., T. Suzuki, K. Aizawa, K. Miyagawa, and R. Nagai. 2010. Ataxia telangiectasia mutated (ATM)-mediated DNA damage response in oxidative stress-induced vascular endothelial cell senescence. *J Biol Chem* 285(38):29662–29670. doi: 10.1074/jbc.M110.125138.

Zhou, K., C. Bellenguez, C. C. Spencer et al. 2011. Common variants near ATM are associated with glycemic response to metformin in type 2 diabetes. *Nat Genet* 43(2):117–120. doi: 10.1038/ng.735.

35 Role of Oxidative Stress in RNA Virus–Induced Cell Death

Mohammad Latif Reshi and Jiann-Ruey Hong

CONTENTS

ABSTRACT

RNA virus infections lead to diseases with broad clinical symptoms and complex pathogenic mechanisms. Studies of reactive oxygen species (ROS) have shown that these reactive molecules are produced in all types of cells and serve as important messengers in cell signaling and various signal transduction pathways. However, excessive oxidant formation is usually a consequence of tissue damage caused by a disease or toxin.

Oxidative stress arises when there is an imbalance between oxidants and antioxidants of diverse origin, which can damage both the structure and function of tissues. A prominent feature of many viral diseases is excessive generation of ROS. An imbalance between pro- and antioxidants appears to be implicated in many aspects of viral infection, including immune function, inflammatory response, virus replication, and apoptosis. Oxidative stress can have profound effects, not only on the host, but also on the pathogen.

The therapeutic use of antioxidants for viral disease therefore has the potential to be applied at many levels and could replace symptomatic therapy. The principal aim of this chapter is to sum up the role of ROS in RNA virus pathogenesis, with particular attention to the human immunodeficiency virus (HIV), hepatitis C virus (HCV), human influenza virus, and lower invertebrate betanodavirus, which mostly infects fishes. Clarifying the role of oxidative stress in apoptosis could lead to the discovery of novel therapeutic strategies and insights into the pathogenesis of different viral diseases.

35.1   INTRODUCTION

Normal cellular metabolism appears to be the main source of endogenous reactive oxygen species (ROS). The oxidants produced are essential to the normal functioning of the body's immune system and take part in signaling pathways and transcription activation [1–3]. However, various external factors, such as viral infections, smoking, and UV radiation, can lead to the generation of excessive ROS. Oxidative stress in biological systems is usually caused by an overload of oxidants, but those of the greatest concern to biological systems are derived from oxygen. Sustained oxidative stress disrupts cellular structures and functions that are usually maintained and mediated by critical oxidation reduction (redox) pathways. The resulting damage to cells and tissues contributes to the pathophysiology of many diseases. ROS and certain free radicals are generated in various human infections caused mostly by viruses, but also bacteria and parasites [4]. ROS have been found to enhance viral replication in various viral infections [5–7]. Increased generation of ROS and reactive nitrogen species (RNS) is a feature of many viral infections and can be caused by the direct effects of viruses on cells and the inflammatory responses of the host [8]. The resulting imbalance in the production of ROS and antioxidants and the body's inability to detoxify ROS is referred to as oxidative stress [9]. Oxidative stress results in the subsequent oxidation of cellular components that activate cytoplasmic/nuclear signal transduction pathways. Recent observations of the multiple pathogenic interactions between ROS and RNA viruses have drawn attention to the possibility that these types of interactions may play a role in the pathogenesis of DNA and RNA viruses [10]. This chapter will consider the accumulated evidence that suggests that patients infected with various RNA viruses are under chronic oxidative stress.

35.2   HOST CELLULAR RESPONSES TO OXIDANT INJURY

The role of oxidants in viral disease is complex because it includes metabolic regulation both of host metabolism and viral replication. There is no doubt that ROS play an important role in various viral diseases. A number of host mechanisms have been shown or are suspected to contribute toward the pathogenesis of viral infection, including excessive cytokines, lipid peroxidation, release of lipid mediators and complement activation [11]. Virus-induced activation of phagocytes is associated with oxidative stress, not only because ROS are released but also because activated phagocytes may release pro-oxidant cytokines such as tumor necrosis factor alpha (TNF-α) [6,12] and interleukin-1, which promote iron uptake through the macrophage system [7,13]. TNF-α is either released from activated phagocytes into circulation or synthesized in infected host cells. In either case, TNF-α can act on host cell mitochondria, producing a pro-oxidant effect; it has been shown to inhibit mitochondrial respiration at complex II, the site of superoxide ($O_2^{-\bullet}$) production [9,14]. This effect can be inhibited by antioxidants such as vitamin E [9,14]. TNF-α also acts to release nuclear transcription factor kappa B (NF-κB) from the cytoplasmic inhibitor protein IκB [3,10]. Following its release, the NF-κB is translocated to the nucleus where it binds to DNA, inducing the transcription of cellular and/or viral genes. NF-κB-induced gene transcription can be inhibited by certain antioxidants such as thiol donors [11,15] (Table 35.1).

TABLE 35.1
Antioxidant Enzymes and Human Viral Diseases

	Disease	Pathogen	Main Key Enzymes	Reference
1.	Infectious disease	Influenza virus	CAT, GPX, SOD	[138]
2.	Infectious disease	HIV	GPX	[139]
3.	Infectious disease	Hepatitis	GPX	[142]

Activated monocytes also release interleukin-1, which stimulates neutrophils to release lyso-somal proteins, including lactoferrin [7,13]. Lactoferrin rapidly binds iron, an effect that explains the hypoferremia of acute inflammatory states; in such conditions, lactoferrin-bound iron accumulates in the macrophage system. If the accumulated iron exceeds the cellular iron-binding capacity, unbound pro-oxidant iron could interact with $O_2^{-\bullet}$ via Fenton's reaction (Figure 35.1) to produce highly reactive hydroxyl radicals ($^{\bullet}OH$). Another aspect of the role of ROS in viral pathogenesis is its positive modulatory role in immune activation, which is important both in eradicating viral infections and immune system–induced cellular injury. The proliferation of T-lymphocytes is a key event in cell-mediated immune responses. Extensive studies by Hunt and coworkers have demonstrated that antioxidants inhibit the proliferation of T-lymphocytes incubated with antigen-presenting cells [12,16]. The role of the host cell pro- and/or antioxidant balance during virally induced apoptosis is currently an area of great interest at the interface between free radical biology, virology, and oncology. Apoptosis, or programmed cell death, an active cell-suicide process with distinct morphologic features, is counteracted by antioxidants, such as N-acetylcysteine and glutathione peroxidase, which can scavenge several peroxides [3,17]. A variety of viruses, including influenza A and B viruses [15,18], produce apoptotic cytotoxicity, a phenomenon that may have evolved to contain viral infections. HIV also produces apoptotic cytotoxicity and HIV gene expression enhances T cell susceptibility to hydrogen peroxide-induced apoptosis [16,17,19]. The antioxidant N-acetylcysteine inhibits HIV-induced apoptosis and viral replication [18,20]. In addition, the host cell, anti-apoptotic gene Bcl-2, inhibits apoptosis via an antioxidant pathway [3,17]. Interestingly, the transfection of host cells with this oncogene results in the conversion of lytic alphavirus infection to a persistent infection [19,21]. Some proteins of both RNA and DNA viruses may actually exert an antioxidant effect that may be linked to the oncogenic potential of the viruses. Viruses can also affect the host cell pro-/antioxidant balance by increasing cellular pro-oxidants such as iron and nitric oxide and by inhibiting the synthesis of antioxidant enzymes such as superoxide dismutase (SOD). ROS could theoretically affect viruses by altering the redox state of the host cell to select certain viral mutants and/or by producing mutations and activating transcription factors such as NF-κB, which enhance viral replication. Given the multitude of pro- and antioxidant effects observed both *in vitro* and *in vivo* during viral infections, and also the efforts to incorporate antioxidants into the therapeutic armamentarium, a variety of viral infections is discussed in this chapter.

35.3 ROS PRODUCTION DURING CERTAIN RNA VIRUS INFECTION

It is well known, and good evidence is present with a large number of published references, that ROS are generated during infection by the following RNA viruses leading to oxidative stress: HIV, HCV, influenza virus, and the fish betanodavirus.

35.3.1 OXIDATIVE STRESS IN THE MOLECULAR PATHOGENESIS OF THE HEPATITIS C VIRUS

The hepatitis C virus (HCV) belongs to the Flaviviridae family of RNA viruses, with a positive strand RNA genome of 9400 bp [22,23]. About 150 million people are infected each year, and, as a result, have an increased chance of developing liver cancer or cirrhosis [24]. The World Health Organization (WHO) has reported that 80% of patients with acute hepatitis C progress to chronic hepatitis, with 2% developing liver cirrhosis and 1%–5% developing liver cancer [25,26]. Researchers first demonstrated the occurrence of oxidative stress during chronic hepatitis C in 1990 [27]. This oxidative stress is associated with hepatic damage, a decrease in glutathione (GSH); an increase in serum malondialdehyde (MDA), 4-hydroxynonenal (HNE) and caspase activity; and a decrease in plasma and hepatic zinc concentrations [28–30]. Zinc therapy increases the functioning of surviving liver tissue [31]. Zinc and selenium deficiencies affect DNA repair and the immune system, increasing the chances of chronicity and malignancy [32]. Although HCV replication takes place in hepatocytes, the virus also potentially attacks and propagates in immune system cells.

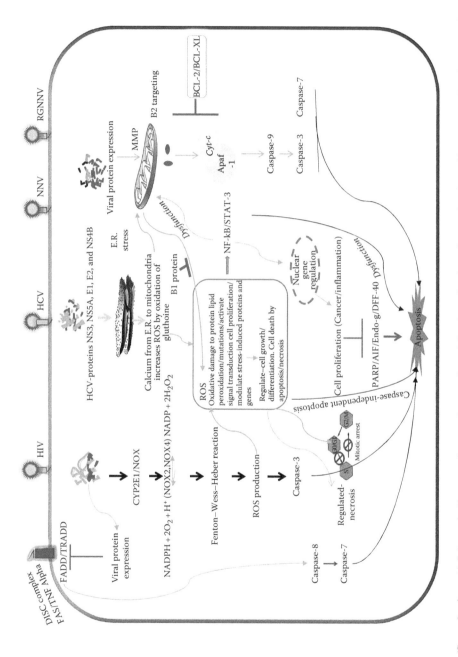

FIGURE 35.1 **(See color insert.)** Schematic diagram of apoptotic cascade and the sites of action of general and specific caspases. The figure also illustrates how ROS produced during viral infection can affect apoptotic cascades. *Abbreviations:* FAS/TNF-α, tumor necrosis factor alpha (death receptors); FADD, Fas-associated death domain; TRADD, tumor necrosis factor receptor–associated death domain; Cas, caspase; AIF, apoptosis-inducing factor; PARP, poly(ADP-ribosyl) polymerase; EndoG, endonuclease G; DFF-40, DNA fragmentation factor 40 kDa; tBID, truncated BID; DISC, death-inducing signaling complex.

Reports have shown that 3% of the world's population is infected with HCV and the rate of infection is increasing alarmingly, especially in developing nations [33]. Although acute hepatitis caused by HCV is naturally cleared in 20%–30% of patients, 70%–80% of cases involve chronic hepatitis [34]. No effective vaccine was available until recently, and the current treatment too is not very effective [35,36]. Reports have shown that HCV gene expression in the host cell increases the level of ROS via mediation of calcium signaling [37]. The release of calcium from the endoplasmic reticulum (ER) results in ER stress. The released calcium is taken up by the mitochondria, resulting in increased ROS production and oxidative stress (Figure 35.1). Oxidative stress is the main contributor to a number of diseases such as cancer, diabetes, and viral infections [38,39]. The livers of patients infected with HCV show elevated levels of ROS and decreased antioxidant levels [40]. It has been reported that the two core proteins, HCV NS3 and NS5A, are responsible for oxidative stress in culture cells [41]. However, the host cell Cox-2 gene, which is the main regulator of prostaglandins, is activated by excess ROS [42]. This activation involves NF-κB, which is present in cells in an inactive form but becomes activated and migrates to the nucleus in HCV-infected cells due to ER stress and ROS [43]. NF-κB controls the expression of the genes responsible for apoptosis and inflammation (Figure 35.1). Elevated ROS also activates another transcriptional factor, STAT-3, which is responsible for cell proliferation, survival, and ontogenesis [42]. This coactivation of NF-κB and STAT-3 in HCV-infected cells is equally important in both acute and chronic liver diseases [39,43,44].

35.3.1.1 HCV Genome in ROS Production

The liver plays an important role in the detoxification and metabolism of harmful substances and is the main target of HCV. HCV replicates in cytoplasm, causing hepatitis, cirrhosis, and hepatocellular carcinoma [45,46]. ROS-induced viral genome heterogeneity has been considered in terms of viral escape from the immune system [47]. The core nucleocapsid protein of HCV is responsible for increasing oxidative stress in the liver [48]. Although this core protein is considered to be the main contributor to oxidative stress, other viral proteins, such as NS3 and NS5A, are also involved in generating oxidative stress [49–53]. Recent studies have shown that many other viral proteins such as E1, E2, and NS4B are also involved [49,51,54–56]. The nonstructural protein NS5A is a membrane integral protein that is important not only for viral replication but also for apoptosis and immune responses, such as interferon resistance and changes in calcium levels [57]. NS5A and NS3 increase calcium uptake and cause glutathione oxidation in mitochondria, thereby increasing ROS production [36,58,59]. The mitochondria activate and translocate the transcription factors NF-κB and STAT3 to the nucleus, leading to oxidative stress. The NS5A activation of NF-κB and STAT3 is repressed by antioxidants [60,61]. NS4B also translocates NF-κB to the nucleus in a PTK-mediated pathway. ROS and nitric oxide not only cause oxidative damage but also affect the DNA repair machinery that leads to cell apoptosis [62–64]. ROS are believed to be the main culprits of liver inflammation in HCV infections [64].

35.3.2 Oxidative Stress in the Molecular Pathogenesis of the Influenza Virus

Infection by different influenza viruses presents different clinical scenarios [65–67]. The influenza A virus is highly active, causing infection of the upper and lower respiratory tract. It has been classified into 16 groups based on hemagglutinin (HA) and neuraminidase (NA) combinations, with HA1, HA2, and HA3, and NA1 and NA2 being prevalent in humans [68]. Studies conducted in 19 countries have shown that H1N1 and H3N2 are the most dominant influenza A viruses prevailing globally [69]. Influenza viruses and parvoviruses have been shown to activate monocytes and poly-morphonuclear monocytes to generate ROS *in vitro* [11]. Activated phagocytes can release not only ROS but also cytokine and TNF-α. The pro-antioxidant effect of TNF-α may be relevant to influenza virus infection because children with Reye's syndrome exhibit increased levels of pro-oxidants and lipid peroxides [70]. Oxidative stress ultimately leads to a decrease in antioxidant levels and indicates a decrease in the functioning of the immune system. Immune system cells generally require a higher concentration of antioxidants than other cells to maintain the system's redox balance and preserve its integrity and function.

The influenza virus induces the production of ROS in host cells that can damage the virus genome [71]. ROS enhance the pathogenesis of infections such as influenza [72,73]. A study on mice infected with the influenza virus showed that although the spread of infection remained confined to the airways and lungs, systemic effects such as weight loss and a decrease in body temperature were clearly apparent [74]. The mice used in the experiment died after 5 or 6 days. The cells taken from the dead mice showed elevated levels of ROS and xanthine oxidase (a ROS generating enzyme), indicating enhanced ROS production [75]. Furthermore, analysis of the antioxidant level revealed an overall decrease in its concentration during the infection. The study suggested that influenza virus infection is associated with oxidative stress. In another study, pyran copolymer–conjugated SOD (PC-SOD) was found to protect mice from the effects of intravenously injected influenza virus. This observation was not immediately apparent because PC-DOD is a well-known antiviral agent [76]. The localized effect of the influenza virus was difficult to detect because the analysis of the amount of ROS in tissues was based on whole tissue homogenates. The infected mice released cytokines and lipid mediators that could have caused the systemic symptoms [77]. To determine the cause of the systemic symptoms, the mice injected with the influenza virus were given cytokine injections (mostly interferon); they showed symptoms resembling influenza [78]. ROS are known for their antiviral activity [79] and can also increase the titer of the influenza virus. The virus carries a glycoprotein known as hemagglutinin on its surface, which is responsible for binding the virus to cells with sialic acid on their membranes, such as cells in the upper respiratory tract and erythrocytes [80]. The hemagglutinin protein is synthesized in an inactive form (HAO) and activated by specific proteases into HA1 and HA2.

The cleavage of HAO into HA1 and HA2 is an important determinant of influenza virulence [80,81]. If the influenza virus released from the cell contains inactive HAO, it may be activated by proteases present in pulmonary surfactants [82]. However, these anti-proteases can be inactivated by ROS, converting a noninfectious influenza virus into an infectious one. Further studies have shown that an oxidant-treated anti-protease is unable to prevent trypsin from converting HAO to HA1 and HA2, resulting in a 10,000-fold increase in virus infection [80]. However, how the influenza virus induces apoptosis is still not clear. ROS production enhances the molecular pathogenesis of the influenza virus. Research has demonstrated that although ROS damage lung parenchyma, the damage can be repaired by taking an appropriate dose of antioxidants [82,83]. ROS are important in the normal development of the whole organism, are important components of adaptive immune responses, and are involved in the normal function of many transcription factors [84,85]. The production of ROS-mediated $O_2^{-\bullet}$ is also an important defense against microbial infections. However, the excessive production of $O_2^{-\bullet}$ associated with infection by the influenza A virus is detrimental. The downregulation of $O_2^{-\bullet}$, achieved by targeting specific enzymes such as NADPH oxidase-2, markedly alleviates lung injuries caused by the influenza virus and viral replication, irrespective of the viral strain [86]. One study showed that influenza infection leads to thymus-specific elevation of the mitochondrial $O_2^{-\bullet}$, which interferes with the normal functioning of T cell lymphocyte damage in influenza A virus infections [87]. A further knockdown of SOD2 indicates that T cells begin the apoptosis process and take on many developmental defects, resulting in an overall weakening of the adaptive immune system and increased susceptibility to the influenza A virus (H1N1). ROS inhibitors and other therapeutic agents targeting ROS may prove useful in controlling influenza [88,89].

35.3.3 Oxidative Stress in the Molecular Pathogenesis of Human Immunodeficiency Virus

Oxidative stress plays a dominant pathogenic role in HIV and hepatitis infections. AIDS, the end phase of HIV infection, is caused by the HIV-1 and HIV-2 groups of cytopathic viruses. Infected patients have increased GSH; decreased cysteine, vitamin C, and SOD; and elevated MDA and HNE levels [90,91]. A decrease in antioxidants indicates the weakening of the immune system, as immune cells require more antioxidants to maintain their function and integrity. The CD4+ T-helper cells that form important components of the immune system are the main targets of the HIV-1 and HIV-2 viruses. The virus production decreases to begin with, as about 5% of the T cells are destroyed and replaced each day via the apoptotic process. This in turn leads to decreases in zinc and vitamin E (antioxidants). Zinc inhibits intracellular virus replication, and selenium slows down the progression of HIV toward AIDS [72]. After suffering from the primary illness, patients may not show any symptoms for more than 10 years, during which time the virus load falls, but it does not stop and continues replication, albeit slowly. A subsequent fall in the CD4+ T cell count ultimately leads to AIDS and the terminal stage of the infection begins. HIV-2 infection progresses slowly compared with HIV-1 infection. In hepatitis patients, HIV preferentially infects CD4+ T-lymphocytes and macrophages.

Oxidative stress has been found to occur in various viral infections and may enhance viral replication [4,92]. *In vitro* studies have shown that oxidative stress promotes HIV replication [6,93,94]. The nuclear transcription factor NF-κB, which is necessary for viral replication, is activated when oxidative stress is present [6,95]. The other role of NF-κB is to activate many of the immune system's inflammatory cytokines [96,97]. Many antioxidants have been examined with a view to determining their antiviral activities. However, effects of antioxidants in different cell cultures have been found to be different and *in vitro* studies demonstrated that no improvement occurs at higher concentrations. HIV-infected AIDS patients exhibit elevated serum hydroperoxides and malondialdehyde, which are the by-products of lipid peroxidation, in addition to membrane damage [98–100]. They also exhibit an increase in resting oxygen consumption, due to the link between free radical formation and oxygen metabolism [100]. These observations are associated with the production of ROS in the neutrophils of HIV-infected patients, whose antioxidant defense systems undergo dramatic changes [101]. Children suffering from HIV infection exhibit decreased SOD levels and activity [102]. Catalase activity increases as AIDS progresses in HIV-infected patients and the level of glutathione peroxidase in red blood cells and plasma decreases, showing that the body's antioxidant system becomes weaker as HIV progresses [17]. The imbalance between oxidation versus antioxidation inside and outside the cell leads to an excess of stored hydrogen peroxide (H_2O_2), which further increases the $^{\bullet}OH$ and lipid peroxide that signal the cell to undergo a programmed cell death [103]. The addition of H_2O_2 and antioxidants results in a respective increase and decrease in apoptosis in cell cultures. AIDS, which is characterized by a decrease in CD4+ lymphocytes, is currently believed to be the main cause of this apoptosis [104,105]. The imbalance in ROS seems to contribute to the progression of AIDS in different ways, including apoptosis of CD4+ cells and changes in the functioning of other immune system components [106].

35.3.3.1 Envelope Glycoprotein gp120 of HIV under Oxidative Stress

HIV-1 uses glycoprotein (gp120) to enter its host cells (T cells and monocytes). Infected monocytes can cross the blood-brain barrier (BBB) and replicate in astrocytes and microglia [107,108]. Recent work has shown that HIV-1 induces oxidative stress in astrocytes and microglia and that gp120 can directly induce apoptosis in neurons [109–112]. It has also been shown that the involvement of cytochrome P450 (CYP) in neurotoxicity may be due to the generation of ROS or other reactive metabolites [110]. Furthermore, gp120 together with the drug methamphetamine involves CYP and NOX pathways in apoptotic cell death (Figure 35.1). Both gp120 and methamphetamine have been found to cause oxidative stress in a time-dependent manner [112,113]. The ROS-mediated BBB damage in HIV-1 infection has been shown to cause lipid peroxidation and a loss of tight junction proteins [111,114,115]. Methamphetamine and gp120 together cause a loss of tight junction proteins in the BBB, making it leaky and thus facilitating the entry of infected monocytes [115]. Methamphetamine increases oxidative stress through dopaminergic and glutamatergic mechanisms, and gp120 increases oxidative stress through glutathione and lipid peroxidation [110,116,117]. A combination of cocaine and gp120 results in excess ROS that in turn activate caspase-3 and NF-κB to force astrocytes to undergo apoptosis [118]. In addition to considering the role of ROS in different diseases, recent reports have shown that oxidative stress is involved in the pathology of neurocognitive disorders associated with HIV [119,120]. The role that CYP plays in different tissues and organs, including the brain, have also been confirmed [121]. Astrocytes have been shown to express CYPs at variable levels, and the roles of CYP2E1 and CYP2A6 in alcohol- and nicotine-mediated oxidative stress have been demonstrated [113,122,123]. Methamphetamine has been shown to increase the expression of CYP2A6, 2B6, and 2D6; gp120 has been shown to increase the expression of CYP2E1, 2B6, and 2D6 [124]. These overall increases suggest that CYPs may be involved in oxidative stress. The interaction between CYP and NADPH is tightly regulated by NOX enzymes, which are currently being used therapeutically in various central nervous system disorders such as Alzheimer's disease and stroke [125,126]. Studies have shown that NOX2 and NOX4 increase oxidative stress in astrocytes [126]. Others have shown that when NOX2 and NOX4 expression is blocked in astrocytes, the level of oxidative stress decreases, indicating that NOX could be used as a therapeutic agent in the treatment of neuro-AIDS. Studies of astrocyte oxidative stress caused by methamphetamine and gp120 have examined the use of antioxidants in HIV-1 pathogenesis and considered the potential of CYP pathways as a target of new drugs.

35.3.4 BETANODAVIRUS AND VIRAL PROTEINS AS MODULATORS OF MITOCHONDRIAL DYSFUNCTION AND ROS PRODUCTION

Betanodavirus is an RNA virus belonging to one of the two genera making up the family Nodaviridae. The virus is an etiological agent of viral nervous necrosis (also known as viral encephalopathy and retinopathy) [127]. The disease is characterized by necrosis of the neural cells of the brain, retina, and spinal cord and can cause up to 100% mortality in larval and juvenile fish and significant losses in older fish. Symptoms commonly shown by infected fish are abnormal swimming behavior, darkening of the skin, and weight loss [127]. The betanodavirus genome encodes proteins alpha and B2, both of which are death inducers (Figure 35.1) [128,129]. Protein B2 causes mitochondria-mediated cell death via a Bax-mediated pathway; it can target mitochondria and induce ROS production and mitochondrial fragmentation via Drp1 recruitment [127,130–132]. Our research shows that the production of ROS during red spotted grouper nervous necrosis virus (RGNNV) infection ultimately leads to mitochondria-mediated cell death [133]. This supports previous research on oxidative stress and cell death during infection with RNA viruses and opens doors for the development of new drugs targeting enzymes or other key factors involved in ROS production. Mitochondria complexes I and II of the electron transport chains are the major sites of ROS production [134,135]. Both the inhibition of complexes I (rotenone) and II (antimycin) and the oxidation of either complex cause increased ROS production [134–137]. However, ROS are also important in the activation of the body's antioxidant enzymes such as SOD and glutathione peroxidase [138,139]. RGNNV-infected cells were found to produce ROS 24 h after infection, with a gradual regulation of catalase and Nrf2 transcription factors [140] and autophagy [141]. However, it remains unclear whether Nrf2 upregulates ROS production. We used antioxidants such as NAC and DPI and overexpressed zebrafish catalase to further explore our hypothesis and found a decrease in RGNNV-induced ROS production and an increase in cell viability. The cell death mechanism influenced by the novel antinecrotic protein B1 remains unknown. To determine which cell death mechanism is influenced by B1, we examined how the betanodavirus nonstructural protein B1 regulates oxidative stress and p53 expression in fish cell lines [140,142].

35.4 CONCLUSION

ROS researchers have described ROS as secondary messengers that influence a number of different molecular processes, including the apoptotic, anti-apoptotic, and pro-apoptotic expression of a number of genes. The physiological role played by ROS is important because viruses depend on the biosynthetic mechanisms of their host cells. The activation of ROS production during viral infections in the absence of antiviral antibodies could play a role in the generation of symptoms and pathologies such as the induction of fever, and it could also lead to hemorrhages in internal organs. Therefore, the main challenge for present-day molecular virologists is to understand the pathophysiological functions of ROS, to gain a deeper understanding of the many aspects of infectious viral diseases. The effect of ROS on the host's immune response is an important aspect of viral pathogenesis and mutation. The toxicity and reactivity of ROS, which are produced in excess by the overreaction of immune responses against the organs or tissues in which viruses replicate, may explain the tissue injury observed in various viral diseases involving immunological interactions.

As discussed in this chapter, most available evidence indicate that free radicals play a complex role in viral diseases, beginning with their influence on the host cell's metabolism and viral replication and extending their desirable inactivation effects on themselves and undesirable toxic effects on host tissues. Therefore, a combination of several molecular mechanisms described in this chapter could be exploited to find new solutions for combating oxidative stress in different viral infections. However, the role of oxidants in viral diseases is complex because it includes regulation of both host metabolism and viral replication. The use of antioxidants in viral disease therapy could therefore be applied at many levels and replace symptomatic therapy, which does not alter viral replication. Any new therapy should also target additional mechanisms that contribute to the symptoms and pathology of viral diseases such as cytokines, lipid peroxidation, and NO$^{\bullet}$. Understanding host pathogen interactions at the molecular level requires the characterization of host-derived small radical molecules, which appear to play an important role in the pathogenesis of viral infection. An energizing concept related to free radicals would contribute insights into the molecular mechanisms of pathological events that occur as a result of the interaction between viruses and their hosts. Deeper and more detailed research must be conducted to better understand the molecular mechanisms and specific apoptotic pathways involved in ROS-mediated cell death and discover how to interfere with their activation or stop their undesired effects. The limitations of interfering with such viral disease mechanisms are similar to those encountered when interfering with oxidant generation, as these pathways are associated with normal host physiology and pathology. It is clear that any useful approach to solving this problem will require a variety of drugs. Clarifying the role of oxidation stress in apoptosis could lead to novel therapeutic strategies and insights into different viral diseases, particularly given that ROS-mediated mechanisms are responsible for apoptosis during viral infections.

ACKNOWLEDGMENT

This work was supported by grants (NSC-99-2321-B-006-010-MY3) awarded to Dr. Jainn-Ruey Hong from the National Science Council, Taiwan, Republic of China.

REFERENCES

1. A. Knebel, H. J. Rahmsdorf, A. Ullrich, and P. Herrlich, Dephosphorylation of receptor tyrosine kinases as target of regulation by radiation, oxidants or alkylating agents. *EMBO Journal*, 15(19): 5314–5325, 1996.
2. R. P. Huang, J.-X. Wu, Y. Fan, and E. D. Adamson, UV activates growth factor receptors via reactive oxygen intermediates. *Journal of Cell Biology*, 133(1): 211–220, 1996.
3. R. Schreck, P. Rieber, and P. A. Baeuerle, Reactive oxygen intermediates as apparently widely used messengers in the activation of the NF-κB transcription factor and HIV-1. *EMBO Journal*, 10(8): 2247–2258, 1991.
4. G. Nabel and D. Baltimore, An inducible transcription factor activates expression of human immunodeficiency virus in T cells. *Nature*, 326(6114): 711–713, 1987.
5. P. A. Baeuerle and D. Baltimore, Activation of DNA-binding activity in an apparently cytoplasmic precursor of the NF-κB transcription factor. *Cell*, 53(2): 211–217, 1988.
6. F. J. T. Staal, M. Roederer, L. A. Herzenberg, and L. A. Herzenberg, Intracellular thiols regulate activation of nuclear factor κB and transcription of human immunodeficiency virus. *Proceedings of the National Academy of Sciences of the United States of America*, 87(24): 9943–9947, 1990.

7. H. U. Simon, S. Yousefi, C. Schranz, A. Schapowal, C. Bachert, and K. Blaser, Direct demonstration of delayed eosinophil apoptosis as a mechanism causing tissue eosinophilia. *Journal of Immunology*, 158(8): 3902–3908, 1997.

8. R. Dringen and B. Hamprecht, Involvement of glutathione peroxidase and catalase in the disposal of exogenous hydrogen peroxide by cultured astroglial cells. *Brain Research*, 759(1): 67–75, 1997.

9. M. Chrobot, A. Szaflarska-Szczepanik, and G. Drewa, Antioxidant defense in children with chronic viral hepatitis B and C. *Medical Science Monitor*, 6(4): 713–718, 2000.

10. M. L. Reshi, Y. C. Su, and J. R. Hong, RNA viruses: ROS-mediated cell death. *International Journal of Cell Biology*, 2014: 467452, 2014.

11. E. Peterhans, Oxidants and antioxidants in viral diseases: Disease mechanisms and metabolic regulation. *Journal of Nutrition*, 127(5): 962S–965S, 1997.

12. D. T. Gloenbock, R. Y. Hampton, R. S. Qusesh, K. Takayom, and C. R. H. Raetz, Lipid Al-like molecules that antagonize the effects of endotoxin on human monocytes. *Journal of Biological Chemistry*, 266: 19490–19498, 1991.

13. M. S. Klempner, C. A. Dinarello, and J. I. Gallis, Human leukocytic pyrogen induces release of specific granule contents from human neutrophils. *Journal of Clinical Investigation*, 61: 1330–1336, 1978.

14. K. Schulze-Osthoff, A. C. Baker, B. Vanhaese broeck, R. Beyaer, W. A. Jacob, and W. Fiers, Cytotoxic activity of tumor necrosis factor is mediated by early damage of mitochondrial functions. *Journal of Biological Chemistry*, 167: 5317–5323, 1992.

15. R. Schreck, B. Meier, D. N. Mintel, W. Drtge, and P. A. Baeuerle, Dithiocarbamates as potent inhibitors of nuclear factor B activation in intact cells. *Journal of Experimental Medicine*, 175: 1181–1194, 1992.

16. N. H. Hunt, D. M. van Reyk, J. C. Fragonas, T. M. Jeimer, and S. D. Goldstone, Redox mechanisms in T cell activation. In: *Oxidative Stress Cell Activation Viral Infection*. Birkhauser Verlag: Basel, Switzerland, 1994, pp. 237–251.

17. D. Hockenbery, Z. Oltvai, X. Yin, C. Milliman, and S. Korsmeyer, Bcl-2 functions in an antioxidant pathway to prevent apoptosis. *Cell*, 75: 241–251, 1993.

18. V. S. Hinshaw, C. W. Olsen, N. Dybahl-Sissoko, and D. Evans, Apoptosis: A mechanism of cell killing by influenza A and B viruses. *Journal of Virology*, 68: 3667–3673, 1994.

19. P. A. Sandstrom, B. Roberts, T. M. Folks, and T. M. Buttke, HIV gene expression enhances T cell susceptibility to hydrogen peroxide-induced apoptosis. *Aids Research and Human Retroviruses*, 9(11): 1107–1113, 1993.

20. W. Malorui, R. Rivabene, M. T. Santini, and G. Donelli, *N*-Acetylcysteine inhibits apoptosis and decreases viral particles in HIV-chronically infected U937 cells. *FEBS Letters*, 327(1):75–78, 1993.

21. M. Rowe, M. Peng-Pilon, D. Huen, R. D. Hardy, Lundgren, and A. B. Rickinson, Upregulation of Bcl-2 by the Epstein-Barr virus latent membrane protein LMPI: A B-cell specific response that is delayed relative to NF-KB activation and to induction of cell surface markers. *Journal of Virology*, 68: 5602–5612, 1994.

22. Q. L. Choo, G. Kuo, A. J. Weiner, L. R. Overby, D. W. Bradley, and M. Houghton, Isolation of a cDNA clone derived from a blood-borne non-A, non-B viral hepatitis genome. *Science*, 244(4902): 359–362, 1989.

23. P. Simmonds, Genetic diversity and evolution of hepatitis C virus—15 years on. *Journal of General Virology*, 85(11): 3173–3188, 2004.

24. F. V. Chisari and C. Ferrari, Hepatitis B virus immunopathogenesis. *Annual Review of Immunology*, 13: 29–60, 1995.

25. G. L. Davis, Treatment of chronic hepatitis C. *British Medical Journal*, 323(7322): 1141–1142, 2001.

26. A. V. Ivanov, B. Bartosch, O. A. Smirnova, M. G. Isaguliants, and S. N. Kochetkov, HCV and oxidative stress in the liver. *Viruses*, 5(2): 439–469, 2013.

27. H. Bantel, A. Lügering, C. Poremba et al. Caspase activation correlates with the degree of inflammatoryliver injury in chronic Hepatitis C virus infection. *Hepatology*, 34(4): 758–767, 2001.

28. P. Boya, A. D. L. Pena, O. Beloqui et al., Antioxidant status and glutathione metabolism in peripheral blood mononuclear cells from patients with chronic hepatitis C. *Journal of Hepatology*, 31(5): 808–814, 1999.

29. N. De Maria, A. Colantoni, S. Fagiuoli et al., Association between reactive oxygen species and disease activity in chronic hepatitis C. *Free Radical Biology and Medicine*, 21(3): 291–295, 1996.

30. G. P. Bianchi, G. Marchesini, M. Brizi et al., Nutritional effects of oral zinc supplementation in cirrhosis. *Nutrition Research*, 20(8): 1079–1089, 2000.

31. P. M. Newberne, S. Broitman, and T. F. Schrager, Esophageal carcinogenesis in the rat: Zinc deficiency, DNA methylation and alkyl transferase activity. *Pathobiology*, 65(5): 253–263, 1997.

32. Anonymous, World Health Organization—Hepatitis C: Global prevalence. *Weekly Epidemiological Record*, 72: 341–344, 1997.
33. H. J. Alter and L. B. Seeff, Recovery, persistence, and sequelae in hepatitis C virus infection: A perspective on long-term outcome. *Seminars in Liver Disease*, 20(1): 17–35, 2000.
34. M. W. Fried, M. L. Shiffman, K. Rajender Reddy et al., Peg interferon alfa-2a plus ribavirin for chronic hepatitis C virus infection. *New England Journal of Medicine*, 347(13): 975–982, 2002.
35. S. J. Hadziyannis, H. Sette Jr., T. R. Morgan et al., Peg interferon-α2a and ribavirin combination therapy in chronic hepatitis C: A randomized study of treatment duration and ribavirin dose. *Annals of Internal Medicine*, 140(5): 346–355, 2004.
36. G. Gong, G. Waris, R. Tanveer, and A. Siddiqui, Human hepatitis C virus NS5A protein alters intracellular calcium levels, induces oxidative stress, and activates STAT-3 and NF-κB. *Proceedings of the National Academy of Sciences of the United States of America*, 98(17): 9599–9604, 2001.
37. H. E. Poulsen, H. Prieme, and S. Loft, Role of oxidative DNA damage in cancer initiation and promotion. *European Journal of Cancer Prevention*, 7(1): 9–16, 1998.
38. K. B. Schwarz, Oxidative stress during viral infection: A review. *Free Radical Biology and Medicine*, 21(5): 641–649, 1996.
39. D. Yadav, H. I. Hertan, P. Schweitzer, E. P. Norkus, and C. S. Pitchumoni, Serum and liver micronutrient antioxidants and serum oxidative stress in patients with chronic hepatitis C. *American Journal of Gastroenterology*, 97(10): 2634–2639, 2002.
40. J. Choi, K. J. Lee, Y. Zheng, A. K. Yamaga, M. M. C. Lai, and J.-H.Ou, Reactive oxygen species suppress hepatitis C virus RNA replication in human hepatoma cells. *Hepatology*, 39(1): 81–89, 2004.
41. G. Waris and A. Siddiqui, Hepatitis C virus stimulates the expression of cyclooxygenase-2 via oxidative stress: Role of prostaglandin E2 in RNA replication. *Journal of Virology*, 79(15): 9725–9734, 2005.
42. G. Waris, A. Livolsi, V. Imbert, J.-F. Peyron, and A. Siddiqui, Hepatitis C virus NS5A and sub genomic replicon activate NF-κB via tyrosine phosphorylation of IκBα and its degradation by calpain protease. *Journal of Biological Chemistry*, 278(42): 40778–40787, 2003.
43. T. Bowman, R. Garcia, J. Turkson, and R. Jove, STATs in oncogenesis. *Oncogene*, 19(21): 2474–2488, 2000.
44. E. R. Jenny-Avital, Hepatitis C. *Current Opinion in Infectious Diseases*, 11(3): 293–299, 1998.
45. T. M. Saito, T. Miyamura, A. Ohbayashi et al., Hepatitis C virus infection is associated with the development of hepatocellular carcinoma. *Proceedings of the National Academy of Sciences of the United States of America*, 87(17): 6547–6549, 1990.
46. X. Forns, R. H. Purcell, and J. Bukh, Quasispecies in viral persistence and pathogenesis of hepatitis C virus. *Trends in Microbiology*, 7(10): 402–410, 1999.
47. K. Koike, Hepatitis C virus contributes to hepato carcinogenesis by modulating metabolic and intracellular signaling pathways. *Journal of Gastroenterology and Hepatology*, 22(1): S108–S111, 2007.
48. V. Ivanov, O. A. Smirnova, O. N. Ivanova, O. V. Masalova, S. N. Kochetkov, and M. G. Isaguliants, Hepatitis C virus proteins activate NRF2/ARE pathway by distinct ROS-dependent and independent mechanisms in HUH7 cells. *PLoS One*, 6(9): e24957, 2011.
49. S. Pal, S. J. Polyak, N. Bano et al., Hepatitis C virus induces oxidative stress, DNA damage and modulates the DNA repair enzyme NEIL1. *Journal of Gastroenterology and Hepatology*, 25(3): 627–634, 2010.
50. C. Bureau, J. Bernad, N. Chaouche et al., Nonstructural 3 protein of hepatitis C virus triggers an oxidative burst in human monocytes via activation of NADPH oxidase. *Journal of Biological Chemistry*, 276(25): 23077–23083, 2001.
51. M. V. Garcia-Mediavilla, S. Sanchez-Campos, P. Gonzalez- Perez et al., Differential contribution of hepatitis C virus NS5A and core proteins to the induction of oxidative and nitrosative stress in human hepatocyte-derived cells. *Journal of Hepatology*, 43(4): 606–613, 2005.
52. F. Thoren, A. Romero, M. Lindh, C. Dahlgren, and K. Hellstrand, A hepatitis C virus-encoded, nonstructural protein (NS3) triggers dysfunction and apoptosis in lymphocytes: Role of NADPH oxidase-derived oxygen radicals. *Journal of Leukocyte Biology*, 76(6): 1180–1186, 2004.
53. C. Garcia-Monzon, P. L. Majano, I. Zubia, P. Sanz, A. Apolinario, and R. Moreno-Otero, Intrahepatic accumulation of nitrotyrosine in chronic viral hepatitis is associated with histological severity of liver disease. *Journal of Hepatology*, 32(2): 331–338, 2000.
54. H. Ming-Ju, H. Yih-Shou, C. Tzy-Yen, and C. Hui-Ling, Hepatitis C virus E2 protein induce reactive oxygen species (ROS)-related fibrogenesis in the HSC-T6 hepatic stellate cell line. *Journal of Cellular Biochemistry*, 112(1): 233–243, 2011.

55. S. Li, L. Ye, X. Yu et al., Hepatitis C virus NS4B induces unfolded protein response and endoplasmic reticulum overload response-dependent NF-κB activation. *Virology*, 391(2): 257–264, 2009.
56. Y. Liang, H. Ye, B. K. Cong, and S. Y. Ho, Domain 2 of nonstructural protein 5A (NS5A) of hepatitis C virus is natively unfolded. *Biochemistry*, 46(41): 11550–11558, 2007.
57. H. H. Yen, K. L. Shih, T. T. Lin, W. W. Su, M. S. Soon, and C. S. Liu, Decreased mitochondrial deoxyribonucleic acid and increased oxidative damage in chronic hepatitis C. *World Journal of Gastroenterology*, 18(36): 5084–5089, 2012.
58. C. Piccoli, G. Quarato, M. Ripoli et al., HCV infection induces mitochondrial bioenergetic unbalance: Causes and effects. *Biochimica et Biophysica Acta*, 1787(5): 539–546, 2009.
59. C. Piccoli, R. Scrima, A. D'Aprile et al., Mitochondrial dysfunction in hepatitis C virus infection. *Biochimica et Biophysica Acta*, 1757(9–10): 1429–1437, 2006.
60. M. Okuda, K. Li, M. R. Beard et al., Mitochondrial injury, oxidative stress, and antioxidant gene expression are induced by hepatitis C virus core protein. *Gastroenterology*, 122(2): 366–375, 2002.
61. M. Jaiswal, N. F. LaRusso, R. A. Shapiro, T. R. Billiar, and G. J. Gores, Nitric oxide-mediated inhibition of DNA repair potentiates oxidative DNA damage in cholangiocytes. *Gastroenterology*, 120(1): 190–199, 2001.
62. K. Machida, G. Mcnamara, K. T.-H. Cheng et al., Hepatitis C virus inhibits DNA damage repair through reactive oxygen and nitrogen species and by interfering with the ATMNBS1/Mre11/Rad50 DNA repair pathway in monocytes and hepatocytes. *Journal of Immunology*, 185(11): 6985–6998, 2010.
63. K. Machida, H. Tsukamoto, J.-C. Liu et al., c-Jun mediates hepatitis C virus hepato carcinogenesis through signal transducer and activator of transcription 3 and nitric oxide-dependent impairment of oxidative DNA repair. *Hepatology*, 52(2): 480–492, 2010.
64. J. Choi and J.-H. J. Ou, Mechanisms of liver injury. III. Oxidative stress in the pathogenesis of hepatitis C virus. *American Journal of Physiology: Gastrointestinal and Liver Physiology*, 290(5): G847–G851, 2006.
65. M. Kaji, A. Watanabe, and H. Aizawa, Differences in clinical features between influenza A H1N1, A H3N2, and B in adult patients. *Respirology*, 8(2): 231–233, 2003.
66. Y. C. Hsieh, T. Z. Wu, D. P. Liu et al., Influenza pandemics: Past, present and future. *Journal of the Formosan Medical Association*, 105(1): 1–6, 2006.
67. B. S. Finkelman, C. Viboud, K. Koelle, M. J. Ferrari, N. Bharti, and B. T. Grenfell, Global patterns in seasonal activity of influenza A/H3N2, A/H1N1, and B from 1997 to 2005: Viral coexistence and latitudinal gradients. *PLoS One*, 2(12): e1296, 2007.
68. E. Peterhans, M. Grob, T. Bürge, and R. Zanoni, Virus-induced formation of reactive oxygen intermediates in phagocytic cells. *Free Radical Research Communications*, 3(1–5): 39–46, 1987.
69. M. S. Cooperstock, R. P. Tucker, and J. V. Baublis, Possible pathogenic role of endotoxin in Reye's syndrome. *The Lancet*, 1(7919): 1272–1274, 1975.
70. O. A. Levander, Nutrition and newly emerging viral diseases: An overview. *Journal of Nutrition*, 127(5): 948S–950S, 1997.
71. T. Hennet, H. J. Ziltener, K. Frei, and E. Peterhans, A kinetic study of immune mediators in the lungs of mice infected with influenza A virus. *Journal of Immunology*, 149(3): 932–939, 1992.
72. T. Akaike, M. Ando, T. Oda et al., Dependence on O_2-generation by xanthine oxidase of pathogenesis of influenza virus infection in mice. *Journal of Clinical Investigation*, 85(3): 739–745, 1990.
73. T. Oda, T. Akaike, T. Hamamoto, F. Suzuki, T. Hirano, and H. Maeda, Oxygen radicals in influenza-induced pathogenesis and treatment with pyran polymer-conjugated SOD. *Science*, 244(4907): 974–976, 1989.
74. S. C. Kunder, L. Wu, and P. S. Morahan, Protection against murine cytomegalovirus infection in aged mice and mice with severe combined immunodeficiency disease with the biological response modifiers poly ribosinicolycytidylic acid stabilized with L-lysine and carboxymethyl cellulose, maleic anhydride divinyl ether and colony stimulating factor I. *Antiviral Research*, 21(3): 233–245, 1993.
75. I. Dvoretzky, Flulike symptoms with interferon. *Journal of the American Academy of Dermatology*, 22(2): 321–322, 1990.
76. M. E. Belding, S. J. Klebanoff, and C. G. Ray, Per oxidase mediated virucidal systems. *Science*, 167(3915): 195–196, 1970.
77. R. J. Russell, P. S. Kerry, D. J. Stevens et al., Structure of influenza hemagglutinin in complex with an inhibitor of membrane fusion. *Proceedings of the National Academy of Sciences of the United States of America*, 105(46): 17736–17741, 2008.
78. R. Rott, H.-D. Klenk, Y. Nagai, and M. Tashiro, Influenza viruses, cell enzymes, and pathogenicity. *American Journal of Respiratory and Critical Care Medicine*, 152(4): S16–S19, 1995.

79. H. Kido, K. Sakai, Y. Kishino, and M. Tashiro, Pulmonary surfactant is a potential endogenous inhibitor of proteolytic activation of Sendai virus and influenza A virus. *FEBS Letters*, 322(2): 115–119, 1993.

80. T. Hennet, E. Peterhans, and R. Stocker, Alterations in antioxidant defences in lung and liver of mice infected with influenza A virus. *Journal of General Virology*, 73(1): 39–46, 1992.

81. J. Geiler, M. Michaelis, P. Naczk et al., *N*-acetyl-ʟ-cysteine (NAC) inhibits virus replication and expression of pro inflammatory molecules in A549 cells infected with highly pathogenic H5N1 influenza A virus. *Biochemical Pharmacology*, 79(3): 413–420, 2010.

82. H. B. Suliman, L. K. Ryan, L. Bishop, and R. J. Folz, Prevention of influenza-induced lung injury in mice overexpressing extracellular superoxide dismutase. *American Journal of Physiology: Lung Cellular and Molecular Physiology*, 280(1): L69–L78, 2001.

83. R. G. Allen and A. K. Balin, Oxidative influence on development and differentiation: An overview of a free radical theory of development. *Free Radical Biology and Medicine*, 6(6): 631–661, 1989.

84. M. J. Hitchler and F. E. Domann, An epigenetic perspective on the free radical theory of development. *Free Radical Biology and Medicine*, 43(7): 1023–1036, 2007.

85. D. J. M. Malvy, M.-J. Richard, J. Arnaud, A. Favier, and O. Amédée-Manesme, Relationship of plasma malondialdehyde, vitamin E and antioxidant micronutrients to human immunodeficiency virus-1 seroptivity. *Clinica Chimica Acta*, 224(1): 89–94, 1994.

86. J. Fuchs, I. Emerit, A. Levy, L. Cernajvski, H. Schofer, and R. Milbradt, Clastogenic factors in plasma of HIV-1 infected patients. *Free Radical Biology and Medicine*, 19(6): 843–848, 1995.

87. E. Peterhans, Reactive oxygen species and nitric oxide in viral diseases. *Biological Trace Element Research*, 56(1):107–116, 1997.

88. J. E. Sprietsma, Zinc-controlled Th1/Th2 switch significantly determines development of diseases. *Medical Hypotheses*, 49(1): 1–14, 1997.

89. R. Vlahos, J. Stambas, and S. Selemidis, Suppressing production of reactive oxygen species (ROS) for influenza A virus therapy. *Trends in Pharmacological Sciences*, 33(1): 3–8, 2012.

90. W. C. L. Ford and A. Harrison, The role of oxidative phosphorylation in the generation of ATP in human spermatozoa. *Journal of Reproduction and Fertility*, 63(1): 271–278, 1981.

91. J. Fuchs, F. Ochsendorf, H. Schofer, R. Milbradt, and H. Rubsamen-Waigmann, Oxidative imbalance in HIV infected patients. *Medical Hypotheses*, 36(1): 60–64, 1991.

92. T. Oda, T. Akaike, T. Hamamoto, F. Suzuki, T. Hirano, and H. Maeda, Oxygen radicals in influenza-induced pathogenesis and treatment with pyran polymer-conjugated SOD. *Science*, 244(4907): 974–976, 1989.

93. E. Peterhans, T. W. Jungi, and R. Stocker, Autotoxicity and reactive oxygen in viral disease. In *Oxy-Radicals in Molecular Biology and Pathology*, I. Fridovich, J. M. McCord, P. A. Cerutti, Eds. New York: Alan R. Liss, Inc., 1988, pp. 543–562.

94. M. J. Lenardo and D. Baltimore, NK-κB: A pleiotropic mediator of inducible and tissue-specific gene control. *Cell*, 58(2): 227–229, 1989.

95. M. A. Brach, S. de Vos, C. Arnold, H.-J. Gruss, R. Mertelsmann, and F. Herrmann, Leukotriene B4 transcriptionally activates interleukin-6 expression involving NK-κB and NF-IL6. *European Journal of Immunology*, 22(10): 2705–2711, 1992.

96. M. Meyer, H. L. Pahl, and P. A. Baeuerle, Regulation of the transcription factors NF-κB and AP-1 by redox changes. *Chemico-Biological Interactions*, 91(2–3): 91–100, 1994.

97. A. Favier, C. Sappey, P. Leclerc, P. Faure, and M. Micoud, Antioxidant status and lipid peroxidation in patients infected with HIV. *Chemico-Biological Interactions*, 91(2–3): 165–180, 1994.

98. J. P. Revillard, C. M. A. Vincent, A. E. Favier, M.-J. Richard, M. Zittoun, and M. D. Kazatchkine, Lipid peroxidation in human immunodeficiency virus infection. *Journal of Acquired Immune Deficiency Syndromes*, 5(6): 637–638, 1992.

99. A. Sonnerborg, G. Carlin, B. Akerlund, and C. Jarstrand, Increased production of malondialdehyde in patients with HIV infection. *Scandinavian Journal of Infectious Diseases*, 20(3): 287–290, 1988.

100. C. Jarstrand and B. Akerlund, Oxygen radical release by neutrophils of HIV-infected patients. *Chemico-Biological Interactions*, 91(2–3): 141–146, 1994.

101. B. Halliwell and J. M. C. Gutteridge, Role of free radicals and catalytic metal ions in human disease: An overview. *Methods in Enzymology*, 186: 1–85, 1990.

102. V. M. Polyakov, A. P. Shepelev, O. E. Kokovkina, and I. V. Vtornikova, Superoxide anion (O_2^-) production and enzymatic disbalance in peripheral blood cells isolated from HIV-infected children. *Free Radical Biology and Medicine*, 16: 15, 1994.

103. J. A. Leff, M. A. Oppegard, T. J. Curiel, K. S. Brown, R. T. Schooley, and J. E. Repine, Progressive increases in serum catalase activity in advancing human immunodeficiency virus infection. *Free Radical Biology and Medicine*, 13(2): 143–149, 1992.

104. G. Laurent-Crawford, B. Krust, S. Muller et al., The cytopathic effect of HIV is associated with apoptosis. *Virology*, 185(2): 829–839, 1991.

105. H. C. Greenspan and O. I. Aruoma, Oxidative stress and apoptosis in HIV infection: A role for plant-derived metabolites with synergistic antioxidant activity. *Immunology Today*, 15(5): 209–213, 1994.

106. W. Droge, H. P. Eck, and S. Mihm, Oxidant-antioxidant status in human immunodeficiency virus infection. In *Oxygen Radicals in Biological Systems*, L. Packer, Ed. San Diego, CA: Academic Press, 1994, pp. 594–601.

107. M. Kaul, J. Zheng, S. Okamoto, H. E. Gendelman, and S. A. Lipton, HIV-1 infection and AIDS: Consequences for the central nervous system. *Cell Death and Differentiation*, 12(1): 878–892, 2005.

108. D. Carroll-Anzinger, A. Kumar, V. Adarichev, F. Kashanchi, and L. Al-Harthi, Human immunodeficiency virus-restricted replication in astrocytes and the ability of gamma interferon to modulate this restriction are regulated by a downstream effector of the Wnt signaling pathway. *Journal of Virology*, 81(11): 5864–5871, 2007.

109. P. V. B. Reddy, N. Gandhi, T. Samikkannu et al., HIV-1 gp120 induces antioxidant response element-mediated expression in primary astrocytes: Role in HIV associated neurocognitive disorder. *Neurochemistry International*, 61: 807–814, 2012.

110. P. T. Ronaldson and R. Bendayan, HIV-1 viral envelope glycoprotein gp120 produces oxidative stress and regulates the functional expression of multidrug resistance protein-1 (Mrp1) in glial cells. *Journal of Neurochemistry*, 106(3): 1298–1313, 2008.

111. P. S. Silverstein, A. Shah, J. Weemhoff, S. Kumar, D. P. Singh, and A. Kumar, HIV-1 gp120 and drugs of abuse: Interactions in the central nervous system. *Current HIV Research*, 10: 369–383, 2012.

112. J. Brown III, C. Theisler, S. Silberman et al., Differential expression of cholesterol hydroxylases in Alzheimer's disease. *Journal of Biological Chemistry*, 279(33): 34674–34681, 2004.

113. A. Shah, S. Kumar, S. D. Simon et al., HIV gp120- and methamphetamine-mediated oxidative stress induces astrocyte apoptosis via cytochrome P450 2E1. *Cell Death & Disease*, 4(10): e850, 2013.

114. M. Park, B. Hennig, and M. Toborek, Methamphetamine alters occludin expression via NADPH oxidase-induced oxidative insult and intact caveolae. *Journal of Cellular and Molecular Medicine*, 16(2): 362–375, 2012.

115. A. Banerjee, X. Zhang, K. R. Manda, W. A. Banks, and N. Ercal, HIV proteins (gp120 and Tat) and methamphetamine in oxidative stress-induced damage in the brain: Potential role of the thiol antioxidant *N*-acetyl cysteine amide. *Free Radical Biology and Medicine*, 48(10): 1388–1398, 2010.

116. P. S. Silverstein, A. Shah, R. Gupte et al., Methamphetamine toxicity and its implications during HIV-1 infection. *Journal of NeuroVirology*, 17(5): 401–415, 2011.

117. J. P. Louboutin and D. S. Strayer, Blood-brain barrier abnormalities caused by HIV-1 gp120: Mechanistic and therapeutic implications. *The Scientific World Journal*, 2012: 482575, 1–15, 2012.

118. H. Yao, J. E. Allen, X. Zhu, S. Callen, and S. Buch, Cocaine and human immunodeficiency virus type 1 gp120 mediate neurotoxicity through overlapping signaling pathways. *Journal of NeuroVirology*, 15(2): 164–175, 2009.

119. K. Dasuri, L. Zhang, and J. N. Keller, Oxidative stress, neurodegeneration, and the balance of protein degradation and protein synthesis. *Free Radical Biology and Medicine*, 62: 170–185, 2012.

120. V. Mollace, H. S. L. M. Nottet, P. Clayette et al., Oxidative stress and neuroAIDS: Triggers, modulators and novel antioxidants. *Trends in Neurosciences*, 24(7): 411–416, 2001.

121. M. Jin, A. Ande, A. Kumar, and S. Kumar, Regulation of cytochrome P450 2e1 expression by ethanol: Role of oxidative stress-mediated pkc/jnk/sp1 pathway. *Cell Death & Disease*, 4: e554, 2013.

122. R. E. Ande, M. Jin et al., An LC-MS/MS method for concurrent determination of nicotine metabolites and the role of CYP2A6 in nicotine metabolite-mediated oxidative stress in SVGA astrocytes. *Drug and Alcohol Dependence*, 125: 49–59, 2012.

123. R. P. Meyer, M. Gehlhaus, R. Knoth, and B. Volk, Expression and function of cytochrome P450 in brain drug metabolism. *Current Drug Metabolism*, 8(4): 297–306, 2007.

124. A. A. Eid, Y. Gorin, B. M. Fagg et al., Mechanisms of podocyte injury in diabetes role of cytochrome P450 and NADPH oxidases. *Diabetes*, 58(5): 1201–1211, 2009.

125. S. Sorce, K. H. Krause, and V. Jaquet, Targeting NOXs enzymes in the central nervous system: Therapeutic opportunities. *Cellular and Molecular Life Sciences*, 69: 2387–2407, 2012.

126. B. Cairns, J. Y. Kim, X. N. Tang, and M. A. Yenari, NOX inhibitors as a therapeutic strategy for stroke and neurodegenerative disease. *Current Drug Targets*, 13(2): 199–206, 2012.

127. K. Bedard and K.-H. Krause, The NOX family of ROS generating NADPH oxidases: Physiology and pathophysiology. *Physiological Reviews*, 87(1): 245–313, 2007.

128. B. L. Munday, J. Kwang, and N. Moody, Betanodavirus infections of teleost fish: A review. *Journal of Fish Diseases*, 25: 127–142, 2002.

129. Y. C. Su, J. L. Wu, and J. R. Hong, Betanodavirus nonstructural protein B2: A novel necrotic death factor that induces mitochondria-mediated cell death in fish cells. *Virology*, 385(1): 143–154, 2009.

130. Y. C. Su and J. R. Hong, Betanodavirus B2 causes ATP depletion-induced cell death via mitochondrial targeting and Complex II inhibition *in vitro* and *in vivo*. *Journal of Biological Chemistry*, 285(51): 39801–39810, 2010.

131. Y. C. Su, H. W. Chiu, J. C. Hung, and J. R. Hong, Beta-nodavirus B2 protein induces hydrogen peroxide production, leading to Drp1-recruited mitochondrial fragmentation and cell death via mitochondrial targeting. *Apoptosis*, 19(10): 1457–1470, 2014.

132. S. P. Chen, H. L. Yang, G. M. Her, H. Y. Lin, M. F. Jeng, J. L. Wu, and J. R. Hong, NNV induces phosphatidylserine exposure and loss of mitochondrial membrane potential in secondary necrotic cells, both of which are blocked by bongkrekic acid. *Virology*, 347: 379–391, 2006.

133. J. F. Turrens and A. Boveris, Generation of superoxide anion by the NADH dehydrogenase of bovine heart mitochondria, *Biochemical Journal*, 191(2): 421–427, 1980.

134. J. F. Turrens, A. Alexandre, and A. L. Lehninger, Ubisemiquinone is the electron donor for superoxide formation by complex III of heart mitochondria. *Archives of Biochemistry and Biophysics*, 237(2): 408–414, 1985.

135. J. St-Pierre, J. A. Buckingham, S. J. Roebuck, and M. D. Brand, Topology of superoxide production from different sites in the mitochondrial electron transport chain. *The Journal of Biological Chemistry*, 277(47): 44784–44790, 2002.

136. M. Ott, V. Gogvadze, and S. Orrenius, Mitochondria, oxidative stress and cell death. *Apoptosis*, 12(5): 913–922, 2007.

137. D Jacoby and A. M. Choi, Influenza virus induces expression of antioxidant genes in human epithelial cells. *Free Radical Biology and Medicine*, 16: 821–824, 1994.

138. K. Banki, E. Hutter, N. J. Gonchoroff, and A. Perl, Molecular ordering in HIV-induced apoptosis. Oxidative stress, activation of caspases, and cell survival are regulated by Trans aldolase. *Journal of Biological Chemistry*, 273, 11944–11953, 1998.

139. L. J. Chen, Y. C. Su, and J.-R. Hong, Betanodavirus nonstructural protein B1: A novel anti-necrotic death factor that modulates cell death in early replication cycle in fish cells. *Virology*, 385(2): 444–454, 2009.

140. C. W. Chang, Y. C. Su, G. M. Her, C. F. Ken, and J. R. Hong, Betanodavirus induces oxidative stress-mediated cell death prevented by anti-oxidants and zfcatalase in fish cells. *PLoS One*, 6(10): e25853, 2011.

141. H.-J. Liao, J.-L. Wu, and J.-R. Hong, Autophagy contributes to betanodavirus infection for viral death expressions, leading to host anti-apoptotic members Bcl-2/Bcl-xL downregulation. In Press, 2016.

142. J. Downey, C. D. Bingle, S. Cottrell, N. Ward, D. Churchman, M. Dobrota, and C. J. Powell, The LEC rat reduced hepatic selenium, contributing to the severity of spontaneous hepatitis and sensitivity to carcinogenesis. *Biochemical and Biophysical Research Communications*, 244: 463–467, 1998.

36 Organophosphate-Induced Oxidative Stress and Neurotoxicity

Alan J. Hargreaves

CONTENTS

ABSTRACT

The widespread agricultural, domestic, and industrial use of organophosphorous compounds has heightened public concern about their toxicity in humans. Those used as insecticides and chemical weapons cause acute toxicity by the irreversible inhibition of acetylcholinesterase (AChE), causing subsequent neuromuscular block, which can be fatal. However, survivors of acute cholinergic toxicity may go on to develop delayed toxicity in the form of intermediate syndrome or various types of delayed neuropathy. Moreover, chronic exposure to some organophosphorous compounds can cause developmental toxicity in the young and cognitive lesions in adults. As there is evidence that antioxidants may attenuate at least some of these toxic effects, the aim of this chapter was to review evidence for the involvement of oxidative stress in the different types of neurotoxicity induced by exposure to organophosphates (OPs). Oxidative stress was found to be a common event in the toxicity of many organophosphorous compounds, suggesting that elevated levels of reactive oxygen species and impaired antioxidant status play a role in the neurotoxicity of these compounds, may have value as prognostic markers and could represent potential targets for therapeutic intervention.

36.1 INTRODUCTION

Organophosphorus compounds (OPs) are widely used as insecticides, herbicides, flame retardants, and lubricants (Chambers and Levi, 1992; Hargreaves, 2012). Many commonly used OPs contain a pentavalent phosphorous atom linked by a double bond to either an oxygen atom (organophosphate [OP]) or a sulfur (organophosphorothioate) atom. Various combinations of aryl and alkyl substituent groups can be ester-linked to the phosphorous atom in the other three positions. The chemical structures of some typical OPs are shown in Figure 36.1. The continued widespread use of OPs has

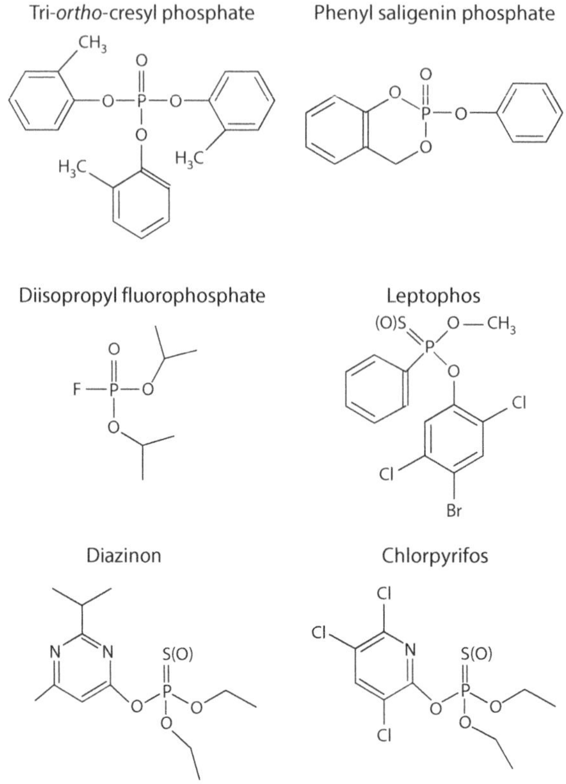

FIGURE 36.1 Chemical structures of some commonly used organophosphorous compounds. Many OPs used as insecticides, nerve agents, and oil additives contain a pentavalent phosphorous atom linked via a double bond to an oxygen atom (in organophosphates) or a sulfur atom (in organophosphorothioates). Two of the three functional groups are typically ethoxy or methoxy groups although others can be present. The third group is an ester-linked aliphatic, homocyclic, or heterocyclic arrangement and may contain halogen groups in some cases, for example, trichlorfon. This is the most easily hydrolyzed group and, as such, becomes the initial "leaving group" that is displaced when an OP binds covalently to the active site of acetylcholinesterase (AChE) or neuropathy target esterase (NTE).

heightened concern over their potential toxicity in humans, which has resulted in bans or restrictions on the use of some OPs over recent years.

Certain OPs, such as the oil additive tricresyl phosphate (TCP) and the insecticide chlorpyrifos (CPF), are able to induce a delayed neuropathic condition known as OP-induced delayed neuropathy (OPIDN), the clinical symptoms of which become apparent 1–3 weeks following exposure (Abou Donia and Lapadula, 1990; Lotti, 1992; Richardson et al., 1993; Abou Donia, 2003). In the case of CPF and other OPIDN-inducing insecticides, the condition normally develops following survival of acute cholinergic crisis caused by deliberate or accidental exposure (Richardson et al., 1993; Soummer et al., 2011; George and Hedge, 2013). The largest recorded outbreak of OPIDN occurred in the United States during the 1930s when alcohol was prohibited. It was caused by the consumption of an alcoholic health "remedy" called Ginger Jake. Increased demand for this remedy as a legal source of alcohol during prohibition led to the adulteration of some batches with TCP when castor oil was in short supply, resulting in partial paralysis in many people who drank it (Bishop and Stewart, 1930; Zeligs, 1938).

TCP is commonly used in aviation fluids and as a plasticiser, lubricant, and flame retardant (Winder and Balouet, 2002; Winder 2006; Murawski and Supplee, 2008). It contains a mixture of *ortho*, *meta*, and *para* isomers, but only tri-*ortho*-cresyl phosphate (TOCP) is associated with the development of OPIDN. Isomers of TCP along with other components of aviation hydraulic fluids have been detected in the cabins and cockpits of many aircrafts, suggesting that OP poisoning may be implicated in the phenomenon of air cabin sickness (CAQPCCA et al., 2002; Winder and Balouet, 2002; CAA, 2004; Winder, 2005, 2006; Murawski and Supplee, 2008). Since TOCP was found to be the most potent isomer in terms of its ability to induce OPIDN (Abou Donia and Lapadula, 1990; Lotti, 1992) the level of this isomer in aviation fluids has been restricted to <1% of total TCP, and the total TCP concentration to less than 3% of the total volume (Harris et al., 1997). However, TOCP toxicity may be underestimated due to the presence of more highly toxic mono- and di-*ortho* cresyl variants and the possibility of interactions with other hydraulic fluid components to exert a toxic effect (Winder, 2005, 2006).

Another OP is the insecticide, diazinon (DZ), the only licensed sheep dip pesticide in the United Kingdom; its chemical structure is presented in Figure 36.1 (Clarke, 2007). Although DZ was judged to be moderately toxic and unable to induce OPIDN (Olsen, 1998), there are concerns for the safety of pesticide handlers and farm workers from exposure to this insecticide (COT, 2007). The use of DZ is therefore either severely restricted or banned in the United States (EPA, 2000, 2006), the European Union (EC, 2007), and Australia (APVMA, 2007; DPI, 2007). Likewise, restrictions were imposed in many countries on the use of CPF due to its association with developmental toxicity (EPA, 2000; Colbourn, 2006).

However, despite these restrictions, there remains a significant public health concern, since exposure to several OPs that are still in use has been linked to various forms of delayed neurotoxicity (Abou Donia and Lapadula, 1990; Lotti, 1992; Nutley and Crocker, 1993; Pilkington et al., 2001; Abou Donia, 2003; Costa, 2006; Karalliedde et al., 2006; Slotkin and Seidler, 2008; Slotkin et al., 2008).

The main types of neurotoxicity following human exposure to OPs can be categorized as (1) acute toxicity, which occurs within hours of exposure, and (2) OP-induced delayed neurotoxicity, which takes several forms, for which symptoms may appear several days, weeks, or even years following exposure.

36.2 ACUTE TOXICITY OF OPS

The organophosphorothioates CPF and DZ (Figure 36.1) are widely employed as insecticides, due to their ability, once bioactivated, to irreversibly inhibit acetylcholinesterase (AChE) in target organisms. Such bioactivation involves the replacement of sulfur by oxygen, resulting in the formation of an oxon derivative, a process which is catalyzed by specific cytochrome P450s (Chambers and Levi, 1992). The immediate consequence of acute exposure to OP pesticides and nerve gases is the irreversible inhibition of AChE, leading to cholinergic crisis caused by hyperstimulation of muscarinic and nicotinic ACh receptors, the associated clinical effects of which are illustrated in Figure 36.2. If death occurs, it is usually due to respiratory failure or cardiac arrest (Chambers and Levi, 1992). The OP binds irreversibly to the hydroxyl group in the side chain of the serine residue at the active site of AChE. It has been suggested that inhibition of nerve AChE activity by more than approximately 70% compared to normal levels will produce elevated levels of ACh in synaptic clefts at neuromuscular junctions, causing neuromuscular block and respiratory failure in severe cases.

Treatment strategies for the alleviation of acute OP toxicity include the use of atropine, which blocks muscarinic AChE receptor activation, cholinergic reactivators (e.g., oximes), and anticonvulsant drugs (Thiermann et al., 1999; Eddleston et al., 2008). However, these treatments do not prevent the development of delayed neurotoxicity in many patients, consistent with the possibility that delayed effects may be triggered by the interaction of OPs with molecular targets other than AChE.

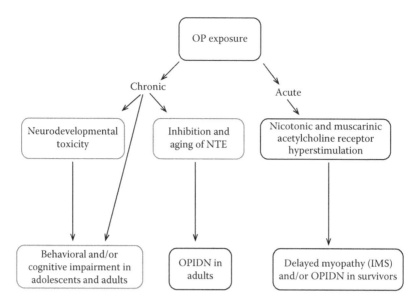

FIGURE 36.2 Pathways to delayed toxicity following exposure to organophosphorous compounds. The diagram is a summary of the main types of delayed neurotoxicity and how they can arise from acute or chronic exposure in terms of AChE inhibition.

36.3 PATHOPHYSIOLOGY OF OP-INDUCED DELAYED NEUROTOXICITY

As indicated in Section 36.1, certain OPs, such as TCP and CPF, are able to induce OPIDN. TOCP and structural congeners of its active neurodegenerative metabolite saligenin cyclic-*o*-tolyl phosphate (e.g., phenyl saligenin phosphate; PSP) are often used in animal studies of OPIDN; the hen being the preferred model for screening purposes due to its high sensitivity to OPIDN induction (Abou Donia and Lapadula, 1990; Lotti, 1992; Abou Donia, 2003). Screening studies, using hens, confirmed that TOCP was a potent inducer of OPIDN and that OP pesticides such as CPF, leptophos, and trichlorfon were also able to induce the condition but only at acutely toxic doses (El Sebae et al., 1977; Richardson et al., 1993).

OPIDN is characterized by delayed onset of an extended period of ataxia and upper motor neurone spasticity arising from single or repeated exposure to OPs (Abou Donia, 2003). OPIDN is a neurodegenerative condition that particularly affects nerves with long fiber tracts in both the central and peripheral nervous systems (Abou Donia and Lapadula, 1990; Lotti 1992; Abou Donia, 2003). A number of case studies on humans exposed to OPs indicate that OPIDN may be preceded by a condition known as intermediate syndrome (IMS), and in some cases, progression to IMS may only be observed. Typically, symptoms of IMS, such as weakness in proximal limb, neck, and respiratory muscles, are observed within a few days of acute exposure (Karalliedde et al., 2006). The pathophysiology of this condition is not well understood but is thought to involve major disruption at neuromuscular junctions and subsequent myopathy. Some patients recovering from OP-induced cholinergic crisis, for example, following fenthion exposure, may develop IMS without progressing to develop OPIDN (Kamayire and Karalliedde, 2004; Karalliedde et al., 2006).

36.4 EPIDEMIOLOGICAL STUDIES

Epidemiological studies indicate that exposure of adults to various levels of OPs can result in long-term neurological and neurobehavioral changes, affecting the CNS to a greater degree than the PNS. For example, a study on pesticide workers occupationally exposed to acute OP pesticide poisoning revealed significant impairment in a range of WHO-approved tests of behavior and cognitive function (Rosenstock et al., 1991). Low-level exposure to DZ in both retired and working sheep dippers showed that subacute exposure was associated with increased risk of neurobehavioral dysfunction (Mackenzie Ross et al., 2010). Subacute levels of CPF were also found to affect cognitive skills of rats exposed during weaning, raising concern about the risk of developmental effects of this pesticide (Jett et al., 2001). This concern was further strengthened by epidemiological studies on children applying pesticides to crops showing significant impairment in tests of neurobehavioral and cognitive skills, the latter becoming worse at longer exposure times (Abdel Rasoul et al., 2008).

These types of lesion differ from both acute toxicity and OPIDN in that they can last for many years during which central effects predominate; they have been collectively referred to as OP-induced chronic neurotoxicity (OPICN) (Abou Donia, 2003). However, it should be noted that this term encompasses a wide range of neurological and neurobehavioral deficits resulting from the neurodegenerative effects of single acutely toxic or repeated subacute doses of OP. Many of these conditions may reflect the interaction of OPs with distinct molecular targets and/or the extent of cell death in specific populations of neurons in the CNS and the level of maturity of the nervous system at the point of exposure.

As indicated in Section 36.4, the neuropathic OP of greatest concern has been TOCP, due to its potent ability to induce OPIDN. Although this isomer is now kept to a minimum level in aviation oils, there are several other OPs in aviation fluids (Winder and Balouet, 2002). It is therefore essential that the ability of all aviation fluid OPs to individually or collectively induce chronic neurotoxicity needs to be investigated.

36.5 MOLECULAR TARGETS OF OP-INDUCED DELAYED NEUROTOXICITY

As delayed neurotoxicity can occur in the absence of cholinergic crisis, it is clear that some delayed neuropathic effects are due to the interaction of OPs with molecular targets other than AChE, such as other serine hydrolases (Casida and Quistad, 2004).

The potential identity of such targets is reviewed in detail elsewhere (Hargreaves, 2012), and here we only give a brief summary. In OPIDN, the primary OP-binding protein is thought to be neuropathy target esterase (NTE) (Johnson, 1974, 1982, 1990; Glynn, 1999a,b, 2000). OPs that are potent inducers of OPIDN are strong inhibitors of NTE and weak inhibitors of AChE (Kropp and Richardson, 2003), which has formed the basis of OP screening in the hen. If such OPs bind to NTE and a phenomenon known as "aging" occurs, which involves increased negative charge after OP binding to the active site serine residue, the inhibition becomes irreversible and OPIDN ensues (Johnson, 1974, 1982, 1990; Glynn, 1999). Biochemical studies with recombinant NTE have shown that it has ion channel and lipid hydrolase activities (Glynn et al., 1999a,b; Glynn, 2000). In this respect, only OPs that induced aging of NTE were able to block ion conductance in vitro (Forshaw et al., 2001), suggesting that this potential role of NTE may be impaired in OPIDN. Its ability to act as a lipid hydrolase using phosphatidyl choline as a substrate (van Teinhoven et al., 2002; Glynn, 2005), suggests that it may play a role in the regulation of phospholipid turnover in cell membranes.

The mutation or deletion of NTE-related genes can result in a central neuropathy, involving glial hyperwrapping, inhibition of lipid hydrolase activity, and neuronal apoptotic cell death in several brain regions, indicating an essential role for NTE and related proteins in neuronal–glial interactions important for nervous system function (Kretzshmar et al., 1997; Glynn, 2000; Akassoglou et al., 2004; Muhlig-Versen et al., 2005). It remains to be established whether the same occurs in OPIDN, which primarily affects the peripheral nervous system. However, the discovery of NTE mutants with disrupted esterase activity in motor neurone disease (Rainier et al., 2008; Hein et al., 2010) is consistent with a vital role for NTE in the maintenance of neurons in adult mammals (Chang and Wu, 2010; Read et al., 2010).

Studies using differentiating neuronal cell lines showed that the inhibition of neurite outgrowth by PSP and CPF was associated with the inhibition of NTE, the former being the most potent inhibitor (Sachana et al., 2001; Hargreaves et al., 2006). However, knock down of NTE had no effect on the extent of neurite outgrowth in differentiating SH-SY5Y neuroblastoma cells, whereas increased NTE expression was associated with an accelerated rate of neurite outgrowth and the inhibition of neuronal cell proliferation, suggesting a more subtle regulatory role for this enzyme in neuronal cell differentiation (Chang and Chen, 2005; Chang et al., 2005, 2006).

TOCP-induced OPIDN was also associated with hyperphosphorylation and abnormal degradation and cellular distribution of cytoskeletal proteins particularly neurofilaments (Suwita et al., 1986; Jensen et al., 1992; Lapadula et al., 1992; Zhao et al., 2004; Song et al., 2009). Similar changes have also been observed in cell culture studies of neurotoxicity (Fowler et al., 2001; Sachana et al., 2001; Hargreaves et al., 2006) and may contribute to the degeneration of axons that occurs in OPIDN. It has also been found that OPs are capable of direct binding to a number of other molecular targets, including cytoskeletal proteins that play a critical role in axon and dendrite development and function (reviewed in Hargreaves, 2012).

Of particular interest to this review is the mounting evidence that mitochondrial disruption and subsequent increases in the cellular levels of reactive oxygen species (ROS) are associated with exposure to anticholinesterase compounds, including OPs, since this can lead to oxidative damage to lipids and DNA, and misfolding and dysfunction of proteins, which can be attenuated by cotreatment with antioxidant compounds (Milatovic et al., 2006; Slotkin and Seidler, 2010).

36.6 OP-INDUCED OXIDATIVE STRESS IN HUMANS

A limited number of clinical studies have been carried out on humans exposed either acutely or chronically to a range of OP pesticides, typically involving the measurement of AChE in peripheral cells or blood fluids. All of these studies showed some degree of oxidative stress in samples taken from human blood, as indicated by increased lipid peroxidation (LPO) and/or altered enzymic and nonenzymic antioxidant levels (Soltaninejad and Abdollahi, 2009). For example, in a study by Ranjbar et al. (2002), it was found that reduced plasma AChE activity correlated well with increased levels of thiobarbituric acid reactive substances (TBARS; indicative of increased LPO), reduced antioxidant capacity, and free thiol levels in blood from workers employed, for at least 1 year, at a plant manufacturing OP pesticides. Shadnia et al. (2005) suggested that longer-term chronic exposure of pesticide workers to OPs could result in DNA damage and enhanced activities of antioxidant enzymes such as superoxide dismutase (SOD), catalase (CAT) and glutathione peroxidase (GPx) in the absence of significant changes in AChE or LPO. A study on 22 individuals with acute symptoms from exposure to a malathion formulation taken orally found high levels of LPO and severely impaired antioxidant defenses in plasma (Ranjbar et al., 2005). Increased oxidative stress and lymphocyte DNA damage were also observed in studies on pesticide applicators and farm workers showing elevated levels of OP metabolites in urine (Muniz et al., 2008; Bayrami et al., 2012). Although one study showed that this was not linked to cognitive impairment (Bayrami et al., 2012), there is clear evidence that long-term chronic exposure to OPs, which is also associated with oxidative damage (Shadnia et al., 2005), can lead to such problems (Rosenstock et al., 1991; Fiedler et al., 1997). Moreover, a systematic review of clinical and experimental findings in patients diagnosed with IMS concluded that, in addition to sustained AChE inhibition, electromyographic disturbance and muscle cell injury, oxidative stress was a key indicator of this syndrome (Abdollahi and Karami-Mohajeri, 2012).

Although there are some inconsistencies between studies with respect to the precise response of antioxidant systems, collectively these studies indicate that both chronic and acute exposure to OPs can cause disruption of antioxidant defenses in human blood, which can result in oxidative damage such as LPO and DNA oxidation depending on the level and duration of exposure. However, as such measurements are made in blood samples, they do not necessarily reflect the level of ROS production and the consequent damage in the nervous system due to oxidative stress. Nevertheless, the results from human studies to date suggest that the measurement of parameters associated with oxidative stress may, in addition to assays of AChE, be useful tools for monitoring the progress of patients suffering from OP poisoning and for monitoring the health of pesticide formulation workers, pesticide applicators, and farm workers for signs of OP-related damage. More extensive studies, however, are needed to understand the role of oxidative stress in OP toxicity more fully and, in particular, how elevated levels of ROS may explain some of the noncholinergic effects associated with OP-induced delayed neurotoxicity.

36.7 OP-INDUCED OXIDATIVE STRESS IN ANIMAL STUDIES

The effects of OP exposure on oxidative stress in nerve tissue have been studied in a number of animal models. In a study with rats exhibiting 80%–90% inhibition of brain AChE (i.e., acute cholinergic toxicity) after a single diisopropyl fluorophosphate (DFP) treatment, neurodegeneration was observed in pyramidal neurons in the CA1 hippocampal area and oxidative stress was detected in brain tissue within 1 h of OP administration; both effects were blocked by pretreatment with antioxidants such as vitamin E and by glutamate receptor antagonists, suggesting that oxidative damage associated with excitotoxicity was a major cause of neurodegeneration after acute exposure to OPs (Zaja-Milatovic et al., 2009). Adult rats treated orally with concentrations of CPF, malathion, and methylparathion causing approximately 50%–60% inhibition of AChE in rat brain (i.e., subacute toxicity) exhibited increased TBARS in brain and a range of other tissues, but coadministration of all three OPs did not increase this effect (Ojha et al., 2011). Reduced activities of the antioxidant enzymes CAT, SOD, and GPx were also observed (Ojha et al., 2011, 2012). Rats administered intraperitoneally with malathion at both acute and chronic doses exhibited increased oxidative stress in certain brain regions but in a non-dose-dependent manner, as indicated by measurement of LPO and elevated protein carbamylation, followed by increased activity of the antioxidant enzymes CAT and SOD (Fortunato et al., 2006). Although increased LPO is a common event, differences in the response of antioxidant enzymes in both studies described may reflect the different dosing regimens and tissue analysis procedures used.

OPs such as dichlorvos and PSP are known to disrupt Ca^{2+} homeostasis via the influx of extracellular Ca^{2+} prior to the development of delayed neuropathy (El-Fawal et al., 1990; Choudhary and Gill, 2001). Impaired energy metabolism, as indicated by reduced activities of NADH dehydrogenase (complex I), succinate dehydrogenase (complex II), and cytochrome oxidase (complex IV) in mitochondria isolated from the brains of rats treated with chronic doses of dichlorvos, was suggested to be the result of increased Ca^{2+} uptake by mitochondria, leading to oxidative stress in mitochondria and caspase 3–mediated apoptotic neuronal cell death (Kaur et al., 2007). Another study in hens administered TOCP found that the levels of LPO increased and antioxidant status was reduced in the cerebrum, sciatic nerve, and spinal cord of treated animals (Zhang et al., 2007). These parameters changed in a time-dependent manner that correlated with increasing clinical severity, suggesting that oxidative stress played an important role in the development of this delayed neuropathy (Zhang et al., 2007). A follow-up study showed that TOCP induced time and dose-dependent increases in mitochondrial damage, particularly vacuolation and fission as monitored by transmission electron microscopy. The same study also demonstrated reduced mitochondrial permeability transition and inhibition of complex II activity in mitochondria isolated from cerebrum and spinal cord of TOCP treated hens (Xin et al., 2011). Mitochondrial dysfunction was also observed in a study on rats induced to develop OPIDN following acute exposure to monocrotophos and dichlorvos (Masoud et al., 2009). Reduced ATP levels and elevated ROS due to mitochondrial dysfunction could be the result, and both could exacerbate the damage to mitochondrial proteins and compromise other ATP-dependent cellular functions. It has also been shown that the myopathic effects of OPs in vivo can be partially blocked by treatment with antioxidants, suggesting that elevated ROS might also contribute to the pathology of IMS (Buyukokuroglu et al., 2008; Abdollahi and Karami-Mohajeri, 2012). The fact that OP-induced mitochondrial dysfunction and oxidative stress can be reduced by treatment with a range of antioxidants such as vitamins E and C, α-tocopherol, electron donors, and by restoring cellular ATP levels further emphasizes the key role played by elevated ROS in OP-induced delayed neurotoxicity and the potential value of antioxidants in its treatment (Ojha and Srivastava, 2012; Karami-Mohajeri and Abdollahi, 2013).

36.8 OP-INDUCED OXIDATIVE STRESS IN DEVELOPMENTAL NEUROTOXICITY

Exposure to CPF and diazinon has been linked to developmental neurotoxicity (Qiao et al., 2005; Slotkin et al., 2005, 2006, 2008; Slotkin and Seidler, 2008, 2010). One study indicated that the generation of elevated levels of TBARS was not elicited when OP was administered at 17–20 days gestation or from 1 to 4 days postnatally (Slotkin et al., 2005). However, administration of CPF in the second postnatal week, at which point neuronal differentiation, synaptogenesis, and metabolic demands reach a peak, resulted in significantly increased levels of TBARs, suggesting that the ability of OPs to induce oxidative stress and developmental neurotoxicity is dependent to some extent on the metabolic demand at the developmental stage at which exposure occurs (Slotkin et al., 2005, 2008). CPF and DZ exposure at an early stage is associated with altered serotonergic and dopaminergic synaptic activity in adults, suggesting that certain aspects of developmental exposure can be manifested at a much later stage in life (Aldridge et al., 2005). A small number of studies with cultured neural cells have shown that OPs such as CPF and diazinon have the ability to inhibit neurite outgrowth and/or proliferation under conditions that induce oxidative stress without cell death (Qiao et al., 2005; Slotkin et al., 2007). Although a clearer picture is beginning to emerge regarding the consequences of OP-induced mitochondrial dysfunction, further work is required to fully understand the role of oxidative stress in OP-induced delayed neurotoxicity. However, CPF was shown to induce oxidative stress–associated, caspase-mediated apoptosis in cultured oligo-dendrocyte progenitor cells, albeit at relatively high concentrations (≥30 μM), suggesting that it may have the ability to induce ROS in a range of neural cell types during development (Saulsbury et al., 2009).

36.9 CONCLUSION

Extensive evidence exists to show that OP exposure can induce oxidative stress and that attenuation of oxidative damage by antioxidant treatment may help to reduce neurological damage following exposure to OPs. Measurement of oxidative stress parameters in human blood–derived samples, in addition to acute toxicity biomarkers such as AChE, may help to monitor OP toxicity more effectively in the clinic. The precise nature of the oxidative imbalance observed in different studies will, to some extent, depend on the level of maturity and metabolic demands at the point of exposure, the dosing regimen, the level and duration of exposure, and the tissue samples used for analyses of oxidative damage. Thus, it is advisable to develop standardized operating procedures across the sector for better comparability.

As indicated in the schematic diagram shown in Figure 36.3, OPs that induce OPIDN may target NTE leading to subsequent increases in intracellular Ca^{2+}, increased mitochondrial uptake of Ca^{2+}, and subsequent mitochondrial dysfunction. Non-OPIDN-induced OPs may cause mitochondrial function through interactions with other protein targets. The resultant increase in ROS levels could further exacerbate these effects, causing reduced levels of ATP, which in turn affect numerous essential cell functions, including axonal transport and proteasome-mediated protein degradation. LPO could lead to membrane disruption and further imbalances in Ca^{2+} homeostasis, DNA oxidation to DNA fragmentation, and protein oxidation to abnormal protein folding and disruption of their proteolytic degradation. Although the precise protein targets for oxidation and abnormal degradation have not been fully characterized, it is highly likely that these will include proteins of the axonal cytoskeleton, which was found to be disrupted in studies of developmental toxicity and OPIDN. Further research is needed in this important area of OP toxicity, as it would not only help to establish the pattern of molecular events following exposure more clearly, but it could also help to

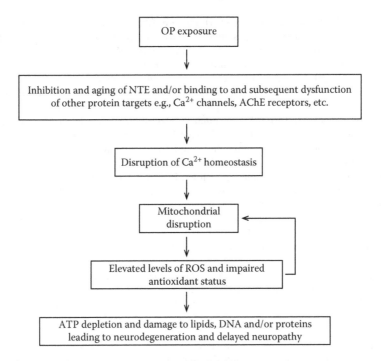

FIGURE 36.3 The role of oxidative stress in organophosphorous compound–induced delayed neurotoxicity. In delayed neurotoxicity, the molecular initiating event is either inhibition and aging of NTE (e.g., in OPIDN) or OP binding to a range of other potential protein targets such as Ca^{2+} channels, neurotransmitter receptors, etc. This causes disruption of Ca^{2+} homeostasis, mitochondrial dysfunction, and elevated levels of ROS. The latter can further affect mitochondrial function by causing oxidation of proteins involved in energy production, leading to ATP depletion and further oxidative damage to other cellular proteins (e.g., those involved in cytoskeletal regulation), lipid peroxidation, and DNA damage.

identify important biomarkers of toxicity for monitoring patient progress and/or for the development of therapeutic strategies to attenuate the effects of OPs that occur due to oxidative damage.

REFERENCES

Abdel Rasoul GM, Abou Salem ME, Mechael AA, Hendy OM, Rohlman DS, Ismail AA. Effects of occupational pesticide exposure on children applying pesticides. *Neurotoxicology* 2008; 29: 833–838.

Abdollahi M, Karami-Mohajeri S. A comprehensive review on experimental and clinical findings in intermediate syndrome caused by organophosphate poisoning. *Toxicol Appl Pharmacol* 2012; 258: 309–314.

Abou Donia MB. Organophosphorus ester induced chronic neurotoxicity. *Arch Environ Health* 2003; 58: 484–497.

Abou Donia MB, Lapadula DM. Mechanisms of organophosphorus ester induced delayed neurotoxicity: Type I and II. *Ann Rev Pharmacol Toxicol* 1990; 30: 405–440.

Akassoglou K, Malester B, Xu J, Tessarollo L, Rosenbluth J, Chao MV. Brain-specific degeneration of neuropathy target esterase/Swiss cheese results in neurodegeneration. *Proc Natl Acad Sci USA* 2004; 101: 5075–5780.

Aldridge JE, Meyer A, Seidler FJ, Slotkin TA. Alterations in central nervous system serotonergic and dopaminergic synaptic activity in adulthood after prenatal or neonatal chlorpyrifos exposure. *Environ Health Perspect* 2005; 113: 1027–1031.

APVMA (Australian Pesticides and Veterinary Medicines Authority). APVMA suspends the use of diazinon for sheep dipping and jetting. Media release, May 2007.

Atkins J, Luthjens LH, Hom ML, Glynn P. Monomers of the catalytic domain of human neuropathy target esterase are active in the presence of phospholipid. *Biochem J* 2002; 361: 119–123.

Bayrami M, Hashemi T, Malekirad AA, Ashayri H, Faraji F, Abdollahi M. Electroencephalogram, cognitive state, psychological disorders, clinical symptoms and oxidative stress in horticulture farmers exposed to organophosphate pesticides. *Toxicol Ind Health* 2012; 28: 90–96.

Bishop EL, Stewart HC. Incidence of partial paralysis. *Am J Pub Health* 1930; 20: 1307–1312.

Bissbort SH, Vermaak WJH, Elias J et al. Novel test and its automation for the determination of erythrocyte acetylcholinesterase and its application to organophosphate poisoning. *Clin Chim Acta* 2001; 303: 139–145.

Buyukokoroglu ME, Cemek M, Tosum M, Yurumez Y. Dantrolene may prevent organophosphate-induced oxidative stress and muscle injury. *Pestic Biochem Physiol* 2008; 92: 156–163.

CAA (Civil Aviation Authority, Safety Regulation Group). Cabin air quality. CAA paper 2004/04.

CAQPCCA (Committee on Air Quality in Passenger Cabins of Commercial Aircraft, Board on Environmental Studies and Toxicology, National Research Council, National Academy of Sciences. The airliner cabin environment and health of passengers and crew. Washington, DC: National Academic Press, 2002.

Casida JE, Quistad GB. Organophosphate toxicology: Safety aspects of nonacetylcholinesterase secondary targets. *Toxicology* 2004; 17: 983–998.

Chambers JE, Levi PE. *Organophosphates: Chemistry, Fate and Effects.* San Diego, CA: Academic Press Inc., 1992.

Chang PA, Wu YJ. Neuropathy target esterase: An essential enzyme for neural development and axonal maintenance. *Int J Biochem Cell Biol* 2010; 42: 573–575.

Chang PA, Wu YJ, Chen R et al. Inhibition of neuropathy target esterase expressing by antisense RNA does not affect neural differentiation in human neuroblastoma (SK-N-SH) cell line. *Mol Cell Biochem* 2005; 272: 47–54.

Chang PA, Chen R, Wu YJ. Reduction of neuropathy target esterase does not affect neuronal differentiation but moderate expression induces neuronal differentiation in human neuroblastoma (SK-N-SH) cell line. *Mol Brain Res* 2005; 141: 30–38.

Chang PA, Liu CY, Chen R, Wu YJ. Effect of over-expression of neuropathy target esterase on mammalian cell proliferation. *Cell Prolif* 2006; 39: 429–440.

Choudhary S, Gill KD. Protective effect of nimodipine on dichlorvos-induced delayed neurotoxicity in rat brain. *Biochem Pharmacol* 2001; 62: 1265–1272.

Clarke Z. Diazinon. In: *X Pharm: The Comprehensive Pharmacology Reference 2007.* Enna SJ, Bylund DB, Eds. Amsterdam, The Netherlands: Elsevier Ltd.

Colbourn T. A case for revisiting the safety of pesticides: A closer look at neurodevelopment. *Environ Health Perspect* 2006; 114: 10–17.

Costa LG. Current issues in organophosphate toxicology. *Clin Chim Acta* 2006; 366: 1–13.

COT (Committee on Toxicology of Chemicals in Food). Consumer products and the environment. Department of Health, London, U.K., 2007.

DPI (Department of Primary Industries). Sheep body lice: Control and eradication. Agriculture notes of the DPI, State of Victoria, 2007; AG1110.

EC—European Commission decision of 6 June 2007 concerning the non-inclusion of diazinon in Annex I to the Council Directive 91/414/EEC and the withdrawal of authorisations for plant protection agents containing that substance. *Off J Eur Union* 2007; 140: 9–10.

Eddleston M, Buckley NA, Eyer P, Dawson AH. Management of acute organophosphorus pesticide poisoning. *The Lancet* 2008; 371: 597–607.

El-Fawal HA, Jortner BS, Ehrich M. Modification of phenyl saligenin phosphate-induced delayed effects by calcium channel blockers: In vivo and in vitro physiological assessment. *NeuroToxicology* 1990; 11: 573–592.

El Sebae AH, Soliman SA, Abo Elamayem M, Ahmed NS. Neurotoxicity of organophosphorus insecticides leptophos and EPN. *J Environ Sci Health* 1977; B12: 269–288.

EPA (Environmental Protection Agency). Annual report, 2000. Office of Pesticide Programs, Washington, DC.

EPA (Environmental Protection Agency). Reregistration eligibility decision for diazinon, Washington, DC, 2006.

Fiedler N, Kipen H, Kelly-McNeil K, Febnske R. Long term use of organophosphates and neuropsychological performance. *Am J Ind Med* 1997; 32: 487–496.

Forshaw PJ, Atkins J, Ray DE, Glynn P. The catalytic domain of human neuropathy target esterase mediates an organophosphate sensitive ion conductance across liposome membranes. *J Neurochem* 2001; 79: 400–406.

Fortunato JJ, Feier G, Vitali AM, Petronilho FC, Dal-Pizzol F, Quevedo J. Malathion induced oxidative stress in rat brain regions. *Neurochem Res* 2006; 31: 671–678.

Fowler MJ, Flaskos J, McLean WG, Hargreaves AJ. Effects of neuropathic and non-neuropathic isomers of tricresyl phosphate and their microsomal activation on the production of axon-like processes by differentiating mouse N2a neuroblastoma cells. *J Neurochem* 2001; 76: 671–678.

George P, Hedge N. Neurotoxic syndromes sequentially occurring after consumption of organophosphorus compound—A case report. *Global J Med Res* 2013; 7: 51–53.

Glynn P. Neuropathy target esterase. *Biochem J* 1999a; 344: 625–631.

Glynn P. Molecular cloning of neuropathy target esterase. *Chem Biol Interact* 1999b; 119–120: 513–517.

Glynn P. Neural development and neuropathy target esterase: Two faces of neuropathy target esterase. *Prog Neurobiol* 2000; 61: 61–74.

Glynn P. Neuropathy target esterase and phospholipid deacylation. *Biochim Biophys Acta* 2005; 1736: 87–93.

Hargreaves AJ. Neurodegenerations induced by organophosphorous compounds. *Adv Exp Med Biol* 2012; 724: 189–204.

Hargreaves AJ, Fowler MJ, Sachana M, Flaskos J, Bountouri M, Coutts IC, Glynn P, Harris W, McLean WG. Inhibition of neurite outgrowth in differentiating mouse N2a neuroblastoma cells by phenyl saligenin phosphate: Effects on neurofilament heavy chain phosphorylation, MAP kinase (ERK 1/2) activation and neuropathy target esterase activity. *Biochem Pharmacol* 2006; 71: 1240–1247.

Harris MO, McLure P, Chessin RL, Corcoran JJ. Toxicological profile for hydraulic fluids. ATDSR, Atlanta, GA; 1997.

Hein ND, Stuckley JA, Rainier SR, Fink JK, Richardson RJ. Constructs of human neuropathy target esterase catalytic domain containing mutations related to motor neuron disease have altered enzymatic properties. *Toxicol Lett* 2010; 196: 67–73.

Jensen KF, Lapadula DM, Knoth Anderson J, Haykal-Coates L. Anomalous phosphorylated neurofilament aggregations in central and peripheral axons of hens treated with tri-*ortho*-cresyl phosphate (TOCP). *J Neurosci Res* 1992; 33: 455–460.

Jett DA, Navoa RV, Beckles RA, McLemore. Cognitive function and cholinergic neurochemistry in weanling rats exposed to chlorpyrifos. *Toxicol Appl Pharmacol* 2001; 174: 89–98.

Johnson MK. The primary biochemical lesion leading to the delayed neurotoxic effects of some organophosphorus esters. *J Neurochem* 1974; 23: 785–789.

Johnson MK. The target for initiation of delayed neurotoxicity by organophosphorus esters: Biochemical studies and toxicological applications. In Hodgson E, Bend JR, Philpot RM, Eds. *Reviews in Biochemical Toxicology*, 1982; Vol. 4, pp. 141–212. Elsevier, New York.

Johnson MK. Organophosphates and delayed neuropathy—Is NTE alive and well? *Toxicol Appl Pharmacol* 1990; 102: 385–389.

Kamanyire R, Karalliedde L. Organophosphate toxicity and occupational exposure. *Occupational Med* 2004; 54: 69–75.

Karalliedde L, Baker D, Marrs TC. Organophosphate-induced intermediate syndrome: Aetiology and relationships with myopathy. *Toxicol Rev* 2006; 25: 1–14.

Karami-Mohajeri S, Abdollahi M. Mitochondrial dysfunction and organophosphorus compounds. *Toxicol Appl Pharmacol* 2013; 270: 39–44.

Kaur P, Radotra B, Minz RW, Gill KD. Impaired mitochondrial metabolism and neuronal apoptotic cell death after chronic dichlorvos exposure in rat brain. *NeuroToxicol* 2007; 28: 1208–1219.

Kretzschmar D, Hasan G, Sharma S, Heisenberg M, Benzer S. The Swiss cheese mutant causes glial hyperwrapping and brain degeneration in *Drosophila*. *J Neurosci* 1997; 17: 7425–7432.

Kropp TJ, Richardson RJ. Relative inhibitory potencies of chlorpyrifos oxon, chlorpyrifos methyloxon and mipafox for acetylcholinesterase versus neuropathy target esterase. *J Toxicol Environ Health A* 2003; 278: 8820–8825.

Lapadula ES, Lapadula DM, Abou-Donia MB. Biochemical changes in sciatic nerve of hens treated with tri-*o*-cresyl phosphate: Increased phosphorylation of cytoskeletal proteins. *Neurochem Int* 1992; 20: 247–255.

Lotti M. The pathogenesis of organophosphate polyneuropathy. *Crit Rev Toxicol* 1992; 21: 465–487.

Lotti M, Moretto R, Zoppellari R et al. Inhibition of neuropathy target esterase predicts the development of OPIDN. *Arch Toxicol* 1986; 59: 176–179.

Lush MJ, Li Y, Read DJ et al. Neuropathy target esterase and a homologous *Drosophila* neurodegeneration mutant protein contain a domain conserved from bacteria to man. *Biochem J* 1998; 332: 1–4.

Mackenzie Ross SJ, Brewin CR, Curran HV, Furlong CE, Abraham-Smith KM, Harrison V. Neuropsychological and psychiatric functioning in sheep farmers exposed to low levels of organophosphate pesticides. *NeuroToxicol Teratol* 2010; 32: 452–459.

Reactive Oxygen Species in Biology and Human Health

Masoud A, Kiran R, Sandhir R. Impaired mitochondrial functions in organophosphate induced delayed neuropathy in rats. *Mol Cell Neurobiol* 2009; 29: 1245–1255.

Milatovic D, Gupta RC, Aschner M. Anti-cholinesterase toxicity and oxidative stress. *Sci World J* 2006; 6: 295–310.

Mühlig-Versen M, Bettencourt de Cruz A, Tschäpe JA, Moser M, Büttner R, Athenstaedt K, Glynn P, Kretzschmar D. Loss of Swiss cheese/neuropathy target esterase activity causes disruption of phosphatidylcholine homeostasis and neuronal and glial cell death in *Drosophila*. *J Neurosci* 2005; 25: 2865–2873.

Muniz JF, McCauley L, Scherer J, Lasarev M, Koshy M, Kow YW, Nazar-Stewart V, Kisby GE. Biomarkers of oxidative stress and DNA damage in agricultural workers: A pilot study. *Toxicol Appl Pharmacol* 2008; 227: 97–107.

Murawski JTL, Supplee DS. An attempt to characterize the frequency, health impact, and operational costs of oil in the cabin and flight deck supply air on U.S. commercial aircraft. *J ASTM Int* 2008; 5: JAI101640.

Nutley BP, Crocker J. Biological monitoring of workers occupationally exposed to organophosphorous pesticides. *Pesticide Sci* 1993; 39: 315–322.

Ojha A, Srivastava N. Redox imbalance in rat tissues exposed with organophosphate pesticides and therapeutic potential of antioxidants. *Ecotoxicol Environ Safety* 2012; 75: 230–241.

Ojha A, Yaduvanshi SK, Srivastava N. Effect of combined exposure to commonly used organophosphate pesticides on lipid peroxidation and antioxidant enzymes in rat tissues. *Pesticide Biochem Physiol* 2011; 99: 148–156.

Olsen KR Ed. *Poisoning and Drug Overdose*. Appleton and Lange, Stamford CT; 2008.

Pilkington A, Buchanan D, Jamal GA, Gillham R, Hansen S, Kidd M, Hurley JF, Soutar CA. An epidemiological study of the relations between exposure to organophosphate pesticides and indices of chronic peripheral neuropathy and neuropsychological abnormalities in sheep farmers and dippers. *Occup Environ Med* 2001; 58: 702–710.

Qiao D, Seidler FJ, Slotkin TA. Oxidative mechanisms contributing to the developmental toxicity of nicotine and chlorpyrifos. *Toxicol Appl Pharmacol* 2005; 206: 17–26.

Rainier S, Bui M, Mark E et al. Neuropathy target esterase gene mutations cause motor neuron disease. *Am J Human Genet* 2008; 82: 780–785.

Ranjbar A, Pasalar P, Abdollahi M. Induction of oxidative stress and acetylcholinesterase inhibition in organophosphorous pesticide manufacturing workers. *Hum Exp Toxicol* 2002; 21: 179–182.

Ranjbar A, Solhi H, Mashayekhi FJ, Susanabdi A, Rezaie A, Abdollahi M. Oxidative stress in human poisoning with organophosphorous pesticides; A case control study. *Environ Toxicol Pharmacol* 2005; 20: 88–91.

Read DJ, Li Y, Chao MV, Cavanagh JB, Glynn P. Neuropathy target esterase is required for adult axon maintenance. *J Neurosci* 2009; 29: 11594–11600.

Richardson RJ, Moore TB, Kayyali US, Randall JC. Chlorpyrifos: Assessment of potential or delayed neurotoxicity by repeated dosing in adult hens with monitoring of brain acetylcholinesterase, brain and lymphocyte neurotoxic esterase and plasma butyrylcholinesterase activities. *Toxicol Sci* 1993; 21: 89–96.

Rosenstock L, Keifer M, Daniell WE et al. Chronic central nervous system effects of acute organophosphate pesticide intoxication. *Lancet* 1991; 338: 223–227.

Sachana M, Flaskos J, Alexaki E, Glynn P, Hargreaves AJ. The toxicity of chlorpyrifos towards differentiating mouse N2a neuroblastoma cells. *Toxicol In Vitro* 2001; 15: 369–372.

Saulsbury MD, Heyliger SO, Wang K, Johnson DJ. Chlorpyrifos induces oxidative stress in oligodendrocyte progenitor cells. *Toxicology* 2009; 259: 1–9.

Shadnia S, Azizi E, Hosseini H, Khoei R, Fouladdel S, Pajoumand A, Jalali N, Abdollahi M. Evaluation of oxidative stress and genotoxicity in organophosphorous insecticide formulators. *Hum Exp Toxicol* 2005; 24: 439–445.

Slotkin TS, MacKillop EA, Ryde IT, Tate CA, Seidler FJ. Screening of developmental toxicity using PC12 cells. Comparisons of organophosphates with a carbamate, an organochlorine and divalent nickel. *Environ Health Perspect* 2007; 115: 93–101.

Slotkin TA, Oliver CA, Seidler FJ. Critical periods for the role of oxidative stress in developmental neurotoxicity of chlorpyrifos and terbutaline alone or in combination. *Dev Brain Res* 2005; 157: 172–180.

Slotkin TA, Ryde TI, Levin ED, Seidler FJ, Developmental neurotoxicity of low dose diazinon exposure of neonatal rats: Effects on serotonin systems in adolescence and adulthood. *Brain Res Bull* 2008; 75: 640–647.

Slotkin TS, Seidler FJ. Developmental neurotoxicants target differentiation into the serotonin phenotype: Chlorpyrifos, diazinon, dieldrin and divalent nickel. *Toxicol Appl Pharmacol* 2008; 233: 211–219.

Slotkin TS, Seidler FJ. Oxidative stress from diverse developmental neurotoxicants: Antioxidants protect against lipid peroxidation without preventing cell loss. *NeuroToxicol Teratol* 2010; 32: 124–131.

Slotkin TS, Tate CA, Ryde IT, Levin ED, Seidler FJ. Organophosphate insecticides target the serotonergic system in developing rat brain regions: Disparate effects of diazinon and parathion at doses spanning the threshold for cholinesterase inhibition. *Environ Health Perspect* 2006; 114: 1542–1546.

Soltaninejad K, Abdollahi M. Current opinion on the science of organophosphate pesticides and toxic stress: A systematic review. *Med Sci Mon* 2009; 15: RA75–RA90.

Song F, Yan Y, Zhao X, Zhang C, Xie K. Neurofilaments degradation as an early molecular event in tri-ortho-cresyl phosphate (TOCP) induced delayed neuropathy. *Toxicology* 2009; 258: 94–100.

Soummer A, Megarbane B, Boroli F, Arbelot C, Saleh M, Moesch C, Fournier E, Rouby JJ. Severe and prolonged neurologic toxicity following subcutaneous chlorpyrifos self-administration: A case report. *Clin Toxicol* 2011; 49: 124–127.

Suwita E, Lapadula DM, Abou-Donia MB. Calcium and calmodulin enhanced in vitro phosphorylation of hen brain cold stable microtubules and spinal cord neurofilament proteins following a single oral dose of tri-*o*-cresyl phosphate. *Proc Natl Acad Sci USA* 1986; 83: 6174–6178.

Thiermann H, Szinicz L, Eyer F, Worek F, Eyer P, Felgenhauer N, Zilker T. Modern strategies in therapy of organophosphate poisoning. *Toxicol Lett* 1999; 107: 233–239.

Van Tienhoven M, Atkins J, Li Y, Glynn P. Human neuropathy target esterase catalyzes hydrolysis of membrane lipids. *J Biol Chem* 2002; 277: 20942–20948.

Winder C, Ed. Contaminated air protection. *Proceedings of the Air Safety and Cabin Air Quality International Aero Industry Conference*. British Airline Pilots Association (BALPA) and the University of New South Wales, Sydney, New South Wales, Australia, 2005.

Winder C. Air monitoring studies for air cabin contamination. *Curr Top Toxicol* 2006; 3: 33–48.

Winder C, Balouet JC. The toxicity of commercial jet oils. *Environ Res* 2002; 89: 146–164.

Xin X, Zeng T, Dou DD, Zhao S, Du JY, Pei JJ, Xie KQ, Zhao XL. Changes of mitochondrial ultrastructures and function in central nervous tissue of hens treated with tri-ortho-cresyl phosphate (TOCP). *Hum Exp Toxicol* 2011; 1062–1072.

Zaja-Milatovic S, Gupta RC, Aschner M, Milatovic D. Protection of DFP-induced oxidative damage and neurodegeneration by antioxidants and NMDA receptor antagonist. *Toxicol Appl Pharmacol* 2009; 240: 124–131.

Zeligs MA. Upper motor neurone sequelae in "Jake" paralysis: A clinical follow up study. *J Nerv Ment Dis* 1938: 87: 464–470.

Zhang LP, Wang QS, Guo X, Zhu YJ, Zhou GZ, Xie KQ. Time dependent changes of lipid peroxidation and antioxidative status in nerve tissues of hen treated with tri-*ortho*-cresyl phosphate (TOCP). *Toxicology* 2007; 239: 45–52.

Zhao XL, Zhu ZP, Zhang TL, Zhang CL, Yu HL, Xe KQ. Tri-ortho-cresyl phosphate (TOCP) decreases the levels of cytoskeletal proteins in hen sciatic nerve. *Toxicol Lett* 2004; 152: 139–147.

37 Reactive Oxygen Species and Morphine Addiction

Zheng Qiusheng and Ma Jun

CONTENTS

37.1 INTRODUCTION

In the World Drug Report 2014, the United Nations Office on Drugs and Crime (UNODC) estimated that in 2012 between 162 and 324 million people aged 15–64 throughout the world had used an illicit drug, and the average of 183,000 (range: 95,000–226,000) deaths of people aged 15–64 were caused by drug abuse. Among all different kinds of drugs, opiates remain the most prevalent abused in Asia and Europe, and cocaine in the Americas. Others contain cannabis and amphetamine-type stimulants. Illicit drug abuse can exert profoundly negative effects on an individual's health, leading to premature death, curtailed quality of life through disability, and infection with HIV and hepatitis B and C. In addition to such outcomes, it also has an extremely severe influence on society and family. Over the period 2003–2012, the annual global crime rates for personal possession and use have increased due to the increase in the total number of drug users, which fluctuate between 3% and 4% all over the world [1]. With the increase of drug abuse (morphine and heroin), drug control and rehabilitation have always been a serious hotspot of society throughout the world, and a big challenge in the field of medical research.

The three terms that are commonly used to describe drug abuse are tolerance, dependence, and addiction. Tolerance refers to an attenuated effect upon repeated exposure to a drug at a constant dose, or the need for an increasing dose or a shortened interval of drug administration to maintain the same effect. Dependence is defined as the desire for sustained drug exposure so as to avoid withdrawal syndrome. Addiction represents a compulsive state character with inappropriate drug use [2].

It is now commonly accepted that drug addiction is one kind of brain disease, which is caused by direct effects of drugs on the brain or neuron adaptions to drugs, and is also influenced by the surrounding environment [3]. Opioid addiction is recognized as a chronic relapsing brain disease resulting from repeated exposure to opioid drugs [4]. Its symptoms are compulsive drug use, persistent desire state, and reduced self-control on drug craving. Once addicted, it can last for the whole lifetime. While it is also well-known that clinical treatment with opioids can produce widespread analgesia, and exert such pharmacologic effects as anti-asthmatic, tranquilizing, antidiarrheal, and hypnotics, either overdose or long-term administration can cause drug dependence and addiction, bringing various toxic effects on the human body. For example, morphine and heroin abuse will impair normal physiological and metabolic activities and destroy immunologic functions, which results in nonaging degenerative change and organic function recession, and even generate various complications in nervous and immunologic systems. Illicit drug use and its prevalence result not only in a high toll on individuals and society through direct adverse effects of drugs and associated healthcare costs, but also in the severe consequences to the world due to loss of vitality and productivity. So these issues brought by drugs abuse actuate people to focus on the pathomechanism of drug addiction and to explore effective withdrawal therapies.

Advancement in the study of opioids addiction provides important insights into the free radicals involved in the process of morphine tolerance, dependence and withdrawal. According to chemical definition, free radical refers to a kind of molecule, atom, or ion that has one or more unpaired valence electrons. Free radicals have a variety of members that mainly possess such active groups of active center as oxygen (O), carbon (C) and nitrogen (N). Oxygen-containing reactive species, known as reactive oxygen species (ROS), commonly possess active chemical properties, which include hydroxyl radical ($\cdot$OH), superoxide radical ($O_2^{\cdot-}$), hydroperoxyl radical ($HO_2\cdot$), hydrogen peroxide (H_2O_2), lipid peroxide radicals ($LOO\cdot$), singlet oxygen ($^1O_2$), nitric oxide (NO), peroxynitrite anion ($ONOO\cdot$) and alkoxy ($RO\cdot$).

The extent of ROS generation relies on the rate of cellular metabolism. A higher metabolic rate is usually observed under oxidative stress, leading to increased ROS production, and a lower rate is happened under normal physiological conditions, implicated in cell proliferation and important to cell viability [5]. Compared with others, ROS and their products exhibit more harmful effects on cells and organs by inducing oxidative damage.

The metabolism of opioids is closely related to ROS, which, along with its metabolic products, can display an important role in opioid dependence, tolerance, and withdrawal. Compared with other opioids, more research attention on morphine and heroin has been drawn into their connection with oxidative stress. First of all, morphine can induce production of high concentration of ROS and inflict oxidative damage to many organs and systems; on the other hand, morphine can lead to the reduction of enzymatic and nonenzymatic antioxidants, such as glutathione, making organisms vulnerable to oxidative injuries. Either the reduced level of ROS or the attenuated antioxidant defense system gives rise to oxidative damage, which in turns drives ROS elevation. These actions can be possibly combined, leading to a vicious cycle of human problems. In addition, some researchers propose the hypothesis that under certain conditions, morphine can exert antioxidant activities via being a direct scavenger [6], the recovery of glutathione levels [7], and/or the inhibition of NADPH oxidase [8]. For example, under the nonenzymatic system, morphine can exert some effects on mitochondria of brain cells where it can decrease the level of lipid peroxide, exhibiting significant antioxidation activity. If oxidative enzyme exists, morphine can increase lipid peroxide.

Khanna and Sharma in 1983 showed that a megadose of vitamin C can withdraw morphine tolerance and dependence in mice [9]. Since then researchers have never stopped contriving ways to figure out the role of free radical in the pathological process of opioids addiction. Morphine administration can lead to opioid receptors activation, which then induces ROS generation [10]. Both heroin and its metabolically produced morphine can interact with various enzymes, receptors, and proteins *in vivo*, which can make piperidine ring of morphine converted into active peridine nitrogen free radicals and other kinds of free radicals, finally leading to ROS reactions and lipid peroxidation [11]. Additionally, it has been found that lipid oxidative damage does not only occur in blood plasma and lymphocytes of heroin addictive patients and liver and brain of addictive mice, but also in such crucial biomolecule as DNA and proteins. DNA damage and protein inactivation can have negative effects on DNA replication, transcription and translation and can even lead to cell death [12,13]. Oxidative stress is widely derived from the process of aging and various neurological lesions, and consequent excess ROS can lead to cellular oxidative damage [14].

37.2 ROS AND MORPHINE ADDICTION

37.2.1 NITRIC OXIDE AND MORPHINE ADDICTION

Nitric oxide (NO) is an important bioactive neurotransmitter, second messenger, and, more notably, a free radical. Due to its various biological activities, NO is closely implicated in many physiological, pathological and pharmacological phenomena, and it mediates various other biological activities. Morphine induces NO generation mainly through a nitric oxide synthase (NOS)-dependent pathway, in which the precursor, L-arginine, is catalyzed by NOS and decomposed to NO. NO is implicated into the process of morphine tolerance and dependence, and NOS activities remarkably increase in such widespread regions of the central nervous system (CNS) as locus coeruleus (LC), cerebellum, amygdala, nucleus accumbens, hippocampus, and medulla oblongata, indicating that NO is one potent contributor to the pathologic mechanism underlying morphine addiction [15–17]. This view is specifically supported by the clinical investigation of blood samples of 137 heroin abuse candidates, suggesting that heroine abuse can destroy antioxidants system and exacerbate oxidative damage, and the longer abuse duration lasts, the more aggravated oxidative stress is [16]. Both NOS activities increase and NOS overexpression results in large amount of NO generation. NO synthesis exceeding the capacity of free radicals scavenging system will deplete cellular antioxidant defenses, resulting in the irreversible lipoperoxidation of various systems and organs. Besides these phenomena, significantly increased levels of NOS and aberrant monoaminergic neurotransmitters contribute to morphine dependence and withdrawal symptoms, which is substantiated in the spinal cord, nucleus accumbens, ventral tegmental area, hippocampus, and the prefrontal cortex in morphine-dependent animal models [17]. The high level of NO can be regarded as the adaptive change to morphine addiction in cells and is implicated in the pathological reaction of opioid in humans.

NO can exert its effects on morphine tolerance and addiction by NO-cyclic guanosine monophosphate (NO-cGMP) and NO-polymers adenosine diphosphate synthase systems [18]. After free diffusion into cells, NO reacts with the active site of c-guanylate cyclase (cGMP) and soluble guanylate cyclase, (sGMP), which is ferrous ion, locating on adjacent presynaptic membrane and subsynaptic membrane, where NO activates and combines with cGMP, leading to intracellular cGMP increase. cGMP, on the one hand, can activate cGMP-dependent protein kinase and regulate various protein functions and physiological activities of neurons. On the other hand, as an essential cell signaling molecule, cGMP indirectly changes the level of cAMP by either activating or inhibiting phosphodiesterase, so the rapid activation in G protein-cAMP system will induce the elevated activities of cAMP-dependent protein kinase, which can alter the synthesis and metabolism of substrate proteins, and exhibit a series of withdrawal symptoms [13,19,20]. Additionally, the N-methyl-D-aspartate (NMDA)-NO-cGMP, as the representative pathway of glutamatergic neurotransmission initiated by glutamate, is involved into morphine dependence and withdrawal symptoms, in which NMDA repressor pathway is responsible for NO generation [21–23]. The administration of glutamate receptor antagonists that act on NMDA receptors can attenuate the analgesic effects of opiates. The NO-cGMP signaling pathway can mediate neurotransmitter release, gene transcription, and neurotoxicity effects, which plays an important role in morphine addiction, withdrawal, and morphine-induced analgesia. Therefore, it is via the indirect reaction that NO exerts its influence on cellular redox systems in the process of morphine dependence and withdrawal.

Actually, the direct reaction of NO with such biomolecules as proteins (except for heme-containing proteins), lipids, and nucleic acids [24,25] is much limited. Morphine can induce cells to release high concentrations of NO not only through directly reacting with various ROS, such as superoxide anion, but also through indirectly oxidizing GSH into GSSG by generating metabolic active nitrogen, such as peroxynitrite ($ONOO^-$) [25]. This causes oxidative damage to lipid, zinc sulfide center, iron-sulfur center, DNA, and protein sulfhydryl, leading to cell and tissue injuries.

37.2.2 SUPEROXIDE RADICALS AND MORPHINE ADDICTION

Morphine can either promote formation of free radicals or reduce activity of different components of antioxidant system in target cells. Thus $O_2^{\bullet-}$ generation elicited by morphine is mainly in the following two ways: first, the inactivation of the antioxidant system will lead to $O_2^{\bullet-}$ scavenge out of control. For example, nitration and inactivation of manganese superoxide dismutase (MnSOD) allows for $O_2^{\bullet-}$ and $ONOO^-$ accumulation during the development and maintenance of central sensitization; second, nicotinamide adenine dinucleotide phosphate (NADPH) oxidase is a noteworthy $O_2^{\bullet-}$ generation system, which is implicated in central sensitization and morphine-induced analgesia and tolerance [26–29]. Upon activation, an $O_2^{\bullet-}$-producing enzyme complex is formed on the membrane, which contains two membrane-bound components and several cytosolic components. The assembly of the active enzyme complex is a principal factor to the posttranslational regulation of NADPH oxidase. NADPH oxidase-derived ROS cause NOS uncoupling and promote ROS formation via mitochondrial electron transport chain. Hence, the interaction among the various sources of ROS may further augment and perpetuate the formation of ROS and oxidative stress [30]. So NADPH oxidase is an important cellular source of ROS under various pathophysiological conditions.

$O_2^{\bullet-}$ increase induced by high concentration of morphine can be observed in liver cells, macrophage, mesangial, and endothelial cells [31]. In a formalin injury test in rats, the enhanced anesthesia effects of morphine caused either by H_2O_2 or $O_2^{\bullet-}$ were withdrawn by naloxone, suggesting that $O_2^{\bullet-}$ is implicated into morphine-induced functions. Morphine and its metabolic products stimulate an increase in $O_2^{\bullet-}$ generation, inducing the disappearance of foot process in glomerular podocyte, the increase of mesangial matrix, and the thickness of the basement membrane, and causing kidney lesion, such as focal segmental glomerulosclerosis [32]. In addition, opioid peptides can activate human polymorphonuclear leukocytes and macrophage to generate $O_2^{\bullet-}$, demonstrating that there exists a close relationship between $O_2^{\bullet-}$ production and immunity [33]. Even at the physiological concentration, opioid peptide can create oxidative effects, so high dose of opioid peptide will cause excessive activation to immune system and inflict various damages to health organs in humans.

Furthermore, $O_2^{\bullet-}$ can interact with H_2O_2 to generate $^\bullet OH$, which is the main way of $^\bullet OH$ formation $in\ vivo$; and there is another way that H_2O_2 can produce $^\bullet OH$ through the Fenton–Haber–Weiss reaction [34]. According to the reduction potential, $^\bullet OH$ is the most reactive oxygen free radical ever known. $^\bullet OH$ can readily interact with almost every molecular component by oxidation, including lipids, proteins, and nucleic acids at an almost diffusion-limited rate. $^\bullet OH$ can also exert the most destructive effects by directly attacking biological macromolecules; for example, it can attack DNA and unsaturated fatty acid (UFA), causing genomic mutation and lipid peroxidation, respectively, and decreasing UFA on the cell membrane; it also changed the phospholipids constitution, cell membrane mobility, cross-linking, permeability, and iron transportation, resulting in autocytolysis. Additionally, $O_2^{\bullet-}$ can interact with NO by a bi-radical reaction to form peroxynitrite ($ONOO^-$), which is a potent oxidant capable of causing cell and tissue injuries. $ONOO^-$ is much more reactive than its parent molecules, $O_2^{\bullet-}$ and NO, and is also able to cross biomembranes, reacting, both directly and indirectly, with most critical biomolecules, such as lipids, proteins, and nucleic acids. Research evidence showed that the system of electron transmission-ROS-oxidative stress

(ET-ROS-OS) is closely related to the receptor transmission pathway in the bioprocess of opioids addiction and other drug addictions [13]. For example, dopamine can produce O-benzoquinone and semiquinone by oxidative metabolism reactions, and its metabolic products are implicated into redox system and finally generate ROS. These evidences will provide effective strategies and ideas to define the mechanism of drug addiction and improve withdrawal therapy.

37.2.3 BIOLOGICAL MOLECULAR OXIDATION AND MORPHINE ADDICTION

Increased endogenous ROS concentration can make certain essential biological molecules, such as DNA, protein, and lipid, more vulnerable to oxidative stress. Oxidative radical reaction plays an important role in metabolic processes in humans, and it is kept in the coordinative and dynamic equilibrium. Such interruptions resulting from morphine or heroin administration, which break the balance, will cause a series of metabolic dysfunctions and impair immunologic functions. The toxic effects of morphine and heroin are attributed to increased lipid peroxidation [35], which can be employed as a marker of the ROS-induced cell damages. Biological membranes contain large amounts of poly-UFAs, which readily react with ROS and cause membrane lipid peroxidation. Research evidence showed that lipid peroxidation is significantly increased in chronic heroin abusers [36], and the levels of malondialdehyde (MDA) and thiobarbituric acid reactive substances (TBARS), as reliable indicators to lipid peroxidation, increased in brain and liver after chronic administration of morphine and heroin [13,25]. Opioids administration can cause increase in ROS, which leads to the chain reaction of the oxidative degradation of UFAs, further causing lipid peroxidation, generating various free radicals, LOO$^\bullet$, LO$^\bullet$, and L$^\bullet$.

Besides these effects, oxidative damage brought about by ROS in morphine addicts is observed in DNA and proteins [13,25]. Oxidative damage in proteins has the following adverse effects: first, it can cause organic injuries by negatively affecting the function of receptors, enzymes, and transport proteins; also, it can lead to the inactivation of DNA repair enzyme and decrease the fidelity of DNA polymerase, and disturb the DNA replication. Finally, it induces long-term damage to biological membrane by causing inactivation and hypofunctioning of biomembrane repair enzymes. Redox regulatory proteins can regulate the specific mRNA and transcription factors and the affinity between ROS and DNA. When ROS overdose destroys the redox system in humans, ROS can detach DNA from proteins and then attack it. As a result of DNA damage, such vital bioprocess as replication, transcription and translation are severely interfered and lead to cell death. Thus it can be concluded that the oxidative damage in lipid, proteins, and DNA induced by morphine undoubtedly brings severe harm to addicts.

37.3 ENHANCEMENT OF METABOLIC FREE RADICALS INDUCED BY MORPHINE

37.3.1 MORPHINE AND PURINE NUCLEOTIDE METABOLISM

Morphine can regulate the activity of xanthine oxidase (XOD) by altering the cellular concentration of calcium ion and inflict oxidative injuries on immune cells and system. XOD has two interconvertible forms, xanthine dehydrogenase and xanthine oxidase. These enzymes catalyze the conversion of hypoxanthine to xanthine and xanthine to uric acid (UA). Both forms of XOD can also catalyze one and two electron reduction of molecular oxygen, respectively, to produce $O_2^{\bullet-}$ and H_2O_2, with xanthine oxidase being more active for generating ROS. The evidence from rat experiments showed that morphine can raise the transcription of XOD in C6 glioma cells, and then increase the xanthine catabolism. It is also observed that the amount of terminal product of xanthine catabolism, UA, is elevated in the serum of morphine-dependent rats [37]. ROS derived from xanthine oxidase and mitochondria further aggravates the oxidative stress burden induced by morphine addiction, leading to disease deterioration.

37.3.2　Morphine and Dopamine

Dopamine (DA) is an important neurotransmitter in the brain, released from substantia nigra of the midbrain area and dopaminergic cells of the ventral tegmental area (VTA). Dopamine mainly regulates neuronal activity by working on γ-aminobutyric acid (GABA)ergic and acetyl cholinergic neurons. It is well known that cell death and ROS generation are principal contributors to neurodegeneration change caused by morphine abuse. Morphine and heroin can bring oxidative damage to neuron cells and even cause neurodegenerative diseases. The DA system is believed to play a crucial role in morphine and heroin addiction behaviors [38–40]. Long-term opioid abuse results in adaptive neurobiological changes in the DA system of brain. And DA system is implicated in the most destructive effects of repeated drug exposure and is a major participant in craving and relapse even after drug-free periods [39]. In the process of its metabolism, dopamine can generate $O_2^{\cdot-}$, $^{\cdot}OH$, H_2O_2, and 3,4-dihydroxy-phenyl acetic acid (DOPAC) by oxidation and degradation, which can be reversed by ascorbate [41]. Dopamine oxidized by monoamine oxidase will produce dopamine quinone and H_2O_2, and it also undergoes autoxidation, especially in the presence of redox-active metal ions, to generate semiquinone radical, superoxide, and electrophilic quinone. Once excessive ROS overwhelm the capability of the cellular defense system, it will inevitably accumulate. This is supported by the research on the injury mechanism of PC12 cells, one dopaminergic cell model, induced by opioids [42]. Opioid abuse has been shown to be associated with decreased densities of dopamine transporters (DAT) and dopamine D2 receptors (D2R) in both animals and humans [43,44]. These neurobiological changes are therefore believed to be the key obstacles in abstinence treatment.

37.4　ROLE OF ROS IN MORPHINE ADDICTION

37.4.1　Opioids Cause Oxidative Injuries by Destroying Redox System

Morphine deregulates the enzymatic and nonenzymatic antioxidant defense system by interferring the redox system and mediating the *in vivo* oxidative stress. The antioxidant enzyme system in cells includes primary antioxidants; secondary antioxidant enzymes, glutathione, glutaredoxin, and thioredoxin systems; and protein and DNA repair enzymes regarded as a part of the antioxidant systems. Primary antioxidants comprise a number of small antioxidant compounds, such as vitamin E, C, and A; SOD with cofactors of selenium, iron, copper, and zinc; catalase (CAT); and glutathione peroxidase (GSH-Px). Secondary antioxidant enzymes consist of glutathione reductase (GR) and glucose-6-phosphate dehydrogenase. The nonenzymatic antioxidant defense system is composed of vitamin E, C, and A; carotenoids; polyphenols; lipoic acid; and other molecule.

ROS is a leading contributor to damages in the opioid receptor signaling pathway. Chronic administration of morphine can generate excessive ROS, and induce the adaptive regulation of gene expression *in vivo*, resulting in addiction memory [45]. As the second messenger, ROS can participate into the opioid receptor–induced signaling process, mediating receptor phosphorylation, changing receptor structure, interfering interaction between morphine and receptor, and altering receptors function. All of these can change gene expression and protein function associated with the opioid receptor signaling pathway and, ultimately, generate a significant effect on cellular cAMP level, which is an important indicator to morphine receptor function change [46]. Thus, ROS indirectly take part in the signaling pathway of opioid tolerance and dependence by influencing cAMP. Also, ROS regulate the effector-cAMP level by interfering with opioid receptors and their coupled G-proteins. But more explorations are needed to find out the exact role of ROS in the opioid receptor signaling pathway.

As a significant component of in vivo redox system, ROS play complex biological roles in cell signal transmission, transcription factors activation, gene expression, cell development, proliferation and apoptosis, Ca^{2+} concentration regulation, oxidation–calcium coupling, and oxidation–phosphorylation coupling. Therefore, when studying the biological effects of ROS, both detrimental and beneficial effects should be taken into considerations. Low levels of ROS can activate signaling pathway and regulate the activities of redox-sensitive transcription factors and mitochondrial enzymes, but a high dose can exert cytotoxicity and interfere with signal transmission, just like its role in opioid addiction.

Heroin can bring severe damage to dynamic balance between intracellular and extracellular ROS, resulting in aggravating ROS reactions and lipid peroxidation, and aberrantly change biological membrane permeability, leading to subcellular structure destruction [13]. ROS and lipid peroxidation interfere biological functions via direct interaction with nucleic acid, protein, lipid and glucose. For example, interaction with lipid causes the dysfunction of lysosomes, mitochondria, and other organelles by disrupting membrane permeability. If this damage occurs in muscle, then it will cause demyelination. Morphine can increase ROS generation by promoting the metabolism of such relevant factors as dopamine and xanthine oxidase. ROS derived from morphine-induced histamine release can generate oxidative stress, and histamine and other inflammatory mediators, such as TNF-α, can induce oxidative stress by ROS elevation. In addition, the oxidative action of morphine lies in its deregulating activities of antioxidant and nonantioxidant enzymes; also receptor mechanism plays an important role in mediating morphine oxidative reactions and regulating redox-sensitive genes expression. Experiment shows that in morphine-induced mouse liver, the levels of antioxidants, SOD, CAT, GSH/GSSH, and peroxidase reduce much lower than in the controls, and cells with DNA injuries, the number of protein carbonyls, and activities of MDA and alanine aminotransferase (ALT) remarkably increase [25]. Similar evidence found in macaque experiments suggested that morphine dependence remarkably increases the level of MDA and 8-isoprostanes and reduces GSH, SOD, GSH-Px, and CAT in blood plasma [47].

37.4.2 ANTIOXIDANTS ENHANCE THE ANALGESIA/ABIRRITATION OF OPIOIDS AND RELIEVE THE SYMPTOM OF OPIOIDS DEPENDENCE AND WITHDRAWAL

As mentioned in Section 37.4.1, antioxidant enzyme systems involve three types of antioxidants. This section discusses several principal antioxidants implicated into antagonizing the oxidative damage induced by morphine addiction.

Glutathione commonly exists in both oxidized (GSSG) and its reduced (GSH) forms. Under the action of glutathione reductase (GR), GSSG can be reduced to GSH, which can scavenge ROS. GSH plays a significant role in the biological system where it stabilizes protein structure and enzyme activities and maintains cell membrane integrality and cytoskeletal structure. At the same time, it works as a primary regulator involved in the redox system and acts as an important antioxidant to protect biological components from oxidative damage.

The effects of morphine on GSH are well established. Experiments on mice suggest that morphine treatment can expose livers to severe oxidative damage, leading to decreased GSH and liver cells apoptosis [45,48]. Heroin significantly reduces the level of GSH in cerebrospinal fluid (CSF); also the levels of 4-hydroxybenzoic acid and 4-hydroxy-3-methoxyphenylacetic acid are remarkably increased [49]. Therefore, any alteration in the concentration of GSH is an indicator to reflect oxidative stress caused by opioids *in vivo*. Once the level of GSH is reduced, the cellular antioxidant defence system will be destroyed, causing oxidative damage to the CNS.

It is still controversial about the mechanism of GSH decrease induced by morphine. The free radical theory proposes that the metabolic product of morphine is ROS, which can *steal* one or more electrons and be endowed with oxidation, and that GSH can work as an electron donor and be oxidized to GSSG, leading to a decrease in GSH content. A second theory is based on a complex because the complex of morphine-GSH is generated in the co-culture of both morphine and liver tissue *in vitro*, which consumes large amounts of GSH, leading to decreased concentration of free GSH in liver cells [50]. The next theory is about GSH depletion. GSH is also a well-known essential antioxidant, playing an indispensable role in the human antioxidant system. Previous studies on morphine addiction emphasized that intracerebroventricular (ICV) morphine can lead to GSH depletion in caudate nucleus cells of rat brain. And ICV morphine in cancer patients can cause rapid decline in GSH in the cerebrospinal fluid, accompanied by a remarkable decrease in 4-hydroxy benzoic acid and homovanillic acid [52]. The depletion of GSH can make the CNS more susceptible to oxidative damage.

Melatonin is another potent candidate antagonizing opioid dependence. Melatonin, N-acetyl-5-methoxy tryptamine, belongs to a neuroendocrine hormone synthesized in the pineal gland. It has extremely wide and effective ROS scavenging effects, and it can exert the most remarkable effects on the strongest oxidant $\cdot$OH with five times stronger scavenging effects than GSH. Thus it is known as the best antioxidant to antagonize oxidative damage, and also as the highly potent scavenger of ROS with promising therapeutic effects on morphine dependence and withdrawal. Morphine-dependent mice usually have decreased immunity due to the declined weight of immune organs, the inhibition of lymphocyte proliferation and the attenuation of macrophage phagocytosis function. Melatonin can remarkably reverse the declined immune functions, significantly improve the phagocytic index of macrophages, and effectively inhibit excessive NO release from peritoneal macrophages. The mechanism of melatonin acting on morphine addiction involves the following facets. First, melatonin can effectively alleviate morphine-induced hyperalgesia and tolerance in the spinal cord. Furthermore, its potential function relies on inhibiting MOR-induced PKCγ activation and NR1 activity, thus decreasing Ca^{2+} influx and reducing neural excitability and gliocytes activity in the spinal cord [53]. Second, melatonin can prohibit the elevated level of NO in blood plasma and brain, which might be related to the ability of melatonin to inhibit NOS activities [54,55]. As mentioned in Section 37.2.1, NOS works as an intermediary to convey the signal from melatonin via regulating NO generation, and finally it exert its influence on oxidative level *in vivo*. Third, the activation of opioid system and benzodiazepine-GABAergic pathway in the spinal cord should be considered as well [56]. Melatonin plays its analgesic effects via directly combing with melatonin receptor MT_2 subtype and modulating $GABA_A$ receptor functions [57]. Also, melatonin shows $GABA_A$ receptor–selective synergism with a benzodiazepine-like agonist action and this pathway seem to be independent for melatonin's receptors [58]. Additionally, there is also evidence suggesting that the analgesic effect of melatonin is exerted by increasing the release of the endogenous β-endorphins in the central nervous system [59,60]. β-endorphin can bind to μ and κ opioid receptors, and then the endogenous opioid system is implicated in the regulation of pain and in the analgesic action of opiate drugs.

In conclusion, oxidative damage is one of the most important pathogenetic contributors to opioids addiction, and exogenous antioxidants could be a feasible therapy to attenuate abstinence syndrome.

37.4.3 OPIOID CAN CAUSE GENETIC AND CELLULAR FACTORS CHANGE AND LEAD TO APOPTOSIS, WHICH IS CLOSELY RELATED TO ROS

Morphine administration can elicit neuronal apoptosis in such important functional regions of brain as the prefrontal cortex, the nucleus accumbens and the hippocampus in neonatal rats [61–63]. The administration of morphine *in vitro* suppresses neural stem cells and glial cells proliferation, inhibits new neurons generation, and promotes apoptosis in a dose-dependent manner, which suggests that potential long-term cellular and behavioral changes are caused by opioids in the developing brain [46].

Apoptosis induced by ROS in brain is regarded as the cause of neurodegenerative disease derived from drugs abuse. Long-term exposure to morphine leads to increased neurons apoptosis, which is mostly observed in the hippocampus of addictive rats, and up-regulating the expressions of three apoptosis-related proteins: Fas, Bcl-2 and caspase-3 [63]. A similar effect is found in the apoptosis of Jurkat cells and newly isolated T-lymphocytes [64]. Morphine can induce apoptosis in these two cells types in a dose-dependent manner, which can be inhibited by CAT that is a heme enzyme and scavenges H_2O_2. ROS can impair the immunity of heroin addicts by directly inducing apoptosis in

T-lymphocytes. Moreover, morphine decreases Bcl-2 expression and activates NF-κB transcription. Morphine may also trigger the mRNA expression of pro-apoptotic receptors in the lymphocytes and in mouse spleen, lung, and heart via activating opioid receptors [66].

The mechanism of morphine-enhanced apoptosis emphasizes that NO upregulation is one contributor to oxidative stress associated with chronic opioid exposure in adult rats, which leads to neuronal apoptosis [67]. High concentration of NO acts as an important initial signaling mediator of nitro-oxidative stress to actuate neuronal necrosis [68], but lower dose of ROS exposure results in delayed neuronal apoptosis [69]. This is supported by the research that morphine induces macrophage apoptosis which is stimulated by $O_2^{\cdot-}$ and NO generation, and antioxidants reverse this effect of morphine [70]. In addition, the receptor system plays an indispensable role in morphine-induced apoptosis. But different mechanism of morphine-induced apoptosis is involved in different kinds of cancer cell. For example, morphine-induced apoptosis in MCF-7 cell (human breast adenocarcinoma cell line) has been shown to be triggered by a novel p53- and caspase-independent apoptotic pathway using a sigma-2 receptor [71], and human epithelial tumor cell line has been shown to undergo apoptosis by the activation of a κ-opioid receptor via the phospholipase C pathway [72]. Delta opioid peptide inhibited the apoptosis of pheochromocytoma cells (PC12) by a delta2 opioid receptor-mediated increase of Fas ligand [73]. Therefore, different receptor systems should be considered in the research on mechanisms underlying different morphine-induced cancer apoptosis.

37.5 CONCLUSION

As discussed, ROS is implicated into the whole bioprocess of morphine addiction. The thorough demonstration of both the role of ROS and its functioning mechanism in morphine addiction is essential to understand the pathophysiological process of morphine addiction. At the same time, the effective intervention of antioxidants provides the strong evidence for the involvement of ROS in morphine addiction and offers the promising future for morphine addiction treatment. The explorations for effective antioxidant therapy and clinical studies about morphine addiction remain to be made.

REFERENCES

1. The United Nations Office on Drugs and Crime (UNODC). 2014. World drug report 2014. Vienna, Austria: United Nations Publication.
2. Nestler EJ. Molecular mechanisms of drug addiction. *Neuropharmacology*. 2004;47(Suppl. 1):24–32.
3. Kreek MJ, Levran O, Reed B, Schlussman SD, Zhou Y, Butelman ER. Opiate addiction and cocaine addiction: Underlying molecular neurobiology and genetics. *J Clin Invest*. 2012;122(10):3387–3393.
4. Leshner AI. Addiction is a brain disease, and it matters. *Science*. 1997;278(5335):45–47.
5. Trachootham D, Lu W, Ogasawara MA, Nilsa RD, Huang P. Redox regulation of cell survival. *Antioxid Redox Signal*. 2008;10(8):1343–1374.
6. Kanesaki T, Saeki M, Ooi Y, Suematsu M, Matsumoto K, Sakuda M, Saito K, Maeda S. Morphine prevents peroxynitrite-induced death of human neuroblastoma SHSY5Y cells through a direct scavenging action. *Eur J Pharmacol*. 1999;372(3):319–324.
7. Gülçin I, Beydemir S, Alici HA, Elmastaş M, Büyükokuroğlu ME. *In vitro* antioxidant properties of morphine. *Pharmacol Res*. 2004;49(1):59–66.
8. Qian L, Tan KS, Wei SJ, Wu HM, Xu Z, Wilson B, Lu RB, Hong JS, Flood PM. Microglia-mediated neurotoxicity is inhibited by morphine through an opioid receptor-independent reduction of NADPH oxidase activity. *J Immunol*. 2007;179(2):1198–1209.
9. Khanna NC, Sharma SK. Megadoses of vitamin C prevent the development of tolerance and physical dependence on morphine in mice. *Life Sci*. 1983;33(Suppl. 1):401–404.
10. Tanaka K, Ludwig LM, Kersten JR, Pagel PS, Warltier DC. Mechanisms of cardioprotection by volatile anesthetics. *Anesthesiology*. 2004;100(3):707–721.
11. Zhou JF, Yan XF, Ruan ZR, Peng FY, Cai D, Yuan H, Sun L, Ding DY, Xu SS. Heroin abuse and nitric oxide, oxidation, peroxidation, lipoperoxidation. *Biomed Environ Sci*. 2000;13(2):131–139.

12. Afify EA, Khedr MM, Omar AG, Nasser SA. The involvement of K(ATP) channels in morphine-induced antinociception and hepatic oxidative stress in acute and inflammatory pain in rats. *Fundam Clin Pharmacol.* 2013;27(6):623–631.

13. Xu B, Wang Z, Li G, Li B, Lin H, Zheng R, Zheng Q. Heroin-administered mice involved in oxidative stress and exogenous antioxidant-alleviated withdrawal syndrome. *Basic Clin Pharmacol Toxicol.* 2006;99(2):153–161.

14. Mattson MP, Liu D. Energetics and oxidative stress in synaptic plasticity and neurodegenerative disorders. *Neuromolecular Med.* 2002;2(2):215–231.

15. Abdel-Zaher AO, Mostafa MG, Farghaly HS, Hamdy MM, Abdel-Hady RH. Role of oxidative stress and inducible nitric oxide synthase in morphine-induced tolerance and dependence in mice. Effect of alpha-lipoic acid. *Behav Brain Res.* 2013;247:17–26.

16. Zhou J, Si P, Ruan Z, Ma S, Yan X, Sun L, Peng F et al. Primary studies on heroin abuse and injury induced by oxidation and lipoperoxidation. *Chin Med J (Engl).* 2001;114(3):297–302.

17. Chen C, Nong Z, Huang J, Chen Z, Zhang S, Jiao Y, Chen X, Huang R. Yulangsan polysaccharide attenuates withdrawal symptoms and regulates the NO pathway in morphine-dependent rats. *Neurosci Lett.* 2014;570:63–68.

18. Zarrindast MR, Karami M, Sepehri H, Sahraei H. Influence of nitric oxide on morphine-induced conditioned place preference in the rat central amygdala. *Eur J Pharmacol.* 2002;453(1):81–89.

19. Granados-Soto V, Rufino MO, Gomes Lopes LD, Ferreira SH. Evidence for the involvement of the nitric oxide-cGMP pathway in the antinociception of morphine in the formalin test. *Eur J Pharmacol.* 1997;340(2–3):177–180.

20. Javadi S, Ejtemaeimehr S, Keyvanfar HR, Moghaddas P, Aminian A, Rajabzadeh A, Mani AR, Dehpour AR. Pioglitazone potentiates development of morphine-dependence in mice: Possible role of NO/cGMP pathway. *Brain Res.* 2013;1510:22–37.

21. Nestler EJ, Aghajanian GK. Molecular and cellular basis of addiction. *Science.* 1997;278(5335):58–63.

22. Gabra BH, Afify EA, Daabees TT, Abou Zeit-Har MS. The role of the NO/NMDA pathways in the development of morphine withdrawal induced by naloxone *in vitro. Pharmacol Res.* 2005;51(4):319–327.

23. Buccafusco JJ, Terry AV Jr, Shuster L. Spinal NMDA receptor—Nitric oxide mediation of the expression of morphine withdrawal symptoms in the rat. *Brain Res.* 1995;679(2):189–199.

24. Sumathy T, Subramanian S, Govindasamy S, Balakrishna K, Veluchamy G. Protective role of *Bacopa monniera* on morphine induced hepatotoxicity in rats. *Phytother Res.* 2001;15(7):643–645.

25. Zhang YT, Zheng QS, Pan J, Zheng RL. Oxidative damage of biomolecules in mouse liver induced by morphine and protected by antioxidants. *Basic Clin Pharmacol Toxicol.* 2004;95(2):53–58.

26. Muscoli C, Mollace V, Wheatley J, Masini E, Ndengele M, Wang ZQ, Salvemini D. Superoxide-mediated nitration of spinal manganese superoxide dismutase: A novel pathway in *N*-methyl-D-aspartate-mediated hyperalgesia. *Pain.* 2004;111(1–2):96–103.

27. Muscoli C, Cuzzocrea S, Ndengele MM, Mollace V, Porreca F, Fabrizi F, Esposito E, Masini E, Matuschak GM, Salvemini D. Therapeutic manipulation of peroxynitrite attenuates the development of opiate-induced antinociceptive tolerance in mice. *J Clin Invest.* 2007;117(11):3530–3539.

28. Salvemini D, Little JW, Doyle T, Neumann WL. Roles of reactive oxygen and nitrogen species in pain. *Free Radic Biol Med.* 2011;51(5):951–966.

29. Ibi M, Matsuno K, Matsumoto M, Sasaki M, Nakagawa T, Katsuyama M, Iwata K, Zhang J, Kaneko S, Yabe-Nishimura C. Involvement of NOX1/NADPH oxidase in morphine-induced analgesia and tolerance. *J Neurosci.* 2011;31(49):18094–18103.

30. Babior BM, Lambeth JD, Nauseef W. The neutrophil NADPH oxidase. *Arch Biochem Biophys.* 2002;397(2):342–344.

31. Lam CF, Liu YC, Tseng FL, Sung YH, Huang CC, Jiang MJ, Tsai YC. High-dose morphine impairs vascular endothelial function by increased production of superoxide anions. *Anesthesiology.* 2007;106(3):532–537.

32. Singhal PC, Pamarthi M, Shah R, Chandra D, Gibbons N. Morphine stimulates superoxide formation by glomerular mesangial cells. *Inflammation.* 1994;18(3):293–299.

33. Sharp BM, Keane WF, Suh HJ, Gekker G, Tsukayama D, Peterson PK. Opioid peptides rapidly stimulate superoxide production by human polymorphonuclear leukocytes and macrophages. *Endocrinology.* 1985;117(2):793–795.

34. Schweikert C, Liszkay A, Schopfer P. Polysaccharide degradation by Fenton reaction—Or peroxidase-generated hydroxyl radicals in isolated plant cell walls. *Phytochemistry.* 2002;61(1):31–35.

35. Lurie E, Soloviova A, Alyabieva T, Kaplun A, Panchenko L, Shvets V. Effect of novel aromatic derivative of GABA on lipid peroxidation in chronically morphinized rats. *Biochem Mol Biol Int.* 1995;36(1):13–19.

36. Panchenko LF, Pirozhkov SV, Nadezhdin AV, Baronets VIu, Usmanova NN. Lipid peroxidation, peroxyl radical-scavenging system of plasma and liver and heart pathology in adolescence heroin users. *Vopr Med Khim.* 1999;45(6):501–506.

37. Liu C, Liu JK, Kan MJ, Gao L, Fu HY, Zhou H, Hong M. Morphine enhances purine nucleotide catabolism *in vivo* and *in vitro*. *Acta Pharmacol Sin.* 2007;28(8):1105–1115.

38. Di Chiara G, Bassareo V. Reward system and addiction: What dopamine does and doesn't do. *Curr Opin Pharmacol.* 2007;7(1):69–76.

39. Melis M, Spiga S, Diana M. The dopamine hypothesis of drug addiction: Hypodopaminergic state. *Int Rev Neurobiol.* 2005;63:101–154.

40. Wu G, Huang W, Zhang H, Li Q, Zhou J, Shu H. Inhibitory effects of processed Aconiti tuber on morphine-induced conditioned place preference in rats. *J Ethnopharmacol.* 2011;136(1):254–259.

41. Rajaei Z, Alaei H, Nasimi A, Amini H, Ahmadiani A. Ascorbate reduces morphine-induced extracellular DOPAC level in the nucleus accumbens: A microdialysis study in rats. *Brain Res.* 2005;1053(1–2):62–66.

42. Oliveira MT, Rego AC, Morgadinho MT, Macedo TR, Oliveira CR. Toxic effects of opioid and stimulant drugs on undifferentiated PC12 cells. *Ann N Y Acad Sci.* 2002;965:487–496.

43. Simantov R. Chronic morphine alters dopamine transporter density in the rat brain: Possible role in the mechanism of drug addiction. *Neurosci Lett.* 1993;163(2):121–124.

44. Kish SJ, Kalasinsky KS, Derkach P, Schmunk GA, Guttman M, Ang L, Adams V, Furukawa Y, Haycock JW. Striatal dopaminergic and serotonergic markers in human heroin users. *Neuropsychopharmacology.* 2001;24(5):561–567.

45. Payabvash S, Beheshtian A, Salmasi AH, Kiumehr S, Ghahremani MH, Tavangar SM, Sabzevari O, Dehpour AR. Chronic morphine treatment induces oxidant and apoptotic damage in the mice liver. *Life Sci.* 2006;79(10):972–980.

46. Ma J, Yuan X, Qu H, Zhang J, Wang D, Sun X, Zheng Q. The role of reactive oxygen species in morphine addiction of SH-SY5Y cells. *Life Sci.* 2015;124:128–135.

47. Pérez-Casanova A, Husain K, Noel RJ Jr., Rivera-Amill V, Kumar A. Interaction of SIV/SHIV infection and morphine on plasma oxidant/antioxidant balance in macaque. *Mol Cell Biochem.* 2008;308(1–2):169–175.

48. Samarghandian S, Afshari R, Farkhondeh T. Effect of long-term treatment of morphine on enzymes, oxidative stress indices and antioxidant status in male rat liver. *Int J Clin Exp Med.* 2014;7(5):1449–1453.

49. Pan J, Zhang Q, Zhang Y, Ouyang Z, Zheng Q, Zheng R. Oxidative stress in heroin administered mice and natural antioxidants protection. *Life Sci.* 2005;77(2):183–193.

50. Kumagai Y, Todaka T, Toki S. A new metabolic pathway of morphine: *In vivo* and *in vitro* formation of morphinone and morphine-glutathione adduct in guinea pig. *J Pharmacol Exp Ther.* 1990;255(2):504–510.

51. Willner D, Cohen-Yeshurun A, Avidan A, Ozersky V, Shohami E, Leker RR. Short term morphine exposure *in vitro* alters proliferation and differentiation of neural progenitor cells and promotes apoptosis via mu receptors. *PLoS One.* 2014;9(7):e103043.

52. Goudas LC, Langlade A, Serrie A, Matson W, Milbury P, Thurel C, Sandouk P, Carr DB. Acute decreases in cerebrospinal fluid glutathione levels after intracerebroventricular morphine for cancer pain. *Anesth Analg.* 1999;89(5):1209–1215.

53. Song L, Wu C, Zuo Y. Melatonin prevents morphine-induced hyperalgesia and tolerance in rats: Role of protein kinase C and *N*-methyl-ᴅ-aspartate receptors. *BMC Anesthesiol.* 2015;15:12.

54. Raghavendra V, Kulkarni SK. Possible mechanisms of action in melatonin reversal of morphine tolerance and dependence in mice. *Eur J Pharmacol.* 2000;409(3):279–289.

55. Yahyavi-Firouz-Abadi N, Tahsili-Fahadan P, Riazi K, Ghahremani MH, Dehpour AR. Melatonin enhances the anticonvulsant and proconvulsant effects of morphine in mice: Role for nitric oxide signaling pathway. *Epilepsy Res.* 2007;75(2–3):138–144.

56. Zurowski D, Nowak L, Machowska A, Wordliczek J, Thor PJ. Exogenous melatonin abolishes mechanical allodynia but not thermal hyperalgesia in neuropathic pain. The role of the opioid system and benzodiazepine-gabaergic mechanism. *J Physiol Pharmacol.* 2012;63(6):641–647.

57. Wan Q, Man HY, Liu F, Braunton J, Niznik HB, Pang SF, Brown GM, Wang YT. Differential modulation of GABAA receptor function by Mel1a and Mel1b receptors. *Nat Neurosci.* 1999;2(5):401–403.

58. Dhanaraj E, Nemmani KV, Ramarao P. Melatonin inhibits the development of tolerance to U-50,488H analgesia via benzodiazepine-GABAAergic mechanisms. *Pharmacol Biochem Behav.* 2004;79(4):733–737.
59. Shavali S, Ho B, Govitrapong P, Sawlom S, Ajjimaporn A, Klongpanichapak S, Ebadi M. Melatonin exerts its analgesic actions not by binding to opioid receptor subtypes but by increasing the release of beta-endorphin an endogenous opioid. *Brain Res Bull.* 2005;64(6):471–479.
60. Yu CX, Zhu B, Xu SF, Cao XD, Wu GC. The analgesic effects of peripheral and central administration of melatonin in rats. *Eur J Pharmacol.* 2000;403(1–2):49–53.
61. Hsiao PN, Chang MC, Cheng WF, Chen CA, Lin HW, Hsieh CY, Sun WZ. Morphine induces apoptosis of human endothelial cells through nitric oxide and reactive oxygen species pathways. *Toxicology.* 2009;256(1–2):83–91.
62. Katebi SN, Razavi Y, Zeighamy Alamdary S, Khodagholi F, Haghparast A. Morphine could increase apoptotic factors in the nucleus accumbens and prefrontal cortex of rat brain's reward circuitry. *Brain Res.* 2013;1540:1–8.
63. Liu LW, Lu J, Wang XH, Fu SK, Li Q, Lin FQ. Neuronal apoptosis in morphine addiction and its molecular mechanism. *Int J Clin Exp Med.* 2013;6(7):540–545.
64. Yin D, Woodruff M, Zhang Y, Whaley S, Miao J, Ferslew K, Zhao J, Stuart C. Morphine promotes Jurkat cell apoptosis through pro-apoptotic FADD/P53 and anti-apoptotic PI3K/Akt/NF- kappaB pathways. *J Neuroimmunol.* 2006;174(1–2):101–107.
65. Jaquet PE, Ferrer-Alcón M, Ventayol P, Guimón J, García-Sevilla JA. Acute and chronic effects of morphine and naloxone on the phosphorylation of neurofilament-H proteins in the rat brain. *Neurosci Lett.* 2001;304(1–2):37–40.
66. Yin D, Mufson RA, Wang R, Shi Y. Fas-mediated cell death promoted by opioids. *Nature.* 1999;397(6716):218.
67. Salvemini D. Peroxynitrite and opiate antinociceptive tolerance: A painful reality. *Arch Biochem Biophys.* 2009;484(2):238–244.
68. Virág L, Szabó E, Gergely P, Szabó C. Peroxynitrite-induced cytotoxicity: Mechanism and opportunities for intervention. *Toxicol Lett.* 2003;140–141:113–124.
69. Niles JC, Wishnok JS, Tannenbaum SR. Peroxynitrite-induced oxidation and nitration products of guanine and 8-oxoguanine: Structures and mechanisms of product formation. *Nitric Oxide.* 2006;14(2):109–121.
70. Bhat RS, Bhaskaran M, Mongia A, Hitosugi N, Singhal PC. Morphine-induced macrophage apoptosis: Oxidative stress and strategies for modulation. *J Leukoc Biol.* 2004;75(6):1131–1138.
71. Wheeler KT, Wang LM, Wallen CA, Childers SR, Cline JM, Keng PC, Mach RH. Sigma-2 receptors as a biomarker of proliferation in solid tumours. *Br J Cancer.* 2000;82(6):1223–1232.
72. Diao CT, Li L, Lau SY, Wong TM, Wong NS. kappa-Opioid receptor potentiates apoptosis via a phospholipase C pathway in the CNE2 human epithelial tumor cell line. *Biochim Biophys Acta.* 2000;1499(1–2):49–62.
73. Hayashi T, Tsao LI, Su TP. Antiapoptotic and cytotoxic properties of delta opioid peptide [D-Ala(2),D-Leu(5)]enkephalin in PC12 cells. *Synapse.* 2002;43(1):86–94.

38 Exercise, Nitric Oxide, and ROS

Leonardo Y. Tanaka, Luiz Roberto G. Bechara, and Paulo R. Ramires

CONTENTS

ABSTRACT

There are several processes by which exercise affects the production of reactive oxygen species (ROS) and nitric oxide (NO). Moreover, the complexity of the association of these issues is accentuated if we consider the specific adjustments caused by physical activity on the different organs and tissues (Kojda and Hambrecht 2005, Powers et al. 2010, Calvert et al. 2011). In this chapter, tissue-specific concerns about NO and ROS during exercise are discussed following a translational view: from the relationship between redox signaling and health promoting effects by chronic exercise to molecular redox targets triggered during acute exercise sessions.

Importantly, the concept of oxidative stress followed here is based on the work by Jones (2006), which describes oxidative stress as a loss of mechanisms controlling redox signaling, in other words, contrary to the balance concept (oxidants vs. antioxidants). The former concept is markedly reinforced by the absence of solid effects by antioxidants treatment during several conditions, including exercise training. Cumulative knowledge has demonstrated that adaptations mediated by regulated ROS production during exercise are not limited to active tissues, for example, skeletal muscle (Kojda and Hambrecht 2005). Therefore, distinct physiological readouts are discussed below such as endothelial function and skeletal/cardiac muscle contraction and hypertrophy.

38.1 EXERCISE, REDOX SIGNALING, AND ENDOTHELIAL FUNCTION

Since the discovery of the fundamental role of the endothelium in controlling acetylcholine-mediated vasodilation (Furchgott and Zawadzki 1980), much attention was given to elucidate the role of endothelial cells on the cardiovascular system, and several studies have elucidated the essential role of the endothelium on counteracting atherosclerosis and its risk factors (Bonetti et al. 2003). The monolayer of endothelial cells is strategically located at the inner side of the vessels acting as a sensor for mechanical and chemical stimulus, and it promptly responds by releasing vasoactive compounds that modulate vasomotricity and prevent leukocyte and platelet adhesion and smooth muscle cells proliferation. Importantly, these effects have largely been shown to be associated with the endothelial release of NO, which is provided from the endothelial nitric oxide synthase (eNOS). Through protein kinase and/or Ca^{2+}/calmodulin-dependent mechanisms, leading to eNOS activation and NO generation from L-arginine, the labile gaseous NO can achieve soluble guanilyl cyclase in vascular smooth muscle cell (VSMC), which converts guanosine triphosphate (GTP) into cyclic guanosine monophosphate (GMP) and promotes a concerted effect resulting in a decrease in Ca^{2+} levels and consequently vasorelaxation. In contrast, the negative regulator of NO, superoxide ($O_2^{-\bullet}$), limits its action by a very fast reaction producing peroxynitrite and limiting NO bioactivity. Thus, a delicate regulation of NO bioavailability is important for endothelial function maintenance (Kojda and Hambrecht 2005).

Indeed, the salutary effect of physical exercise on endothelial function has been observed in experimental models and in humans in health and when associated with several pathological conditions (Jasperse and Laughlin 2006). Collectively, the mechanisms involved in the improvement of endothelial function induced by exercise are associated with NO bioavailability upregulation, such as by increasing its production by eNOS (Tanaka et al. 2015), or limiting its inactivation both by decreasing the expression of the important $O_2^{-\bullet}$ source in vascular cells NADPH oxidase (Nox) and associated subunits (Adams et al. 2005); also by improving the capacity of ROS buffer upregulating the endogenous antioxidant levels. Paradoxically, exercise-induced ROS production is critical for eNOS induction by chronic exercise, as observed by Lauer et al. (2005), who proposed a model of endothelial catalase overexpression increasing in response to exercise training in the absence of eNOS. Indeed, the mechanisms by which redox signaling converges with moderate exercise improvements have been demonstrated to follow the hormesis concept: low or moderate doses of stress triggers adaptive responses resulting in benefits (Radak et al. 2008). The relationship between stress level, such as exercise intensity, and adaptation represents a bell or J-shaped curve. Therefore, the hormetic effect has a critical point, and above this limit, the level of stress overwhelms the endogenous protection capacity, leading to deleterious effects (Radak et al. 2008). Exercise-induced redox signaling is a surrogate statement of hormesis (Ristow et al. 2009).

In vascular cells, the enzymatic complex NADPH oxidase from the Nox family is the most important ROS source for signaling purposes. Several characteristics contribute to compartmentalized ROS production from NADPH oxidases permitting its association with regulated cell signaling, such as specific agonist activation, cell-specific expression, antioxidant system avoiding off-targets, and Nox-regulatory proteins, including polymerase delta-interacting protein 2 (Poldip2) and protein disulfide isomerase (PDI). Endothelial cells express Nox1, Nox2, Nox4, and Nox5, except in rodents, which do not express Nox5 (Lassegue and Griendling 2010). Chronically, physical exercise induces downregulation of NADPH oxidases and its regulatory proteins (Kojda and Hambrecht 2005), but acute exercise triggers $O_2^{-\bullet}$ production by Nox activation, at least in part, by Nox2-dependent mechanisms (Tanaka et al. 2015).

Temporary increase in ROS during exercise not only contributes to chronic adaptations but also converges with immediate benefits on endothelial function. Although unclear, this effect could be mediated by dependent and independent activation of eNOS by hydrogen peroxide (H_2O_2). Of note, this effect is associated with moderate exercise regardless of the oxidative stress induction, while high-intensity exercise may impair endothelium-dependent vasodilation in close association with oxidative stress induction (Goto et al. 2003). Moreover, exercise-induced upregulation of the ROS buffering system has been observed to positively affect NO bioavailability, which is markedly associated with improved endothelium-dependent relaxations (Tanaka et al. 2015) and attenuated constrictor response (Bechara et al. 2008).

The increase in vascular NO levels caused by physical activity have been associated with accentuated mechanical forces imposed by higher blood flow such as shear stress, which also triggers $O_2^{-\bullet}$ production by Nox activation (Duerrschmidt et al. 2006). However, following longer periods of chronic exercise, the capacity of outward vascular remodeling minimizes the impact of shear forces and limits the upregulation of eNOS induction (Maiorana et al. 2003). This mechanism might be important to avoid excessive vasodilation and has highlighted the flexibility of redox regulation cross talking with structural alterations.

38.2 EXERCISE, REDOX SIGNALING, AND SKELETAL MUSCLE ADAPTATIONS

The increased oxygen uptake by exercise-active skeletal muscle makes it one of the main targets of redox processes and has been investigated for several years. Initially, Davies et al. (1982) observed accentuated ROS production by physical exercise. Since then several studies have shown that redox signaling can affect skeletal muscle through certain distinct aspects related to both, acute and chronic exercise.

Cell contraction is strictly controlled by mechanisms regulating Ca^{2+} handling. Of note, ryanodine receptor (RyR), which is essential during excitation–contraction coupling through Ca^{2+} release, has 100 cysteine residues, with about 21 redox-modulated in physiological O_2 tension. The activation of RyR1 (predominant isoform in skeletal muscle) by S-nitrosylation and S-oxidation is a marked example of allosteric activation through redox signaling, with oxidation promoting increased RyR1 activity and overoxidation Ca^{2+} leakage (Stamler et al. 2008). These mechanisms are in concerted action with enhanced ROS production by working muscles and disruption of Ca^{2+} regulation with pathological conditions inducing dystrophy and probably with ROS-induced fatigue. In this context, compartmentalized signaling is possible in the presence of RyR in close proximity with localized neuronal-NOS, providing NO and NADPH oxidase isoforms Nox4 and Nox2, with the former generating H_2O_2 and the latter likely to play a role in tetanic contraction (Sun et al. 2011).

The increase of muscle mass caused by exercise-induced hypertrophy is another important ROS responsive end point. In this sense, both molecular pathways associated with enhanced protein content, including distinct mitogen-activated kinases (MAPKs), and the inhibition of proteolysis have been demonstrated to be redox modulated and participate in hypertrophic responses. In addition, the convergence between oxidative and endoplasmic reticulum stress–mediated signaling has been demonstrated to play a role during exercise-induced hypertrophy. In contrast, a marked paradox in this subject is that there is also loss of muscle content by disuse-induced ROS increase (Powers et al. 2010). Therefore, distinct redox-dependent pathways activated by exercise or disuse have been addressed, and elucidation of this mechanism would further help to synergize with therapies preventing muscle loss and for exercise performance.

38.3 EXERCISE, REDOX SIGNALING, AND CARDIAC ADAPTATIONS

The regulation through redox processes on exercise-adaptations on the heart has been the focus of much attention. Of note, the mechanisms by which ROS influences hypertrophy and cell contraction of striated skeletal and cardiac cells share similarities and will not be described here. Differently, the protective effect mediated by chronic exercise during acute events, such as myocardial infarct, has been reported to be regulated by redox signaling and is another distinct example of NO and ROS cross talk after exercise training. Improvement in antioxidant capacity during chronic exercise combined with the downregulation of ROS generation can enhance myocardial resistance to injury during ischemic periods (Powers et al. 2014). Moreover, stress-responsive pathways, such as adrenergic signaling, activate eNOS during exercise and promote additional protective signaling by increasing NO release and improving its local storage.

There are several ROS sources such as mitochondrial complex I and III, xanthine oxidase, uncoupled NOS, and NADPH oxidases. The relative ROS generation promoted by physical activity varies between the tissue/organ and the presence of preexisting pathology. Here, we highlighted the contribution of NADPH oxidases due to their reported role as master regulators in redox signaling (Brown and Griendling 2009). However, cross talk between NADPH oxidases with mitochondria has been frequently observed (Zinkevich and Gutterman 2011).

Additionally, despite the benefits of exercise in improving antioxidant capacity have been broadly observed, the upregulation of specific antioxidant enzymes (i.e., superoxide dismutase, catalase and glutathione peroxidase) is variable in literature, and this response may also be tissue dependent. However, the redox target Nrf2 has been reported as an important mediator of this effect (Muthusamy et al. 2012).

In this chapter, we described three important exercise targets of physical activity, but other organs/tissues were also evaluated. Moreover, we highlighted studies concerning redox signaling regulation, despite the extensive literature on exercise and alternations in oxidative markers. The growing investigations addressing mechanisms orchestrated from acute to chronic exercise have helped to demystify the association of ROS generation with deleterious processes and have brought valuable information to the physical activity and redox field.

Exercise tolerance has been considered an independent predictor of mortality (Myers et al. 2002), which is different from ancestral humans, in whom the capacity to perform physical activity and survive during periods of starvation was a fundamental factor for human natural selection. Nowadays, however, abundance of food and the low demand for exercise in industrialized nations has turned physical activity into an option for a relaxed quality of life. In fact, the impact of exercise on the body promotes several adjustments, leading to a higher capacity for using the stored energy to produce force and to improve exercise performance, which are intrinsically associated to redox adaptations at different levels, such as oxygen uptake, skeletal fiber, and mitochondrial function (Hawley et al. 2014). The capacity to handle redox processes by improving the compartmentalization of ROS/NO production is fundamental to generate health-promoting effects from moderate and regular physical activity.

REFERENCES

Adams, V., A. Linke, N. Krankel, S. Erbs, S. Gielen, S. Mobius-Winkler, J. F. Gummert et al. 2005. Impact of regular physical activity on the NAD(P)H oxidase and angiotensin receptor system in patients with coronary artery disease. *Circulation* 111(5):555–562.

Bechara, L. R., L. Y. Tanaka, A. M. Santos, C. P. Jordão, L. G. Sousa, T. Bartholomeu, and P. R. Ramires. 2008. A single bout of moderate-intensity exercise increases vascular NO bioavailability and attenuates adrenergic receptor-dependent and -independent vasoconstrictor response in rat aorta. *J Smooth Muscle Res* 44(3–4):101–111.

Bonetti, P. O., L. O. Lerman, and A. Lerman. 2003. Endothelial dysfunction: A marker of atherosclerotic risk. *Arterioscler Thromb Vasc Biol* 23(2):168–175.

Brown, D. I. and K. K. Griendling. 2009. Nox proteins in signal transduction. *Free Radic Biol Med* 47(9):1239–1253.

Calvert, J. W., M. E. Condit, J. P. Aragon, C. K. Nicholson, B. F. Moody, R. L. Hood, A. L. Sindler et al. 2011. Exercise protects against myocardial ischemia-reperfusion injury via stimulation of beta(3)-adrenergic receptors and increased nitric oxide signaling: Role of nitrite and nitrosothiols. *Circ Res* 108(12):1448–1458.

Davies, K. J., A. T. Quintanilha, G. A. Brooks, and L. Packer. 1982. Free radicals and tissue damage produced by exercise. *Biochem Biophys Res Commun* 107(4):1198–1205.

Duerrschmidt, N., C. Stielow, G. Muller, P. J. Pagano, and H. Morawietz. 2006. NO-mediated regulation of NAD(P)H oxidase by laminar shear stress in human endothelial cells. *J Physiol* 576(Pt 2):557–567.

Furchgott, R. F. and J. V. Zawadzki. 1980. The obligatory role of endothelial cells in the relaxation of arterial smooth muscle by acetylcholine. *Nature* 288 (5789):373–376.

Goto, C., Y. Higashi, M. Kimura, K. Noma, K. Hara, K. Nakagawa, M. Kawamura, K. Chayama, M. Yoshizumi, and I. Nara. 2003. Effect of different intensities of exercise on endothelium-dependent vasodilation in humans: Role of endothelium-dependent nitric oxide and oxidative stress. *Circulation* 108(5):530–535.

Hawley, J. A., M. Hargreaves, M. J. Joyner, and J. R. Zierath. 2014. Integrative biology of exercise. *Cell* 159(4):738–749.

Jasperse, J. L. and M. H. Laughlin. 2006. Endothelial function and exercise training: Evidence from studies using animal models. *Med Sci Sports Exerc* 38(3):445–454.

Jones, D. P. 2006. Redefining oxidative stress. *Antioxid Redox Signal* 8(9–10):1865–1879.

Kojda, G. and R. Hambrecht. 2005. Molecular mechanisms of vascular adaptations to exercise. Physical activity as an effective antioxidant therapy? *Cardiovasc Res* 67(2):187–197.

Lassègue, B. and K. K. Griendling. 2010. NADPH oxidases: Functions and pathologies in the vasculature. *Arterioscler Thromb Vasc Biol* 30(4):653–661.

Lauer, N., T. Suvorava, U. Ruther, R. Jacob, W. Meyer, D. G. Harrison, and G. Kojda. 2005. Critical involvement of hydrogen peroxide in exercise-induced up-regulation of endothelial NO synthase. *Cardiovasc Res* 65(1):254–262.

Maiorana, A., G. O'Driscoll, R. Taylor, and D. Green. 2003. Exercise and the nitric oxide vasodilator system. *Sports Med* 33(14):1013–1035.

Muthusamy, V. R., S. Kannan, K. Sadhaasivam, S. S. Gounder, C. J. Davidson, C. Boeheme, J. R. Hoidal, L. Wang, and N. S. Rajasekaran. 2012. Acute exercise stress activates Nrf2/ARE signaling and promotes antioxidant mechanisms in the myocardium. *Free Radic Biol Med* 52(2):366–376.

Myers, J., M. Prakash, V. Froelicher, D. Do, S. Partington, and J. E. Atwood. 2002. Exercise capacity and mortality among men referred for exercise testing. *N Engl J Med* 346(11):793–801.

Powers, S. K., J. Duarte, A. N. Kavazis, and E. E. Talbert. 2010. Reactive oxygen species are signalling molecules for skeletal muscle adaptation. *Exp Physiol* 95(1):1–9.

Powers, S. K., A. J. Smuder, A. N. Kavazis, and J. C. Quindry. 2014. Mechanisms of exercise-induced cardioprotection. *Physiology (Bethesda)* 29(1):27–38.

Radak, Z., H. Y. Chung, E. Koltai, A. W. Taylor, and S. Goto. 2008. Exercise, oxidative stress and hormesis. *Ageing Res Rev* 7(1):34–42.

Ristow, M., K. Zarse, A. Oberbach, N. Kloting, M. Birringer, M. Kiehntopf, M. Stumvoll, C. R. Kahn, and M. Bluher. 2009. Antioxidants prevent health-promoting effects of physical exercise in humans. *Proc Natl Acad Sci USA* 106(21):8665–8670.

Stamler, J. S., Q. A. Sun, and D. T. Hess. 2008. A SNO storm in skeletal muscle. *Cell* 133(1):33–35.

Sun, Q. A., D. T. Hess, L. Nogueira, S. Yong, D. E. Bowles, J. Eu, K. R. Laurita, G. Meissner, and J. S. Stamler. 2011. Oxygen-coupled redox regulation of the skeletal muscle ryanodine receptor-Ca^{2+} release channel by NADPH oxidase 4. *Proc Natl Acad Sci USA* 108:16098–16103.

Tanaka, L. Y., L. R. Grassmann Bechara, A. M. dos Santos, C. P. Jordão, L. G. Oliveira de Sousa, T. Bartholomeu, L. I. Ventura, F. R. Martins Laurindo, and P. R. Ramires. 2015. Exercise improves endothelial function: A local analysis of production of nitric oxide and reactive oxygen species. *Nitric Oxide* 45:7–14.

Zinkevich, N. S. and D. D. Gutterman. 2011. ROS-induced ROS release in vascular biology: Redox-redox signaling. *Am J Physiol Heart Circ Physiol* 301:H647–H653.

39 ROS and Epigenetics

Eric Hervouet, Jianhua Zhang, and Michaël Boyer-Guittaut

CONTENTS

ABSTRACT

Oxidative stress and ROS accumulation are major events involved in the life and death of cells. At low levels, ROS are involved in intracellular pathways but their accumulation can lead to deleterious effects, such as genotoxic stress or metabolic dysregulation, and to initiation or development of pathologies. Therefore, the pathways involved in ROS production and antioxidant response play a major role in the maintenance of cell homeostasis. Gene expression and regulation has been described as a major regulator of the response to oxidative stress, but many questions remain to fully characterize the gene response to oxidative stress. For example, the regulation of ROS production by epigenetic regulation of target genes or the effect of ROS accumulation on epigenetic modifications is still unclear. The link between ROS, mitochondria metabolism, and its effect on nuclear gene regulation are also under investigation. In this chapter, we highlight studies on the role of mitochondrial metabolism in ROS production and the effect of ROS on nuclear epigenetic modifications (DNA methylation and histone modifications). We also discuss the control of ROS production by the epigenetic regulation of genes involved in redox signaling, and the therapeutic potential of epigenetic-targeting drugs.

39.1 INTRODUCTION

Reactive oxygen species (ROS) are a group of molecules, comprising superoxide anion ($O_2^{\bullet-}$), hydroxyl radical ($^{\bullet}OH$), and hydrogen peroxide (H_2O_2), that are synthetized within cells through oxygen metabolism. At low levels, they represent important signaling molecules, but at elevated intracellular levels of ROS can lead to oxidative stress, which is characterized by protein, lipid, and DNA damage or modifications. In healthy cells, mitochondria produce the majority of intracellular ROS during oxidative phosphorylation through the leaking of electrons from the electron transport chain. Elevated intracellular ROS levels can then accentuate mitochondrial dysfunction by inducing an accumulation of high concentration of ROS, an increase in oxidized proteins, lipids, and DNA; a redox imbalance; and oxidative stress.

Recent studies have shown that an alteration of the intracellular redox status can play a pivotal role in the development of numerous pathologies via the deregulation of cellular pathways such as gene regulation, apoptosis, autophagy, or mitochondrial metabolism (Chiurchiù et al. 2016; Guntuku et al. 2016; Hecht et al. 2016; Ratliff et al. 2016). Recently, a growing number of publications have deciphered the mechanism of epigenetic modifications (DNA methylation and/or histone modifications) and their function in gene regulation. More importantly, a deregulation of epigenetic modifications has also been linked to various pathologies such as cancer, AD and PD (Feng et al. 2015; Lardenoije et al. 2015).

DNA methylation, described as the process responsible for the addition of a methyl group on the fifth carbon of cytosine in CpG islands, is catalyzed by the DNMT family. A local increase in DNA methylation in a promoter is frequently associated with the downregulation of the expression of the target gene. DNMT1 is mainly involved in the maintenance of methylation on the newly synthetized DNA strand following DNA replication, whereas both DNMT3A and DNMT3B catalyze de novo methylation on both strands. Besides DNA methylation, posttranslational modification of histones has also been described to regulate the local compaction of chromatin and its accessibility for transcription factors. Numerous modifications of histones have been described, including acetylation, methylation, phosphorylation, and SUMOylation. While some of these modifications relax chromatin (such as acetylation or H3K4me2), others promote the compaction of the chromatin (e.g., H3K9me3, H3K27me3). The local sum of these modifications on a promoter is called the histone code and is a result of the balance between relaxation and compaction signals.

In this chapter, we describe the links between metabolism, more specifically mitochondrial metabolism; ROS and oxidative stress; and the regulation of epigenetic modifications. We discuss how these mechanisms are related to each other and the potentials of targeting these mechanisms as future therapeutic strategies.

39.2 MITOCHONDRIA

Mitochondria are known to generate ATP through oxidative phosphorylation. During this process, ROS are generated through the mitochondrial electron transport chain activity. In addition, mitochondria provide essential substrates required by the epigenetic machinery (Shaughnessy et al. 2014).

39.2.1 MITOCHONDRIA AS PROVIDERS OF ENERGY AND SUBSTRATES FOR THE EPIGENETIC MACHINERY

Mitochondria are a major source of ATP, which is produced by oxidative phosphorylation, α-ketoglutarate (produced by the TCA cycle and described as an activator of TETs), acetyl-CoA (produced from citrate by the ATP citrate lyase), SAM produced from methionine by the methionine adenosyl transferase), and NAD$^+$. ATP is required for epigenetic modification reactions, acetyl-CoA is required for HAT activity, and SAM is required for HMT and DNMT reactions. The role of α-ketoglutarate in epigenetic modification has been demonstrated by the observation that a decrease in this compound due to the hyperactivation of IDHI inhibits DNA demethylases and alters gene expression and global nuclear histone acetylation through an acetyl-CoA-dependent process. NAD$^+$ produced by the oxidative phosphorylation chain can induce the activity of the SIRT1/6 histone deacetylases and therefore nuclear histone deacetylation.

39.2.2 Mitochondria as Targets of Epigenetic Machinery

Recent publications have demonstrated mtDNA as a target for epigenetic modifications (Sadakierska-Chudy et al. 2014) and suggest that these different modifications may serve as biomarkers for various diseases. For example, the mitochondrial genome contains 5mC and 5-hmC, particularly in the D-loop region (Shock et al. 2011), and the detection of specific mtDNA methylation might be used for diagnosis in the future (Iacobazzi et al. 2013). A mitochondrial isoform of DNMT1, DNMT3A, and DNMT3B has been found in mitochondria (Chestnut et al. 2011) and the methylation of mtDNA has been proposed to be involved in the regulation of gene transcription. It has also been shown that the 5hmC proportion is greater in mtDNA than in ncDNA even though the presence of the TET enzymes, responsible for this modification, have not yet been detected in mitochondria. Since mitochondria are sensitive to oxidative stress, different groups studied the effect of ROS-inducing drugs on mitochondrial function in neurons, a cell type particularly dependent on mitochondrial metabolism. Several studies demonstrated that the chronic use of drugs (methamphetamine, cocaine, alcohol, or morphine) can induce oxidative stress and decrease respiratory chain activity, while increasing the expression of nuclear encoded epigenetic enzymes, such as mtDNMT1 and DNMT3A. These changes could then lead to methylation, oxidative damage, or a decrease in the number of mtDNA copies and altered mitochondrial metabolism (Sadakierska-Chudy et al. 2014). Based on these studies, mitochondrial-targeted compounds, MitoQ or Mitoquinone, have been used to protect against the oxidative stress induced by the use of cocaine (Vergeade et al. 2010). Mitochondrial epigenetic modifications have been noted in motor neuron disease and aging (e.g., a decrease in 5hmC in the brains of aging mice) (Dzitoyeva et al. 2012). However, despite the observations that mitochondrial dysfunctions are associated with AD and PD, the epigenetic modification of mitochondrial DNA has not yet been critically investigated in postmortem brains of these patients.

39.3 ROS REGULATE EPIGENETIC MODIFICATIONS AND GENE EXPRESSION

The links between ROS and epigenetic gene expression and DNA and histone modifications have been established. Mitochondria-produced ROS can regulate the transcription factors HIF1-α, p53, and NFκB. In AD, pathology is often associated with reduced mitochondrial function, elevated oxidative stress, and the expression and modification of numerous epigenetic genes. Indeed, it has been shown that altered mitochondrial metabolism can be correlated to alterations of the epigenome, chromatin regulation, and gene regulation (Mastroeni et al. 2015). Therefore, the use of antioxidants, to protect mitochondria against ROS, has been proposed as a new therapeutic treatment in AD, even if their efficiency has not yet been proven *in vivo* (Devall 2014). For example, the treatment of AD-induced cell models (SH-SY5Y cells overexpressing oligomeric amyloid beta peptide, AβO) with antioxidants restored MMP, diminished ROS levels, and maintained ATP levels. More interestingly, these antioxidants restored the normal expression of different epigenetic genes, such as the histone acetyltransferase gene *KAT6B* (K Lysine acetyltransferase 6B) by modulating chromatin opening and accessibility of its promoter (Mastroeni et al. 2015). These data suggested that ROS and mitochondrial homeostasis could be involved in the transcriptional regulation of epigenetic machinery genes.

Similar mechanisms were reported in cancer. For example, *DNMT1* expression and/or *HDAC1* expression and activity were upregulated by H_2O_2 in colorectal or ovarian cancer cells (He et al. 2012; Kang et al. 2012; Zhang et al. 2013). Overexpression of NOX5-S, a variant of NOX5 lacking the calcium-binding domain, and air-pollution-induced ROS increased *DNMT1* expression in Barrett's cells and in normal alveolar epithelial cells, respectively (Soberanes et al. 2012; Hong et al. 2013). In murine macrophages, both ROS treatment and hypoxia decreased JHDMs activity, leading to an increase in repressive markers H3K9me2/3 and H3K36me3, and the subsequent silencing of specific genes (e.g., *CCL2*) (Tausendschon et al. 2011). In CRC, it has been shown that a H_2O_2

treatment decreased TSG *CDX1* expression and protein levels and increased *CDX1* promoter methylation (Zhang et al. 2013). The decrease in *CDX1* expression was partially attenuated by the use of NAC, a GSH precursor, or the inhibitor of DNA methylation, 5-Aza-dC. The effect of H_2O_2 has been shown to be linked to the upregulation of the expression and activity of both DNMT1 and HDAC1 and their increased interaction. A similar model has been described for the TSG *RUNX3* (runt-related transcription factor 3) gene in CRC with an increased expression and activity in DNMT1 and HDAC1 and the silencing of *RUNX3* after H_2O_2 treatment (Zhang et al. 2013).

39.3.1 ROS-Induced DNA Oxidation Regulates DNA Methylation

Oxidative stress and DNA methylation deregulation have been described as hallmarks of cancers even though the link between these mechanisms remains unknown. In most cancers, it has been observed that a hypermethylation of specific loci, usually containing TSGs, and a general hypomethylation are two features often correlated to ROS-induced oxidative stress (Wu and Ni 2015). Indeed, in the presence of oxidative stress or insufficient detoxification, $O_2{}^{\cdot-}$ and NO can react with DNA and lead to the formation of oxidized nucleotides on DNA structures, such as 8-OH-dG or 5hmC, which then regulate the methylation of nearby cytosines (Weitzman et al. 1994; Perillo et al. 2014). On the one hand, 8-OH-dG has been shown to decrease the ability of DNMT1 to methylate cytosine when it is specifically localized on the same strand one or two nucleotides downstream of the methylated target cytosine (Turk et al. 1995). On the other hand, 5hmC can induce DNA methylation to the active strand. Indeed, the conversion of 5mC into 5hmC, then into 5FC, and finally into 5caC is processed by TET family proteins (Vasanthakumar and Godley 2015). The presence of these markers has been observed in many cancers and suggested to be involved in gene dysregulation. It has been shown that ROS can not only function as catalysts of overall DNA methylation, but also induce site-specific hypermethylation due to the induction of DNMTs or the formation of a DNMT-containing complex.

In some cancers, the deregulation of tumor suppressor genes by epigenetic modifications can be linked to overall elevated oxidative stress. For example, the TSG *p16INK4* α expression is induced by a decrease in DNMT1 and a decrease in DNA demethylation in normal human epidermal keratinocytes treated with H_2O_2, which drives these cells toward senescence instead of tumorigenesis (Sasaki et al. 2014). In contrast, it has been described that air pollution–induced ROS production was correlated to *p16INK4* α methylation in murine alveolar epithelial cells, an epigenetic mark frequently observed in lung cancers (Soberanes et al. 2012). Indeed, chronic H_2O_2 exposure can induce the recruitment of the DNMT1/DNMT3B/SIRT1/EZH2 complex to DNA damage–induced foci and their hypermethylation (O'Hagan et al. 2011). Since chronic high amounts of ROS are frequently observed in tumors, these data led to a hypothesis that global DNA hypomethylation, which is frequently reported in cancers, might be explained by a shift of the epigenetic repressive machinery from methylated high repeat elements of the genome to DNA damage loci. Moreover, this shift might also explain the surprising hypermethylation observed in numerous gene promoters in cancer cells linked to the displacement of DNMTs on specific genomic loci previously described as nonmethylated.

During the late stages of cancer development, it has been observed that chronic exposure of HCC to H_2O_2 induced the expression of SNAIL-1, a transcription factor linked to the EMT, and as a consequence, the methylation of *CDH1* coding the E-Cadherin, a hallmark protein of EMT, frequently observed in aggressive carcinoma (Lim et al. 2008). An increase in the biomarkers ERBB2 and ERBB3, linked to cancer aggressiveness, has been quantified in ovarian cancer cells treated with H_2O_2 following the methylation-induced silencing of both *miR-199a* and *miR-125b*, normally involved in *ERBB2* and *ERBB3* gene repression (He et al. 2012). In a model of CRC cells resistant to 5-FU, a compound commonly used in chemotherapies, it was observed that an overexpression of NRF2 and its target HO-1 is correlated with elevated levels of ROS (Kang et al. 2014). Specific silencing of NRF2 or HO-1 decreased tumor growth and restored 5-FU sensitivity. In these resistant

CRC cells, the *NRF2* gene was hypomethylated and DNA demethylase TET-1 levels were upregulated. More interestingly, 5-FU increased ROS levels and TET-1 expression and activity while antioxidants caused an opposite effect. Therefore, the acquisition of 5-FU resistance in CRC cell seems to be linked to *NRF2* gene demethylation following ROS induction and TET-1 activation (Kang et al. 2014).

Another study described an increase in ROS levels following norepinephrine treatment; this involved an increase of NOX1, which resulted in the methylation of the *PKCε* promoter in cardiomyocytes and a decrease in the expression of the cardioprotective protein PKCε (Xiong et al. 2012). But a contradictory result has shown that silencing of *PKCε* following ROS-induced DNA methylation in fetal hearts or in the myocyte cell line H9c2 exposed to hypoxia was in fact independent of the NOX reaction (Patterson et al. 2012).

39.3.2 ROS Regulate Histone Modifications

Oxidative stress may also regulate the balance between an increase *versus* a decrease of histone acetylation (Berthiaume et al. 2006). For example, both HAT activity and H4ac levels were increased in alveolar epithelial cells treated with H_2O_2 or in CD34+ blood cells presenting elevated ROS following thalidomide treatment (Tomita et al. 2003; Aerbajinai et al. 2007). An increase in H3K9ac levels was also observed following an ethanol-linked ROS increase in rat hepatocytes (Choudhury et al. 2010). Another study showed that an H_2O_2 treatment of BEA-2B cells induced the nitration of HDAC2, and its subsequent inhibition (Ito et al. 2004) (Figure 39.1).

In comparison, an induction of ROS levels after a copper (Cu^{2+}) or H_2O_2 treatment in HL-60 leukemia cells decreased histone acetylation, and this inhibition was described to be diminished by the activities of SOD or catalase enzymes (Lin et al. 2005). A similar increase in ROS levels in hepatoma cells treated with nickel provoked a decrease of HAT activities but this effect was annihilated by antioxidant treatment (Kang et al. 2003). ROS have been shown to be necessary for the TGFβ1-dependent increase of HDAC2 activity and histone deacetylation in normal renal rat cells (Noh et al. 2009). In macrophages, hypoxia (pO_2 below 3%) or an increase in ROS led to an increase in the histone markers H3K9me2/me3 and H3K36me3 in specific regions of the promoter of the *CCL2* chemokine gene. This increase in histone modifications associated with a concomitant decrease of JHDMs levels induced the silencing of target genes (Tausendschon et al. 2011).

39.4 OXIDATIVE STRESS SIGNALING IS CONTROLLED BY EPIGENETIC MODIFICATIONS

We have discussed the evidence that ROS can regulate the acquisition and the removal of nuclear epigenetic modifications. Recent studies also demonstrated that epigenetic modifications can regulate the expression of antioxidant or ROS-producing genes to maintain the cellular redox status.

39.4.1 Epigenetic Modulation of Antioxidant Genes

Deregulation of ROS levels have been correlated with several pathologies but further information is still needed to understand whether this increase is an inducer or a consequence of the pathology. For example, it is now clear that high ROS levels have been associated with DNA damage and tumorigenesis (Poillet-Perez et al. 2015). In aging diseases, it has been suggested that mitochondrial dysfunction and increased ROS levels may be an early step of PD or AD (Hroudova and Singh et al. 2014) and that the deregulation of epigenetic modifications might play an important role in the initiation or the development of these pathologies (Landgrave-Gomez et al. 2015). More interestingly, recent data have described that oxidative stress and epigenetic modifications might be linked in this process (Cencioni et al. 2013).

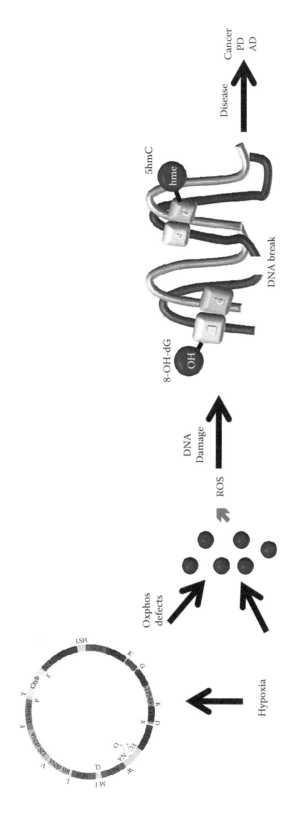

FIGURE 39.1 **(See color insert.)** Effect of oxidative stress on DNA damage and modifications and consequences on disease development. Conditions that induce oxidative stress (e.g., hypoxia, H$_2$O$_2$ treatment, mitochondria dysfunction) increase ROS levels, which in turn can affect mitochondrial function and lead to DNA damage as well as DNA modifications. These alterations will alter gene expression and lead to the development and progression of diseases. AD, Alzheimer's disease; PD, Parkinson's disease; 8-OH-dG, 8-hydroxy-2'-deoxyguanosine; 5hmC, 5-hydroxymethylcytosine.

In tumorigenesis, the role of ROS remains controversial since ROS levels can regulate the balance between pro- and anticancer properties (Hervouet et al. 2007). On the one hand, cancer cells frequently present an accumulation of ROS due to mitochondrial defects, which might be linked to the apparition of DNA mutations and the transformation of normal to cancer cells. On the other hand, it has been described that an increase of ROS production induced by anticancer therapies can lead to decreased cell proliferation and viability.

Nevertheless, one of the hallmarks observed in various cancers is the decreased expression of antioxidant genes correlated with epigenetic changes within their promoters. For example, hypermethylation of the *GPX3* promoter in different ccRCC cell lines has been observed. GPX3 is an extracellular glycosylated enzyme described as a major scavenger of ROS since it can use GSH to reduce H_2O_2, organic hydroperoxide, and lipid peroxides. This gene, which has been described as a TSG in different cancers, presents a decrease of mRNA and protein levels in tumors compared to normal tissues. Moreover, methylation of the GPX3 promoter was significantly higher in tumor tissues, and the use of 5-Aza-dC restored *GPX3* expression. These data suggest that hypermethylation and inhibition of *GPX3* expression may lead to the deregulation of antioxidant response and renal tumorigenesis (Liu et al. 2015). Similar results of the inhibition of *GPX3* expression linked to hypermethylation of its promoter have been described in gastric cancers (Peng et al. 2012) and inflammatory breast carcinoma compared to normal tissues (Chen et al. 2011; Mohamed et al. 2014). More interestingly, GPX3 inhibition by methylation has also been linked to late stages of cancer progression since it has been described to be involved in the apparition of chemoresistance in HNC patients treated with cisplatin (Chen et al. 2011; Peng et al. 2012).

A similar DNA methylation–dependent inhibition of *catalase* expression was also reported in HCC (Min et al. 2010) and *catalase* expression has been also described to be indirectly decreased in H_2O_2-treated HCC cells via the methylation of the *OCT-1* promoter, which codes for a transcription factor essential for *catalase* expression (Quan et al. 2011). The *GSTP1* gene, coding the glutathione S-transferase 1, is also frequently hypermethylated in proliferative inflammatory atrophy lesions of the prostate, which may later lead to carcinogenesis, compared to normal tissues (Millar et al. 1999; Nakayama et al. 2003). The analysis of *GSTP1* promoter methylation has therefore been proposed as a potential diagnosis marker for prostate cancer (Vanaja et al. 2009). A DNA methylation–dependent inhibition of *SOD2* expression, coding the superoxide dismutase MnSOD, was also observed in the human multiple myeloma cell line KAS 6/1 (Hodge et al. 2005) and it has been shown that treatment with the DNA methyltransferase inhibitor zebularine reversed SOD2 inhibition.

A deregulation of the ratio of acetylation versus deacetylation has been described as essential during the development of cardiac hypertrophy. For example, the histone acetylases pCAF and p300 act as hypertrophic factors and, conversely, the histone acetyltransferase *MOF* mRNA and protein levels are decreased in hypertrophic hearts. In a cardiac-specific MOF transgenic mice model, it has been described that MnSOD and catalase expressions increased and inhibited ROS production and the ROS-dependent downstream c-Raf-MEK-ERK pathway, which has been shown to promote hypertrophy. These data suggest that histone modifications may be involved in MnSOD and catalase transcriptional regulation in cardiac hypertrophy models (Qiao et al. 2014) (Figure 39.2).

39.4.2 EPIGENETIC MODULATION OF ROS-PRODUCING GENES

Epigenetics can regulate the expression of antioxidant genes to counteract the deleterious effect of oxidative stress but these modifications have also been linked to the regulation and expression of ROS-producing enzymes. For example, the DUOX1 enzyme, a major contributor of ROS production in the airway, is frequently silenced in human lung cancer. DUOX1 expression was also shown to be reduced in HCC compared to control healthy tissues, and this decrease was correlated with CpG island methylation in its promoter. The restoration of DUOX1 expression, following 5-Aza-dC treatment, decreased cancer cell proliferation and colony formation caused by an increase in ROS levels and a cell cycle arrest in G2/M (Ling et al. 2014).

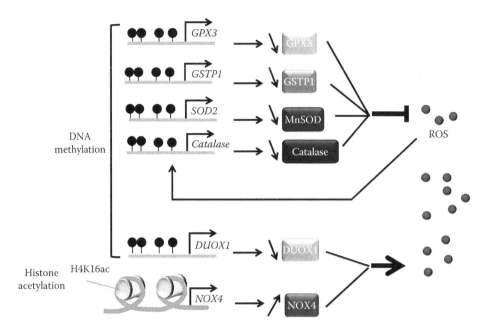

FIGURE 39.2 **(See color insert.)** Effect of epigenetic modifications on the expression of ROS-inhibiting or ROS-producing genes. Epigenetic modifications (DNA methylation or histone post-translational modifications) can alter the expression of numerous genes among which we can find antioxidant (*Catalase*, *SOD*...) or ROS-producing genes (*NOX4*, *DUOX1*...).

Cellular senescence is one of the hallmarks of aging and contributes to the development of age-related diseases. An epigenetic regulation of the ROS-generating enzyme NOX4 has been described as being part of this process. Indeed, in an *in vitro* model of senescent lung fibroblasts, the expression and the activity of the *NOX4* gene is constitutively upregulated. The promoter of the gene is enriched in the H4K16Ac mark and associated with a decrease in the repressive H4K20me3 mark suggesting an open conformation of the chromatin. More interestingly, the silencing of the MOF enzyme, which specifically acetylates H4K16, led to a decreased expression of *NOX4* (Sanders et al. 2015). Similar data obtained in HUVEC demonstrated that the specific invalidation of *HDAC3*, but not *HDAC1* or *HDAC7*, increased histone acetylation of the *NOX4* promoter, but not DNA methylation. Surprisingly, treatment with the HDACi scriptaid inhibited the recruitment of C-Jun despite the open structure of the chromatin. This result was explained by a decrease in RNA pol II recruitment on the *NOX4* promoter probably due to a hyperacetylation of the promoter and a steric inhibition of the transcription factor and enzyme on the gene (Siuda et al. 2012) (Figure 39.2).

39.5 THERAPEUTICS INVOLVING THE INTERPLAY BETWEEN ROS AND EPIGENETIC MODIFICATIONS

As detailed in the previous sections, the link between ROS and epigenetic modifications seems to play a fundamental role in the development of various pathologies. Therefore, with the modern concept of theranostics—"one patient, one diagnosis, one targeted therapeutic"—now emerging, the search for new, efficient drugs or the implementation of new combinations of drugs is rapidly evolving. To target the pathway involving ROS and epigenetics, several therapeutic approaches can be considered. The first one would consist of using epigenetic-targeting drugs in order to regulate overall or specific ROS-regulating gene expression. The second would regulate ROS levels and, as a consequence, regulate epigenetic modifications.

39.5.1 Effects of Epigenetic-Targeting Drugs on ROS

Recent published data have shown that epigenetic-targeting drugs present the ability to regulate DNA or chromatin modifications but, as collateral damage, can also increase ROS levels (Miller et al. 2011; Robert and Rassool 2012). This side effect can either increase cell response to the drug or inhibit the expected therapeutic response (Figure 39.3).

For example, it has been shown that Chaetocin, HMTi targeting SUV39H1, the main methyltransferase for lysine 9 tri-methylation on histone H3, can induce ROS production in a glioma model (Dixit et al. 2014). This increase led to the induction of the YAP-1 (Yes-associated protein 1) and ATM and JNK activation. Increased interaction of YAP-1 with p73- and p300-induced apoptosis, while the activation of JNK led to a glycolytic metabolic switch (increase in glucose uptake, glycolytic enzymes activity, and lactate production). These data are of interest for the development of new therapeutic antiglioma strategies. Similar results were obtained in AML. In the latter, Chaetocin was able to induce apoptosis both *in vitro* and *in vivo* in primary AML cells and inhibit leukemia *in vivo*. Chaetocin increased ROS levels and the transcription of death receptor–related genes to lead to death receptor-linked apoptosis (Chaib et al. 2012). More recently, Lai and collaborators have described a synergistic cytotoxic effect of Chaetocin when used with SAHA, (a histone deacetylase inhibitor) suggesting that a combined inhibition of HDAC and HMT might be a new way to treat AML (Lai et al. 2015). Still in AML, it has been described that DZNep (histone methyltransferase inhibitor) disrupts PRC2 activity leading to the activation of TXNIP (thioredoxin-interacting protein), the inhibition of thioredoxin activity, an increase in ROS levels, and apoptosis. Therefore, DZNep might induce apoptosis in AML cells via the inhibition of HMT EZH2 (Zhou et al. 2011).

Folic acid, which is a precursor of the SAM (S-adenosylmethionine) synthesis pathway, has been previously successfully used to stimulate DNA methylation in low DNA methylation-containing

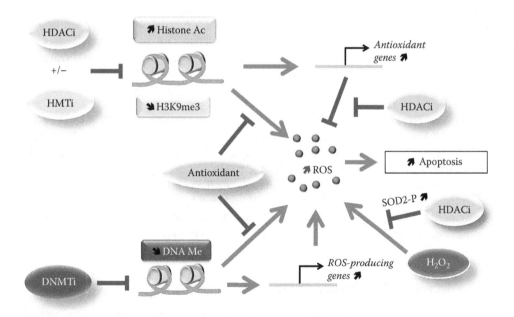

FIGURE 39.3 **(See color insert.)** Interplay between epigenetic-targeting drugs and oxidative stress and its effect on therapeutic efficiency. Inhibitors of epigenetic enzymes (HDACi, HMTi, DNMTi...) used in therapeutics can directly or indirectly regulate the expression of genes involved in the production or degradation of ROS. Depending on the pathology, it might be useful in the future to combine these chemicals with antioxidants or ROS-inducing drugs in order to potentiate their cytotoxic therapeutic effect or inhibit the side effect of ROS production.

glioma cells (Hervouet et al. 2009; Cartron et al. 2012). However, since high folic acid concentrations have been described to induce an oxidative stress, the addition of antioxidants (cocktail of Vitamin C and β-caroten) is required to counterbalance these deleterious effects. An ongoing phase I–II study using L-methylfolate and a supplementation in vitamin C in combination with Bevacizumab and Temozolomide is performed in recurrent high-grade glioma patients (NCT01891747).

In PD, environmental stimuli, such as paraquat or rotenone, formerly used in fertilizers, have been described as risk factors for the development of the pathology. More recently, epigenetic changes have also been described as a potential regulatory factor of the development of PD. A recent study demonstrated that a pre-treatment of PC12 cells with Decitabine (an inhibitor of DNMT), prior to the incubation with paraquat induced an increase in ROS levels and apoptosis (increased BAX and liberation of cytochrome c). These results demonstrated that the inhibition of DNA methylation sensitizes the cells to paraquat-induced ROS levels and mitochondria-linked apoptosis (Kong et al. 2012). Similar results of Decitabine-induced mitochondrial apoptosis through the induction of ROS levels were observed in other models, such as human leukemia cell lines or lymphocytes (Ruiz-Magana et al. 2012; Shin et al. 2012). More interestingly, Fandy and collaborators described that leukemia cells treated with Decitabine presented an induction of apoptosis and a delayed ROS accumulation. This ROS increase was dependent on 5-aza-dC incorporation into DNA. Decitabine also induced the expression of ROS-inducing enzymes such as NADPH oxidase and NOX4 (Fandy et al. 2014).

In human bone marrow–mesenchymal stem cells (hBM-MSCs), increased levels of ROS induced by a H_2O_2 treatment was suppressed by TSA (an inhibitor of HDACs). Indeed, TSA induced the expression of SOD2, which inhibited the lethal effect of ROS in hBM-MSCs. H_2O_2 decreased the levels of phosphorylated FOXO1 and SOD2 but increased P53 levels, while TSA reversed these effects. These data demonstrated that the effect of TSA on ROS modulation goes through the regulation of FOXO1 and SOD2 activities and that this epigenetic drug might prevent ROS accumulation in hBM-MSCs (Jeong and Cho 2015).The use of anticancer drugs has been described to increase ROS levels and induce kidney toxicity. Nadasi et al. showed that TSA and 5-Aza-dC induced mitochondrial ROS in mouse renal proximal tubule cells when used alone or in combination with a mild oxidative stress (e.g., a low dose) and that this increase was abolished by the use of NAC (N-acetylcysteine). The use of this epigenetic drugs also decreased CREB-driven transcription, known to play an important role in nephrotoxicity (Nadasi et al. 2009).

Epigenetic drugs can, therefore, present collateral effects by inducing ROS levels and induce apoptosis, but it has also been described that some epigenetic-targeting drugs can also induce the expression of genes involved in the antioxidant response. For example, in human breast primary tumors and cancer cells lines, vorinostat (SAHA, an HDACi) can induce the expression of genes involved in cell adhesion, but also redox regulation, such as GSH metabolism, which may lead to drug resistance. Indeed, the depletion of GSH induced by buthionine sulphoximine significantly enhanced the vorinostat cytotoxic effect (Chiaradonna et al. 2015). Therefore, inhibiting the antioxidant response and then inducing ROS levels might accentuate the cytotoxic effect of epigenetic drugs.

39.5.2 Effects of ROS-Targeting Drugs on Epigenetic Modifications

We described the effect epigenetic drugs have on the induction of oxidative stress or the antioxidant response, but the question remains whether therapeutics targeting the redox pathway would induce short- or long-term changes in nuclear epigenetic modifications as well.

Some recent data have described the fact that inducing oxidative stress and ROS in target cells might induce cell death in a dependent or independent epigenetic pathway. For example, in a recent study about the possibility of using nanoparticles (NQs) as a therapeutic approach to specifically induce HepG2 cancer cell death, it has been shown that NQs induce an inhibition of cell proliferation together with a decrease in HDAC1 and 2 expression and activity, a decrease in cell cycle regulators Cycline D1 and CDK1, but an increase in the CDK inhibitor, p21^{WAF} (Bishayee et al. 2015). These effects were correlated with an increase in p53-dependent ROS levels and a decrease in MMP leading to the induction of apoptosis (relocalization of Bax to the mitochondria, release of cytochrome c, and activation of Caspase 3). The model presented suggested that NQs-induced ROS would act on mitochondria to induce apoptosis but would also regulate HDACs activities, which in return would induce p53 and p21^{WAF} expression to further activate apoptosis and inhibit cell cycle.

Recently, it has been shown that Cambogin, a drug commonly used in Southeastern Asia for the treatment of cancer, can induce mitochondria-dependent apoptosis in breast cancer cells *in vitro*. This cell death is linked to the increase in pro-apoptotic proteins compared to anti-apoptotic proteins and the release and translocation of AIF to the nucleus (Shen et al. 2015). Moreover, the authors demonstrated, by two-dimensional gel electrophoresis coupled to mass spectrometry, that the ROS pathway was the most deregulated following Cambogin treatment. This increase in ROS levels and the activation of the JNK pathway were also correlated with the induction of ATF-2 (activating transcription factor-2) activity and an increase in the H3K9 mark in the AP-1 binding region of the anti-apoptotic *Bcl-2* gene promoter.

39.6 CONCLUSIONS

In this chapter, we discussed the link between epigenetic modifications and oxidative stress and whether ROS accumulation is a cause or consequence of epigenetic dysregulation. Mitochondria are also a major intermediaries in this interplay since they are the major redox sensors in the cells and the main producers of ROS. Epigenetic modifications occur on mtDNA to regulate mitochondrial gene expression, mitochondrial enzyme activity, and, therefore, ROS production. These findings are fundamental to understand the effect of epigenetic-targeting drugs used for the treatment of various pathologies and already tested in clinical trials. For example, it is now recognized that some epigenetic drugs (HDACi, HMTi, DNMTi) can induce ROS production and apoptosis, which would benefit the treatment if the goal of the drug is to lead to cell death, such as in cancer treatment, but negative if the purpose of the treatment is to lead to gene reexpression. These data would then be essential to determine whether to add an antioxidant to the epigenetic drug to personalize the treatment to the established diagnosis for each patient, a protocol defined as theranostics or personalized medicine and suggested by many as the future of pathology treatment.

ACKNOWLEDGMENTS

MBG and EH were supported by funding from the Ministère de l'enseignement supérieur et de la Recherche and the Ligue contre le Cancer (#007.Y-2014) région Grand-Est (Conférence de Coordination Interrégionale du Grand Est). JZ is supported by NIHR01-NS064090.

ABBREVIATIONS

5-Aza-dc	5-Aza-2′-deoxycytidine
8-OH-dG	8-Hydroxy-2′-deoxyguanosine
5-FU	5-Fluorouracile
5FC	5-Formylcytosine
5caC	5-Carboxylcytosine
5hmC	5-Hydroxymethylcytosine
5mC	5-Methylcytosine

8OH-dG	8-Hydroxy-2′-deoxyguanosine
AD	Alzheimer's disease
AML	Acute myeloid leukemia.
ATF-2	Activating transcription factor-2
ATM	Ataxia telangiectasia mutant
ccRCC	Clear cell renal cell carcinoma
CDK1	Cyclin-dependent kinase 1
CRC	Colorectal cancer
dC	Deoxycytidine
DNMT	DNA methyl transferase
DUOX1	Dual oxidase 1
DZNep	3-Deazaneplanocin A
EMT	Epithelial-mesenchymal transition
ERα	Estrogen receptor α
ERBB2 and ERBB3	ERB-b2 receptor tyrosine kinase 2 and 3
GPX3	Glutathione peroxidase 3
GSH	Glutathione
HAT	Histone acyl transferase
HCC	Hepatocellular carcinoma
HDAC	Histonedeacetylase
HDM	Histone demethylase
HIF-1α	Hypoxia-inducible factor-1 α
HMT	Histone methyltransferase
HMTi	Histone methyltransferase inhibitor
HNC	Head and neck cancer
HO-1	Hemeoxigenase-1
HUVEC	Human umbilical vascular endothelial cells
IDHI	Isocitrate dehydrogenase
JHDM	Jumonji-domain-containing histone demethylase
JNK	C-jun-NH2-kinase
MMP	Mitochondrial membrane potential
MOF	Male absent on the first
mtDNA	Mitochondrial DNA
NAC	N-acetylcysteine
NAD$^+$	Nicotinamide adenine dinucleotide
ncDNA	Nuclear DNA
NFB	Nuclear factor B
NOX4	NADPH oxidase 4
NOX5s	Short NADPH oxidase
NRF2	Nuclear factor-erythroid 2-related factor
NQs	Quercetin-nanoparticles
PD	Parkinson's disease
PKCε	Protein kinase Cε
PRC2	Polycomb-repressive complex 2
ROS	Reactive oxygen species
SAHA	Suberoylanilidehydroxamic acid
SAM	S-adenosylmethionine
TCA	Tricarboxylic acid cycle
TET	Ten-eleven translocations
TSA	Trichostatin A
TSG	Tumor suppressor gene

REFERENCES

Aerbajinai, W., J. Zhu et al. (2007). Thalidomide induces gamma-globin gene expression through increased reactive oxygen species-mediated p38 MAPK signaling and histone H4 acetylation in adult erythropoiesis. *Blood* **110**(8): 2864–2871.

Berthiaume, M., N. Boufaied et al. (2006). High levels of oxidative stress globally inhibit gene transcription and histone acetylation. *DNA Cell Biol* **25**(2): 124–134.

Bishayee, K., A. R. Khuda-Bukhsh et al. (2015). PLGA-loaded gold-nanoparticles precipitated with quercetin downregulate HDAC-Akt activities controlling proliferation and activate p53-ROS crosstalk to induce apoptosis in hepatocarcinoma cells. *Mol Cells* **38**(6): 518–527.

Cartron, P. F., E. Hervouet et al. (2012). Folate supplementation limits the tumourigenesis in rodent models of gliomagenesis. *Eur J Cancer* **48**(15): 2431–2441.

Cencioni, C., F. Spallotta et al. (2013). Oxidative stress and epigenetic regulation in ageing and age-related diseases. *Int J Mol Sci* **14**(9): 17643–17663.

Chaib, H., A. Nebbioso et al. (2012). Anti-leukemia activity of chaetocin via death receptor-dependent apoptosis and dual modulation of the histone methyl-transferase SUV39H1. *Leukemia* **26**(4): 662–674.

Chen, B., X. Rao et al. (2011). GPx3 promoter hypermethylation is a frequent event in human cancer and is associated with tumorigenesis and chemotherapy response. *Cancer Lett* **309**(1): 37–45.

Chestnut, B. A., Q. Chang et al. (2011). Epigenetic regulation of motor neuron cell death through DNA methylation. *J Neurosci* **31**(46): 16619–16636.

Chiaradonna, F., I. Barozzi et al. (2015). Redox mediated suberoylanilide hydroxamic acid (SAHA) sensitivity in breast cancer. *Antioxid Redox Signal* **23**(1):15–29.

Chiurchiù, V., A. Orlacchio et al. (2016). Is modulation of oxidative stress an answer? The state of the art of redox therapeutic actions in neurodegenerative diseases. *Oxid Med Cell Longev* **2016**:7909380. doi: 10.1155/2016/7909380.

Choudhury, M., P. H. Park et al. (2010). Evidence for the role of oxidative stress in the acetylation of histone H3 by ethanol in rat hepatocytes. *Alcohol* **44**(6): 531–540.

Devall, M., J. Mill et al. (2014). The mitochondrial epigenome: A role in Alzheimer's disease? *Epigenomics* **6**(6): 665–675.

Dixit, D., R. Ghildiyal et al. (2014). Chaetocin-induced ROS-mediated apoptosis involves ATM-YAP1 axis and JNK-dependent inhibition of glucose metabolism. *Cell Death Dis* **5**: e1212.

Dzitoyeva, S., H. Chen et al. (2012). Effect of aging on 5-hydroxymethylcytosine in brain mitochondria. *Neurobiol Aging* **33**(12): 2881–2891.

Fandy, T. E., A. Jiemjit et al. (2014). Decitabine induces delayed reactive oxygen species (ROS) accumulation in leukemia cells and induces the expression of ROS generating enzymes. *Clin Cancer Res* **20**(5): 1249–1258.

Feng, Y., J. Jankovic et al. (2015). Epigenetic mechanisms in Parkinson's disease. *J Neurol Sci* **349**(1-2):3–9.

Guntuku, L., V. G. Naidu et al. (2016). Mitochondrial dysfunction in gliomas: Pharmacotherapeutic potential of natural compounds. *Curr Neuropharmacol* (In Press).

He, J., Q. Xu et al. (2012). Reactive oxygen species regulate ERBB2 and ERBB3 expression via miR-199a/125b and DNA methylation. *EMBO Rep* **13**(12): 1116–1122.

Hecht, F., C. F. Pessoa et al. (2016). The role of oxidative stress on breast cancer development and therapy. *Tumour Biol* (In Press).

Hervouet, E., E. Debien et al. (2009). Folate supplementation limits the aggressiveness of glioma via the remethylation of DNA repeats element and genes governing apoptosis and proliferation. *Clin Cancer Res* **15**(10): 3519–3529.

Hervouet, E., H. Simonnet et al. (2007). Mitochondria and reactive oxygen species in renal cancer. *Biochimie* **89**(9): 1080–1088.

Hodge, D. R., B. Peng et al. (2005). Epigenetic silencing of manganese superoxide dismutase (SOD-2) in KAS 6/1 human multiple myeloma cells increases cell proliferation. *Cancer Biol Ther* **4**(5): 585–592.

Hong, J., D. Li et al. (2013). Role of NADPH oxidase NOX5-S, NF-kappaB, and DNMT1 in acid-induced p16 hypermethylation in Barrett's cells. *Am J Physiol Cell Physiol* **305**(10): C1069–C1079.

Hroudova, J., N. Singh et al. (2014). Mitochondrial dysfunctions in neurodegenerative diseases: Relevance to Alzheimer's disease. *Biomed Res Int* **2014**: 175062.

Iacobazzi, V., A. Castegna et al. (2013). Mitochondrial DNA methylation as a next-generation biomarker and diagnostic tool. *Mol Genet Metab* **110**(1–2): 25–34.

Ito, K., T. Hanazawa et al. (2004). Oxidative stress reduces histone deacetylase 2 activity and enhances IL-8 gene expression: Role of tyrosine nitration. *Biochem Biophys Res Commun* **315**(1): 240–245.

Jeong, S. G., G. W. Cho (2015). Trichostatin A modulates intracellular reactive oxygen species through SOD2 and FOXO1 in human bone marrow-mesenchymal stem cells. *Cell Biochem Funct* **33**(1): 37–43.

Kang, J., Y. Zhang et al. (2003). Nickel-induced histone hypoacetylation: The role of reactive oxygen species. *Toxicol Sci* **74**(2): 279–286.

Kang, K. A., M. J. Piao et al. (2014). Epigenetic modification of Nrf2 in 5-fluorouracil-resistant colon cancer cells: Involvement of TET-dependent DNA demethylation. *Cell Death Dis* **5**: e1183.

Kang, K. A., R. Zhang et al. (2012). Epigenetic changes induced by oxidative stress in colorectal cancer cells: Methylation of tumor suppressor RUNX3. *Tumour Biol* **33**(2): 403–412.

Kong, M., M. Ba et al. (2012). 5″-Aza-dC sensitizes paraquat toxic effects on PC12 cell. *Neurosci Lett* **524**(1): 35–39.

Lai, Y. S., J. Y. Chen et al. (2015). The SUV39H1 inhibitor chaetocin induces differentiation and shows synergistic cytotoxicity with other epigenetic drugs in acute myeloid leukemia cells. *Blood Cancer J* **5**: e313.

Landgrave-Gomez, J., O. Mercado-Gomez et al. (2015). Epigenetic mechanisms in neurological and neurodegenerative diseases. *Front Cell Neurosci* **9**: 58.

Lardenoije, R., A. Iatrou et al. (2015). The epigenetics of aging and neurodegeneration. *Prog Neurobiol* **131**:21–64.

Lim, S. O., J. M. Gu et al. (2008). Epigenetic changes induced by reactive oxygen species in hepatocellular carcinoma: Methylation of the E-cadherin promoter. *Gastroenterology* **135**(6): 2128–2140, 2140.e1–8.

Lin, C., J. Kang et al. (2005). Oxidative stress is involved in inhibition of copper on histone acetylation in cells. *Chem Biol Interact* **151**(3): 167–176.

Ling, Q., W. Shi et al. (2014). Epigenetic silencing of dual oxidase 1 by promoter hypermethylation in human hepatocellular carcinoma. *Am J Cancer Res* **4**(5): 508–517.

Liu, Q., J. Jin et al. (2015). Frequent epigenetic suppression of tumor suppressor gene glutathione peroxidase 3 by promoter hypermethylation and its clinical implication in clear cell renal cell carcinoma. *Int J Mol Sci* **16**(5): 10636–10649.

Mastroeni, D., O. M. Khdour et al. (2015). Novel antioxidants protect mitochondria from the effects of oligomeric amyloid beta and contribute to the maintenance of epigenome function. *ACS Chem Neurosci* **6**(4): 588–598.

Millar, D. S., K. K. Ow et al. (1999). Detailed methylation analysis of the glutathione S-transferase pi (GSTP1) gene in prostate cancer. *Oncogene* **18**(6): 1313–1324.

Miller, C. P., M. M. Singh et al. (2011). Therapeutic strategies to enhance the anticancer efficacy of histone deacetylase inhibitors. *J Biomed Biotechnol* **2011**:514261. doi: 10.1155/2011/514261.

Min, J. Y., S. O. Lim et al. (2010). Downregulation of catalase by reactive oxygen species via hypermethylation of CpG island II on the catalase promoter. *FEBS Lett* **584**(11): 2427–2432.

Mohamed, M. M., S. Sabet et al. (2014). Promoter hypermethylation and suppression of glutathione peroxidase 3 are associated with inflammatory breast carcinogenesis. *Oxid Med Cell Longev* **2014**: 787195.

Nadasi, E., J. S. Clark et al. (2009). Epigenetic modifiers exacerbate oxidative stress in renal proximal tubule cells. *Anticancer Res* **29**(6): 2295–2299.

Nakayama, M., C. J. Bennett et al. (2003). Hypermethylation of the human glutathione S-transferase-pi gene (GSTP1) CpG island is present in a subset of proliferative inflammatory atrophy lesions but not in normal or hyperplastic epithelium of the prostate: A detailed study using laser-capture microdissection. *Am J Pathol* **163**(3): 923–933.

Noh, H., E. Y. Oh et al. (2009). Histone deacetylase-2 is a key regulator of diabetes- and transforming growth factor-beta1-induced renal injury. *Am J Physiol Renal Physiol* **297**(3): F729–F739.

O'Hagan, H. M., W. Wang et al. (2011). Oxidative damage targets complexes containing DNA methyltransferases, SIRT1, and polycomb members to promoter CpG islands. *Cancer Cell* **20**(5): 606–619.

Patterson, A. J., D. Xiao et al. (2012). Hypoxia-derived oxidative stress mediates epigenetic repression of PKCepsilon gene in foetal rat hearts. *Cardiovasc Res* **93**(2): 302–310.

Peng, D. F., T. L. Hu et al. (2012). Silencing of glutathione peroxidase 3 through DNA hypermethylation is associated with lymph node metastasis in gastric carcinomas. *PLoS One* **7**(10): e46214.

Perillo, B., A. Di Santi et al. (2014). Nuclear receptor-induced transcription is driven by spatially and timely restricted waves of ROS. The role of Akt, IKKalpha, and DNA damage repair enzymes. *Nucleus* **5**(5): 482–491.

Poillet-Perez, L., G. Despouy et al. (2015). Interplay between ROS and autophagy in cancer cells, from tumor initiation to cancer therapy. *Redox Biol* **4**: 184–192.

Qiao, W., W. Zhang et al. (2014). The histone acetyltransferase MOF overexpression blunts cardiac hypertrophy by targeting ROS in mice. *Biochem Biophys Res Commun* **448**(4): 379–384.

Quan, X., S. O. Lim et al. (2011). Reactive oxygen species downregulate catalase expression via methylation of a CpG island in the Oct-1 promoter. *FEBS Lett* **585**(21): 3436–3441.

Ratliff, B. B., W. Abdulmahdi et al. (2016). Oxidant mechanisms in renal injury and disease. *Antioxid Redox Signal* (In Press).

Robert, C., F. V. Rassool (2012). HDAC inhibitors: roles of DNA damage and repair. *Adv Cancer Res* **116**:87–129.

Ruiz-Magana, M. J., J. M. Rodriguez-Vargas et al. (2012). The DNA methyltransferase inhibitors zebularine and decitabine induce mitochondria-mediated apoptosis and DNA damage in p53 mutant leukemic T cells. *Int J Cancer* **130**(5): 1195–1207.

Sadakierska-Chudy, A., M. Frankowska et al. (2014). Mitoepigenetics and drug addiction. *Pharmacol Ther* **144**(2): 226–233.

Sanders, Y. Y., H. Liu et al. (2015). Epigenetic mechanisms regulate NADPH oxidase-4 expression in cellular senescence. *Free Radic Biol Med* **79**: 197–205.

Sasaki, M., H. Kajiya et al. (2014). Reactive oxygen species promotes cellular senescence in normal human epidermal keratinocytes through epigenetic regulation of p16(INK4a). *Biochem Biophys Res Commun* **452**(3): 622–628.

Shaughnessy, D. T., K. McAllister et al. (2014). Mitochondria, energetics, epigenetics, and cellular responses to stress. *Environ Health Perspect* **122**(12): 1271–1278.

Shen, K., J. Xie et al. (2015). Cambogin induces caspase-independent apoptosis through the ROS/JNK pathway and epigenetic regulation in breast cancer cells. *Mol Cancer Ther* **14**(7): 1728–1737.

Shin, D. Y., Y. S. Park et al. (2012). Decitabine, a DNA methyltransferase inhibitor, induces apoptosis in human leukemia cells through intracellular reactive oxygen species generation. *Int J Oncol* **41**(3): 910–918.

Shock, L. S., P. V. Thakkar et al. (2011). DNA methyltransferase 1, cytosine methylation, and cytosine hydroxymethylation in mammalian mitochondria. *Proc Natl Acad Sci USA* **108**(9): 3630–3635.

Siuda, D., U. Zechner et al. (2012). Transcriptional regulation of Nox4 by histone deacetylases in human endothelial cells. *Basic Res Cardiol* **107**(5): 283.

Soberanes, S., A. Gonzalez et al. (2012). Particulate matter air pollution induces hypermethylation of the p16 promoter Via a mitochondrial ROS-JNK-DNMT1 pathway. *Sci Rep* **2**: 275.

Tausendschon, M., N. Dehne et al. (2011). Hypoxia causes epigenetic gene regulation in macrophages by attenuating Jumonji histone demethylase activity. *Cytokine* **53**(2): 256–262.

Tomita, K., P. J. Barnes et al. (2003). The effect of oxidative stress on histone acetylation and IL-8 release. *Biochem Biophys Res Commun* **301**(2): 572–577.

Turk, P. W., A. Laayoun et al. (1995). DNA adduct 8-hydroxyl-2″-deoxyguanosine (8-hydroxyguanine) affects function of human DNA methyltransferase. *Carcinogenesis* **16**(5): 1253–1255.

Vanaja, D. K., M. Ehrich et al. (2009). Hypermethylation of genes for diagnosis and risk stratification of prostate cancer. *Cancer Invest* **27**(5): 549–560.

Vasanthakumar, A., L. A. Godley (2015). 5-hydroxymethylcytosine in cancer: Significance in diagnosis and therapy. *Cancer Genet* **208**(5): 167–177.

Vergeade, A., P. Mulder et al. (2010). Mitochondrial impairment contributes to cocaine-induced cardiac dysfunction: Prevention by the targeted antioxidant MitoQ. *Free Radic Biol Med* **49**(5): 748–756.

Weitzman, S. A., P. W. Turk et al. (1994). Free radical adducts induce alterations in DNA cytosine methylation. *Proc Natl Acad Sci USA* **91**(4): 1261–1264.

Wu, Q., X. Ni (2015). ROS-mediated DNA methylation pattern alterations in carcinogenesis. *Curr Drug Targets* **16**(1): 13–19.

Xiong, F., D. Xiao et al. (2012). Norepinephrine causes epigenetic repression of PKCepsilon gene in rodent hearts by activating Nox1-dependent reactive oxygen species production. *FASEB J* **26**(7): 2753–2763.

Zhang, R., K. A. Kang et al. (2013). Oxidative stress causes epigenetic alteration of CDX1 expression in colorectal cancer cells. *Gene* **524**(2): 214–219.

Zhou, J., C. Bi et al. (2011). The histone methyltransferase inhibitor, DZNep, up-regulates TXNIP, increases ROS production, and targets leukemia cells in AML. *Blood* **118**(10): 2830–2839.

Index

A

AAFA, *see* Asthma and Allergy Foundation of America
Aceruloplasminemia, 97
Acetylcholinesterase (AChE) inhibition, 488–490, 492–493
Acrosome reaction, 415, 425–427
Acute neurotoxicity, 489
Acyl-CoA:cholesterol acyltransferase (ACAT), 70
AD, *see* Alzheimer's disease
Advanced glycation end products (AGEs), 365
ALS, *see* Amyotrophic lateral sclerosis
Alzheimer's disease (AD)
  amyloid cascade hypothesis, 123
  cholinergic hypothesis, 124–125
  definition, 136
  environmental factors, 161
  genetic factors, 160–161
  OS markers, 159
  oxidative stress
    mitochondria, 126
    neuroprotective strategies, 127–129
    redox active metals, 127
  pathophysiology, 159–160
  Tau hypothesis, 125
Amines, 52
2-amino-1-methyl-6-phenylimidazo[4,5-*b*]pyridine (PhIP), 233–234
Amyloid cascade hypothesis, 123
Amyotrophic lateral sclerosis (ALS)
  Cu,ZnSOD
    A4V, 109
    H43R, 109
    mitochondria, 111
    molecular mechanism, 109
    mutations, 110–111
    structure, 108–109
    targeting therapies, 112–113
  dying-back hypothesis, 110
  etiology of, 106
  FALS and SALS, 106–108
  GSH levels, 113–114
  melatonin, 113
  mitochondrial function, 113
  pharmacological interventions, 112
  prevalence of, 106
  ROS, 108
  SODs, 108
  symptoms, 106
  TDP-43, 114
  VCP mutations, 114
Angiotensin receptor blockers (ARBs), 356
Antioxidant response element (ARE), 341
Antioxidant therapy
  antiaging
    catalase, 448
    glutathione, 448
    metabolic, 449
    natural, 449
    nutrient, 449
    peroxiredoxin, 448
    plants (*see* Phytochemicals)
    SODs, 448
    synthetic, 449
    thioredoxin, 448
  CVD, 314
  DR, 366–367
  FMS, 393–395
  mitochondrial metabolism, 521
  sperm DNA damage, 436
  sperm maturation, 423
Apoptosis
  morphine, 508–509
  RNA virus, 473
  SLE patients, 377–378
  spermatozoa, 416–417
Aspirin, 283, 313
Assisted reproductive technologies (ARTs), 434–435
Astaxanthin, 326
Asthma
  AAFA, 401
  airway hyperreactivity, 190
  allergic/nonallergic, 402
  enzymatic antioxidants, 196–199
  eosinophilic/neutrophilic inflammation, 190
  interleukins, 402, 404
  MOL 294, 404
  nonenzymatic antioxidants, 195–197, 199
  Nox, 402
  oxidative stress, 193–195
  PNRI-299, 404
  primary oxidants, 190–193
  RNS, 402
  secondary oxidants, 191–192
  symptoms, 401
Asthma and Allergy Foundation of America (AAFA), 401
Ataxia telangiectasia mutated (ATM) gene
  activation mechanism, 462
  atherosclerosis, 466
  autophagy regulation, 465–466
  cancer incidence, 463–464
  CVD, 466–467
  cytoplasmic function, 465
  DNA damage, 462–463
  dysfunction, 464–465
  HIF-1 pathway, 465–466
  HSC regulation, 464
  liver disease, 467
  metabolic syndrome, 467
  neural differentiation, 464
  neural stem cells, 464
  neurodegeneration, 464